REVIEW COPY
We would appreciate your sending two copies of your review.

NUMERICAL SOLUTION OF MARKOV CHAINS
(Prpbability: Pure and Applied, A Series of Textbooks and Reference Books/8)

Edited by: William J. Stewart

Published: 1991
728 pages, bound, illustrated

$145.00 (U.S. and Canada)
$166.75 (All other countries)
(Prices subject to change without notice)

marcel dekker, inc.
270 MADISON AVENUE • NEW YORK, NY 10016 • 212-696-9000

NUMERICAL SOLUTION OF MARKOV CHAINS

PROBABILITY: PURE AND APPLIED

A Series of Textbooks and Reference Books

1. Introductory Probability Theory, *by Janos Galambos*
2. Point Processes and Their Statistical Inference, *by Alan F. Karr*
3. Advanced Probability Theory, *by Janos Galambos*
4. Statistical Reliability Theory, *by I. B. Gertsbakh*
5. Structured Stochastic Matrices of M/G/1 Type and Their Applications, *by Marcel F. Neuts*
6. Statistical Inference in Stochastic Processes, *edited by N. U. Prabhu and I. V. Basawa*
7. Point Processes and Their Statistical Inference, Second Edition, Revised and Expanded, *by Alan F. Karr*
8. Numerical Solution of Markov Chains, *edited by William J. Stewart*

Other Volumes in Preparation

NUMERICAL SOLUTION OF MARKOV CHAINS

EDITED BY
WILLIAM J. STEWART
North Carolina State University
Raleigh, North Carolina

Marcel Dekker, Inc. New York • Basel • Hong Kong

Library of Congress Cataloging--in--Publication Data

Numerical solution of Markov chains/edited by William J. Stewart.
p. cm. -- -- (Probability, pure and applied; 8)
Papers presented at a workshop held Jan. 8--10, 1990.
Includes bibliographical references and index.
ISBN 0-8247-8405-7 (alk. paper)
1. Markov processes-- --Congresses. I. Stewart, William J.
II. Series.
QA274.7.N83 1991
519.2'33-- --dc20 91-13877
CIP

This book is printed on acid-free paper.

MARCEL DEKKER, INC.
270 Madison Avenue, New York, New York 10016

Current printing (last digit):
10 9 8 7 6 5 4 3 2 1

PRINTED IN THE UNITED STATES OF AMERICA

Preface

In recent years there has been an enormous increase in interest in Markov chains and their applications. Examples of these applications may be found extensively throughout the biological, physical, mathematical, and social sciences, as well as in business and engineering. There are many people currently working directly or indirectly with Markov chains, and a consensus evolved that this would be an appropriate time to organize a workshop on their numerical solutions. A preliminary letter and questionnaire were sent to about 50 people who are actively carrying out research in this area to determine the extent of their enthusiasm for such a meeting and their willingness to participate. The responses were overwhelming in support of the workshop and it was decided that a three-day workshop with both invited speakers and solicited papers would best serve the needs of the research community.

Requests for a workshop typically came from researchers working on one aspect of the numerical solution of Markov chains who wished to discover what researchers working on different aspects of the same topic have accomplished. There appear to be basically three distinct categories of researchers working in the Markov chain area: applied probabilists, numerical analysts, and system modelers, this latter group consisting of, for example, computer scientists and engineers. Each of these three have their own specific journals and reviews, they organize their own meetings, and, in university environments, they frequently belong to different noninteracting departments. Examples of the lack of cooperation among these groups are evidenced, for example, by the publication in different journals of almost identical results, one clothed in probabilistic terms, the other in terms of matrices and vectors. There has never been a conference or workshop in this area, and consequently there are no proceedings to which a researcher may turn to begin to track down references that may be relevant to his work. The numerical solution of Markov chains is given only perfunctory attention, if any, in textbooks.

The goals of the workshop are thus:

1. To foster cooperation among diverse researchers, to minimize unnecessary duplication, and to identify new long-term research objectives.
2. To bring up-to-date and to inform researchers who work on one aspect of this area, the work being undertaken in different aspects.
3. To provide the opportunity to researchers who work in this area to present their latest research results. The collection of presentations is intended to be an authoritative overview of the field, including its developments, current status, and projections for future directions.
4. To produce a comprehensive set of proceedings, the first ever in this research area. These proceedings are expected to become a significant reference work.

The papers that will be presented cover just about all aspects of solving Markov models numerically. There are papers on matrix generation techniques and generalized stochastic Petri nets; numerous papers on the computation of stationary distributions, including an impressive representation of aggregation/disaggregation approaches, as well as projection type methods and conjugate gradient-based methods; a session is devoted entirely to recursive type methods and another to sensitivity analysis; other sessions are devoted to the computation of transient solutions and to bounds and approximations; a number of papers consider applications, particularly in the area of computer communication models. In addition to the regularly presented papers, eight papers were presented in poster fashion. They were displayed continuously throughout the workshop. A number of software demonstrations were scheduled for presentation during the workshop.

I would like to use this opportunity to express my gratitude to the large number of individuals and organizations that contributed to making this workshop a reality. First I would like to thank the 50-odd people whom I pestered with questionnaires in the early days. Their input was invaluable. Second, I wish to thank everyone who sent in a paper. The response to our call for papers was greater than I had anticipated and the quality of the contributions was high. It was for this reason that the decision was taken to organize a poster session, to enable a greater number of these good papers to be presented. Third, I wish to thank the referees who undertook the often thankless task of insuring the high quality of the papers.

I wish to thank the National Science Foundation and the Army Research Office for the financial support that they provided for this workshop. NSF provided funds to defray the travel expenses of the invited speakers. I believe that this support was crucial in that it enabled the very best researchers in the Markov chain area to commit early to the workshop and by so doing created a nucleus around which the workshop developed. The funds provided by the ARO were used to offset costs such as publishing and printing and enabled us to keep the registration fee modest, particularly for student attendees.

This workshop was organized in cooperation with a number of professional organizations. They are SIAM, IFIP W. G. 7.3, and the ORSA/TIMS Applied Probability Group. Their encouragement was a source of inspiration. Also, at NCSU, the Computer Science Department, the Center for Communication and Signal Processing (CCSP), and the Center for Research in Scientific Computation (CRSC) provided much appreciated support and assistance.

I would like to single our a number of individuals for special thanks. C. D. Meyer (NCSU), R. J. Plemmons (NCSU), K. Trivedi (Duke), and S. Stidham, Jr. (UNC) were always available and willing to lend a hand and/or an ear when the occasion arose. I often called upon them for advice. I especially wish to thank Mladen Vouk (NCSU), who undertook the organizing of the software demonstrations, and Molly Weston from the Computer Science department at NCSU who was instrumental in producing all the printed material, including the call for papers and the program (both preliminary and final) and in general for keeping things moving efficiently. Finally, I would like to acknowledge the great help that Harry Perros (NCSU) has been and to thank him for it. He agreed to handle most of the financial aspects of organizing this workshop and was the person upon whom I most depended when the going got tough.

William J. Stewart

Contents

Contributors

H. H. AMMAR* Department of Electrical and Computer Engineering, Clarkson University, Potsdam, New York

VINCENT A. BARKER Institute for Numerical Analysis, Technical University of Denmark, Lyngby, Denmark

STEVEN BERSON Computer Science Department, University of California at Los Angeles, Los Angeles, California

THOMAS R. BOWE Decision System Associates, Woodside, California

S. CHAKRAVARTHY Department of Science and Mathematics, GMI Engineering and Management Institute, Flint, Michigan

*_Current affiliation_: West Virginia University, Morgantown, West Virginia

GIANFRANCO CIARDO Software Productivity Consortium, Herndon, Virginia

PIERRE-JACQUES COURTOIS Philips Research Laboratory, Louvain-la-Neuve, Belgium

EDWARD J. COYLE School of Electrical Engineering, Purdue University, West Lafayette, Indiana

JOSE J. CRUZ Department of Mechanical/Industrial Engineering, New York Institute of Technology, Old Westbury, New York

JOHN N. DAIGLE Department of Industrial Engineering and Operations Research, Virginia Polytechnic Institute and State University, Blacksburg, Virginia

WILLIAM DAPKUS Decision System Associates, Woodside, California

A. I. ELWALID Department of Electrical Engineering and Center for Telecommunications Research, Columbia University, New York, New York

ANTHONY EPHREMIDES Electrical Engineering Department, University of Maryland, College Park, Maryland

ROBERT GEIST Department of Computer Science, Clemson University, Clemson, South Carolina

WINFRIED K. GRASSMANN Department of Computational Science, University of Saskatchewan, Saskatoon, Saskatchewan, Canada

DONALD GROSS Department of Operations Research, The George Washington University, Washington, D.C.

BINGCHANG GU* Department of Operations Research, George Washington University, Washington, D.C.

LEVENT GÜN† Department of Electrical Engineering, University of Maryland, College Park, Maryland

DANIEL P. HEYMAN Bellcore, Red Bank, New Jersey

Current affiliations

*Chinese Aeronautical Radio Electronics Research Institute, Shanghai, China

†IBM Communications Systems, Research Triangle Park, North Carolina

S. M. R. ISLAM* Department of Electrical and Computer Engineering, Clarkson University, Potsdam, New York

DAVID S. KIM† Department of Industrial and Operations Engineering, University of Michigan, Ann Arbor, Michigan

UDO R. KRIEGER Research Institute of the Deutsche Bundespost, Darmstadt, Germany

GUY LATOUCHE Free University of Brussels, Brussels, Belgium

RUNG-BIN LIN Department of Computer Science, University of Minnesota, Minneapolis, Minnesota

DAVID M. LUCANTONI Department of Teletraffic Theory and System Performance, AT&T Bell Laboratories, Holmdel, New Jersey

JOHN C. S. LUI Department of Computer Science, University of California at Los Angeles, Los Angeles, California

R. B. MATTINGLY Department of Mathematical and Computer Science, Youngstown State University, Youngstown, Ohio

C. D. MEYER Department of Mathematics, North Carolina State University, Raleigh, North Carolina

D. MITRA AT&T Bell Laboratories, Murray Hill, New Jersey

RICHARD R. MUNTZ Department of Computer Science, University of California at Los Angeles, Los Angeles, California

JOGESH MUPPALA Department of Computer Science, Duke University, Durham, North Carolina

A. NAKASIS National Institute for Standards and Technology, Gaithersburg, Maryland

MARCEL F. NEUTS Department of Systems and Industrial Engineering, University of Arizona, Tucson, Arizona

VICTOR F. NICOLA IBM Thomas J. Watson Research Center, Yorktown Heights, New York

---

Current affiliations

*IBM Corporation, Boca Raton, Florida

†General Motors Research Laboratories, Warren, Michigan

BO FRIIS NIELSEN Institute of Mathematical Statistics and Operations Research, Technical University of Denmark, Lyngby, Denmark

KSHEMENDRA N. PAUL Techno-Sciences, Inc., Greenbelt, Maryland

JEFFREY L. POPYACK Department of Mathematics and Computer Science, Drexel University, Philadelphia, Pennsylvania

M. J. M. POSNER Department of Industrial Engineering, University of Toronto, Toronto, Ontario, Canada

K. V. RAVI Department of Electrical and Computer Engineering, GMI Engineering and Management Institute, Flint, Michigan

ANDREW REIBMAN AT&T Bell Laboratories, Holmdel, New Jersey

ROBERT REYNOLDS Department of Computer Science, Clemson University, Clemson, South Carolina

MARIA RIEDERS William E. Simon Graduate School of Business Administration, University of Rochester, Rochester, New York

YOUCEF SAAD* NASA Ames Research Center, Moffett Field, California

ROBIN A. SAHNER Urbana, Illinois

PAUL J. SCHWEITZER William E. Simon Graduate School of Business Administration, University of Rochester, Rochester, New York

PIERRE SEMAL Philips Research Laboratories, Louvain-la-Neuve, Belgium

E. SENETA Department of Mathematical Statistics, University of Sydney, Sydney, New South Wales, Australia

THEODORE J. SHESKIN Industrial Engineering Department, Cleveland State University, Cleveland, Ohio

EUGENE SHRAGOWITZ Department of Computer Science, University of Minnesota, Minneapolis, Minnesota

ROBERT L. SMITH Department of Industrial and Operations Engineering, University of Michigan, Ann Arbor, Michigan

MARK K. SMOTHERMAN Department of Computer Science, Clemson University, Clemson, South Carolina

---

**Current affiliation*: University of Minnesota, Minneapolis, Minnesota

RICHARD M. SOLAND Department of Operations Research, The George Washington University, Washington, D.C.

EDMUNDO DE SOUZA E SILVA Federal University of Rio de Janeiro, Rio de Janeiro, Brazil

T. E. STERN Center for Telecommunications Research, Columbia University, New York, New York

G. W. STEWART Department of Computer Science and Institute for Advanced Computer Studies, University of Maryland, College Park, Maryland

WILLIAM J. STEWART Department of Computer Science, North Carolina State University, Raleigh, North Carolina

SHALER STIDHAM, JR. Department of Operations Research, University of North Carolina, Chapel Hill, North Carolina

USHIO SUMITA William E. Simon Graduate School of Business Administration, University of Rochester, Rochester, New York

J. G. C. TEMPLETON Department of Industrial Engineering, University of Toronto, Toronto, Ontario, Canada

H. C. TIJMS Department of Econometrics, Free University, Amsterdam, The Netherlands

KISHOR S. TRIVEDI Department of Computer Science, Duke University, Durham, North Carolina

NICO M. VAN DIJK Department of Economical Sciences and Econometrics, Free University, Amsterdam, The Netherlands

VICTOR L. WALLACE Department of Computer Science, The University of Kansas, Lawrence, Kansas

JEFFREY E. WIESELTHIER Information Technology Division, Naval Research Laboratory, Washington, D.C.

TAO YANG* Department of Industrial Engineering, University of Toronto, Toronto, Ontario, Canada

JI ZHANG Department of Electrical Engineering, University of Hawaii at Manoa, Honolulu, Hawaii

---

**Current affiliation*: Department of Industrial Engineering, Technical University of Nova Scotia, Halifax, Nova Scotia, Canada

NUMERICAL SOLUTION OF MARKOV CHAINS

1

Numerical Markov Queuing Analysis: A Twenty-Five Year Odyssey

VICTOR L. WALLACE* Department of Computer Science, The University of Kansas, Lawrence, Kansas

ABSTRACT

This meeting occurs just 25 years after the first exploratory work in numerical Markov analysis. Many of the problems perceived then, and a few more, are with us still, though often in different form.

While the work of the 1960s was hindered by weak computers, with tiny memories, and by a system-design world that was almost indifferent to modeling results, it was aided in part by a surfeit of small problems within reach of the techniques of the day. It is fair to say that the early system designers' indifference was engendered in large part by unfamiliarity, tempered by skepticism. Often that skepticism revealed a justifiable fear of the fragility of those early methods.

Today new, highly concurrent architectures are giving rise to new problems that are less and less amenable to classical techniques, both analytical and numerical, and therefore more needful of numerical Markov analytical tools. At the same time, those same new architectures, as well as more economical technologies for older architectures, are presenting modelers with more computing power than their theory now knows how to harness, and a highly performance-competitive industry has simultaneously become much more receptive to modeling as a tool in system design.

This appears an excellent time for a new beginning, kicked off by a "First International Workshop."

1. INTRODUCTION

These are exciting times for work in numerical Markov analysis (NMA). It appears that we have a good "critical mass" of researchers knowledgeable and interested in the subject, and we have a healthy, demanding clientele ready to use our wares. We also have a set of problems, created out of the technology of computing, which will be placing ever more sophisticated demands on our capabilities—demands that, by and large, cannot be served by any alternative approach. Finally, we have built up our storehouse of tools and concepts to the point where we have some hope of meeting the challenge on a consistent basis.

It is an especially attractive time to hold a "First International workshop on the Numerical Solution of Markov Chains." The field appears to have finally matured.

It was not always clear from its birth approximately 25 years ago either that this subject was going to mature so nicely, or that it was going to take so long doing so. In this paper, I would like to address a few remarks about that early childhood and that slow adolescence, to give a perspective on where we are now, and what we have been doing. I think it will help us keep a realistic focus on where the "hot points" are, and how we can achieve the acceptance (not to mention funding) of our science that it deserves.

2. RELATION TO COMPUTER DEVELOPMENTS

NMA is a creature of its applications, and those applications have been predominantly the problems of performance and work flow through computers and computer networks. A science does not usually develop without a source of curiosity welling up from some concrete need. It doesn't always remain with that inspiration, though. In this case, however, this class of application is where the most interesting questions are—questions that cannot be answered by the emaciated models of classical queuing theory, nor by the unabstracted empiricism of stochastic simulations.

It was clear in the beginning that there was a substantial body of questions in computer systems that required a strong mixture of abstraction and attention to the detail of the application. The numerical solution of Markov or Markov renewal (regenerative) models seemed to fit that bill of particulars, and to fill it quite uniquely.

That body of questions appeared, around 1961, with the progression of computer system designs to a new complexity. In order to make the systems of the day interactive, we needed to give the user real-time response. In order to give him real-time response with such an expensive device, we had to exploit concurrency by multiprogramming our systems and by inventing concurrently executing input/output channels, and we had to augment our puny machine memories by ever more sophisticated stores, both disk and drum.

All this gave rise simultaneously to the need to cope with concurrency among unrelated, or randomly related, tasks and to play close attention to such things as response time and resource utilization.

The computer influenced that early thinking in another way, which is less obvious. Science had for centuries come to believe that the only way to abstract a subject was to generate models capable of closed form mathematical analysis. With the beginnings of responsive, interactive computer systems, it also became obvious to some of us that the computer could contribute a new method for abstraction: the computer model. This model, like mathematics, had to be reproducible, and produce reproducible predictions and explanations of cause and effect, but it did not have to be susceptible to closed-form analysis. It only had to be computable and exercisable.

3. EARLY EXPERIENCE

Into this particular mix of developments the Michigan Computer Modelling Project was formed and nurtured, within the Department of Electrical Engineering at the University of Michigan. It was instigated at first by Dr. Harry Goode, the father of the discipline then called "Systems Engineering." Unfortunately, his inspiration was lost to posterity, and to the project, when he died in 1961 in an automobile crash. The project really didn't get underway until a year later.

After that false start, the original work in computer modeling at the University of Michigan was principally stimulated by a research project sponsored by the Rome Air Development Center. Its purpose was, in the words of the task order, "to formulate mathematical models for predicting and controlling the information flow through multiple-element computer systems." The systems in question were multicomputer, multiaccess, interactive, time-shared database-time military systems.

The charter for the project was scientifically enlightened, and included specifications that encouraged the development of new theoretical tools for analysis, control, and "priority determination" in such systems. Participants in the original project included myself, as principal investigator, Drs. Kuei Chuang, Richard V. Evans, and Eugene L. Lawler, and Messrs. Dennis W. Fife, Robert Carlsen, Richard Rosenberg, Robert F. Rosin, and John L. Smith. The latter all took their Ph.D.s at the University of Michigan during this general time frame.

The first report [WALL64] included a remarkable collection of papers, models, and reports, any of which would appear quite at home in this conference.

1. The specification of the "Recursive Queue Analyzer (RQA-0)," a prototype program for the numerical solution of equilibrium state probabilities of large Markov chain models using sparse matrix versions of the power method, and special list-storage representations of matrices and vectors.
2. Applications, using RQA-0, which analyzed (a) multiprogramming with fixed partitions [FIFE64a], (b) multiprogramming with variable partitions, and (c) concurrent arm positioning in a disk store [FIFE64b].
3. A study by Chuang of the sufficient conditions for a controlled Markov chain to have a (stationary) optimum control strategy that is a function of current state only, and for the optimum to be algorithmically determined by Markov renewal programming [JEWE63] and the Howard algorithm [HOWA60].
4. A study of optimal Markov scheduling with random arrivals and deferral costs [FIFE65, LAWL65].
5. An analysis of the statistical properties of the entire throughput of the Michigan Computing Center for a month [ROSI65, WALT67].

The work that followed in successive renewals under the sponsorship by the Rome Air Development Center and, later, ARPA, continued to generate tools and novel approaches to the Markov modeling with an emphasis on the numerical. To cite a few more of the early ones:

1. The development of an improved iterative Markov analysis program (RQA-1) [WALL66a,b], which extended the application to discrete chains (normally generated by embedded Markov chain or semi-Markov chain models), improved the efficiency, and provided tools for the generation of matrices and compact state vectors.
2. Development of a computer-graphic, interactive compiler of Markov chain matrices from queuing flow charts, for input to RQA [IRAN69, WALL69, IRAN71].
3. Use of Markov renewal programming to study optimal foreground-background strategies for scheduling time-sharing [FIFE66].

4. Study of Markov renewal programming and Markov decision analysis for optimization of Markov control strategies when the state is incompletely known to the controller [SMIT70].
5. Development of a technique and theory for solution of Quasi-Birth and Death (QBD) models, including proofs of existence and uniqueness of matrix-geometric solutions [WALL69a].
6. The application of RQA to a large number of system problems that seemed important at the time to the Air Force.

This project was followed by a succession of projects over a number of years under the later leadership of Dr. Keki Irani.

4. THE LESSONS

I cite these things not just to shamelessly call attention to an association and activity of which I am proud, but also to examine some things it teaches about the conditions for science and the conditions for successful transfer to the technology. It was true that we had the field almost to ourselves at the time, but we were actively advertising our work and encouraging others to involve themselves. So perhaps I am not being as provincial as it seems.

What we need to learn is why the period of adolescence has been so long, and the reasons are a harbinger of dangers ahead. Why weren't the NMA techniques more widely used, and why did it take so long to develop the richness of science we see in the program of this workshop?

We have seen that the start was the result of a lucky juxtaposition of technology, its issues and capabilities, and a theoretical capability for abstraction through computers.

I can suggest perhaps five reasons that seemed important at the time. Others can no doubt be raised as well:

1. That the whole idea of computer modeling and Markov modeling was too new and unfamiliar. It was much more "mathematical" and abstract that practitioner-designers were accustomed to. It was also not trusted.
2. Other techniques (product form networks, simulation) were distracting attention because they were more readily understood, and there were still enough problems around that they could solve.
3. The distinction between empirical examination and abstract modeling was not well understood.
4. The NMA techniques tended to be quite fragile. Traps abounded to snare the unwary!
5. Computers were not big enough, or powerful enough, to make the process seem effortless.

There was a strong element of truth in each of these. I can recall a rejection notice from IEEE TEC for one of our papers stating that it was "too mathematical," even though at the time the journal was filled with coding theory, automata theory, and formal language theory. Citations of our work often referred to our modeling analysis as "simulations," with a touch of disdain because the abstractions it presented did not result in closed form.

Two anecdotes can recall the real problems with robustness, and the dangers befalling the unwary.

In the first, although we had been doing NMA models for many years by that time, I can recall the embarrassment in 1970 of encountering a seriously ill-conditioned model for the first time while confidently engaged in consulting. No adjustments seemed to eliminate the problem, and tens of thousands of iterations were getting nowhere. The problem was to model a SCAN strategy on a disk store with a large number of cylinders. While we now have some understanding of why that would produce a problem, there was little experience to draw upon then.

In the second, Jim Foley and I had a project in 1975, using branch-and-bound optimization to automatically determine an optimal configuration for interactive graphics systems [FOLE80]. The program could determine the best choice for workstation processing power, bandwidth of connection to the host supercomputer, the amount of memory and store in the workstation, and even which functional software modules should be migrated to the workstation and which to the host—all as a function of the application statistics and specific market prices for the "commodities." The value function that was being maximized has a Markov analysis embedded in it, and used RQA for evaluation. Unfortunately, as the program modified the Markov model's parameters, it would do things a human modeler never does, like having servers with negligibly short or long service times. This produced nearly decomposable models when we least expected them, and caused a lot of highly technical "corrections" to the algorithm.

Jackson's method (1957 paper) was known to us at the time, and occasionally used. However, the questions we found ourselves addressing didn't lend themselves to it because you had to ignore the key concurrency ideas of blocking and sharing of resources. Although it relies, at least for closed networks, on numerical algorithms for its use, it is only sometimes Markov, and "traditionally" it is not included in NMA.

In any case, my vote for the primary cause for the neglect of such a beautiful tool by the practitioners of the 1960s was probably a combination of user distrust of the theoretical assumptions, and the perception of fragility and unreliability in the algorithm.

5. PROGNOSIS AND CONCLUSIONS

Over the intervening years, we have developed a considerable understanding of technical matters in NMA which were only vaguely understood in the 1960s. We have a lot left to do, but we are no longer so blind. The most important of these are in the category of approximation techniques and error analysis. They include the theory of near-decomposability and aggregation [COUR77], sensitivity analysis [ZARL76], and the various forms of iterative approximation. While our art is not yet completely robust, it is educated.

We have also developed an arsenal of alternative techniques to deal with embarrassing cases. Among those are the direct methods [STEW78], the quasi-birth and death methods [WALL69a, NEUT81], and transient analyses [WHIT73, GRAS77]. There are many additional promising techniques, some of them presented here in this workshop.

Finally, we are attacking the problems of usability by developing "user-friendly" proprietary packages, and even resuming the effort at automatic compilation from network diagrams [IRAN69, WALL69b, IRAN71, WALL72].

We have computers of considerably more power than our current algorithms know how to use effectively, and we have a population of problems generated by parallel systems, distributed systems, visualization systems, multicomputers, and remote window systems. There is an agenda that should keep us coming to Workshops on Numerical Markov Analysis for decades to come!

Given the needs, capabilities, knowledge, and talent now devoted to the problems, the prognosis is for another quarter century of active and fruitful research and practice in NMA.

REFERENCES

[EVAN67] R. V. Evans. Geometric distribution in some two dimensional queuing systems, *Operations Res.*, 5 (September/October 1967).

[FIFE64a] D. W. Fife and R. Rosenberg, Queueing in a memory-shared computer, *Proc. ACM 19th National Conf.*, ACM Publ. (1964).

[FIFE64b] D. W. Fife and J. L. Smith, Transmission capacity of disc storage systems with concurrent arm positioning, *IEEE Trans. Elect. Comp.*, *EC-14*, 4 (August 1965).

[FIFE65] D. W. Fife, Scheduling with random arrivals and linear loss functions, *Manage. Sci.*, *11*, 3 (1965).

[FIFE66] D. W. Fife, An optimization model for time sharing, *Proc. AFIPS Spring Joint Computer Conf.*, *28*, (1966).

[FOLE80] J. D. Foley, Optimization design of two-computer networks, Information Processing 80, IFIP, North Holland, 1980.

[GRAS77] W. K. Grassmann, Transient solutions in Markovian queueing systems, *Comput. Operations Res.*, *4*. (Pergamon Press, 1977).

[HOWA60] Howard, *Dynamic Programming and Markov Processes*, MIT Press, Cambridge (1960).

[IRAN69] K. B. Irani, V. L. Wallace, et al., "SELMA-QAS, Interactive Analysis of Queues," 16-mm sound film, University of Michigan, Ann Arbor (1969).

[IRAN71] K. B. Irani and V. L. Wallace, On network linguistics and the conversational design of computer networks, *J. ACM*, *18*, 4 (October 1971).

[JENN75] A. Jennings and W. J. Stewart, Simultaneous iteration for partial eigensolution of real matrices, *J. IMA*, *15* (1975).

[JEWE63] W. S. Jewell, Markov renewal programming, *Operations Res. 11*, 6 (1963).

[LAWL65] E. L. Lawler, Computational techniques for scheduling problems with deferral costs, *Manage. Sci.*, *11*, 3 (1965).

[NEUT81] M. Neuts, *Matrix Geometric Solutions in Stochastic Models*, Johns Hopkins University Press, Baltimore (1981).

[ROSI65] R. F. Rosin, Determining a computer center environment, *CACM*, *8*, 7 (July 1965).

[SMIT67] J. L. Smith, Multiprogramming under a page on demand strategy, *ACM*, *10*, 10 (October 1967).

[SMIT70] J. L. Smith, Markov decisions on a partitioned state space, *IEEE Trans. on SMC*, *SMC-1*, 1 (1970).

[STEW78] W. J. Stewart, A comparison of numerical techniques in Markov modeling, *CACM*, *21*, 2 (February 1978).

[WALL64] V. L. Wallace, D. W. Fife, and R. F. Rosin, A Study of Information Flow in Multicomputer and Multiconsole Data Processing Systems, *Technical Documentary Report No. RADC-TDR-64-427*, Rome Air Development Center, Griffiss AFB, New York, December 1964.

[WALL66a] V. L. Wallace and R. S. Rosenberg, Markovian models and numerical analysis of computer system behavior, *Proc. AFIPS Spring Joint Computer Conference*, *28*, (New Jersey: AFIPS Press, 1966).

[WALL66b] V. l. Wallace and R. S. Rosenberg, RQA-1, the Recursive Queue Analyzer, *Technical Report No. 2*, Systems Engineering Laboratory, University of Michigan, Ann Arbor (Feb. 1966).

[WALL69a] V. L. Wallace, The solution of Quasi Birth and Death Processes Arising from Multiple Access Computer Systems, Doctoral Dissertation, Dept. of Electrical Engineering, University of Michigan, Ann Arbor, 1969 (University Microfilms).

[WALL69b] V. L. Wallace, On the Representation of Markovian Systems by Network Models, Technical Report No. 42, Systems Engineering Laboratory, University of Michigan, Ann Arbor, Michigan (August 1969).

[WALL72] V. L. Wallace, Toward an algebraic theory of Markovian networks, *Proc. Symp. Computer-Communications Networks and Teletraffic* (New York: Polytechnic press, 1973).

[WALT67] E. S. Walter and V. L. Wallace, Further analysis of a computing center environment, *Comm. ACM*, *10*, 5 (May 1967).

[WHIT73] J. S. Whitlock, Modelling Computer Systems with Time-Varying Markov Chains, Doctoral Dissertation, Department of Computer Science, University of North Carolina, Chapel Hill, 1973 (University Microfilms).

[ZARL76] R. L. Zarling, Numerical Solution of Nearly Decomposable Queueing Networks, Doctoral Dissertation, Department of Computer Science, University of North Carolina, Chapel Hill, 1976 (University Microfilms).

2

A Methodology for the Specification and Generation of Markov Models*

STEVEN BERSON Computer Science Department, University of California at Los Angeles, Los Angeles, California

EDMUNDO DE SOUZA E SILVA Federal University of Rio de Janeiro, Rio de Janeiro, Brazil

RICHARD R. MUNTZ Computer Science Department, University of California at Los Angeles, Los Angeles, California

ABSTRACT

Modelers wish to specify their models in a symbolic high-level language, while analytic Markov process solution techniques require a low-level, numerical representation. The translation between these description levels is a major problem. We describe a general, but surprisingly powerful methodology for specifying system-level models and generating the associated transition rate matrix. The basic approach will be shown to have significant advantages in that it provides the basis for modular, extensible modeling tools. With this methodology, modeling tools tailored to particular application domains can be quickly and easily constructed. An implementation in Prolog of a system based on this methodology is described and some example applications are given.

*This research was supported by a grant from NSF INT-8902183 and CNPq-Brazil.

1. INTRODUCTION

The complexity of the new generation of highly concurrent computer systems that are now being developed has made the use of sophisticated modeling tools for specification and analysis a high-priority enterprise. The variety of architectures and different problems to be analyzed demonstrates the need for general tools, tools that allow specification of general classes of models. There are many examples of these tools in the literature [GOYA86, BAVU87, MAKA82, COST81, CARR86, SAHN87, BUTL86, STEW88]. The usefulness of such tools can be measured in terms of two factors: the sophistication of the underlying analytic and/or simulation techniques used, and the simplicity and power of the user interface. We are particularly concerned with the latter.

We can distinguish two representations of a model: The *modeler's representation* and the *analytic representation*. The analytic representation is the detailed, low-level representation required as input to the analysis modules, that is, some representation of the transition rate matrix. The modeler's representation is the model specification that the user supplies. The modeler's representation is typically in symbolic form and should be in terms of constructs that are natural to the application. For example, a reliability model might be expressed in terms of the number of each component type, their failure modes and rates, etc., while a multiprocessor system might be described in terms of the processors, buses, and memory modules, their physical connection, and the rates of processors accessing memory modules.

Clearly the form of the model specification language determines the ease of defining models, thereby influencing the overall cost of the modeling effort; requiring the user to directly input the transition rate matrix for a model is certainly possible but would severely hinder the application of the tool. A high-level model description language tailored to a particular domain provides the ability to define models that are easy to understand, less time-consuming for the modeler, and less error prone. However, this puts the burden on the modeling tool to provide a *translation* from the modeler's representation to the transition rate matrix.

In our approach, all system models are defined in terms of instances of objects and interactions between objects. The characteristics of specific applications are reflected in the object types that are used in the model specification. By providing libraries of object definitions, the modeling tool can be easily extended to new application domains. System models that combine object types drawn from a library with user-defined object types are accommodated, which permits specialized extensions. Facilities are also provided for general queries as well as custom queries for a specific model. As will be demonstrated by examples that appear later, this has proven to be a powerful modeling paradigm that has easily accommodated a wide variety of applications including reliability models and queuing theoretic models. With this approach one is able to produce tailored interfaces in a matter of hours for reliability modeling and queuing networks. Defining ad hoc models exhibits similar efficiencies. Being able to easily construct and analyze specialized models can be of significant benefit.

Irani and Wallace [IRAN71] investigated the specification of queuing network models using the notion of object types and object interactions. They developed an approach that permits the definition of object types and object interactions. In their system an object is defined in terms of a set of matrices that describe the behavior of that object. In a related paper [WALL72], Wallace describes how the transition rate matrix for a network of interconnected objects can be automatically generated from the matrix descriptions of the individual components.

We have used an approach in which the transition matrix for the complete model is incrementally constructed by searching for reachable states. This technique has several advantages. In many applications it is not feasible to generate the entire transition rate matrix due to the size of the state space. In such situations one may be content to truncate or aggregate the state space (e.g., eliminating states with more than some number of failed components is common in reliability models), particularly if error bounds can be obtained [MUNT89]. Another advantage in the generative approach is that only reachable states are represented in the resulting transition rate matrix. The number of reachable states is typically much smaller than the number of states in the cartesian product of the states of the individual objects due to constraints implicit in object interactions.

It is also interesting to note that in programming terminology our technique uses "lazy evaluation" in the sense that object states and transition rates are only evaluated when needed. This allows the definition of objects with an unbounded number of states, for example, an M/M/1 queue. In the complete models in which these queues exist, there can only be a finite number of states. This is typically accommodated by having only a finite number of customers in the network model.

Other related work includes the METFAC system [CARR86], which provides a general system utilizing production rules to describe system behavior. The production rules operate on the global system state variables. Our approach encourages using only the local object states instead of the global system state. This has the advantage of modularity and leads to natural support for higher-level interfaces in which a particular model is specified in terms of previously defined objects along with simple connections among objects.

Generalized stochastic Petri nets [MARS84] are another general representation for stochastic systems. They have a number of desirable features; in modeling power they are equivalent to Markov chains and they have a natural graphics representation. There are several problems with Petri nets as a modeling paradigm. Since the basic constructs that Petri nets are dealing with, places and transitions, are so primitive, much burden is placed on the modeler. Since all tokens are identical, any objects that have an ordering or priorities can become unwieldy to represent. Two common objects of this nature are first-come first-served queues and priority queues (e.g., priority repair service).

Another approach is relatively specialized tools such as SAVE [GOYA86]. This type of tool has a very convenient interface for specific types of models. The disadvantage is in a lack of flexibility. Models other than those anticipated by the designers of SAVE can be very difficult to accommodate. It is useful to be able to create custom interfaces similar to SAVE, but still retain the flexibility to add or change existing modeling constructs. Our system, in addition to allowing a modeler to add to the basic system, allows for the creation of custom interfaces.

In Section 2 we describe the system modeling paradigm. In Section 3 we describe the organization of the system that has been constructed to implement this approach. Section 4 contains several example applications, and Section 5 presents our conclusions.

2. MODELING PARADIGM

As noted by Irani and Wallace [IRAN71], a modeling system consists of a large amount of software including command interpreters, analysis packages, and query and display facilities. This software represents a significant investment, and one would like it to be applicable to a wide range of application areas, such as queuing network models, reliability models, etc. The objective is to develop an approach that allows a modeling tool to be tailored to different application domains while requiring that most of the software be reusable across application domains.

The approach we advocate is based on the observation that models are typically very modular in that the model consists of a set of component "objects" that are "connected" in some manner. This is exactly what comes to mind if one thinks of queuing network models, the objects being the queues. The approach then is to build a system that allows object types and object interactions to be defined in some standard manner that is interpreted by the (invariant) remainder of the software. Particular application domains are characterized by the object types that are applicable to that domain. Such a system is extensible since new object types can be declared and used in conjunction with previously defined object types.

We begin with a short informal description and then follow with a more formal description of the modeling paradigm. With the view that we have adopted, a system model is composed of a set of interacting components called *objects*. The interactions among objects are represented via a message-based mechanism, as will be described shortly.

Each object is an entity that has an internal state, which can evolve over time. The state of an object can change due either to (1) an *event* that the object itself generates or (2) the receipt of a message from another object. The state of an object will determine the types of events it can generate and the rates at which they occur. An event may simply cause the object generating it to change state with no effect on other objects, but in general, an event will cause other objects to react in some manner. This is modeled by allowing objects to generate *messages* to be *sent* to other objects, notifying them of an event. The specification of an object therefore also includes a definition of how it reacts (i.e., changes state and sends messages) to received massages.

The state of the system is given by the set of states of the component objects and the list of "undelivered" messages in the system. Note that messages are an abstraction introduced to model the way objects interact and are "delivered" and reacted to in zero time. Thus some of the states are transitory in that they have zero holding time. Using stochastic Petri net terminology, a state is called *vanishing* if it has zero holding time, and *tangible* otherwise. In our system, the vanishing states are those with one or more undelivered messages in the system and the tangible states are those with no undelivered messages in the system.

For the purposes of analysis, we are only interested in transitions between tangible states. We define a vanishing sequence to be a sequence of state transitions that starts and ends in a tangible state where all intermediate states are vanishing. The actions that may happen in response to an event or a message determine sets of vanishing sequences. Once all the sets of vanishing sequences are determined, the transition rates for each pair of tangible states can be easily found. These rates are collected to form the state transition rate matrix, which is then fed to the Markov chain solver.

The above informal description should provide an intuitive basis on which to interpret the more formal definition which follows. Formally, we say that an object $\mathcal{O}$ is defined as

$$\mathcal{O} \triangleq (\mathcal{I}, \mathcal{S}, \mathcal{S}_0, \mathcal{E}, \mathcal{M}^r, \mathcal{M}^s, \mathcal{R}, \delta', \delta)$$

where

$\mathcal{I}$ = the unique name of the object

S = the set of possible states for the object
$S_0 \in \mathcal{S}$ = the initial state for the object
$\mathcal{E}$ = the set of events that can be generated by the object
$\mathcal{M}^r$ = the set of messages that can be received by the object
$\mathcal{M}^s$ = the set of messages that can be sent by the object
$\mathcal{R}$ = the rate function of the object

$$\mathcal{R} : \mathcal{S} \times \mathcal{E} \rightarrow \Re^+$$

δ' = the event function of the object

$\delta' : \mathcal{S} \times \mathcal{E} \rightarrow 2^{(0,1] \times \mathcal{S} \times [M^s]}$, where $[M^s]$ is an ordered list of messages each belonging to $\mathcal{M}^s$

δ = the message function of the object

$$\delta : \mathcal{S} \times \mathcal{M}^r \rightarrow 2^{(0,1] \times \mathcal{S} \times [\mathcal{M}^s]}$$

The function δ', given a state and an event, returns a set of 3-tuples representing the possible responses to an event. Each 3-tuple consists of a probability, a new state, and a list of outgoing messages. Associated with the event function, there is a rate function $\mathcal{R}(s,e)$ that gives the rate at which event e occurs in state s. The function δ is analogous to δ'.

We say that a state S for an ordered set of N objects $\mathcal{O}_i$, $1 \leq i \leq N$, is a vector of object states, one for each object, and a list of messages.

$$S = \langle s_0, s_1, \ldots, s_N; [m_1, m_2, \ldots, m_k] \rangle$$

We define a system transition relation γ in terms of the object transition functions δ and δ'.

$$\gamma : (\{\langle s_1, \ldots, s_i, \ldots, s_N; [m_1, \ldots, m_k] \rangle\}) \longrightarrow (\langle s_1, \ldots, s_i', \ldots, s_N; [m_2, \ldots, m_k, m_{k+1}, \ldots, m_{k+n}] \rangle)$$

if message m_1 has as a destination object $\mathcal{O}_i$ and where m_{k+1}, ..., m_{k+n} are messages generated by object $\mathcal{O}_i$ as a response to message m_1, that is, $(p, s_i', [m_{k+1}, \ldots, m_{k+n}]) \in \delta_i(s_i, m_1)$

$$\gamma : (\langle s_1, \ldots, s_i, \ldots, s_N; [] \rangle) \longrightarrow (\langle s_1, \ldots, s_i', \ldots, s_N; [m_1, \ldots, m_k] \rangle)$$

if there exists an event e generated by $\mathcal{O}_i$ such that $(p, s_i', [m_1, \ldots, m_k]) \in \delta_i'(s_i, e)$

We define a state S to be *reachable* if there is a sequence γ_0, γ_1, ..., γ_M such that $S = \gamma_M \circ \gamma_{M-1} \circ \cdots \circ \gamma_0(S_0)$. That is, a state S is reachable if there is a valid sequence of states from the initial state S_0 to S. A reachable state with no outstanding messages, $(\langle s_1, \ldots, s_N; [] \rangle)$, is said to be tangible.

We define a *vanishing sequence* to be a sequence of transitions initiated by an event e, $V' \xrightarrow{p'} V_0 \xrightarrow{p_0} \cdots V_K \xrightarrow{p_K} V''$ such that

1. These transitions are allowed by γ and occur with probability p' or p_i, $0 \leq i \leq K$.
2. V' and V'' are tangible states.
3. $V_0, \ldots, V_K$ are vanishing states.

Let v be a vanishing sequence. Then the transition rate $\alpha(v)$ from tangible state V' to tangible state V'' corresponding to sequence v is given by

$$\alpha(v) = \mathcal{R}(V',e) \times p' \times \prod_{i=0}^{K} p_i$$

where $\mathcal{R}$ is the rate function as defined previously. Let $\mathcal{V}(V',V'')$ be the set of all vanishing sequences from V' to V'', and let $R(V',V'')$ be the transition rate from V' to V''. Then

$$R(V',V'') = \sum_{v \in \mathcal{V}(V',V'')} \alpha(v)$$

The transition rate matrix is $R[V',V'']$ for all reachable tangible states V' and V''. Note that $\mathcal{V}(V',V'')$ is not necessarily finite. However as long as the Markov chain is ergodic, the sum will exist. This is covered in more detail in Section 3.3.2.

3. IMPLEMENTATION ORGANIZATION

We have developed a prototype written in Prolog based on the object-oriented model presented in the previous section. This prototype allows users to easily define, solve, and query Markov models at a high level of abstraction. This section concentrates on the implementation organization, which is largely independent of the use of Prolog. However, since Prolog was perhaps an unusual choice, we will briefly present our reasons for this choice before describing the organization of this prototype.

3.1 Implementation Language

The basic operation in building the transition rate for a model is finding all the reachable states from a given initial state, according to rules that describe the behavior of the model. These rules specify preconditions on the state of an object and the actions to be taken if the preconditions are satisfied. Three features that Prolog provides make it appropriate for this prototype. First,

Prolog allows the use of untyped complex data structures, which allows the user full freedom in the description of object states. Second, Prolog provides *unification*, a powerful form of pattern matching. The pattern matching is used in the preconditions of the rules to determine which actions can be taken. This allows very general rules to be written quite simply, as will be seen in the examples. Third, Prolog has backtracking search as a basic feature of the language. As more than one rule may have its preconditions satisfied at the same time, the backtracking automatically applies all possible such rules to find all reachable states. These features allow rapid development both of complex models and complex modeling domains.

The one potential problem with using Prolog is some sacrifice of speed. In our current prototype, a typical model with 1099 states and 4282 transitions is generated in 54 seconds (on a Sun 3/60 workstation), which is not as fast as custom-complied tools. However we feel that the flexibility and the clear and concise representation of models more than make up for the slower speed. With better compliers for Prolog becoming available, speed will become even less of an issue.

3.2 Prototype Organization

In order to facilitate the use of the tool and aid in tailoring it to particular applications we have distinguished four different "layers" in its organization, as shown in Figure 1.

Core Layer

The core of the system provides a generic interface, which takes descriptions of objects, events, messages, and an initial state and generates the Markov state description. The "interface" to the core is the basis on which everything else is built. The core expects the following "schema" or format.

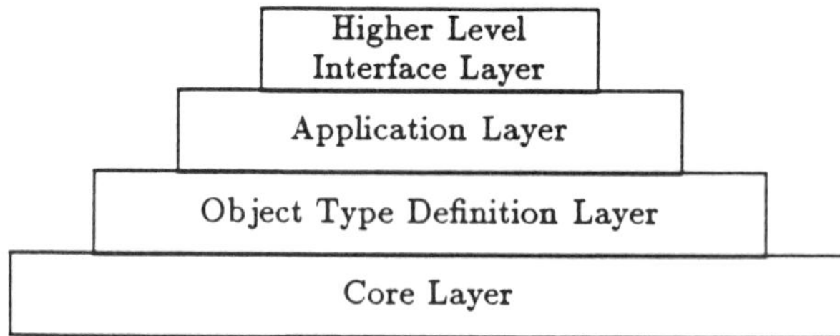

Figure 1 System layers.

1. Each object has a unique name.
2. Each object has an initial state.
3. Each object has a set of events it can generate, the rates at which they occur, and a list of corresponding actions.
4. Each object has a set of messages it can accept and their corresponding actions.

The core operates on this description by searching for the states reachable from the given initial state. A standard breadth first search procedure is used that, for each reachable state S, determines the states reachable from S by some event.

Object Type Definition Layer

An obvious extension to the basic paradigm presented in the previous section is the notion of object types. The notion of object types is important since it allows economy in specifying a model. It is expected that object type definitions will be parameterized so that instances of the objects can be declared and parameters specified for each instance. In this section, we introduce by way of example the language used to define object types. The example used is a reliability model consisting of a set of cpus, a set of memories that fail at some rate, and a repair service that fixes the cpus and memories at some rate. This is to be modeled as three objects, a cpu object, a memory object, and a repair service object. The cpu object represents a "pool" of identical available cpus, and similarly the memory object represents a "pool" of memories. The cpu and memory objects will generate one event, *failure*, and will be able to receive one message, *repaired*. The repair-service will generate one event, a *repair*, and will be able to receive one message, *fail*. In this section, the necessary object types will be defined: specifically a repairable-component-type for the cpu and memories objects and a repairman-type for the repair service object.

Figure 2 shows the two object type definitions. First, the *repairable-component-type* object type is defined. The OBJECT NAME statement defines a formal variable that, when instantiated, will be the name of this object. The OBJECT TYPE statement declares the name of the type of object being defined. Next the behavior of an object of this type is defined. The EVENT statement says that this object can generate a failure event. When it does, the object state at the start of the event is N and the object state at the end of the event is $N1$. The CONDITION statement gives the conditions under which this event can occur. In this case, a failure can occur when the number of working components, N, is greater than zero. The ACTION statement describes the actions the object takes in response to the event. When

```
OBJECT TYPE: repairable-component-type
OBJECT NAME: component-name
EVENT failure: N—> N1
        CONDITION: N > 0
        ACTION: N1 is N - 1,
                send(repair-service,fail(component-name))
RATE: N*failure-rate(component-name)
MESSAGE repaired: N—> N1
        ACTION: N1 is N + 1

OBJECT TYPE: repairman-type
OBJECT NAME: repair-name
EVENT repair: [H|T]—> T
        ACTION: send(H,repaired)
RATE: repair-rate(repair-name)
MESSAGE fail(Component): L—> L1
        ACTION: append(L,[Component],L1)
```

Figure 2 Examples of object type definition.

there is a failure, then number of operational components is decremented and a message is sent to the repair service. If a response were probabilistic instead of deterministic, a PROBABILITY statement would be used. This event occurs at a rate described in the RATE statement. In this case N is the number of operational components and failure-rate (component-name) is a value defined in the application layer. The description for MESSAGE repaired is very similar.

The OBJECT TYPE repairman-type defines the repair service. The major difference between repairman-type and repairable-component-type objects is that repairman-type objects use first-come first-serve service discipline so the state is represented as an ordered list rather than a number. When an object is repaired, it is taken off the front of the list. Following Prolog syntax the $[H \mid T]$ refers to the input state being a list with head H, and tail T, while the $\longrightarrow T$ indicates that the output state is the tail of the input state. When an object is repaired a repair message is sent to it. The append function creates a new list by merging the old list L with the list containing the name of the newly failed component.

There are now two object types that are defined that can be used by an end user in building a model. Note that nowhere in the object type definition is the state space explicitly defined. As mentioned earlier, this makes defining general objects easier (Section 4.2 gives one example where this is useful). These object types can be combined with others to form libraries. A user can then define instances of object types by simply referring to the library definitions, as will be shown in the following application layer subsection.

Application Layer

At the application layer, an "end user" can define a model by instantiating objects from an object types library and declaring the required parameters. These required parameters are used to customize individual instances.

```
type(cpu,repairable-component-type).
type(memory,repairable-component-type).
type(repair-service,repairman-type).

initial(cpu,2).
initial(memory,3).
initial(repair-service,[]).

failure-rate(cpu,pfc).
failure-rate(memory,pfm).
repair-rate(cpu,prc).
repair-rate(memory,prm).

value(pfc,0.001).
value(pfm,0.01).
value(prc,0.25).
value(prm,1.0).
```

Continuing with the previous example, the instances of objects can now be created. The type(cpu, repairable-component-type) and type(memory, repairable-component-type) statements declare two components to be of object type "repairable-component-type." Similarly the type(repair-service, repairman-type) statement declares the repair-service to be of object type "repairman-type." The initial(cpu,2), initial(memory,3) and initial(repair-service,[]) statements declare that initially in the model there are two operating cpus, three operating memories and an empty repair queue. It remains only to fill in the rate parameters. The failure-rate(cpu,pfc) gives a symbolic value for the cpu failure rate parameter, similarly with the failure-rate(memory,pfm), repair-rate(cpu,prc), and repair-rate(memory,prm) statements. The state transition rate matrix can be generated at this point in terms of the symbolic parameters. Only when the matrix is actually solved are the numerical values necessary. These values are specified by the value(pfc,0.001), value(pfm, 0.01), value(prc,0.25), and value(prm,1.0) statements. The model is complete and ready to evaluate. The results can be examined by a high-level query language described in a later section.

Higher-Level Interface Layer

The highest layer is where sophisticated user interfaces (e.g., English-like or graphical) can be defined. A graphical interface has been developed that allows a modeler to define icons for object types. These icons have a pictorial representation of the object plus zero or more *ports*. These ports represent possible interactions between objects, typically sending or receiving a message. Once these icons are defined, a model can be created by making instances of these object types and connecting the ports (with colored lines), the connections representing messages and message routes. In addition submodels can be created allowing (and encouraging) hierarchical modeling. The modeler can zoom in to see the detailed construction of a model or can zoom out to see the higher-level model structure. Our interface also provides animated (and nonanimated) simulation. Animated simulation in combination with the ability to zoom in and out provides a very powerful way to debug models.

Specialized graphical interfaces can easily be constructed using this framework. A translation must be written from these icons and connections to the objects and parameters of the application layer, but this is typically straightforward. We have already constructed two interfaces of this nature, one for reliability systems and another for queuing networks. Further applications are under development. Figure 3 shows a simple model done in the queuing network modeling domain. A detailed discussion of this example is presented in Section 4.1.

3.3 Other Implementation Issues

This section discusses some issues that were omitted from the general description.

Conditions and Functions

There are many times when the behavior of one object is dependent on the state of another. In reliability models, for example, an object can go dormant if certain combinations of components fail. If a disk controller D fails, any disks attached to D become dormant. An object will often have a different failure rate when it is dormant (often zero) than when it isn't. If the disk is modeled as not being able to fail when dormant, the disk object's behavior is dependent on the state of the disk controller. To interrogate the state of the controller, the disk object could send a message to the controller and change to an intermediate state waiting for a response from the controller. The controller would send back a message indicating whether it was operational or not. With this information the disk would know whether it could fail. From

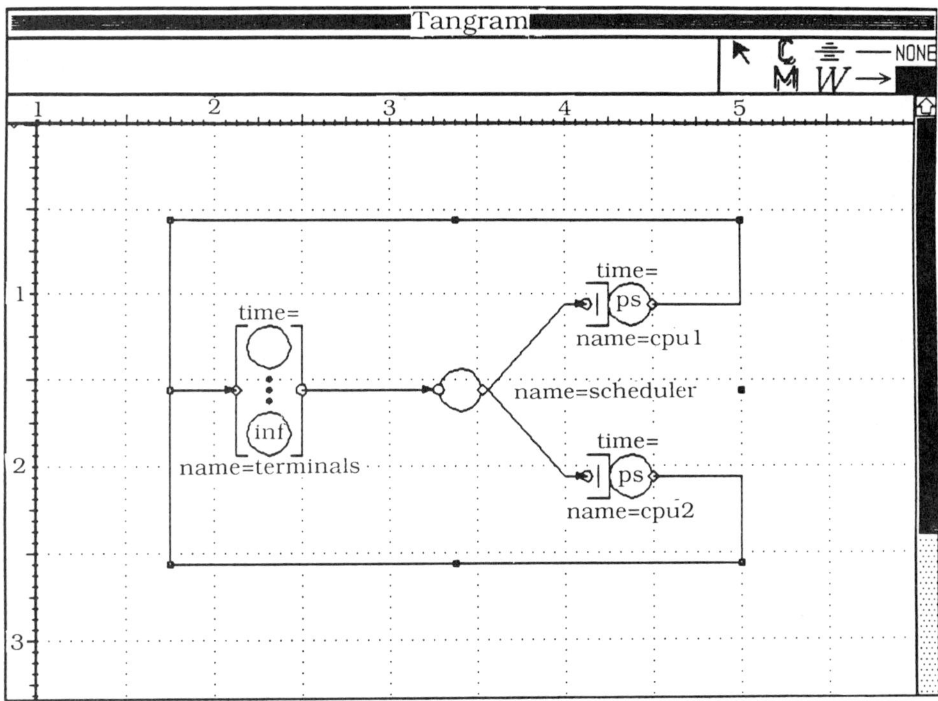

Figure 3 Queuing network example.

our experience, it has proven easier to let the disk inquire about the controller object state without resorting to the message mechanism. We allow functions to be declared, as part of an object's specification, that return information about that object's state. The inquiring object uses the evaluate(object, function, Return) predicate, where object is the destination object, function is the service requested from object and Return is the return value. The object responding to the inquiry must have declared the function.

Loops

Similar to generalized stochastic Petri nets (GSPNs), discrete event simulations and other modeling systems that have "zero-time" events, a model can be created that has "infinite" loops of zero-time events.

As an example, consider a queuing system consisting of an infinite server (representing terminals) and two cpus with no waiting room. If a customer arrives at one of the cpus and it is busy, the customer is rerouted to the other cpu with probability p_c and to the terminals with probability p_t. Departures from the terminals are routed with equal probability to either cpu. This model has an infinite loop. If both cpus are busy and a customer arrives at one of them, it is routed to the other cpu with probability p_c. The second cpu may reroute it back again with probability p_c (total probability p_c^2). With some probability the customer can keep bouncing from one cpu to the other. More specifically, if there are three customers in the system and the system state is represented as the number of customers in the terminals and cpus respectively (plus messages in transit), then the state transitions would be:

$$\langle 1,1,1\rangle \longrightarrow \langle 0,1,1\rangle\,[cpu_1\ arrival] \xrightarrow{p_c} \langle 0,1,1\rangle\,[cpu_2\ arrival] \xrightarrow{p_c}$$
$$\langle 0,1,1\rangle\,[cpu_1\ arrival] \xrightarrow{p_c} \cdots$$

As all possible transition sequences must be examined in the direct implementation, an infinite number of sequences have to be examined. This problem can be resolved by noticing that the state of the system when the customer arrives at the first cpu from the terminals is identical to the state of the system when the customer is rerouted from the second cpu back to the first. When an event causes a transition to a vanishing state, a transition probability matrix is constructed containing every state reachable in zero time. This transition probability matrix can be solved to determine the rates between tangible states. See [MARS84] for more details.

Ordering of Messages

The order in which messages are delivered can be significant, as illustrated in the following example of a computer system model in which components can fail and be repaired. In addition to failing themselves, the failure of one component may cause another component to fail. The model consists of cpu and disk components where a disk failure causes a cpu to fail. The repair service discipline is first-come first-serve. When a disk fails, a message is sent to the cpu to tell it to fail and a message is sent from the disk to the repair service to request repair. When the cpu receives the fail message, it also sends a message requesting repair. These massages are all sent in zero time. Even though the messages from the cpu and from the disk requesting repair both arrive at the repair service in zero time, if the message from the cpu comes to the repair service first, then the cpu would get repaired first; otherwise the disk would get repaired first. The order of messages determines the order that the components are repaired, even though the messages are nominally delivered at the same time.

In terms of understanding the behavior of a model, this is the weakest point of our paradigm. However, this problem is not unique to our system but rather is shared with every other modeling system with zero time events.

If needed, this ambiguity is detectable automatically by the system, but in general will be very expensive. To detect ambiguity, all permutations of messages must be tried. If two permutations of messages derive different tangible states, there is an ambiguity. However, many standard types of objects are not affected by the order in which messages arrive, for example, a processor sharing server is not affected by the order in which customers arrive, and neither is a priority server. In addition, as long as no object that is sensitive to message order receives more than one message in zero time, there can be no ambiguity. This will usually be the case. In those other cases where there is an ambiguity, it should be noted that any ordering of messages can be specified by the modeler.

4. EXAMPLES

In this section we present several examples that further illustrate use of the system. In the first example, we show the ease of using predefined library objects along with user-defined objects in a model. The second example is a processor-bus-memory example drawn from the Petri net literature. The third example is of a more complex condition that could be used in a model. A final example shows an example query of analysis results. These examples show some of the power of the approach.

4.1 Queuing Network Example

In this example we demonstrate the simplicity of building new object types and using them in conjunction with object types from a library. We emphasize the ease with which new objects can be created and used in conjunction with predefined objects. This model is of a simple load balancing system as shown in Figure 3. The model is in two parts; the first part instantiates objects from predefined types, and the second defines a new object. The type statements establish the basic objects, a set of terminals using the infinite server queuing discipline, and two cpus using the processor sharing queue discipline. The *initial* statements, *route* statements, and *departure-rate* statements fill in the necessary parameters for the objects. The other object needed is the scheduler. Assuming that no such object type exists in the library, the scheduler must be defined specifically for this model.

The scheduler tries to maintain an up to date view of the system by getting new state information from the two cpus. To do this, it samples the queue lengths of the two cpus at intervals that are exponentially distributed with mean $1/ur$. When an arrival comes to the scheduler from the terminals, it is sent to the cpu the scheduler believes has the lowest load. If the scheduler believes the loads are equal, it chooses one cpu or the other with probability ½.

The scheduler has a state that is a 2-tuple (n_1,n_2), where n_i is the queue length last supplied from cpu_i. The initial state has zero customers in each cpu. The only event that the scheduler generates is the update event, which gets the currents state information from both cpus. The scheduler also receives arrival messages from the terminals. Upon receipt of an arrival, the scheduler compares the queue lengths of the two cpus in its most recent load update. If they are the same, one cpu of the other is chosen with probability ½, and an arrival message is sent to that cpu. If they are different, an arrival is sent to the cpu that apparently has the smallest number of customers.

```
type(terminals,inf).
type(cpu1,ps).
type(cpu2,ps).

initial(terminals,3).
initial(cpu1,0).
initial(cpu2,0).

route(terminals,scheduler,1.0).
route(cpu1,terminals,1.0).
route(cpu2,terminals,1.0).

departure-rate(terminals,tr).
departure-rate(cpu1,cr).
departure-rate(cpu2,cr).

OBJECT: scheduler
INITIAL STATE: [0,0]
EVENT update: DONT-CARE → [N1,N2]
        ACTION: evaluate(cpu1,state,N1),
                evaluate(cpu2,state,N2)
        RATE: ur
MESSAGE terminal-arrival:[M,N] → [M,N]
        CONDITION: M = N
        ACTION: send(cpu1,arrival)
        PROBABILITY: 1/2
        CONDITION: M = N
        ACTION: send(cpu2,arrival)
        PROBABILITY: 1/2
        CONDITION: M < N
        ACTION: send(cpu1,arrival)
        CONDITION: M > N
```

ACTION: send(cpu2,arrival)

4.2 Multiprocessor System Model

This next example is from [MARS84] and is a generalized stochastic Petri net (GSPN) model of a multiprocessor system with multiple buses and multiple common memories. We describe this model using our methodology and compare it with the GSPN model. This particular example is a five processor, three common memory, two bus system as in Figure 4. The GSPN model for this system is shown in Figure 5.

The following description of the system is also from [MARS84]:

> In this model, the processors execute in their private memory for an exponentially distributed random time with mean $1/\lambda$ before issuing an access request directed to one of the common memories in the system. The request may not be immediately served, either because there is no bus available or because the addressed memory is busy. The durations of access to common memories are independent, exponentially distributed, random variables with mean $1/\mu$.
>
> The following bus arbitration policy is assumed: if a processor, say A, is in the queue for a memory already accessed by another processor, say B, then at the end of the access the bus is not released but is immediately given to processor A.

Using our methodology, the system is modeled as three objects, CPU representing the processor, CM the common memories, and B the buses. The

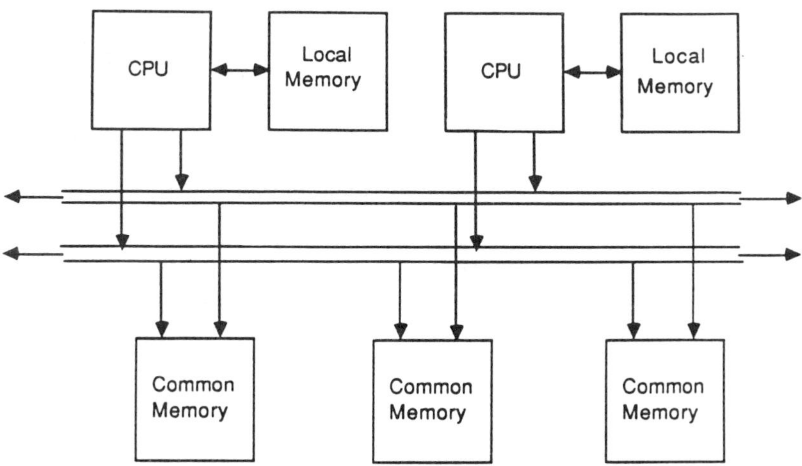

Figure 4 Multiprocessor system model.

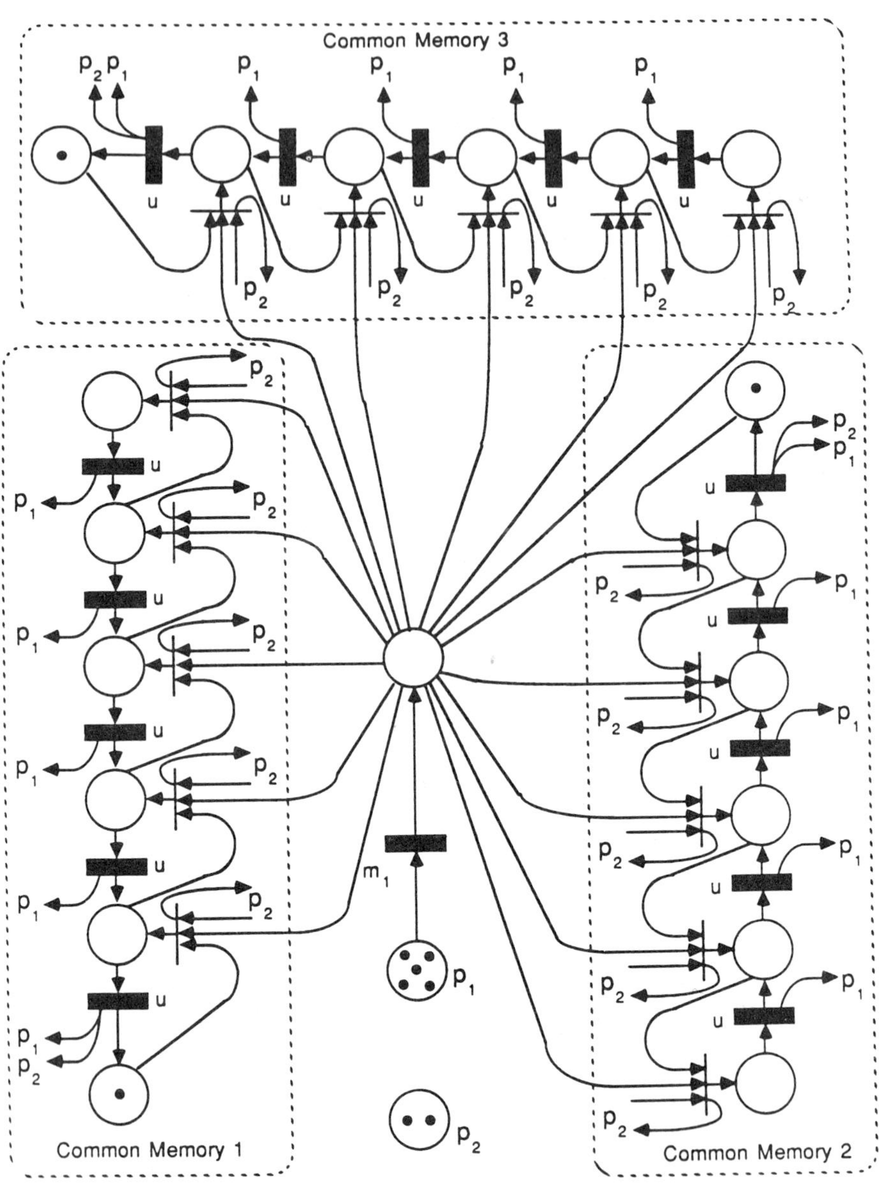

Figure 5 GSPN model.

CPU object state is just the number of cpus executing in their private memory. The bus state is the number of free buses when positive, the number of customers waiting for buses when negative. The state of the common memory is a 3-tuple representing the number of processors waiting for or accessing each memory. Initially all five cpus are executing in their local memory, both buses are free, and no cpu is trying to access a common memory.

The description of the CPU object is very straightforward. When there are $N > 0$ cpus executing in their private memory, they make an access to common memory at a rate of $N * \lambda$. A request to acquire the bus is sent and the state of the cpu object is decremented. The cpu object can receive one message, *process*, the result of a cpu finishing its accessing of a common memory. The state is then incremented as the processor goes into its private memory execution phase.

The bus object implements the arbitration policy. It generates no events but handles two messages, *acquire* and *release*. The handling of the acquire message depends on whether there is a free bus, $N > 0$ or not, $N \leq 0$. If a bus is free, the request is immediately granted by decrementing the number of free buses and sending the *access* message to the memories. If no bus is free, the number of free buses is decremented representing the number of common memory access requests waiting for a bus. Handling the *release* message is similar to *acquire*. If there are requests waiting for a bus, $N < 0$; when one is released, the number waiting is decremented and an *access* is sent to the memories.

The memories are described in two different ways. The event *release* is handled in an initially simpler fashion, explicitly spelling out the transitions. The *access* message is handled in a more general way, but a way that seems more confusing at first. Normally *access* and *release* would both be described in the general way.

The memories object generates a release event at rate μ for each memory being accessed. If there are no more cpus queued for that memory, ($N = 1, O = 1$, or $P = 1$), that memory is decremented, the cpu is sent the process message, and the bus is sent the release message. If on the other hand, other cpus are queued for that memory, ($N > 1, O > 1$, or $P > 1$), the queue at that memory is still decremented and the cpu is still sent the process message, but the bus is not released.

The *access* message is very similar. It is handled by using a substitute (subst( )) that nondeterministically picks one element in the memories state such that it meets the rest of the condition. If there are already cpus queued at the selected memory ($N > 1$), the queue size is incremented and the bus is released. If the memory is free ($N = 0$), the number of cpus at that memory is set to 1. Not only is this more general, it is also more compact.

OBJECT: cpu
INITIAL STATE: 5
EVENT access-common: $N \rightarrow N1$
CONDITION: $N > 0$
ACTION: N1 is N−1, send(bus,acquire)
RATE: $N * \lambda$
MESSAGE process: $N \rightarrow N1$
ACTION: N1 is N+1

OBJECT: bus
INITIAL STATE: 2
MESSAGE acquire: $N \rightarrow N1$
CONDITION: $N \leq 0$
ACTION: N1 is N−1
CONDITION: $N > 0$
ACTION: N1 is N−1, send(memories,access)
MESSAGE release: $N \rightarrow N1$
CONDITION: $N < 0$
ACTION: $N1$ is $N + 1$, send(memories,access)
CONDITION: $N \geq 0$
ACTION: N1 is N+1

OBJECT: memories
INITIAL STATE: [0,0,0]
EVENT release: [N,O,P] → [N1,O1,P1]
CONDITION: $N > 1$
ACTION: N1 is N−1, O1=O, P1=P, send(cpu,process)
CONDITION: $O > 1$
ACTION: N1=N, O1 is O−1, P1=P, send(cpu,process)
CONDITION: $P > 1$
ACTION: N1=N, O1=O, P1 is P−1, send(cpu,process)
CONDITION: $N = 1$
ACTION: N1 is N−1, O1=O, P1=P, send(cpu,process), send(bus,release)
CONDITION: $O = 1$
ACTION: N1=N, O1 is O−1, P1=P, send(cpu,process), send(bus,release)
CONDITION: $P = 1$
ACTION: N1=N, O1=O, P1 is P−1, send(cpu,process), send(bus,release)
RATE: μ
MESSAGE access: $M \rightarrow M1$
CONDITION: subst(N,M,N1,M1), $N > 0$

ACTION: N1 is N+1, send(bus,release)
CONDITION: subst(N,M,N1,M1), $N = 0$
ACTION: N1 is N+1

We see two main advantages in our approach over GSPNs. The first is the modularity in describing each object separately and only letting objects interact in one stylized way. This has been discussed earlier. The second advantage is in the generality of the models.

To accommodate an increase in the number of common memories in the processor-bus-memory GSPN model, additional places and transitions must be added. The number of places and transitions needed for this model is linear in the number of memories. As is shown in [MARS84] this linear growth of places and transitions with respect to the number of memories can be avoided by skillfully reorganizing the model. But this is done by making the number of places and transitions linear in the number of buses in the model. In addition, changing from a memory-dependent model to a bus-dependent one is nontrivial and in our opinion prone to error. In our model, changing the number of bus, cpu, or memory components involves only changing the initial state. For example, a 10 cpu, six memory, four bus system, would only require changing the cpu initial state from 5 to 10, the bus state from 2 to 4, and the memory state from $[0, 0, 0]$ to $[0, 0, 0, 0, 0, 0]$.

4.3 Complex Conditions

Another illustration of the flexibility of our system concerns the manner in which complex conditions and control schemes can be incorporated into a model. For example, in reliability models, a user might have a repair policy that gives priority to any single component that if repaired would make the system operational. If no single component being repaired would make the system operational, the repair policy would be a user-defined priority. This kind of behavior can be very difficult to model in other systems because of its ad hoc nature. Note that any one specific example could be incorporated in a system—the real trick is to allow ad hoc rules to be specified. In this section are presented two examples to illustrate these points.

The following two examples are written in Prolog. Prolog gives the user much more expressiveness than the stylized interface language. These predicates written in Prolog can then be used directly by the high level language in the CONDITION statement or ACTION statement.

Data Availability Modeling Example

This simple example should serve to illustrate the ease of representing a useful, nontrivial system model feature in Prolog. Suppose we have a distributed

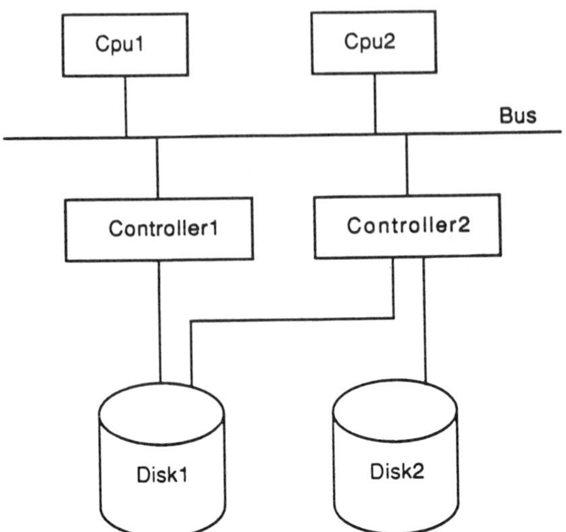

Figure 6 Distributed architecture model.

architecture model as illustrated in Figure 6. The "connectivity" between components can be represented by a set of Prolog "facts."

```
connected(State, cpu1, bus) :- up(State, cpu1), up(State, bus).
connected(State, cpu2, bus) :- up(State, cpu2), up(State, bus).
connected(State, bus, controller1) :- up(State, bus), up(State, controller1).
connected(State, bus, controller2) :- up(State, bus), up(State, controller2).
connected(State, controller1, disk1) :- up(State, controller1), up(State,
   disk1).
connected(State, controller2, disk1) :- up(State, controller2), up(State,
   disk1).
connected(State, controller2, disk2) :- up(State, controller2), up(State,
   disk2).
```

These rules state that a direct data path exists between the named components if both components are operational ("up") in the current state.

A rule that defines the existence of a data path between two components can be given by the following rules.

```
path(State,X,Y) :- connected(State,X,Y).
path(State,X,Y) :- connected(State,X,Z),path(State,Z,Y).
```

The first rule states that there is a path from "X" to "Y" if "X" and "Y" are directly connected. The second rule states that there is a path from "X" to

"Y" if "X" is directly connected to some component "Z" and there is a path from "Z" to "Y."

Now suppose that critical data are replicated as described by the following facts:

```
copy(d1, disk1).     % copy of data item d1 on disk1
copy(d1, disk2).     % copy of data item d1 on disk2
copy(d2, disk2).     % copy of data item d2 on disk2
```

A rule can be used to define the availability of a data path to at least one copy of each data item:

```
data-available(State) :- data-available(State,d1), data-available(State,d2).
data-available(State,d1) :- copy(d1,Disk), cpu(CPU), path(State,CPU,Disk).
data-available(State,d2) :- copy(d2,Disk), cpu(CPU), path(State,CPU,Disk).

operational(State) :- data-available(State).
```

The first rule states that all data are accessible if both data items "d1" and "d2" are accessible. The next two rules state the conditions for availability of each data item. For example, "d1" is available if there is a disk containing a copy of "d1" and a processor with a path from the processor to the disk. The last rule says that the system is operational if all the data are available. The operational predicate would typically be used in a condition statement.

The above description is one example of how relationships and rules of behavior can be represented relatively simply in Prolog.

"Trap States"

There are cases for which it is desirable to define trap states for the model. For example, in availability models one would want to define the conditions under which the system is considered to have failed and have one trap state associated with a failed system. It is also convenient to be able to truncate the state space by not allowing more than some number of failures. These cases are handled by allowing the definition of trap states. The trap predicate allows the user to define what system states correspond to trap conditions. If we want to trap on states with two failures in the examples in Section 3 we would add:

trap(State) :- member([repair-facility, $[R_1, R_2]$], State).

While searching the state space, a transition to a state S for which trap(S) is true can be automatically changed to a transition to a state *trap-state*.

4.4 Querying the Analysis Results

So far, we have concentrated on generating the Markov chain. Another feature of our prototype is the query language. If interested in equilibrium state information, a user is generally interested in a more compact answer such as the mean queue length at a particular object, or the marginal probability that an object will be in a particular state. To answer this type of query, we allow a user to associate a "reward" with each state. These states are specified by rules similar to "ACTION" rules. To calculate the result, the system simply sums up the product of the state probabilities and the associated rewards.

As an example, consider the load-balancing scheduler system model of Section 4.1. Suppose we are interested in the percentage of time that the system is in an unbalanced state, that is, states in which the number of customers at one cpu has at least two customers more than the other cpu. A reward of 1.0 is associated with each state that has this property. This is specified as follows.

QUERY: unbalanced
OBJECT: cpu1
STATE: N1
OBJECT: cpu2
STATE: N2
CONDITION: $N1 - N2 > 1$
REWARD: 1.0
CONDITION: $N2 - N1 > 1$
REWARD: 1.0

The actual query would be:

query(unbalanced, Result).

For particular models of libraries, a user might have many queries predefined in a library similar to an object library.

5. CONCLUSIONS AND FUTURE WORK

We have described a methodology for the specification and generation of Markov models. This methodology allows a modeler to define models in a convenient and powerful way. Further, it allows high-level language tools to be easily built for specific modeling applications. Finally, we use a generative approach so that only reachable states are generated.

The extensibility of the systems is easily seen. A model can easily be defined that incorporates object definitions from a library of predefined objects with new objects that the modeler wishes to define.

The system is also easy to extend. It took only a few hours to define the "objects" for availability modeling of the same order of sophistication as the SAVE system (System AVailability Estimator, a state-of-the-art availability modeling package from IBM). The same level of effort was needed to define a set of objects for extended queuing network models. In addition we have used the system to define various ad hoc models (e.g., a priority queuing system in which the low-priority customers received service after waiting for some number of high-priority customers). It has also been used to generate the Markov chain for the models in [MUNT89]. To our knowledge, the ease with which the interface was tailored to these application domains is not possible with other systems.

Future work consists of expanding the types of matrix techniques available (e.g., allowing heuristics to determine where to truncate a Markov chain, allowing a user to specify how a model to aggregate a model) and expanding the types of numerical techniques available (e.g., transient analysis and embedded Markov chains).

REFERENCES

[BAVU87] S. J. Bavuso, J. B. Dugan, K. S. Trivedi, E. M. Rothmann, and W. E. Smith, Analysis of typical fault-tolerant architectures using HARP, *IEEE Trans. Reliability*, Vol. R-36, No. 1, June 1987, 176–185.

[BUTL86] R. W. Butler, An abstract language for specifying Markov reliability Models, IEEE *Trans. Reliability*, Vol. R-35, No. 5, December 1986, 595–601.

[CARR86] J. A. Carrasco and J. Figueras, METFAC: Design and implementation of a software tool for modeling and evaluation of complex fault-tolerant computing systems, *Proc. FTCS-16*, pp. 424–429, July 1986.

[COST81] A. Costes, J. E. Doucet, C. Landrault, and J. C. Laprie, SURF: A program for dependability evaluation of complex fault-tolerant computing systems, *Proc. FTCS-11*, pp. 72–78, June 1981.

[GOYA86] A. Goyal, W. C. Carter, E. de Souza e Silva, S. S. Lavenberg, and K. S. Trivedi, The system availability estimator, *Proc. FTCS-16*, pp. 84–89, July 1986.

[IRAN71] K. B. Irani and V. L. Wallace, On network linguistics and the conversational design of queueing networks, *J. ACM*, Vol. 18, No. 4, pp. 616–629, October 1971.

[MAKA82] S. V. Makam and A. Avizienis ARIES 81: A Reliability and life-cycle evaluation tool for fault tolerant systems, *Proc. FTCS-12*, pp. 276–274, June 1982.

[MARS84] A. M. Marsan, G. Conte, and G. Balbo, A class of generalized stochastic Petri nets for the performance evaluation of multiprocessors systems, *ACM Trans. Computer Systems*, pp. 93–122, May 1984.

[MUNT89] R. R. Muntz, E. de Souza e Silva, and A. Goyal, Bounding availability of repairable computer systems, *Proc. Sigmetrics*, pp. 29–38, May 1989.

[SAHN87] R. A. Sahner and K. S. Trivedi, Reliability Modeling using SHARPE, *IEEE Trans. Reliability*, Vol. R-36, No. 2, June 1987, 186–193.

[STEW88] W. J. Stewart, MARCA: MARkov Chain Analyzer, *North Carolina State University Technical Report TR-88-32*, October 1988.

[WALL72] V. L. Wallace, Toward an algebraic theory of Markovian networks, *Symp. Computer-Communications Networks and Teletraffic*, pp. 397–407, April 1972.

3

MARCA: Markov Chain Analyzer, A Software Package for Markov Modeling

WILLIAM J. STEWART Department of Computer Science, North Carolina State University, Raleigh, North Carolina

1. BASIC CONCEPTS AND TERMINOLOGY

MARCA (MARkov Chain Analyzer) is a software package designed to facilitate the generation of large Markov chains and to compute transient probability distributions of the chain at different times as well as its stationary probability vector. It was originally developed in 1973 while the author was completing his Ph.D. dissertation at the Queen's University of Belfast in Northern Ireland. Later, while the author was at the University of Rennes in France, a modified version of the package was developed and incorporated into QNAP (Queuing Network Analysis Package), a sophisticated package that incorporates many different types of methods (exact analytical methods, approximate methods, simulation, etc.) for the solution of queuing models. MARCA still forms the basis for the numerical solution approach in the most recent versions of QNAP.

When developing Markov chain models, it is important to have a framework in which the system being studied may be represented. One commonly used approach is to represent the system as a network of queues and servers whose state at any moment describes the state of the system at that moment. For maximum flexibility, MARCA employs the concept of *balls and buckets*.

Consider, for example, the case of several processes competing for a set of resources. The resources are effectively on loan to the process, to be returned when asked for by the operating system, or when the process has finished needing them. Unused resources are kept in a free resource pool. In terms of balls and buckets, a unit of resource may be represented as a ball, and a process or resource pool as a bucket. The movement of balls among the buckets represents the behavior of the resources as they are allocated and deallocated among the processes. At any instant the state of the model is determined by the number of balls in each bucket, just as the state of the computer system is, at any instant, given by the amount of resource that each process possesses, and the amount that remains idle in the resource pool.

Alternatively, a process may be represented as a ball, and a unit of resource as a bucket. In this case the number of balls in a bucket represents the number of processes using, or queuing to use the resource. Depending on the formulation of the problem and on the particular results required, other representations are possible. It is due to this flexibility that the ball and bucket concept has been chosen to provide the basis of the modeling techniques employed by MARCA. Using a queuing network formalism is too restrictive, for not all Markov models can be conveniently structured as queues and servers. However, once a state description for a Markov chain has been developed, it is relatively simple to transcribe this description in terms of balls and buckets. All systems that are analyzed by the software package MARCA must be formulated in terms of balls and buckets.

The movement of a ball from one bucket to another is called a *transition* from a *source bucket* to a *destination bucket*, and the *rate of transition* is defined as the rate at which the source bucket loses balls to the destination bucket. The rates will obviously vary from system to system and from bucket to bucket. A transition of particular interest is an *instantaneous transition*, which means, as its name implies, that the actual movement takes no time at all. An example of this may be that as soon as a bucket becomes empty of balls, one is immediately taken from another bucket and given to it. This instantaneous transition occurs at the instant the bucket becomes empty. Such transitions usually serve to implement scheduling policies.

The *state of the system* is represented in row vector form, the ith element of which denotes the number of balls in the ith bucket at the moment of interest. Thus, at any particular moment, the model whose state is given by (a,b,c,d) has a balls in bucket 1, b balls in bucket 2, c balls in bucket 3, and d balls in bucket 4. Suppose that while in state (a,b,c,d), a transition involving one ball occurs from bucket 2 to bucket 4, then state $(a,b-1,c,d+1)$ is called the destination state, state (a,b,c,d) being the source state. The rate of transition from source state to destination state is just the rate at which bucket 2 loses balls to bucket 4. Further, suppose that state $(a,b-1,c,d+1)$ occasions an instantaneous transition from bucket 1 to bucket 2;

$$(a,b,c,d) \longrightarrow (a,b-1,c,d+1) \xrightarrow{\infty} (a-1,b,c,d+1).$$

then $(a-1,b,c,d+1)$ becomes the destination state, and the rate of transition from state (a,b,c,d) to state $(a-1,b,c,d+1)$ equals the rate at which bucket 2 loses balls to bucket 4.

2. THE FUNCTIONAL COMPONENTS OF MARCA

The software package MARCA is written in Fortran and may be considered as consisting of three parts or segments;

1. A master segment, which controls the operation of the package.
2. A section that is responsible for generating the states of the system and the infinitesimal generator matrix.
3. A section that derives and analyzes the stationary probability vector and the transient probability distributions.

2.1 The Master Segment

The master segment has several purposes. Array sizes are specified here by means of parameter statements. Dummy arguments are employed in the remaining subroutines so that changes to array sizes need only be made in this segment. It is this segment that reads the input data, checks it for inconsistencies, and also writes it to the output file. If errors are detected, the checking procedure is applied to the remaining data before the package aborts the run. If the master segment detects no errors, it then initiates other subroutines to generate and analyze the model. The first routine it calls is STATEMATX, which generates the system states and transition probability matrix.

2.2 Generation of System States and Transition Matrix

It is the purpose of the subroutine STATEMATX to generate all the states of the system, to store them in a one-dimensional array called LIST, and to construct the infinitesimal generator, the matrix of transition rates. This is the matrix whose ijth element denotes the rate of transition from state i to state j. Due to the nature of this matrix (very large and very sparse), it is necessary, and indeed efficient, to store it in condensed form. The nonzero elements of the infinitesimal generator matrix are stored by *rows* in a one-dimensional real array, A. The column position of each nonzero element is stored in an integer array, JA; the kth element of JA denotes the column position of the nonzero element that is stored in the kth position of A. Finally, a third integer array, IA, keeps track of the beginning positions of the elements of each row in the arrays A and JA; the kth element of IA denotes the position in A and JA at which the nonzero elements of the kth row begin.

The list of states is built as follows. An initial feasible state supplied by the user is examined to determine which states it can reach in a single step transition. For each destination state, *newstate*, a subroutine written by the user and called RATE is used to determine the rate at which this transition occurs. If the transition rate is zero or negative, the destination state is ignored. On the other hand, if the transition rate is strictly positive, a second user-written subroutine called INSTANT is called to see if an instantaneous transition(s) may be made from *newstate*. If so, the result of the instantaneous transition becomes *newstate*, so that it is now a single step transition from *currentstate*. The list of states is now searched, by a subroutine called SEARCHLIST, to determine if it already contains *newstate*. If *newstate* is not in the list of states, then it is added to the end of LIST and a pointer, LIMIT, which is set equal to the number of states in the list, is incremented by one. The rate of transition obtained by RATE is now stored in the array A and its position in arrays JA and IA.

When all destination states that emanate from *currentstate* have been so treated, the diagonal element is computed as the negated sum of the rates and is also stored. The next state in LIST, which is flagged by a variable MARKER, becomes *currentstate* and the process repeated. When the value of MARKER exceeds that of LIMIT, all the states will have been generated.

The list of states and the arrays that contain the transition matrix are now returned to the master calling routine. Error messages will be issued by the subroutine if insufficient storage has been allocated to the array that is used to store the list of states and/or the arrays that store the nonzero elements of the transition rate matrix. Some information is also printed to help the user estimate how much more storage is likely to be needed. Finally, MARCA will print a warning message if it detects a state from which there is no transition.

2.3 Computation and Interpretation of the Numerical Solutions

After the states of the system and the matrix of transition rates have been generated, MARCA tests the system for near-decomposability. This is accomplished by temporarily setting to zero those nonzero elements of the stochastic transition probability matrix that are less than some threshold, γ, and then applying an algorithm to determine the strongly connected components of the resulting matrix. MARCA performs this for a range of values of γ and produces a table that displays the number of disjoint groups, their average size, and their standard deviation. If the number of groups is not large (less then 50), it will also list the size of each group. This information may be useful to the user in choosing a particular solution method.

Control is now passed to one of several numerical solution procedures for the determination of the stationary probability vector and transient distributions. Currently, in MARCA there is a variety of different types of algorithm for determining the stationary probability vector: direct methods, fixed-point iteration methods, subspace iteration methods, and combinations of these. A particular method must be chosen by the user. In future implementations, it is envisaged that this choice could be made automatically by MARCA, but for the moment this is still one of our research objectives.

When the stationary probability vector has been generated, control in MARCA is passed to a subroutine, called PROBAB, to determine the probability distribution of balls in each of the buckets. It also calculates the mean number of balls in each bucket and the standard deviation of the distribution.

After computing the marginal probability distributions, MARCA enters a user-supplied subroutine called SELECTPI, which gives the user access to the stationary probability vector and other model parameters. The user may wish to manipulate this data to compute model characteristics other than the marginal probabilities just described.

The numerical method used to compute the transient distribution is the randomization method developed by Grassmann [GRA77]. The user is requested to input a time at which the probability distribution is required, a beginning state, and an accuracy parameter. The transient distribution is computed at that time, and from it, the marginal probabilities. The user may request the probabilities of different states, or the transient solution at a different point of time. When this is finished, MARCA stops executing.

3. INTERACTING WITH MARCA

To analyze a particular model, the software package must be supplied with the following information

1. The number of buckets, L.
2. The maximum numbers of balls that each bucket can hold.
3. An initial state, (i.e., an initial value for each bucket).
4. A list of all the transitions that can occur among the buckets.
5. The information necessary to determine the rates of transition among all states. This is supplied in the form of a Fortran subroutine called RATE.

6. The information necessary to detect a state that yields an instantaneous transition, and to determine to what state the transition is destined. This is supplied in the form of a Fortran subroutine called INSTANT.
7. Common data that might be needed during execution of the subroutines RATE and INSTANT.
8. The type of output listing required for the state space.
9. Information pertaining to the choice of a numerical solution method.

In order to be able to supply this information, the user must have a clear idea of the model of his system expressed in the concepts of *balls and buckets*. Given this information, MARCA generates the state space and the infinitesimal generator matrix. From this matrix it computes the stationary probability vector and derives the marginal probability distribution of each of the buckets. This, however, may not be sufficient information for the user. A user may wish to examine or manipulate the stationary probabilities belonging to a very specific subset of states. A facility is included in MARCA to allow such an interaction. Before terminating, MARCA calls a subroutine called SELECTPI. This subroutine gives the user access to the list of states, the stationary probability vector, the hashing functions used to find the position of a given state in the list of states, and both double-precision and integer work arrays. The user is responsible for writing this subroutine in order to extract whatever information he requires. Examples in the use of SELECTPI are provided later.

3.1 The Input Data File

All of the above items, with the exception of 5 and 6, are read directly by MARCA from an input file called fort.12. We shall consider them separately.

1. The number of buckets, L, is the first quantity to be input, since this determines the number of values the package should expect in each of the remaining sections of the input date; for example, if there are nine buckets, then the package expects nine maximum values to be input, initial values for each of nine buckets, etc. The parameter statements in the current version of MARCA are such that the maximum number of buckets that can be handled, LMAX, is 20. This may, of course, be altered by the user. A check is performed by MARCA to ensure that $1 \leq L \leq \text{LMAX}$.
2. The maximum value of each bucket is input to a one-dimensional integer array called MAX, the ith element of which contains the maximum value of the ith bucket.
3. The third quantity to be input is an initial state. This is obtained by assigning to each bucket a specific value that is within the capacity of that bucket, and such that, when the L values are taken together, they satisfy

all the constraints of the system; in other words, the initial state must be a realizable state of the system. The L values are stored in the first L positions of a one-dimensional integer array LIST. A test is carried out to ensure that $0 \leq \text{LIST}(i) \leq \text{MAX}(i)$ for $i = 1, 2, \ldots, L$.

4. The interbucket transitions are input in the following manner. Each bucket is considered in turn as a source bucket, and the destinations of all transitions which emanate from it (i.e., the destination buckets) are listed. This list, however, must be preceded by the number of such transitions. Thus if transitions are possible from bucket i to each of three buckets j, k, and l, this information must be input as

Source bucket	Number of transitions	Destination buckets
i	3	$j\ k\ l$

It is not necessary for j, k, or l to be any particular order, nor for the source buckets to be in order. However, from the user's point of view it is probably better to maintain some measure of order to facilitate checking, etc. Checks are performed by the package to test that no bucket has transitions to more than L buckets and that the indices of source bucket and destination bucket lie between 1 and L. When no transitions occur from a source bucket (we shall see that this is a useful device in certain modeling problems), it is best to specify that, from the source bucket, a single transitions back to the source bucket again may occur.

7. It is believed that in certain cases it may be useful for the user to be able to have access to common data items in the writing of the subroutines RATE and INSTANT. MARCA provides a mechanism to facilitate such access. Two arrays are placed in *common blocks*: one, a double-precision array called RCOMM, is kept in a common block called CBLCK1, and the other, an integer array called ICOMM, is kept in CBLCK2. The user needs to specify the number of elements in each array. This means that two zeros should be entered if no common data are required. On the following lines, the user should enter the data items for which he requires access in RATE or INSTANT. The list of double-precision data items should be given first. An example in the use of the common block storage is provided later.
8. Three options for listing the generated state space are available. Level 0 simply lists the first 10 states generated by the package. Level 1 lists all

the states that are generated. Level 2, in addition to listing all the states, also outputs, for each source state, the destination states and the rate of transition.

9. The particular numerical method chosen for the computation of the stationary distribution should depend on the insight that the user has on his system being modeled. Certain parameters pertinent to the solution methods must also be assigned numerical values.

The methods for the stationary distribution that are currently available in MARCA are as follows:

1. SOR (Successive Over Relaxation).
2. SSOR (Symmetric SOR).
3. POWER.
4. ARNOLDI.
5. GE (compact Gaussian elimination).
6. MA28 (a sparse direct method from Harwell).
7. ILUGMR (GMRES with incomplete factorization).
8. SORGMR (GMRES with SOR acceleration).
9. SSORGMR (GMRES with SSOR acceleration).
10. FXPTIT (fixed-point iterations with preconditioning).
11. CGS (conjugate gradient squared).
12. NCDSOLVER (iterative aggregation/disaggregation).

To select a method, the user inputs the number of the method. Thus inputting a 1 results in the SOR method being chosen, inputting a 5 results in the compact Gaussian elimination method being chosen, etc. It is hoped that in later versions of MARCA, the choice of a method can be left to the program itself. This approach is under active study at the moment.

Once the method has been chosen, a number of additional parameters must be specified. These are (the variable names are noted alongside):

a. The precision desired of iterative methods (TOL).
b. The maximum number of iterations permitted (ITMAX).
c. An omega value for all SOR-related methods (OMEGA).
d. The size of the subspace for subspace iteration methods (M).
e. The type of incomplete factorization (ILU) to be used (ILUTYPE).
f. The threshold for threshold-based ILU methods (THRESH).
g. The methods to be used for individual blocks (NCDSOLVER).

Note that not all methods require all these parameters. The direct methods require none at all, SOR requires only the first three, etc. However, MARCA expects values to be input for all six parameters. This is to facilitate moving from one numerical solution method to another, just by changing a minimum number of parameters in fort.12. One exception is that the parameter for the NCDSOLVER may be either input interactively or read from the file. Most of the parameters are self explanatory. Some typical values are

$$\text{TOL} = 10^{-10}, \qquad \text{ITMAX} = 400, \qquad \text{OMEGA} = 1.2,$$

$$\text{M} = 10, \qquad \text{THRESH} = 10^{-3}$$

There are three possible choices for ILUTYPE. The first is called ILU0 and is an incomplete factorization in which the structure of the LU decomposition is forced to be identical to the structure of the infinitesimal generator matrix. The second is called ILUTH, and in this decomposition elements are kept only if they are larger than a specified threshold. The third is ILUK, and here only a fixed number (K) of elements are kept per row.

This is not an appropriate place for an extensive discussion of these methods. Here we shall be content to give some information of a general nature. Further information may be found in [PSS89].

If the number of states generated by the model is not large, less than 1000, say, then it has been observed that method 5, compact Gaussian elimination, performs very efficiently. Above this size, the amount of fill-in generated often becomes too large and an iterative method should be employed. If computer memory is at a premium, then SOR with a reasonably good omega value should be satisfactory. (A good value of omega can often be deduced from tests conducted on similar but smaller models.) In most other cases, it has been found that methods 7 and 8 perform well. The MA28 algorithm was developed by Iain Duff at Harwell and at the moment requires users to contact Harwell for permission to use it. Consequently, the code is not incorporated into MARCA. Rather an interface is in place so that a user who obtains the MA28 Fortran source can simply link in with MARCA and then use this method. However, in many tests it has been found that GE performs better than MA28 because it has been designed specifically for Markov chain problems whereas MA28 has been developed for a very extensive application base. GE is more efficient for Markov chain problems because it does not include the bells and whistles, which may be effective in applications other than Markov modeling. For problems that are nearly completely decomposable (NCD), the method NCDSOLVER has shown itself to be very effective. Finally, the conjugate gradient method, despite the success it has achieved in other domains, has not proven to be very reliable in Markov chain models, and unless the user has special needs this method should be avoided.

3.2 The Fortran Subroutine RATE

This routine is called when, in a certain state, a source bucket L1 and a destination bucket L2 have been found such that

1. the content of L1 is strictly greater than zero.
2. the content of L2 is strictly less that its maximum value.
3. a transfer if permissible from L1 to L2. In other words, such a transition has been defined in part 4 of the input data.

Under such circumstances a transition is possible, and it is the purpose of the subroutine RATE to determine the rate at which this transition occurs and the destination state. It must necessarily be written by the user himself, since only he will be aware of what the transitions are and upon what they depend. A certain transition may, for example, depend upon the state the system currently occupies, or more simply, just upon the number of balls which the source bucket possesses. Instantaneous transitions need not be considered, since they are catered for in a separate subroutine, INSTANT.

In all cases, the initial statements of the subroutine RATE should be

```
SUBROUTINE RATE (L1, L2, ISTATE, INTO, L, R)
DOUBLE PRECISION R
DIMENSION ISTATE(L), INTO(L)
```

where

L1 is the source bucket.

L2 is the destination bucket.

ISTATE is a one-dimensional integer array that contains the current state of the system.

INTO is a one-dimensional integer array that, on exit, contains the destination state for a nonzero transition rate.

L is the number of buckets in the system.

R denotes, on exit, the required rate of transition.

The parameters L1, L2, ISTATE, and L should be left unaltered by the routine. The parameters INTO and R are undefined on entry. Since R is declared to be a double-precision variable in the calling program, it should also be declared as double-precision in this subroutine. If the user requires access to common block data, the subroutine should also contain one or more of the statements:

```
COMMON /CBLCK1/ RCOMM(1)
COMMON /CBLCK2/ ICOMM(1)
```

The common data items input from the file fort.12 are available in the arrays RCOMM and ICOMM. Note that it is only necessary to specify the first element of these arrays in the common statement.

The body of the subroutine may be structured as L different sections, the ith of which considers transitions from the ith bucket. Further, the ith section should again be divided into as many subsections as there are transitions from the ith bucket. Thus when the subroutine is entered, the parameter L1 will determine the section, and parameter $L2$ the subsection to which control should be passed. This may be programmed efficiently by means of computed GOTO statements. The structure of RATE will then be as follows:

```
      SUBROUTINE RATE (L1, L2, ISTATE, INTO, L, R)
      DOUBLE PRECISION R
      DIMENSION ISTATE(L), INTO(L)
      GOTO(1, 2, ....  )  L1

  1          GOTO(11, 12, ....  )  L2
 11          ............
             ............
                    ............
 12          ............
             ............
             ............
  2          GOTO(21, 22, ....  )  L2
 21          ............
             ............
             ............
 22          ............
```

Section L1, subsection L2 contains the programming that determines the rate of transition from source bucket L1 to destination bucket L2. It also stores the required destination state into INTO, and finally returns control to the calling routine.

Note that the above structure for the subroutine RATE should be used only as a guide. In many cases it may be considerably simplified.The computed GOTO statements with integer variable L2 are only useful if there are several possible transitions from a particular source bucket. It is sometimes found that different buckets have the same possible transitions and rates of transition and may therefore be considered together. Also, it is possible to perform some computation that is common to all transitions before sending control to a particular section and subsection. Each model is distinct and must be treated as such. It is, after all, for this reason that the user must write the routine himself.

3.3 The Fortran Subroutine INSTANT

This subroutine is called each time a destination state is obtained by the subroutine RATE. Its purpose is to examine this state and determine if an instantaneous transition occurs from it to any other state. Instantaneous transitions will usually occur as the result of certain scheduling policies imposed upon the system and that the user wishes to consider as taking no time at all to implement. For example, a user may use INSTANT to model the case in which possession of the CPU is instantaneously taken from a low-level process to be given to a higher-priority process that has just requested it. In this case it is known from the state of the system (i.e., the low-priority process computing and the high-priority process waiting), that an instantaneous transition will occur; further, the state to which the instantaneous transition is destined may be easily determined. Consider, however, the case where the CPU has just finished performing a task for a high priority job and finds that not just one, but several jobs having the same priority are waiting for it. It is known in advance that an instantaneous transition will occur—for the CPU will be allocated to one of the waiting processes—but sufficient information may not be available to determine to which process. In this case it is necessary for the subroutine INSTANT to return to the list of all possible destination states (which should not be large) and associate with each a probability that it will be the resultant of the instantaneous transition.

In reality, of course, there are no instantaneous transitions, since even the simplest operations take a very small but still a finite amount of time to perform. The advantage of taking such transitions to be of zero duration is that it considerably simplifies the analysis. In many cases, and perhaps even most, the subroutine RATE is capable of handling all possible transitions. Nevertheless, there are cases in which the use of INSTANT can greatly facilitate the task of constructing a model.

In all cases, the initial statements of the subroutine INSTANT should be

```
SUBROUTINE INSTANT (L, INTO, IIN, PROB, IC)
DOUBLE PRECISION PROB
DIMENSION INTO(IIN), PROB(L)
```

where

L is the number of buckets in the system. It should be returned unaltered by the subroutine.

INTO is a one-dimensional integer array. On entry it contains in the first L positions the state to be tested for an instantaneous transitions. If such a transition is possible, then on exit INTO must contain the resulting state or states.

IIN is an integer quantity, the declared length of the array INTO. It should be returned unaltered by the subroutine.

PROB is a one-dimensional, double-precision array whose ith element must denote on exit the probability that the instantaneous transition will lead to the ith state stored in INTO.

IC is an integer quantity that on exit denotes the number of states stored in INTO.

Once again, if the user requires access to common block data, the subroutine should also contain one or more of the statements:

```
COMMON /CBLCK1/ RCOMM(1)
COMMON /CBLCK2/ ICOMM(1)
```

The common data items input from the file *fort.12* are available in the arrays RCOMM and ICOMM. Note that it is only necessary to specify the first element of these arrays in the common statement.

Perhaps the best way to program INSTANT is for the user to write down the conditions that a state must satisfy before an instantaneous transition can occur. If a state filters through these conditions, then no instantaneous transitions occurs, and in this case IC should be set to zero and control returned to the calling routine. No alterations should be made to the array INTO. If, on the other hand, a state satisfies one of the conditions, then control should be passed to another section of the subroutine where the resulting state (or states) is determined. IC should be set equal to the number of states which can result from the instantaneous transition and these states should then be stored contiguously in the first $L \times$ IC positions of INTO. This implies that the state that originated the instantaneous transition is overwritten. Finally, the respective probabilities of the destination states should be stored in the first IC positions of PROB. Note that if IC = 1, then PROB(1) = 1.0.

3.4 The Fortran Subroutine SELECTPI

As mentioned previously, this subroutine provides the user with access to the states of the system and to the stationary probability vector. This information may be manipulated in any way the user desires. The initial statements should be as follows:

```
SUBROUTINE SELECTPI (L, MAX, N, LIST, LMAX, PI, MHASH,
       NHASH, MMO, ISTATE, MIW, DWORK, MDW )
INTEGER MAX(L), LIST(LMAX), MHASH(0:MMO), NHASH(N),
       ISTATE(MIW)
DOUBLE PRECISION PI(N), DWORK(MDW)
```

where

L is the number of buckets.

MAX stores the maximum value of each bucket.

N is the number of states.

LIST stores the states in contiguous locations.

LMAX is the declared length of LIST.

PI is the stationary probability vector.

MHASH and NHASH are arrays used by the hashing function.

ISTATE is an integer work array.

MIW is the declared length of ISTATE.

DWORK is a double-precision work array.

MDW is the declared length of DWORK.

The array length indicators LMAX, MIW, and MDW are parameters that are set by MARCA at program initiation and may be changed by the user. The contents of the two work arrays ISTATE and DWORK are unimportant. In fact, the user may choose to use more meaningful names for these arrays if that is felt to be appropriate. The remaining variables do contain relevant information on entry. However, since MARCA terminates immediately on exit from SELECTPI, their content on exit is unimportant.

When searching through the list of states, the user may wish to take advantage of the hashing function. To do so, it suffices to execute the statement

```
CALL SEARCHBIS (L, LIST, LMAX, N, ISTATE, MHASH, NHASH,
        MMO, IR)
```

On exit, IR will denote the index of the state contained in the first L positions of ISTATE. It then follows that PI(IR) contains the stationary probability of the state. If the state cannot be found among the list of states, then SEARCHBIS returns the value IR = 0.

3.5 Output from MARCA

After the user has prepared his data (in fort.12), written the Fortran subroutines RATE, INSTANT, and SELECTPI, and submitted the package for execution, then, provided no errors occur, the following information is written to a file called fort.13:

1. The input data.
2. The states generated by MARCA.

3. Information relating to the numerical solution method chosen.
4. The marginal probability (stationary) distributions.
5. Results specified by the user in SELECTPI.
6. The marginal probability (transient) distributions.

The Input Data

This information is output in the same order as that given by the user. It provides a check that the problem solved is the intended problem and that incorrect data has not been supplied. Care should be taken to verify this output, since MARCA cannot detect all input errors and inconsistencies. This data may also be used as a file index, in the case when more than one problem of a similar nature is being solved, as for example when different service rates or numbers of customers are used in otherwise identical models. This output serves as a useful means of distinguishing the different cases at a later date.

The States Generated by MARCA

As stated earlier, three different levels of output are available for the states, ranging from a list of the first 10 states generated to a complete list of all of the states together with all possible states and transition rates to these states from each source state. The states are printed as and when they are generated. The first state is always the initial state supplied by the user.

If the user has already debugged his model and is confident that his interfaces with MARCA are working correctly, then it is best to limit the number of states printed. On the other hand, if the user is developing a new model and wishes to be sure that his subroutines RATE and INSTANT are functioning correctly, it is more appropriate to output the maximum amount of information. During this verification stage, the user should include as few customers as possible to keep the generated output to a manageable quantity.

At this point it is appropriate to discuss the state space generated for the case of completely decomposable systems. Suppose that a system is decomposable into two subsystems A and B, so that there are no transitions from any state of A to any state of B. If the initial state chosen in the input data belongs to A, then the only states generated by MARCA will be the states of the subsystem A. Even if transitions are possible from B to A (but not the converse), and the initial state is in A, none of the states of B will be generated. However, exact decomposability is almost certainly a result of incorrect statement of the input, a very inefficient scheduling algorithm, etc. The user should use the list of states to check that:

All states generated are reasonable.
No reasonable state has been omitted.

If these fail, there has been in all probability an error in the statement of the problem. When the system is known to be decomposable, it is advisable to treat the different parts separately. The reasons for this are threefold. First, there is no difference in the actual results obtained; second, there are no problems with linear dependence to adversely affect the numerical solution procedure; and third, less memory is required to store arrays, and less time is needed to analyze the model, resulting in a more efficient operation of the package.

Information Relating to the Numerical Solution Method

This information depends to a large extent on the method used to compute the stationary probability vector. In all cases, however, MARCA will specify the method that was used along with any parameters associated with the method. For example, with iterative methods, MARCA will always print out the precision requested and the maximum number of iterations permitted. Once the solution has been computed, MARCA will output other relevant information, like the amount of fill-in for direct methods, the accuracy achieved by iterative methods, and so on.

The Marginal Probability Distributions

Each bucket is taken in turn, and the distribution of balls in the bucket, the average number of balls in the bucket, and the standard deviation of the distribution are printed. In most cases this is the information the user requires from MARCA, for with them, he has a quantitative measure of the performance of his system. By careful examination he may detect in which buckets most of the balls spend their time. If this is in buckets representing resources, then the user may deduce high resource utilization. Perhaps, however, one or two buckets monopolize the system, possessing for a large proportion of the time all, or most, of the balls. Such a situation represents a bottleneck, introduced by the scheduling algorithm, or poor performance of a component vis-á-vis the remainder of the system. Buckets that for a large proportion of the time are empty or possess only a few balls obviously indicate that little use is being made of the system components they represent.

4. SOME MODELING STRATEGIES

In this section, we shall consider ways to incorporate certain modeling features in MARCA, features such as Coxian servers and queues with blocking. We wish to stress that we present only one possible way to model these features. The user may wish to experiment with other possibilities.

4.1 Modeling Coxian Servers in MARCA

When modeling a station with a single Coxian server, the two parameters that completely characterize the station are the number of customers at the station and the current phase of service. The most convenient way to do this in MARCA is to associate a bucket with each. Thus the number of balls in the first bucket will denote the number of customers in the station, and the number in the second will indicate the phase of service. We shall call the first bucket the *customer population bucket*, and the second, the *phase indicator bucket*.

The number of balls in the phase indicator bucket is usually unimportant when there are zero customers in the station, but by convention we shall insist that both buckets must be simultaneous empty and simultaneously nonempty. In MARCA, it is not necessary for the number of balls that circulate among the buckets of the model to be a constant value. Balls can be made to appear and disappear at will, for it is the user who describes the destination states that arise from source states, and it is completely within his ability to augment the number of balls in a bucket without having to take a ball from a different bucket. Thus the total number of balls in a model can vary from state to state. This is the case in modeling a Coxian server, for it is important to be able to augment the number of balls, indicating a transition from phase i to phase $i + 1$ without stealing a ball from somewhere else. To handle this in the input data, it is necessary to indicate that a transition can occur from the phase indicator bucket to itself. Transitions should be specified from the first bucket, the customer population bucket, to all possible destination buckets.

An alternative approach is to introduce a bucket for each phase and let a ball propagate through the phase buckets, just as a customer propagates through the different service phases. This however, has the disadvantage of making the Markov chain state description very large and has a detrimental effect on storage requirements.

Note that it is unnecessary to introduce *instantaneous* transitions when modeling Coxian servers. In the subroutine RATE, we need to capture the interphase transitions that can occur in addition to the usual interstation transitions. For an interphase transition, the destination state is obtained from the source state by simply incrementing the phase indicator bucket by 1. MARCA will ensure that this does not occur if the bucket contains its maximum number of balls (i.e., if the server is at its last service phase). The rate of transition should be set to $a_i \mu_i$. In a typical Coxian server it is possible to exit the server from any phase i [with rate $(1-a_i)\mu_i$] and proceed to the next station. When a departure occurs from a Coxian server, it is important to remember to set the phase indicator to 1 if the station is not empty, and to 0 otherwise. If the

Coxian Server

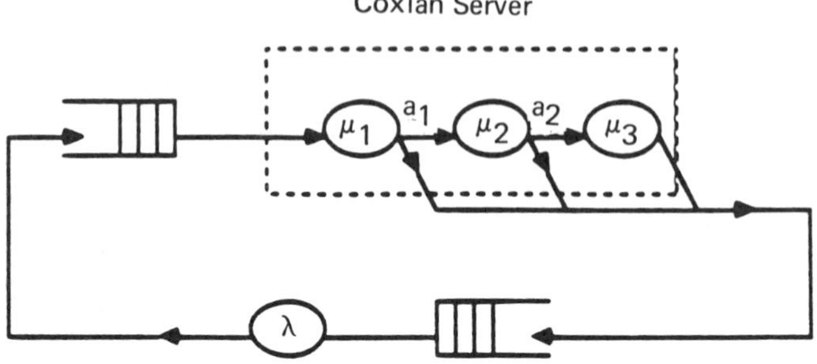


```
      subroutine rate(l1,l2,istate,into,l,r)
      implicit double precision (a-h,o-z)
      dimension istate(1),into(1)
      double precision lambda, mu(3), a(3)
c
c
c         ***   SET UP MODEL PARAMETERS   ***
c
      lambda = 0.5d0
      mu(1) = 1.0d0
      mu(2) = 2.0d0
      mu(3) = 10.0d0
      a(1) = 0.5d0
      a(2) = 0.1d0
      a(3) = 0.0d0
c
c
c         ***   INITIALIZE DESTINATION STATE   ***
c
      do 10 i=1,l
         into(i)=istate(i)
   10    continue
      r=0.0d0
c
c
      goto (1,2,3)l1
      return
c
c
c         ***   DEPARTURE FROM COXIAN SERVER   ***
c
    1 r = (1.0d0 - a(into(2))) * mu(into(2))
      into(1)=into(1)-1
      into(2) = 0
      if(into(1).gt.0)into(2) = 1
      into(3)=into(3)+1
      return
c
c
c         ***   COXIAN INTERPHASE TRANSITION   ***
c
    2 r = a(into(2)) * mu(into(2))
      into(2)=into(2)+1
      return
c
```

```
c
c        ***   DEPARTURE FROM EXPONENTIAL SERVER    ***
c
    3 r = lambda
      into(1)=into(1)+1
      if(into(1).eq.1)into(2) = 1
      into(3)=into(3)-1
      return
c
c
      end


      subroutine instant(l,into,iin,prob,ic)
      implicit double precision (a-h,o-z)
      dimension into(iin),prob(l)
      ic=0
      return
      end
```

Coxian is in fact an Erlang server (with $a_i = 1$ for all i), then a transition cannot occur to the next station until the last phase of service has been reached. In this case, the rate of transition must be set to zero if the service phase is not equal to k in a k-phase Erlang server. We now provide a sample subroutine RATE for the following simple two-station network, one station of which is a Coxian of order three.

4.2 Modeling Hyperexponential Type Servers in MARCA

In servers of Coxian type such as those discussed in Section 4.1, the service process is always begun in phase one. In this section we shall consider the case in which the initial service phase is chosen randomly as in a hyperexponential server.

Again, the two important aspects that must be taken into consideration are the number of customers in the station and the phase of service. As before, we shall employ two buckets, a population size bucket and a phase indicator bucket, to represent such a station. We shall assume that there is only one hyperexponential server in the station. In the input data we again specify transitions from the population size bucket to other buckets according to the model description. We should also include a single transition from the phase indicator bucket back to itself.

In this server there are no interphase transitions. As soon as service is completed at a phase, the customer departs and the next customer in the queue begins service at any one of the possible phases. A convenient way to incorporate this in MARCA is by use of instantaneous transitions as follows. In the subroutine RATE, whenever a customer leaves the station from phase

Hyperexponential—type Station

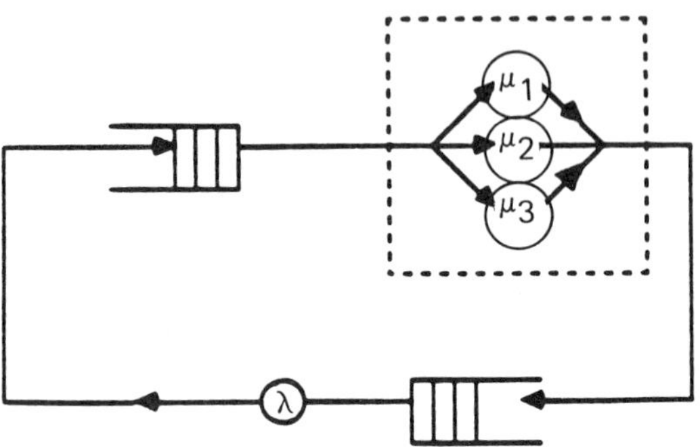


```
      subroutine rate(l1,l2,istate,into,l,r)
      implicit double precision (a-h,o-z)
      dimension istate(l),into(l)
      double precision lambda, mu(3)
c
c
c        ***    SET UP MODEL PARAMETERS    ***
c
      lambda = 0.5d0
      mu(1) = 1.0d0
      mu(2) = 2.0d0
      mu(3) = 10.0d0
c
c
c        ***    INITIALIZE DESTINATION STATE    ***
c
      do 10 i=1,l
         into(i)=istate(i)
   10    continue
      r=0.0d0
c
c
      goto (1,2,3)l1
      return
c
c
c        ***    DEPARTURE FROM HYPEREXPONENTIAL    ***
c
    1 r = mu(into(2))
      into(1)=into(1)-1
      into(2) = 0
      into(3)=into(3)+1
      return
c
c
c        ***    NO INTERPHASE TRANSITIONS POSSIBLE    ***
c
    2 return
c
c
c        ***    DEPARTURE FROM EXPONENTIAL SERVER    ***
c
    3 r = lambda
      into(1)=into(1)+1
      into(3)=into(3)-1
      return
c
c
      end
```

```
      subroutine instant(l,into,iin,prob,ic)
      implicit double precision (a-h,o-z)
      dimension into(iin),prob(l)
c
c
c          ***    CHECK FOR AN INSTANTANEOUS TRANSITION    ***
c
      ic=0
      if(into(1).eq.0)return
      if(into(2).ne.0)return
c
c
c          ***    SET UP PARAMETERS    ***
c
      k = 3
      p1 = 0.1d0
      p2 = 0.6d0
      p3 = 0.3d0
c
c
c          ***    GENERATE THE THREE DESTINATION STATES    ***
c
      ic = k
      do 1 i = 1,k
         into(i+k) = into(i)
         into(i+2*k) = into(i)
    1    continue
      into(2) = 1
      into(5) = 2
      into(8) = 3
c
c
c          ***    ASSIGN PROBABILITIES    ***
c
      prob(1) = p1
      prob(2) = p2
      prob(3) = p3
c
c
      return
      end
```

i, say, the number of balls in the population size bucket is decremented by 1; in addition, the number of balls in the phase indicator bucket (i) should be modified to represent the phase in which the next customer is to be served. However, this is not convenient to do in the subroutine RATE because each of the phases is a possible starting phase. Instead, we suggest that the number of balls in the phase indicator bucket be set to zero by the subroutine RATE and the subroutine INSTANT used to generate a destination state for each possible phase. The probability of choosing a particular phase must also be determined by INSTANT.

Consider as an example a two-station network in which one of the stations has a hyperexponential type of server with three possible phases. A graphical representation of the model as well as possible implementations of RATE and INSTANT are presented below. Note that in INSTANT, each time a state is found in which there is at least one customer in the population size bucket and the phase indicator bucket is zero, instantaneous transitions generate three destinations states and the originating state is overwritten.

4.3 Modeling Open Queuing Networks in MARCA

It is sometimes necessary to construct models in which customers arrive from outside the system, circulate among the stations of the network, and eventually leave the network. Theoretically, the number of states generated may be infinite and special steps have to be taken to obtain a finite representation, which is quite naturally an approximation. In MARCA, this is achieved by associating a bucket with the exterior of the network. Arrivals to the network are represented by transitions from this *exterior bucket* to the bucket or buckets that represent the station or stations that normally receive the arriving customers. Departures from the network are represented by transitions from the stations to the external bucket. In this case a fixed number of balls representing customers circulate among the stations of the network and the external bucket.

If the parameters of the model are such that the probability of having zero balls in the external bucket is very small, of the order of 10^{-10}, then we may assume that the arrival process is adequately represented. On the other hand, if the probability of having zero balls in the external bucket is not small, this means that the arrival process will be shut off with nonnegligible probability and is thus not accurately modeled. In this case it is necessary to increment the number of balls in the model until the probability of having zero at the external bucket becomes sufficiently small.

Suppose we wish to model an open network consisting of a station with a single hyperexponential server. The arrival process is assumed to be Poisson with rate λ. In this case we may use the model of Section 4.2, in which $\lambda = 0.5$. With 10 customers circulating in the network we find the probability that the external bucket is empty is $0.53 * 10^{-5}$. This gives us insufficient accuracy. By increasing the number of customers to 20, we obtain a probability of $0.14 * 10^{-9}$, while with 30 customers, the probability that the arrival process is shut off is only $0.39 * 10^{-14}$. This is more than sufficiently accurate to model the arrival process.

Note that the number of customers that must be included in the closed network to adequately represent an open network depends on the parameters of the network. If the service process is very much faster than the arrival process, then a much smaller number of customers suffices. If the arrival rate is taken to be $\lambda = 0.05$ in the previous example, then the same accuracy is achieved with only 10 customers circulating in the network. With

10 customers we find the probability the external bucket is empty is $0.68 * 10^{-14}$.

4.4 Modeling Queues with Blocking in MARCA

Finally we consider the case when the model contains a server that may be blocked. There are many different types of blocking possibilities. For an extensive review, the interested reader should consult some of the papers of Perros and his co-workers. In this example we consider the case of *transfer blocking*, in other words, the server becomes blocked after it has completed service of a customer and finds itself unable to transfer the customer to its desired destination because the waiting room of the destination station is full. The server suspends service until the transfer can be completed.

In MARCA, a server that can become blocked may be represented by two buckets, one bucket to denote the number of customers at the station and a second to indicate if the server is blocked or not. This latter bucket shall be called the *blocking indicator bucket*. It will contain a single ball if the server is blocked and zero balls if it is not blocked. If the server provides service that is not exponentially distributed, for example a Coxian server, then an additional (third) bucket is required to indicate the current phase of service, just as described in Sections 4.1 and 4.2 above.

There are many ways in which the subroutines RATE and INSTANT may be written to handle this situation. For example, in RATE it suffices to verify that the blocking indicator bucket is zero before permitting a transition to occur. If there is a ball in the blocking indicator bucket, the rate of transition should simply be set to zero. In INSTANT, it suffices to check that each time the blocking indicator bucket is full, the destination bucket is also at its maximum value. If the destination bucket is not full, indicating that a transition has occurred out of this bucket thereby freeing a slot for the customer in the blocked station, the function that should be performed by the subroutine INSTANT is to set the contents of the blocking indicator bucket to zero and to transfer a ball from the previous blocked bucket to the destination bucket.

Consider as an example a tandem network consisting of an exponential server with infinite waiting room and a second with finite waiting room. A graphical description of the network follows along with sample RATE and INSTANT subroutines.

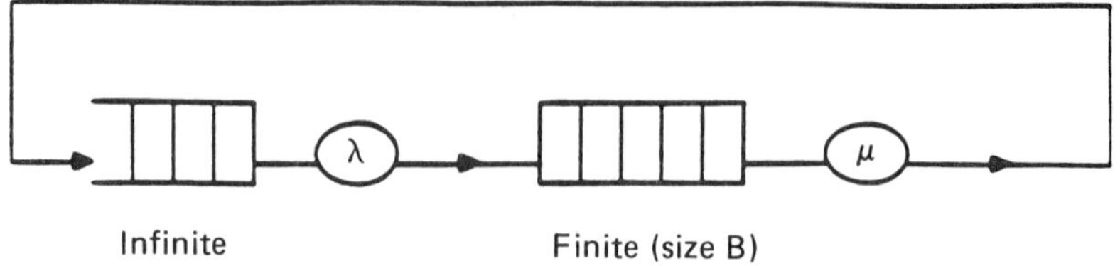


```
      subroutine rate(l1,l2,istate,into,l,r)
      implicit double precision (a-h,o-z)
      dimension istate(l),into(l)
      double precision lambda, mu
c
c
c         ***    SET UP MODEL PARAMETERS AND      ***
c         ***    INITIALIZE DESTINATION STATE     ***
c
      lambda = 0.5d0
      mu = 1.0d0
c
      do 10 i=1,l
         into(i)=istate(i)
   10    continue
      r=0.0d0
c
c
      goto (1,2,3)l1
      return
c
c
c         ***    TRANSITION FROM INFINITE QUEUE     ***
c                ***    IF NOT BLOCKED    ***
c
    1 if(into(2).eq.1)return
      r = mu
      into(1)=into(1)-1
      into(2) = 1
      return
c
c
c         ***    BLOCKING INDICATOR BUCKET    ***
c         ***    NO TRANSITION POSSIBLE    ***
c
    2 return
c
c
c         ***    DEPARTURE FROM FINITE QUEUE     ***
c
    3 r = lambda
      into(1)=into(1)+1
      into(3)=into(3)-1
      return
c
c
      end
```

```
      subroutine instant(l,into,iin,prob,ic)
      implicit double precision (a-h,o-z)
      dimension into(iin),prob(l)
c
      b = 4
c
c
c         ***    TEST FOR AN INSTANTANEOUS TRANSITION    ***
c
      ic = 0
      if(into(2).eq.0)return
      if(into(3).eq.b)return
c
c
c         ***    AN INSTANTANEOUS TRANSITION OCCURS    ***
c
      ic = 1
      into(2) = 0
      into(3) = into(3) + 1
c
c
      return
      end
```

5. CONCLUSIONS

In this paper, we have given a brief overview of the software package MARCA. More information on using the package, including some extensive examples, may be found in the technical report [STE88]. For further information on the numerical methods for the stationary solution, the interested reader should consult, [PSS89]. A description of the transient approach that is used in MARCA may be found in [AS89].

REFERENCES

[AS89] Allouba, H., and Stewart, W. J. Stationary and Transient Numerical Methods for Telecommunication Problems. Technical Report 89–29, Dept. of Computer Science, North Carolina State University, Raleigh, N.C. 27695, 1989

[PSS89] Philippe, B., Saad, Y., and Stewart, W. J. Numerical Methods in Markov Chain Models. Technical Report 89–21, Dept. of Computer Science, North Carolina State University, Raleigh, N.C. 27695, 1989.

[GRA77] Grassmann, W. Transient solutions in Markovian queues. Eur. J. Operations Res., 1, pp. 396–402. 19777.

[STE88] Stewart, W. J. MARCA: Markov Chain Analyzer. Technical Report 88–32, Dept. of Computer Science, North Carolina State University, Raleigh, N.C. 27695, 1989.

4

A Survey of Aggregation–Disaggregation in Large Markov Chains

PAUL J. SCHWEITZER William E. Simon Graduate School of Business Administration, University of Rochester, Rochester, New York

ABSTRACT

This paper surveys results since 1984 on aggregation–disaggregation for large Markov chains. The topics covered include alternative forms of the aggregate equations; alternative forms of the disaggregation equations; alternative schemes for iterative aggregation–disaggregation; aggregation for generalized birth-and-death (row-continuous) processes; and aggregation for nearly completely decomposable systems. Implementation issues are described, including overrelaxation, Gauss–Seidel variants, and block scaling. Many topics for further research are identified, including adaptive aggregation, multilevel aggregation, parallel computation, bounds, and applications.

1. INTRODUCTION

This paper surveys recent methodology for aggregation and disaggregation in large Markov chains. It is an update of the author's 1984 survey [Schweitzer,

1984] and overlaps the recent survey in [Schweitzer, 1989a]. It is unavoidable that such a survey reflects the author's interests, and cannot claim completeness. The author apologizes in advance for any omissions of relevant material.

The central problem addressed here is computation of the equilibrium distribution $\pi \equiv \{\pi_i, i \in S\}$ satisfying

$$\pi = \pi P, \qquad \sum_{i \in S} \pi_i = 1 \tag{1}$$

for a discrete-time time-homogeneous Markov chain. Here $S \equiv \{1, 2, \ldots, n\}$ is the finite state space, n is large, and for simplicity the $n \times n$ transition probability matrix (tpm) $P = [P_{ij}]$ is assumed to be ergodic (all states communicate and are aperiodic). The vector π will be unique and strictly positive. The main results have straightforward extensions when transient states occur, and to continuous-time Markov chains.

Recall that aggregation involves five key concepts:

1. *Partitioning* of the state space into say $\bar{n}$ blocks, where usually $\bar{n} \ll n$:

$$S = S(1) + S(2) + \cdots + S(\bar{n}) \tag{2}$$

2. Defining *aggregate probabilities* for each block $S(\alpha)$:

$$\bar{\pi}_\alpha \equiv \sum_{i \in S(\alpha)} \pi_i = \Pr[in\ S(\alpha)] \qquad 1 \leq \alpha \leq \bar{n} \tag{3}$$

3. Specifying the *aggregate equations*:

$$\bar{\pi}_\alpha = \sum_{\beta=1}^{\bar{n}} \bar{\pi}_\beta \bar{P}_{\beta\alpha} \qquad 1 \leq \alpha \leq \bar{n} \tag{4a}$$

$$\sum_{\alpha=1}^{\bar{n}} \bar{\pi}_\alpha = 1 \tag{4b}$$

where

$$\bar{P}_{\beta\alpha} \equiv \frac{\sum_{j \in S(\beta)} \sum_{i \in S(\alpha)} \pi_j P_{ji}}{\sum_{k \in S(\beta)} \pi_k} \qquad 1 \leq \alpha, \beta \leq \bar{n} \tag{5}$$

is the aggregate transition probability from block $S(\beta)$ to block $S(\alpha)$.

4. Specifying the *disaggregation equations* for each block $S(\alpha)$

$$\pi_i = \sum_{\beta=1}^{\bar{n}} \bar{\pi}_\beta P^{\#}_{\beta i} \qquad i \in S(\alpha) \tag{6a}$$

or

$$\pi_i - \sum_{j \in S(\alpha)} \pi_j P_{ji} = \sum_{\substack{\beta=1 \\ \beta \neq \alpha}}^{\bar{n}} \bar{\pi}_\beta P^{\#}_{\beta i} \qquad i \in S(\alpha) \tag{6b}$$

where

$$P^{\#}_{\beta i} \equiv \frac{\sum_{j \in S(\beta)} \pi_j P_{ji}}{\sum_{k \in S(\beta)} \pi_k} \qquad 1 \leq \beta \leq \bar{n}, i \in S \tag{7}$$

is the disaggregate transition probability from block $S(\beta)$ to state i.

5. Devising an *iterative scheme* to solve the fourfold set of equations (4,b, 5, 6, 7) for the fourfold set of unknowns ($\pi_i, \bar{\pi}_\alpha, \bar{P}_{\beta\alpha}$ and $P^{\#}_{\beta i}$).

Recall from Schweitzer and Kindle [1986 (Theorem 3)] that solution of the fourfold set of equations is *equivalent* to solving equation (1). Note that one must distinguish between two concepts of aggregation, the first being the *exact* fourfold set of equations (4, 5, 6, 7) into which equation (1) is transformed and the second being the *iterative procedure* chosen in step 5 to solve this fourfold set.

Recall that Takahashi [1975] was the first to note that the matrices $\bar{P}$ and $P^{\#}$ in equations (5) and (7) depend only upon the *conditional* probabilities (*relative* probabilities)

$$q_j \equiv \pi_j / \sum_{k \in S(\beta)} \pi_k \qquad j \in S(\beta), \quad 1 \leq \beta \leq \bar{n} \tag{8}$$

of the various states within each block. This implies that iterative aggregation can be implemented so as to be *completely insensitive to scale* (within each block), and also to exhibit *one-step convergence* if the relative probabilities are known perfectly. The last property is central to the one-pass algorithms for birth-and-death processes and other time-reversible Markov chains [Sumita and Rieders, 1989d], for aggregation in product-form queuing networks, [Balsamo and Iazeolla, 1984, 1985; Conway and Georganas, 1986] and for processes with other structural features, such as only one entrance or exit state per block [Kim and Smith, 1989; Sumita and Rieders, 1989d; Feinberg and Chiu, 1987].

Aggregation predicts the *one-step* transition probabilities correctly, but not the *multistep* transition probabilities: if $X(k)$ is the state at time k,

$$\lim_{k \to \infty} Pr[X(k+1) \in S(\beta) \mid X(k) \in S(\alpha)] = \bar{P}_{\alpha\beta}$$

$$\lim_{k \to \infty} Pr[X(k+m) \in S(\beta) \mid X(k) \in S(\alpha)] \neq [(\bar{P})^m]_{\alpha\beta} \qquad m \geq 2.$$

The latter implies that the lumped process has *lost* the Markov property, while the former implies that the equilibrium distribution $\{\bar{\pi}_\alpha\}$ of the lumped process can nevertheless be computed correctly (while pretending the lumped process is Markovian), because $\{\bar{\pi}_\alpha\}$ depends only upon the one-step transition probabilities, which are predicted correctly.

2. INCOMPLETE SPECIFICATION

2.1 Need for Empirical Experimentation

Note that some *lack of complete specification* occurs both in the choice of the exact fourfold equations and in the choice of iterative procedure. Several *alternative specifications* are possible, and only *empirical experimentation* can determine the preferable choices. Iterative aggregation is currently in this stage of numerical experimentation with various choices [e.g., Haviv, 1987].

2.2 Alternate Specifications of the Exact Aggregation Equations

The *exact aggregation equations* from combining equations (4a) and (5) may be recast as

$$\bar{\pi}_\alpha = \sum_{\beta=1}^{\bar{n}} \left[\frac{\bar{\pi}_\beta}{\sum_{k \in S(\beta)} \pi_k} \right]^b NUM_{\beta\alpha} \qquad 1 \le \alpha \le \bar{n} \tag{9}$$

where $NUM_{\beta\alpha}$ is the numerator appearing on the right side of equation (5). The original formulation has $b = 1$ but, since the term in brackets is unity, *any positive value of* b may be employed instead. If (say) successive substitution is used to solve equation (9) for $\{\bar{\pi}_\alpha, 1 \le \alpha \le \bar{n}\}$, then it becomes an empirical question which values of b perform best. [More generally, the bracketed term in equation (9) may be replaced by any function of $\bar{\pi}_\beta$ and $\sum_{k \in S(\beta)} \pi_k$ that is unity when these are equal.]

2.3 Alternate Specifications of the Exact Disaggregation Equations

Two alternate formulations of the *exact disaggregation equations* are given by equations (6a) and (6b), called *point disaggregation* and *block disaggregation*, respectively. Block disaggregation, which was introduced by Takahashi [1975], is a *splitting method*. It offers potentially higher accuracy per iteration, with the disadvantage of having to solve $|S(\alpha)|$ linear equations. Point disaggregation, employed by Vakhutinsky, Dudkin, and Ryvkin [1979], has the advantage of avoiding solution of $|S(\alpha)|$ linear equations, and it alone permits solution of extremely large systems. For example, Seelen [1986] solved 140,000-state systems via point disaggregation. (For discussion of block-iterative versions of Jacobi and Gauss–Seidel schemes, without aggregation, see Courtois and Semal [1986b].)

In iterative aggregation schemes, point disaggregation always looks like a *single successive substitution step*. Block disaggregation, in which the $|S(\alpha)|$ linear equations are solved by an iterative scheme, will look like a *sequence of successive substitution steps*. Near the *end* of the algorithm, where good estimates of π_i are available, this sequence will have only a few members (ultimately one member) so that point and block aggregation will behave similarly.

Three additional alternative formulations of the disaggregation equations are due to Kim and Smith [1989], Sumita and Rieders [1989a], and Feinberg and Chiu [1987]. In Kim and Smith [1989], a modified process is constructed with observations on $S(\alpha)$ alone, and with a new tpm on $S(\alpha)$ given by

$$\hat{P}_{ij} = P_{ij} + \left[\sum_{k \in S/S(\alpha)} P_{ik} \right] w_{\alpha j}, \qquad i,j \in S(\alpha) \tag{10}$$

where $A/B \equiv \{i \mid i \in A, i \notin B\}$. Note we require $w_{\alpha j} \geq 0$, $\sum_{j \in S(\alpha)} w_{\alpha j} = 1$. Here $w_{\alpha j}$ may be interpreted as the probability of reappearance in $S(\alpha)$ at j, if the system leaves $S(\alpha)$. If the w's are chosen properly, namely,

$$w_{\alpha j} = \frac{\sum_{i \in S/S(\alpha)} \pi_i P_{ij}}{\sum_{i \in S/S(\alpha)} \sum_{k \in S(\alpha)} \pi_i P_{ik}} \qquad j \in S(\alpha) \tag{11}$$

the equilibrium distribution of $\hat{P}$ on $S(\alpha)$ will yield perfect estimates of the conditional probabilities $\{q_j, j \in S(\alpha)\}$ on $S(\alpha)$, which is all that aggregation theory requires.

A similar idea is employed in Sumita and Rieders [1989a] in which $\hat{P}$, the tpm on $S(\alpha) \times S(\alpha)$ is again an estimate (or modification) of the *stochastic complement* of $S(\alpha)$. This estimate involves inversion of a matrix of size only $\bar{n} - 1$ (rather than the hopeless task of inverting a matrix of size $n - |S(\alpha)|$ for the exact stochastic complement) and must be iterated because only an estimate of π is available. (For generalized birth-and-death processes, as in Section 3, the estimate completely avoids solution of $\bar{n} - 1$ linear equations [Sumita and Rieders, 1989b, 1989c].)

The *stochastic complement* of $S(\alpha)$ [Meyer, 1989] is the tpm

$$P^{S(\alpha) \times S(\alpha)} + P^{S(\alpha) \times S/S(\alpha)} [I - P^{S/S(\alpha) \times S/S(\alpha)}]^{-1} P^{S/S(\alpha) \times S(\alpha)}$$

on $S(\alpha) \times S(\alpha)$, irreducible if P is, where $P^{A \times B}$ denotes $[P_{ij}]_{i \in A, j \in B}$. It describes the transition probabilities for the original process if only states within $S(\alpha)$ are observed. It is known that the unique stationary distribution of the stochastic complement is $\{q_i, i \in S(\alpha)\}$, where q_i is defined by equation (8), namely, a scaled version of $\{\pi_i, i \in S(\alpha)\}$. Hence *disaggregation on $S(\alpha)$ can be accomplished after computing the stochastic complement.*

Alternative disaggregation schemes correspond to different ways of estimating or modifying the stochastic complement. For example, Kim and Smith [1989] use a rank-1 correction in the second term equation (10) while Sumita and Rieders [1989a, 1989b] uses a replacement process on the $\bar{n} - 1$ blocks, leading to a rank-2 correction in the GBD case. Estimation or modification of the stochastic complement seems to be the most promising approach for modifying the disaggregation equations, and also allows the possibility of parallel computation [Mattingly, 1988].

A related approach in Feinberg and Chiu [1987, Theorem 2] *simultaneously disaggregates all blocks* by solving the $n + \bar{n}$ linear equations

$$\pi_j = \sum_{i \in S(\alpha)} \pi_i P_{ij} + \bar{\pi}^*_\alpha w_{\alpha j} \qquad j \in S(\alpha), 1 \leq \alpha \leq \bar{n} \tag{12}$$

$$\bar{\pi}^*_\alpha = \sum_{i \in S/S(\alpha)} \sum_{j \in S(\alpha)} \pi_i P_{ij} \qquad 1 \leq \alpha \leq \bar{n} \tag{13}$$

$$\sum_{i \in S} \pi_i = 1 \tag{14}$$

for $\{\pi_i, i \in S\}$ and $\{\bar{\pi}^*_\alpha, 1 \leq \alpha \leq \bar{n}\}$. Here $w_{\alpha j}$ in equation (12) is given by the right-hand side of (11). By exploiting the structure of these equations, inversions of only $\bar{n} \times \bar{n}$ and $|S(\alpha)| \times |S(\alpha)|$ matrices are required, as in the Takahashi approach. These may be interpreted as follows: $\bar{\pi}^*_\alpha$ is the probability of being outside $S(\alpha)$ (which is assumed to last only *one* period, hence $\bar{\pi}^*_\alpha \neq 1 - \bar{\pi}_\alpha$) while $w_{\alpha j}$ is, as in Kim and Smith, the conditional probability that the return to $S(\alpha)$ will occur at state j. By inserting equation (13) into the right side of (12), we see that $\bar{\pi}^*_\alpha$ cancels the denominator in $w_{\alpha j}$, so that equation (12) reduces to the usual Kolmogoroff equation

$$\pi_j = \sum_{i \in S(\alpha)} \pi_i P_{ij} + f_j \qquad j \in S(\alpha) \tag{15}$$

where

$$f_j \equiv \sum_{i \in S/S(\alpha)} \pi_i P_{ij} \qquad j \in S(\alpha) \tag{16}$$

denotes the flow rate from $S/S(\alpha)$ to state j.

[For all three of these alternative disaggregations, the property that the components of π are now improperly normalized is not a problem, since the *conditional probabilities* q_i are correct. The absolute probabilities may be subsequently obtained from $\pi_j = q_j \bar{\pi}_\beta$ for $j \in S(\beta)$.]

A related idea for disaggregating block $S(\alpha)$ is to note that

$$\hat{\pi} \equiv [\{\pi_i, i \in S(\alpha)\} \mid \{\bar{\pi}_\beta, \beta \in L\}]$$

where

$$L \equiv \{1, 2, \ldots, \bar{n}\} / \{\alpha\}$$

is the solution to

$$\sum_i \hat{\pi}_i \equiv \sum_{i \in S(\alpha)} \pi_i + \sum_{\beta \in L} \bar{\pi}_\beta = 1 \tag{17a}$$

$$\pi_i = \sum_{j \in S(\alpha)} \pi_j P_{ji} + \sum_{\beta \in L} \bar{\pi}_\beta P^{\#}_{\beta i} \qquad i \in S(\alpha) \tag{17b}$$

$$\bar{\pi}_\beta = \sum_{\delta \in L} \bar{\pi}_\delta \bar{P}_{\delta\beta} + \sum_{i \in S(\alpha)} \pi_i \sum_{j \in S(\beta)} P_{ij} \qquad \beta \in L \tag{17c}$$

Here equations (17b) and (17c) follow from (6b) and (4a), respectively, while equation (17a) follows from (3) and (4b). The interpretation of equation (17) is that $\hat{\pi}$ is the stationary distribution for a Markov chain with state space $S(\alpha) \cup L$ containing $|S(\alpha)| + \bar{n} - 1$ states. Each block $S(\beta)$ with $\beta \neq \alpha$ has been collapsed into a singleton, which acts as an artificial source for transitions from $S(\beta)$ back to $S(\alpha)$. Since the transition probabilities $\bar{P}$ and $P^{\#}$ in equation (17) for these artificial sources depend, in a nonlinear way, upon the unknown π, they must be estimated iteratively. The replacement process decomposition [Sumita and Rieders, 1989a] uses a similar approach, solving the $\bar{n} - 1$ linear equations in (17c) for $[\bar{\pi}_\beta]_{\beta \in L}$ and substituting into (17b).

Finally, note that the original Takahashi [1975] paper showed that disaggregation of $S(\alpha)$ could be accomplished with a Markov chain Q of size $|S(\alpha)| + 1$:

$$Q = \begin{bmatrix} P^{S(\alpha) \times S(\alpha)} & G \\ H & J \end{bmatrix} \tag{18}$$

where

$$G_{i1} \equiv \sum_{j \in S/S(\alpha)} P_{ij} \qquad i \in S(\alpha)$$

$$= \text{transition probability from } i \text{ to } S/S(\alpha)$$

$$H_{1i} \equiv \frac{1}{1-\bar{\pi}_\alpha} \sum_{j \in S/S(\alpha)} \pi_j P_{ji} \qquad i \in S(\alpha)$$

$$= \text{expected transition probability from } S/S(\alpha) \text{ to } i$$

$$J_{11} \equiv 1 - \sum_{i \in S(\alpha)} H_{1i}$$

$$= \text{expected transition probability from } S/S(\alpha) \text{ to } S/S(\alpha)$$

The tpm Q has a unique equilibrium distribution

$$x = xQ \qquad \sum x_i = 1$$

given by

$$x = [\{\pi_i, i \in S(\alpha)\} \mid 1 - \bar{\pi}_\alpha]$$

Here all of $S/S(\alpha)$ has been collapsed into *one* state. This has lower dimensionality than the Feinberg and Chiu [1987] approach or equation (17).

The four versions of disaggregation: equations (12) to (14) with $n + \bar{n}$ states; equation (17) with $|S(\alpha)| + \bar{n} - 1$ states; equation (18) with $|S(\alpha)| + 1$ states; equation (10) with $|S(\alpha)|$ states—may be thought of as *hierarchical levels of aggregation.* In particular, we note that equation (18) may be obtained from (17) by applying the Takahashi theory of equations (3) to (7) whereby L is aggregated into a single state. Also, equations (10) and (11) are obtained from (18) by extracting the stochastic complement of $S(\alpha)$ from Q.

Another variant of the disaggregation equations is to rewrite the pair of equations (6a) and (7) as

$$\pi_i = \sum_{\beta=1}^{\bar{n}} (E_\beta)^b \sum_{j \in S(\beta)} \pi_j P_{ji} \tag{19}$$

where

$$E_\beta \equiv \bar{\pi}_\beta / \sum_{j \in S(\beta)} \pi_j \qquad 1 \le \beta \le \bar{n} \tag{20}$$

with minor changes if equation (6b) is used. Just as in equation (9), the choice $b = 1$ is not mandatory, and E_β could be replaced by any function of π and $\bar{\pi}$ that equals unity at the exact solution. Further discussion of equation (19) is given in Section 3.5.

Author	Estimate of $f_j, j \in S(\alpha)$
Takahashi, equations (6), (17b)	$\sum_{\substack{\beta=1 \\ \beta\neq\alpha}}^{\bar{n}} \bar{\pi}_\beta P^{\#}_{\beta j}$
Feinberg and Chiu, equation (12)	$\bar{\pi}^*_\alpha w_{\alpha j}$
Kim and Smith, equation (10)	$\sum_{i\in S(\alpha)} \pi_i(\hat{P}_{ij} - P_{ij}) = [\sum_{i\in S(\alpha)} \sum_{k\in S/S(\alpha)} \pi_i P_{ik}] w_{\alpha j}$
Takahashi, equation (18)	$(1 - \bar{\pi}_\alpha) H_{1j}$

An alternate viewpoint for the above disaggregation schemes is that all start from the exact equation (15) and employ different estimates of the flow rate f_j, defined by equation (16), to state $j \in S(\alpha)$ from $S/S(\alpha)$ (the equivalence between the Feinberg and Chiu and the Kim and Smith approaches follows from the flow balance between $S(\alpha)$ and $S/S(\alpha)$ at equilibrium):

$$\sum_{i\in S(\alpha)} \sum_{k\in S/S(\alpha)} \pi_i P_{ik} = \sum_{i\in S/S(\alpha)} \sum_{k\in S(\alpha)} \pi_i P_{ik}. \tag{21}$$

In all cases, f_j can only be *estimated* because it (through $P^{\#}, w_{\alpha j}$, or H_{1j}) depends upon the unknown π. Hence all the schemes are *iterative*, involving updating the estimates of f_j whenever estimates of π are updated. It is noteworthy that none of the iterative schemes can guarantee convergence, although convergence can be forced via fallback to the power method [Schweitzer, 1984; Schweitzer and Kindle, 1986; Sumita and Rieders, 1989a].

The encouraging numerical results by Sumita and Rieders [1989b, 1989c], Feinberg and Chiu [1987, Section 5], and Seelen (see Section 3 below) demonstrate the value of modifying the disaggregation equations, and suggest that future research in this area will be fruitful.

2.4 Alternate Specifications of the Iterative Procedure

An extremely large class of iterative procedures can be devised to solve the fourfold set of exact equations. Some versions are as follows:

Straightforward Aggregation–Disaggregation

One cycle is: enter with estimate of π. Use equations (5) and (7) to estimate $\bar{P}$ and $P^{\#}$. Use equation (4) (with estimate of $\bar{P}$) to estimate $\bar{\pi}$. Use equation (6) (with estimates of $P^{\#}$ and $\bar{\pi}$) to reestimate π.

Gauss–Seidel Variant

Similar to above but the latest values of $\{\pi_j, j \in S(\alpha)\}$ [obtained by disaggregation of block $S(\alpha)$] are used either to revise estimates of $P^{\#}_{\alpha i}$ for disaggregation of subsequent blocks [Takahashi, 1975], or else the entire aggregation procedure of equations (4) and (5) is redone after each block is disaggregated (see Section 3.5). Also, Gauss–Seidel can be used *within* the disaggregation of block $S(\alpha)$, if equation (6b) is solved by an iterative procedure. (This was done, e.g., by Seelen [1986].)

Over-Relaxation

If the one cycle aggregation–disaggregation in the straightforward or Gauss–Seidel variant converts an estimate π^{old} into a revised estimate π^{new}, the estimate $w\pi^{\text{new}} + (1 - w)\pi^{\text{old}}$ with $w > 1$ is sometimes even better. Care is needed in choosing the over-relaxation factor w so that the algorithm is accelerated, not degraded or even rendered divergent. At the same time, it is important to avoid over-investment of computer time searching for the optimal w.

A successful version of adaptive over-relaxation, whereby w is updated occasionally, has been described by Seelen [1986] and Tijms [1986, Appendix D]. It appears that much of the impressive success in Seelen's implementation, described below, is due to the combination of Gauss–Seidel and adaptive over-relaxation.

Strategic Issues in Choosing an Iterative Procedure

These include how to partition; how to make an initial guess for π (see Schweitzer, 1984); whether iterative aggregation is used as the algorithm to actually compute π or whether it is invoked occasionally as an acceleration step [Chatelin, 1984; Chatelin and Miranker, 1982; Haviv, 1987; Koury, McAllister and Stewart, 1984] for some other primary algorithm such as the power method; choice of the disaggregation equations [(6a) or (6b) or others] and whether such linear equations are to be solved exactly (LU decomposition) or by an iterative scheme: implementation of *nested* (*multilevel*) *aggregation* [Chow and Tsitsiklis, 1988; Mandel, McCormick and Ruge, 1988; Schweitzer and Kindle, 1986; Shima and Haimes, 1984], *dynamic aggregation* [Bertsekas and Castanon, 1989], or *parallel computation* on the separate blocks [Lubachevsky and Mitra, 1986; Mattingly, 1988]. Relatively little is known for making these decisions.

In particular, practically all published analyses are of *static* schemes, without separate analyses of begin-game (initial partition, starting point, initial tolerances), middle-game (dynamic partitioning, levels of nesting, midgame tolerances, acceleration steps), and end-game (termination criterion, reduction to successive substitution, error bounds, etc.).

Choice of Convergence Criteria

These include acceleration of convergence by, for example, the Aitken δ^2 method [Schweitzer, 1984] or multilevel aggregation, estimation of errors, identification of blocks $S(\alpha)$ where π_i or $\bar{\pi}_\alpha$ has apparently converged and no longer requires further computations [Sumita and Rieders, 1989b], and termination criteria.

Also included are identifying and exploiting deviations to improve convergence; for example, $\bar{\pi}_\alpha$ versus $\sum_{i \in S(\alpha)} \pi_i$ in Takahashi's algorithm and the difference between the left and right sides of equations (21) in all algorithms, especially Kim and Smith and Feinberg and Chiu.

Preprocessing the Transition Probability Matrix

These replace P by a new matrix, and may improve the rate of convergence, although possibly with loss of sparsity. They include replacement of P by P^2, the Jacobi and Gauss–Seidel schemes, over-relaxation [replace P by $wP + (1 - w)I$], and exploitation of special structure to express all $\{\pi_i\}$ as positive combinations of a subset of the $\{\pi_i\}$.

Simultaneous versus Alternate Aggregation

In the Feinberg and Chiu [1987, Theorem 2] approach, the $n + \bar{n}$ linear equations (12, 13, 14) must be solved. Their scheme solves *all* $n + \bar{n}$ linear equations *simultaneously*, in a practical way, with $[w_{\alpha j}]$ being updated between iterations. This may be considered as *simultaneous aggregation* (for all $\bar{\pi}^*_\alpha$) and *disaggregation* (for every block). An alternative exists, similar to Takahashi [1975], in which aggregation and disaggregation (one block at a time) are done *alternately*: enter with estimate of π; estimate $\{w_{\alpha j}\}$ from right-hand side of equation (11); aggregation step: compute $\{\bar{\pi}^*_\beta\}$ from equation (13); disaggregation of each block $S(\alpha)$: solve the $|S(\alpha)|$ linear equations in equation (12); exit with revised estimate of π.

3. BLOCK SCALING AND ITERATIVE AGGREGATION FOR GENERALIZED BIRTH-AND-DEATH PROCESSES

3.1 GBD Processes

A generalized birth-and-death (GBD) process is a Markov chain such that states in block $S(\alpha)$ can make transitions only to blocks $S(\alpha - 1)$, $S(\alpha)$,

$S(\alpha + 1)$. A Markov chain has this structure if and only if the tpm P is block tridiagonal. GBDs occur in queuing processes where a typical state is a pair $i = (i_1, i_2)$ with i_1 denoting the customer population and i_2 denoting all remaining information. Since each transition causes i_1 to change by -1, 0, or $+1$, the queuing process with $S(\alpha) = \{(i_1, i_2) \mid i_1 = \alpha\}$ is GBD. Conversely, any state $i \in S(\alpha)$ in a GBD process can be labeled as a pair (i_1, i_2) where $i_1 = \alpha$ and i_2 denotes the relative position of this state within $S(\alpha)$. GBD processes are better known as *row-continuous* Markov chains [Sumita and Rieders, 1989b, 1989c] because the "row index" i_1 is skip-free. The alternate name GBD is used here to emphasize that the allowed transitions between blocks (i.e., the aggregate tpm $\bar{P}$) look like an ordinary birth-and-death process.

Takahashi [1975] was the first to exploit the GBD structure with an iterative aggregation algorithm. He showed how to compute the conditional probabilities $\{q_j, j \in S(\alpha)\}$ exclusively from the conditional probabilities $\{q_j, j \in S(\alpha \pm 1)\}$. The resulting algorithm for the q's alone looks like exclusively disaggregation. The terms $\{\bar{\pi}_\alpha\}$ may then be obtained from

$$\frac{\bar{\pi}_{\alpha+1}}{\bar{\pi}_\alpha} = \frac{\sum_{i \in S(\alpha)} \sum_{j \in S(\alpha+1)} q_i P_{ij}}{\sum_{i \in S(\alpha+1)} \sum_{j \in S(\alpha)} q_i P_{ij}} \qquad 1 \le \alpha \le \bar{n} - 1, \quad \sum_{\alpha=1}^{\bar{n}} \bar{\pi}_\alpha = 1$$

and π from $\pi_i = \bar{\pi}_\alpha q_i$ for $i \in S(\alpha)$. This algorithm was successfully employed in Takahashi and Takami [1976]. The later algorithms in Seelen [1986] and Sumita and Rieders [1989b, 1989c] are reported to outperform it.

3.2 The Seelen Algorithm

Seelen [1986] has presented an iterative algorithm for GBD processes (in continuous time) with impressive convergence properties for queuing problems: typically 3–10 iterations to achieve 1% accuracy; one problem with more that 140,000 states took only 53 iterations and 110 CPU-seconds on a CYBER 170-750 computer. The number of iterations was remarkably insensitive to the starting point and to the number of states.

The Seelen algorithm assumes that the blocks are updated in ascending order [first $S(1)$, then $S(2)$, etc.] and that states within each block are also updated in ascending order (i before j if $i < j$). The specific update from π^{old} to π^{new} is

$$\begin{aligned} \bar{\pi}_i^{\text{new}} = &\sum_{j \in S(\alpha-1)} \pi_j^{\text{new}} P_{ji} + \sum_{\substack{j \in S(\alpha) \\ j<i}} \pi_j^{\text{new}} P_{ji} + f_\alpha \sum_{\substack{j \in S(\alpha) \\ j>i}} \pi_j^{\text{old}} P_{ji} \\ &+ g_\alpha \sum_{j \in S(\alpha+1)} \pi_j^{\text{old}} P_{ji} \qquad i \in S(\alpha), \quad 1 \le \alpha \le \bar{n} \end{aligned} \tag{22}$$

$$\pi_i^{\text{new}} = w\tilde{\pi}_i^{\text{new}} + (1 - w)\pi_i^{\text{old}} \tag{23}$$

(the first [last] sum is ignored if $\alpha = 1[\bar{n}]$). Here it is assumed that $P_{ii} = 0$ and that when computing π_i^{new} for $i \in S(\alpha)$, π_j^{new} is available for $j \in S(\alpha - 1)$ and for $j \in S(\alpha)$ with $j < i$. Equation (22) is a *point disaggregation scheme* with a *Gauss–Seidel variant* (π_i^{old} will be overwritten by π_i^{new}), and *over-relaxation* given by equation (23), where w is adjusted dynamically. Note that the values of π_j^{new} are treated as if scaled properly, while the values of $\{\pi_j^{\text{old}}, j \in S(\alpha)\}$ and $\{\pi_j^{\text{old}}, j \in S(\alpha + 1)\}$ are *scaled up* by f_α and g_α, respectively, where

$$f_1 \equiv \frac{1}{\sum_{i \in S} \pi_i^{\text{old}}}$$

$$f_\alpha \equiv \frac{\sum_{i \in S(\alpha-1)} \sum_{j \in S(\alpha)} \pi_i^{\text{new}} P_{ij}}{\sum_{i \in S(\alpha)} \sum_{j \in S(\alpha-1)} \pi_i^{\text{old}} P_{ij}} \qquad 2 \le \alpha \le \bar{n} \tag{24}$$

$$g_\alpha \equiv f_\alpha \frac{\sum_{i \in S(\alpha)} \sum_{j \in S(\alpha+1)} \pi_i^{\text{old}} P_{ij}}{\sum_{i \in S(\alpha+1)} \sum_{j \in S(\alpha)} \pi_i^{\text{old}} P_{ij}} \qquad 1 \le \alpha \le \bar{n} - 1 \tag{25}$$

to force conformity with the *aggregate balance equation*

$$\sum_{i \in S(\alpha)} \sum_{j \in S(\alpha+1)} \pi_i P_{ij} = \sum_{i \in S(\alpha+1)} \sum_{j \in S(\alpha)} \pi_i P_{ij} \tag{26}$$

namely, flow from $S(\alpha)$ to $S(\alpha + 1)$ = flow from $S(\alpha + 1)$ to $S(\alpha)$, and similar aggregate balance between $S(\alpha - 1)$ and $S(\alpha)$.

Much of the impressive behavior of the Seelen algorithm is due to the *adaptive over-relaxation* and to the *Gauss–Seidel* variant of disaggregation (both between blocks and within each block). For clarity of exposition, we shall omit both these features and describe below the relationship of Seelen's *block scaling* to both the *power method* [Koury, McAllister and Stewart, 1984; Barker and Plemmons, 1986]

$$\pi^{\text{new}} = \pi^{\text{old}} P \tag{27}$$

and to Takahashi's [1975] iterative aggregation algorithm. (Seelen's algorithm may be viewed as an extremely effective modification of each of these methods. Although Seelen's algorithm gives the overt appearance of pure disaggregation, the aggregation aspect is actually present in the f's and g's.) We shall also see that *the birth-and-death structure is not essential for implementation of block scaling*, although in more general structures it will be more laborious to calculate the generalization of the f's and g's.

3.3 The Power Method and Block Scaling

With respect to the power method, the key observation is that equation (27) produces much faster convergence of the *relative* probabilities q_j defined by equation (8) than of the *absolute* probabilities $\bar{\pi}_\alpha$ of each block [Simon and Ando, 1961; McAllister, Stewart and Stewart, 1984; Meyer, 1989; Peponides and Kokotovic, 1983]. Consequently, one is led to modify equation (27) as

$$\begin{aligned}\pi_i^{\text{new}} &= A_\alpha \sum_{j\in S(\alpha-1)} \pi_j^{\text{old}} P_{ji} + B_\alpha \sum_{j\in S(\alpha)} \pi_j^{\text{old}} P_{ji} \\ &\quad + C_\alpha \sum_{j\in S(\alpha+1)} \pi_j^{\text{old}} P_{ji} \qquad i \in S(\alpha), \quad 1 \le \alpha \le \bar{n}\end{aligned} \tag{28}$$

(the first [last] sum is ignored if $\alpha = 1[\bar{n}]$). This amounts to rescaling all π_j^{old} within the same block. The original power method, with $A_\alpha = B_\alpha = C_\alpha = 1$ does no scaling at all. For *any* choice of $\{A_\alpha, B_\alpha, C_\alpha\}$, equation (28) is an easily programmed modification of the power method, with noticeable acceleration if good choices are made.

The aggregate balance condition of equation (26) between $S(\alpha - 1)$ and $S(\alpha)$ leads to

$$B_\alpha / A_\alpha = \frac{\sum_{i\in S(\alpha)} \sum_{j\in S(\alpha-1)} \pi_j^{\text{old}} P_{ji}}{\sum_{i\in S(\alpha-1)} \sum_{j\in S(\alpha)} \pi_j^{\text{old}} P_{ji}} \tag{29}$$

while the aggregate balance condition of equation (26) between $S(\alpha)$ and $S(\alpha + 1)$ leads to

$$C_\alpha / B_\alpha = \frac{\sum_{i\in S(\alpha+1)} \sum_{j\in S(\alpha)} \pi_j^{\text{old}} P_{ji}}{\sum_{i\in S(\alpha)} \sum_{j\in S(\alpha+1)} \pi_j^{\text{old}} P_{ji}} \tag{30}$$

Thus equation (28) fixes $\{\pi_i^{\text{new}}, i \in S(\alpha)\}$ up to one common scale factor A_α, which is estimated in the next iteration. Upon identification of B_α/A_α with f_α and of $C_\alpha/A_\alpha = (C_\alpha/B_\alpha)(B_\alpha/A_\alpha) = f_\alpha(C_\alpha/B_\alpha)$ with g_α [cf. equations (24) and (25)], the relationship of equation (28) with (22) is clear: $\{\pi_i^{\text{old}}, i \in S(\alpha - 1)\}$ are treated as if scaled correctly ($A_\alpha = 1$), while $\{\pi_i^{\text{old}}, i \in S(\alpha)\}$ and $\{\pi_i^{\text{old}}, i \in S(\alpha + 1)\}$ are scaled up by f_α and g_α, respectively.

3.4 Iterative Aggregation and Block Scaling

Consider the point disaggregation equation (6a), which assumes the form of equation (28) with $(A_\alpha, B_\alpha, C_\alpha) = (E_{\alpha-1}, E_\alpha, E_{\alpha+1})$ where

$$E_\alpha \equiv \bar{\pi}_\alpha^{\text{old}} \Big/ \sum_{j \in S(\alpha)} \pi_j^{\text{old}} \qquad 1 \leq \alpha \leq \bar{n}$$

Hence knowledge of the $\{E_\alpha\}$ fixes the scaling. From the aggregate birth-and-death process solved in (4a),

$$\bar{\pi}_\alpha^{\text{old}} \bar{P}^{\text{old}}_{\alpha,\alpha+1} = \bar{\pi}^{\text{old}}_{\alpha+1} \bar{P}^{\text{old}}_{\alpha+1,\alpha} \qquad 1 \leq \alpha \leq \bar{n}-1$$

which in turn implies

$$E_\alpha N_{\alpha,\alpha+1} = E_{\alpha+1} N_{\alpha+1,\alpha} \qquad 1 \leq \alpha \leq \bar{n}-1 \tag{31}$$

where $N_{\beta\alpha}$ is the numerator in equation (5) evaluated at $\pi = \pi^{\text{old}}$. Thus equation (31) gives all of the scale factors (E's) up to one proportionality factor, which drops out when π is renormalized.

Note that the scale factors

$$f_\alpha = B_\alpha / A_\alpha = E_\alpha / E_{\alpha-1} = N_{\alpha-1,\alpha} / N_{\alpha,\alpha-1}$$
$$g_\alpha = f_\alpha (C_\alpha / B_\alpha) = f_\alpha E_{\alpha+1} / E_\alpha = f_\alpha N_{\alpha,\alpha+1} / N_{\alpha+1,\alpha}$$

agree with those derived in equations (29) and (30), showing that the correct block scaling is done *automatically* by the Takahashi theory with point disaggregation *without* having to resort to ad hoc balance arguments such as equation (26).

3.5 Block Scaling in the General Case

Note that the *general form* of equation (6a) is the power method with *block scaling*, namely,

$$\begin{aligned} \pi_i^{\text{new}} &= \sum_{\beta=1}^{\bar{n}} \left[\frac{\bar{\pi}_\beta^{\text{old}}}{\sum_{k \in S(\beta)} \pi_k^{\text{old}}} \right] \sum_{j \in S(\beta)} \pi_j^{\text{old}} P_{ji} \\ &= \sum_{\beta=1}^{\bar{n}} E_\beta \sum_{j \in S(\beta)} \pi_j^{\text{old}} P_{ji}. \end{aligned} \tag{32}$$

where the expression $\pi_j^{\text{old}} / \sum_{k \in S(\beta)} \pi_k^{\text{old}} \equiv q_j^{\text{old}}$ contained therein is expected to converge quickly to the conditional probability q_j. Thus equation (32) is *efficient whenever there is a fast, accurate way to solve the aggregate equations* (4) for $\{\bar{\pi}_\beta^{\text{old}}\}$. This was the case when the aggregate process was birth-and-death, but may hold for other structures as well. For example, it holds when the aggregate Markov chain has a tree structure (rather than a linear structure) because it is time-reversible and has a product-form solution.

Summarizing, the block scaling procedure from the point disaggregation equations in Takahashi's algorithm produces the *same* block scaling as Seelen's algorithm; both appear essentially as equation (32), namely, the power method with separate scaling E_β of each block $S(\beta)$. Block scaling may be applied to problems without the birth-and-death structure, provided only that the aggregation equations are easily solved and that the conditional probabilities converge quickly.

Restoring the Gauss–Seidel version into equation (32) adds the following complication. After *each* block is disaggregated, the *entire* set of aggregate equations (4) must be resolved to supply *revised* values of $\{\bar{\pi}_\beta\}$ and $\{E_\beta\}$ for equation (32) before the next block is disaggregated. In the special GBD case where the aggregate equations have birth-and-death form or more generally possess product-form solution, revision of $\{E_\beta\}$ reduces to simple formulas like equations (24) and (25) based upon block-balance considerations. However, for more general structures where revising $\{E_\beta\}$ is more onerous, the revisions could be done occasionally rather than after disaggregation of each block.

3.6 Replacement Process Approach for GBD

Sumita and Rieders have specialized their replacement process approach [Sumita and Rieders, 1989a] to the GBD case [Sumita and Rieders, 1989b, 1989c], exploiting the property that the stochastic complement of $S(\alpha)$ can be easily estimated, without any need for matrix inversions, from estimates of $\{\pi_i, i \in S(\alpha - 1)\}$ and $\{\pi_i, i \in S(\alpha + 1)\}$. A rank-2 correction to $P^{S(\alpha)\times S(\alpha)}$ is involved.

Their computational experiments show that for large problems, the replacement process approach significantly out-performs both the modified Takahashi [1975] and the modified Van de Heyden [Haviv, 1987] algorithms. (The latter two modified algorithms are explained in the paper, and tailored to the GBD structure.) It would be of great interest to compare the Seelen (point Gauss–Seidel disaggregation) with the replacement process approach, and investigate further syntheses. For example, the solution of $|S(\alpha)|$ equations when disaggregating $S(\alpha)$ in the replacement process approach, currently implemented by successive approximations, could be modified or accelerated by point Gauss–Seidel or over-relaxation.

4. NEARLY COMPLETELY DECOMPOSABLE SYSTEMS

NCD systems have tpm $P(\epsilon) \equiv A + \epsilon B$ where A is a block-diagonal tpm and $|\epsilon|$ is small. An extensive literature has appeared on such systems, including computation of the equilibrium distribution by aggregation. For reasons of space, we mention only a few recent results.

Schweitzer [1986b] showed that the equilibrium distribution $\pi(\epsilon)$ of $P(\epsilon)$ has the simple form $\pi(\epsilon) = \pi(0^+)[I - \epsilon BZ^{(0)}]^{-1}$ where $Z^{(0)}$ is the coefficient of ϵ^0 in the Laurent expansion of the fundamental matrix of $P(\epsilon)$.

Stewart [1983] provides a simple one-pass algorithm for generating an estimate of the equilibrium distribution $\pi(\epsilon)$ in terms of the Perron left and right eigenvectors of the $\bar{n}$ matrices $\{[P_{ij}]_{ij \in S(\alpha)}, 1 \leq \alpha \leq \bar{n}\}$. A prior error bound on the estimate is also provided. Courtois and Semal [1984a, 1984b, 1986a] also provide a one-pass estimate of $\pi(\epsilon)$, with a prior error bound that is apparently smaller than Stewart's. Additional one-pass estimates are given in Haviv and Ritov [1986] and Haviv [1986], and an additional error bound is given in Haviv and Van der Heyden [1984]. See also Haviv [1989].

Koury, McAllister, and Stewart [1984] describe several iterative schemes to compute $\pi(\epsilon)$, including (1) the power method, (2) the point and block Gauss–Seidel, and (3) the $0(\epsilon^2)$ Vantilborgh [1981] method (one-pass and iterative versions), and others. Methods (1) and (2) include occasional acceleration steps using the Courtois aggregation procedure. Numerical computations suggest that, in general, block Gauss–Seidel (with and without aggregation) had the fewest number of iterations and outperforms point Gauss–Seidel. Occasional aggregation steps are recommended to accelerate convergence, and frequently reduced the number of iterations significantly. Haviv [1987] describes use of an occasional aggregation step to accelerate the power method.

McAllister, Stewart, and Stewart [1984] supply a rapidly convergent iterative scheme to compute $\pi(\epsilon)$ exactly. It combines the power method with a Rayleigh–Ritz refinement technique. Special attention is paid to the $\bar{n}$ eigenvalues of $P(\epsilon)$ that are at or near unity.

Vantilborgh [1985] has provided an algorithm [Vantilborgh, 1981] that computes the equilibrium distribution with error $0(\epsilon^2)$ rather than the usual $0(\epsilon)$ in one-pass aggregation. This involves aggregating all states not of interest into a single fictitious state, and has a probabilistic interpretation in terms of taboo sets for a Markov chain. The analysis involves finding a sufficiently accurate estimate of the stochastic complement of $S(\alpha)$. (This algorithm was simplified in Cao and Stewart [1985].)

Cao and Stewart [1985] examine three iterative schemes for computing the stationary distribution for NCD systems: Takahashi [1975] with block disaggregation and Gauss–Seidel; Koury, McAllister, and Stewart [1984] with block disaggregation and Gauss–Seidel; and Vantilborgh [1981]. In all three methods, each iteration reduces the error by a factor of order ϵ, so that excellent local convergence can be expected—k iterations give an error of order $0(\epsilon^k)$.

5. TOPICS FOR FUTURE RESEARCH

The following areas are suggested as fruitful ones for further research on computational methods for aggregation.

1. All the topics listed as *strategic issues* in Section 2.4, plus all of the incomplete specifications mentioned in Sections 2.2, 2.3, and 2.4. In particular, the issues of *multilevel aggregation*, *parallel computation*, and *adaptive aggregation* appear promising for extremely large systems.
2. Integration of iterative aggregation with results on prior or posterior *bounds* on the equilibrium distribution [e.g., Balsamo and Pandolfi, 1988; Courtois and Semal, 1984a, 1984b, 1986a, 1986c; Funderlic and Meyer, 1986; Haviv, 1986, 1987; Haviv and Ritov, 1986; Haviv and Van der Heyden, 1984; Schweitzer, 1984, 1986a, 1987; Stewart 1983; Van der Wal and Schweitzer, 1987; Vantilborgh, 1985]. The goal is to incorporate a dynamic error analysis.
3. Integration with iterative aggregation of recent insights into Gauss–Seidel schemes [e.g., Cao and Stewart, 1985; Greenberg and Vanderbei, 1988; Koury, McAllister and Stewart, 1984; Mitra and Tsoucas, 1988; Van der Wal and Schweitzer, 1987]. The goal is to improve convergence of the iterative schemes.
4. Iterative aggregation for quantities besides the equilibrium distribution, for example, aggregation for transient solutions [Bobbio and Trivedi, 1986; Coderch, Willsky, Sastry, and Castanon, 1983; Gross and Miller, 1984; Rohlicek and Willsky, 1988; Trivedi, Reibman, and Smith, 1988] or first-passage quantities [Sumita and Rieders, 1988] or the fundamental matrix, or tail probabilities or moments, or busy periods in queuing networks [Daduna, 1988]; also aggregation in Markov decision processes [e.g., Bean, Birge, and Smith, 1987; Mendelssohn, 1982; Schweitzer, 1990; Schweitzer, Sumita and Ohno, 1989] and in Markov renewal processes.
5. Convergence properties of various iterative schemes. This immediately involves an investigation of problem structures (e.g., of approximate

lumpability [Schweitzer, 1984], of matrix density, etc.) since convergence rates are known to be sensitive to structure. This investigation should identify which classes of problems would benefit from aggregation, and suggest partitioning schemes.

One structural issue arising in the disaggregation step is whether the $|S(\alpha)|$ linear equations (6b) should be solved by an *iterative* scheme such as Gauss–Seidel (or one of the newer iterative schemes for large, sparse, nonsymmetric linear systems), or an *exact* scheme such as LU decomposition. Gauss–Seidel is attractive if $P^{S(\alpha)\times S(\alpha)}$ is sparse (the case in Seelen [1986]) or if $\rho(P^{S(\alpha)\times S(\alpha)}) \ll 1$, which may hold if $S(\alpha)$ is small. LU decomposition is attractive if $P^{S(\alpha)\times S(\alpha)}$ is dense, especially since the decomposition can be done once, stored, and used during subsequent iterations. Alternatively, these equations may themselves be solved by aggregation–disaggregation (multilevel aggregation).

6. The analysis of non-product-form networks of queues, including queues with blocking, is an important area for approximations. Recent approaches have included those explicitly involving *aggregation* ideas [Brandwajn, 1985; Brandwajn and Jow, 1988; DeKoster, 1987; Schweitzer, 1989b; Schweitzer and Altiok, 1988, 1989; Seelen, 1986; Takahashi, 1985, 1989] as well as some excellent *heuristic* approaches [e.g., Alkaff and Muth, 1987; Altiok and Perros, 1987; Gershwin, 1987; Lee and Pollock, 1987; Perros and Snyder, 1989; Pollock, Birge and Alden, 1985]. Integration of these approaches appears very promising, either via heuristics supplying the parameters $\bar{P}_{\alpha\beta}$ needed in the aggregate model, or else aggregation suggesting the type (and functional dependence) of variables needed to be estimated by the heuristics.
6. Implementation issues such as adaptive over-relaxation, Gauss–Seidel, and block scaling were shown by Seelen to have dramatic importance. These and the above-mentioned strategic issues offer great promise in reducing computer times.

ACKNOWLEDGMENTS

Partial support for this study was provided by the IBM Program of Support for Education in the Management of Information Systems and by the Center for Manufacturing and Operations Management at the William E. Simon Graduate School of Business Administration of the University of Rochester. I thank Harry Groenevelt for useful observations on LU decompositions, and the reviewers for helpful suggestions.

REFERENCES

Alkaff, A., and E. J. Muth (1987), The throughput rate of multistation production lines with stochastic servers, *Probability Eng. Info. Sci.*, 1, 309–326.

Altiok, T., and H. G. Perros (1987). Approximate analysis of arbitrary configurations of open queueing networks with blocking, Ann. Operations Res., 9, 481–509.

Balsamo, S., and G. Iazeolla (1984). Aggregation and disaggregation in queueing networks; The principle of product-form synthesis, in G. Iazeolla, P. J. Courtois, and A. Hordijk (eds.), *Mathematical Computer Performance and Reliability*, North-Holland, Amsterdam, 95–109.

Balsamo, S., and G. Iazeolla (1985), Product-form synthesis of queueing networks, IEEE Trans. Software Eng., SE-11, 194–199.

Balsamo, S., and B. Pandolfi (1988), Bounded aggregation in Markovian networks, in G. Iazeolla, P. J. Courtois, and O. J. Boxma (eds.), *Computer Performance and Reliability*, North-Holland, Amsterdam, 73–92.

Barker, G. P., and R. J. Plemmons (1986), Convergent iterations for computing stationary distributions of Markov chains, SIAM J. Alg. Disc. Meth., 7, 390–397.

Bean, J. C., J. R. Birge, and R. L. Smith (1987), Aggregation in dynamic programming, Operations Res., 35, 215–220.

Bertsekas, D. P., and D. A. Castanon (1989), Adaptive aggregation methods for infinite horizon dynamic programming, IEEE Trans. Automatic Control, 34, 589–598.

Bobbio, A., and K. S. Trivedi (1986), An aggregation technique for the transient analysis of stiff Markov chains, IEEE Trans. Comput., C-35, 803–814.

Brandwajn, A. (1985), Equivalence and decomposition in queueing systems—A unified approach, Perform. Eval., 5, 175–186.

Brandwajn, A., and Y. L. Jow (1988), An approximation method for tandem queues with blocking, Operations Res., 36, 73–83.

Cao, W. L., and W. J. Stewart (1985), Iterative aggregation/disaggregation techniques for nearly uncoupled Markov chains, J. Assoc. Comput. Mach., 32, 702–719.

Chatelin, F. (1984), Iterative aggregation/disaggregation methods, in G. Iazeolla, P. J. Courtois, and A. Hordijk (eds.), *Mathematical Computer Performance and Reliability*, North-Holland, Amsterdam, 199–207.

Chatelin, F., and W. L. Miranker (1982), Acceleration by aggregation of successive approximation methods, Linear Algebra Appl., 43, 17–47.

Chow, C.-S., and J. N. Tsitsiklis (1988), An Optimal Multigrid Algorithm for Continuous State Discrete Time Optimal Control, Technical Report No. LIDS-P-1751. M.I.T. Laboratory for Information and Decision Systems.

Coderch, M., A. S. Willsky, S. S. Sastry, and D. A. Castanon (1983), Hierarchical aggregation of singularly perturbed finite state Markov processes, *Stochastics*, 8, 259–289.

Conway, A. E., and N. D. Georganas (1986), Decomposition and aggregation by class in closed queueing networks, IEEE Trans. Software Eng., SE-12, 1025–1040.

Courtois, P. J., and P. Semal (1984a), Bounds for the positive eigenvectors of nonnegative matrices and their approximations by decomposition, J. Assoc. Comput. Mach., 31, 804–825.

Courtois, P. J., and P. Semal (1984b), Error bounds for the analysis by decomposition of nonnegative matrices, in G. Iazeolla, P. J. Courtois, and A. Hordijk (eds.), *Mathematical Computer Performance and Reliability*, North-Holland, Amsterdam, 209–224.

Courtois, P. J., and P. Semal (1986a), Bounds on conditional steady state distributions in large Markovian and queueing models, in O J. Boxma, J. W. Cohen, and H. C. Tijms (eds.), *Teletraffic Analysis and Computer Performance Evaluation*, Elsevier-North-Holland, Amsterdam, 499–520.

Courtois, P. J., and P. Semal (1986b), Block iterative algorithms for stochastic matrices, Linear Algebra Appl., 76, 59–70.

Courtois, P. J., and P. Semal (1986c), Computable error bounds for conditional steady-state probabilities in large Markov chains and queueing models, IEEE J. Selected Areas in Communications, SAC-4, 926–937.

Daduna, H. (1988). Busy periods for subnetworks in stochastic networks: Mean value analysis, J. Assoc. Comput. Mach., 35, 668–674.

DeKoster, M. B. M. (1987), Estimation of line efficiency by aggregation, Int. J. Prod. Res., 25, 615–626.

Feinberg, B. N., and S. S. Chiu (1987), A method to calculate steady-state distributions of large Markov chains by aggregating states, Operations Res., 35, 282–290.

Funderlic, R. E., and C. D. Meyer, Jr. (1986), Sensitivity of the stationary distribution vector for an ergodic Markov chain, Linear Algebra Appl., 76, 1–17.

Gershwin, S. B. (1987), An efficient decomposition method for approximate evaluation of production lines with finite storage space, Operations Res., 35, 291–305.

Greenberg, A. G. and R. J. Vanderbei (1988), Quicker Convergence for Iterative Numerical Solutions to Stochastic Problems: Probabilistic Interpretations, Ordering Heuristics, and Parallel Processing, AT&T Bell Laboratories, Murray Hill, New Jersey, (unpublished manuscript, December 1988).

Gross, D., and D. R. Miller (1984), The randomization technique as a modelling tool and solution procedure for transient Markov processes, Operations Res., 32, 343–361.

Haviv, M. (1986), An approximation to the stationary distribution of a nearly completely decomposable Markov chain and its error analysis, SIAM J. Alg. Disc. Meth., 7, 589–593.

Haviv, M. (1987), Aggregation/disaggregation methods for computing the stationary distribution of a Markov chain, SIAM J. Numer. Anal., 24, 952–966.

Haviv, M. (1989), More on a Rayleigh–Ritz refinement technique for nearly uncoupled stochastic matrices, SIAM J. Matrix Anal. Appl., 10, 287–293.

Haviv, M., and Y. Ritov (1986), An approximation to the stationary distribution of a nearly completely decomposable Markov chain and its error bound, SIAM J. Alg. Disc. Meth., 7, 583–588.

Haviv, M., and L. Van der Heyden (1984), Perturbation bounds for the stationary probabilities of a finite Markov chain, Adv. Appl. Probability, 16, 804–818.

Kim, D. S., and R. L. Smith (1989), An Exact Aggregation Algorithm for a Special Class of Markov Chains, Technical Report No. 89-2, Dept. of Industrial and Operations Eng., University of Michigan, Ann Arbor.

Koury, J. R., D. F. McAllister, and W. J. Stewart (1984), Iterative methods for computing stationary distributions for nearly completely decomposable Markov chains, SIAM J. Alg. Disc. Meth., 5, 164–186.

Lee, H., and S. M. Pollock (1987), Approximate Analysis of Open Exponential Queueing Networks with Blocking: General Configuration, Technical Report No. 87-13, Dept. of Industrial and Operations Eng., University of Michigan, Ann Arbor. [To appear in Operations Res.]

Lubachevsky, B. D., and D. Mitra (1986), A chaotic asynchronous algorithm for computing the fixed point of a nonnegative matrix of unit spectral radius, J. Assoc. Comput. Mach., 33, 130–150.

Mandel, J., S. McCormick, and J. Ruge (1988), An algebraic theory for multigrid methods for variational problems, SIAM J. Numer. Anal., 25, 91–110.

Mattingly, R. B. (1988), Vector and Parallel Algorithms for Computing the Stationary Distribution Vector of an Irreducible Markov Chain, unpublished Ph.D. thesis, North Carolina State University, Raleigh, North Carolina.

McAllister, D. F., G. W. Stewart, and W. J. Stewart (1984), On a Rayleigh–Ritz refinement technique for nearly uncoupled stochastic matrices, Linear Algebra Appl., 60, 1–25.

Mendelssohn, R. (1982), An iterative aggregation procedure for Markov decision processes, Operation Res., 30, 62–73.

Meyer, C. D. (1989), Stochastic complementation, uncoupling Markov chains, and the theory of nearly reducible systems, SIAM Rev., 31, 240–272.

Mitra, D., and P. Tsoucas (1988), Convergence of relaxations for numerical solutions of stochastic problems, in G. Iazeolla, P. J. Courtois, and O. J. Boxma (eds.), *Computer Performance and Reliability*, North-Holland, Amsterdam, 119–133.

Peponides, G. M., and P. V. Kokotovic (1983), Weak connections, time scales, and aggregation of nonlinear systems, IEEE Trans. Systems, Man, Cybernet., SMC-13. 527–532.

Perros, H. G., and P. M. Snyder (1989), A computationally efficient approximation algorithm for feed-forward open queueing networks with blocking, Perform. Eval., 9, 217–224.

Pollock, S. M., J. R. Birge, and J. M. Alden (1985), Approximation analysis for Open Tandem Queues with Blocking: Exponential and General Service Distribution, Technical Report No. 85–30, Dept. of Industrial and Operations Engineering, University of Michigan, Ann Arbor.

Rohlicek, J. R., and A. S. Willsky (1988), The reduction of perturbed Markov generators: An algorithm exposing the role of transient states, J. Assoc. Comput. Mach., 35, 675–696.

Schweitzer, P. J. (1984), Aggregation methods for large Markov chains, in G. Iazeolla, P. J. Courtois, and A. Hordijk (eds.), *Mathematical Computer Performance and Reliability*, North-Holland, Amsterdam, 275–286.

Schweitzer, P. J. (1986a), Posterior bounds on the equilibrium distribution of a finite Markov chain, Commun. Statist.-Stochastic Models, 2, 323–338.

Schweitzer, P. J. (1986b), Perturbation series expansions for nearly completely-decomposable Markov chains, in O. J. Boxma, J. W. Cohen, and H. C. Tijms (eds.), *Teletraffic Analysis and Computer Performance Evaluation*, Elsevier-North-Holland, Amsterdam, 319–328.

Schweitzer, P. J. (1987), Dual bounds on the equilibrium distribution of a finite Markov chain. *J. Math. Anal. Appl.*, 126, 478–482.

Schweitzer, P. J. (1989a), Survey of Aggregation Methods, presented at CORS/TIMS/ORSA Joint National Meeting, May 8–10, 1989, Vancouver, B.C.

Schweitzer, P. J. (1989b), Aggregative modelling of queueing networks, in M. Bonatti (ed.), *Teletraffic Science for New Cost-Effective Systems, Networks and Services, ITC-12*, Proceedings of the 12th International Teletraffic Congress, Turino, Italy, June 1–8, 1988, Elsevier-North-Holland, Amsterdam, 1522—1528.

Schweitzer, P. J. (1991), Aggregation in Markov decision processes, Forthcoming.

Schweitzer, P. J., and T. Altiok (1988), Aggregate modelling of tandem queues with blocking, in G. Iazeolla, P. J. Courtois, and O. Boxma (eds.), *Computer Performance and Reliability*, North-Holland, Amsterdam, 135–149.

Schweitzer, P. J., and T. Altiok (1989), Aggregate modelling of tandem queues without intermediate buffers, in H. G. Perros and T. Altiok (eds.), *Queueing Networks with Blocking*, Elsevier-North-Holland, Amsterdam, 47–72.

Schweitzer, P. J., and K. W. Kindle (1986), An iterative aggregation–disaggregation algorithm for solving linear equations, Appl. Math. Comput., 18, 313–353.

Schweitzer, P. J., U. Sumita, and K. Ohno (1989), A replacement process decomposition for discounted Markov renewal programs, To appear in Ann. Oper. Res.

Seelen, L. P. (1986), An algorithm for Ph/Ph/c queues, Eur. J. Operations Res., 23, 118–127.

Shima, T. and Y. Y. Haimes (1984), The convergence properties of hierarchical overlapping coordination, IEEE Trans. Systems, Man, Cybernet., SCM-14, 74–87.

Simon, H., and A. Ando (1961), Aggregation of variables in dynamic systems, Econometrica, 29, 111–138.

Stewart, G. W. (1983), Computable error bounds for aggregated Markov chains, J. Assoc. Comput. Mach., 30, 271–285.

Sumita, U., and M. Rieders (1988), First passage times and lumpability of semi-Markov processes, J. Appl. Prob., 25, 675–687.

Sumita, U., and M. Rieders (1989a), A New Algorithm for Computing the Ergodic Probability Vector for Large Markov Chains: Replacement Process Approach, Working Paper Series No. QM-88-08, University of Rochester, William E. Simon Graduate School of Business Administration. Revised, July 1989. To appear in Probability Eng. Info. Sci., 4 (1990) 89–116.

Sumita, U., and M. Rieders (1989b), Application of the Replacement Process Approach for Computing the Ergodic Probability Vector of Large Scale Row-Continuous Markov Chains, Working Paper Series No. QM-88-10, University of Rochester, William E. Simon Graduate School of Business Administration, Revised, April 1989.

Sumita, U., and M. Rieders (1989c), Numerical Comparison of the Replacement Process Approach with the Aggregation-Disaggregation Algorithm for Row-Continuous Markov Chains, Working Paper Series No. QM-89-09, University of Rochester, William E. Simon Graduate School of Business Administration.

Sumita, U., and M. Rieders (1989d), Lumpability and time reversibility in the aggregation-disaggregation method for large Markov chains, *Commun. Statist.-Stochast. Models*, 5, 63–81.

Takahashi, Y. (1975), A Lumping Method for Numerical Calculations of Stationary Distributions of Markov Chains, Research Report No. B-18, Department of Information Sciences, Tokyo Institute of Technology, Tokyo, Japan.

Takahashi, Y. (1985), A new type aggregation method for large Markov chains and its application to queueing networks, in M. Akiyama (ed.), *Proceedings of ITC 11 (Kyoto, Japan 1985)*, North-Holland. Amsterdam, 3.4.A.1.1–1.4.

Takahashi, Y. (1989), Aggregate approximation for acyclic queueing networks with communication blocking, in H. G. Perros and T. Altiok (eds.), *Queueing Networks with Blocking*, Elsevier-North-Holland, Amsterdam, 33–46.

Takahashi, Y., and Y. Takami (1976), A numerical method for the steady-state probabilities of a GI/G/c queueing system in a general class, J. Operations Res. Soc. Jpn., 19, 147–157.

Tijms, H. (1986), *Stochastic Modelling and Analysis, A Computational Approach*, John Wiley and Sons, New York.

Trivedi, K., A. Reibman, and R. Smith (1988), Transient analysis of Markov and Markov reward models, in G. Iazeolla, P. J. Courtois, and O. J. Boxma (eds.), *Computer Performance and Reliability*, North-Holland, Amsterdam, 535–545.

Vakhutinsky, I. Y., L. M. Dudkin, and A. A. Ryvkin (1979), Iterative aggregation—A new approach to the solution of large-scale problems, Econometrica, 47, 821–841.

Van der Wal, J., and P. J. Schweitzer (1987), Iterative bounds on the equilibrium distribution of a finite Markov chain, Probability Eng. Info. Sci., 1, 117–131.

Vantilborgh, H. (1981), The Error of Aggregation. A Contribution to the Theory of Decomposable Systems and Applications, unpublished Ph.D. thesis, These de Docteur en Sciences Appliquees, Faculte des Sciences Appliquees, Universite Catholique de Louvain, Louvain-la-Neuve, Belgium.

Vantilborgh, H. (1985), Aggregation with an error $0(\epsilon^2)$, J. Assoc. Comput. Mach., 32, 162–190.

5

An Exact Aggregation–Disaggregation Algorithm for Mandatory Set Decomposable Markov Chains

DAVID S. KIM* and **ROBERT L. SMITH** Department of Industrial and Operations Engineering, University of Michigan, Ann Arbor, Michigan

ABSTRACT

Many Markov chain models have very large state spaces, making the computation of stationary probabilities very difficult. Often the structure and numerical properties of the Markov chain allows for more efficient computation through state aggregation and disaggregation. In this paper we develop an efficient exact single-pass aggregation–disaggregation algorithm that exploits structural properties of large finite irreducible mandatory set decomposable Markov chains. Mandatory set decomposable structure is a generalization of several other Markov chain structures for which exact aggregation–disaggregation algorithms exist.

---

*Current affiliation: General Motors Research Laboratories, Warren, Michigan.

1. INTRODUCTION

Markov chain modeling is one of the most powerful tools for analyzing complex stochastic systems. Markov chains have been used to model computer and manufacturing systems and have been applied in the social and biological sciences in such diverse areas as manpower planning and genetics. The models are usually developed to understand the long-run behavior of some complex stochastic system. Unfortunately, when modeling many real systems, the state space is often very large and moreover no simple analytic solution for the stationary probabilities exists (in fact the research to be described in this paper was motivated by the development of a finite Markov chain model of an automated guided vehicle system). With no simple analytic solution available, the stationary probabilities can be found by solving a large system of linear equations. As the number of states and hence equations grows, computation of the stationary probabilities becomes impractical in terms of computer memory and computation time.

Often the structure and numerical properties of a Markov chain model can be exploited, in an approximate or exact procedure, to reduce the computational loan and memory requirements. In this paper we develop an efficient exact single-pass aggregation–disaggregation (AD) algorithm that exploits a structure we call "mandatory set decomposable." This structure requires that the states can be partitioned into mutually exclusive and exhaustive sets such that all transitions into a set of states must pass through some state contained in a fixed subset before exiting the set and, for each state in the subset, the conditional stationary probabilities, given that the state is in the mandatory set, are known. This subset is called the "mandatory set."

Exact Markov chain AD algorithms are not new. Either special exact algorithms have been developed or existing iterative AD algorithms converge exactly for Markov chains with special structures. This includes the work of Feinberg and Chiu [1987], Schweitzer [1984] and Guardabassi and Rinaldi [1970], and Sumita and Reiders [1988,1989]. For a general discussion of Markov chain AD algorithms refer to Schweitzer [1984], Cao and Stewart [1985], and Haviv [1987].

In Section 2, we present notation and definitions of various Markov chain structures. Section 3 establishes general results. Finally, Section 4 applies these results to mandatory set decomposable Markov chains and also presents some actual applications. Throughout this paper, consideration will be restricted to discrete time finite irreducible Markov chains.

2. DEFINITIONS

In this section notation and various Markov chain structures, which are special cases of the mandatory set structure, will be defined. Let $\mathbf{P} = [P_{ij}]$ be the transition probability matrix of a finite irreducible Markov chain $\{x_t, t = 0, 1, 2, \ldots\}$ of order n with stationary probability row vector π. Let $A = \{A_1, A_2, \ldots, A_N\}$ be a partition of the states into mutually exclusive and exhaustive sets. Each set A_I, $I = 1, 2, \ldots, N$ makes up an aggregate state which will be referred to as a *macrostate*. The states within a macrostate (or set) will be referred to as *microstates*. Thus we have N macrostates with macrostate I, $I = 1, 2, \ldots, N$ comprised of n_I microstates and $\sum_{I=1}^{N} n_I = n$.

DEFINITION *Exact lumpability* [Schweitzer 1984]. A Markov chain $\{x_t = 0, 1, 2, \ldots\}$ is said to be *exactly lumpable* with respect to the partition A if the sum $\sum_{i \in A_I} P_{ij}$ is the same for every $j \in A_j$. For a Markov chain that is exactly lumpable with respect to partition A, the stationary probabilities for all microstates within a macrostate are equal.

DEFINITION *Weak lumpability* [Kemeny and Snell 1960]. A Markov chain $\{x_t, t = 0, 1, 2, \ldots\}$ is said to be *weakly lumpable* with respect to partition A if the macrostate process (the N state process where each macrostate is viewed as a single state) is a Markov chain for some initial probability vector π^0.

DEFINITION *Single input state decomposable* [Feinberg and Chiu 1987]. A Markov chain $\{x_t, t = 0, 1, 2, \ldots\}$ is said to be *single input state decomposable* with respect to partition A if each macrostate can only be entered through one microstate.

DEFINITION *Single exit state decomposable* (new definition). A Markov chain $\{x_t, t = 0, 1, 2, \ldots\}$ is said to be *single exit state decomposable* with respect to partition A if it is possible to move out of each macrostate only through a single state.

DEFINITION *Mandatory set decomposable* (new definition). A Markov chain $\{x_t, t = 0, 1, 2, \ldots\}$ is said to be *mandatory set decomposable* with respect to partition A if for each macrostate there exists a subset of the macrostate such that once the macrostate is entered, some microstate in this subset must be visited before departure from the macrostate is possible (see Figure 1) and that for each state in the mandatory set, the conditional stationary probabilities given that the state is in the mandatory set, are known. If there is more than one possible mandatory set then the largest such subset is the mandatory

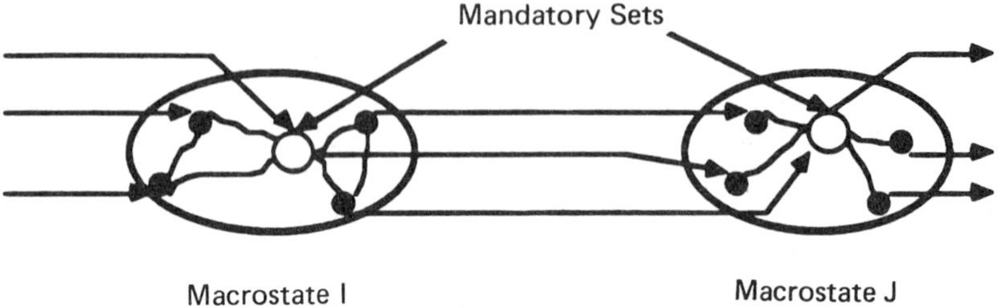

Figure 1 State transition diagram for a mandatory set decomposable Markov chain.

set. If there is no single largest such subset then any of these largest subsets may be the mandatory set.

All of the previous structures defined are special cases of the mandatory set structure. Note that the mandatory set may consist of only a single state as in the single exit and single input cases, or it may consist of all the microstates in the macrostate as in the exactly and weakly lumpable case. Note also that all Markov chains have a trivial partition of the states into singletons, which satisfies the definition of mandatory set decomposability.

3. STOCHASTIC PROCESSES DERIVED FROM AN IRREDUCIBLE MARKOV CHAIN

In this section some stochastic processes derived from a finite irreducible Markov chain will be identified and some general results presented. The results presented in this section form the general framework of the mandatory set algorithm presented in the next section. We will use t as a time index for all processes defined. As each process is presented, the differences in the interpretation of t will be explained. We will also often refer to the Markov chain $\{x_t, t = 0, 1, 2, \ldots\}$ as the *original* Markov chain.

3.1 Macrostate Process

For a finite irreducible Markov chain we can define an N-state stochastic process $\{X_t, t = 0, 1, 2, \ldots\}$, whose order corresponds to the number of macrostates defined by partition A. The following proposition, proved in Meyer [1989], shows that we can define a transition probability matrix for this process so that it is an irreducible Markov chain. Moreover, the stationary probabilities for each state in this defined process will equal the sum of the stationary probabilities of all microstates contained in a macrostate in the original Markov chain.

PROPOSITION 1 Let A be a partition of the state space of $\{x_t, t = 0, 1, 2, \ldots\}$ into N macrostates. Define the order N matrix $\underline{\mathbf{P}} = [\underline{P}_{IJ}]$ as

$$\underline{P}_{IJ} = \sum_{i \in A_I} \left(\frac{\pi_i P_{iJ}}{\sum_{l \in A_I} \pi_l} \right) \tag{1}$$

where $P_{iJ} = \sum_{j \in A_J} P_{ij}$. Then $\underline{\mathbf{P}}$ is stochastic and irreducible. Moreover, if $\mathbf{\Pi}$ is the stationary probability row vector for $\underline{\mathbf{P}}$ then $\mathbf{\Pi}_I = \sum_{i \in A_I} \pi_i$ for $I = 1, 2, \ldots, N$.

From the previous proposition, we see that the transition probabilities and stationary probabilities of each state in the $\{X_t, t = 0, 1, 2, \ldots\}$ process are directly related to the macrostates of the original Markov chain. For this reason the $\{X_t, t = 0, 1, 2, \ldots\}$ process will be called the *macrostate Markov chain* with transition probability matrix $\underline{\mathbf{P}} = [\underline{P}_{IJ}]$ and N-dimensional stationary probability row vector $\mathbf{\Pi}$. The time index t for this process is the same as that for the original Markov chain $\{x_t, t = 0, 1, 2, \ldots\}$.

3.2 Microstate Processes

We can also define N different stochastic processes $\{x_t^I, t = 0, 1, 2, \ldots\}$ by viewing the microstates within macrostate I, $I = 1, 2, \ldots, N$ in isolation. When viewing the microstates in macrostate I in isolation, a transition leaving macrostate I via microstate i and reentering the macrostate through microstate j is defined to be a transition from state i to state j.

We will define transition probabilities for these microstate processes as a function of π and show that these processes have some very useful properties. This will be presented in the following proposition, but first the intuition used in defining these processes will be described. To better understand the intuition behind a microstate process $\{x_t^I, t = 0, 1, 2, \ldots\}$ it will help to consider probability as physical flows. Consider transitions out of microstate i in macrostate I. The next transition will be into microstate j in macrostate I with probability P_{ij}. With probability $\sum_{l \notin A_I} P_{il}$ the next transition will be into a state outside of macrostate I. The transitions out of macrostate I, from all the microstates in macrostate I, will make up the flow of probability out of macrostate I (see Figure 2). The idea is to consider the Markov chain $\{x_t, t = 0, 1, 2, \ldots\}$ to have state distribution π and redirect this probability flow back into macrostate I by distributing it over the microstates appropriately. If we let $\mathbf{P^I} = [P_{ij}^I]$ be the transition probability matrix for the microstate

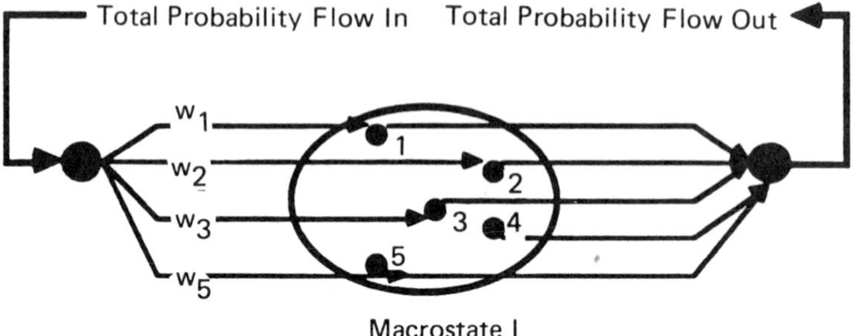

Figure 2 Redirection of transitions back into the macrostate.

process, this suggests that the transition probabilities for $\{x_t^I, t = 0, 1, 2, \ldots\}$, $I = 1, 2, \ldots, N$ be defined as

$$P_{ij}^I = P_{ij} + \left(1 - \sum_{k \in A_I} P_{ik}\right) w_j^I \tag{2}$$

where $w_j^I = \mathbf{Pr}\{x_{t+1} = j \mid x_t \notin A_I, x_{t+1} \in A_I\}$ (i.e. w_j^I is the probability macrostate I is reentered through microstate $j \in A_I$ at time $t + 1$, given that we just entered macrostate I from some state not in macrostate I). Assuming the original Markov chain has state distribution π at time t it can be shown that

$$w_j^I = \frac{\sum_{l \notin A_I} \pi_l P_{lj}}{\sum_{i \notin A_I} \sum_{k \in A_I} \pi_i P_{ik}} \tag{3}$$

We will call these values the *reentering probabilities*. The following proposition is proved in Kim and Smith [1989].

PROPOSITION 2 If the transition probabilities for the microstate process $\{x_t^I, t = 0, 1, 2, \ldots\}$, $I = 1, 2, \ldots, N$, are computed as in equation (2) with the reentering probabilities computed as in equation (3) then the microstate processes are finite irreducible Markov chains. Moreover, if we let $\boldsymbol{\pi}^\mathbf{I}$ be the stationary probability row vector of $\mathbf{P^I}$, then for all $i \in A_I$, $\pi_i^I = \pi_i / \sum_{j \in A_I} \pi_j$.

The stochastic processes $\{x_t^I, t = 0, 1, 2, \ldots\}$, $I = 1, 2, \ldots, N$ will be called *microstate Markov chains* with transition probability matrices $\mathbf{P^I}$ and stationary row vectors $\pi^\mathbf{I}$. We can view the time index t being incremented each time a transition in A_I occurs in the original Markov chain.

$\mathbf{P^I}$ is different from the stochastic complement of A_I in Meyer [1989] and is also different from the replacement process on A_I in Sumita and Reiders [1988]. In general $\mathbf{P^I}$ will be more dense since for a $\{x_t^I, t = 0, 1, 2, \ldots\}$

process, we ignore the fact that we may have left macrostate I from some particular microstate and assume that the chain has state distributions π.

Note that computing w_j^I exactly requires knowledge of π. Fortunately, for certain structures the w_j^I can be computed efficiently and exactly without knowledge of π. One case is the single input state structure. For this case, $w_j^I = l$ if j is the single input state and zero otherwise. We will show that the w_j^I can also be found exactly and efficiently for mandatory set decomposable Markov chains.

3.3 Embedded Processes

Now let $\mathbf{P} = [P_{ij}]$ be the transition probability matrix of a finite irreducible *mandatory set decomposable* Markov chain. Let $A = \{A_1, A_2, \ldots, A_N\}$ be a partition that separates the states of the chain into N mandatory set decomposable macrostates. In general we can assume that $N < n$, so that our partition is not the trivial mandatory set partition. All results that follow also hold when A is the trivial partition.

Results that follow also hold for the case when there are strict subsets of each macrostate that satisfy all the criteria for mandatory sets except that we do not know the conditional stationary probabilities of the states contained in these subsets. If these subsets are the whole macrostates, and we do not know the conditional stationary probabilities of the states contained in the macrostates, then the following results are exactly the same as the previous microstate and macrostate Markov chain results.

To describe the embedded processes we decompose each macrostate into two mutually exclusive and exhaustive subsets. For $I = 1, 2, \ldots, N$ let $A_I = \{B_I, C_I\}$ be a partition of macrostate A_I where $C_I = A_I \setminus B_I$. We define B_I and C_I as follows:

B_I = the states in macrostate I that may be visited prior to visiting a state in the mandatory set

C_I = the states in macrostate I that include the mandatory set and all states that may be visited only after a visit to some state in the mandatory set occurs

An embedded stochastic process $\{z_t, t = 0, 1, 2, \ldots\}$ can be identified if the Markov chain $\{x_t, t = 0, 1, 2, \ldots\}$ is viewed only at those points in time when transitions into and throughout the set $\bigcup_{I=1}^{N} C_I$ occur. The time index t for the embedded process is only incremented when transitions in the set $\bigcup_{I=1}^{N} C_I$ occur in the original Markov chain. We have the following proposition proved in Kim and Smith [1989].

PROPOSITION 3 Suppose $\{x_t, t = 0, 1, 2, \ldots\}$ is a finite irreducible mandatory set decomposable Markov chain and $\{z_t, t = 0, 1, 2, \ldots\}$ is the embedded process corresponding to partition A. Then $\{z_t, t = 0, 1, 2, \ldots\}$ is a finite irreducible Markov chain.

The process $\{z_t, t = 0, 1, 2, \ldots\}$ will be called the *embedded Markov chain*. By itself, there is nothing special about the embedded chain and it can be viewed as simply another finite irreducible Markov chain, and thus propositions 1 and 2 can be applied to $\{z_t, t = 0, 1, 2, \ldots\}$. The macrostates with respect to the original partition A are now sets C_I, $I = 1, 2, \ldots, N$. As with the original Markov chain, we can identify embedded microstate processes $\{z_t^I, t = 0, 1, 2, \ldots\}$, $I = 1, 2, \ldots, N$, and the embedded macrostate process $\{Z_t, t = 0, 1, 2, \ldots\}$.

The embedded microstate processes $\{z_t^I, t = 0, 1, 2, \ldots\}$, $I = 1, 2, \ldots, N$, will be called the *embedded microstate Markov chains* with transition probability matrices $\mathcal{P}^I$ and stationary probability row vectors τ^I. The elements of $\mathcal{P}^I$ are defined as follows:

$$\mathcal{P}_{ij}^I = P_{ij} + \left(1 - \sum_{l \in C_I} P_{il}\right) w_j^{C_I} \tag{4}$$

where

$$w_j^{C_I} = \frac{\sum_{l \notin C_I} \pi_l P_{lj}}{\sum_{i \notin C_I} \sum_{k \in C_I} \pi_i P_{ik}}, \qquad \text{for} \quad j \in C_I.$$

Note that the summations for the $w_j^{C_I}$ include all states in the original Markov chain. The time index t for the embedded process is only incremented when transitions in the set C_I occur in the original Markov chain. We know from proposition 2 that for $i \in C_I$, $I = 1, 2, \ldots, N$,

$$\tau_i^I = \frac{\pi_i}{\sum_{j \in C_I} \pi_j} \tag{5}$$

The embedded process $\{Z_t, t = 0, 1, 2, \ldots\}$ will be called the *embedded macrostate Markov chain* with transition probability matrix $\mathcal{P} = [\mathcal{P}_{IJ}]$ and stationary probability row vector τ. $\mathcal{P}$ is defined as follows

$$\mathcal{P}_{IJ} = \sum_{i \in C_I} \sum_{j \in A_J} \tau_i^I P_{ij} \qquad \text{for} \quad I \neq J \tag{6}$$

$$\mathcal{P}_{II} = 1 - \sum_{J=1}^{N} \mathcal{P}_{IJ} \tag{7}$$

In Kim and Smith [1989] it is shown that when the τ^I satisfy equation (5) then for $I = 1, 2, \ldots, N$,

$$\tau_I = \frac{\sum_{i \in C_I} \pi_i}{\sum_{J=1}^{N} \sum_{j \in C_J} \pi_j} \tag{8}$$

4. EXACT ALGORITHM FOR A MANDATORY SET DECOMPOSABLE MARKOV CHAIN

We now develop the exact algorithm for a mandatory set decomposable Markov chain. If the reentering probabilities are known, then we can proceed as follows:

1. Compute the microstate Markov chain transition probabilities and solve for the stationary probabilities.
2. Using the stationary probabilities, compute the transition probabilities for the macrostate Markov chain and then solve for the macrostate stationary probabilities.
3. Find the stationary probabilities of the original chain by multiplying the microstate stationary probabilities by the corresponding macrostate stationary probability.

This is the same basic algorithm used in Feinberg and Chiu [1987] and in Sumita and Reiders [1988].

The mandatory set structure allows for efficient computation of the reentering probabilities. This will now be developed for the case when the mandatory sets are singletons (the nonsingleton case is treated in the Appendix). When the mandatory sets are singletons, the sets C_I defined in section 3.3 are single input macrostates and we can analyze the embedded microstate Markov chain associated with this macrostate as done in Feinberg and Chiu [1987]. The reentering probabilities for each embedded microstate Markov chain are trivial since we can only reenter through the singleton mandatory sets. From the embedded process results, we can find the conditional stationary probabilities of microstates in the macrostates $C_I, I = 1, 2, \ldots, N$.

We can now take advantage of the embedded macrostate Markov chain to compute the w_j^I exactly for the singleton mandatory set case. We find the w_j^I as follows (assuming the original Markov chain has state distribution π at time t):

$$w_j^I = \mathbf{Pr}\{x_{t+1} = j \mid x_t \notin A_I, x_{t+1} \in A_I\}$$

$$= \sum_{K \neq I} (\mathbf{Pr}\{x_{t+1} = j \mid x_t \in A_K, x_{t+1} \in A_I\}$$
$$\times \mathbf{Pr}\{x_t \in A_K \mid x_t \notin A_I, x_{t+1} \in A_I\}) \tag{9}$$

Here, because of the mandatory set structure, t is a time when the Markov chain is in some set C_K, $K \neq I$, and moves into macrostate I in the next step. Viewing the chain with state distribution π we have, for the first probability in equation (9),

$$\mathbf{Pr}\{x_{t+1} = j \mid x_t \in A_K, x_{t+1} \in A_I\} = \frac{\sum_{i \in C_K} \tau_i^K P_{ij}}{\sum_{i \in C_K} \sum_{j \in A_I} \tau_i^K P_{ij}} \tag{10}$$

For the second probability in equation (9) it can be shown that for $K \neq I$,

$$\mathbf{Pr}\{x_t \in A_K \mid x_t \notin A_I, x_{t+1} \in A_I\} = \frac{\mathbf{Pr}\{x_t \in A_K \mid x_{t+1} \in A_I\}}{\sum_{L \neq I} \mathbf{Pr}\{x_t \in A_I \mid x_{t+1} \in A_I\}}$$

We apply Bayes's rule to compute the probabilities on the right-hand side of the previous expression

$$\begin{aligned}
\mathbf{Pr}\{x_t \in A_K \mid x_{t+1} \in A_I\} &= \mathbf{Pr}\{Z_T = K \mid Z_{T+1} = I\} \\
&= \frac{\mathbf{Pr}\{Z_T = K, Z_{T+1} = I\}}{\mathbf{Pr}\{Z_{T+1} = I\}} \\
&= \frac{\mathbf{Pr}\{Z_{T+1} = I \mid Z_T = K\} * \mathbf{Pr}\{Z_T = K\}}{\mathbf{Pr}\{Z_{T+1} = I\}} \\
&= \frac{\mathcal{P}_{KI} \tau_K}{\tau_I} \qquad (11)
\end{aligned}$$

Note that T denotes the different time index for the embedded macrostate Markov chain. Substituting equations (10) and (11) back into equation (9) gives, after some algebra,

$$w_j^I = \frac{\sum_{K \neq I} \left(\sum_{i \in C_K} \tau_i^K P_{ij}\right) \tau_K}{\tau_I (1 - \mathcal{P}_{II})} \tag{12}$$

Note that we can compute w_j^I without knowing π. Also, since we know the reentering probabilities for the embedded microstate Markov chains, the τ^I, $I = 1, 2, \ldots, N$ and τ satisfy equations (5) and (8) respectively. The mandatory set algorithm is presented next.

MANDATORY SET ALGORITHM

1. If any mandatory sets are not singletons, collapse them into singletons as described in the Appendix.

2. For $I = 1, 2, \ldots, N$ compute the embedded microstate Markov chain transition probability matrices $\mathcal{P}^I$ as in equation (4). Solve for τ^I.
3. Form the $\mathcal{P}$ matrix for the embedded macrostate Markov chain as in equations (6) and (7) and solve for τ.
4. For $I = 1$ to N compute w_j^I for all microstates j in macrostate I as in equation (12).
5. For all macrostates I, compute the microstate Markov chain transition probability matrices $\mathbf{P}^I$ as in equation (2). Solve for $\pi^{\mathbf{I}}$.
6. Compute the $\underline{\mathbf{P}}$ matrix for the macrostate Markov chain as in equation (1) and solve for $\mathbf{\Pi}$.
7. $\pi = [\Pi_1\pi^1, \Pi_2\pi^2, \ldots, \Pi_N\pi^N]$.

Note that since we have mandatory set structure, there may exist subsets of microstates in each macrostate that terminate transitions from other macrostates and also subsets of microstates from which exits out of the macrostate are possible. We can make some of the above computations more efficient by restricting computations to these subsets. The following theorem, proved in Kim and Smith [1989], shows that the mandatory set algorithm computes the exact stationary probabilities.

THEOREM 1 For a mandatory set decomposable Markov chain $\{x_t, t = 0, 1, 2, \ldots\}$ with singletons for all mandatory sets and transition probability matrix $\mathbf{P}$, π produced by the mandatory set algorithm satisfies the system $\pi = \pi\mathbf{P}$, $\sum_{i=1}^{n} \pi_i = 1$.

Under the assumption that the number of macrostate N equals $\sqrt{n}$ and the number of microstates contained in each macrostate equals $\sqrt{n}$ the computational complexity of the algorithm is $O(n^2)$. Note that this algorithm is also very efficient in terms of storage since there will be no fill-in of the original transition probability matrix outside of the main diagonal blocks, as would occur with Gaussian elimination (fill-in occurs when a zero element of the matrix becomes nonzero). Note also that this algorithm is ideally suited for parallel processing, since computationally the microstate Markov chains can be considered as separate systems.

4.1 Applications

In this section we begin by describing some existing models that have the mandatory set structure or for which the mandatory set concept may be of some use. The main purpose of these examples is to illustrate some general types of models where the mandatory set structure may occur.

The first model is a model of social mobility [Henry et al. 1971]. In this model the states of the Markov chain are defined in terms of occupational groups and the length of stay in the group. This state definition is meant to reflect the declining chance of moving to another class as the length of stay in the current class increases and can also be applied to human migration over different geographical locations. In the social science literature these models are known as "cumulative inertia" models, and such models can easily have very large state spaces. The transition probability matrix for this model is shown in Figure 3.

Note that the states of this Markov chain can be partitioned into sets that correspond to each occupational group and that each set forms a single input state macrostate that is a special case of the mandatory set structure.

The second example is a model for police patrols in rural areas [Pollock et al. 1982]. In this model a rural area is broken up into regions and given that a patrol car is in a region, it can be in any of eight possible states. The eight states for region i are:

1. Patrol in region i.
2. Travel from i to a routine or unfounded call.
3. Travel from i to an emergency call.
4. Service of a routine call in i with no calls waiting.
5. Service of a routine call in i with one call waiting.
6. Service of an emergency call in i with no calls waiting.
7. Service of an emergency call in i with one call waiting.
8. Service of an unfounded call in i.

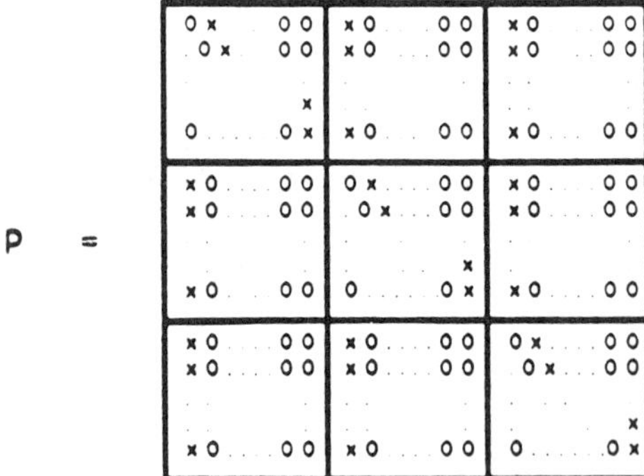

Figure 3 Transition probability matrix for the occupational or geographical mobility model. The x's denote nonzero transition probabilities. [From Bartholomew 1973.]

Now if states two and three above can be combined into a single state or if the probabilities of the two types of travel are known relative to each other, then for a single patrol car this model has mandatory set structure. In particular, if states two and three are combined then the structure is the single exit state decomposable structure.

The last example illustrates the type of structure where the mandatory set algorithm may be the most useful. This is when there exist subsets of the macrostates that satisfy all the requirements of a mandatory set except that the conditional stationary probabilities are not known. It may then be possible to get a good estimate of these conditional stationary probabilities from a limited amount of data or to develop a much smaller model concerning only those states in a particular subset.

Acknowledgments:: We would like to thank all the referees for their feedback. We are very grateful to Paul Schweitzer for his many helpful comments, suggestions, and corrections, which considerably improved this paper.

REFERENCES

Bartholomew, D. J. (1973), *Stochastic Models for Social Process*, 2nd Edition, John Wiley & Sons, New York.

Cao, W., and W. J. Stewart. (1985), Iterative aggregation/disaggregation techniques for nearly uncoupled Markov chains, *J. Assoc. Comput. Machinery*, Vol. 32, No. 3, pp. 702–719.

Feinberg, B. N., and S. S. Chiu. (1987), A method to calculate steady state distributions of large Markov chains, *Operations Res.*, Vol. 35, No. 2, pp. 282–290.

Guardabassi, G., and S. Rinaldi. (1970), Two problems in Markov chains: A topological approach, *Operations Res.*, Vol. 18, No. 2, pp. 324–333.

Haviv, M. (1987), Aggregation/disaggregation methods for computing the stationary distribution of a Markov chains, *SIAM J. Numer. Anal.*, Vol. 24, No. 4, pp. 952–966.

Henry, N. W., R. McGinnis, and H. W. Tegtmeyer. (1971), A finite model of mobility, *J. Math. Sociol.*, Vol. 1, pp. 107–116.

Kemeny, J. G., and J. D. Snell. (1960), *Finite Markov Chains*, D. Van Nostrand, Princeton, N.J.

Kim, D. S., and R. L. Smith. (1989), An Exact Aggregation Algorithm for a Special Class of Markov Chains, Department of Industrial and Operations Engineering, University of Michigan, Ann Arbor, Technical Report 89–2.

Meyer, C. D. (1989), Stochastic complementation, uncoupling Markov chains, and the theory of nearly reducible systems, *SIAM Rev.*, Vol. 31, No. 2, pp. 240–272.

Pollock, S. M., J. R. Birge, and W. J. Hopp. (1982), A Model for Wide Area Patrol Policy Evaluation, Department of Industrial and Operations Engineering, The University of Michigan.

Schweitzer, P. J. (1984), Aggregation methods for large Markov chains, In *Mathematical Computer Performance and Reliability*, G. Iazeolla, P. J. Courtois, and A. Hordijk (eds.), Elsevier Science Publishers B.V. North Holland, Amsterdam.

Sumita, U., and M. Reiders. (1988), Application of the Replacement Process Approach for Computing the Ergodic Probability Vector of Large Row Continuous Markov Chains, Working Paper Series No. QM88–10, William E. Simon Graduate School of Business Administration, University of Rochester.

Sumita, U., and M. Reiders. (1989), Lumpability and time reversibility in the aggregation disaggregation method for large Markov chains, *Commun. Statist.-Stochastic Models*, Vol. 5, No. 1, pp. 63–81.

APPENDIX: GENERAL MANDATORY SET

We now extend our mandatory set result to the case when the mandatory set is not a singleton. When the mandatory set is not a singleton, we can no longer decompose a macrostate into smaller macrostates such that we get a single input macrostate. However, by definition we know the conditional stationary probabilities of each microstate in the mandatory set. We use this information to collapse the mandatory set into a single state and then apply our previous results for singleton mandatory sets.

It is clear that the system with the mandatory set collapsed into a single state will be a finite irreducible Markov chain, as long as all transitions out of states in the mandatory set will still occur out of the single collapsed state. This can be guaranteed by defining the transition probabilities properly. Let $\mathbf{P}^c$ be the transition probability matrix of this collapsed system of order n^c with stationary probability row vector π^c. Let I^* be the singleton mandatory set in the collapsed system and let A_{I^*} denote the nonsingleton mandatory set in the noncollapsed system. The transition probabilities of this collapsed system are defined as follows:

$$P_{ij}^c = P_{ij} \qquad \text{for} \quad i,j \neq I^*$$

$$P_{iI^*}^c = \sum_{j \in A_{I^*}} P_{ij} \qquad \text{for all} \quad i, i \neq I^*$$

$$P_{I^*j}^c = \sum_{k \in A_{I^*}} \left(\frac{\pi_k}{\sum_{i \in A_{I^*}} \pi_i} \right) P_{kj} \qquad \text{for all} \quad j, j \neq I^*$$

$$P_{I^*I^*}^c = \sum_{k \in A_{I^*}} \left(\frac{\pi_k}{\sum_{i \in A_{i^*}} \pi_i} \right) \sum_{j \in A_{I^*}} P_{kj}$$

We have the following proposition proved in Kim and Smith [1989].

PROPOSITION 4 In the collapsed system π^c satisfies

$$\pi_i^c = \pi_i \qquad \text{for} \quad i \neq I^*$$

$$\pi_{I^*}^c = \sum_{i \in A_{I^*}} \pi_i$$

6

On the Sensitivity of Nearly Uncoupled Markov Chains

G. W. STEWART Department of Computer Science and Institute for Advanced Computer Studies, University of Maryland, College Park, Maryland

ABSTRACT

Nearly uncoupled Markov chains (also known as nearly completely decomposable Markov chains) arise in a variety of applications, where they model loosely coupled systems. In such systems it may be difficult to determine the transitions probabilities with high accuracy. This paper investigates the sensitivity of the limiting distribution of the chain to perturbations in the transition probabilities. The conclusion is that nearly uncoupled Markov chains are quite sensitive to such perturbations but the perturbation of the limiting distribution is not arbitrary.

1. INTRODUCTION

A nearly uncoupled Markov chain (NUMC) is a discrete chain whose matrix $\mathbf{P}$ of transition probabilities is almost block diagonal. More precisely, the states of a NUMC with k blocks can be ordered so that its transition matrix assumes the form

$$\mathbf{P} = \mathbf{D} + \mathbf{E} \equiv \begin{bmatrix} \mathbf{D}_{11} & \mathbf{E}_{12} & \cdots & \mathbf{E}_{1k} \\ \mathbf{E}_{21} & \mathbf{D}_{22} & \cdots & \mathbf{E}_{1k} \\ \vdots & \vdots & & \vdots \\ \mathbf{E}_{k1} & \mathbf{E}_{k2} & \cdots & \mathbf{D}_{kk} \end{bmatrix} \tag{1}$$

where the elements of the off-diagonal blocks $\mathbf{E}_{ij}$ are small. Chains of this kind are used to model systems in which certain groups of states are loosely coupled to one another. Since the system spends a relatively large amount of time in a group before passing on to another, it seems natural that a NUMC would achieve steady states within groups rather quickly and a steady state between groups more slowly. This behavior was first noted by Simon and Ando [11] and has been the subject of much subsequent research [e.g., 1, 2, 7, 14].

In addition, Simon and Ando showed that an approximation to the steady-state probability vector π^T could be pieced together from the left Perron vectors* of the diagonal blocks $\mathbf{D}_{ii}$ and an easily formed coupling matrix—a technique that is sometimes called aggregation. When aggregation is combined with iterative techniques, such as block relaxation, the result is a family of powerful algorithms for solving NUMCs.

One practical difficulty with NUMCs is that it is difficult to determine the off-diagonal elements accurately, at least empirically. This is because transitions between blocks are rare events. Consequently, the behavior of the chain must be observed over a long period to estimate the elements of $\mathbf{E}$. It is therefore important to have some idea of the sensitivity of π^T to perturbation in the elements of $\mathbf{P}$. The purpose of this paper is describe the factors that make π^T sensitive to such perturbation.

In the next section we will review the perturbation theory for the left Perron vector of an irreducible Markov chain. The following section consists of a review the theory of NUMCs. We will then apply the results of the two sections to determine the sensitivity of π^T to perturbations in $\mathbf{P}$.

Throughout this paper, we will assume that $\mathbf{P}$ is an irreducible stochastic matrix partitioned as in equation (1). For simplicity we will take $k = 3$ in displayed formulas, the general case being an obvious extension. The vector π^T will denote the unique positive left Perron vector of $\mathbf{P}$ normalized so that

---

* We will use the term Perron vectors to refer to the positive eigenvectors, left and right, of an irreducible nonnegative matrix. Since a NUMC is *a fortiori* aperiodic, it has only one eigenvalue of magnitude one and its left Perron vector is a steady-state vector.

its components sum to one. The corresponding right eigenvector of $\mathbf{P}$, whose elements are all one, will be written $\mathbf{1}$.*

The symbol $\|\cdot\|$ will denote the Euclidean vector norm and the subordinate matrix norm defined by

$$\|\mathbf{A}\| = \max_{\|\mathbf{x}\|=1} \|\mathbf{Ax}\|$$

Since we will be concerned with the behavior of NUMCs as $\mathbf{E}$ approaches zero, we will set

$$\epsilon = \|\mathbf{E}\|$$

2. PERTURBATION THEORY FOR IRREDUCIBLE MARKOV CHAINS

In this section we will be concerned with the following problem. Let $\mathbf{P}$ be an irreducible stochastic matrix (here we do not assume that the associated chain is a NUMC) with left Perron vector $\boldsymbol{\pi}^{\mathrm{T}}$. Let

$$\tilde{\mathbf{P}} = \mathbf{P} + \mathbf{F}$$

be an irreducible, stochastic matrix with left Perron vector $\tilde{\boldsymbol{\pi}}^{\mathrm{T}}$. The problem is to find a bound on $\|\tilde{\boldsymbol{\pi}}^{\mathrm{T}} - \boldsymbol{\pi}^{\mathrm{T}}\|$.

The perturbation theory for the steady state vectors of Markov chains was initiated by Schweitzer [10], and has been developed in various forms since [e.g., 3–6, 9]. Here we present it in a form that emphasizes its interaction with the transient behavior of the chain.

2.1 A Useful Lemma

The following result will be used repeatedly throughout this paper.

LEMMA 2.1 Let $\mathbf{v}^{\mathrm{T}}\mathbf{u} = 1$. Then there are matrices $\mathbf{U}$ and $\mathbf{V}$ with $\|\mathbf{V}\| = 1$ and $\|\mathbf{U}\| = \|\mathbf{u}\|\|\mathbf{v}\|$ such that

$$[\mathbf{u}\ \mathbf{U}]^{-1} = \begin{bmatrix} \mathbf{v}^{\mathrm{T}} \\ \mathbf{V}^{\mathrm{T}} \end{bmatrix}$$

---

*For historical reasons stochastic matrices are defined so that the steady state vector (if it has one) is the left Perron vector of $\mathbf{P}$. This is unfortunate, since the usual convention in matrix algebra is that an unadorned vector is a column vector. In this paper we will keep the right and left sides of our matrices distinct by always transposing the vectors and matrices associated with the latter.

In fact, the matrix $\mathbf{V}$ may be taken to be any matrix with orthonormal columns whose column space is orthogonal to $\mathbf{u}$. The importance of this result is not so much that it shows that there is a pair of inverse matrices that have $\mathbf{u}$ and $\mathbf{v}^T$ for their first column and row, but that these matrices are well behaved if the angle between $\mathbf{u}$ and $\mathbf{v}$ is small. Let us turn to one such case now.

2.2 A Decomposition of P

In the above lemma take $\mathbf{u} = \mathbf{1}$ and $\mathbf{v}^T = \boldsymbol{\pi}^T$. Then $\mathbf{P}$ is similar to the matrix

$$\begin{bmatrix} \boldsymbol{\pi}^T \\ \mathbf{V}^T \end{bmatrix} \mathbf{P}[\mathbf{1} \quad \mathbf{U}] = \begin{bmatrix} \boldsymbol{\pi}^T\mathbf{P1} & \boldsymbol{\pi}^T\mathbf{PU} \\ \mathbf{V}^T\mathbf{P1} & \mathbf{V}^T\mathbf{PU} \end{bmatrix} = \begin{bmatrix} \boldsymbol{\pi}^T\mathbf{1} & \boldsymbol{\pi}^T\mathbf{U} \\ \mathbf{V}^T\mathbf{1} & \mathbf{V}^T\mathbf{PU} \end{bmatrix}$$
$$\equiv \begin{bmatrix} 1 & \mathbf{0} \\ \mathbf{0} & \mathbf{B} \end{bmatrix} \tag{2}$$

Thus we have reduced $\mathbf{P}$ to block diagonal form, in which the eigenvalue one is isolated from the others. These remaining eigenvalues are the eigenvalues of $\mathbf{B}$, and are all less than one in magnitude.

The decomposition of equation (2) has an equivalent form, which, with a suitable generalization, we will use to represent NUMCs. Specifically,

$$\mathbf{P} = [\mathbf{1}\ \mathbf{U}]\,\mathrm{diag}[1, \mathbf{B}] \begin{bmatrix} \boldsymbol{\pi}^T \\ \mathbf{V}^T \end{bmatrix} = \mathbf{1}\boldsymbol{\pi}^T + \mathbf{UBV}^T \equiv \mathbf{S} + \mathbf{T}$$

It is easy to verify that the σth power of $\mathbf{P}$ is given by

$$\mathbf{P}^\sigma = \mathbf{S} + \mathbf{T}^\sigma$$

The nonzero eigenvalues of $\mathbf{T}$ are the same as the eigenvalues of $\mathbf{B}$. If the chain is aperiodic, these eigenvalues are all less than one in magnitude. Hence $\lim_{\sigma\to\infty} \mathbf{T}^\sigma = 0$, and the rate at which it converges determines the rate at which the chain converges to the *steady-state matrix* $\mathbf{S}$. For this reason, we call $\mathbf{T}$ the *transient matrix* of the system. Note that both $\mathbf{S}$ and $\mathbf{T}$ are independent of the choice of $\mathbf{U}$ and $\mathbf{V}$.

2.3 The Perturbation of π

Now let us see what happens when we apply the transformations from the last subsection to the matrix $\tilde{\mathbf{P}} = \mathbf{P} + \mathbf{F}$. Since $\tilde{\mathbf{P}}$ is a stochastic matrix, we have

$$\mathbf{1} = \tilde{\mathbf{P}}\mathbf{1} = \mathbf{P1} + \mathbf{F1} = \mathbf{1} + \mathbf{F1}$$

which implies that $\mathbf{F1} = \mathbf{0}$. Hence

$$\begin{bmatrix} \pi^T \\ \mathbf{V}^T \end{bmatrix} \tilde{\mathbf{P}}[\mathbf{1}\ \mathbf{U}] = \begin{bmatrix} 1 & \pi^T\mathbf{FU} \\ \mathbf{0} & \mathbf{V}^T(\mathbf{P}+\mathbf{F})\mathbf{U} \end{bmatrix} \equiv \begin{bmatrix} 1 & \mathbf{h}^T \\ \mathbf{0} & \tilde{\mathbf{B}} \end{bmatrix} \tag{3}$$

Since $\tilde{\mathbf{P}}$ is an irreducible stochastic matrix, no eigenvalue of $\tilde{\mathbf{B}}$ is one, and $\mathbf{I} - \tilde{\mathbf{B}}$ is nonsingular.

Now let us look for a right eigenvector of the matrix (3) in the form $[1\ \mathbf{p}^T]$. We must have

$$[1\ \mathbf{p}^T] \begin{bmatrix} 1 & \mathbf{h}^T \\ \mathbf{0} & \tilde{\mathbf{B}} \end{bmatrix} = [1\ \mathbf{p}^T]$$

from which it follows that $\mathbf{h}^T + \mathbf{p}^T\tilde{\mathbf{B}} = \mathbf{p}^T$ or

$$\mathbf{p}^T = \mathbf{h}^T[\mathbf{I} - \tilde{\mathbf{B}}]^{-1}$$

In terms of the original coordinate system,

$$\tilde{\pi}^T = \pi^T + \mathbf{p}^T\mathbf{V}^T = \pi^T + \pi^T\mathbf{FU}[\mathbf{I} - \tilde{\mathbf{B}}]^{-1}\mathbf{V}^T \tag{4}$$

(since $\mathbf{V}^T\mathbf{1} = 0$, the vector $\tilde{\pi}^T$ is properly normalized). Hence

$$\frac{\|\tilde{\pi}^T - \pi^T\|}{\|\pi^T\|} \leq \|\mathbf{U}[\mathbf{I} - \tilde{\mathbf{B}}]^{-1}\mathbf{V}^T\|\|\mathbf{F}\| \tag{5}$$

For $\mathbf{F}$ sufficiently small, $[\mathbf{I} - \tilde{\mathbf{B}}]^{-1} \cong [\mathbf{I} - \mathbf{B}]^{-1}$. Hence if we set

$$\mathbf{T}^\natural = \mathbf{U}[\mathbf{I} - \mathbf{B}]^{-1}\mathbf{V}^T$$

(Schweitzer [10] calls this the *fundamental matrix*), we may assert that

The condition of π^T is $\|\mathbf{T}^\natural\|$

The bound (5) will always be somewhat less than sharp. The reason is that $\mathbf{F}$ is not arbitrary but must satisfy $\mathbf{F1} = \mathbf{0}$. Consequently we do not have the freedom to adjust $\mathbf{F}$ so that equality is attained in the inequality $\|\mathbf{FU}[\mathbf{I} - \tilde{\mathbf{B}}]^{-1}\mathbf{V}^T\| \leq \|\mathbf{F}\|\|\mathbf{U}[\mathbf{I} - \tilde{\mathbf{B}}]^{-1}\mathbf{V}^T\|$.

The matrix $\mathbf{T}^\natural$ is the same as the Drazin generalized inverse $[\mathbf{I}-\mathbf{P}]^{\#}$, whose norm is also the condition number for the problem. However, the use of $\mathbf{T}^\natural$ in our bounds is more natural here because it exhibits the matrix $\mathbf{T}$ explicitly. For NUMCs the matrix $\mathbf{T}$ can be further decomposed into a slow transient and a fast transient, and it is the slow transient that controls the sensitivity of π. We now turn to establishing this decomposition.

3. THE THEORY OF NUMCS

3.1 A Typical NUMC

One difficulty in writing about NUMCs is that their behavior is easy to describe qualitatively but requires a good deal of notation to state precisely. The qualitative statement is that for a typical NUMC and an initial state vector $\mathbf{s}_0$ the state vector $\mathbf{s}_t^{\mathrm{T}} = \mathbf{s}_0^{\mathrm{T}}\mathbf{P}^{\sigma}$ at time σ can be decomposed uniquely as the sum of three vectors:

$$\mathbf{s}_{\sigma}^{\mathrm{T}} = \boldsymbol{\pi}^{\mathrm{T}} + \mathbf{y}_{\mathrm{s}}^{(\sigma)\mathrm{T}} + \mathbf{y}_{\mathrm{f}}^{(\sigma)\mathrm{T}} \tag{6}$$

As $\sigma \to \infty$, the vectors $\mathbf{y}_{\mathrm{s}}^{(\sigma)}$ and $\mathbf{y}_{\mathrm{f}}^{(\sigma)}$ approach zero. But $\mathbf{y}_{\mathrm{s}}^{(\sigma)}$ approaches zero at a rate that becomes slower as $\mathbf{E}$ approaches zero; hence the subscript "s," which stands for "slow transient." On the other hand, $\mathbf{y}_{\mathrm{f}}^{(\sigma)}$ is a fast transient, which goes to zero at a rate that is essentially independent of $\mathbf{E}$.

The qualitative statement leaves three questions unanswered.

1. In what subspaces do the vectors $\mathbf{y}_{\mathrm{s}}^{(\sigma)}$ and $\mathbf{y}_{\mathrm{f}}^{(\sigma)}$ lie?
2. What is the relation of these subspaces to the block decomposition (1) of $\mathbf{P}$?
3. Under what conditions does this decomposition remain stable as $\mathbf{E}$ approaches zero?

We will answer each of these questions in the following subsections.

3.2 A Decomposition

In the last section we saw that a stochastic matrix could be reduced to block diagonal form in which one block corresponds to the steady state of the system and the other to the transient. The basic fact about a typical NUMC is that the transient matrix can be further decomposed into a slow transient and a fast transient. Specifically, we can find matrices

$$\mathbf{X} = [\mathbf{1}\ \mathbf{X}_{\mathrm{s}}\ \mathbf{X}_{\mathrm{f}}]$$

and

$$\mathbf{Y}^{\mathrm{T}} = \begin{bmatrix} \boldsymbol{\pi}^{\mathrm{T}} \\ \mathbf{Y}_{\mathrm{s}}^{\mathrm{T}} \\ \mathbf{Y}_{\mathrm{f}}^{\mathrm{T}} \end{bmatrix}$$

with $\mathbf{X}^{-1} = \mathbf{Y}^{\mathrm{T}}$ such that

$$\mathbf{Y}^{\mathrm{T}}\mathbf{P}\mathbf{X} = \mathrm{diag}[1, \mathbf{B}_{\mathrm{s}}, \mathbf{B}_{\mathrm{f}}]$$

Here $\mathbf{X}_{\mathrm{s}}$ and $\mathbf{Y}_{\mathrm{s}}$ have $k - 1$ columns, where k is the number of blocks in the decomposition of $\mathbf{P}$.

There are two important consequences of this diagonalization. First, as in the last section we can show that

$$\mathbf{P}^{\sigma} = \mathbf{1}\boldsymbol{\pi}^{\mathrm{T}} + \mathbf{X}_{\mathrm{s}}\mathbf{B}_{\mathrm{s}}^{\sigma}\mathbf{Y}_{\mathrm{s}}^{\mathrm{T}} + \mathbf{X}_{\mathrm{f}}\mathbf{B}_{\mathrm{f}}^{\sigma}\mathbf{Y}_{\mathrm{f}}^{\mathrm{T}} \equiv \mathbf{S} + \mathbf{T}_{\mathrm{s}}^{\sigma} + \mathbf{T}_{\mathrm{f}}^{\sigma} \tag{7}$$

This means that we can determine the properties of the powers of $\mathbf{P}$ by examining the properties of the smaller matrices $\mathbf{B}_{\mathrm{s}}$ and $\mathbf{B}_{\mathrm{f}}$.

But we can do more. A little matrix algebra shows that any probability vector $\mathbf{s}_0^{\mathrm{T}}$ can be written in the form

$$\mathbf{s}_0^{\mathrm{T}} = \boldsymbol{\pi}^{\mathrm{T}} + \mathbf{g}_{\mathrm{s}}^{\mathrm{T}}\mathbf{Y}_{\mathrm{s}}^{\mathrm{T}} + \mathbf{g}_{\mathrm{f}}^{\mathrm{T}}\mathbf{Y}_{\mathrm{f}}^{\mathrm{T}}$$

where

$$\mathbf{g}_{\mathrm{i}}^{\mathrm{T}} = \mathbf{s}_0^{\mathrm{T}}\mathbf{X}_{\mathrm{i}} \qquad \mathrm{i} = \mathrm{s}, \mathrm{f}$$

From this and equation (7), we see that

$$\mathbf{s}_{\sigma}^{\mathrm{T}} = \boldsymbol{\pi}^{\mathrm{T}} + \mathbf{g}_{\mathrm{s}}^{\mathrm{T}}\mathbf{B}_{\mathrm{s}}^{\sigma}\mathbf{Y}_{\mathrm{s}}^{\mathrm{T}} + \mathbf{g}_{\mathrm{f}}^{\mathrm{T}}\mathbf{B}_{\mathrm{f}}^{\sigma}\mathbf{Y}_{\mathrm{f}}^{\mathrm{T}}$$

Thus the vectors in the decomposition (6) are just

$$\mathbf{y}_{\mathrm{i}}^{(t)\mathrm{T}} = \mathbf{g}_{\mathrm{i}}^{\mathrm{T}}\mathbf{B}_{\mathrm{i}}^{t}\mathbf{Y}_{\mathrm{i}}^{\mathrm{T}} \qquad \mathrm{i} = \mathrm{s}, \mathrm{f}$$

In other words they lie respectively in the subspaces spanned by the rows of $\mathbf{Y}_{\mathrm{s}}^{\mathrm{T}}$ and $\mathbf{Y}_{\mathrm{f}}^{\mathrm{T}}$, and their behavior is controlled by the powers of the matrices $\mathbf{B}_{\mathrm{s}}$ and $\mathbf{B}_{\mathrm{f}}$. It turns out that for a typical NUMC the eigenvalues of $\mathbf{B}_{\mathrm{s}}$ are very near one, while the eigenvalues of $\mathbf{B}_{\mathrm{s}}$ are bounded away from one—hence the slow and fast transients. The matrix $\mathbf{T} = \mathbf{T}_{\mathrm{s}} + \mathbf{T}_{\mathrm{f}}$ is the transient matrix introduced in the last section.

Although the existence of this decomposition is of theoretical interest, its practical importance lies in the fact that we can approximate it in terms of the blocks of the partition (1). We now turn to the details of the approximation.

3.3 How to Dissect a NUMC

In this subsection we will give a step-by-step recipe for approximating the decomposition described is the last subsection. We will begin with the first term of the decomposition $\mathbf{1}\boldsymbol{\pi}^{\mathrm{T}}$. Of course we know this term, since we assume we know $\boldsymbol{\pi}^{\mathrm{T}}$. However, a little creative reexpression will set us up for what follows.

Let us write

$$\boldsymbol{\pi}^{\mathrm{T}} = (\boldsymbol{\pi}_1^{\mathrm{T}}\ \boldsymbol{\pi}_2^{\mathrm{T}}\ \boldsymbol{\pi}_3^{\mathrm{T}})$$

where the partitioning is conformal with equation (1). Since $\boldsymbol{\pi}^{\mathrm{T}} > 0$, none of the components in the above partitioning is zero. Hence we may set

$$\boldsymbol{\pi}^{\mathrm{T}} = (\nu_1\bar{\boldsymbol{\pi}}_1^{\mathrm{T}}\ \nu_2\bar{\boldsymbol{\pi}}_2^{\mathrm{T}}\ \nu_3\bar{\boldsymbol{\pi}}_3^{\mathrm{T}}) \tag{8}$$

where the ν_i are chosen so that the sums of the components of the $\bar{\pi}_i$ are one. Note that $\nu_1 + \nu_2 + \nu_3 = 1$. We will call the numbers ν_i the *coupling coefficients* of the NUMC.

Let

$$\mathbf{Q} = \begin{bmatrix} \mathbf{1} & \mathbf{0} & \mathbf{0} \\ \mathbf{0} & \mathbf{1} & \mathbf{0} \\ \mathbf{0} & \mathbf{0} & \mathbf{1} \end{bmatrix}$$

and

$$\mathbf{R}^{\mathrm{T}} = \begin{bmatrix} \bar{\pi}_1^{\mathrm{T}} & \mathbf{0}^{\mathrm{T}} & \mathbf{0}^{\mathrm{T}} \\ \mathbf{0}^{\mathrm{T}} & \bar{\pi}_2^{\mathrm{T}} & \mathbf{0}^{\mathrm{T}} \\ \mathbf{0}^{\mathrm{T}} & \mathbf{0}^{\mathrm{T}} & \bar{\pi}_3^{\mathrm{T}} \end{bmatrix} \tag{9}$$

Then it is easy to verify that

1. $\mathbf{1} = \mathbf{Q1}$
2. $\pi^{\mathrm{T}} = \mathbf{v}^{\mathrm{T}}\mathbf{R}^{\mathrm{T}}$

These two equations are the expressions for the right and left eigenvectors of $\mathbf{P}$.

Let us now produce an approximation for the slow transient. To do this, we introduce the *coupling matrix*

$$\mathbf{C} = \mathbf{R}^{\mathrm{T}}\mathbf{PQ} = \begin{bmatrix} \pi_1^{\mathrm{T}}\mathbf{D}_{11}\mathbf{1} & \pi_1^{\mathrm{T}}\mathbf{E}_{12}\mathbf{1} & \pi_1^{\mathrm{T}}\mathbf{E}_{13}\mathbf{1} \\ \pi_2^{\mathrm{T}}\mathbf{E}_{21}\mathbf{1} & \pi_2^{\mathrm{T}}\mathbf{D}_{22}\mathbf{1} & \pi_2^{\mathrm{T}}\mathbf{E}_{23}\mathbf{1} \\ \pi_3^{\mathrm{T}}\mathbf{E}_{31}\mathbf{1} & \pi_3^{\mathrm{T}}\mathbf{E}_{32}\mathbf{1} & \pi_3^{\mathrm{T}}\mathbf{D}_{33}\mathbf{1} \end{bmatrix}$$

It is easy to see that $\mathbf{C}$ is stochastic:

$$\mathbf{C1} = \mathbf{R}^{\mathrm{T}}\mathbf{PQ1} = \mathbf{R}^{\mathrm{T}}\mathbf{P1} = \mathbf{R}^{\mathrm{T}}\mathbf{1} = \mathbf{1}$$

Moreover, its steady-state vector is $\mathbf{v}^{\mathrm{T}}$ (the vector of coupling coefficients):

$$\mathbf{v}^{\mathrm{T}}\mathbf{C} = \mathbf{v}^{\mathrm{T}}\mathbf{R}^{\mathrm{T}}\mathbf{PQ} = \pi^{\mathrm{T}}\mathbf{PQ} = \pi^{\mathrm{T}}\mathbf{Q} = \mathbf{v}^{\mathrm{T}}$$

Since $\mathbf{v}^{\mathrm{T}}\mathbf{1} = 1$, we can find matrices $\mathbf{U}$ and $\mathbf{V}$ such that

$$[\mathbf{1}\ \mathbf{U}]^{-1} = \begin{bmatrix} \mathbf{v}^{\mathrm{T}} \\ \mathbf{V}^{\mathrm{T}} \end{bmatrix}$$

Then the matrix

$$\begin{bmatrix} \mathbf{v}^{\mathrm{T}} \\ \mathbf{V}^{\mathrm{T}} \end{bmatrix} \mathbf{C}[\mathbf{1}\ \mathbf{U}] = \begin{bmatrix} \mathbf{v}^{\mathrm{T}}\mathbf{C1} & \mathbf{v}^{\mathrm{T}}\mathbf{CU} \\ \mathbf{V}^{\mathrm{T}}\mathbf{C1} & \mathbf{V}^{\mathrm{T}}\mathbf{CU} \end{bmatrix} = \begin{bmatrix} \mathbf{v}^{\mathrm{T}}\mathbf{1} & \mathbf{v}^{\mathrm{T}}\mathbf{U} \\ \mathbf{V}^{\mathrm{T}}\mathbf{1} & \mathbf{V}^{\mathrm{T}}\mathbf{CU} \end{bmatrix} \equiv \mathrm{diag}(1, \tilde{\mathbf{B}}_s)$$

is similar to the coupling matrix. With this notation we may write down the approximations to the slow transient:

1. $\mathbf{X}_s \cong \mathbf{QU}$
2. $\mathbf{Y}_s^T \cong \mathbf{V}^T\mathbf{R}$
3. $\mathbf{B}_s \cong \tilde{\mathbf{B}}_s$

Finally, let us describe the approximation to the fast transient. Since $\bar{\pi}_i^T\mathbf{1} = 1$, we can find matrices $\mathbf{J}_i$ and $\mathbf{K}_i$ such that

$$[\mathbf{1}\ \mathbf{J}_i]^{-1} = \begin{bmatrix} \bar{\pi}_i^T \\ \mathbf{K}_i^T \end{bmatrix}$$

Let

$$\mathbf{J} = \begin{bmatrix} \mathbf{J}_1 & \mathbf{0} & \mathbf{0} \\ \mathbf{0} & \mathbf{J}_2 & \mathbf{0} \\ \mathbf{0} & \mathbf{0} & \mathbf{J}_3 \end{bmatrix}$$

and

$$\mathbf{K}^T = \begin{bmatrix} \mathbf{K}_1^T & \mathbf{0}^T & \mathbf{0}^T \\ \mathbf{0}^T & \mathbf{K}_2^T & \mathbf{0}^T \\ \mathbf{0}^T & \mathbf{0}^T & \mathbf{K}_3^T \end{bmatrix}$$

and set

$$\tilde{\mathbf{B}}_f = \mathbf{K}^T\mathbf{PJ}$$

Then

1. $\mathbf{X}_f \cong \mathbf{J}$,
2. $\mathbf{Y}_f^T \cong \mathbf{K}^T$
3. $\mathbf{B}_f \cong \tilde{\mathbf{B}}_f$

3.4 Assessment of the Approximations

We now turn to the problem of how good are the approximations described above. We will be concerned with the behavior as ϵ approaches zero. Unfortunately, the hypothesis that $\mathbf{P}$ is a NUMC is not sufficient to guarantee that the approximations become increasingly accurate: we must impose additional regularity conditions. We will now give an informal derivation of the conditions.

Let us begin by backing off from the full decomposition and consider only the decomposition into the coupling matrix and the fast transient. By construction,

$$[\mathbf{Q}\ \mathbf{J}]^{-1} = \begin{bmatrix} \mathbf{R}^{\mathrm{T}} \\ \mathbf{K}^{\mathrm{T}} \end{bmatrix}$$

Let

$$\begin{bmatrix} \mathbf{R}^{\mathrm{T}} \\ \mathbf{K}^{\mathrm{T}} \end{bmatrix} \mathbf{P}[\mathbf{Q}\ \mathbf{J}] = \begin{bmatrix} \mathbf{C} & \mathbf{H}^{\mathrm{T}} \\ \mathbf{G} & \tilde{\mathbf{B}}_{\mathrm{f}} \end{bmatrix} \tag{10}$$

Each of the blocks in this matrix inherits a block structure from the structures of $\mathbf{Q}$, $\mathbf{R}^{\mathrm{T}}$, $\mathbf{J}$, and $\mathbf{K}^{\mathrm{T}}$. For example,

$$\mathbf{G} = \begin{bmatrix} \mathbf{K}_1^{\mathrm{T}}\mathbf{D}_{11}\mathbf{1} & \mathbf{K}_1^{\mathrm{T}}\mathbf{E}_{12}\mathbf{1} & \mathbf{K}_1^{\mathrm{T}}\mathbf{E}_{13}\mathbf{1} \\ \mathbf{K}_2^{\mathrm{T}}\mathbf{E}_{21}\mathbf{1} & \mathbf{K}_2^{\mathrm{T}}\mathbf{D}_{22}\mathbf{1} & \mathbf{K}_2^{\mathrm{T}}\mathbf{E}_{23}\mathbf{1} \\ \mathbf{K}_3^{\mathrm{T}}\mathbf{E}_{31}\mathbf{1} & \mathbf{K}_3^{\mathrm{T}}\mathbf{E}_{32}\mathbf{1} & \mathbf{K}_3^{\mathrm{T}}\mathbf{D}_{e3}\mathbf{1} \end{bmatrix}$$

It is clear from this that all the off-diagonal blocks of $\mathbf{G}$ are of order ϵ. The same is true of the off-diagonal blocks of $\mathbf{H}^{\mathrm{T}}$, $\mathbf{C}$, and $\tilde{\mathbf{B}}_{\mathrm{f}}$. Thus the matrix (10) has the form

$$\begin{bmatrix} c_{11} & * & * & \mathbf{h}_{11}^{\mathrm{T}} & * & * \\ * & c_{22} & * & * & \mathbf{h}_{22}^{\mathrm{T}} & * \\ * & * & c_{33} & * & * & \mathbf{h}_{33}^{\mathrm{T}} \\ \mathbf{g}_{11} & * & * & \mathbf{B}_{11}^{(\mathrm{f})} & * & * \\ * & \mathbf{g}_{22} & * & * & \mathbf{B}_{22}^{(\mathrm{f})} & * \\ * & * & \mathbf{g}_{33} & * & * & \mathbf{B}_{33}^{(\mathrm{f})} \end{bmatrix} \tag{11}$$

where an asterisk indicates a block of order ϵ.

It is important that the matrices $\mathbf{G}$ and $\mathbf{H}^{\mathrm{T}}$ approach zero with ϵ, since this insures that the fast transient becomes uncoupled from the slow transient. From the matrix (11) above we see that we need only be concerned with the diagonal blocks of $\mathbf{G}$ and $\mathbf{H}^{\mathrm{T}}$. To derive a condition under which they approach zero, consider the equation

$$\pi_1^{\mathrm{T}}\mathbf{D}_{11} + \pi_2^{\mathrm{T}}\mathbf{E}_{21} + \pi_3^{\mathrm{T}}\mathbf{E}_{31} = \pi_1^{\mathrm{T}}$$

which follows from the fact that $\pi^{\mathrm{T}}\mathbf{P} = \pi^{\mathrm{T}}$. Since by construction $\pi_1^{\mathrm{T}}\mathbf{J}_1 = 0$, we have

$$\mathbf{h}_{11}^{\mathrm{T}} = \bar{\pi}_1^{\mathrm{T}}\mathbf{D}_{11}\mathbf{J}_1 = -\nu_1^{-1}(\pi_2^{\mathrm{T}}\mathbf{E}_{21} + \pi_1^{\mathrm{T}}\mathbf{E}_{31})$$

If ν_1 remains bounded below as $\epsilon \to 0$, then the right-hand side of the above equation is of order ϵ. A similar argument shows that under similar conditions the diagonal blocks of **G** are of order ϵ. Thus we are lead to our first condition on the behavior of **P** as $\epsilon \to 0$.

REGULARITY CONDITION 1 There is a constant $M_1 > 0$ such that

$$\nu_i \equiv \pi_i^{\mathrm{T}}\mathbf{1} \geq M_1 \qquad i = 1, 2, \ldots, k$$

The second condition we must impose insures, among other things, that the fast transient is really fast. To derive it, note that as $\epsilon \to 0$, the matrix **P** must have at least k eigenvalues near one, since each of the diagonal blocks $\mathbf{D}_{ii}$ is near a stochastic matrix. Now the coupling matrix **C** is a $k \times k$ stochastic matrix with off-diagonal elements that approach zero. Hence **C** approaches the identity matrix, and its eigenvalues account for the k eigenvalues near one. What we need is a condition that insures that the remaining eigenvalues stay away from unity. A natural condition is to require that the eigenvalues of the matrices $\mathbf{B}_{ii}^{(\mathrm{f})}$ be bounded away from unity. But we actually require a stronger condition.

REGULARITY CONDITION 2 There is a constant $M_2 > 0$ such that

$$\|(\mathbf{I} - \mathbf{B}_{ii}^{(\mathrm{f})})^{-1}\| \leq M_2 \qquad i = 1, 2, \ldots, k$$

Since to any eigenvalue λ of $\mathbf{B}_{ii}^{(\mathrm{f})}$ there corresponds an eigenvalue $(1-\lambda)^{-1}$ of $[\mathbf{I} - \mathbf{B}_{ii}^{(\mathrm{f})}]^{-1}$, the second regularity condition implies that $|(1-\lambda)^{-1}| \leq M_2$ or $|(1-\lambda)| \geq M_2^{-1}$; i.e., the eigenvalues of $\mathbf{B}_{ii}^{(\mathrm{f})}$ are really bounded away from unity. But the condition also allows us to ignore the matrices **G** and $\mathbf{H}^{\mathrm{T}}$. In fact, using standard perturbation theory techniques [12,13], we can prove the following theorem.

THEOREM 3.1 Under regularity conditions 1 and 2, there are matrices $\mathbf{\Pi}$ and $\mathbf{\Xi}$ of order ϵ such that

1. $\mathbf{1} = \mathbf{Q1}$
2. $\pi^{\mathrm{T}} = \mathbf{v}^{\mathrm{T}}\mathbf{R}^{\mathrm{T}}$
3. $\mathbf{X}_{\mathrm{s}} = \mathbf{QU} + \mathbf{J\Pi} + O(\epsilon^2)$
4. $\mathbf{Y}_{\mathrm{s}}^{\mathrm{T}} = \mathbf{V}^{\mathrm{T}}\mathbf{R} + \mathbf{\Xi}^{\mathrm{T}}\mathbf{K}^{\mathrm{T}} + O(\epsilon^2)$
5. $\mathbf{B}_{\mathrm{s}} = \tilde{\mathbf{B}}_{\mathrm{s}} + O(\epsilon^2)$
6. $\mathbf{X}_{\mathrm{f}} = \mathbf{J} - \mathbf{QU\Xi}^{\mathrm{T}} + O(\epsilon^2)$
7. $\mathbf{Y}_{\mathrm{f}}^{\mathrm{T}} = \mathbf{J}^{\mathrm{T}} - \mathbf{\Xi V}^{\mathrm{T}}\mathbf{R}^{\mathrm{T}} + O(\epsilon^2)$
8. $\mathbf{B}_{\mathrm{f}} = \tilde{\mathbf{B}}_{\mathrm{f}} + O(\epsilon^2)$

Although we will not pursue the matter in this paper, the foregoing development suggests a computational procedure approximating π^{T}: namely, obtain approximations to the vector $\bar{\pi}_i^{\mathrm{T}}$ and use them in the above recipe. This is the aggregation procedure alluded to in the introduction. There are many ways of obtaining approximations to the $\bar{\pi}_i^{\mathrm{T}}$, but perhaps the simplest is to compute them as the left Perron vectors of the $\mathbf{D}_{ii}$. Unfortunately, this procedure works only once: any attempt to use the new approximation to π^{T} in the recipe gives back the same approximation. This has led to composite methods that combine aggregation with iterative methods, such as block Gauss–Seidel [1,15].*

4. THE PERTURBATION OF NUMCS

We are now ready to combine the results from the last two sections. Specifically, let $\tilde{\mathbf{P}} = \mathbf{P} + \mathbf{F}$ be a perturbation of a NUMC that satisfies regularity conditions one and two. From equation (4), we see that the problem of assessing the effects of $\mathbf{F}$ on the steady-state vector amounts to finding the matrix $\mathbf{T}^{\natural}$.

We have seen that for a typical NUMC the transient matrix decomposes into a slow transient and a fast transient. Specifically, from equation (7) with $\sigma = 1$, we find that

$$\mathbf{T} = \mathbf{T}_{\mathrm{s}} + \mathbf{T}_{\mathrm{f}}$$

Moreover, it is easily verified from the definitions of $\mathbf{T}_{\mathrm{s}}$ and $\mathbf{T}_{\mathrm{f}}$ that if we set

$$\mathbf{T}_{\mathrm{i}}^{\natural} = \mathbf{X}_{\mathrm{i}}(\mathbf{I} - \mathbf{T}_{\mathrm{i}})^{-1}\mathbf{Y}_{\mathrm{i}} \qquad \mathrm{i} = \mathrm{s}, \mathrm{f}$$

then

$$\mathbf{T}^{\natural} = \mathbf{T}_{\mathrm{s}}^{\natural} + \mathbf{T}_{\mathrm{f}}^{\natural}$$

Consequently the condition of π^{H} is given by

$$\|\mathbf{T}^{\natural}\| \leq \|\mathbf{T}_{\mathrm{s}}^{\natural}\| + \|\mathbf{T}_{\mathrm{f}}^{\natural}\| \tag{12}$$

Since from (11)

$$\mathbf{T}_{\mathrm{f}}^{\natural} \cong \mathbf{X}_{\mathrm{f}} \operatorname{diag}[(\mathbf{I} - \mathbf{B}_{11}^{(\mathrm{f})})^{-1}, (\mathbf{I} - \mathbf{B}_{22}^{(\mathrm{f})})^{-1}, (\mathbf{I} - \mathbf{B}_{33}^{(\mathrm{f})})^{-1}]\mathbf{Y}_{\mathrm{f}}^{\mathrm{T}}$$

it follows from the second regularity condition that the second term in the inequality (12) is bounded as $\epsilon \to 0$.

---

*It is worth noting that the analysis of such methods requires the introduction of a third regularity condition to insure that the slow transient is not too slow. Specifically, we must assume that there is a constant M_3 such that $\|(I - B_s)^{-1}\| \leq \epsilon^{-1} M_3$.

The first term is another story. It is equal to $\|\mathbf{X}_s(\mathbf{I} - \mathbf{B}_s)^{-1}\mathbf{Y}_s^T\|$. Now we have noted in the last section that $\mathbf{B}_s$ is equal to $\mathbf{I} + O(\epsilon)$. It follows that

$$\|\mathbf{T}_s^{\natural}\| \geq O(\epsilon^{-1})$$

In other words,

> The condition of $\boldsymbol{\pi}^T$ increases at least in inverse proportion to the size of $\mathbf{E}$

This negative result is perhaps disappointing, but it accords with common sense. If the condition were bounded, we could find a value of $\|\mathbf{F}\|$ for which $\boldsymbol{\pi}^T$ is satisfactorily accurate no matter what the value of ϵ. In particular, if ϵ were less than this value of $\|\mathbf{F}\|$, we could set $\mathbf{E}$ to zero and still get an accurate steady-state vector—which is obvious nonsense.

It should be stressed that the perturbations introduced by the slow transient are by no means arbitrary. From equation (4) and the foregoing, we can write

$$\tilde{\boldsymbol{\pi}}^T \cong \boldsymbol{\pi}^T + \boldsymbol{\pi}^T\mathbf{F}\mathbf{X}_s(\mathbf{I} - \mathbf{B}_s)^{-1}\mathbf{Y}_s^T$$

so that the perturbation in $\boldsymbol{\pi}^T$ is along the row space of $\mathbf{Y}_s^T$. From Item 2 in Theorem 3.1 and the definition [equation (9)] of the matrix $\mathbf{R}$, we see that the perturbation is along the directions of the pieces $\boldsymbol{\pi}_i$ of the steady-state vector. To put it another way, if in analogy with equation (8) we write

$$\tilde{\boldsymbol{\pi}}^T = (\tilde{\nu}_1\bar{\mathbf{r}}\tilde{\bar{\boldsymbol{\pi}}}_1^T \ \tilde{\nu}_2\bar{\mathbf{r}}\tilde{\bar{\boldsymbol{\pi}}}_2^T \ \tilde{\nu}_3\bar{\mathbf{r}}\tilde{\bar{\boldsymbol{\pi}}}_3^T)$$

then the vectors $\bar{\mathbf{r}}\tilde{\bar{\boldsymbol{\pi}}}_1^T$ are quite stable, whereas the coupling coefficients $\tilde{\nu}_i$ are sensitive to perturbations in $\mathbf{P}$. Again this accords with our intuition about NUMCs.

The numerical assessment of the condition of $\boldsymbol{\pi}^T$ is not difficult. Most aggregation procedures require one to compute the an approximation $\tilde{\mathbf{C}}$ to the coupling matrix $\mathbf{C}$. From there it is a small step to compute an approximation to $\mathbf{B}_s$ and estimate the norm of $(\mathbf{I} - \mathbf{B}_s)^{-1}$, which can be done by a number of well-known techniques [8].

There is only one caveat. Since the eigenvalues of $\mathbf{I} - \mathbf{B}_s$ are $O(\epsilon)$, the matrix $\tilde{\mathbf{C}}$ must be an $o(\epsilon)$ approximation to $\mathbf{C}$. Fortunately, it is easy to see that if the approximations to $\bar{\boldsymbol{\pi}}_i$ used to compute $\tilde{\mathbf{C}}$ are accurate to $O(\epsilon)$ then $\tilde{\mathbf{C}} = \mathbf{C} + O(\epsilon^2)$.

ACKNOWLEDGMENT

This work was supported in part by the Air Force Office of Sponsored Research under grant AFOSR-82-0078.

REFERENCES

[1] W. L. Cao and W. J. Stewart (1985). Iterative aggregation/disaggregation techniques for nearly uncoupled Markov chains. J. Assoc. Comput. Mach., 32, 702–719.

[2] P. J. Courtois (1977). *Decomposability.* Academic Press, New York.

[3] P. J. Courtois and P. Semal (1984). Error bounds for the analysis by decomposition of non-negative matrices. In G. Iazeolla, P. J. Courtois, and A. Hordijk, editors, *Mathematical Computer Performance and Reliability*, pages 287–302, North Holland, Elsevier.

[4] R. E. Funderlic (1986). Sensitivity of the stationary distribution vector for an ergodic Markov chain. Linear Algebra Appl. 76, 1–17.

[5] G. H. Golub and C. D. Meyer (1985). Using the QR factorization and group inversion to compute, differentiate, and estimate the sensitivity of stationary probabilities for Markov chains. SIAM J. Algebraic Discrete Methods, 7, 273–281.

[6] M. Haviv and L. van der Heyden (1984). Perturbation bounds for the stationary probabilities of a finite Markov chain. Adv. Appl. Probability, 16, 804–818,

[7] M. Haviv (1987). Aggregation/disaggregation methods for computing the stationary distribution of a Markov chain. SIAM J. Numer. Anal., 24, 952–966.

[8] N. J. Higham (1987). A survey of condition number estimation for triangular matrices. SIAM Rev., 29, 575–596.

[9] C. Meyer and G. W. Stewart (1988). Derivatives and perturbations of eigenvectors. SIAM J. Number Anal., 25, 679–691.

[10] P. J. Schweitzer (1968). Perturbation theory and finite Markov chains. J. Appl. Probability, 5, 401–413.

[11] H. A. Simon and A. Ando (1961). Aggregation of variables in dynamic systems. Econometrica, 29, 111–138.

[12] G. W. Stewart (1973). Error and perturbation bounds for subspaces associated with certain eigenvalue problems. SIAM Rev., 15, 727–764.

[13] G. W. Stewart (1983). Computable error bounds for aggregated Markov chains. J. Assoc. Comput. Mach., 30, 271–285.

[14] G. W. Stewart (1984). On the structure of nearly uncoupled Markov chains. In G. Iazeolla, P. J. Courtois, and A. Hordijk, editors, *Mathematical Computer Performance and Reliability*, pages 287–302, North Holland, Elsevier.

[15] G. W. Stewart, W. J. Stewart, and D. F. McAllister (1984). A Two-Stage Iteration for Solving Nearly Uncoupled Markov Chains. Technical Report TR-1384, Department of Computer Science, University of Maryland.

7

Sensitivity Analysis, Ergodicity Coefficients, and Rank–One Updates for Finite Markov Chains

E. SENETA* Department of Mathematical Statistics, University of Sydney, Sydney, New South Wales, Australia

ABSTRACT

We show that a natural measure of relative sensitivity of the stationary distribution under perturbation of the transition matrix is an ergodicity coefficient of the corresponding fundamental matrix **Z**. Problems of bounding this coefficient in terms of simply computed quantities, and of iterative computation of **Z**, are also addressed. The paper seeks to simplify and unify the work (1986) of Funderlic and Meyer, and of Hunter, by beginning with results of Schweitzer (1968) and use of Bartlett's identity.

---

*Work done in part while on leave (1988–1989) at the University of Virginia, Charlottesville, Virginia.

1. INTRODUCTION

Suppose $\mathbf{P} = \{p_{ij}\}$ is an $(n \times n)$ stochastic matrix containing a single irreducible set of indices, so that there is a unique stationary distribution vector $\pi^T[\pi^T(\mathbf{I}-\mathbf{P}) = \mathbf{0}^T, \pi^T\mathbf{1} = 1]$. Let $\overline{\mathbf{P}}$ be any other $(n \times n)$ stochastic matrix with this structure (the irreducible sets need not coincide), and $\overline{\pi}^T$ its unique stationary distribution vector.

In this paper we focus on two aspects of the perturbation of $\mathbf{P}$ to $\overline{\mathbf{P}}$. Firstly, we consider measurement of the relative sensitivity of the corresponding perturbation of π to $\overline{\pi}$. Such a measurement, in terms of a condition number, has recently been discussed by Funderlic and Meyer (1986). We consider the question in terms of coefficients of ergodicity; the central ideas, with numerical examples, have been briefly presented in Seneta (1988a,b). We also set out the relation of the ergodicity coefficients involved to the eigenvalue structure, inasmuch as the structure of the spectrum is closely related to relative stability under perturbation.

Second, we consider the effect of taking $\mathbf{E} = \overline{\mathbf{P}} - \mathbf{P}$ to be of rank 1. This question, within the context of a sequence of such sequential rank 1 updates that transform the matrix $\mathbf{P}$ to a matrix $\mathbf{P}_A$ of similar structure, with a view to computationally simple sequential modifications of π^T to the corresponding distribution π_A^T of $\mathbf{P}_A$, has been discussed by Hunter (1986).

Our intention is to simplify, unify, and extend the work of Funderlic and Meyer (1986) and Hunter (1986). Further, we are able to do so by beginning with results already present in the early important paper on perturbation of Schweitzer (1968), and making repeated use of Bartlett's identity (a nonstatistical name sometimes used is the Sherman–Morrison identity):

$$\left(\mathbf{A} + \mathbf{a}\mathbf{b}^T\right)^{-1} = \mathbf{A}^{-1} - \frac{(\mathbf{A}^{-1}\mathbf{a})(\mathbf{b}^T\mathbf{A}^{-1})}{1 + \mathbf{b}^T\mathbf{A}^{-1}\mathbf{a}}$$

providing $\mathbf{A}^{-1}$ exists and $\mathbf{b}^T\mathbf{A}^{-1}\mathbf{a} \neq -1$. The usefulness of this identity in the rank 1 perturbation setting has been perceived by Hunter (1986).

Finally, note that our assumptions on $\mathbf{P}$ and $\overline{\mathbf{P}}$ are those of Schweitzer (1968); it is more restrictive, and not necessary, to assume either $\mathbf{P}$ or $\overline{\mathbf{P}}$ irreducible, as is sometimes done.

2. BACKGROUND RESULTS

It is known that under the assumptions on $\mathbf{P}$ and $\overline{\mathbf{P}}$, the corresponding *fundamental matrices* (Kemeny and Snell, 1960) $\mathbf{Z}$ and $\overline{\mathbf{Z}}$ exist, where $\mathbf{Z} = (\mathbf{I} - \mathbf{P} + \mathbf{1}\pi^T)^{-1}$. The following is a restatement of Theorem 2 of Schweitzer (1968).

THEOREM 1 Under our prior conditions on $\mathbf{P}$ and $\overline{\mathbf{P}}$,

$$\overline{\pi}^T = \pi^T \mathbf{H} \tag{1}$$

$$\overline{\mathbf{Z}} = (\mathbf{I} - \mathbf{1}\overline{\pi}^T \mathbf{EZ})\,\mathbf{ZH} \tag{2}$$

where $\mathbf{E} = \overline{\mathbf{P}} - \mathbf{P}$, and $\mathbf{H} = [\mathbf{I} - \mathbf{EZ}]^{-1}$.

The invertibility of $\mathbf{I} - \mathbf{EZ}$ is shown in Schweitzer's (1968) Theorem 1.

If $\mathbf{E} = \mathbf{ab}^T$, $\mathbf{a} \neq \mathbf{0}$, then $\mathbf{I} - \mathbf{EZ} = \mathbf{I} - \mathbf{ac}^T$, where $\mathbf{c}^T = \mathbf{b}^T\mathbf{Z}$. Now if $\mathbf{c}^T\mathbf{a} = 1$, there is clearly a right eigenvector $\mathbf{a}$ of $\mathbf{I} - \mathbf{ac}^T$ corresponding to eigenvalue 0. Since a zero eigenvalue would contradict the invertibility of $\mathbf{I} - \mathbf{EZ}$, it follows $\mathbf{c}^T\mathbf{a} \neq 1$, and so, by Bartlett's identity,

$$\mathbf{H} = (\mathbf{I} - \mathbf{ab}^T\mathbf{Z})^{-1} = \mathbf{I} + \frac{\mathbf{ab}^T\mathbf{Z}}{1 - \mathbf{b}^T\mathbf{Za}} \tag{3}$$

which we shall use in the sequel.

A matrix whose use is often more advantageous than that of $\mathbf{Z}$ is the *group generalized inverse* $\mathbf{A}^{\#}$ of $\mathbf{A} = \mathbf{I} - \mathbf{P}$. In our setting, $\mathbf{A}^{\#} = \mathbf{Z} - \mathbf{1}\pi^T$ (Meyer, 1975).

For any $(m \times n)$ real matrix $\mathbf{B} = \{b_{ij}\}$, an l_p *coefficient of ergodicity* can be defined by

$$\tau_p(\mathbf{B}) = \sup_{\substack{\|\delta^T\|_p = 1 \\ \delta^T \mathbf{1} = 0}} \|\delta^T \mathbf{B}\|_p \tag{4}$$

(see Seneta, 1984), and in the case when $p = 1$,

$$\tau_1(\mathbf{B}) = \tfrac{1}{2} \max_{i,j} \sum_{s=1}^{n} |b_{is} - b_{js}|$$

If $\mathbf{B}$ is $(n \times n)$, and there is a real eigenvalue of r of $\mathbf{B}$ with corresponding (real) left and right eigenvectors $\mathbf{v}^T, \mathbf{w}$ which can be normed so that $\mathbf{v}^T\mathbf{w} = 1$, then

$$\rho(\mathbf{B} - r\mathbf{wv}^T) \leq \tau_1(\mathbf{B})$$

where $\rho(\mathbf{A})$ is the spectral radius of $\mathbf{A}$. Thus, if $\mathbf{B}$ has all its row sums equal (to b, say), then since $\mathbf{B1} = b\mathbf{1}$, it follows that if there is a left eigenvector $\mathbf{v}^T$ corresponding to b such that $\mathbf{v}^T\mathbf{1} = 1$, then

$$\rho(\mathbf{B} - b\mathbf{1v}^T) \leq \tau_1(\mathbf{B}) \tag{5}$$

This last therefore holds for the matrices $\mathbf{B} = \mathbf{P}, \mathbf{Z}, \mathbf{I} - \mathbf{P}, \mathbf{A}^{\#}$, with $\mathbf{v}^T = \pi^T$, and $b = 1, 1, 0, 0$, respectively. (Recall that $\pi^T\mathbf{Z} = \pi^T$, $\mathbf{Z1} = \mathbf{1}$.)

It is worth noting that if all row sums of $\mathbf{B}$ are equal, then the expressions for $\tau_1(\mathbf{B})$ can be rewritten

$$\tau_1(\mathbf{B}) = b - \min_{i,j} \sum_{s=1}^{n} \min(b_{is}, b_{js}) \tag{6}$$

It is clear from equation (6) that $0 \leq \tau_1(\mathbf{S}) \leq 1$ for any stochastic matrix $\mathbf{S}$. For a detailed discussion of $\tau_1(\mathbf{S})$ for stochastic $\mathbf{S}$, see Seneta (1981). In particular, if $\tau_1(\mathbf{S}) < 1$, $\mathbf{S}$ (then called *scrambling*) contains only a single irreducible set of states which is aperiodic. Thus if the irreducible set of $\mathbf{P}$ of our previous section is periodic, then $\tau_1(\mathbf{P}) = 1$. Since the property of the kind $\tau_1(\mathbf{S}) < 1$ will be seen from the sequel to be desirable, we can sometimes achieve it for our $\mathbf{P}$ through the transformation

$$\mathbf{P}^* = \tfrac{1}{2}(\mathbf{I} + \mathbf{P})$$

which eliminates any periodicity present. $\mathbf{P}^*$ has the same irreducible set, and same stationary distribution vector $\boldsymbol{\pi}^T$, as $\mathbf{P}$. Further, since

$$\begin{aligned} \mathbf{I} - \mathbf{P} + \mathbf{1}\boldsymbol{\pi}^T &= 2(\mathbf{I} - \mathbf{P}^* + \mathbf{1}\boldsymbol{\pi}^T - (\mathbf{1}\boldsymbol{\pi}^T)/2) \\ &= 2((\mathbf{Z}^*)^{-1} - (\mathbf{1}\boldsymbol{\pi}^T)/2) \end{aligned}$$

it follows from Bartlett's identity that

$$\mathbf{Z} = \tfrac{1}{2}(\mathbf{Z}^* + \mathbf{1}\boldsymbol{\pi}^T) \tag{7}$$

since $\mathbf{Z}^*\mathbf{1} = \mathbf{1}$, $\boldsymbol{\pi}^T\mathbf{Z}^* = \boldsymbol{\pi}^T$. It follows from the definition (4) that

$$\tau_1(\mathbf{Z}) = \tau_1(\mathbf{A}^{\#}) = \tau_1(\mathbf{Z}^*)/2 \tag{8}$$

Of course, a more general transformation than $\mathbf{P}^*$ of $\mathbf{P}$, that is, $\alpha\mathbf{I} + (1 - \alpha)\mathbf{P}$, $0 < \alpha < 1$, could be used to achieve similar ends.

3. COEFFICIENTS OF ERGODICITY AND RELATIVE SENSITIVITY

The relative effect on $\boldsymbol{\pi}^T$ of the perturbation $\mathbf{E}$ to $\mathbf{P}$ is measured by the quantity

$$\frac{\|\bar{\boldsymbol{\pi}}^T - \boldsymbol{\pi}^T\|_1 / \|\boldsymbol{\pi}^T\|_1}{\|\mathbf{E}\|_1 / \|\mathbf{P}\|_1} = \frac{\|\bar{\boldsymbol{\pi}}^T - \boldsymbol{\pi}^T\|_1}{\|\mathbf{E}\|_1} \tag{9}$$

the equality following since $\|\boldsymbol{\pi}^T\|_1 = 1 = \|\mathbf{P}\|_1$. (With an l_p norm, $1 \leq p \leq \infty$, where $p \neq 1$, this simplification would not obtain.)

THEOREM 2 Under our prior conditions on $\mathbf{P}$ and $\overline{\mathbf{P}}$,

$$\frac{\|\overline{\boldsymbol{\pi}}^T - \boldsymbol{\pi}^T\|_1/\|\boldsymbol{\pi}^T\|_1}{\|\mathbf{E}\|_1/\|\mathbf{P}\|_1} \leq \tau_1(\mathbf{Z}) \tag{10}$$

Moreover

$$\tau_1(\mathbf{Z}) = \tau_1(\mathbf{C}(\mathbf{u},\mathbf{v})) \leq \|\mathbf{C}(\mathbf{u},\mathbf{v})\|_1 \tag{11}$$

where for any $\mathbf{v}$, and any $\mathbf{u}$ such that $\mathbf{u}^T\mathbf{1} \neq 0$,

$$\mathbf{C}(\mathbf{u},\mathbf{v}) = (\mathbf{I} - \mathbf{P} + \mathbf{1}\mathbf{u}^T)^{-1} - \mathbf{1}\mathbf{v}^T \tag{12}$$

Proof. From equation (1) of Theorem 1, it follows that

$$\overline{\boldsymbol{\pi}}^T - \boldsymbol{\pi}^T = \overline{\boldsymbol{\pi}}^T\mathbf{E}\mathbf{Z}$$

[which is the starting point of Funderlic and Meyer (1986)]. Taking inverses,

$$(\overline{\boldsymbol{\pi}}^T - \boldsymbol{\pi}^T)(\mathbf{I} - \mathbf{P} + \mathbf{1}\overline{\boldsymbol{\pi}}^T) = \overline{\boldsymbol{\pi}}^T\mathbf{E}$$

that is,

$$(\overline{\boldsymbol{\pi}}^T - \boldsymbol{\pi}^T)(\mathbf{I} - \mathbf{P} + \mathbf{1}\mathbf{u}^T) = \overline{\boldsymbol{\pi}}^T\mathbf{E}$$

for any $\mathbf{u}$, since $(\overline{\boldsymbol{\pi}}^T - \boldsymbol{\pi}^T)\mathbf{1} = 0$; and if $\mathbf{u}^T\mathbf{1} \neq 0$, $(\mathbf{I} - \mathbf{P} + \mathbf{1}\mathbf{u}^T)$ is invertible, so

$$\overline{\boldsymbol{\pi}}^T - \boldsymbol{\pi}^T = \overline{\boldsymbol{\pi}}^T\mathbf{E}\left\{(\mathbf{I} - \mathbf{P} + \mathbf{1}\mathbf{u}^T)^{-1} - \mathbf{1}\mathbf{v}^T\right\}$$

since $\overline{\boldsymbol{\pi}}^T\mathbf{E}\mathbf{1} = 0$, since $\mathbf{E}\mathbf{1} = \mathbf{0}$. Thus

$$\overline{\boldsymbol{\pi}}^T - \boldsymbol{\pi}^T = \overline{\boldsymbol{\pi}}^T\mathbf{E}\mathbf{C}(\mathbf{u},\mathbf{v})$$

Taking norms,

$$\|\overline{\boldsymbol{\pi}}^T - \boldsymbol{\pi}^T\|_1 \leq \|\overline{\boldsymbol{\pi}}^T\mathbf{E}\|_1 \|(\overline{\boldsymbol{\pi}}^T\mathbf{E}/\|\overline{\boldsymbol{\pi}}^T\mathbf{E}\|_1)\mathbf{C}(\mathbf{u},\mathbf{v})\|_1$$

whenever $\|\overline{\boldsymbol{\pi}}^T\mathbf{E}\|_1 \neq 0$;

$$\|\overline{\boldsymbol{\pi}}^T - \boldsymbol{\pi}^T\|_1 \leq \|\overline{\boldsymbol{\pi}}^T\mathbf{E}\|_1\tau_1(\mathbf{C}(\mathbf{u},\mathbf{v}))$$

and so, since $\|\overline{\boldsymbol{\pi}}^T\mathbf{E}\|_1 \leq \|\overline{\boldsymbol{\pi}}_1^T\|_1\|\mathbf{E}\|_1$, using equation (9), we obtain as upper bound of the relative sensitivity, $\tau_1(\mathbf{C}(\mathbf{u},\mathbf{v}))$. In particular, equation (10) obtains. Further, from equation (12),

$$\mathbf{C}(\mathbf{u},\mathbf{v}) = \left(\mathbf{Z}^{-1} + \mathbf{1}(\mathbf{u} - \boldsymbol{\pi})^T\right)^{-1} - \mathbf{1}\mathbf{v}^T$$

and since $(\mathbf{u} - \boldsymbol{\pi})^T\mathbf{1} = \mathbf{u}^T\mathbf{1} - 1 \neq -1$, it follows from Bartlett's identity that

$$\begin{aligned}\mathbf{C}(\mathbf{u},\mathbf{v}) &= \mathbf{Z} - \frac{\mathbf{Z}\mathbf{1}(\mathbf{u} - \boldsymbol{\pi})^T\mathbf{Z}}{1 + (\mathbf{u} - \boldsymbol{\pi})^T\mathbf{Z}\mathbf{1}} \\ &= \mathbf{Z} - \frac{\mathbf{1}(\mathbf{u} - \boldsymbol{\pi})^T\mathbf{Z}}{\mathbf{u}^T\mathbf{1}}\end{aligned}$$

whence from the definition (4) of τ, $\tau_1(\mathbf{C}(\mathbf{u},\mathbf{v})) = \tau_1(\mathbf{Z})$. The last inequality in equation (11) is obvious, from the definition (4).

Theorem 2 suggests that a natural *condition number* [to measure sensitivity of perturbation in our setting, following the lead of Funderlic and Meyer (1986)], through the use of matrix norms, is $\tau_1(\mathbf{Z})$, or equivalently, $\tau_1(\mathbf{A}^{\#})$ [noting $\mathbf{C}(\boldsymbol{\pi},\mathbf{0}) = \mathbf{Z}$, $\mathbf{C}(\boldsymbol{\pi},\boldsymbol{\pi}) = \mathbf{A}^{\#}$]. The use of rank 1 modification procedures, as embodied in equation (10) through use of arbitrary $\mathbf{u}$ and $\mathbf{v}$, is a frequently used device in perturbation theory [see, e.g., the references in Seneta (1988a)]. It is occasionally useful in this setting also via equation (11) through $\|\mathbf{C}(\mathbf{u},\mathbf{v})\|_1$ (which does depend on $\mathbf{u},\mathbf{v}$) if $\mathbf{C}(\mathbf{u},\mathbf{v})$ can be put into propitious form through relevant choice of $\mathbf{u},\mathbf{v}$ (Seneta, 1988a).

An easily computable bound on $\tau_1(\mathbf{Z})$ is also obtainable from the following [note from equation (6) that $\tau_1(\mathbf{P}) < 1 \Rightarrow \tau_1(\mathbf{P}^*) < 1$].

THEOREM 3 Under our conditions on $\mathbf{P}$ and $\overline{\mathbf{P}}$, if $\tau_1(\mathbf{P}) < 1$, then $\tau_1(\mathbf{Z}) \leq (1-\tau_1(\mathbf{P}))^{-1}$, and if $\tau_1(\mathbf{P}^*) < 1$, then

$$\tau_1(\mathbf{Z}) \leq \tfrac{1}{2}(1-\tau_1(\mathbf{P}^*))^{-1}$$

Proof. The result for $\tau_1(\mathbf{P}^*)$ follows from that for $\tau_1(\mathbf{P})$ through the use of equation (8). When $\tau_1(\mathbf{P}) < 1$, since from equation (4)

$$\tau_1(\mathbf{Z}) = \sup_{\substack{\|\boldsymbol{\delta}^T\|_1=1 \\ \boldsymbol{\delta}^T\mathbf{1}=0}} \|\boldsymbol{\delta}^T(\mathbf{I}-\mathbf{P}+\mathbf{1}\boldsymbol{\pi}^T)^{-1}\|_1$$

if we follow the development of Section 2 of Seneta (1988b) with $\boldsymbol{\delta}$ replacing α, we obtain

$$\tau_1(\mathbf{Z}) \leq \|\boldsymbol{\delta}^T\|_1 \sum_{k=0}^{\infty} (\tau_1(\mathbf{P}))^k$$

whence the required follows.

The relation of Theorem 3 to the results of Funderlic and Meyer (1968) has been discussed in Seneta (1988a,b).

To conclude this section, we note by example that equality may obtain in equation (10). If we take $\mathbf{P}$ (respectively $\overline{\mathbf{P}}$) to have its first (respectively, second) column to be $\mathbf{1}$, and remaining columns zero, then $\boldsymbol{\pi}^T = \mathbf{f}_1^T, \overline{\boldsymbol{\pi}}^T = \mathbf{f}_2^T$, where $\mathbf{f}_i^T$ is the vector with unit in the ith position and zeros elsewhere, we find $\|\overline{\boldsymbol{\pi}}^T - \boldsymbol{\pi}^T\|_1 = 2$, $\|\mathbf{E}\|_1 = 2$ so the left-hand side of equation (10) is unity, while $\tau_1(\mathbf{P}) = 0$, so from Theorem 3, $\tau_1(\mathbf{Z}) \leq 1$, whence from equation (10), $\tau_1(\mathbf{Z}) = 1$. (Note the equality in the bound of Theorem 3 also, for this example.)

Deeper optimality questions of theoretical interest, which we do not address here, relate to the value of the maximum of equation (9) when optimized over all suitable $\overline{\mathbf{P}}$ for a given $\mathbf{P}$, and of the minimum of $\|\mathbf{C}(\mathbf{u}, \mathbf{v})\|_1$ in equation (11) minimized over all $\mathbf{u}$ and $\mathbf{v}$.

4. EIGENVALUES AND RELATIVE SENSITIVITY

Eigenvalues of $\mathbf{P}$ close to unity are a typical cause of instability of $\boldsymbol{\pi}^T$ following a small change of $\mathbf{P}$ to $\overline{\mathbf{P}}$. This can be perceived when $\mathbf{P}$ has almost more than one irreducible set (then there will be a second eigenvalue close to unity). Also, if the irreducible set of $\mathbf{P}$ is periodic with large periodicity d, another empirically observed case of instability, the eigenvalues of $\mathbf{P}$ will include $\exp(2\pi ij/d), j = 1, \ldots, d-1$, several of which will be close to unity. The transformation to $\mathbf{P}^* = \frac{1}{2}(\mathbf{I} + \mathbf{P})$ makes the irreducible set aperiodic, the above eigenvalues transforming to $\frac{1}{2}(1 + \exp(2\pi ij/d))$. The corresponding modulus is $\{(1 + \cos(2\pi j/d))/2\}^{1/2}$, so, although these transformed eigenvalues are no longer on the unit circle, the closeness to unity of, say the cases $j = 1, j = d - 1$ persists. It is therefore necessary to investigate briefly the relationship between eigenvalues of $\mathbf{P}$, $\mathbf{Z}$, $\mathbf{I} - \mathbf{P}$, $\mathbf{A}^{\#}$, and how the closeness to unity of eigenvalues of $\mathbf{P}$ affects $\tau_1(\mathbf{Z})$.

Suppose $\lambda \neq 1$ is an eigenvalue of $\mathbf{Z}$ (since $\mathbf{Z}$ is invertible, $\lambda \neq 0$) and $\mathbf{x}$ ($\neq \mathbf{0}$) a corresponding right eigenvector; since $\boldsymbol{\pi}^T$ is a left eigenvector corresponding to a different eigenvalue (that is, unity), it follows that $\boldsymbol{\pi}^T\mathbf{x} = 0$, whence

$$\mathbf{x} = \lambda\mathbf{Z}^{-1}\mathbf{x} = \lambda(\mathbf{I} - \mathbf{P} + \mathbf{1}\boldsymbol{\pi}^T)\mathbf{x} = \lambda(\mathbf{I} - \mathbf{P})\mathbf{x}$$

so $1/\lambda$ is an eigenvalue of $\mathbf{I} - \mathbf{P}$ (clearly $\lambda \neq 0$), whence $1 - \lambda^{-1} (\neq 0, 1)$ is an eigenvalue of $\mathbf{P}$. Conversely, if $\rho \neq 0, 1$ is an eigenvalue of $\mathbf{P}$, then $(1 - \rho)^{-1}$ is an eigenvalue of $\mathbf{Z}$. If λ is an eigenvalue of $\mathbf{Z}$, $\lambda - 1$ is the corresponding eigenvalue of $\mathbf{A}^{\#}$. Thus in terms of ergodicity coefficients, for any eigenvalue $\rho \neq 0, 1$ of $\mathbf{P}$, from equation (5)

$$\frac{1}{\tau_1(\mathbf{I} - \mathbf{P})} \leq \frac{1}{|1 - \rho|} \leq \tau_1(\mathbf{Z}) \equiv \tau_1(\mathbf{A}^{\#})$$

Thus, as expected, the presence of a $\rho \neq 1$ but close to unity gives a large value of $\tau_1(\mathbf{Z})$. It is conceivable that $\tau_1(\mathbf{Z})$ may be large even if there is no such eigenvalue; whether this can occur, and, if so, whether instability of $\boldsymbol{\pi}^T$ is indicated, will not be addressed here.

5. SEQUENTIAL RANK-ONE UPDATES

Although Theorem 3 provides a simply computed upper bound for $\tau_1(\mathbf{Z})$, in general we are faced with the problem of prior computation of $\mathbf{Z}$, or $\mathbf{A}^{\#}$. Inversion of a matrix is in general an expensive operation that one might wish to avoid (Meyer, 1975; Funderlic and Meyer, 1986). A reduction of the problem emerges from the following.

In the case that $\mathbf{E} = \overline{\mathbf{P}} - \mathbf{P} = \mathbf{a}\mathbf{b}^T$, $\mathbf{a} \neq \mathbf{0}$, it follows from equations (1), (2), and (3) that

$$\overline{\pi}^T = \pi^T \left(I + \frac{\mathbf{a}\mathbf{b}^T\mathbf{Z}}{1 - \mathbf{b}^T\mathbf{Z}\mathbf{a}} \right) \tag{13}$$

$$\overline{\mathbf{Z}} = (I - (\overline{\pi}^T\mathbf{a})\,\mathbf{1}\mathbf{b}^T\mathbf{Z}) \left(\mathbf{Z} + \frac{\mathbf{Z}\mathbf{a}\mathbf{b}^T\mathbf{Z}}{1 - \mathbf{b}^T\mathbf{Z}\mathbf{a}} \right) \tag{14}$$

Suppose we have available the matrix $\mathbf{P}(= \mathbf{P}_0)$ (containing a single irreducible class of indices), its stationary distribution π_0^T, and corresponding $\mathbf{Z} = (\mathbf{Z}_0)$; and $\mathbf{P}$ is perturbed by a sequence of m rank 1 updates

$$\mathbf{P}_{k+1} = \mathbf{P}_k + \mathbf{a}_{k+1}\mathbf{b}_{k+1}^T \qquad k = 0, \ldots, m-1$$

so that each $\mathbf{P}_{k+1}$ still contains a single irreducible class, and $\mathbf{P}_m = \mathbf{P}_A$ is a matrix for which we desire to know the stationary distribution π_A^T (Hunter, 1986) and/or the corresponding $\mathbf{Z}$ matrix $\mathbf{Z}_A$. It is clear that π_A^T and $\mathbf{Z}_A$ can be obtained iteratively by repeating equation (14) after equation (13) in succession to obtain from π_0^T, $\mathbf{Z}_0$, π_k^T, $\mathbf{Z}_k$, $k = 1, \ldots, m$ (with obvious notation), by replacing in equations (13) and (14) π^T, $\mathbf{Z}$, $\overline{\pi}^T$, and $\overline{\mathbf{Z}}$ by π_{k-1}^T, $\mathbf{Z}_{k-1}$, π_k^T, and $\mathbf{Z}_k$, respectively. Hunter's (1986) proposed iterative schemes involve sequential construction of more quantities at each state, and are more involved.

It is clear that if one's primary interest is in π_A^T and/or $\mathbf{Z}_A$, as ours is in the $\mathbf{Z}$ of our earlier sections, one would wish to start with a $\mathbf{P}_0$ that is sufficiently simple that π_0^T and $\mathbf{Z}_0$ can be written down immediately. Apart from the simplicity of π_0^T and $\mathbf{Z}_0$, care must be taken with the rank one updates so that $\mathbf{P}_{k+1}$, $k = 0, \ldots, m-1$, retain the property of containing a single irreducible class of indices. The following choices achieve these aims.

Take

$$\mathbf{P}_0 = \mathbf{1}\mathbf{1}^T\mathbf{P}_A/n \tag{15}$$

That is, $\mathbf{P}_0$ is of rank one, with every row being the average of the rows of the "final" matrix $\mathbf{P}_A$. Since $\mathbf{P}_0$ has positive entries in at least the same positions as $\mathbf{P}_A$, it likewise must contain at most one irreducible class. Thus

$$\pi_0^T = \mathbf{1}^T\mathbf{P}_A/n \qquad \mathbf{Z}_0 = (\mathbf{I} - \mathbf{P}_0 + \mathbf{1}\pi_0^T)^{-1} = \mathbf{I} \tag{16}$$

For the updates, replace each row in the current approximating matrix $\mathbf{P}_k$ in turn by the corresponding one of $\mathbf{P}_A$ to go to $\mathbf{P}_{k+1}$. That is:

$$\mathbf{a}_{k+1} = \mathbf{f}_{k+1} \qquad \mathbf{b}_{k+1}^T = \mathbf{f}_{k+1}^T(\mathbf{P}_A - \mathbf{P}_0) \qquad k = 0, \ldots, m-1 \tag{17}$$

This will result in π_A^T, $\mathbf{Z}_A$ in $m = n$ iterative steps.

Obvious modifications of equations (15)–(17) suggest themselves. If $\mathbf{P}_A$ has a strictly positive row, a better choice of $\mathbf{P}_0$ may be the matrix with every row equal to this strictly positive one. This will save on at least one update.

Acknowledgment: Section 4 owes much to a stimulating discussion with Carl D. Meyer.

REFERENCES

Funderlic, R. E., and Meyer, C. D. (1986). Sensitivity of the stationary distribution vector for an ergodic Markov chain. *Linear Algebra Appl.*, 76, 1–17.

Hunter, J. J. (1986). Stationary distributions of perturbed Markov chains. *Linear Algebra Appl.*, 82, 201–214.

Kemeny, J. G., and Snell, L. J. (1960). *Finite Markov Chains*. Van Nostrand, New Jersey.

Meyer, C. D. (1975). The role of the group generalized inverse in the theory of finite Markov chains. *SIAM Rev.*, 17, 443–464.

Schweitzer, P. J. (1968). Perturbation theory and finite Markov chains. *J. Appl. Prob.*, 5, 401–413.

Seneta, E. (1981). *Non-Negative Matrices and Markov Chains*. 2nd ed. Springer-Verlag, New York.

Seneta, E. (1984). Explicit forms for ergodicity coefficients and spectrum localization. *Linear Algebra Appl.*, 60, 187–197.

Seneta, E. (1988a). Sensitivity to perturbation of the stationary distribution: Some refinements. *Linear Algebra Appl.*, 108, 121–126.

Seneta, E. (1988b). Perturbation of the stationary distribution measured by ergodicity coefficients. *Adv. Appl. Prob.*, 20, 228–230.

8

Iterative Methods for Determining Derivatives of Stationary Distributions of Finite Markov Chains

JOSE J. CRUZ Department of Mechanical/Industrial Engineering, New York Institute of Technology, Old Westbury, New York

SHALER STIDHAM, JR. Department of Operations Research, University of North Carolina, Chapel Hill, North Carolina

ABSTRACT

We present algorithms for obtaining derivatives of steady-state probability distributions of finite Markov chains. Our focus is on a conceptually simple approach for developing such algorithms from established methods of computing steady state probability distributions. We briefly discuss our prototype (an application of the approach to the power method) and present the results of applying the approach to LOPSI (lopsided iteration) and Neut's matrix geometric methods.

This material is based upon work supported by the National Science Foundation under grant DDM-8719825. The government has certain rights in this material. Any opinions, findings, and conclusions or recommendations expressed in this material are those of the author(s) and do not necessarily reflect the views of the National Science Foundation.

1. INTRODUCTION

Consider an aperiodic irreducible finite Markov chain with transition matrix P and steady-state probability distribution π; and assume that the entries of P are differentiable with respect to a design parameter h with $h \in H \subseteq R$, the feasible region for h. We let a prime (′) denote the operator $\partial(\quad)/\partial h$. It has been shown in a number of papers (see [2], [3], and [18]) that, given π, its derivative π' can be determined from the equation

$$\pi' = \lim_{r\to\infty} \pi P' \sum_{k=0}^{r} P^k$$

where P' denotes the matrix where P'_{ij} is the derivative of the (i,j) entry of P.

The identity

$$\pi P' \sum_{k=0}^{r} P^k \equiv \pi P' + \left(\pi P' \sum_{k=0}^{r-1} P^k\right) P$$

suggests the following algorithm (where s'_i denotes the ith estimate of π'):

Algorithm A

(i) Initial condition:

$$s'_0 = 0$$

(ii) rth iteration:

$$s'_r = (s_{r-1})'P + \pi p'$$

We refer to the above method as the *sequential* method since π must be estimated ahead of π'.

By contrast, consider the following alternative series form expression for π',

$$\pi' = \lim_{r\to\infty} a \sum_{k=0}^{r-1} P^k P' P^{r-k-1}$$

where a is an arbitrary probability vector. This expression suggests a *concurrent* iterative method of estimating π and π'. This method, attributed to Evans ([4] and [5]), is formulated as follows:

Algorithm B (Evans's Concurrent Method)

(i) Initial conditions:

$$s_0 = a, \quad s'_0 = 0$$

(ii) rth iteration:

$$s_r = s_{r-1}P$$
$$s'_r = s'_{r-1}P + s_{r-1}P'$$

(A discussion of circumstances where concurrent methods have definite advantages over sequential methods is beyond the scope of this paper. Interested readers are referred to [2] and [5].)

Theorem 5.2 of [2] formally states the conditions of convergence of Evans's concurrent method and is restated below.

THEOREM 1 Let P be the $n \times n$ transition matrix for an irreducible and aperiodic Markov chain. Let π be the stationary distribution corresponding to P, let a be an initial estimate of π, and s_r the rth estimate of π (according to the power method). Assuming the entries of P are differentiable functions of parameter h, let $P' = \partial P/\partial h$, $\pi' = \partial\pi/\partial h$, and

$$s'_r = \partial s_r/\partial h = a \sum_{m=0}^{r-1} P^m P' P^{r-1-m}$$

Then

$$\lim_{r\to\infty} s'_r = \lim_{r\to\infty} a \sum_{m=0}^{r-1} P^m P' P^{r-1-m} = \pi'$$

A rigorous proof of the above theorem is given in [2]. We note that, because of the possibility that some of the subdominant eigenvalues of the transition probability matrix P are of multiple type, the differentiability of these eigenvalues and their associated eigenvectors is not guaranteed. The complexity of the proof is due mainly to this possible nondifferentiability of certain eigenvalues and eigenvectors.

We shall consider the special case where all eigenvalues of p are simple so they and the corresponding eigenvectors are differentiable (assuming the matrix elements of P are). The convergence proof under this condition is relatively simple and yields information concerning the convergence rate of Evans's concurrent method. This convergence rate will be our basis of comparison when we develop a similar algorithm based on LOPSI.

1.1 Convergence Proof of Evans's Concurrent Method for a Simplified Case

Let the eigenvalues of the transition probability matrix P be $\lambda_1, \lambda_2, \ldots, \lambda_n$ and the associated eigenvectors $q_1, q_2, q_3, \ldots, q_n$.

As mentioned earlier, the concurrent method (of determining π and π') is based on the expression

$$\pi' = \lim_{r\to\infty}\left[a\sum_{m=0}^{r-1} P^m P' P^{r-m-1}\right]$$

where the expression within the brackets is obtained by applying $\partial(\ \ )/\partial h$ to aP^r. To prove the convergence of $\partial(aP^r)/\partial h$, we first express the initial estimate a of π in terms of the eigenvectors of P:

$$a = q_1 + \sum_{j=2}^{n} \alpha_j q_j$$

Postmultiplying this equation by P^r yields

$$s_r = aP^r = q_1 + \sum_{j=2}^{n} \alpha_j \lambda_j^r q_j$$

so the derivative of s_r with respect to h is

$$s_r' = q_1' + \sum_{j=2}^{n} [\alpha_j' \lambda_j^r q_j + \alpha_j r \lambda_j' \lambda_j^{r-1} q_j + \alpha_j \lambda_j^r q_j']$$

Since $|\lambda_j| < 1$ for $n \geq j \geq 2$, all three terms within the brackets vanish as $r \to \infty$. For an arbitrary j, the dominant term among the three is the second, i.e.,

$$[\alpha_j \lambda_j' q_j]\, r \lambda_j^{r-1}$$

which exhibits a convergence rate of

$$r/[(r+1)|\lambda_j|]$$

Since the eigenvalues are arranged in decreasing order of magnitude, the most dominant among all vanishing terms will be

$$[\alpha_2 \lambda_2' q_2] r \lambda_2^{r-1}$$

The upper bound for the convergence rate of s_r' to π' is then a variable $r/[(r+1)|\lambda_2|]$ approaching $|\lambda_2^{-1}|$ as $r \to \infty$.

1.2 Computational Efficiencies of Algorithms Based on the Power Method

The convergence rate of the power method is $|\lambda_1|/|\lambda_2|$, which for an irreducible stochastic matrix will be $1/|\lambda_2|$ where $|\lambda_2|$ is less than one. Evidently, for $|\lambda_2|$ close to one the convergence will be very slow. An algorithm for computing derivatives based on the power method, such as the sequential and concurrent methods discussed above, is likely to exhibit the same slow convergence. Our analysis of the concurrent method, applied to this special class of transition matrices where all eigenvalues are simple, shows this. It has a variable convergence rate, which is, in fact, slower than but asymptotically approaches that of the power method.

It seems logical then to develop a method (for determining π') based on a faster converging algorithm (for determining π) and see if it leads to a more efficient algorithm. LOPSI, developed by Stewart and Jennings [19], appears to be a first likely choice as such basis algorithm since it is a natural extension of the power method. While the latter iterates on a single vector and has a convergence rate of $|\lambda_1|/|\lambda_2|$, LOPSI simultaneously operates on a number (say m) of vectors and exhibits a convergence rate of $|\lambda_1|/|\lambda_{m+1}|$. We shall briefly describe LOPSI and then develop an algorithm based on this method for obtaining π'.

2. LOPSI (LOPSIDED ITERATION ALGORITHM)

At each iteration, LOPSI estimates a number, say m, of the most dominant eigenvalues of any real nonsymmetric square matrix P of, say, dimension n and the corresponding eigenvectors. The following notation will be used to present LOPSI. Let

$$Q_a = [q_1, q_2, \ldots, q_m]^t$$

be the m eigenvectors corresponding to the m most dominant eigenvalues of matrix P, arranged in decreasing order. Let

$$Q_b = [q_{m+1}, q_{m+1}, \ldots, q_n]^t$$

be the remaining eigenvectors, and

$$U = [u_1, u_2, \ldots, u_m]^t$$

be the initial estimate of Q_a. Then u may be expressed as a linear combination of the full set of eigenvectors Q_a, Q_b as follows:

$$U = [C_a \mid C_b] \left[\frac{Q_a}{Q_b} \right] = C_a Q_a + C_b Q_b$$

where C_a and C_b are matrix coefficients of sizes $(m \times m)$ and $m \times (n - m)$.

The procedure for the rth iteration of LOPSI is summarized below:

Step 1: $V(r) = U(r)P$

Step 2: $G(r) = U(r)U^t(r)$

$H(r) = V(r)U^t(r)$
$B(r)G(r) = H(r)$ Solve for $B(r)$.
Step 3: $D(r)B(r) = D(r)$ Obtain the eigensolution (left eigenvectors D and the eigenvalues Λ) of this equation
Step 4: $W(r) = D(r)V(r) \equiv U(r + 1)$
Step 5: Go back to step 1 using the new estimate $U(r + 1)$.

The implementation of the algorithm actually involves more than the above five basic steps. For instance, it may be desirable, in implementing some convergence test, to maintain the eigenvector estimates in decreasing order (according to the magnitude of their corresponding eigenvalues). Thus, if the eigenvalue routine used produces eigenvalue estimates that are not necessarily in this order, it will be necessary to sort them and perform a corresponding permutation on the columns of matrix P. Also, steps 2 to 4, otherwise known as the reorientation process, may be omitted in some iterations to speed up the "washing out effect" of step 1 on lower eigenvector components. Assuming that the "washing out effect" of step 1 has indeed reduced the lower eigenvector components, it can be shown that step 4, which provides the starting matrix U to be used for the next iteration, yields the following:

$$W = D = \Lambda_a Q_a + \left(C_a^{-1} C_b\right) \Lambda_b Q_b \equiv U$$

where Λ_a and Λ_b are, respectively, the matrices of eigenvalues for Q_a and Q_b. Note that the new matrix coefficient of Q_a is now Λ_a. Assuming this to be the case after the first iteration, it can then be shown by induction (see Appendix A6.1 of [2]) that after $(r + 1)$ iterations, LOPSI yields

$$U(r + 1) = \Lambda_a Q_a + \left(\Lambda_a^{-1}\right)^{k-1} \left(C_a^{-1} C_b\right) \Lambda_b^r Q_b \tag{1}$$

Decoupling this matrix equation (see Appendix A6.2 of [2]) into vector equations yields

$$u_j(r + 1) = \lambda_j q_j + \sum_{k=1}^{n-m} M_{j,k} q_{m+k} \lambda_j \left(\lambda_{m+k}/\lambda_j\right)^r$$

for $j = 1, 2, \ldots, m$ and where $M = (C_a^{-1} C_b)$. Since

$$|\lambda^{m+1}|/|\lambda_j| < 1$$

then

$$\lim_{r\to\infty} u_j(r + 1) = \lambda_j q_j$$

(Note that when P is an irreducible stochastic matrix, then

$$\lim_{r\to\infty} u_1(r + 1) = q_1$$

since $\lambda_1 = 1$ and $|\lambda_j| < 1$ for $j \geq 2$.)

2.1 A LOPSI-Based Algorithm for Determining Derivatives

We now develop an algorithm, based on LOPSI, for determining the derivatives $\partial u_j/\partial h$. This algorithm will be referred to as *Lopsider.*

Recall that Evans's concurrent method for estimating π' is based on the power method,

$$s_{r+1} = s_r P$$

where $s_0 = a$, an initial estimate of π.

A concurrent estimate of π' is obtained by differentiating the preceding equation with respect to h:

$$s'_{r+1} = s'_r P + s_r P'$$

These two equations, taken together, will then provide estimates that converge to π and π'.

We shall develop a similar concurrent method based on LOPSI. Assuming the latter simultaneously iterates on m trial eigenvectors, our concurrent method will be applicable if the m most dominant eigenvalues are simple and we can prove that the method converges if all eigenvalues of the transition matrix P are simple. The explanation for the qualifying conditions will be discussed later.

The approach for developing an algorithm that concurrently determines $U(r)$ and $U'(r)$ is basically the same as that used by Evans in deriving his concurrent method from the power method. The procedure for arriving at $U'(r)$, however, will be much more complicated. While the iterative form of the power method consists of a single step, the iterative form of LOPSI consists of four steps with each, particularly steps 3 and 4, being more complicated than the single step of the power method. To obtain $U'(r)$, we need to perform differentiation starting from the very first step of LOPSI up to the fourth step, as will now be shown. Before proceeding, we note that the main objective of developing Lopsider is to illustrate an extension of Evans's concurrent method for determining π and π'. To obtain eigenvalues and their derivatives, we shall use the conceptually simplest methods as they are best suited for illustration purposes.

The first step of LOPSI is

$$V(r) = U(r)P$$

so the corresponding step in Lopsider is

$$V'(r) = U'(r)P + U(r)P'$$

where the primes indicate differentiation with respect to h. Similarly, for the other steps, the corresponding equations are

Step 2a′: $G(r) = U'(r)U^t(r) + U(r)[U^t(r)]'$
Step 2b′: $H'(r) = V'(r)U^t(r) + V(r)[U^t(r)]'$
Step 2c′: $B'(r)G(r) + B(r)G'(r) = H'(r)$ or
$B'(r)G(r) = H'(r) - B(r)G'(r)$ Solve for B'
Step 3a′: $D'(r)B(r) + D(r)B'(r) = \Lambda'(r)D(r) + \Lambda(r)D'(r)$

Differentiability of the m eigenvalues in $\Lambda(r)$ and their corresponding eigenvectors is guaranteed if none of the eigenvalues are multiple. Assuming this to be the case, we can, in principle, obtain $\Lambda'(r)$ by differentiating the characteristic equation with respect to h. The result,

$$|B - \lambda I|' = 0$$

is an equation with two unknowns (λ and λ'). Thus, substituting each value of λ_j (obtained from, say, the characteristic equation) in the above differential equation yields an equation in one unknown (λ'), which can then be solved for the corresponding derivative λ_j'. Having obtained Λ' in this manner, the rearranged equation

$$D'(r)B(r) - \Lambda(r)D'(r) = \Lambda'(r)D(r) - D(r)B'(r)$$

can now be solved for $D'(r)$.

Step 4a′: $W'(r) = D'(r)V(r) + D(r)V'(r) \equiv U'(r+1)$
Step 5a′: Go back to step 1a′ using the new estimates $U(r+1)$ and $U'(r+1)$.

It can be shown by induction (see Appendix A6.3 of [2]) that

$$U'(r+1) = \Lambda_a' Q_a + \Lambda_a Q_a' + \left[\Lambda_a^{-1}(r) C_a^{-1} C_b \Lambda_b Q_a\right]' \tag{2}$$

or

$$U'(r+1) = \left[\Lambda_a Q_a + \Lambda_a^{-1}(r) C_a^{-1} C_b \Lambda_b Q_a\right]'$$

Thus, if we decouple equation (2), the expression obtained for $u_j'(n+1)$ will be that obtained by differentiating the expression for $u_j(n+1)$ [obtained by decoupling equation (1)]. The result is

$$u_j'(r+1) = \lambda_j' q_j + \lambda_j q_j' + \sum_{k=1}^{n-m} M_{j,k}' q_{m+k} (\lambda_{m+k}/\lambda_j)^r \lambda_j + \sum_{k=1}^{n-m} M_{j,k} q_{m+k}' (\lambda_{m+k}/\lambda_j)^r \lambda_j$$

$$+ \sum_{k=1}^{n-m} M_{j,k} q_{m+k} r \left(\lambda_{m+k}/\lambda_j\right)^{r-1} \lambda'_{m+k}$$

$$- \sum_{k=1}^{n-m} M_{j,k} q_{m+k} (r+1) \left(\lambda_{m+k}/\lambda_j\right)^{r} \lambda'_j \tag{3}$$

Since $(\lambda_{m+k}/\lambda_j) < 1$,

$$\lim_{r\to\infty} u_j(r+1) = \lambda'_j q_j + \lambda_j q'_j$$

For an irreducible Markov chain, $\lambda_1 = 1$ so

$$u'_1(r+1) = q'_1 + \sum_{k=1}^{n-m} M'_{1,k} q_{m+k} \left(\lambda_{m+k}\right)^r$$

$$+ \sum_{k=1}^{n-m} M_{1,k} q'_{m+k} \left(\lambda_{m+k}\right)^r$$

$$+ \sum_{k=1}^{n-m} M_{1,k} q_{m+k} r \left(\lambda_{m+k}\right)^{r-1} \lambda'_{m+k} \tag{4}$$

and, since $|\lambda_j| < 1$ for $j \geq 2$,

$$\lim_{r\to\infty} u_1(r+1) = q'_1$$

Lopsider converges more quickly (i.e., it requires fewer iterations) than Evans's concurrent algorithm. The convergence rate for Lopsider is bounded above at the nth iteration by $r/[(r+1)|\lambda_{m+1}|]$; the corresponding quantity for Evans's concurrent method is $r/[(r+1)|\lambda_2|]$. However, for each iteration, the computational complexity of Lopsider is definitely greater than that of Evans. To compare computational efficiencies, we should then consider total computational complexity for convergence as our criterion.

More specifically, our comparison method is as follows. For a given $|\lambda_2|$ and integer m, we compute the maximum value of $|\lambda_{m+1}|$ such that Lopsider is more computationally efficient than Evans's concurrent algorithm. Since both methods are iterative and with variable convergence rates, this threshold value of $|\lambda_{m+1}|$, denoted as $|\lambda^*_{m+1}|$, will depend on what we consider as an acceptable estimate error (denoted as ϵ).

2.2 Application of Comparison Method to a Computer Model

We applied our comparison method to a system architecture model of a computer with several terminals (discussed by Stewart [20]). The number of

Table 1 Threshold Values for $|\lambda^*_{m+1}|$ for $n = 20$ and $\lambda_2 = 0.999071$

	$\lvert\lambda^*_{m+1}\rvert$				
m	$\epsilon = 0.1$	$\epsilon = 0.01$	$\epsilon = 0.001$	$\epsilon = 0.0001$	$\epsilon = 0.00001$
3	0.99613	0.99602	0.99595	0.99595	0.99590
4	0.99393	0.99370	0.99354	0.99354	0.99343
5	0.98976	0.98923	0.98887	0.98887	0.98860
6	0.98601	0.98516	0.98458	0.98458	0.98416
7	0.98155	0.98028	0.97941	0.97941	0.97877
8	0.97636	0.97459	0.97331	0.97331	0.97240
9	0.97039	0.96791	0.96618	0.96618	0.96495
10	0.96361	0.96029	0.95807	0.95807	0.95637

SOURCE: Based on data from Stewart [20].

possible states of the system varies with the number of terminals. Our numerical results for the case of 3 terminals, yielding 20 possible states, are shown in Table 1. This is extracted from Appendix A6.4 of [2], which provides numerical results for a variety of cases.

3. ALGORITHMS BASED ON MATRIX GEOMETRICAL METHODS

In Chapter 6 of [2], the approach for developing concurrent algorithms was applied to a class of matrix geometric methods. In particular, algorithms were presented for the general quasi-birth-and-death (QBD) process and two special cases (bounded and unbounded M/PH/1 queues). To illustrate the approach as applied to matrix geometric methods, we shall consider the bounded M/PH/1 class of queues and perform a detailed numerical analysis of the M/PH/1/3 model. But first, we provide a brief background for probability distributions of phase type.

Consider a Markov process on the states $\{1, \ldots, m+1\}$ with transition rate matrix

$$\bar{Q} = \begin{bmatrix} T & T_0 \\ \underline{0} & 0 \end{bmatrix}$$

where $T_{ij} \geq 0$ for $i = j$ and $T_{ii} \leq 0$. The initial probability vector is given by $\underline{\alpha}$, α_{m+1}. We assume that the states $1, \ldots, m$ are all transient, so that absorption into the state $m+1$, from any initial state, is certain. It can be shown that the probability distribution $F(\cdot)$ of the time t until absorption in the state $m+1$, corresponding to the initial vector $(\underline{\alpha}, \alpha_{m+1})$, is given by

$$F(t) = 1 - \underline{\alpha} \exp(Tt)\underline{e},$$

where

$$\exp(Tt) = I + Tt + (Tt)^2/2 + (Tt)^3/3! + \cdots$$

and e is an m-column vector of 1's and $t \geq 0$. A probability distribution $F(\cdot)$ on $[0, \infty)$ is said to be a distribution of phase type (PH-distribution) if and only if it is the distribution of time until absorption in a finite Markov process of the type defined above. The pair $(\underline{\alpha}, T)$ is called a representation of $F(\cdot)$.

If the phase-type service-time distribution of an M/PH/1 queue has representation $(\underline{\alpha}, T)$, then the mean service time for the queue will be

$$\mu_M = -\underline{\alpha} T^{-1} \underline{e}$$

Denoting the Poisson arrival rate by λ, we assume that $\lambda \mu_M \equiv \rho < 1$ and, for ease of notation, that $\alpha_{m+1} = 0$.

The bounded M/PH/1 queue (with capacity $\underline{K} + 1$) can likewise be viewed as a QBD process with the following transition rate matrix:

$$\underline{Q} = \begin{bmatrix} -\lambda & \lambda\underline{\alpha} & 0 & 0 & \cdots & 0 & 0 \\ T_0 & (T - \lambda I) & \lambda I & 0 & \cdots & 0 & 0 \\ 0 & T_0\alpha & (T - \lambda I) & \lambda I & 0 & \cdots & \cdots \\ \vdots & \vdots & \vdots & \vdots & \vdots & \vdots & \vdots \\ 0 & 0 & 0 & 0 & 0 & T_0\underline{\alpha} & (T - \lambda I) \end{bmatrix}$$

The stationary distribution vector, denoted by $\underline{x}$, is partitioned into vector components, denoted by $\underline{x}_i$, which are m-component row vectors for $i \geq 2$; x_0 is a scalar. These component vectors are expressed in terms of $\underline{\alpha}$, $R = \lambda(\lambda I - \lambda B^{00} - T)^{-1}$ where $B^{00} = \underline{e\alpha}$, and x_0, which is obtained by normalization. The geometric form of $\underline{x}_i$ is evident in the following expressions:

$$\underline{x}_i = x_0 \underline{\alpha} R^i \qquad \text{for} \quad 1 < i \leq K,$$

or

$$x_1 = x_0 \underline{\alpha} R \qquad \text{and} \qquad \underline{x}_i = \underline{x}_{i-1} R \qquad \text{for} \quad 2 < i \leq K$$

and

$$\underline{x}_{K+1} = x_0 \underline{\alpha} R^K \left(-\lambda T^{-1}\right) = \underline{x}_K \left(-\lambda T^{-1}\right)$$

while x_0 (obtained by normalization) is given by

$$x_0 = \left\{ \underline{\alpha} \left[\sum_{i=0}^{K} R^i - \lambda R^K T^{-1} \right] \underline{e} \right\}^{-1}$$

To obtain the derivatives concurrently, we first compute R' using the equation

$$R' = -R \left[R^{-1}\right]' R$$

where the derivative of $R^{r-1} = [I - B^{00} - (1/\lambda)T]$ is easily determined. To obtain x_0', we start by differentiating the expression

$$\left[\sum_{i=0}^{K} R^i - \lambda R^K T^{-1} \right]$$

$$\left[\sum_{i=0}^{K} R^i - \lambda R^K T^{-1} \right]' = \sum_{i=0}^{K} \sum_{r=0}^{i-1} R^r R' R^{i-r-1} - \lambda \sum_{n=0}^{K-1} R^r R' R^{K-r-1} T^{-1}$$
$$+ \lambda R^K \left(T^{-1} T' T'^{-1}\right)$$

With the above result, we can compute

$$y = \left[x_0^{-1}\right]' = \underline{\alpha} \left[\sum_{i=0}^{K} R^i - \lambda R^K T^{-1} \right]' \underline{e}, \qquad \text{a scalar}$$

It follows that

$$x_0' = y x_0^2$$

The other derivatives can then be readily computed:

$$\underline{x}_2' = x_0' \underline{\alpha} R + x_0 \underline{\alpha} R'$$
$$\underline{x}_{i+1}' = \underline{x}_i' R + \underline{x}_i R' \qquad \text{for} \quad 2 \le i \le K$$
$$\underline{x}_{K+1}' = x_0' \underline{\alpha} R^K (-\lambda T^{-1}) + x_0 \underline{\alpha} \left[\sum_{n=0}^{K-1} R^r R' R^{K-r-1} \right] (-\lambda T^{-1})$$
$$+ x_0 \underline{\alpha} R^K (\lambda T^{-1} T' T^{-1})$$

3.2 Numerical Analysis of an M/PH/1/3 Queue

Consider the M/PH/1/3 queue of Figure 1. The service facility of the system consists of two phases with the mean service rate for the first phase being μ_1 and for the second phase μ_2. The system has two waiting slots. Since

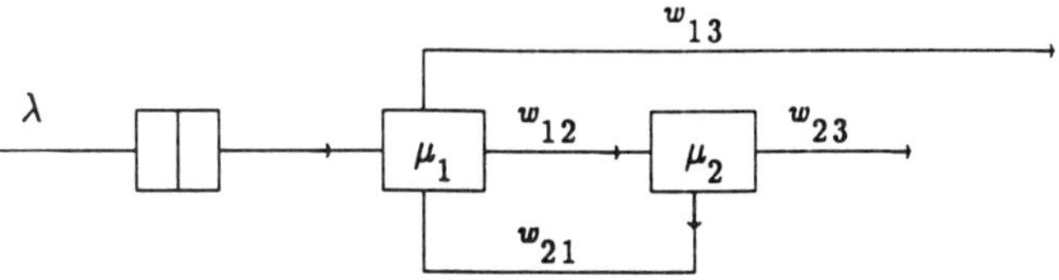

Figure 1 An M/PH/1/3 model.

the service facility can accommodate only a single customer at a time, the system's total capacity is three. The state of the system is described by an ordered pair (i,j) where i gives the total number of customers in the system (i.e., including the customer in the service facility and those in the waiting slots) and j indicates whether the customer in the service facility is in the first or second phase. Considering that the system has capacity three and a two-phase service facility, it follows that the system will have seven possible states:

$$(0,0),\ (1,1),\ (1,2),\ (2,1),\ (2,2),\ (3,1),\ \text{and}\ (3,2)$$

which, for convenience, will be denoted, respectively, as states 0 to 6.

The model assumes Poisson arrivals with mean rate λ and exponential service rates with a mean of μ_1 for the first phase and μ_2 for the second phase. Upon completion of the first service phase, a customer in service can exit with probability w_{13} or proceed to the second phase with probability w_{12}. A customer leaving the second service phase can exit with probability w_{23} or move back to the first phase with probability w_{21}.

3.3 The Matrix Geometric Approach

The stationary probability vector is partitioned as follows:

$$\underline{x} = (x_0, \underline{x}_1, \underline{x}_2, \underline{x}_3)$$

where

$$x_0 = \pi_0$$
$$\underline{x}_1 = (\pi_1, \pi_2)$$
$$\underline{x}_2 = (\pi_3, \pi_4)$$
$$\underline{x}_3 = (\pi_5, \pi_6)$$

Using the notations for the general M/PH/1/K + 1 model, we have

$$T = \begin{bmatrix} -\mu_1 & \mu_1 w_{12} \\ \mu_2 w_{21} & -\mu_2 \end{bmatrix} \qquad T_0 = \begin{bmatrix} \mu_1 & w_{13} \\ \mu_2 & w_{23} \end{bmatrix}$$

$$\underline{\alpha} = [0, 1] \qquad \underline{e} = \begin{bmatrix} 1 \\ 1 \end{bmatrix}$$

$$B^{00} = \underline{e\alpha} = \begin{bmatrix} 1 \\ 1 \end{bmatrix} [1\ 0] = \begin{bmatrix} 1 & 0 \\ 1 & 0 \end{bmatrix}$$

so the transition rate matrix can be written as follows:

$$Q = \begin{bmatrix} -\lambda & \lambda\underline{\alpha} & 0 & 0 \\ T_0 & (T - \lambda I) & \lambda I & 0 \\ 0 & T_0\underline{\alpha} & (T - \lambda I) & \lambda I \\ 0 & 0 & T_0\underline{\alpha} & (T - \lambda I) \end{bmatrix}$$

3.4 Numerical Results for the Matrix Geometric Model (MGM)

Our computational procedure was applied using the following parameter values:

$$\mu_1 = 0.6000 \qquad \mu_2 = 0.7000 \qquad w_{12} = 0.5000 \qquad w_{13} = 0.5000$$
$$w_{21} = 0.5000 \qquad w_{23} = 0.5000 \qquad \lambda = 0.3000$$

A comparison of computational complexities (considering only multiplications and divisions) for obtaining π and π' shows that, when applied to this particular M/PH/1/3 model, the matrix geometric method is, by far, the most computationally efficient. A summary of the results follows:

Method	Computational complexity
Matrix geometric method	135
Sequential method	1868.86
Concurrent method	2265.76

An acceptable solution for either the sequential or concurrent method has been defined as one where both estimates of the steady-state probability distribution and its derivative differ from their previous estimates by no more than 0.0001 in magnitude.

The computational complexities of the above methods will vary as the size of the problem grows, and, most likely, so will their relative computational efficiencies.

Recall that the state of our M/PH/1/3 model is specified by an ordered pair (i,j), which indicates that i customers are in the system and the customer being served is in phase number j. Assuming the capacity of the system is K and the number of phases Ph, the total number of states will be $(K\ Ph + 1)$.

The computational complexity for any of the three methods depends, possibly among other things, on three factors: K, Ph, and a sparsity of certain matrices. The number of phases in the service process is the same as the dimension f of submatrices $(T - \lambda I)$ and λI in the block partitioned transition rate matrix. It has been shown that the computational complexity for MGM is cubic in f. This cubic term, together with lower-order terms, comes from obtaining x_0 and its derivative x_0'. The computational complexities for obtaining $\underline{x}_i$ and its derivative $\underline{x}_i'$, for $i \geq 1$, are quadratic in f. For the other methods, the computational complexities are quadratic in n, the dimension of the transition rate matrix. Increasing Ph by ΔPh will increase f by the same amount and n by $\Delta Ph\ K$, while increasing the capacity K by ΔK will only affect n, increasing it by $\Delta K\ Ph$. Sparsity may change depending on how the problem grows. We expect that the computational efficiency of the MGM, as compared to the others, will improve as K is increased and deteriorate as Ph is increased because its computational complexity is cubic in f.

REFERENCES

[1] Clint, M., and A. Jennings (1971). A simultaneous iteration method for the unsymmetric eigenvalue problem, J. Inst. Math. Applic. 8, 111–121.

[2] Cruz, Jose J. (1989). Perturbation Analysis and Design of Network of Queues, Ph.D. dissertation, North Carolina State University at Raleigh.

[3] Evans, R. E. (1971). Programming problems and changes in the stable behavior of a class of Markov chains, J. Appl. Prob. 8, 543–550.

[4] Evans, R. E. (1975). An Optimization Algorithm and Its Applications in Queueing, Faculty Working Papers, #270, College of Commerce and Business Administration, University of Illinois at Urbana-Champaign.

[5] Evans, R. E. (1981). Markov chain design problems, Operations Res. 29, No. 5, 957–970.

[6] Funderlic, R. E., and C. D. Meyer, Jr. (1986). Sensitivity of the stationary distribution vector for ergodic Markov chain, Linear Algebra Applic. 76, 1–17.

[7] Golub, G. H., and C. D. Meyer, Jr. (1986). Using the QR factorization and group inversion to compute, differentiate, and estimate the sensitivity of stationary probabilities for Markov chains, SIAM J. Discrete Math. 7, No. 2.

[8] Golub, G. H., and C. F. Van Loan (1974). *Matrix Computations*, Johns Hopkins University Press, Baltimore.

[9] Gross, D., and C. M. Harris (1974). *Fundamentals of Queueing Theory*, John Wiley and Sons, New York.

[10] Jennings, A., and W. J. Stewart (1975). Simultaneous iteration for partial eigensolution of real matrices, J. Inst. Math. Applic. 15, 351–361.

[11] Kemeny, J. G., and J. L. Snell (1960). *Finite Markov Chains*, Van Nostrand Reinhold, New York.

[12] Meyer, C. D., Jr., and G. W. Stewart (1988). Derivatives and perturbations of eigenvectors, SIAM J. Numer. Anal. 25, No. 3, 1–13.

[13] Neuts, M. F. (1981). *Matrix Geometric Solutions in Stochastic Models: An Algorithmic Approach*, Johns Hopkins University Press, Baltimore.

[14] Noble, B. (1969). *Applied Linear Algebra*, Prentice-Hall, Englewood Cliffs, N.J.

[15] Schweitzer, P. J. (1968). Perturbation theory and finite Markov chains, J. Appl. Prob. 5, 401–413.

[16] Seneta, E. (1973). *Non-negative Matrices*, John Wiley and Sons, New York.

[17] Steinberg, D. I. (1974). *Computational Matrix Algebra*, McGraw-Hill, New York.

[18] Stewart, W. J., and A. Goyal (1985). Matrix Methods in Large Dependability Models, Technical Report, IBM Research Center, Yorktown Heights, N.Y.

[19] Stewart, W. J., and A. Jennings (1981). A simultaneous iteration algorithm for real matrices, ACM Trans. Math. Software 7, No. 2, 184–198.

[20] Stewart, W. J. (1978). A comparison of numerical techniques in Markov modeling, Commun. ACM 21, No. 2.

[21] Stewart, G. W. (1972). *Introduction to Matrix Computations*, Academic Press, Orlando, Fla.

[22] Wilkinson, J. H. (1965a). *The Matrix Eigenvalue Problem*, Clarendon Press, Oxford.

9

The Joint Distribution of Arrivals and Departures in Quasi-Birth-and-Death Processes

MARCEL F. NEUTS Department of Systems and Industrial Engineering, University of Arizona, Tucson, Arizona

ABSTRACT

In queues modeled as quasi-birth-and-death processes, the transitions toward higher (lower) levels often correspond to arrivals to (departures from) the queue. By using matrix-analytic methods, we obtain a joint vector transform for the numbers of right and left transitions during an interval $(0, t)$ in a stationary quasi-birth-and-death process. That transform may be used to study a number of mathematical descriptors of the arrival and departure processes. The implementation of these descriptors presents challenging problems in numerical analysis.

This research was supported in part by grants ECS-8803061 from the National Science Foundation and AFOSR-88-0076 from the Air Force Office of Scientific Research.

1. INTRODUCTION

The appeal of the matrix-analytic methods, of which detailed presentations may be found in references [2] and [3], lies in the fact that unifying matrix analogues of many results known for elementary stochastic models may be obtained. As these matrix analogues are derived by using structural properties of certain two-dimensional random walks, but do not otherwise depend on specific distributional assumptions, they are of much broader generality and methodological depth than their elementary counterparts. In the oral presentation at the conference, the relationship between the Markov renewal processes amenable to analysis by the matrix analytic methods and random walks on semi-infinite two-dimensional lattices was discussed in greater generality. In the present paper, which deals with a specific problem, we derive the joint transform of the numbers of left and right transitions during an interval of time $(0,t)$ in a stationary quasi-birth-and-death process. When applied to queues described by quasi-birth-and-death processes, the resulting formulas provide much information on the relationship between the arrival and departure processes.

In Chapter 3 of [2], a comprehensive treatment of quasi-birth-and-death processes may be found. To the extent possible, we shall use notation and terminology that is introduced there. In brief, a quasi-birth-and-death process (QBD–process) is a Markov process whose state space consists of the pairs (i,j) with $i \geq 1, 1 \leq j \leq m$, and an additional set of boundary states $(0,j)$ with $1 \leq j \leq m_1$. The parameters of m and m_1 are positive integers. The set of states $\{(i,j), 1 \leq j \leq m\}$ is called the *level* i, and the level 0 is correspondingly defined for the m_1 boundary states.

The infinitesimal generator $\tilde{Q}$ of the Markov process has the block tridiagonal form

$$\tilde{Q} = \begin{bmatrix} B & C & & & & \cdots \\ D & A_1 & A_0 & & & \cdots \\ & A_2 & A_1 & A_0 & & \cdots \\ & & A_2 & A_1 & A_0 & \cdots \\ & & & A_2 & A_1 & \cdots \\ \cdot & \cdot & \cdot & \cdot & \cdot & \end{bmatrix}$$

where the diagonal elements of B and A_1 are negative and all other elements of the displayed blocks are nonnegative. The matrices A_0, A_1, A_2, are square and of order m, while the matrices B, C, and D are respectively of dimensions $m_1 \times m_1$, $m_1 \times m$, and $m \times m_1$. We shall assume that the infinitesimal generator $\tilde{Q}$ is *irreducible*. In order to avoid cases that arise only in rare applications but require more detailed discussion, we further assume that the matrix

$$A = A_0 + A_1 + A_2$$

is an irreducible generator. The stationary probability vector of A is denoted by π. It is known [2, Thm. 3.1.1], that the QBD-process is *positive recurrent* if and only if

$$\pi A_2 \mathbf{e} > \pi A_0 \mathbf{e} \tag{1}$$

The stationary probability vector $\mathbf{x}$ of $\tilde{Q}$ is partitioned as $[\mathbf{x}_0, \mathbf{x}_1, \mathbf{x}_2, \ldots]$ and has the *matrix-geometric form*:

$$\mathbf{x}_i = \mathbf{x}_1 R^{i-1}, \qquad \text{for} \quad i \geq 1 \tag{2}$$

The pair consisting of the m_1-vector $\mathbf{x}_0$ and the m-vector $\mathbf{x}_1$ is the unique solution to the system of linear equations

$$\begin{aligned} &\mathbf{x}_0 B + \mathbf{x}_1 D = \mathbf{0} \\ &\mathbf{x}_0 C + \mathbf{x}_1 (A_1 + RA_2) = \mathbf{0} \\ &\mathbf{x}_0 \mathbf{e} + \mathbf{x}_1 (I - R)^{-1} \mathbf{e} = 1 \end{aligned} \tag{3}$$

and the matrix R is the minimal nonnegative solution to the equation

$$R^2 A_2 + RA_1 + A_0 = 0 \tag{4}$$

Techniques for the numerical computation of R, as well as the significance and the mathematical properties of that matrix, are discussed in reference [2].

2. THE FUNDAMENTAL PERIOD

In the derivations in this paper, we shall use techniques from the matrix-geometric analysis [2] and from that for Markov chains of $M/G/1$ type [3]. Because of the tridiagonal structure of the generator $\tilde{Q}$, QBD-processes belong to the intersection of the two classes of two-dimensional random walks analyzed in these books. The matrix-analytic properties of QBD-processes are therefore particularly tractable.

The *fundamental period* is defined as the first-passage time from a level $\mathbf{i+1}$ to the level $\mathbf{i}$ for $i \geq 1$. A brief discussion of the semi-Markov matrix which describes joint transition probabilities of the duration of the fundamental period and of the first state visited in the level $\mathbf{i}$, given the initial state in the level $\mathbf{i+1}$, may be found in Section 3.3 of reference [2]. In the interest of brevity, we shall list statements only, for those results whose proofs are an immediate adaptation of arguments given there. In studying the first passage from $\mathbf{i+1}$ to the level $\mathbf{i}$, we shall also keep track of the numbers of *right transitions* and *left transitions*. These are respectively the transitions in which the path moves to the next upward level and those in which the path moves one level downward. We shall consider the transform matrix $G(z_1,z_2,s)$ where the variables z_1 and z_2 refer respectively to the right and the left transitions, while s is the transform variable referring to the duration (in real time) of the fundamental period. We note that the matrix defined in formula (3.3.1) of reference [2] encodes on the left transitions.

By arguments that are now classical, we obtain that the matrix $G(z_1,z_2,s)$ satisfies the matrix quadratic equation

$$G(z_1,z_2,s) = (sI - A_1)^{-1} z_1 A_0 G^2(z_1 z_2 s) + (sI - A_1)^{-1} z_2 A_2 \tag{5}$$

The existence and uniqueness properties of the appropriate solution of that equation are discussed in greater generality in Chapter 2 of reference [3].

We also record corresponding formulas for the transform matrices that describe the first-passage probabilities for the first passage from the level **1** to the boundary level **0** and for the recurrence time of the level **0**. A more general discussion of such passage times is given in Chapters 2 and 3 of reference [3]. We denote those matrices by $L(z_1,z_2,s)$ and $K(z_1,z_2,s)$ and stress that the transform variables similarly correspond to the right and left transitions and the time duration involved. The matrices L and K are respectively of dimensions $m \times m_1$ and $m_1 \times m_1$. The standard first-passage arguments lead to the equations

$$\begin{aligned} L(z_1,z_2,s) = {} & (sI - A_1)^{-1} z_1 A_0 G(z_1,z_2,s) L(z_1,z_2,s) \\ & + (sI - A_1)^{-1} z_2 D \end{aligned} \tag{6}$$

and

$$K(z_1,z_2,s) = (sI - B)^{-1} z_1 C L(z_1,z_2,s) \tag{7}$$

In a positive recurrent QBD-process, the matrix $G(1,1,s)$ is the Laplace–Stieltjes transform of a *stochastic* semi-Markov matrix and is therefore a substochastic matrix for $s > 0$. That observation and a simple comparison argument imply that the matrix $T(z_1,z_2,s)$, defined by

$$T(z_1,z_2,s) = [sI - A_1 - z_1 A_0 G(z_1,z_2,s)]^{-1} \tag{8}$$

exists for all triples (z_1,z_2,s) of complex numbers satisfying $|z_1| < 1$, $|z_2| \leq 1$, and $\mathrm{Re}\, s \geq 0$, or $|z_1| \leq 1$, $|z_2| < 1$, and $\mathrm{Re}\, s \geq 0$, or $|z_1| \leq 1$, $|z_2| \leq 1$, and $\mathrm{Re}\, s > 0$.

Using the matrix $T(z_1,z_2,s)$, the formulas (6) and (7) may be rewritten as

$$L(z_1,z_2,s) = T(z_1,z_2,s) z_2 D \tag{9}$$

and

$$K(z_1,z_2,s) = (sI - B)^{-1} z_1 C T(z_1,z_2,s) z_2 D \tag{10}$$

Moreover, (5) and (8) readily lead to the equations

$$[T(z_1,z_2,s)]^{-1} G(z_1,z_2,s) = z_2 A_2 \tag{11}$$

and

$$[I - K(z_1,z_2,s)]^{-1} = [sI - B - z_1 C T(z_1,z_2,s) z_2 D]^{-1} (sI - B) \tag{12}$$

The preceding formulas, which are used in simplifying some of the calculations that follow, are merely matrix analogues of simple relations between the transforms of the distributions of the duration, the numbers of departures, and the numbers of arrivals during the busy cycle of the $M/M/1$ queue. In the sequel this will be illustrated by the considerations of that and other special cases. In particular, the matrix $K(z_1,z_2,s)$ is the analogue of the Laplace–Stieltjes transform of the duration of the *busy cycle* jointly with the generating function of the numbers of departures and arrivals during it.

3. THE COUNTS DURING THE INTERVAL $(0, t)$

The stationary version of the QBD-process is obtained by choosing its initial state probabilities according to the stationary probability vector $\mathbf{x}$ of $\tilde{Q}$. With that choice, a variety of derived point processes, such as that of the right transitions (arrivals, in queuing applications), and of left transitions (departures) are also stationary. In this section we obtain an equation for the transform of the joint distribution of:

The state (i,j) visited at time t.
The number of right transitions in $(0,t)$.
The number of left transitions in $(0,t)$.

That transform involves four variables, s with respect to the time variable t, z_1 and z_2 to account for the numbers of right and left transitions respectively, and a variable w for the generating function with respect to the index i of the level visited at time t. The index j, which plays a different role for the level $\mathbf{0}$ than for the other levels, arises as the component index in the transform, which is a row vector of dimensions $m_1 + m$. Specifically, we shall write the transform as

$$\Psi^*(z_1,z_2,s,w) = [\Psi_0(z_1,z_2,s), \Psi(z_1,z_2,s,w)] \tag{13}$$

where the vectors on the right-hand side are respectively of dimensions m_1 and m. The m_1-vector $\Psi(z_1,z_2,s)$ corresponds to the case where, at time t, the

QBD-process is in the set of states 0; the m-vector $\Psi(z_1, z_2, s, w)$ corresponds to the cases where at t, the QBD-process is in one of the levels **i**, with $i \geq 1$. The generating function evaluated with respect to i has the corresponding variable w and for $w = 0$, the second generating function is clearly identically zero.

THEOREM 1 The transform $\Psi^*(z_1, z_2, s, w)$ is given by the equations

$$\Psi_0(z_1, z_2, s) = \left[x_0 + x_1 \sum_{i=0}^{\infty} R^i G^i(z_1, z_2, s) T(z_1, z_2, s) z_2 D \right] \times [sI - B - z_1 C T(z_1, z_2, s) D z_2]^{-1} \tag{14}$$

$$\Psi(z_1, z_2, s)\,[sI - A_1 - z_1 A_0 G(z_1, z_2, s) - w z_1 A_0] = w z_1 \Psi_0(z_1, z_2, s) C + w \mathbf{x}_1 (I - wR)^{-1} \sum_{i=0}^{\infty} R^i G^i(z_1, z_2, s) \tag{15}$$

Proof. The proof is by an application of the law of total probability, followed by a judicious and elaborate simplification of the expressions obtained. In the interest of brevity, we shall immediately state the terms contributed by the various alternatives in transformed form.

Case 1: There are no level changes in $(0, t)$. This case contributes the term

$$\mathbf{x}_0 (sI - B)^{-1} \tag{16}$$

to Ψ_0 and the term

$$\sum_{i=1}^{\infty} \mathbf{x}_i (sI - A_1)^{-1} w^i = w \mathbf{x}_1 (I - wR)^{-1} (sI - A_1)^{-1} \tag{17}$$

to the generating function Ψ.

Case 2: There are some level changes in $(0, t)$. That alternative needs to be decomposed by considering the times of the first and the last level changes to occur in $(0, t)$ and further according to the number of transitions to occur between those two epochs. The resulting formulas are quite transparent once we introduce the transform $H(z_1, z_2, s)$ of the semi-Markov process embedded at successive level changes. That matrix is given by:

$$H(z_1, z_2, s) = \begin{bmatrix} 0 & U_0(z_1, s) & & & & \cdots \\ D_0(z_2, s) & 0 & U(z_1, s) & & & \cdots \\ & D(z_2, s) & 0 & U(z_1, s) & & \cdots \\ & & D(z_2, s) & 0 & U(z_1, s) & \cdots \\ & & & D(z_2, s) & 0 & \cdots \\ \cdot & \cdot & \cdot & \cdot & \cdot & \cdots \end{bmatrix}$$

where the matrices $U(z_1,s)$, $D(z_2,s)$, $U_0(z_1,s)$, and $D_0(z_2,s)$ stand for

$$U(z_1,s) = (sI - A_1)^{-1} z_1 A_0, \qquad D(z_2,s) = (sI - A_1)^{-1} z_2 A_2$$
$$U_0(z_1,s) = (sI - B)^{-1} z_1 C, \qquad D_0(z_2,s) = (sI - A_1)^{-1} z_2 D$$

In order to account for all transitions up to the last level change in $(0,t)$, we form the infinite dimensional vector

$$\theta^*(z_1,z_2,s) = \mathbf{x}H(z_1,z_2,s)[I - H(z_1,z_2,s)]^{-1} \tag{18}$$

where $\mathbf{x}$ is the stationary probability vector of the generator $\tilde{Q}$. The vector $\theta^*(z_1,z_2,s)$ is partitioned into an m_1-vector $\theta_0(z_1,z_2,s)$ and a sequence of m-vectors $\theta_\nu(z_1,z_2,s)$, $\nu \geq 1$.

By a "last exit" argument familiar from the theory of Markov renewal processes, we see that case 2 contributes the term

$$\theta_0(z_1,z_2,s)(sI - B)^{-1} \tag{19}$$

to the generating function Ψ_0, and the term

$$\sum_{\nu=1}^{\infty} \theta_\nu(z_1,z_2,s) w^\nu (sI - A_1)^{-1} \tag{20}$$

to the generating function Ψ. This last equation arises from the case where at some time v in $(0,t)$, there is a transition that takes the QBD-process to enter the level ν and during the remaining time interval (v,t) only transitions within the level ν may occur.

Equation (18) is equivalent to the recursive equations

$$\theta_0(z_1,z_2,s) = [\theta_1(z_1,z_2,s) + \mathbf{x}_1] D_0(z_2,s) \tag{21}$$

$$\begin{aligned}\theta_1(z_1,z_2,s) = {} & [\theta_0(z_1,z_2,s) + \mathbf{x}_0] U_0(z_1,s) \\ & + [\theta_2(z_1,z_2,s) + \mathbf{x}_2] D(z_2,s)\end{aligned} \tag{22}$$

and for $\nu \geq 2$,

$$\begin{aligned}\theta_\nu(z_1,z_2,s) = {} & [\theta_{\nu-1}(z_1,z_2,s) + \mathbf{x}_{\nu-1}] U(z_1,s) \\ & + [\theta_{\nu+1}(z_1,z_2,s) + \mathbf{x}_{\nu+1}] D(z_2,s)\end{aligned} \tag{23}$$

The next steps are aimed at eliminating the vectors $\theta_\nu(z_1,z_2,s)$ from the terms in the expressions (19) and (20). These steps are crucial in deriving the final expressions, which involve only the matrices R and $G(z_1,z_2,s)$, which are known (at least in principle).

We begin by forming the vector generating function

$$\theta^{**}(z_1,z_2,s,w) = \sum_{\nu+1}^{\infty} \theta_\nu(z_1,z_2,s) w^\nu \tag{24}$$

It follows from the equations (22) and (23) that

$$w\theta^{**}(z_1,z_2,s,w) = w^2\,[\theta_0(z_1,z_2,s) + \mathbf{x}_0]\,U_0(z_1,s) + w^2\left[\theta^{**}(z_1,z_2,s,w) + \sum_{\nu=1}^{\infty}\mathbf{x}_\nu w^\nu\right]U(z_1,s) + \left[\theta^{**}(z_1,z_2,s,w) + \sum_{\nu=2}^{\infty}\mathbf{x}_\nu w^\nu - w\theta_1(z_1,z_2,s)\right]D(z_2,s) \tag{25}$$

After rearranging terms, that equation yields

$$\begin{aligned}
&\theta^{**}(z_1,z_2,s,w)\,(sI - A_1)^{-1}\left[swI - wA_1 - w^2z_1A_0 - z_2A_2\right]\\
&= w^2\,[\theta_0(z_1,z_2,s) + \mathbf{x}_0]\,U_0(z_1,s)\\
&\quad + w\mathbf{x}_1\,(I - wR)^{-1}\,(sI - A_1)^{-1}\left[w^2z_1A_0 + z_2A_2\right]\\
&\quad - w\mathbf{x}_1\,(sI - A_1)^{-1}z_2A_2 - w\theta_1(z_1,z_2,s)\,(sI - A_1)^{-1}z_2A_2
\end{aligned} \tag{26}$$

The right-hand side of that equation still contains the vectors $\theta_0(z_1,z_2,s)$ and $\theta_1(z_1,z_2,s)$. They are eliminated by the following step, which is rather subtle. We multiply the equation (23) of index ν on the right by the matrix $G^{\nu-1}(z_1,z_2,s)$ and form the sum of (22) and the resulting equations over all positive ν. This yields the equality

$$\begin{aligned}
&\sum_{\nu=1}^{\infty}\theta(z_1,z_2,s)G^{\nu-1}(z_1,z_2,s)\\
&= [\theta_0(z_1,z_2,s) + \mathbf{x}_0]\,U_0(z_1,s) + [\theta_1(z_1,z_2,s) + \mathbf{x}_1]\\
&\quad \times U(z_1,s)G(z_1,z_2,s) + \sum_{\nu=2}^{\infty}[\theta_\nu(z_1,z_2,s) + \mathbf{x}_\nu]\\
&\quad \times \left[D(z_1,s) + U(z_1,s)G^2(z_1,z_2,s)\right]G^{\nu-2}(z_1,z_2,s)
\end{aligned}$$

But now we notice that by equation (5),

$$D(x_2,s) + U(z_1,s)G^2(z_1,z_2,s) = G(z_1,z_2,s)$$

which results in a major cancellation and leads to the essential auxiliary equation

$$\begin{aligned}
\theta_1(z_1,z_2,s) = {} & [\theta_0(z_1,z_2,s) + \mathbf{x}_0]\,U_0(z_1,s)\\
& + [\theta_1(z_1,z_2,s) + \mathbf{x}_1]\,U(z_1,s)G(z_1,z_2,s)
\end{aligned}$$

$$+\sum_{\nu=2}^{\infty}\mathbf{x}_\nu G^{\nu-1}(z_1,z_2,s) \tag{27}$$

By using the matrix $T(z_1,z_2,s)$ defined in (8), that equation may be rewritten as

$$\begin{aligned}[]
&[\theta_1(z_1,z_2,s)+\mathbf{x}_1](sI-A_1)^{-1}\\
&=\Bigg\{[\theta_0(z_1,z_2,s)+\mathbf{x}_0]U_0(z_1,s)\\
&\quad+\sum_{\nu=1}^{\infty}\mathbf{x}_1R^{\nu-1}G^{\nu-1}(z_1,z_2,s)\Bigg\}T(z_1,z_2,s)
\end{aligned} \tag{28}$$

From the equations (21) and (28), we now obtain that

$$\begin{aligned}
\theta_0(z_1,z_2,s) &= \Bigg\{[\theta_0(z_1,z_2,s)+\mathbf{x}_0]U_0(z_1,s)\\
&\quad+\mathbf{x}_1\sum_{\nu=0}^{\infty}R^{\nu}G^{\nu}(z_1,z_2,s)\Bigg\}T(z_1,z_2,s)z_2D
\end{aligned}$$

Using formula (12), that equation may be rewritten in terms of $K(z_1,z_2,s)$, to yield that

$$\begin{aligned}
&\theta_0(z_1,z_2,s)+\mathbf{x}_0\\
&=\left[\mathbf{x}_0+\mathbf{x}_1\sum_{\nu=0}^{\infty}R^{\nu}G^{\nu}(z_1,z_2,s)T(z_1,z_2,s)z_2D\right]\\
&\quad\times[I-K(z_1,z_2,s)]^{-1}
\end{aligned} \tag{29}$$

By adding the term in equation (16) and using formula (12), we first calculate $\Psi_0(z_1,z_2,s)$ and obtain the expression stated in equation (14).

In order to evaluate $\Psi(z_1,z_2,s,w)$ we now eliminate $\theta_0(z_1,z_2,s)$ and $\theta_1(z_1,z_2,s)$ from the right-hand side of equation (26). After some further simplifications, the right-hand side of (26) reduces to

$$\begin{aligned}
&w\left[\mathbf{x}_0+\mathbf{x}_1\sum_{\nu=0}^{\infty}R^{\nu}G^{\nu}(z_1,z_2,s)T(z_1,z_2,s)z_2D\right]\\
&\times[I-K(z_1,z_2,s)]^{-1}(sI-B)^{-1}z_1C\,[wI-G(z_1,z_2,s)]\\
&+w\mathbf{x}_1(I-wR)^{-1}(sI-A_1)^{-1}\left[w^2z_1A_0+z_2A_2\right]\\
&-w\mathbf{x}_1\sum_{\nu=0}^{\infty}R^{\nu}G^{\nu}(z_1,z_2,s)T(z_1,z_2,s)z_2A_2
\end{aligned} \tag{30}$$

We now add the term in equation (17) to form $\Psi(z_1,z_2,s,w)$ and we simplify. We recognize that the expression for $\Psi_0(z_1,z_2,s)$ arises in the right-hand side, which leads to the following equation:

$$\begin{aligned}&\Psi(z_1,z_2,s,w)\left[swI - wA_1 - w^2z_1A_0 - z_2A_2\right]\\&= wz_1\Psi_0(z_1,z_2,s)C\left[wI - G(z_1,z_2,s)\right]\\&\quad + w^2\mathbf{x}_1(I - wR)^{-1} - w\mathbf{x}_1\sum_{i=0}^{\infty}R^iG^{i+1}(z_1,z_2,s)\end{aligned}$$

The final simplification is quite remarkable. It is well known that the singularities in the unit disk $|w| \leq 1$ of the matrix in the left-hand side of equation (30) arise at the eigenvalues of the matrix $G(z_1,z_2,s)$ and that these are all *removable singularities*. In the present case, these singularities can be explicitly removed by noting that, using equation (5) for $G(z_1,z_2,s)$, we may, by straightforward calculation, derive the equality

$$\begin{aligned}&\left[sI - A_1 - z_1A_0G(z_1,z_2,s) - wz_1A_0\right]\left[wI - G(z_1,z_2,s)\right]\\&= swI - wA_1 - w^2z_1A_0 - z_2A_2\end{aligned} \tag{31}$$

The last two terms in equation (30) may be rewritten as

$$w\mathbf{x}_1(I - wR)^{-1}\sum_{i=0}^{\infty}R^iG^i(z_1,z_2,s)\left[wI - G(z_1,z_2,s)\right]$$

and by canceling the matrix $wI - G(z_1,z_2,s)$, we obtain the stated formula (15).

The matrix $sI - A_1 - z_1A_0G(z_1,z_2,s) - wz_1A_0$ is nonsingular in the domain of interest of the transform variables. To see this, it suffices to note that the matrix

$$(sI - A_1 - A_0)^{-1}A_0G(1,1,s)$$

is the transform of a (substochastic) semi-Markov matrix, so that the stated nonsingularity follows by a standard comparison argument.

REMARK Equation (31) corresponds to an elementary factorization in the case of the $M/M/1$ queue and it was found by examining to what extent an analogous factorization might hold in the matrix case. It corresponds to an explicit matrix Wiener–Hopf factorization and, given the rarity of these, the search for additional equalities of this type may prove to be rewarding.

4. SOME SPECIAL CASES

4.1 A Simple Boundary

The case where $B = A_1 + A_2$, $C = A_0$, and $D = A_2$, results in the simplest boundary equations. Examples of that boundary behavior are the $PH/M/1$ queue and the $M/M/1$ queue in a Markovian environment, both discussed in Neuts [2]. For QBD-processes with a simple boundary, the vector $\mathbf{x}$ is given by the "pure" matrix-geometric solution

$$\mathbf{x}_i = \pi(I - R)R^i \qquad \text{for} \quad i \geq 0 \tag{32}$$

Furthermore, the vector Ψ_0 and Ψ are of the same dimension, so that we may evaluate the generating function $\Psi^{**}(z_1,z_2,s,w)$, which is the sum of these two vectors. Applying formulas (14) and (15) for this particular case, we obtain after simplification of the resulting expressions that

$$\begin{aligned}
&\Psi^{**}(z_1,z_2,s,w)\,[sI - A_1 - z_1A_0G(z_1,z_2,s) - wz_1A_0] \\
&= \pi\,(I - R)\,(I - wR)^{-1}\,[wI \\
&\quad - (w - z_2)\sum_{i=0}^{\infty} R^iG^i(z_1,z_2,s)\,[z_2I - G(z_1,z_2,s)]^{-1}
\end{aligned} \tag{33}$$

4.2 The $M/M/1$ Queue

The simplest case of a QBD-process with a simple boundary is the $M/M/1$ queue that corresponds to $A_0 = \lambda$, $A_2 = \mu$, and $A_1 = -(\lambda + \mu)$, where λ and μ are the arrival and service rates. The matrix R reduces to the scalar $\rho = \lambda\mu^{-1}$ and $x_0 = 1 - \rho$.

The transform $G(z_1,z_2,s)$ is then the solution in (0,1) of the quadratic equation

$$\lambda z_1G^2(z_1,z_2,s) - (s + \lambda + \mu)\,G(z_1,z_2,s) + \mu z_2 = 0 \tag{34}$$

The solution is readily evaluated and expanded in an explicit power series [8].

For the particular case of the $M/M/1$ queue, formula (33) may, after lengthy elementary manipulations, be written as:

$$\begin{aligned}
\Psi^{**}(z_1,z_2,s,w) &= \frac{1-\rho}{1-\rho w}\,\frac{z_2 - w\rho G(z_1,z_2,s)}{z_2 - z_1w\rho G(z_1,z_2,s)} \\
&\quad \times \frac{1}{1-\rho G(z_1,z_2,s)}\,\frac{1}{s + \lambda - \lambda z_1G(z_1,z_2,s)} \\
&= \frac{1-\rho}{1-\rho w}\,\frac{z_2 - w\rho G(z_1,z_2,s)}{z_2 - z_1w\rho G(z_1,z_2,s)}
\end{aligned}$$

$$\times \frac{1}{s + \lambda(1 - z_1)G(z_1, z_2, s) + \lambda(1 - z_2)} \tag{35}$$

An immediate consequence of (35) is the equation

$$\Psi^{**}(1, z_2, s, w) = \frac{1 - \rho}{1 - \rho w} \frac{1}{s + \lambda(1 - z_2)}$$

which shows that, in the stationary $M/M/1$ queue, the queue length at time t is *independent* of the number of departures in the interval $(0, t)$. That result is also readily established by a time-reversal argument; see Kelly [1].

5. IMPLICATIONS OF THEOREM 1

By application of Theorem 1, we can evaluate and compare such descriptors of the arrival and departure processes as their variance-time curves and their square-wave spectra [4]. Particularly for the second-order properties, the required analytic calculations remain quite involved. These require two differentiations with respect to z_1 or z_2, after which both these variables are set equal to one. After lengthy manipulations, we obtain transform equations in s, which can be inverted to yield a system of equations involving matrix integro-differential and integral equations of convolution type, whose numerical analysis and computer implementation are subjects for future research. This subject requires a lengthy treatment, which cannot be presented here. We shall only discuss some examples of the highly modular steps that are involved.

In order to evaluate the inverse $V(x)$ of the transform matrix

$$V(s) = \sum_{i=0}^{\infty} R^i G^i(1, 1, s)$$

that equation is written as

$$V(s) = I + RV(s)G(1, 1, s)$$

which, upon inversion, corresponds to the integral equation

$$V(x) = I + R\int_0^x V(x-u)G'(1, 1, u)\,du, \qquad \text{for} \quad x \geq 0$$

As shown in reference [2], the semi-Markov matrix $G(x) = G(1, 1, x)$ is itself the solution to the nonlinear integro-differential equation

$$G'(x) = A_2 + A_1 G(x) + A_0 \int_0^x G(x-u)G'(u)\,du, \qquad \text{for} \quad x \geq 0$$

The other equations arising in the moment calculations are linear, but analytically more complex. Their general study, possibly by such promising tools as the matrix Laguerre transform [7], offers significant numerical analytic challenges. For particular QBD-processes, such as that describing the $PH/PH/1$ queue, the expansions obtained by Ramaswami and Lucantoni [5,6] should result in significant algorithmic simplifications.

REFERENCES

[1] Kelly, F. P. *Reversibility and Stochastic Networks*. New York: Wiley &Sons, 1979.

[2] Neuts, M. F. *Matrix-Geometric Solutions in Stochastic Models: An Algorithmic Approach*. Baltimore: The Johns Hopkins University Press, 1981.

[3] Neuts, M. F. *Structured Stochastic Matrices of $M/G/1$ Type and Their Applications*. New York: Marcel Dekker, 1989.

[4] Neuts, M. F., and Sitaraman, H. The square-wave spectral density of a stationary renewal process. *J. Appl. Math. Simul.*, 2, 117–130, 1989.

[5] Ramaswami V., and Lucantoni, D. Stationary waiting time distributions in queues with phase type service and quasi-birth-and-death processes. *Stoch. Models*, 1, 125–136, 1985.

[6] Ramaswami V., and Lucantoni, D. Algorithms for the multi-server queue with phase type service. *Stoch. Models*, 1, 393–417, 1985.

[7] Sumita, U. The matrix Laguerre transform. *Appl. Math. Comp.*, 15, 1–28, 1984.

[8] Takács, L. *Introduction to the Theory of Queues*. New York: Oxford University Press, 1962.

10

Queuing Systems Having Phase-Dependent Arrival and Service Rates

JOHN N. DAIGLE Department of Industrial Engineering and Operations Research, Virginia Polytechnic Institute and State University, Blacksburg, Virginia

DAVID M. LUCANTONI Department of Teletraffic Theory and System Performance, AT&T Bell Laboratories, Holmdel, New Jersey

ABSTRACT

The behavior of many systems of practical interest in communications and other areas is well modeled by a single-server exponential queuing system in which the arrival and service rates are dependent upon the state of a Markov chain, the dynamics of which are independent of the queue length. Formal solutions to such models based on Neuts's matrix geometric approach have appeared frequently in the literature. A major problem in using the matrix geometric approach is the computation of the rate matrix, which requires the solution of a matrix polynomial. In particular, computational times appear to be unpredictable and excessive for many problems of practical interest, especially under heavy traffic loads. In this paper, we present an efficient method of determining the rate matrix that performs well at all traffic loads. We then show how the rate matrix may be applied to determine not only the joint and marginal queue distributions, but also many other quantities of practical interest, such as the virtual and stochastic equilibrium waiting time distributions. We present numerical examples that illustrate the application of our techniques.

1. INTRODUCTION

The behavior of many systems of practical interest in communications and other areas is well modeled by a single-server exponential queuing system in which the arrival and service rates are dependent upon the state of a Markov chain, the dynamics of which is independent of the queue length. For example, consider the following scenario:

> A communication line services both circuit telephone calls and packet switched data. There are a finite number, say, M telephone subscribers, each of which has exponentially distributed on-hook and off-hook times. Data packets arrive according to a Poisson process and their lengths are assumed to be approximated well by an exponentially distributed random variable. The communication line has a transmission capacity of 1.544 megabits per second, and each active telephone call consumes 64 kilobits per second. A maximum of, say, $N \leq \min\{M, 23\}$ active telephone subscribers are allowed to have calls in progress at any given time. The transmission capacity not used in servicing telephone calls is used to transmit data packets.

In the above scenario, data packets arrive according to a Poisson process, the rate of which is independent of everything, but the service rate for the data packets at any given time is dependent upon the number of telephone calls in progress. The latter process is governed by a birth-death process, the dynamics of which are independent of the number of packets awaiting transmission; we shall refer to this process as the *phase process*. In this paper, we limit our discussion to the special case in which the service rates for all phases are strictly positive. Chapter 6 of Neuts [1981] is devoted to the analysis of more general problems of this nature, which he refers to as the *M/M/1 queue in a random environment*.

Many models for such systems have recently appeared in the literature. For the most part, solutions to the models have been based upon Neuts's matrix geometric approach [Neuts 1981]. Use of this method requires the computation of the so-called *rate matrix*, which is the minimum nonnegative solution of a matrix quadratic equation. A problem with using the matrix geometric approach that has been consistently cited in the literature is that the computation time of the rate matrix is unpredictable and possibly excessive, the latter being especially true for high traffic utilizations. Indeed, some researchers [cf. Stewart and Cao 1987] concluded that it is more computationally efficient to solve the equilibrium balance equations directly than to compute the rate matrix.

Although computation of the rate matrix is recognized as a potential problem, matrix geometric methods offer more than compact presentation of results in the form of the rate matrix. Indeed, the rate matrix is a fundamental parameter of the underlying stochastic process, and it can be used to determine quantities of interest other than the stochastic equilibrium joint and marginal occupancy distributions. In particular, the rate matrix is a determining factor for the occupancy distribution as seen by an arbitrary arrival and the virtual and stochastic equilibrium waiting time distributions. Thus, there is adequate motivation to compute the rate matrix even if its computation is difficult.

Generating function approaches to solving models in the class under consideration here have also received poor reviews in the literature. In fact, the belief that transform approaches are numerically difficult has motivated many researchers to choose the matrix geometric approach. However, the experience of the first named author using transform methods for problems of this nature, dating back prior to the time matrix geometric methods were widely known, provides no evidence of difficulty in inverting transforms when the solution methodology is formulated in terms of a set of eigenproblems. Indeed, in that author's own experience, solutions via transform/eigenanalysis approaches appear to be "blindingly fast" when compared to any other technique, including both direct solution of the balance equations and the matrix geometric approach.

Thus, there is an apparent dilemma: one can compute occupancy distributions "blindingly fast" via transform methods and not have all the nice results available via matrix geometric techniques, or one can suffer through the long computational times required by the matrix geometric technique. The obvious solution to this dilemma is to have it all: compute the rate matrix "blindingly fast" via a transform/eigenanalysis approach!

We begin the formal part of our presentation in Section 2 by formulating the balance equations for the system in matrix form. We then apply transform techniques to obtain an expression for the probability generating function (PGF) for the joint occupancy distribution, and we discuss inversion of this PGF using classical eigenanalysis. Next, in Section 3, we obtain the same PGF in terms of the matrix geometric rate matrix, and we briefly discuss the usual techniques for solving for the rate matrix. We then show the relationship between the rate matrix and the eigensystem analyzed in using the transform approach. In Section 4, we provide numerical examples that illustrate the computational properties of our approach. Then, in Section 5, we describe the application of the rate matrix in computing waiting time distributions. Finally, we summarize our findings in Section 6.

2. TRANSFORM SOLUTION

As we have stated in the introductory section, arrival and service rates in our system are determined by the state of a birth-death process that operates independently of the queue occupancy distribution. We shall refer to this process as the *phase process*. We denote the phase process for the system by $\{\tilde{\wp}(t), t \geq 0\}$ and its infinitesimal generator (Cohen [1969], Chapter 3) by $\mathcal{Q}$. Our convention is that $\{\tilde{\wp}(t), t \geq 0\}$ takes on integer values between 0 and K so that

$$\begin{aligned}\frac{d}{dt}&[P\{\tilde{\wp}(t)=0\}P\{\tilde{\wp}(t)=1\}\cdots P\{\tilde{\wp}(t)=K\}] \\ &= [P\{\tilde{\wp}(t)=0\}P\{\tilde{\wp}(t)=1\}\cdots P\{\tilde{\wp}(t)=K\}]\mathcal{Q}\end{aligned} \tag{1}$$

The state diagram for the phase process is shown in Figure 1. As can be seen from this diagram, the birth rate while the phase process is in state i, $i = 0, 1, \ldots, K-1$ is equal to β_i, and the rate at which the process transitions from state i to state $i-1$, for $i = 1, 1, \ldots, K$, is given by δ_i. Thus, we find that

$$\mathcal{Q} = \begin{bmatrix} -\beta_0 & \beta_0 & 0 & \cdots & & & & \\ \delta_1 & -(\beta_1+\delta_1) & \beta_1 & 0 & \cdots & & & \\ 0 & \delta_2 & -(\beta_2+\delta_2) & \beta_2 & 0 & \cdots & & \\ \cdots & \cdots & \cdots & \cdots & \cdots & \cdots & \cdots & \cdots \\ \cdots & \cdots & \cdots & \cdots & \cdots & \cdots & \cdots & \cdots \\ \cdots & \cdots & \cdots & \cdots & \cdots & \delta_{K-1} & -(\beta_{K-1}+\delta_{K-1}) & \beta_{K-1} \\ \cdots & \cdots & \cdots & \cdots & \cdots & \cdots & \delta_K & -\delta_K \end{bmatrix}$$

Whenever $\{\wp(t), t \geq 0\}$ is in phase i, $0 \leq i \leq K$, the arrival rate of units to the server is λ_i and the service rate is μ_i. For convenience, we define the

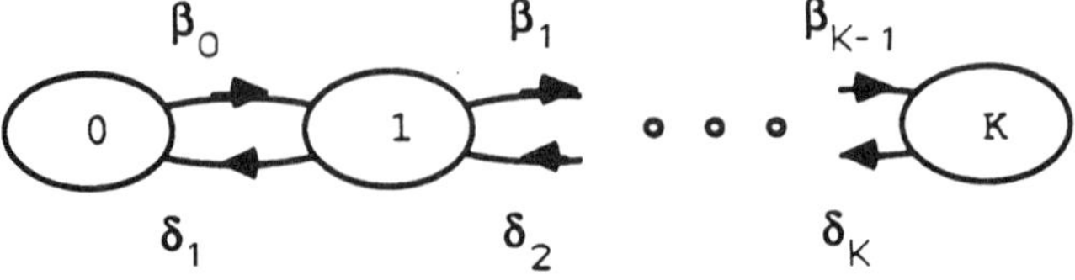

Figure 1 State diagram for phase process.

arrival rate matrix and the service rate matrix, respectively, as follows:

$$\Lambda = \text{diag}(\lambda_0, \lambda_1, \ldots, \lambda_K) \tag{3}$$

and

$$\mathcal{M} = \text{diag}(\mu_0, \mu_1, \ldots, \mu_K) \tag{4}$$

As we have stated in the previous section, our discussion is limited to the special case in which $\mu_i > 0$ for $0 \leq i \leq K$. Modifications required to handle the more general case are under development, and will be discussed elsewhere.

A typical state for our process is specified by the point (n, i), where the first component specifies the current system occupancy and the second component specifies the current phase. Thus, our process is a quasi-birth-death (QBD) process [Neuts 1981] on the state space $\{(n,i), n \geq 0, 0 \leq i \leq K\}$. Let

$$P_{n,i} = \lim_{t \to \infty} P\,\{\tilde{n}(t) = n, \tilde{\wp}(t) = i\} \tag{5}$$

Figure 2 shows a partial state diagram for a queuing system having phase-dependent arrival and service rates. From the state diagram, we find for a typical state on the interior of the diagram, that is, with $n \geq 1$ and $0 < i < K$, that

$$(\lambda_i + \mu_i + \beta_i + \delta_i)P_{ni} = \lambda_i P_{n-1,1} + \beta_{i-1}P_{n,i-1} + \delta_{i+1}P_{n,i+1} + \mu_i P_{n+1,i} \tag{6}$$

From this balance equation, the balance equations for the states that are not interior to the state diagram are readily determined. We simply specialize

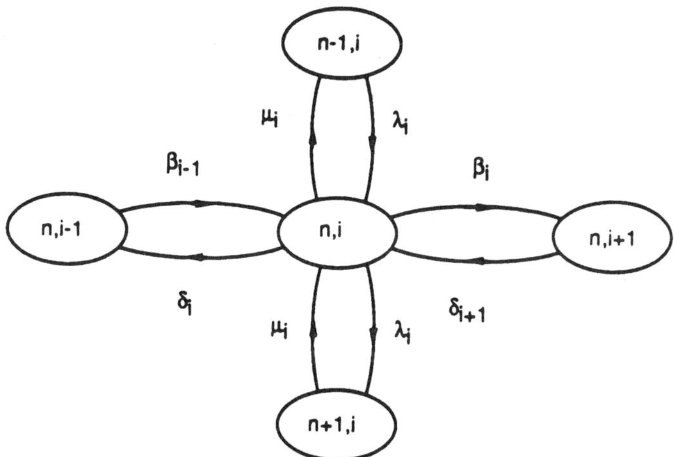

Figure 2 Partial state diagram for system having phase-dependent arrival and service rates.

the above equation to account for the changes due to the boundary conditions. First, we consider the case $n = 0$ and $i = 0$ for which there are no transitions from state $(0, 0)$ due to "deaths," no transitions into state $(0, 0)$ due to "births," and no transitions into state $(0, 0)$ due to new arrivals. Thus, we find the balance equations for state $(0, 0)$ is

$$(\lambda_0 + \beta_0)P_{00} = \delta_1 P_{01} + \mu_0 P_{10} \tag{7}$$

Next, we consider the case $n = 0$, $0 < i < K$. On this boundary, there are no transitions from state $(0, i)$ due to service completions, and no transitions into state $(0, i)$ due to new arrivals. Thus, we obtain the following set of equations:

$$(\lambda_i + \beta_i + \delta_i)P_{0i} = \beta_{i-1}P_{0,i-1} + \delta_{i+1}P_{0,i+1} + \mu_i P_{1,i} \tag{8}$$

Finally, we consider the case $n = 0$ and $i = K$. On this boundary, there are no transitions from state $(0, K)$ due to service completions or "births," and no transitions into state $(0, K)$ due to either new arrivals or "deaths." Thus, the appropriate equation for this boundary is

$$(\lambda_K + \delta_K)P_{0K} = \beta_{K-1}P_{0,K-1} + \mu_K P_{1K} \tag{9}$$

Equations (7)–(9) can be rewritten in more compact form by using matrix notation. Toward this end, we define

$$\pi_n = [P_{n0} \quad P_{n1} \quad \cdots \quad P_{nK}]$$

Then, upon rearranging equations (7)–(9), we find

$$\pi_0(\Lambda - \mathcal{Q}) - \pi_1 \mathcal{M} = 0 \tag{10}$$

The matrix equation (10) summarizes all of the required information for the states in which $n = 0$. A similar set of equations is required for the states in which $n > 0$. These equations can also be obtained from equation (6). For $i = 0$, we find

$$(\lambda_0 + \mu_0 + \beta_0)P_{n0} = \lambda_0 P_{n-1,0} + \delta_1 P_{n1} + \mu_0 P_{n+1,0} \tag{11}$$

Finally, for $i = K$ with $n > 0$, we find, by setting $P_{n,K+1} = 0$ in equation (6) and noting that the birth rate while in state (n, K) is equal to 0, that

$$(\lambda_K + \mu_K + \delta_K)P_{nK} = \lambda_K P_{n-1,K} + \beta_{K-1}P_{n,K-1} + \mu_K P_{n+1,K} \tag{12}$$

Upon rewriting equations (6), (11), and (12) in matrix form, it is readily seen that

$$\pi_{n-1}\Lambda - \pi_n(\lambda - \mathcal{Q} + \mathcal{M}) + \pi_{n+1}\mathcal{M} = 0 \qquad \text{for} \quad n \geq 1 \tag{13}$$

where all of the terms have been previously defined.

Define

$$\mathcal{G}(z) = \sum_{n=0}^{\infty} z^n \pi_n \tag{14}$$

or equivalently,

$$\mathcal{G}(z) = [\, G_0(z) \quad G_1(z) \quad \cdots \quad G_K(z) \,] \tag{15}$$

where

$$G_i(z) = \sum_{n=0}^{\infty} z^n P_{n,i} \qquad \text{for} \quad 0 \le i \le K.$$

Then, upon multiplying both sides of equation (13) by z^n, summing from 1 to ∞, and rearranging terms, we find

$$\mathcal{G}(z)\Lambda z - [\mathcal{G}(z) - \pi_0](\Lambda - \mathcal{Q} + \mathcal{M}) + [\mathcal{G}(z) - \pi_1 z - \pi_0]\mathcal{M}\frac{1}{z} = 0$$

After rearranging terms, the resulting equation is

$$\begin{aligned} &\mathcal{G}(z)\Lambda z - \mathcal{G}(z)(\Lambda - \mathcal{Q} + \mathcal{M}) + \mathcal{G}(z)\mathcal{M}\frac{1}{z} \\ &\qquad = \pi_0\mathcal{M}\left(\frac{1}{z} - 1\right) - [\pi_0(-\mathcal{Q} + \Lambda) - \pi_1\mathcal{M}] \end{aligned} \tag{16}$$

Upon comparison of the last term on the right-hand side of equation (16) to equation (10), we see that this bracketed term is equal to zero. Thus equation (16) reduces to

$$\mathcal{G}(z)\Lambda z - \mathcal{G}(z)(\Lambda - \mathcal{Q} + \mathcal{M}) + \mathcal{G}(z)\mathcal{M}\frac{1}{z} = \pi_0\mathcal{M}\left(\frac{1}{z} - 1\right) \tag{17}$$

or equivalently,

$$\mathcal{G}(z)[\Lambda z^2 - (\Lambda - \mathcal{Q} + \mathcal{M})z + \mathcal{M}] = (1 - z)\pi_0\mathcal{M} \tag{18}$$

To simplify notation, define

$$\mathcal{A}(z) = \Lambda z^2 - (\Lambda - \mathcal{Q} + \mathcal{M})z + \mathcal{M} \tag{19}$$

Then, equation (18) can be rewritten as

$$\mathcal{G}(z)\mathcal{A}(z) = (1 - z)\pi_0\mathcal{M} \tag{20}$$

Upon solving equation (20), we find

$$\mathcal{G}(z) = \frac{1 - z}{\det \mathcal{A}(z)} \pi_0\mathcal{M} \operatorname{adj} \mathcal{A}(z) \tag{21}$$

In principle, equation (21) specifies the generating function for the joint phase-occupancy distribution. However, the form of the solution has several problems. First, equation (21) contains an unknown vector of coefficients, π_0. Second, equation (21) involves the specification of the adjoint of $\mathcal{A}(z)$, which is a second order λ-matrix [Lancaster 1966]; the specification of this matrix is incredibly tedious, at best. Third, equation (21) involves the specification of the determinant of $\mathcal{A}(z)$. This determinant is a polynomial in z of order $2(K + 1)$, and its specification is very tedious except when K is very small.

With regard to the unknown vector π_0, we observe that the vector function $\mathcal{G}(z)$ is a vector of marginal probability generating functions and is therefore bounded at least for $|z| \le 1$ [Hunter 1983]. This means that if the denominator of the right-hand side equation of (21) is zero at z_i such that $|z_i| \le 1$, then there is also a zero of the numerator of the right-hand side equation of (21) at z_i. We show in the appendix (see also Daigle [1977]) that all of the zeros of $\det \mathcal{A}(z)$ are real and positive and that there are exactly K zeros of $\det \mathcal{A}(z)$ in the interval (0, 1) provided that $\det \mathcal{M} \neq 0$. These K zeros lead to K linear equations in the $K + 1$ unknowns of π_0.

From equation (19), we see that $\mathcal{A}(1) = \mathcal{Q}$, since $\mathcal{Q}$ is an infinitesimal generator, $\mathcal{Q}e = 0$; thus, each row of $\mathcal{A}(1)$ sums to zero. Therefore, $\det \mathcal{A}(1) = 0$ so that $\det \mathcal{A}(z)$ has a $(1 - z)$ factor, which cancels the same factor in the numerator. This fact, when coupled with the fact that the probabilities must sum to unity leads to one additional equation in the $K + 1$ unknowns of π_0.

An alternate equation involving the zero of det $\mathcal{A}(z)$ at $z = 1$, which also relates to system stability, can be obtained as follows. By differentiating equation (20), taking the limit as $z \to 1$, and postmultiplying by e, where e is a column vector of 1's, we find that

$$\mathcal{G}(1) \frac{d}{dz} \mathcal{A}(z) \bigg|_{s=1} \mathbf{e} + \frac{d}{dz} \mathcal{G}(z) \bigg|_{s=1} \mathcal{A}(1)\mathbf{e} = -\pi_0 \mathcal{M}\mathbf{e} \tag{22}$$

but, $\frac{d}{dz} \mathcal{A}(z)|_{s=1}\mathbf{e} = (\Lambda + \mathcal{Q} - \mathcal{M})\mathbf{e}$ so that equation (22) reduces to

$$\mathcal{G}(1)(\Lambda + \mathcal{Q} - \mathcal{M})\mathbf{e} = -\pi_0 \mathcal{M}\mathbf{e} \tag{23}$$

Now, from equation (18), we find that $\mathcal{G}(1)\mathcal{A}(1) = \mathcal{G}(1)\mathcal{Q} = 0$ so that equation (23) becomes

$$\mathcal{G}(1)(\mathcal{M} - \Lambda)\mathbf{e} = \pi_0 \mathcal{M}\mathbf{e} \tag{24}$$

We also note in passing that $\mathcal{G}(1)$ is the stationary probability vector for the phase process. Thus, the rearrangement of equation (24), $[\mathcal{G}(1) - \pi_0]\mathcal{M}\mathbf{e} = \mathcal{G}(1)\Lambda\mathbf{e}$, states simply that the overall average rate at which customers are serviced by the system is equal to the overall rate at which customers arrive to the system. The term π_0 is subtracted from the equilibrium phase probability vector in specifying the overall average service rate because there are no customers serviced when the system is empty. In addition, we note that at least one element of π_0 must be greater than zero in order that a stochastic equilibrium solution exist, else all π_n for $n \geq 0$ would be zero. Thus, a condition for stochastic equilibrium that is readily apparent from equation (24) is that

$$\mathcal{G}(1)(\mathcal{M} - \Lambda)\mathbf{e} \geq 0$$

Finally, we point out that we may obtain $\mathcal{G}(1)$ by normalizing the left eigenvector of $\mathcal{Q}$, or equivalently, $\mathcal{A}(1)$, corresponding to its zero eigenvalue; that is, if we denote the left eigenvector of $\mathcal{Q}$ corresponding to its zero eigenvalue by ψ_K, then

$$\mathcal{G}(1) = \frac{\psi_K}{\psi_K \mathbf{e}} \tag{25}$$

With regard to the zeros of $\det \mathcal{A}(z)$, we define $\mathcal{Z}^{(0,\infty)}$ to be the set of all z such that $\det \mathcal{A}(z) = 0$. We partition the set $\mathcal{Z}^{(0,\infty)}$ into three sets: $\mathcal{Z}^{(0,1)}$ and $\mathcal{Z}^{(1,\infty)}$ contains those elements of $\mathcal{Z}^{(0,\infty)}$ less then and greater than unity, respectively, and the third set contains only the unity element. Thus, $\mathcal{Z}^{(0,\infty)} = \mathcal{Z}^{(0,1)} \cup \{1\} \cup \mathcal{Z}^{(1,\infty)}$. The elements of $\mathcal{Z}^{(0,1)}$ are then labeled $z_0, z_1, \ldots, z_{K-1}$; the unit element is referred to as z_K; and the elements of $\mathcal{Z}^{(1,\infty)}$ are labeled $z_{K+1}, \ldots, z_n$ where n is the total number of elements in $\mathcal{Z}^{(0,\infty)}$. We assume the elements of $\mathcal{Z}^{(0,\infty)}$ are distinct,* and for convenience, we order the indexing such that $z_i < z_j$ if $i < j$.

Let $\mathcal{F}_n(z) = \sum_{i=0}^{K} G_i(z)$ or in vector notation, $\mathcal{F}_n(z) = \mathcal{G}(z)e$. Then, from equation (21), we find

$$\mathcal{F}_n(z) = \frac{1-z}{\det \mathcal{A}(z)} \pi_0 \mathcal{M} \operatorname{adj} \mathcal{A}(z)\mathbf{e} \tag{26}$$

Then, for each $z_i \in \mathcal{Z}^{(0,1)}$, we find from equation (26) and the boundedness of $\mathcal{F}_n(z)$ inside and on the unit circle of the complex plane that

$$0 = \pi_0 \mathcal{M} \operatorname{adj} \mathcal{A}(z_i)\mathbf{e} \qquad \text{for} \quad z_i \in \mathcal{Z}^{(0,1)} \tag{27}$$

Equations (24) and (27) then form a system of $K + 1$ linear equations through which π_0 may be determined. The result may then be substituted into equation (21) to obtain the marginal PGFs or into equation (26) to obtain the total PGF.

---

*We make no claim that the elements of $\mathcal{Z}^{(0,\infty)}$ are always distinct, but in our computational experience we have not encountered any cases in which they are not. Alternate techniques would be required to handle the case of nondistinct values, but this case is not addressed here.

It should be noted that one would usually not want to write explicit expressions for adj $\mathcal{A}(z)$ and det $\mathcal{A}(z)$ in order to formulate the linear system of equations. The entire problem can be formulated in terms of the eigenvalues and eigenvectors of a matrix that can be specified directly from inspection of $\mathcal{A}(z)$, and the expressions for $\mathcal{F}_n(z)$ and $\mathcal{G}(z)$ and their corresponding probabilities can be specified in a convenient manner without the need for direct manipulation of $\mathcal{A}(z)$, as will be explained in the remainder of this section.

Before describing our numerical approach to specifying the marginal and joint occupancy distributions, we state a few definitions and theorems related to λ-matrices that may be unfamiliar to many readers and that are directly needed in our development.

DEFINITION: λ-MATRIX A λ-matrix is a matrix, $\mathcal{A}(z)$, the (scalar) elements of which are polynomials in its argument, z.

DEFINITION: NULL VALUE Let $\mathcal{A}(z)$ be a λ-matrix. Then a value z_i, such that $\det \mathcal{A}(z_i) = 0$ is called a null value of $\mathcal{A}(z)$. For example, $(\lambda I - A)$ is a λ-matrix, and the eigenvalues of A are null values of the λ-matrix $(\lambda I - A)$.

DEFINITION: (RIGHT) NULL VECTOR Let $\mathcal{A}(z)$ be a λ-matrix, and let $X(z_i)$ be a nontrivial column vector such that $\mathcal{A}(z_i)X(z_i) = 0$. Then $X(z_i)$ is called a *null vector* of the λ-matrix $\mathcal{A}(z)$ corresponding to the null value z_i. For example, $(\lambda I - A)$ is a λ-matrix, and the eigenvectors of the matrix A are null vectors of the λ-matrix $(\lambda I - A)$.

DEFINITION: LEFT NULL VECTOR Let $\mathcal{A}(z)$ be a λ-matrix, and let $X(z_i)$ be a nontrivial row vector such that $X(z_i)\mathcal{A}(z_i) = 0$. Then $X(z_i)$ is called a *left null vector* of the λ-matrix $\mathcal{A}(z)$ corresponding to the null value z_i.

THEOREM 1 Let $\mathcal{A}(z)$ be a λ-matrix having distinct null values, and let $X(z_i)$ be a null vector of $\mathcal{A}(z)$ corresponding to the null value z_i. Then, the columns of adj $\mathcal{A}(z_i)$ are proportional to $X(z_i)$.

Proof. If $\det \mathcal{A}(z) \neq 0$, then we find

$$\mathcal{A}(z)[\mathcal{A}(z)]^{-1} = I$$

where I is the identity matrix. But,

$$[\mathcal{A}(z)]^{-1} = \operatorname{adj} \mathcal{A}(z) \frac{1}{\det \mathcal{A}(z)}$$

Thus,

$$\mathcal{A}(z) \operatorname{adj} \mathcal{A}(z) = \det \mathcal{A}(z) I \tag{28}$$

But the λ-matrix, and consequently its adjoint and determinant, are continuous functions of its argument. Therefore, the limits of the left- and right-hand sides of equation (28) as $z \to z_i$, exist and the limits of each of the terms exists individually. Thus,

$$\lim_{z \to z_i} \mathcal{A}(z) \lim_{z \to z_i} \operatorname{adj} \mathcal{A}(z) = \lim_{z \to z_i} \det \mathcal{A}(z) I \tag{29}$$

or, equivalently,

$$\mathcal{A}(z_i) \operatorname{adj} \mathcal{A}(z_i) = 0 \tag{30}$$

But, since z_i is a null value of $\mathcal{A}(z)$, equation (30) shows that each column of $\operatorname{adj} \mathcal{A}(z_i)$ is a null vector of $\mathcal{A}(z)$ corresponding to z_i. Since the null values are distinct by hypothesis, all null vectors corresponding to z_i are proportional to each other. Consequently, all columns of $\operatorname{adj} \mathcal{A}(z_i)$ are proportional to each other and to $X(z_i)$.

As a consequence of the above theorem, since the left-hand side of equation (27) is zero, we may replace $\operatorname{adj} \mathcal{A}(z_i)\mathbf{e}$ in equation (27) by $X(z_i)$ where $X(z_i)$ is the null vector of $\mathcal{A}(z)$ corresponding to z_i. This null vector is, in turn, simply the eigenvector off $\mathcal{A}(z_i)$ corresponding to the zero eigenvalue of $\mathcal{A}(z_i)$. Thus, if the null values of $\mathcal{A}(z)$ are known exactly, then computation of the corresponding null vectors is trivial.*

The null values and vectors of $\mathcal{A}(z)$ can be obtained from standard eigenvalue–eigenvector routines. Toward this end, consider the system $\mathcal{A}(\sigma)X_\sigma = 0$ where σ is any null value of $\mathcal{A}(z)$, and X_σ is the corresponding null vector. Define $Y_\sigma = \sigma X_\sigma$. Then, from the definition of $\mathcal{A}(z)$ given by equation (19), we find

$$\sigma \Lambda Y_\sigma - \sigma(\Lambda + \mathcal{Q} + \mathcal{M})X_\sigma + \mathcal{M} X_\sigma = 0$$

and, by definition,

$$Y_\sigma - \sigma X_\sigma = 0$$

Combining these two systems, we find that

$$\left\{ \begin{bmatrix} \mathcal{M} & 0 \\ 0 & I \end{bmatrix} - \sigma \begin{bmatrix} \Lambda - \mathcal{Q} + \mathcal{M} & -\Lambda \\ I & 0 \end{bmatrix} \right\} \begin{bmatrix} X_\sigma \\ Y_\sigma \end{bmatrix} = 0$$

---

*This is true in general as can be seen by the argument following equation (41). In this particular case, $\mathcal{A}(z_i)$ is tridiagonal, and the elements of its null vector can be computed recursively by setting the first component to 1 and then obtaining the remaining elements by solving the linear system $\mathcal{A}(z_i)X(z_i)$ row by row.

or, equivalently,

$$\left\{\begin{bmatrix} I & 0 \\ 0 & I \end{bmatrix} - \sigma \begin{bmatrix} (\Lambda - \mathcal{Q} + \mathcal{M})\mathcal{M}^{-1} & -\Lambda\mathcal{M}^{-1} \\ I & 0 \end{bmatrix}\right\} \begin{bmatrix} \mathcal{M}X_\sigma \\ \mathcal{M}Y_\sigma \end{bmatrix} = 0 \tag{31}$$

From equation (31), we see that the null values of $\mathcal{A}(z)$ are equivalent to the inverses of the eigenvalues of the matrix

$$\mathcal{A}_E = \begin{bmatrix} (\Lambda - \mathcal{Q} + \mathcal{M})\mathcal{M}^{-1} & -\Lambda\mathcal{M}^{-1} \\ I & 0 \end{bmatrix} \tag{32}$$

where we have used the subscript E to denote *expanded.* The null vectors of $\mathcal{A}(z)$ are proportional to $\mathcal{M}^{-1}$ times the upper and lower $(K+1)$-subvectors of the eigenvectors of this same $2(K+1)$-dimensional square matrix. Thus, let $\phi_i = X_{z_i}$, $i = 0, \ldots, K-1$ where $z_i \in \mathcal{Z}^{(0,1)}$. Then, we find that we may form the linear system of equations (24) and (27) as follows:

$$\begin{aligned} 0 &= \pi_0\phi_0 \\ 0 &= \pi_0\phi_1 \\ &\vdots \\ 0 &= \pi_0\phi_{K-1} \\ \mathcal{G}(1)(\mathcal{M} - \Lambda)\mathbf{e} &= \pi_0\mathcal{M}\mathbf{e} \end{aligned} \tag{33}$$

The system (33) may then be solved for π_0.

Having solved for π_0, we have a formal solution for $\mathcal{G}(z)$ in equation (21). However, the form of $\mathcal{G}(z)$ in its present state is not suitable for manipulation, and algebraic manipulation of $\mathcal{A}(z)$ would be required to complete the specification. A reasonable approach at this point may be to expand $\mathcal{G}(z)$ using partial fraction expansions. At first glance, this would appear a formidable task, but a little further investigation will show that this is not the case.

As a starting point, we repeat equation (21), putting the $(1-z)$ factor in the denominator:

$$\mathcal{G}(z) = \frac{1}{\frac{\det \mathcal{A}(z)}{(1-z)}} \pi_0\mathcal{M} \operatorname{adj} \mathcal{A}(z) \tag{34}$$

Now, because Λ is not required to have full rank, the polynomial $\det \mathcal{A}(z)$ may have order less than $2(K+1)$.* If so, there will be a corresponding

*More about this will be discussed in the next section.

number of eigenvalues of $\mathcal{A}_E$ that are zero. We therefore find that

$$\begin{aligned}\det \mathcal{A}(z) &= \sum_{i=0}^{n} a_i z^i \\ &= a_n \prod_{z_i \in \mathcal{Z}^{(0,\infty)}} (z - z_i)\end{aligned} \tag{35}$$

where n is the number of eigenvalues of $\mathcal{A}_E$ having nonzero values, and we recall that $\mathcal{Z}^{(0,\infty)}$ is the set of null values of $\mathcal{A}(z)$ corresponding to those n eigenvalues of $\mathcal{A}_E$. Thus,

$$\det \mathcal{A}(z) \mid_{z=0} = (-1)^n a_n \prod_{z_i \in \mathcal{Z}^{(0,\infty)}} z_i \tag{36}$$

But

$$\begin{aligned}\det \mathcal{A}(z) \mid_{z=0} &= \det \mathcal{M} \\ &= \prod_{i=0}^{K} \mu_i\end{aligned}$$

so that

$$a_n = (-1)^n \prod_{z_i \in \mathcal{Z}^{(0,\infty)}} z_i^{-1} \prod_{i=0}^{K} \mu_i \tag{37}$$

Upon substituting equation (37) into (35), we find

$$\frac{\det \mathcal{A}(z)}{(1-z)} = \prod_{i=0}^{K} \mu_i \prod_{z_i \in \mathcal{Z}^{(0,1)} \cup \mathcal{Z}^{(1,\infty)}} (1 - z_i^{-1} z) \tag{38}$$

Substitution of equation (38) into (34) then yields

$$\mathcal{G}(z) = \frac{1}{\prod_{i=0}^{K} \mu_i \prod_{z_i \in \mathcal{Z}^{(0,1)} \cup \mathcal{Z}^{(1,\infty)}} (1 - z^{-1} z)} \pi_0 \mathcal{M} \operatorname{adj} \mathcal{A}(z) \tag{39}$$

We note that the zeros of $\det \mathcal{A}(z)$ in the interval $(0, 1)$ are canceled by the choice of π_0 so that the remaining zeros are the ones in the interval $(1, \infty)$. Recall that $\mathcal{Z}^{(1,\infty)}$ denotes the set of null values of $\mathcal{A}(z)$ in $(1, \infty)$. Thus, expressing equation (39) using partial fraction expansion results in

$$\mathcal{G}(z) = C_0 + \sum_{z_{K+i} \in \mathcal{Z}^{(1,\infty)}} \frac{1}{(1 - z_{K+i}^{-1} z)} A_i \tag{40}$$

where C_0 is a row vector of constants reflecting the fact that the numerator polynomials may have degree larger than that of the denominator polynomial, and A_i is a row vector representing the residue of $\mathcal{G}(z)$ corresponding to

$z_{K+i} \in \mathcal{Z}^{(1,\infty)}$ [Hunter 1983]. Upon multiplying both sides of equations (39) and (40) by $(1 - z_{K+i}^{-1} z)$ and taking limits as $z \to z_{K+i}^{-1}$, we find

$$A_i = \frac{1}{\prod_{j=0}^{K} \mu_i \prod_{z_i \in \mathcal{Z}^{(0,1)} \cup \mathcal{Z}^{(1,\infty)} \setminus \{z_{K+i}\}} (1 - z_j^{-1} z_{K+i})} \pi_0 \mathcal{M} \operatorname{adj} \mathcal{A}(z_{K+i}) \tag{41}$$

At this point, it is worthwhile to contemplate the difficulties of computation of A_i. At first glance, this computation may appear difficult because it involves evaluating $\operatorname{adj} \mathcal{A}(z_i)$. However, an LUD decomposition approach [Press et al. 1988] that takes into account the fact that $\mathcal{A}(z_i)$ is both tridiagonal and singular leads to a very simple algorithm for obtaining $\operatorname{adj} \mathcal{A}(z_i)$ as the outer product of two vectors that are easily obtained.* Specifically, if we solve the linear equation

$$\mathcal{A}(z_i) B = I$$

for the unknown matrix B, we find

$$B = \mathcal{A}(z_i)^{-1} = \frac{1}{\det \mathcal{A}(z_i)} \operatorname{adj} \mathcal{A}(z_i)$$

Now, analytically, $\det \mathcal{A}(z_i) = 0$, so that B does not exist. However, numerically, we can substitute a very small number for the zero value obtained in the LUD algorithm, and consequently compute a value for B. We may then compute the value implicitly used for the determinant of $\mathcal{A}(z_i)$ by taking the product of the terms on the major diagonal of the *upper matrix*. Upon multiplying the terms of B by the determinant, a very close approximation to $\operatorname{adj} \mathcal{A}(z_i)$ results. Once a numerical representation for $\operatorname{adj} \mathcal{A}(z_i)$ is obtained, one can obtain a scaling vector that gives the multiplying factors between the computed null vector of $\mathcal{A}(z_i)$ and the columns of $\operatorname{adj} \mathcal{A}(z_i)$. Then, $\operatorname{adj} \mathcal{A}(z_i)$ can be stored as the outer product of the null vector of $\mathcal{A}(z_i)$ and the scaling vector.

Now, from equation (40), we find that

$$\begin{aligned} \mathcal{G}(z) &= C_0 + \sum_{z_{K+i} \in \mathcal{Z}^{(1,\infty)}} A_i \sum_{j=0}^{\infty} (z_{K+i}^{-1} z)^j \\ &= C_0 + \sum_{z_i \in \mathcal{Z}^{(1,\infty)}} A_i + \sum_{z_{K+i} \in \mathcal{Z}^{(1,\infty)}} A_i \sum_{j=1}^{\infty} (z_{K+i}^{-1} z)^j \end{aligned}$$

---

*Press et al. [1988] do not describe this method of finding adjoints. On the other hand, the very enlightening discussions presented in that reference with regard to solving linear systems of the form $Ax = b$ where A is singular matrix and b is an arbitrary column vector have directly led to our approach.

$$= \pi_0 + \sum_{z_{K+i} \in \mathcal{Z}^{(1,\infty)}} \sum_{j=1}^{\infty} (z_{K+i}^{-1} z)^j \tag{42}$$

Thus, we need not obtain C_0 explicitly in order to compute the state probabilities. We find simply that

$$\pi_j = \sum_{z_{K+i} \in \mathcal{Z}^{(1,\infty)}} A_i z_{K+i}^{-j} \qquad \text{for} \quad j \geq 1 \tag{43}$$

The marginal probabilities are obtained by simply summing the joint probabilities. We find that

$$\begin{aligned} p_n &= \pi_n e \\ &= \sum_{z_{K+i} \in \mathcal{Z}^{(1,\infty)}} A_i \mathbf{e} z_{K+i}^{-n} \qquad \text{for} \quad j \geq 1 \end{aligned} \tag{43}$$

where $p_n = P\{\tilde{n} = n\}$ is the equilibrium probability that the occupancy is n.

We now present a simple example to illustrate our technique. Larger examples will be provided later.

EXAMPLE A computer accesses a transmission line via a statistical multiplexor or packet switch. The computer acts as a source of traffic; in this case, arrivals of packets from the computer are analogous to arrivals of customers to a queue. The computer alternates between idle and busy periods, which have exponential durations with parameters β and δ, respectively. During busy periods, the computer generates packets at a Poisson rate λ. The service times on the transmission line form a sequence of iid exponential random variables with parameter μ. Compare the state probabilities and the occupancy distribution for the parameter values $\lambda = \beta = \delta = \mu = 1$.

Solution. In referring back to the model, we find $\beta_0 = \beta$, $\delta_1 = \delta$, $\lambda_0 = \lambda$, $\lambda_1 = \lambda$, and $\mu_0 = \mu_1 = \mu$. Thus, we find

$$\mathcal{Q} = \begin{bmatrix} -\beta & \beta \\ \delta & -\delta \end{bmatrix} = \begin{bmatrix} 1 & -1 \\ -1 & 1 \end{bmatrix}$$

$$\mathcal{M} = \begin{bmatrix} \mu & 0 \\ 0 & \mu \end{bmatrix} = \begin{bmatrix} 1 & 0 \\ 0 & 1 \end{bmatrix}$$

$$\Lambda = \begin{bmatrix} 0 & 0 \\ 0 & \lambda \end{bmatrix} = \begin{bmatrix} 0 & 0 \\ 0 & 1 \end{bmatrix}$$

From the definition of $\mathcal{A}_E$, which is given by equation (32), we find

$$\mathcal{A}_E = \begin{bmatrix} 2 & -1 & 0 & 0 \\ -1 & 2 & 0 & -1 \\ 1 & 0 & 0 & 0 \\ 0 & 1 & 0 & 0 \end{bmatrix}$$

Then, from a standard eigenanalysis routine, we find the eigenvalues of $\mathcal{A}_E$ to be $\{3.414214, 1.0, 0.585786, 0\}$. Thus, there are three nonzero eigenvalues of $\mathcal{A}_E$ and we find $\mathcal{Z}^{(0,\infty)} = \{0.292893, 1.0, 1.707107\}$. Following our indexing scheme, $z_0 = 0.292893$, $z_1 = 1$, and $z_2 = 1.707107$.

The null vectors of $\mathcal{A}(z)$ corresponding to z_0 and z_1 are found by partitioning the matrix of eigenvectors of $\mathcal{A}_E$; the results are as follows:

$$\phi_0 = \begin{bmatrix} -0.414214 \\ 0.585786 \end{bmatrix}$$

and

$$\phi_1 = \begin{bmatrix} -0.5 \\ -0.5 \end{bmatrix}$$

respectively. From equation (25), we find

$$\mathcal{G}(1) = \begin{bmatrix} 0.5 \\ 0.5 \end{bmatrix}$$

Thus, from equation (33), we find

$$[\,0 \quad 0.5\,] = \pi_0 \begin{bmatrix} -0.414214 & 1.0 \\ 0.585786 & 1.0 \end{bmatrix}$$

from which we find

$$\pi_0 = [\,0.292893 \quad 0.207107\,]$$

Evaluation of $\mathcal{A}(z_2)$ yields

$$\mathcal{A}(1.707107) = \begin{bmatrix} -2.414214 & 1.707107 \\ 1.707107 & -1.207107 \end{bmatrix}$$

from which we find

$$\operatorname{adj} \mathcal{A}(1.707107) = \begin{bmatrix} -1.207107 & -1.707107 \\ -1.707107 & -2.414214 \end{bmatrix}$$

From equation (41), we find

$$A_1 = \frac{1}{(1 - z_0^{-1} z_2)} \pi_0 \mathcal{M} \operatorname{adj} \mathcal{A}(z_2)$$

$$A_1 = \frac{1}{(1 - 3.414214 \times 1.707107)} \begin{bmatrix} 0.292893 \\ 0.207107 \end{bmatrix}^T$$

$$\times \begin{bmatrix} -1.207107 & -1.707107 \\ -1.707107 & -2.414214 \end{bmatrix}$$

$$= [\, 0.146447 \quad 0.207107 \,]$$

where the notation "T" denotes the matrix transpose operator. Thus, from equation (42), we find

$$\mathcal{G}(z) = \pi_0 + A_1 \sum_{n=1}^{\infty} z_2^{-n} z^n$$

$$= [\, 0.292893 \quad 0.2207107 \,] + [\, 0.146447 \quad 0.207107 \,] \sum_{n=1}^{\infty} 0.585786^n z^n$$

Upon postmultiplication of the both sides of this expression by **e**, we readily find

$$\mathcal{F}(z) = 0.5 + 0.353554 \sum_{n=1}^{\infty} 0.585786^n z^n$$

so that $p_0 = 0.5$ and $p_n = 0.353554 \times 0.585786^n$ for $n \geq 1$.

3. MATRIX GEOMETRIC TRANSFORM SOLUTION

Neuts [1981] has shown that the system of equations (10) and (13) has a matrix-geometric solution; that is, there exists a matrix $\mathcal{R}$ such that

$$\pi_n = \pi_{n-1} \mathcal{R} \quad \forall \quad n \geq 1 \tag{44}$$

Then, we find by successive substitutions into equation (44) that

$$\pi_n = \pi_0 \mathcal{R}^n \quad \forall \quad n \geq 0 \tag{45}$$

The key to solving a matrix geometric system is to specify the matrix $\mathcal{R}$, which we shall address below.

Following our probability generating function approach, we find from equation (44) that

$$\sum_{n=1}^{\infty} z^n \pi_n = \sum_{n=1}^{\infty} z^n \pi_{n-1} \mathcal{R}$$

so that

$$\mathcal{G}(z) - \pi_0 = z\mathcal{G}(z)\mathcal{R}$$

or

$$\mathcal{G}(z)[I - z\mathcal{R}] = \pi_0 \tag{46}$$

and

$$\mathcal{G}(z) = \pi_0[I - z\mathcal{R}]^{-1} \tag{47}$$

Thus,

$$\lim_{z \to 1} \mathcal{G}(z) = \pi_0[I - \mathcal{R}]^{-1} \tag{48}$$

and

$$\pi_0 = \mathcal{G}(1)[I - \mathcal{R}] \tag{49}$$

Also, we have from equation (13) that

$$\pi_{n-1}\Lambda - \pi_n(\Lambda - \mathcal{Q} + \mathcal{M}) + \pi_{n+1}\mathcal{M} = 0 \qquad n \geq 1$$

Hence, upon substituting equation (44) into the above equation, we find

$$\pi_{n-1}\Lambda - \pi_{n-1}\mathcal{R}(\Lambda + \mathcal{Q} + \mathcal{M}) + \pi_{n-1}\mathcal{R}^2\mathcal{M} = 0 \qquad n \geq 1$$

so that

$$\pi_{n-1}[\Lambda - \mathcal{R}(\Lambda - \mathcal{Q} + \mathcal{M}) + \mathcal{R}^2\mathcal{M}] = 0 \qquad n \geq 1 \tag{50}$$

Clearly, a sufficient condition for equation (50) to hold is that $\mathcal{R}$ satisfy

$$\Lambda - \mathcal{R}(\Lambda - \mathcal{Q} + \mathcal{M}) + \mathcal{R}^2\mathcal{M} = 0 \tag{51}$$

Thus, if one could solve equation (51) for $\mathcal{R}$, one could then use equation (49) to solve for π_0, having previously computed $\mathcal{G}(1)$ by normalizing the left eigenvector of $\mathcal{Q}$ corresponding to its zero eigenvalue as described in the previous section, equation (25).

Obtaining a solution for $\mathcal{R}$ of equation (51) is not necessarily an easy task. One possibility is to first solve equation (51) for $\mathcal{R}$ and then use successive approximations to obtain $\mathcal{R}$ numerically as follows:

$$\mathcal{R} = \Lambda(\Lambda - \mathcal{Q} + \mathcal{M})^{-1} + \mathcal{R}^2\mathcal{M}(\Lambda + \mathcal{Q} + \mathcal{M})^{-1}$$

and

$$\mathcal{R}_j = \Lambda(\Lambda - \mathcal{Q} + \mathcal{M})^{-1} + \mathcal{R}_{j-1}^2\mathcal{M}(\Lambda + \mathcal{Q} + \mathcal{M})^{-1} \qquad \text{for} \quad j \geq 1 \qquad (52)$$

The idea is to start with $\mathcal{R} = 0$ and then compute successive approximations to $\mathcal{R}$ using equation (52). Neuts [1981] has shown that the sequence $\{\mathcal{R}_0, \mathcal{R}_1, \mathcal{R}_2, \ldots\}$ is a monotonically increasing sequence that converges to the minimal nonnegative solution to equation (51), and that this solution is the solution that uniquely provides the rate matrix $\mathcal{R}$ that satisfies equation (44).

Another possibility is to specify an arbitrary contraction map [Hewitt and Stromberg [1969] on $\mathcal{R}$ based on equation (51), and then use successive approximations to obtain $\mathcal{R}$. One way to specify a contraction map is to first multiply both sides of equation (51) by some positive number τ and then add $\mathcal{R}$ to both sides of the result. This procedure yields

$$\mathcal{R} = \tau\mathcal{R}^2\mathcal{M} + \mathcal{R}[I - \tau(\mathcal{M} - \mathcal{Q} + \Lambda)] + \tau\Lambda \qquad (53)$$

We then set

$$\mathcal{R}_i = \tau\mathcal{R}_{i-1}^2\mathcal{M} + \mathcal{R}_{i-1}[I - \tau(\mathcal{M} - \mathcal{Q} + \Lambda)] + \tau\Lambda \qquad (54)$$

and iterate on i, starting with some suitable value for $\mathcal{R}_0$, until $\mathcal{R}_i$ converges. One possibility is to set $\mathcal{R}_0 = \text{diag}(1/(K+1), 1/(K+1), \ldots, 1/(K+1))$.

This approach is called the *matrix-geometric* approach, and it is elegantly described in Neuts [1981]. Approaches based on such iterative schemes with special variations [cf. Williams and Leon–Garcia 1984] for certain problems having special structures are the typical way in which $\mathcal{R}$ is determined in the literature. Such approaches suffer from an inability to predict computation times except in very special cases. Indeed, excessive computational times have prompted several researchers to develop special variations to solve special cases in which they are interested.

Matrix geometric techniques are very powerful, and their use is not limited to the analysis of quasi-birth-and-death (QBD) processes. The literature contains many applications of matrix geometric techniques to the solution of problems. An application of this method to a non-Markovian queuing system is presented in Daigle and Langford [1986], and applications of this method to analysis of ethernet-based local area networks are presented in Coyle and Liu [1985] and Beuerman and Coyle [1988]. In addition, the results that may be obtained via matrix-geometric techniques are not limited to occupancy distributions. Ramaswami and Lucantoni [1985] discuss the application of matrix geometric techniques to obtaining stationary waiting time distributions in QBD processes and other systems. Thus, it is very worthwhile to pursue alternate techniques for computing the rate matrix $\mathcal{R}$.

We now turn our attention to a discussion of the relationship between the transform approach described in the previous section and the matrix geometric approach. To begin our discussion, let $\nu \neq 0$ be an eigenvalue of $\mathcal{R}$, and let V_ν be a left eigenvector of $\mathcal{R}$ corresponding to ν. Then, by the defining relationship between eigenvalues and eigenvectors,

$$V_\nu \mathcal{R} = V_\nu \nu \tag{55}$$

Also, upon premultiplying equation (51) by V_ν, we find that

$$V_\nu[\Lambda - \mathcal{R}(\Lambda - \mathcal{Q} + \mathcal{M}) + \mathcal{R}^2\mathcal{M}] = 0$$

Substitution of equation (55) into the previous equation yields

$$V_\nu[\Lambda - \nu(\Lambda - \mathcal{Q} + \mathcal{M}) + \nu^2\mathcal{M}] = 0 \tag{56}$$

Then, upon substituting $\sigma = 1/\nu$ into equation (56) and multiplying both sides by σ^2, we find

$$V_\nu[\Lambda\sigma^2 - (\Lambda - \mathcal{Q} + \mathcal{M})\sigma + \mathcal{M}] = 0 \tag{57}$$

But this latter expression is exactly

$$V_\nu \mathcal{A}(\sigma) = 0 \tag{58}$$

This means that if V_ν is a left eigenvector of $\mathcal{R}$ corresponding to ν, then V_ν is a left null vector of $\mathcal{A}(z)$ corresponding to its null value $\frac{1}{\nu}$. Since it is obvious from equation (45) that, for a stable system, all of the eigenvalues of $\mathcal{R}$ have magnitudes less than unity, the null values of $\mathcal{A}(z)$ that are of interest are those that have magnitudes greater than unity, that is, those in the set $\mathcal{Z}^{(1,\infty)}$. Analogous to the case of the previous section, these eigenvalues and left eigenvectors can be found via a standard eigenanalysis of the matrix

$$\mathcal{A}'_E = \begin{bmatrix} (\Lambda - \mathcal{Q}^T + \mathcal{M})\mathcal{M}^{-1} & -\Lambda\mathcal{M}^{-1} \\ I & 0 \end{bmatrix} \tag{59}$$

where the only difference between equations (32) and (59) is that the matrix $\mathcal{Q}$ is transposed.

Now, the matrix $\mathcal{A}'_E$ may have zero eigenvalues. In fact, it is easy to show, following arguments similar to those of Appendix A, that the number of zero eigenvalues of this matrix is exactly the same as the number of terms that are zero on the major diagonal of Λ. As is shown in the appendix, following an elementary transformation to transform $\mathcal{A}(z)$ into a symmetric λ-matrix, the quadratic form of the transformed symmetric λ-matrix, which has the same null values as the original matrix, can be written as

$$V_\nu \hat{\mathcal{A}}(\sigma) V_\nu^T = l\sigma^2 - (l - p + m)\sigma + m = 0$$

where l, m, p are nonnegative, positive, and nonpositive, respectively. The discriminant of the solution of this quadratic equation is readily found to be $(l - m)^2 - 2p(l + m) + p^2$, which is always nonnegative. Therefore, all null values of $\mathcal{A}(z)$ are real. It is also easy to see that the null values of $\mathcal{A}(z)$ are nonnegative. But as the value of l becomes smaller and smaller, the value of the null value corresponding to V_ν becomes larger and larger, and its inverse becomes smaller and smaller. In the limit, the null value becomes infinity and its inverse becomes zero. In effect, the matrix $\mathcal{A}$ has one less null value, but the inverse of this "null value at infinity" shows up as a zero eigenvalue of $\mathcal{A}_E$.

From equation (52), it is easy to see that if $\lambda_i = 0$ for some i, then the corresponding row of the $\mathcal{R}$ matrix will be zero; this can be seen by simply doing successive substitutions in equation (52) starting with $\mathcal{R}_0 = 0$. Thus, if we define the $(K + 1) \times 1$ column vector whose ith element is 1 with all other elements being 0 by $\mathbf{e}_i$, then it is easy to see that $\mathbf{e}_i$ is a left eigenvector of $\mathcal{R}$.

Now, suppose there are n values of i for which $\lambda_i = 0$. Then there will be n rows of $\mathcal{R}$ that will be identically 0. Define $\mathcal{T}$ to be the elementary transformation such that $\mathcal{T}\Lambda\mathcal{T}$ is a diagonal matrix in which the 0 values of λ_i appear as the first n diagonal elements. Then the first n rows of the matrix $\mathcal{T}\mathcal{R}\mathcal{T}$ will be identically 0, and the rows vectors $\mathbf{e}_i$, $0 \leq i \leq n$ will be left eigenvectors of this matrix. Next, we denote the matrix formed by the collection of the remaining left eigenvectors of $\mathcal{T}\mathcal{R}\mathcal{T}$ by $\hat{\mathcal{V}}$, and partition this matrix into the matrix $[\hat{\mathcal{V}}_1\hat{\mathcal{V}}_2]$, where $\hat{\mathcal{V}}_1$ contains the first n columns of $\hat{\mathcal{V}}$. Then it is readily verified that

$$\begin{bmatrix} I & 0 \\ \hat{\mathcal{V}}_1 & \hat{\mathcal{V}}_2 \end{bmatrix} \mathcal{T}\mathcal{R}\mathcal{T} = \begin{bmatrix} 0 & 0 \\ 0 & \hat{\mathcal{N}} \end{bmatrix} \begin{bmatrix} I & 0 \\ \hat{\mathcal{V}}_1 & \hat{\mathcal{V}}_2 \end{bmatrix} \tag{60}$$

where $\hat{\mathcal{N}}$ is the diagonal matrix of the nonzero eigenvalues of $\mathcal{R}$. The form of equation (60) indicates that if the matrix of left eigenvalues spans the $K + 1$ dimensional eigenspace, then the matrix $\hat{\mathcal{V}}_2$ is nonsingular. Thus, we find

$$\mathcal{R} = \mathcal{T} \begin{bmatrix} 0 & 0 \\ \hat{\mathcal{V}}_2^{-1}\hat{\mathcal{N}}\hat{\mathcal{V}}_1 & \hat{\mathcal{V}}_2^{-1}\hat{\mathcal{N}}\hat{\mathcal{V}}_2 \end{bmatrix} \mathcal{T} \tag{61}$$

The implications of the above is that a zero-valued eigenvalue of $\mathcal{R}$ having multiplicity greater than one simplifies, rather than complicates, computation of $\mathcal{R}$.

The computation of the matrix of left eigenvectors of $\mathcal{R}$ is quite straightforward using standard eigenanalysis packages. First, we formulate the matrix $\mathcal{A}_E'$ and obtain its eigenvalues and corresponding eigenvectors. We then select the set of eigenvalues that are less than unity together with their corresponding eigenvectors. The last $K + 1$ elements of the eigenvector of $\mathcal{A}_E'$ corresponding to each eigenvalue of $\mathcal{A}_E'$ that is less than unity are then transformed by $\mathcal{M}^{-1}$ to yield the elements of the left eigenvector of $\mathcal{R}$ corresponding to the same eigenvalue. If the diagonal matrix whose diagonal elements are the eigenvalues of $\mathcal{R}$ is denoted by $\mathcal{N}$, and the matrix of corresponding left eigenvectors is denoted by $\mathcal{V}$, then we compute $\mathcal{R}$ from

$$\mathcal{R} = \mathcal{V}^{-1}\mathcal{N}\mathcal{V} \tag{62}$$

Once $\mathcal{R}$ is known, we may solve for x_0 using equation (49) and then compute x_n for $n \geq 1$ via equation (45). We note that the computational effort required to compute x_0 via equations (62) and (49) is roughly equal to that required to compute x_0 via equation (33). We also note that computation of a particular power of $\mathcal{R}$ is readily accomplished by making use of the identity

$$\mathcal{R}^n = \mathcal{V}^{-1}\mathcal{N}^n\mathcal{V}$$

4. NUMERICAL EXAMPLES

In this section, we present a limited number of numerical examples that illustrate the computational properties and the viability of the algorithms presented in the previous two sections. In particular, we present some numerical results relative to the example problem that we described in the introduction. For continuity, we restate the problem below and briefly describe the parameters for the system under consideration.

Among the numerical results presented below are average queue length and both conditional and marginal complementary distribution functions, or survivor functions. Before proceeding to our numerical examples, per se, we shall briefly discuss techniques for computing these quantities. With regard to the survivor functions, define the joint (occupancy, phase) survivor function as

$$\Sigma_n = \sum_{m=n+1}^{\infty} \pi_m \tag{63}$$

Then due to equation (45), we find

$$\begin{aligned} \Sigma_n &= \sum_{m=n+1}^{\infty} \pi_0\mathcal{R}^m \\ &= \pi_0[I - R]^{-1}\mathcal{R}^{n+1} \\ &= \mathcal{G}(1)\mathcal{R}^{n+1} \end{aligned} \tag{64}$$

Thus, it is readily seen that the values of the vector Σ_n for successive values of n can be obtained via a simple postmultiplication by $\mathcal{R}$. The marginal survivor function for the queue occupancy can then be obtained by summing the elements of Σ_n, or equivalently, postmultiplication by $\mathbf{e}$. Also, the terms of the conditional survivor functions for the queue occupancy can be obtained by dividing the ith element of Σ_n by the ith element of $\mathcal{G}(1)$.

By using the final form of equation (64) and the well-known result that the expected value of a nonnegative random variable is given by the integral of its survivor function, we find

$$\begin{aligned} E[\tilde{n}] &= \sum_{n=0}^{\infty} \Sigma_n \mathbf{e} \\ &= \mathcal{G}(1)[I - \mathcal{R}]^{-1}\mathcal{R}\mathbf{e} \end{aligned}$$

But, since our technique for computing $\mathcal{R}$ yields the eigenvalues and eigenvectors, the expectation can be computed using these quantities. In particular, by using equation (62) and modest algebraic manipulation, we find

$$E[\tilde{n}] = \mathcal{G}(1)\mathcal{V}^{-1}\operatorname{diag}\left[\frac{\nu_0}{1-\nu_0} \quad \frac{\nu_1}{1-\nu_1} \quad \cdots \quad \frac{\nu_k}{1-\nu_k}\right]\mathcal{V}\mathbf{e} \tag{65}$$

Computational forms for higher moments can be easily derived along the same lines. The resulting formulas are as follows:

$$\begin{aligned} \mathcal{G}^{(n)}(1)\mathbf{e} &= \left.\frac{d^n}{dz^n}\mathcal{G}(z)\right|_{z=1}\mathbf{e} \\ &= \mathcal{G}(1)\mathcal{V}^{-1}\operatorname{diag}\left[\frac{\nu_0}{1-\nu_0}^{n} \quad \frac{\nu_1}{1-\nu_1}^{n} \quad \cdots \quad \frac{\nu_k}{1-\nu_k}^{n}\right]\mathcal{V}\mathbf{e} \end{aligned} \tag{66}$$

Formulas for the above quantities based on the partial expansion representation of the occupancy distribution can be readily developed, the most complicated operation being the summing of a geometric series. These formulas have been developed, and in fact, all numerical results presented below have been computed using both representations, except for accuracy which refers only to the partial fraction expansion technique.

We now turn to the discussion of our sample problem. A communication line services both circuit switched telephone calls and packed switched data. There are a finite number, M, of telephone subscribers, each of which has exponentially distributed on-hook and off-hook times, the latter having parameter r and the rate for the former being dependent upon the offered load a, which is given in Erlangs. In particular, the rate for the on-hook distribution is given by the quantity $a/(Mr)$. Data packets arrive according to a Poisson process and their lengths are assumed to be approximated well by an exponentially distributed random variable having mean 8000, including packet overhead bits.

The communication line has a transmission capacity of 1.544 megabits per second of which 8000 bits per second are used for synchronization. Thus, at full line capacity, the line can transmit 192 packets per second. Each active telephone call consumes 64 kilobits per second. A maximum of min $\{M, 23\}$ active telephone subscribers are allowed to have calls in progress at any given time. The transmission capacity not used in servicing telephone calls is used to transmit data packets. Thus, if there are i active callers, then the service rate for the packets is $(1 - i/24) \times 192$.

Based on the above description, it is readily seen that the parameters for the model are as follows:

$$\begin{aligned}
K &= 23, \\
\beta_i &= \frac{a}{Mr}(M - i) \qquad \text{for} \quad 0 \le i < 23 \\
\delta_i &= ir \qquad \text{for} \quad 1 \le i \le 23 \\
\mu &= (1 - i/24) \times 192 \qquad \text{for} \quad 0 \le i \le 23 \\
\lambda_i &= 192 \times \rho_d
\end{aligned}$$

where ρ_d is the proportion of the full line capacity that is devoted to data transmission.

For our first example, we fix the offered load for the voice traffic at 18 Erlangs, offered data traffic at 15%, and call holding time at 300 s. We then consider system behavior as a function of the calling population size, M. Table 1 shows performance results for values of M between 64 and 65,536 by powers of 2. From Table 1, we see that as the population size increases, the call blocking probability tends to become constant, as would be the case for Poisson arrivals. Also, the probability of finding an empty data queue decreases with increasing M, the rate of decrease tending to zero as M becomes large. The latter is due to the fact that as the voice call arrival process tends to Poisson, the equilibrium distribution of the number of active calls tends to stabilize; this equilibrium distribution is the one seen by arriving data packets since they arrive according to a Poisson process. Similar observations hold for the average queue occupancy and the data occupancy distribution. The maximum eigenvalue across the entire range of M is very close to unity and changes very little as M is increased.

Table 1 also lists a measure of performance of the partial fraction expansion version of the computational procedure called *accuracy*. This term represents the difference between unity and the number obtained by summing the joint probabilities obtained form the computation. That is, the partial fraction expansion is completed via solving for the null values, null vectors, and adjoints, and then the results are used to compute the total probability mass. From the table, it is seen that the probability mass sums to within one part in 10^{14} of unity in all cases. One would then conclude that the tail probabilities could be computed to a reasonably high degree of accuracy.

Table 1 Algorithm and System Performance as a Function of Calling Population Size with Offered Voice Traffic at 18 Erlangs and $\rho_d = 0.15$

Population size	Call blocking probability	$P\{\text{empty queue}\}$	Average data occupancy	Maximum eigenvalue	Accuracy
64	3.330120e−003	5.859597e−001	3.557980e+001	9.989110e−001	1.927386e−015
128	1.627150e−002	4.829850e−001	1.714576e+002	9.992334e−001	1.666365e−015
256	2.967740e−002	4.193942e−001	3.335587e+002	9.993705e−001	4.038653e−016
512	3.849125e−002	3.856325e−001	4.541037e+002	9.994337e−001	3.610393e−016
1024	4.343366e−002	3.684486e−001	5.271198e+002	9.994641e−001	4.718990e−016
2048	4.603680e−002	3.598060e−001	5.672587e+002	9.994789e−001	2.360417e−015
4096	4.737094e−002	3.554752e−001	5.882985e+002	9.994863e−001	1.217559e−016
8192	4.804608e−002	3.533078e−001	5.990694e+002	9.994899e−001	3.011588e−015
16384	4.838567e−002	3.522237e−001	6.045187e+002	9.994918e−001	8.893168e−016
32768	4.855596e−002	3.516815e−001	6.072594e+002	9.994927e−001	6.577312e−016
65536	4.864123e−002	3.514104e−001	6.086338e+002	9.994931e−001	1.542440e−015

Table 2 shows results similar to those of Table 1, except that the data traffic intensity has been increased. The primary purpose of this table is to demonstrate that the computational method does not break down under high traffic loads. For example, when $M = 65{,}536$, the probability of finding an empty queue at an arbitrary point in time is 4×10^{-6}, yet the probabilities sum to within 5 parts in a million of unity. This appears to be quite remarkable in that the maximum eigenvalue is larger than 0.9999999, and the solution depends upon solving for 48 eigenvectors and eigenvalues and repeatedly solving systems of 24 linear equations based on these eigenvalues and eigenvectors. For the case of $M = 32{,}768$, the probability of an empty data queue is 2.5×10^{-4}, which is also very close to saturation. Yet the probabilities sum to within about 1 part in 10^{10} of unity.

In the previous two cases, the maximum system eigenvalue changed very little with increasing M. In the third example, we fix the calling population at 512 and the average off-hook time at 100 seconds and then vary the data traffic intensity from very low to very high load. The results are shown in Table 3. We observe that with no data traffic,* the probability of an empty queue is unity as expected. The maximum eigenvalue ranges from zero, as shown in the previous section, to above 0.999999, with average queue length ranging from zero to about 38 million. From the table, it is seen that the numerical stability of the method does not appear to break down even under very heavy load.

One of the factors affecting numerical stability is the magnitude of the eigenvalues relative to each other. Table 4 shows the complete set of eigenvalues for the third example for three values of traffic intensity: one low, one medium, and one high. Notice that at high traffic intensities, the eigenvalues do not all bunch together near unity as might be the case for some problems. This may account in some part for the seemingly amazing stability of the method. On the other hand, the eigenvalues are not all that far apart either. For example, the three largest eigenvalues are within 0.01 of each other at the highest traffic intensity.

Figure 3 shows conditional and marginal queue length distributions for the $p_d = 0.16$ case of the third example. This figure simply indicates the kind of results that can be obtained. One important point to note from the figure is that the tails of all distributions, conditional, joint, and marginal, decay according to the maximum eigenvalue of $\mathcal{R}$. The decay probabilities begin at roughly the same occupancy levels, but of course, the ordinate values at the point at which the geometric decay begins may differ drastically.

---

*We literally set all the λ terms to zero.

Table 2 Algorithm and System Performance as a Function of Calling Population Size with Offered Voice Traffic at 18.244 Erlangs and $\rho_d = 0.28$

Population size	Call blocking probability	$P\{\text{empty queue}\}$	Average data occupancy	Maximum eigenvalue	Accuracy
64	3.765679e−003	2.575932e−001	1.172273e+003	9.997999e−001	1.055362e−015
128	1.799335e−002	1.325646e−001	6.568136e+003	9.999153e−001	3.089814e−015
256	3.244355e−002	6.551767e−002	1.974862e+004	9.999610e−001	1.936889e−014
512	4.184344e−002	3.227063e−002	4.738760e+004	9.999813e−001	1.393586e−013
1024	4.708666e−002	1.589722e−002	1.038203e+005	9.999909e−001	4.064648e−013
2048	4.984103e−002	7.793980e−003	2.196087e+005	9.999956e−001	2.318497e−012
4096	5.125085e−002	3.765687e−003	4.626801e+005	9.999979e−001	5.990437e−012
8192	5.196384e−002	1.757668e−003	1.000001e+006	9.999990e−001	6.097013e−011
16384	5.232234e−002	7.552270e−004	2.337514e+006	9.999996e−001	1.433846e−010
32768	5.250209e−002	2.544028e−004	6.954315e+006	9.999999e−001	1.031161e−010
65536	5.259210e−002	4.090362e−006	4.330027e+008	1.000000e+000	5.320682e−006

Table 3 Algorithm and System Performance as a Function of ρ_d with Calling Population of 512 and Offered Voice Traffic = 18.244 Erlangs

Data traffic intensity	Call blocking probability	P {empty queue}	Average data occupancy	Maximum eigenvalue	Accuracy
0.000	4.184344e−002	1.000000e+000	0.000000e+000	0.000000e+000	1.857238e−015
0.010	4.184344e−002	9.495388e−001	5.607618e−002	2.315753e−001	2.642743e−015
0.025	4.184344e−002	8.741457e−001	1.774883e−001	5.645127e−001	3.541113e−015
0.050	4.184344e−002	7.533826e−001	1.010752e+000	9.247615e−001	1.634055e−015
0.075	4.184344e−002	6.448787e−001	6.150972e+000	9.833023e−001	1.269601e−015
0.100	4.184344e−002	5.462015e−001	2.418806e+001	9.937754e−001	7.527616e−016
0.120	4.184344e−002	4.731717e−001	5.779082e+001	9.965915e−001	1.503138e−015
1.140	4.184344e−002	4.048339e−001	1.218496e+002	9.979463e−001	7.174166e−016
0.160	4.184344e−002	3.408049e−001	2.366025e+002	9.986871e−001	1.615461e−017
0.180	4.184344e−002	2.807915e−001	4.362660e+002	9.991314e−001	1.750553e−015
0.200	4.184344e−002	2.245312e−001	7.831238e+002	9.994176e−001	7.183924e−016
0.220	4.184344e−002	1.717838e−001	1.404345e+003	9.996130e−001	4.749510e−015
0.240	4.184344e−002	1.223174e−001	2.607990e+003	9.997536e−001	2.339275e−015
0.260	4.184344e−002	7.590597e−002	5.389706e+003	9.998598e−001	2.294551e−015
0.280	4.184344e−002	3.233099e−002	1.581604e+004	9.999440e−001	9.407839e−015
0.290	4.184344e−002	1.154133e−002	4.911922e+004	9.999805e−001	2.696379e−013
0.295	4.184344e−002	1.384957e−003	4.301674e+005	9.999977e−001	3.367907e−012
0.296	4.184344e−002	1.572177e−005	3.814735e+007	1.000000e+000	8.276163e−008

Table 4 Eigenvalues of $\mathcal{R}$ for Various Values of Data Traffic Intensity

Eigenvalue rank	$\rho_d = 0.01$	$\rho_d = 0.16$	$\rho_d = 0.29568$
0	0.231575	0.998687	1.000000
1	0.116740	0.981114	0.997261
2	0.078630	0.953036	0.991985
3	0.059261	0.876432	0.985788
4	0.047543	0.741692	0.979111
5	0.039693	0.627486	0.966280
6	0.034067	0.541263	0.930370
7	0.029838	0.475229	0.855700
8	0.026543	0.423328	0.772968
9	0.023903	0.381549	0.700337
10	0.021740	0.347228	0.638919
11	0.019937	0.318547	0.586939
12	0.018409	0.294227	0.542575
13	0.017100	0.273348	0.504342
14	0.015964	0.255230	0.471087
15	0.014969	0.239361	0.441912
16	0.014091	0.225347	0.416119
17	0.013311	0.212881	0.393157
18	0.012612	0.201721	0.372588
19	0.011983	0.191672	0.354058
20	0.011414	0.182576	0.337279
21	0.010896	0.174303	0.322015
22	0.010424	0.166747	0.308070
23	0.009990	0.159819	0.295281

With regard to computation time, we found that computational effort for the partial fraction expansion approach and the matrix geometric transform approach are roughly equivalent. The time required to determine the rate matrix via the transform approach for this particular problem is roughly 30 s, the computations being performed on an Apple Macintosh II executing Lightspeed C.† It is worthwhile to compare the times required to compute

†The MC68881 floating-point processor is standard equipment for the Apple Macintosh II.

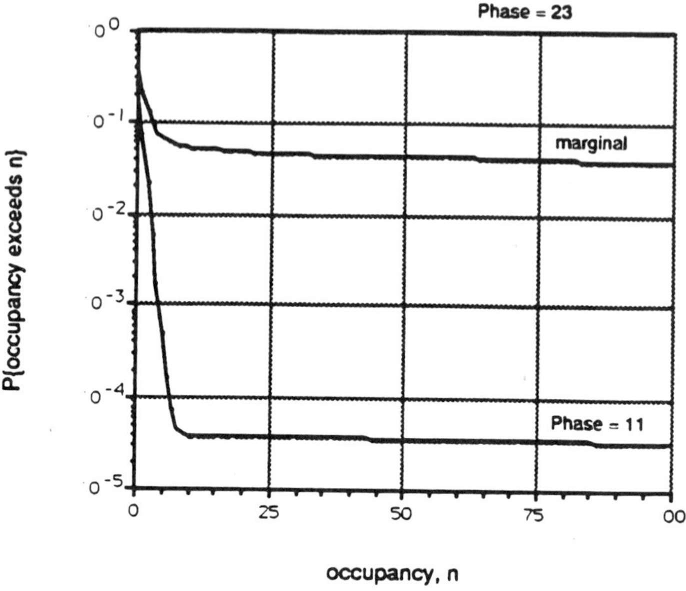

Figure 3 Example conditional and marginal survivor functions for ρ_d = 0.15, calling population of 64, offered voice traffic at 18 Erlangs, and call holding time of 100 s.

the rate matrix via the iterative approach of equation (52) and the transform method described above as well as their accuracies. Several trial comparisons were made, but the time required for the iterative procedure to converge in the examples tried was too large to admit extensive comparison on the target machine. In particular, for a 16-phase system in which the overall traffic intensity was 0.98 and the maximum eigenvalue of $\mathcal{R}$ was 0.999, computation via the transform approach required on the order of 18 s while the iterative procedure required over 4 h to obtain the rate matrix accurate to within six significant figures of that obtained via the transform approach. Since the iterative procedure generates an increasing sequence, and the iterative scheme appears to converge in the direction of the rate matrix obtained via the transform approach, we conclude that the transform approach appears to be quite accurate.

Figure 4 shows a curve that illustrates the rate of conversion of the iterative procedure to the rate matrix computed via transform techniques. In particular, the figure provides a plot of the sum of the absolute values of the differences between the terms of the rate matrix computed via transform tech-

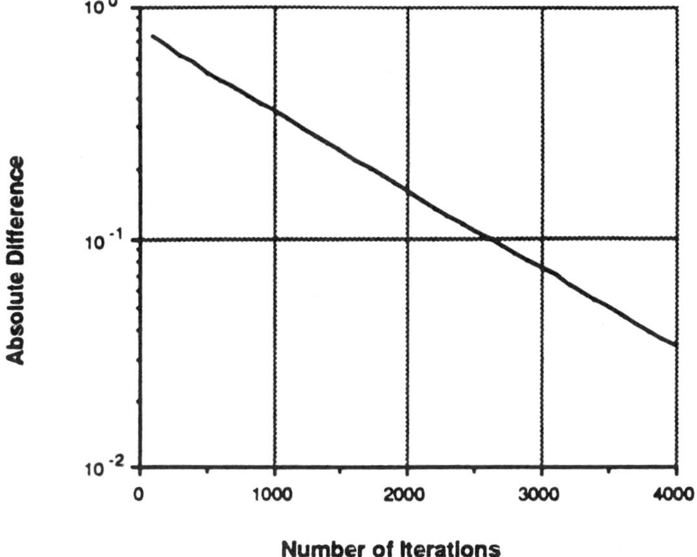

Figure 4 Sum of the absolute values of the difference between the elements of $\mathcal{R}$ computed via transform techniques and $\mathcal{R}_i$ of the iterative technique as a function of the number of iterations for $\rho_d = 0.15$, calling population of 64, offered voice traffic at 18 Erlangs, and call holding time of 100 s.

niques and terms of $\mathcal{R}$, computed according to equation (52) as a function of i. As can be seen from the figure, the rate of convergence is approximately geometric, the geometric rate being roughly 0.926 per 100 iterations. At this rate of convergence, the total number of iterations required for the iterative scheme to converge to within one part in 10^6 of the value achieved with the transform method would be 17,500 iterations. Each 100 iterations requires approximately 105 s, so that total computation time required for this level of accuracy would be over 5 h. By contrast, computation of the rate matrix for this particular size matrix via transform techniques requires slightly less than 30 s. Thus, the transform technique performs faster by a factor of about 600 at this level of accuracy. Since modern machines and languages, such as the Apple Macintosh executing Lightspeed C, store floating-point numbers using 18 significant digits and carry out floating-point calculations using even more digits, the degree of accuracy which may be achieved is remarkable. Thus, the time required to compute $\mathcal{R}$ to within machine accuracy via iterative techniques would be enormous for this particular problem.

It should be noted that convergence of the iterative method is highly sensitive to the magnitude of the maximum eigenvalue of $\mathcal{R}$. However, the transform method is relatively insensitive to this quantity. Thus, in cases in which the traffic intensity is high, or the maximum eigenvalue of $\mathcal{R}$ is close to unity because of the structure of the problem even under low traffic loads as is the case here, then the transform approach will likely perform even better by comparison.

5. THE WAITING TIME DISTRIBUTIONS

In Section 3 we described an efficient procedure for computing the rate matrix $\mathcal{R}$. As we mentioned earlier, $\mathcal{R}$ contains much information about the underlying stochastic process, and many quantities of interest can be computed in terms of $\mathcal{R}$. As an example, we will describe a simple algorithmic procedure for computing the actual and virtual waiting time distributions in terms of $\mathcal{R}$. The derivation of the actual waiting time distribution is presented below. It follows closely the analysis in Ramaswami and Lucantoni [1985] for the virtual waiting time of a general QBD process. The derivation of the virtual waiting time distribution will be omitted, and we refer to Ramaswami and Lucantoni [1985] for the details.

We have already computed $\boldsymbol{\pi}_n$ whose ith component is the stationary probability that there are n customers in the system while the phase process in state i. When all λ_i's are not equal the arrival process is not Poisson, and hence, the distribution of the number in the system and the phase at arrivals will not be the stochastic equilibrium distribution given by $\boldsymbol{\pi}_n$. Let $\mathbf{y}_n$ be a vector whose ith component is the stationary probability that an *arrival* sees n customers in the system and the phase process in state i. Then it is easy to see that

$$\mathbf{y}_n = \frac{1}{\mathcal{G}(1)\Lambda\mathbf{e}} \pi_n \Lambda = (\mathcal{G}(1)\Lambda\mathbf{e})^{-1} \mathcal{G}(1)(I - \mathcal{R})\mathcal{R}^n \Lambda, \quad \forall \quad n \geq 0 \tag{67}$$

Note that when $\lambda_i = \lambda \; \forall \; i$, then $\mathbf{y}_n = \boldsymbol{\pi}_n$ as expected. If a customer arrives at the queue to find n customers in the system and the phase process in state i, then the waiting time of that customer is equal to the sum of n consecutive service times. This is equivalent to the time until absorption in a Markov process, as can be seen as follows. Consider the state space $\{(n,i), n \geq 0, 0 \leq i \leq K\}$ where n is the number in the system and i is the state of the phase process. We list the states in lexicographic order and define the level $\mathbf{n} = \{(n,0), \ldots, (n,K)\}$.

The waiting time of an arriving customer is the time until absorption in a Markov process on the above state space with infinitesimal generator

$$\tilde{Q} = \begin{bmatrix} 0 & 0 & 0 & 0 & \cdots & \cdots \\ \mathcal{M} & \mathcal{Q}-\mathcal{M} & 0 & 0 & \cdots & \cdots \\ 0 & \mathcal{M} & \mathcal{Q}-\mathcal{M} & 0 & \cdots & \cdots \\ 0 & 0 & \mathcal{M} & \mathcal{Q}-\mathcal{M} & \cdots & \cdots \\ \cdots & \cdots & \cdots & \cdots & \cdots & \cdots \\ \cdots & \cdots & \cdots & \cdots & \cdots & \cdots \end{bmatrix}$$

and initial probability vector $y = (y_0, y_1, \ldots)$. Note that the set of states $(0,0)$, $\ldots$, $(0,K)$ are the absorbing states and that starting in a level $\mathbf{n}$, $n \geq 1$, the time to reach level $\mathbf{0}$ is the sum of n consecutive service times. The probability that a customer does not wait is given by

$$P\{\tilde{w} = 0\} = \mathbf{y}_0\mathbf{e} = 1 - \frac{\mathcal{G}(1)\mathcal{R}\Lambda\mathbf{e}}{\mathcal{G}(1)\Lambda\mathbf{e}} \tag{68}$$

We derive a computational procedure for computing the absorption time distribution by uniformization [Ross 1985]. In particular, choose $\theta = \max_{0 \leq i \leq K} \{(\mathcal{M} - \mathcal{Q})_{ii}\}$. Then the evolution of the Markov process with infinitesimal generator $\tilde{Q}$ is equivalent to the following construction. Start a Poisson process with rate θ. At each Poisson epoch choose the next state according to the discrete time Markov chain with transition probability matrix $S = I + \theta^{-1}\tilde{Q}$, that is,

$$S = \begin{bmatrix} 0 & 0 & 0 & 0 & \cdots & \cdots \\ \theta^{-1}\mathcal{M} & I + \theta^{-1}(\mathcal{Q}-\mathcal{M}) & 0 & 0 & \cdots & \cdots \\ 0 & \theta^{-1}\mathcal{M} & I + \theta^{-1}(\mathcal{Q}-\mathcal{M}) & 0 & \cdots & \cdots \\ 0 & 0 & \theta^{-1}\mathcal{M} & I + \theta^{-1}(\mathcal{Q}-\mathcal{M}) & \cdots & \cdots \\ \cdots & \cdots & \cdots & \cdots & \cdots & \cdots \\ \cdots & \cdots & \cdots & \cdots & \cdots & \cdots \end{bmatrix}$$

It is clear that "downward" transitions, that is, transitions from level $\mathbf{i}+1$ to $\mathbf{i}$, $i \geq 0$, correspond to service completions. In the uniformized process, each Poisson epoch may or may not correspond to a service completion. Therefore. if a customer arrives to find n customers in the system, then the probability that the customer's waiting time is less than or equal to x is equal to the probability that some number $j \geq n$ of Poisson epochs occur in time $\leq x$ and at least n of these correspond to service completions. Conversely, the complementary waiting time distribution $S_{\tilde{w}}(x) = P\{\tilde{w} > x\}$ is the probability that

$j \geq 0$ Poisson epochs occur in $(0,x]$ and at most $n-1$ of these correspond to service completions.

Define the matrices $K_n^{(j)}, j \geq 0, n \geq 0$, whose (i,k)-th element is the conditional probability that starting in some level **l**, $l \geq j$, and the phase process in state i, $0 \leq i \leq K$ in j transitions of the Markov chain with transition matrix S, exactly n of these are downward transitions, and after the jth transition, the phase process is in state k, $0 \leq k \leq K$. It is easy to see that these matrices satisfy $K_0^{(0)} = I$, $K_n^{(j)} = 0$, $n > j$, and for $j+1 \geq n$ with $j \geq 0$,

$$K_n^{(j+1)} = K_n^{(j)}[I + \theta^{-1}(\mathcal{Q} - \mathcal{M})] + \theta^{-1}K_{n-1}^{(j)}M \tag{70}$$

with $K_{-1}^{(j)} = 0$.

The above discussion on the relationship between the complementary waiting time distribution and the uniformized process leads to

$$\begin{aligned} S_{\tilde{w}}(x) &= \sum_{i=1}^{\infty} y_i \sum_{j=0}^{\infty} e^{-\theta x} \frac{(\theta x)^j}{j!} \sum_{n=0}^{i-1} K_n^{(j)}\mathbf{e} \\ &= \sum_{j=0}^{\infty} d_j e^{-\theta x} \frac{(\theta x)^j}{j!} \end{aligned} \tag{71}$$

where $S_{\tilde{w}}(x) = P\{\tilde{w} > x\}$, and

$$d_j = \frac{\mathcal{G}(1)(I-\mathcal{R})}{\mathcal{G}(1)\Lambda\mathbf{e}} \sum_{i=1}^{\infty} \mathcal{R}^i \Lambda \sum_{n=0}^{i-1} K_n^{(j)}\mathbf{e} \tag{72}$$

Now,

$$\begin{aligned} &\sum_{i=1}^{\infty} \mathcal{R}^i \Lambda \sum_{n=0}^{i-1} K_n^{(j)}\mathbf{e} \\ &= \sum_{i=1}^{j+1} \mathcal{R}^i \Lambda \sum_{n=0}^{i-1} K_n^{(j)}\mathbf{e} + \sum_{i=j+2}^{\infty} \mathcal{R}^i \Lambda \sum_{n=0}^{j} K_n^{(j)}\mathbf{e} \\ &= \sum_{n=0}^{j} \sum_{i=n+1}^{j+1} \mathcal{R}^i \Lambda K_n^{(j)}\mathbf{e} + (I-\mathcal{R})^{-1}\mathcal{R}^{j+2}\Lambda\mathbf{e} \\ &= \sum_{n=0}^{j} (I-\mathcal{R})^{-1}(\mathcal{R}^{n+1} - \mathcal{R}^{j+2})\Lambda K_n^{(j)}\mathbf{e} + (I-\mathcal{R})^{-1}\mathcal{R}^{j+2}\Lambda\mathbf{e} \\ &= \sum_{n=0}^{j} (I-\mathcal{R})^{-1}\mathcal{R}^{n+1}\Lambda K_n^{(j)}\mathbf{e}, \end{aligned} \tag{73}$$

where we have used the fact that

$$\sum_{n=0}^{j} K_n^{(j)}\mathbf{e} = \mathbf{e}$$

which is readily verified using equation (70). Substitution of equation (73) into (72) yields

$$d_j = \frac{1}{\mathcal{G}(1)\Lambda\mathbf{e}}\mathcal{G}(1)\sum_{n=0}^{j}\mathcal{R}^{n+1}\Lambda K_n^{(j)}\mathbf{e} = \frac{1}{\mathcal{G}(1)\Lambda\mathbf{e}}\mathcal{G}(1)\mathcal{L}_j\mathbf{e} \tag{74}$$

where $\{\mathcal{L}_0, \mathcal{L}_1, \ldots\}$ is a sequence of matrices defined by the following recursion:

$$\mathcal{L}_0 = \mathcal{R}\Lambda \tag{75}$$

$$\begin{aligned}
\mathcal{L}_{j+1} &= \sum_{n=0}^{j+1}\mathcal{R}^{n+1}\Lambda K_n^{(j+1)} \\
&= \sum_{n=0}^{j+1}\mathcal{R}^{n+1}\Lambda K_n^{(j)}[I + \theta^{-1}(\mathcal{Q} - \mathcal{M})] + \sum_{n=0}^{j+1}\mathcal{R}^{n+1}\Lambda K_{n-1}^{(j)}[\theta^{-1}\mathcal{M}] \\
&= \mathcal{L}_j[I + \theta^{-1}(\mathcal{Q} - \mathcal{M})] + \theta^{-1}\mathcal{R}\mathcal{L}_j\mathcal{M}
\end{aligned} \tag{76}$$

We summarize these results in the following theorem:

THEOREM 2 The stationary complementary waiting time distribution at arrival times is given by

$$S_{\tilde{w}}(x) = \sum_{j=0}^{\infty} d_j e^{-\theta s}\frac{(\theta x)^j}{j!}$$

where $\theta = \max\{\mathcal{M}_{ii} - \mathcal{Q}_{ii}\}$, $d_j = [\mathcal{G}(1)\Lambda\mathbf{e}]^{-1}\mathcal{G}(1)\mathcal{L}_j\mathbf{e}$, $\mathcal{L}_0 = \mathcal{R}\Lambda$, and $\mathcal{L}_{j+1} = \mathcal{L}_j[I + \theta^{-1}(\mathcal{Q} - \mathcal{M})] + \theta^{-1}\mathcal{R}\mathcal{L}_j\mathcal{M}$ for $j \geq 0$.

We note that the sequence $\{d_j\}$ is computed only once. Then equation (71) is used to evaluate $S_{\tilde{w}}(x)$ for different values of x. In computing successive d_j's, only the previous matrix $\mathcal{L}_j$ needs to be stored in memory. Also, for the special case in which $K = 0$, the system reduces to an $M/M/1$ queue. In this case, $\mathcal{R} = \rho = \lambda/\mu$, $\theta = \mu$, $\mathcal{L}_j = \rho^{j+1}\lambda$, and $d_j = \rho^{j+1}$. Therefore, equation (71) reduces to the familiar formula

$$S_{\tilde{w}}(x) = \rho e^{-\mu(1-\rho)s}$$

A little thought reveals that the virtual waiting time distribution is equal to the time until absorption in the Markov process having infinitesimal generator $\tilde{Q}$ and initial probability vector $\pi = (\pi_0, \pi_1, \ldots)$. A derivation analogous to that presented above leads to the computational algorithm for this quantity summarized by the following theorem.

THEOREM 3 The stationary complementary virtual waiting time distribution is given by

$$S_{\tilde{v}}(x) = \sum_{j=0}^{\infty} \hat{d}_j e^{-\theta s} \frac{(\theta x)^j}{j!} \tag{77}$$

where $\tilde{v}$ is the virtual waiting time, $S_{\tilde{v}}(x) = P\{\tilde{v} > x\}$, $\theta = \max\{\mathcal{M}_{ii} - \mathcal{Q}_{ii}\}$, $d_j = \mathcal{G}(1)\hat{\mathcal{L}}_j\mathbf{e}$, $\hat{\mathcal{L}}_0 = \mathcal{R}$, and $\hat{\mathcal{L}}_{j+1} = \hat{\mathcal{L}}_j[I + \theta^{-1}(\mathcal{Q} - \mathcal{M})] + \theta^{-1}\mathcal{R}\hat{\mathcal{L}}_j\mathcal{M}$ for $j \geq 0$.

We note in conclusion that if both the virtual waiting time and the waiting time at arrivals are to be computed, then the similarity of the above algorithms can be exploited, and they can be computed simultaneously with many common ingredients.

6. CONCLUSIONS

In this paper, we have presented computationally efficient methods through which to analyze the queuing behavior, both occupancy and waiting time, of systems that are will modeled by a single-server exponential queuing system in which the arrival and service rates are dependent upon the state of a Markov chain, the dynamics of which are independent of the queue length. We have shown how transform and eigenanalysis techniques can be applied to efficiently compute occupancy probabilities via partial fraction expansions and the rate matrix required in the popular matrix geometric solution methodology of Neuts [1981]. Our techniques appear to be at least 100 to 1000 times faster than a direct iterative procedure that produces results of comparable quality for a system having 24 phases at moderate traffic loads. Given the drastic difference in performance at moderate loads, we saw no reason to investigate the performance of the iterative procedure at higher loads, given its inherent disadvantages. We have demonstrated that our techniques do not break down under heavy load, at least in the class of problem under discussion here. In our computational experience, we have not encountered any numerical difficulties.

Although we have limited discussion here to the case in which the phase process is a birth-and-death process, none of the standard library routines used in our computations are limited to these types of processes. Neither are any of the key steps in our development dependent upon our system having real or nonnegative eigenvalues. Thus, we anticipate that our approach will be applicable to more general settings, including general QBD processes. Extensions of our approach to more general environments are currently under investigation.

Acknowledgments: We are grateful to Dr. A. E. Eckberg of AT&T Bell Laboratories for the many useful ideas he has shared with us, relative to this work, for more than 10 years. We also thank the referees who provided many suggestions for improving our presentation.

APPENDIX

In this appendix, we show that the null values of the matrix $\mathcal{A}(z)$ are all real and positive and that exactly K of these null values are in the interval $(0, 1)$. In order to prove that the null values are all positive, we first transform the λ-matrix $\mathcal{A}(z)$ into a symmetric λ-matrix $\hat{\mathcal{A}}(z)$, then form the quadratic form of $\hat{\mathcal{A}}(z)$. This quadratic form results in a scalar quadratic equation, one of the roots of which is a null value of $\mathcal{A}(z)$ and the other, if any, being extraneous. However, both roots are positive given the form of $\mathcal{A}(z)$. To show that there are exactly K null values of $\mathcal{A}(z)$ in the interval $(0, 1)$, we first establish $\det \mathcal{A}(z)$ as $(1-z) \det \mathcal{B}(z)$. We then show, using a modified version of $\mathcal{B}(z)$, $\hat{\mathcal{B}}(z)$, that there are exactly K null values of $\mathcal{A}(z)$ in the interval $(0, 1)$. We then show that it is impossible for any of these null values to migrate across the point $z = 1$ as the elements of $\hat{\mathcal{B}}(z)$ are modified to become $\mathcal{B}(z)$.

To begin our development, we repeat the definition of $\mathcal{A}(z)$. We have

$$\mathcal{A}(z) = \Lambda z^2 - (\Lambda - \mathcal{Q} + \mathcal{M})z + \mathcal{M} \tag{A.1}$$

where

$$\Lambda = \text{diag}(\lambda_0, \lambda_1, \ldots, \lambda_K) \tag{A.2}$$

$$\mathcal{M} = \text{diag}(\mu_0, \mu_1, \ldots, \mu_k) \tag{A.3}$$

and

$$\mathcal{Q} = \begin{bmatrix} -\beta_0 & \beta_0 & 0 & \cdots & & & & \\ \delta_1 & -(\beta_1+\delta_1) & \beta_1 & 0 & \cdots & & & \\ 0 & \delta_2 & -(\beta_2+\delta_2) & \beta_2 & 0 & \cdots & & \\ \cdots & \cdots & \cdots & \cdots & \cdots & \cdots & \cdots & \cdots \\ \cdots & \cdots & \cdots & \cdots & \cdots & \cdots & \cdots & \cdots \\ \cdots & \cdots & \cdots & \cdots & \cdots & \delta_{K-1} & -(\beta_{K-1}+\delta_{K-1}) & \beta_{K-1} \\ \cdots & \cdots & \cdots & \cdots & \cdots & \cdots & \delta_K & -\delta_K \end{bmatrix} \tag{A.4}$$

Now, since $\mathcal{Q}$ is tridiagonal and the off-diagonal terms have the same sign, it is possible to transform $\mathcal{Q}$ to a symmetric matrix via a diagonal similarity transformation as is done to examine the eigenvalues of a specific $\mathcal{Q}$ matrix by Karlin and McGregor [1965]. In particular, we define

$$W_{ij} = \begin{cases} \prod_{j=0}^{i} l_j & \text{for } i = j = 0, 1, \ldots, K \\ 0, & \text{else} \end{cases}$$

where $l_0 = 1$ and

$$l_j = \sqrt{\frac{\delta_j}{\beta_{j-1}}} \qquad \text{for} \quad 1 \leq j \leq K$$

Then it is easy to verify that

$$\tilde{\mathcal{Q}} = W^{-1}\mathcal{Q}W$$

where $\tilde{\mathcal{Q}}$ is the following tridiagonal symmetric matrix:

$$\tilde{\mathcal{Q}} = \begin{bmatrix} -\beta_0 & \sqrt{\beta_0\delta_1} & 0 & \cdots & & & & \\ \sqrt{\beta_0\delta_1} & -(\beta_1+\delta_1) & \sqrt{\beta_1\delta_2} & 0 & \cdots & & & \\ 0 & \sqrt{\beta_1\delta_2} & -(\beta_2+\delta_2) & \sqrt{\beta_2\delta_3} & 0 & \cdots & & \\ \cdots & \cdots & \cdots & \cdots & \cdots & \cdots & \cdots & \cdots \\ \cdots & \cdots & \cdots & \cdots & \cdots & \cdots & \cdots & \cdots \\ \cdots & \cdots & \cdots & \cdots & \cdots & \sqrt{\beta_{K-2}\delta_{K-1}} & -(\beta_{K-1}+\delta_{K-1}) & \sqrt{\beta_{K-1}\delta_K} \\ \cdots & \cdots & \cdots & \cdots & \cdots & \cdots & \sqrt{\beta_{K-1}\delta_K} & -\delta_K \end{bmatrix} \tag{A.4}$$

It is then straightforward to show that $\hat{\mathcal{Q}}$ is negative semidefinite. Also, since the determinate of a product of matrices is equal to the product of the individual determinants, and W is nonsingular, it follows that

$$\det \hat{\mathcal{A}}(z) = \det W^{-1}\mathcal{A}(z)W \tag{A.5}$$

Further, since Λ, $\mathcal{M}$, and W are all diagonal, it follows that

$$\det \hat{\mathcal{A}}(z) = \det[\Lambda z^2 - (\Lambda - \hat{\mathcal{Q}} + M)z + M] \tag{A.6}$$

Now, suppose $\det \mathcal{A}(\sigma) = 0$. Then $\det \hat{\mathcal{A}}(\sigma) = 0$, and there exists X_σ, nontrivial, such that

$$\hat{\mathcal{A}}(\sigma)_\sigma = 0$$

Therefore,

$$\langle X_\sigma, \hat{\mathcal{A}}(\sigma)X_\sigma \rangle = 0 \tag{A.7}$$

where $\langle \cdot, \cdot \rangle$ denotes the inner product. But all of the matrices of equation (A.6) are semidefinite, and therefore we find that

$$\langle X_\sigma, \hat{A}(\sigma) X_\sigma \rangle = l\sigma^2 - (l - p + m)\sigma + m = 0 \tag{A.9}$$

where

$$\begin{aligned} l &= \langle X_\sigma, \Lambda X_\sigma \rangle \geq 0 \\ m &= \langle X_\sigma, \mathcal{M} X_\sigma \rangle > 0 \\ p &= \langle X_\sigma, \hat{Q} X_\sigma \rangle \leq 0 \end{aligned}$$

The discriminant of the quadratic equation for equation (A.9) is given by

$$(l - p + m)^2 - 4lm = (l - m)^2 - 2p(l + m) + p^2$$

Clearly, the right-hand side of the above equation is the sum of nonnegative terms so that the value of σ must be real: this establishes the fact that all null values of $A(z)$ are real.

With regard to the value of σ, there are two possibilities: $l = 0$, or $l > 0$. In the former case, we find from equation (A.9) that

$$\sigma = \frac{m}{(-p + m)} > 0$$

In the latter case, from the quadratic equation, we find

$$\sigma = \frac{1}{2l}\left[(l - p + m) \pm \sqrt{(l - p + m)^2 - 4lm}\right] \tag{A.10}$$

Clearly, $(l - p + m) > \sqrt{(l - p + m)^2 - 4lm} > 0$. Therefore, both roots of equation (A.9) are positive, so that all of the null values of $A(z)$ are positive.

We now turn our attention to the proof of the fact that there are exactly K null values of $A(z)$ in the interval $(0, 1)$. We first note that $\det \mathcal{A}(z)$ is equal to the determinant of the matrix obtained from $\mathcal{A}(z)$ by replacing its last column by $\mathcal{A}(z)\mathbf{e}$. Now, it is easy to show that

$$\mathcal{A}(z)\mathbf{e} = \begin{bmatrix} \lambda_0 z^2 - (\lambda_0 + \mu_0)z + \mu_0 \\ \lambda_1 z^2 - (\lambda_1 + \mu_1)z + \mu_1 \\ \vdots \\ \lambda_K z^2 - (\lambda_K + \mu_K)z + \mu_K \end{bmatrix} = \begin{bmatrix} (1 - z)(\mu_0 - \lambda_0 z) \\ (1 - z)(\mu_1 - \lambda_1 z) \\ \vdots \\ (1 - z)(\mu_K - \lambda_K z) \end{bmatrix}$$

Therefore, $\det \mathcal{A}(z) = (1 - z)\det \mathcal{B}(z)$, where $\mathcal{B}(z)$ is the matrix obtained by replacing the last column of $\mathcal{A}(z)$ by the column vector $(\mathcal{M} - \Lambda z)e$.

Now, we may compute the determinant of $\mathcal{B}(1)$ by expanding about its last column. In doing this, we find by use of Theorem 1 (Section 2) that the vector of cofactors corresponding to the last column is proportional to ψ_K, the left eigenvector of $\mathcal{Q}$ corresponding to its zero eigenvalue. Since

$$G(1) = \frac{\psi_K}{\psi_K \mathbf{e}}$$

we then see that

$$\det \mathcal{B}(1) = \alpha G(1)[\mathcal{M} - \Lambda]\mathbf{e} \tag{A.11}$$

for some $\alpha = \psi_K e \neq 0$. Thus, by comparing equation (A.11) to (24), we see that $\det \mathcal{B}(1)$ is nonzero unless $\pi_0 = 0$. That is, if an equilibrium solution exists, then $\mathcal{B}(1)$ is nonsingular. Equivalently, if an equilibrium solution exists, then $\mathcal{B}(z)$ has no null value at $z = 1$.

We note that the null values of a λ-matrix are continuous functions of the parameters of the matrix. Thus, we may choose to examine the behavior of the null values of $\mathcal{B}(z)$ as a function of the *death rates* of the phase process while keeping the *birth rates* positive. Toward this end, we define $\mathcal{B}(z) = \hat{\mathcal{B}}(z, \delta_1, \delta_2, \ldots, \delta_K)$.

First, consider $\det \mathcal{B}(z, 0, 0, \ldots, 0)$. Since the lower subdiagonal of this λ-matrix is zero, the determinant of the matrix is equal to the product of the diagonal elements, which are as follows:

$$\lambda_i z^2 - (\lambda_i + \beta_i + \mu_i)z + \mu_i \qquad 0 \leq i < K$$

$$\mu_K - \lambda_K z$$

The null values of $\hat{\mathcal{B}}(z, 0, 0, \ldots, 0)$ may therefore be found by setting each of the above terms equal to zero. Thus, the null values of $\hat{\mathcal{B}}(z, 0, 0, \ldots, 0)$ may be obtained by solving the following equations for z:

$$\lambda_i z^2 - (\lambda_i + \beta_i + \mu_i)z + \mu_i = 0 \qquad 0 \leq i < K \tag{A.13}$$

and

$$\mu_K - \lambda_K z = 0 \tag{A.14}$$

As before, when solving equation (A.12), two possibilities must be considered: $\lambda_i = 0$, and $\lambda_i > 0$. In the first case,

$$z = \frac{\mu_i}{\beta_i + \mu_i}$$

so for each i, the corresponding null value is between zero and one. In the latter case, we find that

$$\lim_{s \to 0} \lambda_i z^2 - (\lambda_i + \beta_i + \mu_i)z + \mu_i = \mu_i > 0$$

$$\lim_{s \to 1} \lambda_i z^2 - (\lambda_i + \beta_i + \mu_i)z + \mu_i = -\beta_i < 0$$

and

$$\lim_{s \to \infty} \lambda_i z^2 - (\lambda_i + \beta_i + \mu_i)z + \mu_i$$

tends to $+\infty$. Therefore, each quadratic equation of (A.12) has one root in $(0, 1)$ and one root in $(1, \infty)$ if $\lambda_i > 0$. In summary, the equations (A.12) yield exactly K null values of $\hat{\mathcal{B}}(z, 0, 0, \ldots, 0)$ in $(0, 1)$ and between 0 and K null values of $\hat{\mathcal{B}}(z, 0, 0, \ldots, 0)$ in $(1, \infty)$, the latter quantity being equal to the number of positive values of λ_i.

One additional null value of $\hat{\mathcal{B}}(z, 0, 0, \ldots, 0)$ is obtained from equation (A.13). Since the phase process in this case is a pure birth process, the equilibrium state of the phase process is K. Thus, in order that the queuing process have an equilibrium solution, we require $\mu_K > \lambda_K$ so that the null value of $\hat{\mathcal{B}}(z, 0, 0, \ldots, 0)$ corresponding to equation (A.14) is in the interval $(1, \infty)$. In summary, there are exactly K null values of $\hat{\mathcal{B}}(z, 0, 0, \ldots, 0)$ in $(0, 1)$ and between 0 and $K + 1$ null values of $\hat{\mathcal{B}}(z, 0, 0, \ldots, 0)$ in $(1, \infty)$, the latter quantity being one more than the number of positive values of λ_i.

Due to equation (A.11), there exists no $(\delta_1, \delta_2, \ldots, \delta_K)$ such that

$$\det \hat{\mathcal{B}}(1, \delta_1, \delta_2, \ldots, \delta_K) = 0$$

In addition, all of the null values of $\hat{\mathcal{B}}(z, \delta_1, \delta_2, \ldots, \delta_K)$ are real if $\mathcal{Q}$ is a generator for an ergodic birth-and-death process. Therefore, it is impossible for any of the null values of $\hat{\mathcal{B}}(z, \delta_1, \delta_2, \ldots, \delta_K)$ in the interval $(0, 1)$ to assume the value 0 for some value of $(\delta_1, \delta_2, \ldots, \delta_K)$. Hence, the number of null values of $\hat{\mathcal{B}}(1, \delta_1, \delta_2, \ldots, \delta_K)$ in $(0, 1)$ is independent of the *death rates*. Therefore, $\hat{\mathcal{B}}(z)$, and consequently $\mathcal{A}(z)$, has exactly K null values in the interval $(0, 1)$.

REFERENCES

Beuerman, S. L., and E. J. Coyle. 1988. The delay characteristics of CSMA/CD networks. *IEEE Trans. Commun.* COM-36:5, 553–563.

Cohen, J. W. 1969. *The Single Server Queue*. Wiley Interscience, New York.

Coyle, E. J., and B. Liu. 1985. A matrix representation of CSMA/CD networks. *IEEE Trans. Commun.* COM-33:1, 53–64.

Daigle, J. N. 1977. *Queueing Analysis of a Packet Switching Node in a Data Communication System*. Doctoral dissertation, Columbia University, New York.

Daigle, J. N., and J. D. Langford. 1986. Models for analysis of packet voice communication systems. IEEE J. Selected Areas Commun. SAC-4:6, 847–855.

Hewitt, E., and K. Stromberg. 1969. *Real and Abstract Analysis*. Springer Verlag, New York.

Hunter, J. J. 1983. *Mathematical Techniques in Applied Probability*, Vol. 1, *Discrete Time Models: Basic Theory*. Academic Press, New York.

Karlin, S., and J. McGregor. 1965. Ehrenfest urn models. *J. Appl. Probability*, 2, 352–376.

Lancaster, P. 1966. *Lambda Matrices and Vibrating Systems*. Pergamon Press Ltd., London.

Neuts, M. F. 1981. *Matrix Geometric Solutions in Stochastic Models*. The Johns Hopkins University Press, Baltimore.

Press, W. H., Flannery, B. P., Teukolsky, S. A., and Vetterling, W. T. 1988. *Numerical Recipes in C—The Art of Scientific Computing*. Cambridge University Press, Cambridge.

Ramaswami, V., and D. M. Lucantoni. 1985. Stationary waiting time distributions in queues with phase-type service and in quasi-birth and death processes. *Stochastic Models* 1:2, 125–134.

Ross, S. M. 1985. *Introduction to Probability Models*, 3rd. ed. Academic Press, New York.

Stewart, W. J., and W.-L. Cao. 1987. Queueing models, block Hessenberg matrices, and the method of Neuts. *Ann. Operations Res.* 8, 265–284.

Williams, G. F., and A. Leon-Garcia. 1984. Performance analysis of integrated voice and data hybrid-switched links. *IEEE Trans. Commun.* COM-32:6, 695–706.

11

A Generalized Recursive Technique for Finite Markov Processes

TAO YANG*, M. J. M. POSNER, and J. G. C. TEMPLETON
Department of Industrial Engineering, University of Toronto, Toronto, Ontario, Canada

ABSTRACT

In this paper, the concept of the ith-order recursive computation for state probabilities of finite Markov processes is introduced. As a special case, the conventional recursive computation of Markov processes turns out to be the first-order recursive computation. Structures of Markov processes are systematically analyzed and the set of all finite Markov processes is partitioned according to their structures into classes M_1, M_2, ..., etc. It is shown that any Markov process in M_i can be evaluated with an ith order recursive computation whose complexity is $O(2^{i-1}n)$, where n is the number of states in the process. Examples and comparisons with some other techniques are also presented.

1. INTRODUCTION

We consider an irreducible finite Markov process $\{X(t), t \in T\}$ taking on integer values from the set $S = \{1, 2, \ldots, n\}$ $(n < \infty)$. The time parameter t can

*Current affiliation: Department of Industrial Engineering, Technical University of Nova Scotia, Halifax, Nova Scotia, Canada

be either discrete (where $T = \{0, 1, 2, \dots\}$) or continuous (where $T = [0, \infty)$). It is well known (see, for example, Bhat, 1971) that, for an irreducible finite Markov process, there is a unique positive solution to its balance equations

$$(1 - p_{i,i})p(i) = \sum_{j \neq i} p_{j,i}p(j) \qquad i = 1, 2, \dots, n \tag{1a}$$

for the discrete case, or

$$\sum_{j \neq i} r_{i,j}p(i) = \sum_{j \neq i} r_{j,i}p(j) \qquad i = 1, 2, \dots, n \tag{1b}$$

for the continuous case, together with the normalizing equation

$$\sum_{i=1}^{n} p(i) = 1 \tag{2}$$

where $p_{i,j}$, satisfying $p_{ij} \geq 0$ and $\sum_{j=1}^{n} p_{ij} = 1$, is called the transition probability from state i to state j, $r_{ij} \geq 0$ is called the transition rate from state i to state j, and $p(i)$ is the steady-state probability that the process is in state i if it is aperiodic or the long-run proportion of time that the process is in state i if it is periodic. For convenience of discussion, we shall simply call it the state probability that the process is in state i. Theoretically, state probabilities can always be evaluated by solving simultaneous linear equations (1) and (2) using Gauss elimination. This general method, however, is good only for problems of small size and becomes impractical when the number of states is substantially large.

It is well known that balance equations of some Markov processes can be arranged in a certain order so that if one state probability is known, then the others can be evaluated in a recursive fashion by applying the balance equations in that order. This evaluation procedure is widely known as recursive computation of state probabilities. Whether a Markov process can be recursively computed depends on its structure; in most cases, such a structure can be easily identified after a simple analysis of either its balance equations or its state transition diagram. Examples of these processes include the finite birth-death process, the simple random walk with two reflecting barriers, etc.

For Markov processes whose state probabilities cannot be recursively computed, there has been a large body of research work on computational methods, and numerous ingenious approaches have been proposed. Examples can be found in Guardabassi and Rinaldi (1970), Herzog, Woo, and Chandy (1975), Takahashi (1975), Courtois (1978), Morrison (1980), Morris (1981), Carroll, van de Liefvoort, and Lipsky (1982), Gershwin and Schick (1983), Schweitzer (1984), Grassmann, Taksar, and Heyman (1985), Buzacott and Kostelski (1987), and Feinberg and Chiu (1987). However, there seems to be a gap in the literature between the conventional recursive technique mentioned earlier and the existing techniques proposed so far. More specifically, there are open questions such as: Are there any Markov processes whose state probabilities cannot be recursively computed but can be computed in a recursive form by applying the balance equations twice instead of once as in the conventional recursive computation? If yes, then, what are the structural characteristics of such processes and how can they be systematically identified? Are there any Markov processes whose structures allow other forms of recursive computation that require more than two applications of the balance equations?

The major objective of this paper is to demonstrate results of our recent research in an attempt to fill in this gap in some sense. We will present a systematic analysis on structures of finite Markov processes on a general basis. We note that, in a conventional recursive computation, balance equations of the Markov process concerned are applied only once. If we call such a recursive computation the *first-order* recursive computation, then we show that the set of all finite, irreducible Markov processes can be divided into subsets M_1, $M_2, \ldots$ such that those in subset M_1 can be evaluated with a first-order recursive computation and those in M_i can be evaluated with an ith-order recursive computation that is equivalent to 2^{i-1} first-order recursive computations in terms of the amount of computation involved. That is, in an ith-order recursive computation, balance equations are applied up to 2^{i-1} times. Hence the complexity of the ith-order recursive computation is approximately $O(2^{i-1}n)$ for a Markov process with n states.

The paper is organized as follows. In the next section we introduce necessary notations and basic concepts that play important roles in our discussion. In Section 3 we present in detail the concept of second-order recursive computation, which is an important step toward the understanding of the principle of the more general ith-order recursive computation. In Section 4, the concept of ith-order recursive computation is introduced and classification of Markov processes is discussed. In Section 5 we present some applications of our approach and compare it with some other existing techniques. Section 6 concludes our results and findings.

2. NOTATIONS AND BASIC CONCEPTS

We focus on structures of Markov processes. In other words, we pay attention only to whether there is a transition from state i to state j and ignore the specific value of the transition rate or transition probability. A very useful representation of the structure of a Markov process is its state transition diagram with the values of transition rates or transition probabilities being suppressed. An example of such a representation for a simple 10-state Markov chain is given in Figure 1(a).

Let S_1 be a subset of S. We denote by $P(S_1)$ the set of all state probabilities of states in S_1. That is, $P(S_1) = \{p(s) : s \in S_1\}$. For each state $s \in S$, we define $A(s)$ as the set of states having transitions into s. The ith equation in equation (1) is called the equation associated with state i. Then the equation associated with state s contains variables $p(s)$ and $P(A(s))$ only. If all but one state probability in $\{p(s)\} \cup P(A(s))$ are known, then it is clear that the unknown state probability can be computed by solving a single-variable linear equation that can be easily derived from the equation associated with state s. In this case, we say that the unknown probability can be recursively computed from the known ones with *one application* of the balance equation associated with the state s.

For example, in Figure 1(a), we have $A(10) = \{1, 2, 4, 9\}$. Hence, we can see that $p(10)$ can be recursively computed from $P(A(10))$ with one application of the equation associated with state 10. Similarly, one may also say that $p(1)$ can be recursively computed from $p(2)$, $p(4)$, $p(9)$ and $p(10)$ with one application of the same balance equation.

Let $S_1 \subset S$ and s_1 be a state not in S_1. We define $R(s_1, S_1)$ as the set of *all* states in $S - S_1$, including s_1, whose state probabilities can be recursively computed with *one application* of the balance equations associated with states in $R(s_1, S_1)$, given that $p(s_1)$ and $P(S_1)$ are known. The set $R(s_1, S_1)$ is called the *recursive group* induced by s_1 and based on S_1. The state s_1 and the set S_1 are called the *inducing state* and the *base*, respectively, of the recursive group.

A few remarks should be made here. First, for any given s_1 and S_1, $R(s_1, S_1)$ is uniquely determined. Second, a recursive group is always a nonempty set for it contains at least the inducing state s_1. Third, balance

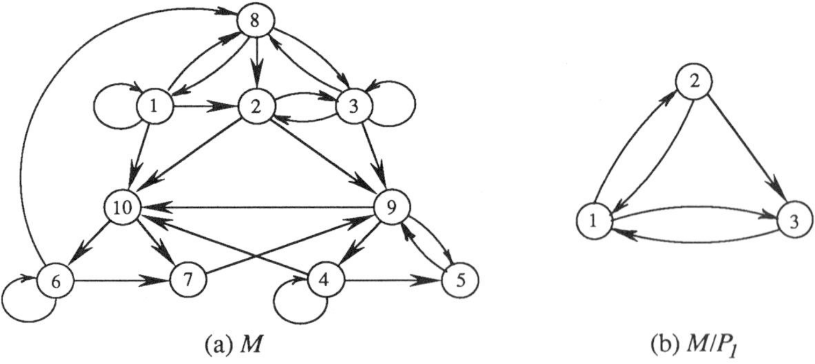

Figure 1 A 10-state Markov chain: M.

equations that are used to recursively compute $P(R(s_1, S_1))$ are restricted to those whose associated states are in $R(s_1, S_1)$. Finally, assuming that $R(s_1, S_1)$ contains j states, since $p(s_1)$ is assumed to be known there are only $j - 1$ unknowns in $P(R(s_1, S_1))$. Therefore, the recursive computation of $P(R(s_1, S_1))$ will use exactly $j - 1$ equations and there exists one state h in $R(s_1, S_1)$ whose associated equation *will not be used*. An interesting property of h is that if there are any transitions from some states in $S - S_1 - R(s_1, S_1)$ to some states in $R(s_1, S_1)$, then it is necessary that *all* these transitions must be directed to the single state h. Hence we shall call h the *sink* of $R(s_1, S_1)$.

Returning to the Markov process shown in Figure 1(a), we have $R(6, \varnothing) = \{6, 10, 7\}$ because if $p(6)$ is known we can compute $p(10)$ by applying the balance equation associated with state 6 and then $p(7)$ by applying the balance equation associated with state 7. Clearly, state 10 is the sink of $R(6, \varnothing)$. Similarly, we have $R(2, \varnothing) = \{2\}$ since no other state probabilities can be recursively computed if $p(2)$ is the only known state probability. However, if $S_1 = \{8\}$ we then have $R(2, S_1) = \{2, 3, 1\}$. We point out that, although $p(6), p(10)$, and $p(7)$ can also be recursively computed, they do not belong to $R(2, S_1)$ because the evaluation of $p(6)$ requires one application of the balance equation associated with state 8 that is definitely *not* in $R(2, S_1)$ (note that $8 \in S_1$).

Let $S_1 \subset S$ and $s_1 = \min\{s : s \in S - S_1\}$. We define $CR(s_1, S_1)$ as the set of all states in $S - S_1$, including s_1, that satisfy the following conditions:

1. Their state probabilities can be recursively computed *in increasing order of their index numbers* with one application of the balance equations associated with the states in $CR(s_1, S_1)$.
2. If h, called the *sink* of $CR(s_1, S_1)$, is the state whose associated equation will not be used in the recursive computation of $P(CR(s_1, S_1))$, and $H = A(h) - S_1 - CR(s_1, S_1)$ is not empty, then the index number of h is less than those of any states in H.

The set $CR(s_1, S_1)$ so defined is called the *coherent recursive group* induced by s_1 and based on S_1. Similarly, s_1 and S_1 are called the inducing state and the base of the coherent recursive group. The second condition imposed on the coherent recursive group $CR(s_1, S_1)$ is quite easy to understand, while the first one may seem obscure. To explain it, let us assume that $CR(s_1, S_1) = \{s_1, s_2, \ldots, s_j\}$ and that state probabilities (except $p(s_1)$) can be recursively computed in the order $p(s_2)$, $p(s_3)$, $\ldots$, $p(s_j)$. Then the first condition says that $s_1 < s_2 < \cdots < s_j$.

Clearly, the definition of $CR(s_1, S_1)$ is similar to that of $R(s_1, S_1)$ except $CR(s_1, S_1)$ is more restricted. Hence we have, for any given S_1 and $s_1 = \min\{s : s \in S - S_1\}$, $CR(s_1, S_1) \subseteq R(s_1, S_1)$, and $CR(s_1, S_1)$ is always a nonempty, uniquely determined set since, in the worst case, it contains at least s_1.

A partition $(S_1, S_2, \ldots, S_k)$ of S is called an *r-partition* if $S_1 = R(s_1, \varnothing)$, $S_2 = R(s_2, S_1)$, …, and $S_k = R(s_k, \cup_{j=1}^{k-1} S_j)$, where s_j $(j = 1, 2, \ldots, k)$ is the inducing state of S_j as defined earlier. An r-partition is said to be *coherent* if $S_i = CR(s_i, \cup_{j=1}^{i-1} S_j)$ for $i = 1, 2, \ldots, k$. It is clear that each Markov process has a *unique* coherent r-partition. Further, r-partitions have an important property that states that for a finite, irreducible Markov process M with n states, the number of subsets in any of its r-partitions including the coherent r-partition is always less than n. A rigorous proof of this property can be found in Yang (1989).

Let us consider again the example shown in Figure 1(a). Let $s_1 = 6$, $s_2 = 1$, and $s_3 = 4$. Then we have $R(s_1, \varnothing) = \{6, 10, 7\}$, $R(s_2, S_1) = \{1, 8, 3, 2\}$, and $R(s_3, S_1 \cup S_2) = \{4, 9, 5\}$, where $S_1 = R(s_1, \varnothing) = \{6, 10, 7\}$ and $S_2 = R(s_2, S_1) = \{1, 8, 3, 2\}$. Let $S_3 = R(s_3, S_1 \cup S_2) = \{4, 9, 5\}$; we have that (S_1, S_2, S_3) is an r-partition of the 10-state Markov chain. The unique coherent r-partition of the structure shown in Figure 1(a) is given by $\{S_1, S_2, \ldots, S_8\}$, where $S_i = \{i\}$ for $i = 1, 2, \ldots, 5$ and $i = 7$, $s_6 = \{6, 8\}$, and $S_8 = \{9, 10\}$.

We consider process M with state space $S = \{1, 2, \ldots, n\}$ and let $P = (S_1, S_2, \ldots, S_k)$ be a partition of S. We construct the process M/P, called the *quotient process of M relative to partition P*, in such a way that the state space of M/P is $S^1 = \{1, 2, \ldots, k\}$ and that there is a transition from state i to state j $(i \neq j)$ in process M/P if and only if, in process M, there are transitions from some states in S_i to some in S_j (transitions from one state to itself are *not* considered in quotient processes). Quotient processes have such a property that M/P is always irreducible if M is irreducible.

Let P be the r-partition mentioned earlier of the 10-state Markov chain. Then the quotient process M/P relative to P is a three-state process whose structure is shown in Figure 1(b). Clearly, we see that M/P is also irreducible.

3. THE SECOND-ORDER RECURSIVE COMPUTATION

As mentioned in Section 1, there are Markov processes whose state probabilities can be evaluated recursively by applying each balance equation once if one state probability is known. Since in this conventional recursive computation each balance equation is applied only *once*, it will be referred later as a *first-order recursive computation*. In this section we will identify a class of Markov processes whose state probabilities cannot be evaluated through a first-order recursive computation, but may be recursively calculated by applying each balance equation at least once and some twice. This recursive computation is said to be of *second order* since some balance equations are used *twice*.

In terms of structures, we can say that a Markov process can be evaluated with a first-order recursive computation if and only if there exists a state $s \in S$ such that the recursive group $R(s, \varnothing)$ induced by s contains all states in S. Suppose we are to evaluate a Markov process M with state space $S = \{1, 2, \ldots, n\}$ and it does not belong to the class of Markov processes that can be evaluated with a first order recursive computation. First, we find a state $s_1 \in S$ such that s_1 maximizes $|R(s, \varnothing)|$ for $s \in S$ and let $S_1 = R(s_1, \varnothing)$. We then find a state $s_2 \in S - S_1$ such that s_2 maximizes $|R(s, S_1)|$ for $s \in S - S_1$ (where $|R|$ is the number of elements in the set R) and let $S_2 = R(s_2, S_1)$. Repeat in a similar manner to find S_3, S_4, …, S_k so that $S_1 \cup S_2 \cup \cdots \cup S_k = S$. Clearly, $P = (S_1, S_2, \ldots, S_k)$ is an r-partition of S and will be called a *largest r-partition* of M. A very interesting result is that if state probabilities of the quotient process M/P can be evaluated, given that $p(1)$ is known, in the order $p(2), p(3)$, …, and $p(k)$ with a *first-order recursive computation*, then state probabilities of process M can be evaluated with a *second-order recursive computation* if $p(s_1)$ is known. Mathematically, we have the following.

THEOREM 1 Let M be a Markov process with state space $S = \{1, 2, \ldots, n\}$ and $P = (S_1, S_2, \ldots, S_k)$, $k > 1$, be a largest r-partition of M. Also, let M/P be the quotient process of M related to the partition P. If $CR(1, \varnothing)$ of M/P contains all states of M/P, then state probabilities of process M can be evaluated with a second-order recursive computation if $p(s_1)$ is known, where s_1 is the inducing state of S_1.

Proof. Let s_i and h_i be the inducing state and the sink of S_i, respectively, for $i = 1, 2, \ldots, k$. Since $p(s_1)$ is known and S_1 is a recursive group induced by s_1 and based on an empty set, then, by definition of recursive group, we have that state probabilities of other states (if any) in S_1 can be evaluated with a first-order recursive computation.

Since S_2 is a recursive group induced by s_2 and based on S_1, if $p(s_2)$ and $P(S_1)$ are known, then state probabilities of other states (if any) in S_2 can be evaluated with a first-order recursive computation. This, however, implies that each state probability in $P(S_2)$ can be expressed as a linear function of $p(s_2)$. That is, we have $p(i) = c_i p(s_2) + d_i$ for each $i \in S_2$. Since $P(S_1)$ are known, if we set $p(s_2) = 1$ and carry out the first-order recursive computation for states in S_2, we will obtain $c_i + d_i$ for each state $i \in S_2$. Similarly, if we set $p(s_2) = 0$ and carry out the same first-order recursive computation, we will obtain d_i for each state $i \in S_2$. At this point, c_i can be obtained with a simple subtraction whose computation is negligible in comparison with a first-order recursive computation.

We further note that the implication of the assumption that $CR(1, \varnothing)$ contains all states of M/P is that, in process M/P, $p(2)$ can be computed from $p(1)$ by applying the equation associated with either state 1 or state 2. Suppose the equation associated with state 1 must be used, then we have that the set of variables involved in the equation associated with state h_1, the sink of S_1, is a subset of $P(S_1) \cup P(S_2)$. Since $P(S_1)$ are known and $p(i) = c_i p(s_2) + d_i$ for each $i \in S_2$, where c_i's and d_i's are known, $p(s_2)$ can be evaluated with one application of the equation associated with h_1. Similarly, if, in the quotient process M/P, the equation associated with state 2 must be used, then $p(s_2)$ can be determined by using the equation associated with h_2, the sink of S_2. Once $p(s_2)$ is known, the computation for other state probabilities in $P(S_2)$ is negligible due to $p(i) = c_i p(s_2) + d_i$.

Evaluation of $P(S_3)$, $P(S_4)$, ..., and $P(S_k)$ can be carried out one after another in the same way as we did for $P(S_2)$. Because the major part of the computation is due to the two first-order recursive computations required for the determination of c_i's and d_i's, we conclude that state probabilities of process M can be evaluated with a second-order recursive computation in which each of the balance equations is applied at most twice. □

We should point out that in Theorem 1 $p(s_1)$ is assumed to be known. In general, this is not the case. However, as usual, one can always set $p(s_1) = 1$ and carry out the second-order recursive computation as indicated in the proof of Theorem 1. Then the numbers obtained can be normalized into probabilities by dividing each of them by their sum. We further point out that if, in process $M/P, p(j)$ is evaluated by applying the equation associated with state l, then, in process M, $p(s_j)$ can be evaluated by applying the equation associated with state h_l after c_i's and d_i's have been obtained for $i \in S_j$. Finally, we note that the proof of Theorem 1 actually gives the algorithm for the computation of state probabilities of M and we can see that the specific computational procedure is determined by the r-partition P of M.

EXAMPLE Let us consider the 10-state Markov chain discussed in Feinberg and Chiu (1987). The transition diagram is shown in Figure 2(a). In terms of structure, it is the same as the one discussed in the previous section. Therefore, some of the results can be used here. From Section 2 we have $P = (S_1, S_2, S_3)$ is an r-partition, where $S_1 = \{6, 10, 7\}$, $S_2 = \{1, 8, 3, 2\}$, and $S_3 = \{4, 9, 5\}$. It is not difficult to verify that P is also a largest r-partition. The quotient process M/P is shown in Figure 2(b). From the structure of M/P, we have $CR(1, \varnothing) = \{1, 2, 3\}$ which contains all states of M/P. Thus, state probabilities of M can be evaluated with a second-order recursive computation.

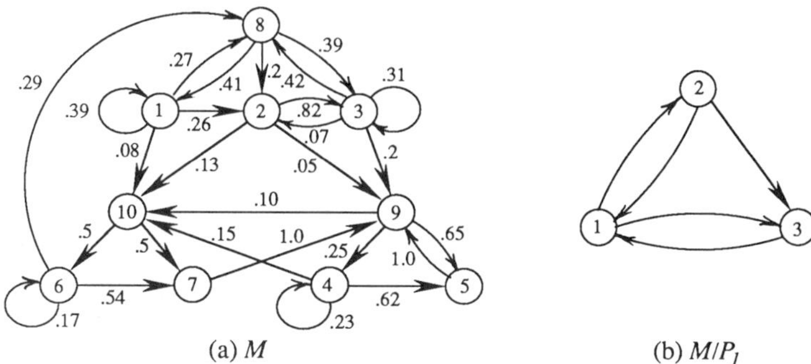

(a) M (b) M/P_1

Figure 2 A 10-state Markov chain.

To start, we set $p(6) = 1$ and apply equations associated with states 6 and 7 to compute $p(10)$ and $p(7)$. Carrying out this first-order recursive computation, we obtain $p(10) = 1.66$ and $p(7) = 1.37$. This completes our computation for $P(S_1)$.

To compute $P(S_2)$, we first set $p(1) = 1$ and apply equations associated with states 1, 8 and 3 to compute $c_8 + d_8$, $c_3 + d_3$, and $c_2 + d_2$. Carrying out this first-order recursive computation, we have $c_8 + d_8 = 1.488$, $c_3 + d_3 = 2.210$, and $c_2 + d_2 = 1.152$. We then set $p(1) = 0$ and carry out the same first-order recursive computation, which gives $d_8 = 0$, $d_3 = -0.690$, and $d_2 = -0.581$. From these we have $c_8 = 1.488$, $c_3 = 2.90$, and $c_2 = 1.733$. Since in the quotient process M/P, $p(2)$ is computed from $p(1)$ by using the equation associated with state 2, then in process M we should use the equation associated with state 2, the sink of S_2, to compute $p(1)$. From Figure 2(a), equation 2 reads $p(2) = 0.26p(1) + 0.07p(3) + 0.2p(8)$. Replacing $p(2), p(3)$, and $p(8)$ by their linear expressions in terms of $p(1)$ and solving for $p(1)$, we have $p(1) = 0.548$, which in turn gives $p(8) = 0.815$, $p(3) = 0.899$, and $p(2) = 0.368$. This completes the computation for $P(S_2)$.

Similarly, we set $p(4) = 1$ and apply equations associated with states 4 and 5 to obtain $c_9 + d_9 = 3.08$ and $c_5 + d_5 = 2.622$. Then we set $p(4) = 0$ and carry out the same first-order recursive computation, and we have $d_9 = d_5 = 0$. Since, in process $M/P, p(3)$ can be computed from $p(1)$ and $p(2)$ by applying either the equation associated with state 1 or the equation associated with state 3, hence, in process M, we can use either the equation associated with state 10 (the sink of S_1) or the equation associated with state 9 (the sink of S_3) to determine $p(4)$. Suppose we apply the equation associated with state 9, we then obtain $p(4) = 3.424$, which gives $p(9) = 10.546$ and $p(5) = 8.977$. This completes the computation for $P(S_3)$.

Normalizing these numbers into probabilities we obtain the state probability vector for the 10-state Markov chain as (.0185, .0124, .0304, .1156, .3032, .0338, .0463, .0275, .3562, .0561).

4. HIGHER-ORDER RECURSIVE COMPUTATIONS

Let us recall that a Markov process M can be evaluated with a *first-order recursive computation* if there exists a state s of M such that $R(s, \varnothing)$ contains all states of M. Analogously, a Markov process can be evaluated with a *second-order recursive computation* if $CR(1, \varnothing)$ contains all states of M/P, where P is a largest r-partition of process M. In other words, we have that a Markov process M can be evaluated with a first-order recursive computation if P_1, a largest r-partition of M, contains only one subset (the state space of M), or otherwise with a second-order recursive computation if P_2, the coherent r-partition of M/P_1, contains only one subset (the state space of M/P_1), where P_1 is a largest r-partition of M. Here the pattern becomes so obvious that we can easily generalize the idea to higher-order recursive computations. For example, we might imagine that a Markov process M could be evaluated with a third-order recursive computation if P_3, a coherent r-partition of process $M/(P_1P_2)$ contains only one subset [the state space of $M/(P_1P_2)$], where P_1 is a largest r-partition of M, P_2 is a coherent r-partition of M/P_1, and $M/(P_1P_2)$ is the quotient process of M/P_1 related to P_2.

We note that a two-state Markov process can always be evaluated with a first-order recursive computation and hence its coherent r-partition always contains only one subset. Further, the number of states in a quotient process is always less than that of the original process if the partition is an r-partition or a coherent r-partition. Therefore, for any Markov process M, there exists a positive integer j such that the coherent (or largest, if $j = 1$) r-partition P_j of process $M/(P_1P_2\cdots P_{j-1})$ (or M, if $j = 1$) contains only one subset, the state space of the process $M/(P_1P_2\cdots P_{j-1})$ (or M, if $j = 1$), where P_1 is a largest r-partition of M, and $P_2, \ldots$ and P_{j-1} are, respectively, coherent r-partitions of processes $M/P_1, \ldots$ and $M/(P_1\cdots P_{j-2})$. The sequence $(P_1, P_2, \ldots, P_j)$ is called a *partition sequence* of M.

THEOREM 2 Let M be an irreducible Markov process with state space $S = \{1, 2, \ldots, n\}$. Let $P_1 = (S_1^1, S_2^1, \ldots, S_{k_1}^1)$, $k_1 \geq 1$, be a largest r-partition of M and $(P_1, P_2, \ldots, P_j)$ be a partition sequence of M. Then state probabilities of process M can be evaluated with a jth-order recursive computation that requires at most 2^{j-1} applications of balance equations of M if $p(s_1)$ is known, where s_1 is the inducing state of S_1^1.

A strict mathematical proof is rather lengthy and therefore is omitted here (interested readers may refer to Yang, 1989). However, the idea is quite obvious. Let us, for example, consider the case when $j = 3$. Let $S^1 = \{1,2,\ldots,k_1\}$ $(k_1 > 2)$ be the state space of the quotient process M/P_1 and $S^2 = \{1,2,\ldots,k_2\}$ $(k_2 > 1)$ the state space of $M/(P_1P_2)$, where $P_2 = (S_1^2, S_2^2, \ldots, S_{k_2}^2)$ is the coherent r-partition of M/P_1. Since (P_1, P_2, P_3) is a partition sequence of M, we have that the coherent recursive group $CR(1, \varnothing)$ of $M/(P_1P_2)$ contains all its states. That is, $CR(1, \varnothing) = S^2 = \{1,2,\ldots,k_2\}$. For each state $s \in S^2$, we define $X(s) = \cup_{i \in S_s^2} S_i^1$, $\beta(s)$ as the inducing state of S_i^2, where i is the inducing state of S_s^2 and $\mu(s)$ as the sink of S_i^1, where i is the sink of S_s^2. Clearly, $P = (X(1), X(2), \ldots, X(k_2))$ is a partition of M and M/P has the same structure as that of $M/(P_1P_2)$. Further, we have $\beta(1) = s_1$.

Since P_2 is the coherent r-partition of M/P_1, S_1^2 is the coherent recursive group (of M/P_1) induced by 1 and based on an empty set. With an approach similar to that used in the proof of theorem 1, it is not difficult to see that $P(X(1))$ can be evaluated with a *second-order* recursive computation since $p(\beta(1))$ $(= p(s_1))$ is known. Following the same line, it can be shown that if $P(X(1))$ and $p(\beta(2))$ are known, then $P(X(2))$ can also be evaluated with a second-order recursive computation. This, however, implies that each state probability in $P(X(2))$ can be expressed as a linear function of $p(\beta(2))$, that is, $p(i) = c_i p(\beta(2)) + d_i$ for each $i \in X(2)$. Since $P(X(1))$ are already known, we can set $p(\beta(2)) = 1$ and carry out the second-order recursive computation to obtain $c_i + d_i$ for each $i \in X(2)$. Then we set $p(\beta(2)) = 0$ and carry out the same second-order recursive computation to evaluate d_i. After c_i and d_i are evaluated for each $i \in X(2)$, we then can use the equation associated with either $\mu(1)$ or $\mu(2)$ to obtain $p(\beta(2))$ depending on whether the equation associated with state 1 or state 2 should be applied to compute $p(2)$ from $p(1)$ in the quotient process $M/(P_1P_2)$. The computation of $P(X(3))$, ..., and $P(X(k_2))$ follows in the same fashion.

We note that the major part of the third-order recursive computation is due to the evaluation of coefficients c_i's and d_i's, which requires two second-order recursive computations, each of which in turn requires *at most* two applications of associated balance equations. Hence evaluating state probabilities for the Markov process M requires *at most* four applications of associated equations. For cases when $j > 3$, it can be shown in a similar way that a jth-order recursive computation is equivalent to two $(j-1)$st recursive computations and the balance equations are applied at most 2^{j-1} times. Therefore, the complexity of a jth-order recursive computation is $O(2^{j-1}n)$, assuming that the basic operation is one application of a balance equation.

From the above arguments, we see that the computational procedure of a jth-order recursive computation is determined by the partition sequence $(P_1, P_2, \ldots, P_j)$. To improve computational efficiency, we need to find a partition sequence of M that contains the smallest number of partitions among all possible partition sequences of M. Such a partition sequence will be called an *optimal partition sequence*. Since, for any Markov process, the coherent r-partition is unique, there exists exactly one partition sequence corresponding to each largest r-partition of M. Hence, the determination of an optimal partition sequence depends on the choices of the largest r-partition P_1 of process M. For example, there are two largest r-partitions for the 10-state Markov chain given in Section 2. We have met one of them in both Sections 2 and 3, and know that it gives a partition sequence that contains only two partitions. The other is $P_1 = (S_1, S_2, S_3)$ in which $S_1 = R(4, \varnothing) = \{4, 9, 5\}$, $S_2 = R(6, S_1) = \{6, 10, 7\}$, and $S_3 = R(1, S_1 \cup S_2) = \{1, 8, 3, 2\}$. It can be verified, however, that this second largest r-partition of M leads to a partition sequence that contains three partitions. Therefore, the largest r-partition given in Section 2 is better than P_1. Given an arbitrary Markov process, the problem of finding its optimal partition sequence seems very complicated. Fortunately, in many practical cases, Markov processes possess certain regularity that allows one to determine the optimal partition sequence with a simple visual analysis of the structure of the Markov process concerned. Examples given in the next section may help explain this.

At this point, we can see that for any finite Markov process there is an optimal partition sequence. The number of partitions in the optimal partition sequence in some sense indicates how well the process is structured and therefore it will be called the *order* of the Markov process. The set of all finite Markov processes is then partitioned into classes: M_1, M_2, M_3, etc. Class M_i contains those of order i whose state probabilities can be evaluated with an ith-order recursive computation in which balance equations are applied no more than 2^{i-1} times.

5. APPLICATIONS AND COMPARISONS

5.1 The $M/G/1/\lambda_i$ Queue

To demonstrate their computational procedure, Herzog, Woo, and Chandy (1975) considered the $M/G/1/\lambda_i$ queuing system in which the input process is Markovian with state-dependent rate λ_i (where i is the number of customers in the system) and there are $N - 1$ waiting positions and a single server with general service time distribution. The general service time is then approximated by a Coxian process of K stages. This results in a finite Markov process with continuous time parameter. The structure of the process (as usual, we denote it by M) is shown in Figure 3(a). To simplify our discussion, let us denote by s_{ij} the state of the process in which there are i customers in the system and the service process is in stage j. It

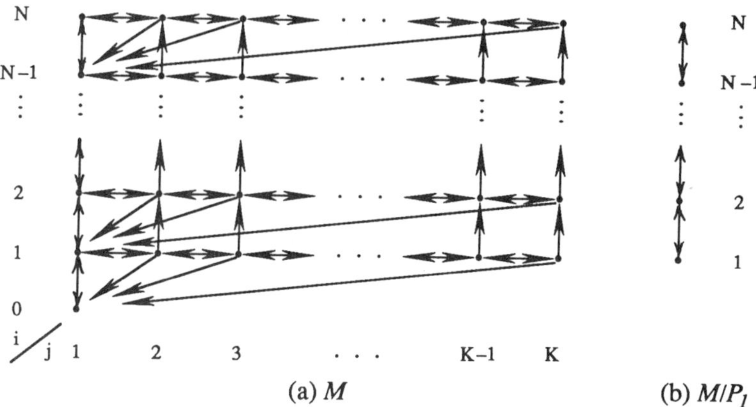

Figure 3 The $M/G/1/\lambda_i$ system.

is not difficult to see that $R(s_{1,K}, \varnothing) = \{s_{1,K}, s_{1,K-1}, \ldots, s_{1,1}, s_{0,1}\}$. Let $S_1 = R(s_{1,K}, \varnothing)$; we have $R(s_{2,k}, S_1) = \{s_{2,K}, s_{2,K-1}, \ldots, s_{2,1}\}$. In general, we have $S_i = R(s_{i,K}, \cup_{k=1}^{i-1} S_k) = \{s_{i,K}, s_{i,K-1}, \ldots, s_{i,1}\}$ for $i = 2, 3, \ldots, N$. The partition $P_1 = (S_1, S_2, \ldots, S_N)$ is obviously a largest r-partition of M. The related quotient process M/P_1 is shown in Figure 3(b). From the structure of M/P_1, we see that $CR(1, \varnothing) = \{1, 2, \ldots, N\}$ contains all states of M/P_1. By Theorem 1, M belongs to M_2 and can be evaluated with a second-order recursive computation with a complexity of $O(2NK)$. With the approach proposed by Herzog, Woo, and Chandy (1975), the selected boundary states are $s_{0,1}$, $s_{1,1}, \ldots,$ and $s_{N,1}$. In principle, for each boundary state, a first-order recursive computation should be carried out to evaluate the corresponding coefficients. However, after making use of some special property of this particular process, the complexity of their algorithm is approximately $O(N^2K/2)$, which is still considerably less efficient than our second-order recursive computation, especially when N is very large.

5.2 A Multiprocessor System

Our second example concerns a multiprocessor system that was first studied by Mitrani (1987). The system consists of N_1 primary processors and N_2 secondary ones. These processors are not always available. Each of them goes through alternating periods of being operative and inoperative, independently of others. These periods are distributed exponentially. Normally, only the primary processors are used, with the secondary processors being used when *all* primary ones are inoperative. When the secondary processors

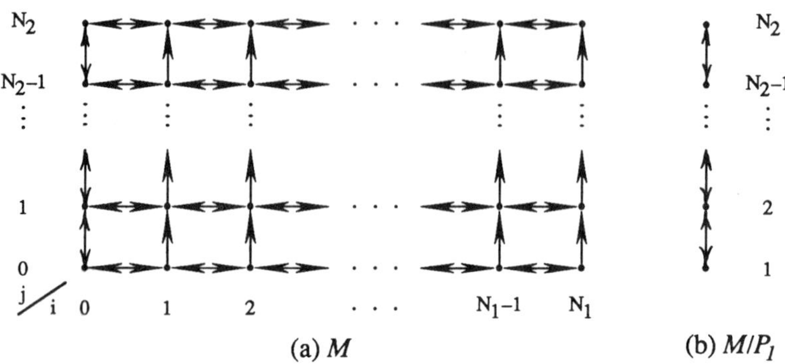

Figure 4 A multiprocessor system.

are not used, they cannot break down. Again, we have a finite Markov process with continuous time parameter, and its structure is given in Figure 4(a).

Let $s_{i,j}$ be the state in which there are i operative primary processors and j secondary ones in the system. Then, it is not difficult to find a largest r-partition for the Markov process (M): $P_1 = (S_1, S_2, \ldots, S_{N_2})$, where $S_i = \{s_{N_1,i-1}, s_{N_1-1,i-1}, \ldots, s_{0,i-1}\}$ for $i = 1, 2, \ldots, N_2 - 1$ and $S_{N_2} = \{s_{N_1,N_2-1}, \ldots, s_{0,N_2-1}, s_{0,N_2}, \ldots, s_{N_1,N_2}\}$. The structure of the quotient process M/P_1 is given in Figure 4(b), from which we have $CR(1, \varnothing) = \{1, 2, \ldots, N_2\}$, which contains all states of M/P_1. Hence, process M belongs to M_2 and can be evaluated with a second-order recursive computation whose complexity is approximately $O(2N_1N_2)$.

Mitrani (1987) presented a very interesting theoretical study of this system. He chose to use analytic methods to obtain the joint steady-state probability $p_{i,j}$ that there are i primary and j secondary processors operative. By solving a first-order partial differential equation he found an explicit expression for the two-variable generating function $G(x,y) = \sum_{i=0}^{N_1} \sum_{j=0}^{N_2} p_{i,j} x^i y^j$. Then the method to invert $G(x,y)$ is given. Analysis of equations (32) and (33) in Mitrani (1987) shows that the complexity of the inversion of $G(x,y)$ is $O(N_1^2 N_2)$, which is less efficient than a second-order recursive computation when N_1 is large.

5.3 The $C_a/M/s/m$ Retrial Queue

Let us consider a service station in which there are s identical, independent exponential servers and m waiting positions. Customers arrive according to some renewal process with interarrival time distribution $A(t)$ being of Coxian type with a stages. An arriving customer will receive immediate service if there are idle servers. If all servers are busy and there are free waiting positions, the customer will join the queue and wait for service. If all servers and waiting positions are occupied, he will decide either to retry for service later (with probability α) or to leave the system forever (with probability $\bar{\alpha} = 1-\alpha$). The time until his next retrial is exponentially distributed and is independent of any other stochastic processes in the system. Customers in the process of retrial are said to be in *orbit* and are called *orbiting customers*. On retrial, an orbiting customer is treated at the service station in the same way as when he first arrived.

This is clearly a Markov process with a countably infinite number of states. To apply our approach, we need to truncate it into a finite process. One way to do this is to assume that the capacity of the orbit is finite, say N. That is, any customer joining the orbit who finds N customers in it will leave the system forever. With an N sufficiently large, the truncated finite process can approximate the infinite process with any degree of accuracy. Let $s_{i,j,k}$ denote the state of the truncated process (we shall call it process M) in which there are i customers in the service station, j customers in the orbit, and the arrival process is in stage k. The transition diagram of M is given in Figure 5(a) (note that for convenience not all transitions are shown).

The structure of process M looks more complicated than our previous examples. However, with little effort we find that $S_1 = R(s_{c,N,a}, \emptyset) = \{s_{c,N,a}, s_{c,N,a-1}, \ldots, s_{c,N,1}\}$ and $S_2 = R(s_{c-1,N,a}, S_1) = \{s_{c-1,N,a}, s_{c-1,N,a-1},$

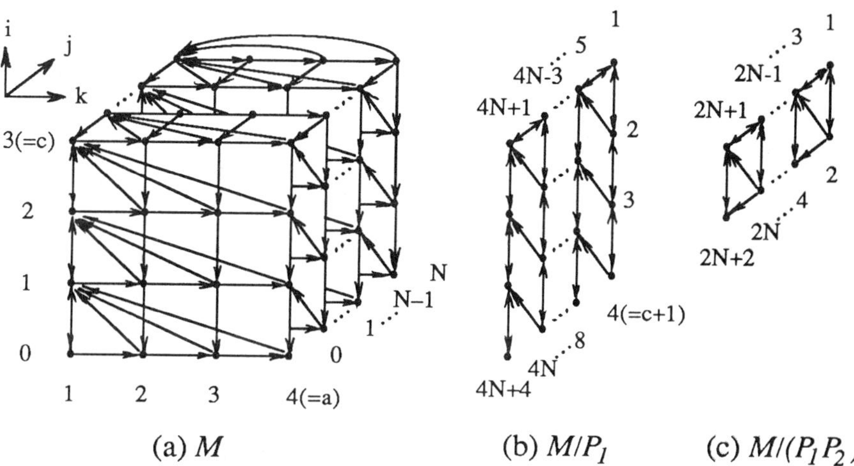

Figure 5 The $C_a/M/s/m$ retrial queue.

$\ldots, s_{c-1,N,1}\}$. In general, we have $S_{(c+1)(N+1-j)-i} = \{s_{ij,a}, s_{ij,a-1}, \ldots, s_{ij,1}\}$ for $i = 0, 1, \ldots, c$ and $j = 0, 1, \ldots, N$, where $c = s + m$. The partition $P_1 = (S_1, S_2, \ldots, S_{(c+1)(N+1)})$ is a largest r-partition and the quotient process M/P_1 is shown in Figure 5(b). Then, the coherent r-partition of M/P_1 is easily derived as $P_2 = (S_1^1, S_2^1, \ldots, S_{2(N+1)}^1)$, with $S_i^1 = \{s_i\}$ if i is odd and $S_i^1 = \{t_i + 1, t_i + 2, \ldots, t_i + c\}$ if i is even, where $s_i = (i-1)(c+1)/2 + 1$ and $t_i = s_{i-1}$. The corresponding quotient process $M/(P_1P_2)$ is shown in Figure 5(c). In process $M/(P_1P_2)$, we have that $CR(1, \varnothing) = \{1, 2, \ldots, 2(N+1)\}$ contains all its states. Therefore, we conclude that M belongs to M_3 and can be evaluated with a third-order recursive computation. Since the total number of states in process M is $a(c+1)(N+1)$, the complexity of the third-order recursive computation for M is approximately $O(4a(c+1)(N+1))$.

A special case of $C_a/M/s/m$ is the $M/M/s/m$ retrial queue, which has drawn attention of some researchers in recent years (see, for example, Greenberg and Wolff, 1986; Stepanov, 1987; Neuts and Rao, 1991). Interested readers may find that the truncated $M/M/s/m$ retrial queue is a second-order Markov process. Thus, evaluation of its state probabilities requires only a second-order recursive computation with a complexity of $O(2(c+1)(N+1))$. A widely used numerical method for this type of retrial queue is the Gauss–Seidel iteration (see, for example, Keilson, Cozzolino, and Young 1968; Ridout, 1984; Neuts and Rao, 1991). To compare it with our approach, we note that each iteration requires the same amount of computation as would be needed with a first-order recursive computation since, in each iteration, the new probabilities are computed from the old ones by one application of each of the balance equations. Therefore a second-order recursive computation is equivalent to the computation needed by two iterations, a third-order recursive computation is equivalent to the computation required by four iterations, etc. It is quite unlikely that Gauss-Seidel iteration will converge in two iterations, and in many cases tens, even hundreds of iterations are needed for its convergence. Therefore, in general, the second-order recursive computation is much more efficient than Gauss–Seidel iteration for the $M/M/s/m$ retrial queue.

6. CONCLUSIONS

Given an arbitrary finite Markov process M, we first find one of its largest r-partitions P_1. If P_1 contains more than one subset, we then construct the related quotient process M/P_1 and proceed to find the coherent r-partition P_2 of M/P_1. Repeat this procedure to find coherent r-partitions $P_3, P_4, \ldots$ of quotient processes $M/(P_1P_2), M/(P_1P_2P_3), \ldots$ until we reach the point where P_j, the coherent r-partition of $M/(P_1P_2 \cdots P_{j-1})$, contains a single subset. The sequence $(P_1, P_2, \ldots, P_j)$ is called a partition sequence of process M. The one with the smallest number of partitions is called an optimal partition sequence of M, and the number of partitions in it is called the order of process M. State probabilities of a Markov process of order i can be evaluated with an ith-order recursive computation whose complexity is $O(2^{i-1}n)$.

The determination of the optimal partition sequence of a given Markov process, in general, is a complicated combinatoric problem. Fortunately, many Markov processes of practical interest possess certain regularity that allows us to identify their optimal partition sequences quite easily. It is therefore recommended that whenever possible the order of a Markov process be determined before one applies a numerical method to evaluate its state probabilities, since the ith-order recursive computation is very efficient for processes of extremely low orders (for example, an order that is much less than $\log_2 n$).

Finally, we should point out that so far we have only discussed theoretical aspects of the generalized recursive technique. It is well known that for a numerical method to be practically useful, it should be numerically stable. Stability of the generalized recursive technique depends, in general, on the order of the Markov process concerned and the implementation of the technique. From the nature of the technique, we expect that the technique may become numerically unstable for Markov processes of high orders. Our computational experience with the second-order and the third-order recursive computations indicates that they are numerically very stable when the state probabilities are evaluated roughly in an increasing order of their values. Considering the computational efficiency of the generalized technique, it is worthwhile to identify the conditions under which higher-order recursive computations are unstable and to introduce remedies for these cases.

Acknowledgments: We would like to express our gratitude to the Natural Sciences and Engineering Research Council of Canada and to the government of Ontario for the support of this work through grants A5639, A4374, and a graduate scholarship.

REFERENCES

U. N. Bhat, *Elements of Applied Stochastic Processes*, John Wiley and Sons, New York, 1971.

J. A. Buzacott and D. Kostelski, Matrix-geometric and recursive solution of a two-stage unreliable flow line, IIE Trans., Vol. 19(4), 429–438 (1987).

J. L. Carroll, A. van de Liefvoort, and L. Lipsky, Solution of $M/G/1//N$-type loops with extensions to $M/G/1$ and $GI/M/1$ queues, Operations Res., Vol. 30(3), 490 (1982).

P. J. Courtois, *Decomposability: Queueing and Computer Applications*, Academic Press, New York, 1977.

B. N. Feinberg and S. S. Chiu, A method to calculate steady-state distributions of large Markov chains by aggregating states, Operations Res., Vol. 35, No. 2, 282 (1987).

S. B. Gershwin and I. C. Schick, Modelling and analysis of three-stage transfer lines with unreliable machines and finite buffers, Operations Res., Vol. 31, No. 2, 354 (1983).

W. K. Grassmann, M. I. Taksar, and D. P. Heyman, Regenerative analysis and steady state distributions of Markov chains, Operations Research, Vol. 33, 1107 (1985).

B. S. Greenberg and R W. Wolff, An upper bound on the performance of queues with returning customers, J. Appl. Prob. 24, 466–475 (1987).

G. Guardabassi and S. Rinaldi, The problems in Markov chains; A topological approach, Operations Res., Vol. 18, 324 (1970).

U. Herzog, L. Woo, and K. M. Chandy, Solution of queueing problems by a recursive technique, IBM J. Res. Develop., Vol. 19, No. 3, 295 (1975).

J. Keilson, J. Cozzolino, and H. Young, A service system with unfilled requests repeated, Operations Res., Vol. 16, 1126 (1968).

I. Mitrani, Multiprocessor systems with reserves and preferences, Queueing Systems, Vol. 2, No. 3, 245 (1987).

R. J. T. Morris, Priority queueing networks, Bell System Tech. J., Vol. 59, No. 8, 1745 (1980).

J. A. Morrison, Analysis of some overflow problems with queueing, Bell System Tech. J., Vol. 59, No. 8, 1427 (1980).

M. F. Neuts and B. M. Rao Numerical Investigation of a Multi-Server Retrial Model, Queueing Systems, Vol. 7, No. 2, to appear (1991).

G. E. Ridout, A Study of Retrial Queueing Systems with Buffers, M.A.Sc. thesis, Department of Industrial Engineering, University of Toronto (1984).

P. J. Schweitzer, Aggregation methods for large Markov chains. In *Mathematical Computer Performance and Reliability*, edited by G. Iaseolla, P. J. Courtois, and A. Hordijk, Elsevier Science, Amsterdam, 1984.

S. N. Stepanov, Increase the efficiency of numerical methods for models with repeat calls, Problems Info. Transmission, 22(4), 78 (1987).

Y. Takahashi, A Lumping Method for Numerical Calculations of Stationary Distributions of Markov chains, Research Report B-18, Department of Information Sciences, Tokyo Institute of Technology, Tokyo, Japan, 1975.

T. Yang, A Study On Retrial Queues, Ph.D. thesis, Department of Industrial Engineering, University of Toronto (1990).

12

A Theory of Statistical Multiplexing of Markovian Sources: Spectral Expansions and Algorithms

A. I. ELWALID* Department of Electrical Engineering and Center for Telecommunications Research, Columbia University, New York, New York

D. MITRA AT&T Bell Laboratories, Murray Hill, New Jersey

T. E. STERN Center for Telecommunications Research, Columbia University, New York, New York

ABSTRACT

This paper considers the problem of exactly calculating the stationary-state distribution of a system that statistically multiplexes the output of K sources. Each source is an N-state Markov chain and its Poisson rate of packet generation is determined by its state. Since Broadband ISDN (B-ISDN) will statistically multiplex many bursty sources and the burstiness can be modeled by Markov chains, the efficient analyses of large examples of the model considered here are an essential requirement for B-ISDN design. An algorithm is given for the spectral expansion of the state distribution whose complexity is $O(K^{3(N-1)})$ for N fixed and K large. The efficiency of the algorithm is derived from an algebraic theory that gives the (exact) decomposition of the eigenvalue problem of the entire system into many small eigenvalue problems, and

---

*Part of this work was done while at AT&T Bell Laboratories, Murray Hill, New Jersey, under the University Relations Program.

also gives the Kronecker product form to the eigenvectors; another important element is a recursive algorithm for (exact) aggregation.

1. INTRODUCTION

In this paper we give a theory and efficient computational algorithms for the system sketched in Figure 1. The system is composed of K independent sources, each an N-state continuous-time, irreducible Markov chain with generator $\mathbf{M}$. When a source is in state i it generates discrete units of information, say packets, in a Poisson stream at a rate $\lambda_i (1 \leq i \leq N)$. The packets are buffered if necessary and transmitted on a trunk. The service time of each packet is an independent, exponentially distributed random variable with mean $1/\mu$.

In this paper we shall assume that the sources are time-reversible, that is, $\mathbf{M}$ is symmetrizable [1, 2]. This assumption allows the exposition to be simpler; however, it is not essential, a point that should become obvious to the reader.

The model is an established paradigm and yet its practical importance in communications is even growing today. Various approaches have been taken for its analysis [3–13]. Our approach is based on computing the *spectral expansion* of the equilibrium state distribution of the system. This representation requires explicit knowledge of eigenvalues and eigenvectors, which typically is prohibitively expensive to obtain. We have obtained various explicit decompositions of the eigensystem from an algebraic theory, and these are essential elements of a provably efficient algorithm, which we give for exactly computing the system's state distribution.

More specifically, the analysis has two phases: the first phase analyzes the *unaggregated system*, where the description of the state of the sources is oblivious of the fact that the K sources are statistically identical; in the second

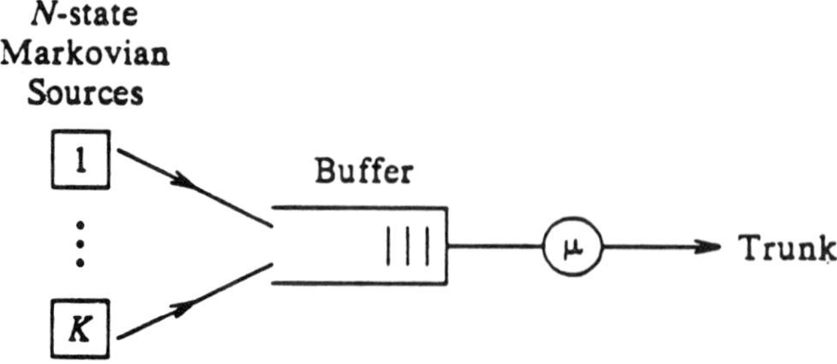

Figure 1 A buffered statistical multiplexing system with K statistically identical N-state Markovian sources.

phase a transition is made to the *aggregated system*, in which the homogeneity of the sources is taken into account to give a minimal state description.

The output of the first phase is the spectral representation

$$\boldsymbol{\pi}(n) = \sum_k a_k z_k^n \boldsymbol{\phi}_k, \qquad (n = 0, 1, 2, \ldots) \tag{1}$$

where n is the buffer content and $\boldsymbol{\pi}(n) = \{\pi(n; \mathbf{s})\}$ is the N^K-tuple of equilibrium probabilities in which $\mathbf{s}$ is the unaggregated state of the sources.

The output of the second phase of aggregation is

$$\boldsymbol{\pi}^{(A)}(n) = \sum_\tau b_\tau z_\tau^n \boldsymbol{\xi}_\tau \qquad (n = 0, 1, 2, \ldots) \tag{2}$$

where $\boldsymbol{\pi}^{(A)} = \{\pi(n, \boldsymbol{\sigma})\}$ in which $\boldsymbol{\sigma}$ is an aggregate state of the sources and therefore $\boldsymbol{\pi}^{(A)}(n)$ is of dimension L, where

$$L \triangleq \binom{N + K - 1}{N - 1} \tag{3}$$

Two key decomposition results are proven in the first phase in the context of the unaggregated system. It is shown that any eigenvector $\boldsymbol{\phi}$ has the Kronecker product form $\mathbf{u}_1 \otimes \mathbf{u}_2 \otimes \cdots \otimes \mathbf{u}_K$ where each $\mathbf{u}_i$ is N-dimensional. A second decomposition shows that the eigenvalues $\{z_k\}$ are obtained by solving a system of K coupled eigenvalue problems where each is of dimension N.

Proceeding to the second phase, first we prove an important decomposition result, which states that the eigenvalues of the aggregated source-buffer system, $\{z_\tau\}$ in equation (2), are the roots of a family of $L/N!$ polynomials in which each polynomial is of degree $2N!$. Second, a simple recursive algorithm is devised to obtain exactly the aggregate eigenvector $\boldsymbol{\xi}$ from $\boldsymbol{\phi}$. Finally, the coefficients $\{b_\tau\}$ are obtained by solving a system of L normalization equations.

The complexity of the entire computational procedure is $O(L^3)$. For the case of particular practical interest where N is small and K large, the complexity is $O(K^{3(N-1)})$, that is, polynomial in the large parameter K.

The model for statistical multiplexing given above has been investigated extensively in the past. In particular, Neuts's matrix-geometric methods have been shown to apply [3]. Another technique closely coupled to the matrix-geometric method is due to Heffes and Lucantoni [4], who approximate the source process in striving for computational tractability. A different approach has been used by Sriram and Whitt [5] in which the approximation is by renewal processes and their approximate characterization is by two moments. Yet another approach is due to Halfin [6], who develops a recursive technique. Asymptotic analysis of a queuing system like the one considered here is due to Burman and Smith [7]. A recent application of the method of spectral expansions to a queuing system is by Mitra and Mitrani [8].

The direct antecedents of the present work are the results in Anick, Mitra, and Sondhi [9], Kosten [10], Mitra [11], and Stern and Elwalid [12] on stochastic fluid models, which are obtained from queuing models by a limiting process. These papers established that the method of spectral expansions is a powerful one due primarily to algebraic techniques that explicitly factor the characteristic polynomial and give the Kronecker product form to the eigenvectors. These techniques have evolved since their introduction and have been shown to apply to systems of increasing complexity. The present paper develops the methodology, hitherto restricted to fluid models, for queuing systems. We mention that certain results such as the one on the factorization (Theorem 3.1) and the recursive aggregation algorithm (Theorem 3.2) are new and should now be exported to fluid models. We also mention that the asymptotic treatments of fluid models due to Weiss [13] and Morrison [14] are particularly interesting when the number of sources is large.

New impetus for the analysis of statistical multiplexing is coming from the imminence of B-ISDN, which proposes to have sources of varying degrees of burstiness share the bandwidth [15]. Specifically, the access rules for either accepting or blocking calls with bursty traffic requires analysis of the kind performed here [16, 17]. Zukerman and Kirton [18] point out the limitations of the matrix-geometric method in the context of the problem sizes and number of sources expected in B-ISDN. The need arising from burstiness characteristics for considering individual Markovian sources in which the number of states exceeds two has been stated by Ramaswami and Latouche [19]. Our own work on advanced switches subject to random routings of calls carrying bursty traffic has provided some of the motivation for considering high-order Markovian sources.

2. THE UNAGGREGATED SYSTEM

2.1 Eigenvalues and Eigenvectors

Let the state of source i be denoted by $s(i)$ where $s(i) \in \{1,2,\ldots,N\}$. The unaggregated state of the sources is given by $\mathbf{s} = (s(1), s(2), \ldots, s(K))$. We let the state space of the unaggregated source process be

$$\mathcal{H}_{N,K} \triangleq \{\mathbf{k} \mid \mathbf{k} \in \mathcal{Z}^K, 1 \le k(i) \le N (1 \le i \le K)\} \tag{4}$$

Note $|\mathcal{H}_{N,K}| = N^K$. Let $\boldsymbol{\pi}(n)$ denote the lexicographic arrangement of $\{\boldsymbol{\pi}(n;\mathbf{s})\}$, that is; $\boldsymbol{\pi}(n) = \{\pi(n;1,\ldots,1), \pi(n;1,\ldots,1,2), \ldots, \pi(n;N,\ldots,N)\}$.

The generator of the unaggregated source process is

$$\mathbf{Q} = \mathbf{M} \oplus \mathbf{M} \oplus \cdots \oplus \mathbf{M}. \tag{5}$$

a K-fold Kronecker sum on $\mathbf{M}$. (The Kronecker sum $\mathbf{A} \oplus \mathbf{B} = \mathbf{A} \otimes \mathbf{I} + \mathbf{I} \otimes \mathbf{B}$ [2, 3].)

The diagonal elements of the diagonal *rate* matrix $\mathbf{R}$ are the (Poisson) rates for each of the states of the unaggregated source process, that is,

$$\mathbf{R} = \mathbf{\Lambda} \oplus \Lambda \oplus \cdots \oplus \mathbf{\Lambda} \tag{6}$$

another K-fold Kronecker sum in which $\mathbf{\Lambda} = \text{diag}(\lambda_1, \lambda_2, \ldots, \lambda_N)$.

We can now give the *balance equations* of the composite source and buffer system.

$$\begin{aligned} 0 &= \boldsymbol{\pi}(n)[\mathbf{Q} - \mathbf{R}] + \mu\boldsymbol{\pi}(n+1) & (n = 0) \\ &= \boldsymbol{\pi}(n-1)\mathbf{R} + \boldsymbol{\pi}(n)[\mathbf{Q} - (\mathbf{R} + \mu)] + \mu\boldsymbol{\pi}(n+1) & (n \geq 1) \end{aligned} \tag{7}$$

The independent solutions of equation (7) have the form

$$\boldsymbol{\pi}(n) = z^n \boldsymbol{\phi}$$

where z is a root of the characteristic equation

$$|\mathbf{C}(z)| = 0 \tag{8}$$

and

$$\boldsymbol{\phi}\mathbf{C}(z) = 0 \tag{9}$$

where

$$\mathbf{C}(z) \triangleq \mu z^2 + z[\mathbf{Q} - (\mathbf{R} + \mu)] + \mathbf{R} \tag{10}$$

In equation (9) consider the following particular form:

$$\boldsymbol{\phi} = \mathbf{u}_1 \otimes \mathbf{u}_2 \otimes \cdots \otimes \mathbf{u}_K \tag{11}$$

On substituting it and the expressions in equations (5) and (6) for $\mathbf{Q}$ and $\mathbf{R}$ in equation (9) we get

$$\mu z(z-1)\mathbf{u}_1 \otimes \cdots \otimes \mathbf{u}_K + \sum_{i=1}^{K} \mathbf{u}_1 \otimes \cdots \otimes \mathbf{u}_i[z(\mathbf{M} - \mathbf{\Lambda}) + \mathbf{\Lambda}] \otimes \cdots \otimes \mathbf{u}_K = 0 \tag{12}$$

If now we have a set of K numbers $\nu_1, \nu_2, \ldots \nu_K$ such that

$$\sum_{i=1}^{K} \nu_i = \mu \tag{13}$$

then from equation (12) we obtain

$$\sum_{i=1}^{K} \mathbf{u}_1 \otimes \cdots \otimes \mathbf{u}_i[\nu_i z(z-1) + z\mathbf{M} - (z-1)\mathbf{\Lambda}] \otimes \cdots \otimes \mathbf{u}_K = 0 \tag{14}$$

In particular, therefore, the original equation (9) is satisfied if

$$\left.\begin{array}{ll} \mathbf{u}_i\mathbf{A}(z,\nu_i) = 0, & (1 \le i \le K) \\ \displaystyle\sum_{i=1}^{K} \nu_i = \mu & \end{array}\right\} \tag{15}$$

where

$$\mathbf{A}(z,\nu) \triangleq \nu z^2 + z[\mathbf{M} - (\mathbf{\Lambda} + \nu)] + \mathbf{\Lambda} \tag{16}$$

The above represents an important decomposition of the original eigenvalue problem since it is a system of K *coupled eigenvalue problems* where each is N-dimensional. The coupling (13) can be interpreted as a partitioning of the outgoing trunk of the multiplexing system and this is reminiscent of a similar interpretation in reference [11]. Notice too the striking similarity in structure (and dissimilarity in dimension) of $\mathbf{A}(z,\nu)$ and $\mathbf{C}(z)$ in equations (16) and (10).

In the coupled eigenvalue problem of equation (15),

$$|\mathbf{A}(z,\nu)| = \{z(z-1)\}^N \left|\nu\mathbf{I} - \frac{1}{1-z}\mathbf{M} - \frac{1}{z}\mathbf{\Lambda}\right| \tag{17}$$

It will be useful to view the equation $|\mathbf{A}(z,\nu)| = 0$ as the following equation in ν with parameter z:

$$|\nu\mathbf{I} - \mathbf{B}(z)| = 0 \tag{18a}$$

where

$$\mathbf{B}(z) \triangleq \frac{1}{1-z}\mathbf{M} + \frac{1}{z}\mathbf{\Lambda} \tag{18b}$$

The function in the left side of equation (18a) is of course none other than the characteristic polynomial (of degree N) in ν of the matrix $\mathbf{B}(z)$. Denote the solutions in ν by

$$g_j(z) \qquad (1 \le j \le N) \tag{19}$$

These functions are continuous and have continuous derivatives except at singularities located at 0 and 1. A great deal is known about their form, asymptotic behavior and derivatives. A sketch is given in Figure 2. For the moment consider these N functions as given. Information on related functions that arise in fluid models is in reference [12].

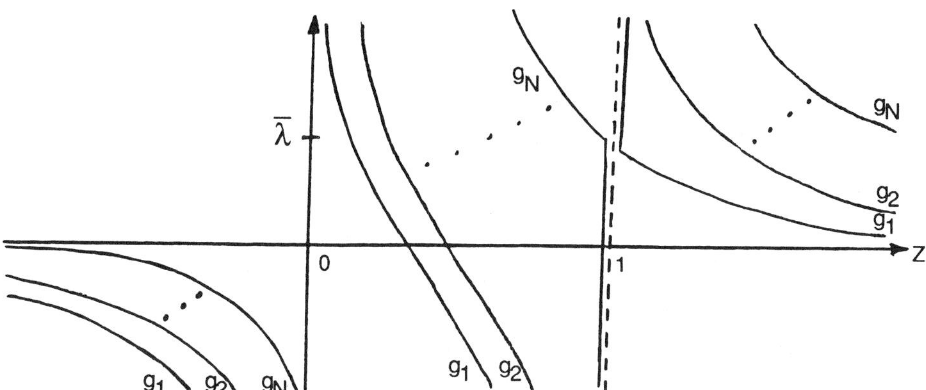

Figure 2 A sketch of the functions $g_j(z)$ $(1 \leq j \leq N)$ in which $\bar{\lambda} = \langle \mathbf{w}, \boldsymbol{\lambda} \rangle$ where $\mathbf{w}$ is the equilibrium probability vector of an individual source.

For each $\mathbf{k} \in \mathcal{H}_{N,K}$ the solutions z of the equation*

$$\sum_{i=1}^{K} g_{k(i)}(z) = \mu \tag{20}$$

are solutions to the coupled eigenvalue problem in equation (15). Let us denote these solutions by $z_k^{(1)}, z_k^{(2)}, \ldots$. We now show that in fact there are exactly two solutions and, furthermore,

$$0 < z_k^{(1)} \leq 1 \leq z_k^{(2)} \tag{21}$$

To see this, observe from Figure 2 that for any j, $g_j(z)$ is monotonically decreasing in $(0, 1)$ and $(1, \infty)$, and $g_j(1^-) = -\infty$, $g_j(1^+) = \infty$, and $g_j(z) \to 0$ as $z \to \infty$. It follows that the quantity on the left side of equation (20), which is a sum of such functions, also has those properties. Hence the solutions of equation (20) for any positive μ satisfy equation (21).

We now introduce the *ergodicity condition*, which allows us to be specific about the number of solutions $z_k^{(\cdot)}$ in $(0, 1)$. Let $\mathbf{w}$ be the stationary probability vector of an individual source, that is,

$$\mathbf{wM} = \mathbf{0}, \qquad \langle \mathbf{w}, \mathbf{1} \rangle = 1 \tag{22}$$

---

*We attempt to conform to the practice of denoting the elements of a vector $\mathbf{k}$ by $k(1)$, $k(2), \ldots$. The subscript on a vector is reserved for indexing the vector.

where $\langle \cdot, \cdot \rangle$ represents the inner product of vectors and $\mathbf{1}$ is the vector in which all elements are unity. Then the ergodicity condition is

$$K\bar{\lambda} < \mu, \tag{23}$$

where, $\bar{\lambda} \triangleq \langle \mathbf{w}, \boldsymbol{\lambda} \rangle$.

The role of $\bar{\lambda}$ is clear from inspecting Figure 2.

THEOREM 2.1 (i) If the ergodicity condition is satisfied then

$$0 < z_{\mathbf{k}}^{(1)} < 1 \qquad (\mathbf{k} \in \mathcal{H}_{N,K}) \tag{24a}$$

$$z_{k}^{(2)} = 1 \qquad (\mathbf{k} = [1, 1, \dots, 1]) \tag{24b}$$

$$1 < z_{\mathbf{k}}^{(2)} \qquad (\mathbf{k} \in \mathcal{H}_{N,K} \text{ and } \mathbf{k} \neq 1) \tag{24c}$$

(ii) If the inequality in the ergodicity condition is reversed, then

$$0 < z_{\mathbf{k}}^{(1)} < 1 \qquad (\mathbf{k} \in \mathcal{H}_{N,K} \text{ and } \mathbf{k} \neq [N, N, \dots, N]) \tag{25a}$$

$$z_{\mathbf{k}}^{(1)} = 1, \qquad (\mathbf{k} = (N, N, \dots, N]) \tag{25b}$$

$$1 < z_{\mathbf{k}}^{(2)} \qquad (\mathbf{k} \in \mathcal{H}_{N,K}) \tag{25c}$$

Observe that when the ergodicity condition is satisfied there are N^K real eigenvalues in the interval $(0, 1)$ and with the reversal of the inequality in the ergodicity condition there is one less eigenvalue in $(0, 1)$ Also observe that there are typically many $\mathbf{k}$ (later to be aggregated into one class) with the same numerical value for $z_{\mathbf{k}}^{(\cdot)}$.

Henceforth we shall assume that the ergodicity condition is satisfied and focus on the eigenvalues in equation (24a) that are in $(0, 1)$. It is now convenient to drop the superscript and to denote these eigenvalues by

$$z_{\mathbf{k}} \qquad (\mathbf{k} \in \mathcal{H}_{N,K}) \tag{26}$$

Associated with each $\mathbf{k} \in \mathcal{H}_{N,K}$ and $z_{\mathbf{k}}$ there is an eigenvector $\phi_{\mathbf{k}}$ with the structure

$$\phi_{\mathbf{k}} = \mathbf{u}_{k(1)} \otimes \mathbf{u}_{k(2)} \otimes \cdots \otimes \mathbf{u}_{k(K)} \tag{27}$$

This follows from equations (11), (15), and (20).

The spectral expansion for the equilibrium unaggregated source-buffer probabilities is

$$\pi(n) = \sum_{k \in \mathcal{H}_{N,K}} a_k z_k^n \phi_k \qquad (n \geq 0) \tag{28}$$

3. THE AGGREGATED SYSTEM

3.1 Aggregate Source State

We begin by giving a formal definition of the aggregate states of the source process. Let $\boldsymbol{\sigma}$ denote an aggregate source state and $\mathcal{S}_{N,K}$ the state space, that is,

$$\mathcal{S}_{N,K} \triangleq \left\{ \boldsymbol{\sigma} \mid \boldsymbol{\sigma} \in \mathbb{Z}^N, 0 \leq \sigma(i)(1 \leq i \leq N), \text{ and } \sum_{i=1}^{N} \sigma(i) = K \right\} \tag{29}$$

Note that $|\mathcal{S}_{N,K}| = L$ which has been given in equation (3). In interpreting $\boldsymbol{\sigma}$ to be an aggregate source state we let $\sigma(i)$ denote the number of sources in state $i(1 \leq i \leq N)$. It is also useful to have an explicit characterization of $\mathcal{A}(\boldsymbol{\sigma})$, the set of unaggregated source states that are lumped into a particular aggregate source state $\boldsymbol{\sigma}$:

$$\mathcal{A}(\boldsymbol{\sigma}) = \left\{ \mathbf{k} \in \mathcal{H}_{K,N} \mid \sum_{i=1}^{K} 1(k(i) = j) = \sigma(j)\ (1 \leq j \leq N) \right\} \tag{30}$$

where $1(\cdot)$ is the indicator function whose value is either 0 or 1. The number of elements in $\mathcal{A}(\boldsymbol{\sigma})$ is

$$c(\boldsymbol{\sigma}) = \frac{K!}{\sigma(1)!\sigma(2)!\cdots\sigma(N)!} \tag{31}$$

3.2 Eigenvalues of the Aggregate Source–Buffer System

Our starting point is equation (20). We note that for any $\boldsymbol{\sigma} \in \mathcal{S}_{N,K}$, all $\mathbf{k} \in \mathcal{A}(\boldsymbol{\sigma})$ have a common left side in equation (20) and therefore also a common set of roots. Hence the aggregated counterpart to equation (20) is

$$\sum_{j=1}^{N} \sigma(j) g_j(z) = \mu \qquad (\boldsymbol{\sigma} \in \mathcal{S}_{N,K}) \tag{32}$$

We denote the roots of equation (32) by

$$z_{\sigma}^{(1)}, z_{\sigma}^{(2)} \qquad (\boldsymbol{\sigma} \in \mathcal{S}_{N,K}) \tag{33}$$

Since the ergodicity condition is assumed to hold, it follows from Theorem 2.1 that only one of the roots is in the interval (0, 1), and we denote it by z_{σ}.

We present two algorithmic approaches for solving equation (32) for the eigenvalues. The first is a Newton-type numerical procedure for finding the roots of equation (32) viewed as a nonlinear equation:

$$z^{(n+1)} = z^{(n)} - \left\{ \sum_{j=1}^{N} \sigma(j) g_j(z^{(n)}) - \mu \right\} \Big/ \sum_{j=1}^{N} \sigma(j) g_j'(z^{(n)}) \tag{34}$$

where $z^{(n)}$ is the nth iterate of a process converging to an eigenvalue z. Whenever the procedure calls for an evaluation of $\{g_1(z), g_2(z), \ldots, g_N(z)\}$ for given z, the value returned are the N solutions of the following eigenvalue problem in ν:

$$|\nu \mathbf{I} - \mathbf{B}(z)| = 0 \tag{35}$$

where $\mathbf{B}(z)$ is defined in equation (18b). Similarly when the procedure calls for the derivatives at z, $\{g_1'(z), g_2'(z), \ldots, g_N'(z)\}$, these are calculated as follows [20]:

$$g_j'(z) = \frac{1}{(1-z)^2} \langle \tilde{\mathbf{u}}_j, \tilde{\mathbf{u}}_j \tilde{\mathbf{M}} \rangle - \frac{1}{z^2} \langle \tilde{\mathbf{u}}_j, \tilde{\mathbf{u}}_j \mathbf{\Lambda} \rangle \tag{36}$$

where $\tilde{\mathbf{M}}$ and $\{\tilde{\mathbf{u}}_j\}$ are simply obtained from the process of symmetrization, that is,

$$\tilde{\mathbf{M}} \triangleq \mathbf{W}^{1/2} \mathbf{M} \mathbf{W}^{-1/2} \qquad \text{and} \qquad \tilde{\mathbf{u}}_j \triangleq \mathbf{u}_j \mathbf{W}^{-1/2} \tag{37}$$

where $\mathbf{W} \triangleq \text{diag}(\mathbf{w})$.

The second approach for finding the eigenvalues $\{z_\sigma\}$ for $\boldsymbol{\sigma} \in \mathcal{S}_{N,K}$ consists of finding the roots of a large family of polynomials $\{P(z;\boldsymbol{\sigma})\}$ parameterized by $\boldsymbol{\sigma}$ in which each polynomial is of low degree. These polynomials in z are defined as follows:

$$P(z;\boldsymbol{\sigma}) \triangleq \{z(1-z)\}^{N!} \prod_{\omega \in \Omega_N} \{\langle \omega(\boldsymbol{\sigma}), \mathbf{g}(z) \rangle - \mu\} \tag{38}$$

where Ω_N is the set of all permutations ω on N-tuples. Clearly, $|\Omega_N| = N!$.

THEOREM 3.1 The polynomials defined in equation (38) have the following properties:

1. The set of roots of $P(z;\boldsymbol{\sigma})$ is the union of the roots of $N!$ equations where each equation is obtained from a distinct permutation of $\boldsymbol{\sigma}$ in equation (32).
2. $P(z;\boldsymbol{\sigma})$ is a polynomial in z of degree $2N!$.
3. $P(z;\boldsymbol{\sigma}) = P(z;\omega(\boldsymbol{\sigma}))$ for all $\omega \in \Omega_N$, that is, P is symmetric in $\boldsymbol{\sigma}$. Hence there are only $L/N!$ such distinct polynomials.
4. When the ergodicity condition is satisfied $P(z;\boldsymbol{\sigma})$ has $N!$ roots in the interval $(0,1)$ and $N!$ roots in $[1,\infty)$.

The proof of the polynomial structure of $P(z;\boldsymbol{\sigma})$ is as follows. The right side of equation (38) is a symmetric polynomial in g_1, g_2, ..., g_N. From the *Fundamental Theorem on Symmetric Polynomials* [21], any symmetric polynomial can be expressed as a polynomial in the "elementary symmetric polynomials." Now, the elementary symmetric polynomials of $g_1, g_2, \ldots,$ g_N may be expressed in z by noting that they are also the coefficients of the powers of ν in the characteristic polynomial of $\mathbf{B}(z)$ which is in the left side of equation (35). Finally, the factor $\{z(1-z)\}^{N!}$ in equation (38) allows the elementary symmetric polynomials in $g_1, g_2, \ldots, g_N$ to be expressed as polynomials in z. This proof is also the basis for a complete algorithmic procedure for obtaining the coefficients of the powers of z in $P(z;\boldsymbol{\sigma})$. Later we give $P(z;\boldsymbol{\sigma})$ explicitly for $N = 2$.

Notice that in effect what has been achieved is a closed-form factorization of the characteristic polynomial of the aggregate source-buffer system, which is of degree $2L$, into $L/N!$ factors where each factor is a polynomial of degree $2N!$ For fixed N and $K \to \infty$, the complexity of obtaining all the eigenvalues of the composite buffer system is $O(L)$.

3.3 Eigenvectors of the Aggregated System

For each $\boldsymbol{\sigma} \in \mathcal{S}_{N,K}$ and the eigenvalue $z_{\boldsymbol{\sigma}}$ of the aggregated system there is a set of N-dimensional vectors $\{\mathbf{u}_1(z_{\boldsymbol{\sigma}}), \ldots, \mathbf{u}_N(z_{\boldsymbol{\sigma}})\}$ that are solutions of equation (15). With appropriate multiplicities these vectors define the unaggregated eigenvectors indexed by $\mathbf{k}$, for all $\mathbf{k} \in \mathcal{A}(\boldsymbol{\sigma})$. Specifically, if

$$\phi_{\mathbf{k}} = \phi_{k(1)} \otimes \phi_{k(2)} \otimes \cdots \otimes \phi_{k(K)} \tag{39}$$

then

$$\phi_{k(i)} \in \{\mathbf{u}_1(z_\sigma), \ldots, \mathbf{u}_N(z_\sigma)\} \qquad (1 \le i \le K) \tag{40}$$

From equation (32) it follows that $\mathbf{u}_j(z_{\boldsymbol{\sigma}})$ has multiplicity $\sigma(j)$ $(1 \le j \le N)$. Hence, the set of unaggregated eigenvectors corresponding to various $\mathbf{k}$ in $\mathcal{A}(\boldsymbol{\sigma})$ are Kronecker products of a common set of distinct eigenvectors $\{\mathbf{u}_j(z_\sigma)\}$ with a common set of multiplicities $\{\sigma(j)\}$. Any pair of unaggregated eigenvectors in the same aggregate class differ only in the order in which the Kronecker product in equation (39) is taken. The above is the basis for the following result.

PROPOSITION 3.1 For each $\boldsymbol{\sigma} \in \mathcal{S}_{N,K}$,

$$\sum_{l \in \mathcal{A}(\boldsymbol{\tau})} \phi_{k}(l) = \xi_{\sigma}(\boldsymbol{\tau}) \qquad \text{for all} \quad \mathbf{k} \in \mathcal{A}(\sigma) \tag{41}$$

The vectors $\{\boldsymbol{\xi}_\sigma\}$ are the eigenvectors of the aggregate source–buffer system. We give below an efficient recursive algorithm for the exact calculation of the element $\xi_{\boldsymbol{\sigma}}(\boldsymbol{\tau})$ that exploits the Kronecker product structure.

3.4 Recursive Algorithm for Exactly Computing the Aggregated Eigenvectors $\{\xi_\sigma\}$

From equations (31) and (41), it is clear that typically there are may terms in the composition of $\xi_{\boldsymbol{\sigma}}(\boldsymbol{\tau})$. Therefore the efficiency of our method is dependent on the following algorithm for aggregation.

For $\boldsymbol{\sigma} \in \mathcal{S}_{N,K}$, the eigenvalue $z_{\boldsymbol{\sigma}}$, and the set from equation (15), $\{\mathbf{u}_1, \ldots, \mathbf{u}_N\}$. To evaluate $\boldsymbol{\xi}_{\boldsymbol{\sigma}}$, first define its multivariate generating function,

$$\mathcal{G}(\boldsymbol{\sigma};\mathbf{y}) \triangleq \sum_{\boldsymbol{\tau} \in \mathcal{G}_{N,K}} y_1^{\tau_1} \cdots y_N^{\tau_N} \xi_\sigma(\boldsymbol{\tau}) \tag{42}$$

Note that alternatively we have from equation (39) and Proposition 3.1,

$$\mathcal{G}(\boldsymbol{\sigma};\mathbf{y}) = \langle \mathbf{y} \otimes \mathbf{y} \otimes \cdots \otimes \mathbf{y}, \phi_{k(1)} \otimes \phi_{k(2)} \otimes \cdots \phi_{k(K)} \rangle \tag{43}$$

where $\mathbf{k}$ is any element of $\mathcal{A}(\boldsymbol{\sigma})$. From equation (43),

$$\begin{aligned}\mathcal{G}(\boldsymbol{\sigma};\mathbf{y}) &= \prod_{i=1}^{K} \langle \mathbf{y}, \phi_{k(1)} \rangle \\ &= \prod_{j=1}^{N} \langle \mathbf{y}, \mathbf{u}_j(z_{\boldsymbol{\sigma}}) \rangle^{\sigma(j)} \end{aligned} \tag{44}$$

from equation (40) and the discussion on multiplicities following the equation. In principle we now have a procedure for calculating $\xi_{\boldsymbol{\sigma}}(\boldsymbol{\tau})$: expand equation (44) as in equation (42), and obtain the appropriate coefficient. However, this procedure will be inefficient for $N > 2$.

Hence proceed by defining the multivariate generating function of $\mathcal{G}(\boldsymbol{\sigma};\mathbf{y})$:

$$\mathcal{H}(\mathbf{x};\mathbf{y}) \triangleq \sum_{\boldsymbol{\sigma} \geq \mathbf{0}} \mathbf{x}^{\boldsymbol{\sigma}} \mathcal{G}(\boldsymbol{\sigma};\mathbf{y}) \tag{45}$$

where we have abbreviated $x(1)^{\sigma(1)} x(2)^{\sigma(2)} \cdots x(N)^{\sigma(N)}$ by $\mathbf{x}^{\sigma}$. Now from equation (44),

$$\mathcal{H}(\mathbf{x};\mathbf{y}) = \sum_{\boldsymbol{\sigma} \geq \mathbf{0}} \prod_{j=1}^{N} \{x(j) \langle \mathbf{y}, \mathbf{u}_j \rangle\}^{\sigma(j)} \tag{46}$$

The right-hand quantity may be evaluated to give

$$\mathcal{H}(\mathbf{x};\mathbf{y}) = \prod_{j=1}^{N} \frac{1}{1 - x(j)\langle \mathbf{y}, \mathbf{u}_j \rangle} \tag{47}$$

Now define

$$\mathcal{H}_i(\mathbf{x};\mathbf{y}) \triangleq \prod_{j=1}^{i} \frac{1}{1 - x(j)\langle \mathbf{y}, \mathbf{u}_j \rangle} \qquad (1 \le i \le N) \tag{48}$$

so that the generating function in equation (47) is also $\mathcal{H}_N(\mathbf{x};\mathbf{y})$. Our algorithm recursively calculates $H_i(\mathbf{s},\mathbf{t})$ where

$$\mathcal{H}_i(\mathbf{x};\mathbf{y}) = \sum_{\sigma \ge 0} \mathbf{x}^{\sigma} \sum_{\tau \in \mathcal{S}_{N,|\sigma|}} \mathbf{y}^{\tau} H_i(\sigma;\tau) \tag{49}$$

where $|\sigma| = \langle \sigma, \mathbf{1} \rangle$.

THEOREM 3.2 For each $\sigma \in \mathcal{S}_{N,K}$,

$$\xi_{\sigma}(\tau) = H_N(\sigma;\tau) \qquad (\tau \in \mathcal{S}_{N,K}) \tag{50}$$

where

$$H_i(\mathbf{s};\mathbf{t}) = \sum_{j=1}^{N} H_i(\mathbf{s} - \mathbf{e}_i; \mathbf{t} - \mathbf{e}_j) u_i(j) + H_{i-1}(\mathbf{s};\mathbf{t}) \qquad (1 \le i \le N) \tag{51}$$

$\mathbf{e}_i$ is the unit vector with unity for its ith element and zero elsewhere, and the initial conditions are

$$\begin{aligned} H_0(\mathbf{s};\mathbf{t}) &= 1 \qquad \text{if } \mathbf{s} = \mathbf{t} = 0 \\ &= 0 \qquad \text{otherwise} \end{aligned} \tag{52}$$

For each i, the recursion in equation (51) is executed for $\mathbf{s} \le \sigma$ and $|\mathbf{t}| = |\mathbf{s}|$.

3.5 Normalizations, Coefficients of the Spectral Expansion

Recall that $\mathbf{w}$ defined in equation (22) is the stationary probability vector of the source Markov chain. It follows directly from the Kronecker sum structure of $\mathbf{Q}$, the generator of the unaggregated source process, that

$$\mathbf{pQ} = 0, \qquad \langle \mathbf{p}, \mathbf{1} \rangle = 1 \tag{53}$$

where

$$\mathbf{p} = \mathbf{w} \otimes \mathbf{w} \otimes \cdots \otimes \mathbf{w} \tag{54}$$

a K-fold Kronecker product. Moreover, the elements of the stationary probability vector of the aggregated source process is obtained easily by aggregating $\mathbf{p}$:

$$p^{(A)}(\boldsymbol{\sigma}) \triangleq c(\boldsymbol{\sigma}) \{w(1)\}^{\sigma(1)} \{w(2)\}^{\sigma(2)} \dots \{w(N)\}^{\sigma(N)} \qquad (\boldsymbol{\sigma} \in \mathcal{S}_{N,K}) \quad (55)$$

where $c(\boldsymbol{\sigma})$ is given in equation (31).

Now consider the spectral expansion of the stationary probability vectors of the aggregated source-buffer system $\{\boldsymbol{\pi}^{(A)}(n)\}$:

$$\pi^{(A)}(n; \boldsymbol{\sigma}) = \sum_{\boldsymbol{\tau} \in \mathcal{S}_{N,K}} b_{\boldsymbol{\tau}} z_{\boldsymbol{\tau}}^n \xi_{\boldsymbol{\tau}}(\boldsymbol{\sigma}) \qquad (n \geq 0; \boldsymbol{\sigma} \in \mathcal{S}_{N,K}) \quad (56)$$

Clearly these probabilities must satisfy the normalization conditions,

$$p^{(A)}(\boldsymbol{\sigma}) = \sum_{n=0}^{\infty} \pi^{(A)}(n; \boldsymbol{\sigma}) \qquad (\boldsymbol{\sigma} \in \mathcal{S}_{N,K}) \quad (57)$$

That is, from equation (56),

$$p^{(A)}(\boldsymbol{\sigma}) = \sum_{\boldsymbol{\tau} \in \mathcal{S}_{N,K}} b_{\boldsymbol{\tau}} (1 - z_{\boldsymbol{\tau}})^{-1} \xi_{\boldsymbol{\tau}}(\boldsymbol{\sigma})$$

In matrix form:

$$\mathbf{p}^{(A)} = \mathbf{b}[\mathbf{I} - \mathbf{Z}]^{-1} \boldsymbol{\Xi} \quad (58)$$

where $\mathbf{p}^{(A)}$ is the vector with elements $p^{(A)}(\boldsymbol{\sigma})$, the rows of the $L \times L$ matrix $\boldsymbol{\Xi}$ are $\{\boldsymbol{\xi}_{\boldsymbol{\tau}}\}$, and $\mathbf{Z} = \text{diag}\,\{z_{\boldsymbol{\sigma}}\}$. Inverting equation (58) gives

$$\mathbf{b} = \mathbf{p}^{(A)} \boldsymbol{\Xi}^{-1} [\mathbf{I} - \mathbf{Z}] \quad (59)$$

Thus, our procedure requires the systems of L linear equations in equation (58) to be numerically solved.

The algorithms for computing all the elements of the spectral expansion in equation (56) have all been described.

4. MARKOVIAN SOURCES WITH TWO STATES

All Markov chains in equilibrium are reversible when $N = 2$. Let

$$\mathbf{M} = \begin{bmatrix} -\alpha & \alpha \\ \beta & -\beta \end{bmatrix} \quad \text{and} \quad \boldsymbol{\Lambda} = \begin{bmatrix} \lambda_1 & \\ & \lambda_2 \end{bmatrix} \quad (60)$$

The family of $(K+1)$ polynomials $\{P(z; \boldsymbol{\sigma})\}$ whose roots give the eigenvalues of the aggregated system are as follows: for $\sigma_1 \geq 0$, $\sigma_2 \geq 0$, $\sigma_1 + \sigma_2 = K$.

$$P(z; \boldsymbol{\sigma}) \triangleq \mu^2 \{z(1 - z)\}^2$$

$$
\begin{aligned}
&- \mu(\sigma_1 + \sigma_2)z(1-z)\left\{(\lambda_1 + \lambda_2)(1-z) - (\alpha + \beta)z\right\} \\
&+ \sigma_1\sigma_2\left\{(\lambda_1 + \lambda_2)(1-z) - (\alpha + \beta)z\right\}^2 \\
&- (\sigma_1 - \sigma_2)^2(1-z)\left\{(\lambda_1\lambda_2 + \alpha\lambda_2 + \beta\lambda_1)z - \lambda_1\lambda_2\right\} \qquad (61)
\end{aligned}
$$

Note that $P(z; \sigma_1, \sigma_2) = P(z; \sigma_2, \sigma_1)$. Hence only half of the $(K + 1)$ polynomials are distinct. The ergodicity condition is

$$K(\alpha\lambda_2 + \beta\lambda_1) < \mu(\alpha + \beta) \qquad (62)$$

When this condition is satisfied, each of the $(K + 1)/2$ distinct polynomials $P(z; \boldsymbol{\sigma})$ has two roots in the interval $(0, 1)$ and two roots in $[1, \infty]$.

REFERENCES

[1] J. Keilson, *Markov Chain Models—Rarity and Exponentiality*, Springer-Verlag, New York, 1979.

[2] R. Bellman, *Introduction to Matrix Analysis*, 2nd ed., McGraw-Hill, New York, 1970.

[3] M. F. Neuts, *Matrix Geometric Solutions in Stochastic Models*, John Hopkins University Press, Baltimore, 1981.

[4] H. Heffes and D. Lucantoni, A Markov modulated characterization of packetized voice and data traffic and related statistical multiplexer performance, *IEEE J. SAC*, 4 (6), pp. 856–868, 1986.

[5] K. Sriram and W. Whitt, Characterizing superposition arrival processes in packet multiplexers for voice and data, *IEEE J. SAC*, 4 (6), 833–846, 1986.

[6] S. Halfin, The backlog of data in buffers with variable input and output rates, in *Performance of Computer-Communication Systems*, eds. H. Rudin and W. Bux, Elsevier, Amsterdam, pp. 307–319, 1984.

[7] D. Y. Burman and D. R. Smith, An asymptotic analysis of a queueing system with Markov-modulated arrivals, *Operations Res.*, 34, (1), 1986.

[8] D. Mitra and I. Mitrani, Analysis of a Kanban discipline for cell coordination in production lines, II: Stochastic demands, *Operations Res.*, in press.

[9] D. Anick, D. Mitra, and M. M. Sondhi, Stochastic theory of a data-handling system with multiple sources, *Bell System Tech. J.*, 61, 1982, pp. 1871–1894.

[10] L. Kosten, Stochastic theory of data-handling systems with groups of multiple sources, in *Performance of Computer-Communication Systems*, eds. H. Rudin and W. Bux, Elsevier, Amsterdam, pp. 321–331, 1984.

[11] D. Mitra, Stochastic theory of a fluid model of producers and consumers coupled by a buffer, *Adv. Appl. Probability*, 20, 1988, pp. 646–676.

[12] T. E. Stern and A. I. Elwalid, Analysis of separable Markov-modulated rate models for information-handling systems, to appear in *Adv. Appl. Probability*, March 1991.

[13] A. Weiss, A new technique for analyzing large traffic systems, *Adv. Appl. Probability*, 18, 1986, pp. 506–532.

[14] J. A. Morrison, Asymptotic analysis of a data-handling system with many sources, *SIAM J. Appl. Math.* 49 (2) 1989, pp. 617–637.

[15] J. Filipiak, Multi-layer analysis of packet delay and blocking in statistical multiplexing, University of Adelaide report TRC 3/89, 1989, submitted for publication.

[16] A. E. Eckberg, D. T. Luan, and D. M. Lucantoni, Bandwidth management: A congestion control strategy for broadband packet networks—Characterizing the throughput-burstiness filter, Proc. ITC Specialist Seminar, Adelaide, paper no. 4.4, 1989.

[17] G. Woodruff, R. Kositpaiboon, G. Fitzpatrick, and P. Richards, Control of ATM statistical multiplexing performance, Proc. ITC Specialist Seminar, Adelaide, paper no. 17.2, 1989.

[18] M. Zukerman and P. Kirton, Applications of matrix-geometric solutions to the analysis of the bursty data queue in a B-ISDN switching system, Proc. GLOBECOM '88, pp. 1635–1639.

[19] V. Ramaswami and G. Latouche, Modeling packet arrivals from asynchronous input lines, Proc. 12th International Teletraffic Congress, Torino, Italy, June 1988, pp. 41A.5.1–41A.5.7.

[20] J. H. Wilkinson, *The Algebraic Eigenvalue Problem*, Clarendon Press, Oxford, pp. 67–69, 1965.

[21] G. Birkhoff and S. Maclane, *A Survey of Modern Algebra*, rev. ed., Macmillan, New York, pp. 145–147, 1953.

13

Some Markov Chain Problems in the Evaluation of Multiple-Access Protocols

JEFFREY E. WIESELTHIER Information Technology Division, Naval Research Laboratory, Washington, D.C.

ANTHONY EPHREMIDES Electrical Engineering Department, University of Maryland, College Park, Maryland

ABSTRACT

In this paper we review models for several variants of a class of multiple-access protocols in communication systems, each of which is characterized by an underlying Markov chain of infinite dimension. The performance evaluation of these protocols requires an exact, or a reasonably accurate, estimate of the steady-state vectors of these Markov chains. For each case we discuss the form of the Markov chains and the computational techniques of truncation and approximation that we have used. These chains have special structure that may not be fully exploited by our techniques; thus improvements may be possible.

1. INTRODUCTION

In the application area of performance evaluation of multiple access protocols for a variety of communication systems, it is customary to use models that involve Markov chains. Often, these chains possess special structure that may be exploited to achieve efficient computation and high accuracy in obtaining their steady-state vectors. The steady-state vector is usually indispensable in computing the performance, as it is for the protocols considered here. Sometimes the first moment of the stationary vector suffices. In most cases these chains are of infinite or very large dimensionality.

In this paper, we first describe the basic model of the communication system and the Markov chain it leads to. Then we examine variants of it (that are motivated by the special needs of the application) and the corresponding versions of the Markov chain models. In each case, we outline our approach in evaluating the stationary vectors. Truncation is needed in all cases for numerical evaluation. Some of the protocols considered lead to problems of extremely large dimensionality; in these cases we exploit special features of the system to reduce complexity, without requiring approximations other than truncation. In one case, an approximation is developed that exploits the difference in time scales of two classes of traffic in the network. There is no conclusive evaluation of these techniques as yet. In fact, we conclude by pointing out that improvements are desirable and, perhaps, possible.

2. THE COMMUNICATION SYSTEM

We consider M ground-based users (terminals) that communicate among themselves via a transponder, which is typically satellite-based. The transponder broadcasts all messages it receives to all members of the user population. A method is needed to determine how the users should schedule their packet transmissions to avoid destructive "collisions," which occur when two or more terminals transmit simultaneously, and to maintain high efficiency. This is known as the channel-access problem.

In much of this paper we assume that the transponder is located on a geosynchronous satellite, which results in a substantial round-trip propagation time of approximately 0.27 s. Since we consider fixed-length packets as units of transmission (each requires one time "slot" for transmission), this delay may be expressed in terms of packet length as R slots. The value of R is an important system parameter that has a great impact on protocol operation and performance; typically, R is of the order of 10–12 slots, but for ground-based transmissions R can be significantly smaller.

It is assumed that each user has an infinite buffer in which it stores the arriving packets, which are assumed to form a Bernoulli process with rate λ. The total arrival rate is, therefore, $M\lambda$ packets per slot, which is equal to the throughput rate under stable operation.

Out of the numerous protocols that have been proposed in the literature for such a system, we single out one class, called the interleaved-frame flush out (IFFO) protocols [1, 2], which is especially designed and well suited for bursty traffic in satellite-based systems. We also consider some new variants of these protocols. Most other types of protocols are modeled in ways similar to the ones that we describe for the IFFO protocols (e.g., [3]). They all use Markov chains with different special structural features.

3. A MARKOV CHAIN MODEL FOR THE IFFO PROTOCOLS

Although the basic model structure and analysis for the IFFO protocols has been presented in [1], here we confine ourselves to a brief discussion of them to facilitate the presentation of subsequently discussed new protocols, and to display the numerical problems associated with their analysis.

The basic frame structure for the IFFO protocols, as shown in Figure 1, is a reservation structure, in which the unreserved slots may be used for transmission on a contention basis. The first slot of each frame, which consists of M "minislots" that are exclusively allocated to the M terminals in (contention-free) time-division multiple access (TDMA) fashion, is known as the status slot; it is used by each of the terminals to reserve a transmission slot for each of the packets that were generated in the previous frame.

It is assumed that all reservation minipackets are received successfully by all terminals following the round-trip propagation delay of R slots. We define:

R_k = total number of reserved slots in frame k

N_k = total number of contention (unreserved) slots in frame k

L_k = total length (in slots) of frame k

Thus the status slot is followed by R_k reserved slots. It is required that each frame have a length of at least R slots to ensure that the reservation information generated at the beginning of the frame is received before the start of the next frame. If $R_k < R - 1$, the remaining $N_k = R - 1 - R_k$ slots in the frame are used for transmission on a contention basis, as in the so-called slotted-ALOHA protocol. If $R_k > R - 1$, additional slots are added to accommodate all of the reservations. Therefore,

$$N_k = \max(R - 1 - R_k, 0)$$

and

$$L_k = \max(R_k + 1, R) = 1 + R_k + N_k$$

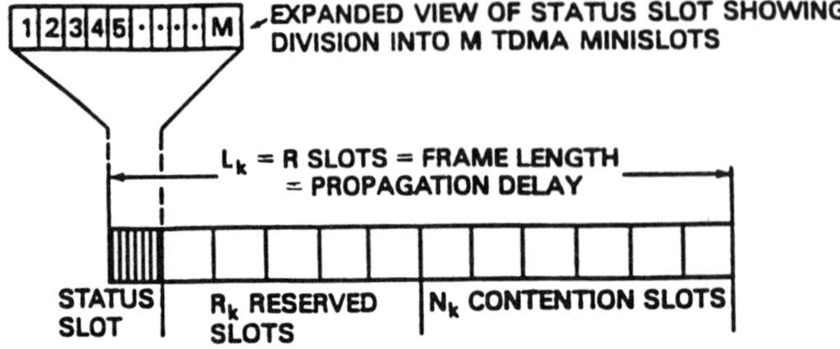

CASE 1: THE CASE WITH CONTENTION SLOTS, i.e., $R_k \leq R-2$

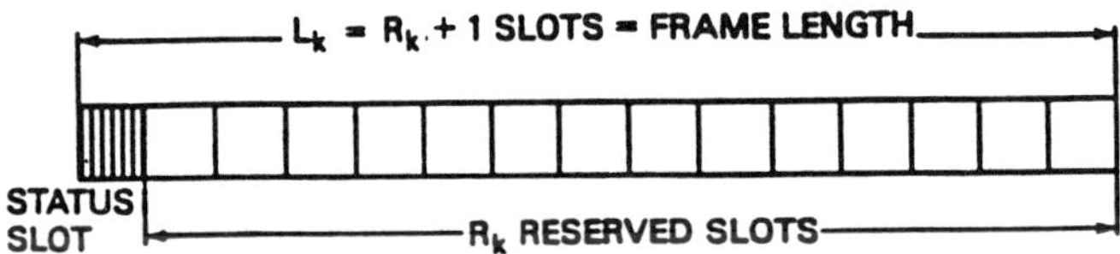

CASE 2: THE CASE WITH NO CONTENTION SLOTS, i.e., $R_k \geq R-1$

Figure 1 Frame structure for the IFFO protocols.

The quantity of interest that needs to be tracked is R_k, which evolves as a first-order Markov chain. A complete discussion of the dynamics is presented in [1]. Here we summarize the derivation and show the resulting equations.

The operation of the IFFO protocols is illustrated in Figure 2. Each packet arriving in frame k is known as a *k-packet.* Reservation minipackets for all k-packets will be transmitted in the first slot of frame $k + 1$. Although reservations are transmitted for all k-packets, some of these reservations may not be needed, owing to the possibility of successful transmission in contention slots. It is easy to cancel the unneeded reservations because knowledge of the outcome of all contention transmissions in frame k will be available to all users not later than the beginning of frame $k + 2$.

Let A_k be the total number of packet arrivals in frame k, summed over all terminals, and S_k be the number of successful contention transmissions in frame k. Then,

$$R_{k+2} = A_k - S_k$$

Note that R_{k+2} is totally independent of A_{k+1}, L_{k+1}, and R_{k+1}. It depends only on R_k and on what happens during frame k. Thus we can split

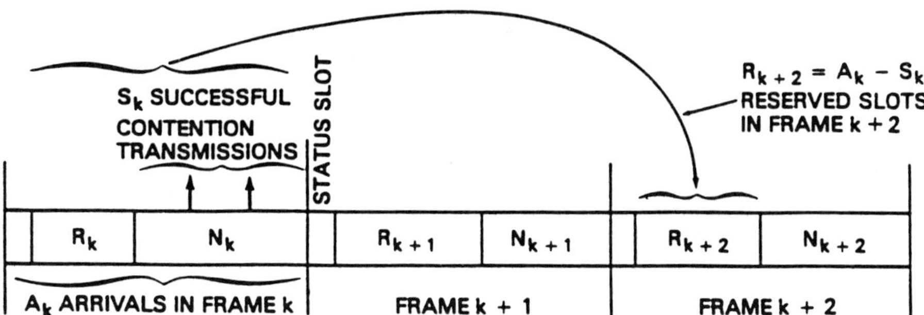

Figure 2 Operation of the IFFO protocols, showing transmission procedure for frame-k arrivals.

the process $\{R_k\}$ into two interleaved Markov chains, which may be denoted $\{R_{2j}\}$ and $\{R_{2j+1}\}$. The reserved slot process in even-numbered frames is independent of the reserved slot process in odd-numbered frames. These processes have identical statistics and may be analyzed separately. In Section 5 we consider modifications of the IFFO protocols under which the even- and odd-numbered slots are no longer independent.

There are several versions of the IFFO protocols, each of which is characterized by a different transmission procedure in the unreserved slots:

1. Pure Reservation IFFO (PR-IFFO): The unreserved slots are not used for contention; they simply remain idle and wasted. All packets that arrive in frame k are transmitted in the reserved slots of frame $k + 2$.
2. Fixed Contention IFFO (F-IFFO): The transmission policy depends on the slot number in which packets arrive.

    a. A packet arriving in slot n, for $n \in [R_{k+1}, R-1]$, will be transmitted in slot $n + 1$, that is, in each contention slot, each terminal will transmit the packet that may have arrived in the previous slot. Each colliding packet will be retransmitted in a reserved slot in frame $k + 2$.
    b. All packets arriving during the first R_k slots of frame k will not be allowed to contend because of the high risk of collision, and will be assigned reserved slots for transmission during frame $k + 2$.
    c. All packets arriving during slot R (i.e., the last slot in the frame) will be assigned reserved slots in frame $k + 2$.

3. Controlled Contention IFFO (C-IFFO): In this version, the transmission procedure is state-dependent. A complete description and an approximate analysis of this protocol are presented in [1].

3.1 The Transition Probability Matrices for IFFO

For PR-IFFO, since $R_{k+2} = A_k$, it is easy to see that the elements of the transition probability matrix for R_k can be written as

$$p_{ij} \triangleq \Pr(R_{k+2} = j \mid R_k = i) = \begin{cases} \binom{MR}{j} \lambda^j (1-\lambda)^{MR-j} & 0 \le i \le R-1 \\ \binom{M(i+1)}{j} \lambda^j (1-\lambda)^{M(i+1)-j} & i \ge R-1 \end{cases}$$

The evaluation of the combinatorial expressions in the above equation raises an interesting computational problem. For large values of M (values up to 50 have been considered) and i (as great as 700) the values of the combinatorial expressions become extremely large, while the values of the exponential expressions become extremely small. To avoid the resulting computational problems, we recognize that, since the arrival process in each slot is an independent sequence, the probability mass function (pmf) of the number of arrivals in $i+1$ slots is simply the convolution of $i+1$ pmf's of the number of arrivals in a single slot. Thus we can perform the computation by convolving expressions that have less extreme values. We have

$$a(j) = \Pr(j \text{ arrivals in one slot}) = \binom{M}{j} \lambda^j (1-\lambda)^{M-j}$$

Therefore, for $i \ge R-1$,

$$p_{ij} = a(j) * a(j) * \cdots * a(j)$$

where $*$ represents the convolution operator and the expression is the convolution of $i+1$ pmf's of the form $a(j)$. Similarly, for $i < R-1$, p_{ij} is the convolution of R $a(j)$'s.

For F-IFFO we can easily show that (see [1])

$$p_{ij} \triangleq \Pr(R_{k+2} = j \mid R_k = i) = \binom{M(i+1)}{j} \lambda^j (1-\lambda)^{M(i+1)-j} * c_{N_k}(j)$$

where

$$c_{N_k}(j) = c(j) * c(j) * \cdots * c(j)$$

is the convolution of N_k (which depends on $R_k = i$) pmf's of the form $c(j)$, where

$$c(0) = (1-\lambda)^{M-1}[(1-\lambda) + M\lambda]$$
$$c(1) = 0$$

$$c(j) = \binom{M}{j} \lambda^j (1 - \lambda)^{M-j}, \qquad 2 \leq j \leq M$$

All convolutions are performed numerically. Whenever $R_k \geq R - 1$, the convolution vanishes because $N_k = 0$, and the p_{ij} are the same as those for PR-IFFO. Thus the performance of F-IFFO is closely bounded by that of PR-IFFO in the limit of high input rates.

3.2 On Obtaining Equilibrium Results for the IFFO Protocols

The equilibrium pmf for the number of reserved slots per frame is needed to evaluate the two steady-state performance indexes that have been considered for the IFFO protocols, that is, the expected time spent in the system per packet, and the expected number of reserved slots per frame. To evaluate performance we need to compute the stationary vector of the chain R_k. Expressions for expected packet delay are presented in [4] and [2].

An equilibrium pmf for R_k does, in fact, exist under the IFFO protocols, as long as the input rate is less than one packet per slot. In this case, R_k, which is irreducible and aperiodic, is easily shown to be ergodic by Foster's theorem [5].

Note that the chain that describes R_k has an unbounded (hence, infinite) number of states. To obtain a numerical solution, we truncate the probability vector and transition probability matrix to some finite dimension N. The N can be chosen sufficiently large so that the effect of truncation error is small. An $N \times N$ transition probability matrix permits the modeling of up to $N - 1$ reserved slots per frame (since the N elements in the probability vector range from 0 to $N - 1$). The truncation operation involves setting $\Pr(j = N - 1 \mid i) = 1 - \Pr(j \leq N - 2 \mid i)$ to maintain conservation of probability. Typical values of N can range from 20 to 400, depending on the throughput rate.

Method Used to Determine the Equilibrium pmf $\boldsymbol{\pi}$

The equilibrium pmf $\boldsymbol{\pi}$ (a row vector) must satisfy the following matrix equation:

$$\boldsymbol{\pi} = \boldsymbol{\pi}\mathbf{P}$$

where $\mathbf{P}$ is the transition probability matrix with elements $p_{ij} \triangleq \Pr(R_{k+2} = j \mid R_k = i)$.

Instead of solving the N equations in N unknowns, it is computationally preferable to use the iterative procedure of relaxation, that is,

$$\boldsymbol{\pi} = \lim_{n \to \infty} \boldsymbol{\pi}(0)\mathbf{P}^n$$

where $\pi(0)$ is an arbitrary initial pmf. In all of our numerical examples we assumed that $\pi(0)$ is uniformly distributed between 0 and 9. A scaling operation, which ensures that the elements of the pmf sum to 1 at each iteration, must be included to minimize the effects of computer roundoff error.

The iteration was stopped when the following criterion was satisfied:

$$|\pi_j(n+1) - \pi_j(n)| \leq 0.5 \times 10^{-6} \qquad 0 \leq j \leq N-1$$

Table 1 shows the size of the arrays used for PR-IFFO as a function of throughput, as well as the number of iterations needed for convergence for $M = 10$ (the number is somewhat smaller for smaller values of M and somewhat larger for larger values). To provide an indication of the behavior of the tail of the distribution, we also show the largest value of R_k that has a probability mass greater than 10^{-9}, which is denoted as $R_k(10^{-9})$. To verify that the array sizes are sufficiently large so that the effects of truncation are insignificant, additional computer runs have been made for larger array sizes. For the array sizes shown, all of the performance criteria considered [e.g., expected value and variance of R_k, expected delay, expected frame length, number of iterations required, and $R_k(10^{-9})$] are unchanged to at least the fourth dec-

Table 1 Size of Array Used, Number of Iterations Needed, and $R_k(10^{-9})$ as a Function of Throughput under PR-IFFO for $M = 10$

Throughput	Size of array	Iterations	$R_k(10^{-9})$
0.00001	20	2	2
0.1	20	2	12
0.2	20	2	16
0.3	20	3	19
0.4	30	4	22
0.5	30	5	25
0.6	30	7	30
0.7	50	10	39
0.8	60	19	55
0.85	80	29	71
0.9	120	52	104
0.92	150	70	128
0.94	200	100	167
0.95	230	123	199
0.96	300	157	246
0.97	400	209	323

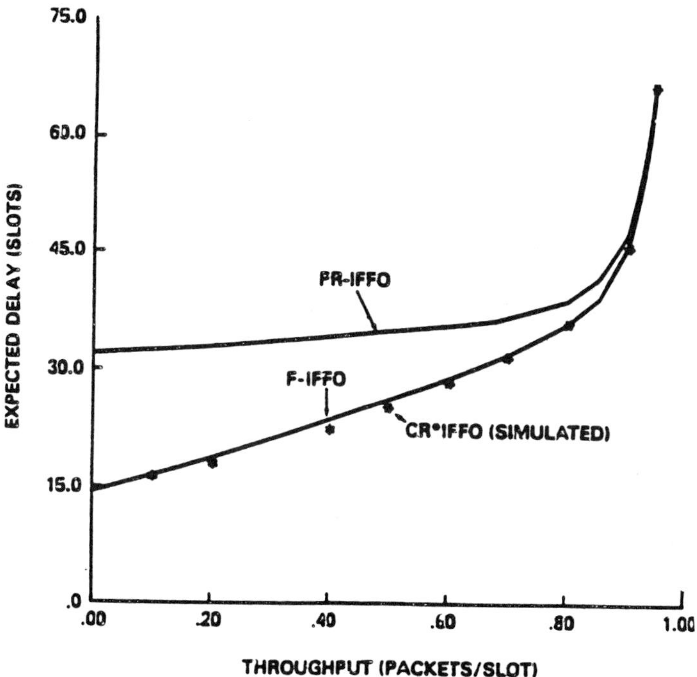

Figure 3 Delay vs. throughput curves for PR-IFFO, F-IFFO, and CR * IFFO ($R = 12$, $M = 10$).

imal place when larger array sizes are used. In general, convergence and acceptable results are achieved for somewhat smaller array sizes.

3.3 Performance Results

Extensive performance results for the IFFO protocols, including comparisons with other multiple-access schemes, are presented in [1] and [4]. We now summarize some of these results. Figure 3 shows delay-throughput curves for the three IFFO protocols that have been studied. At low to moderate throughput rates, F-IFFO provides considerable performance improvement over PR-IFFO; the ability to transmit in contention slots (thus avoiding the delay associated with the reservation process) has a significant impact on system operation. At high throughput rates, few packets are able to take advantage of the contention mode of operation, and PR-IFFO provides a close upper bound on expected delay under F-IFFO that becomes increasingly tight as throughput increases, as discussed earlier. Simulation results for PR-IFFO and F-IFFO were virtually identical to the numerically computed results; this was expected because the mathematical model used to characterize these protocols is exact, except for the truncation of the pmf's. The CR∗IFFO protocol (a special case of C-IFFO under which the transmission probability in contention slots depends on the system state and is chosen to optimize performance) provides only slightly better performance than F-IFFO. Simulation results are shown for CR∗IFFO because they are more accurate than the numerically computed results, since approximations were needed in the derivation of the transition probability matrix.

For all of the IFFO protocols, performance is quite insensitive to the number of terminals for M greater than about 5; thus these curves for $M = 10$ are representative of higher values as well. Figure 4 compares the delay-throughput performance of F-IFFO with that of slotted ALOHA and several hybrid protocols that also combine random access with reservations. Of these protocols, F-IFFO performs best over a wide range of throughput (from 0.33 to 0.93 packets/slot).

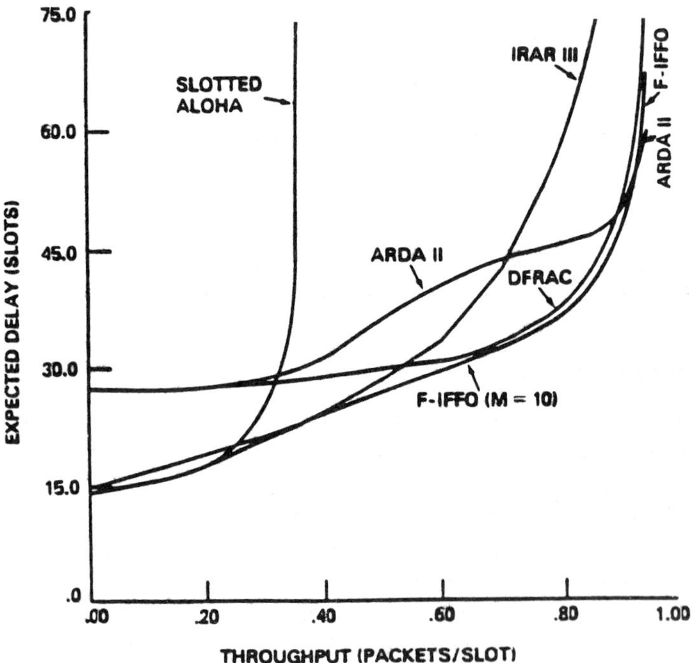

Figure 4 Delay vs. throughput curves for F-IFFO, ARDAII, IRARIII, DFRAC, and slotted ALOHA ($R = 12$).

4. THE INTERLEAVED-FRAME FIXED-LENGTH (IFFL) SCHEMES

The IFFO protocols are characterized by a frame length that adapts to channel traffic. This adaptive feature guarantees that all k-packets are successfully transmitted not later than the end of frame $k + 2$; this is the flush-out feature of these schemes that, indeed, motivated them and that results in very high efficiency and excellent delay performance. In this section we consider a variation of these protocols under which the frame length is kept fixed at R slots. In certain applications (e.g., voice/data integration, which is discussed in Section 6) it is desirable to keep a constant frame length, although doing so reduces the efficiency of the protocol.

We call these schemes the interleaved-frame fixed-length (IFFL) schemes. Operation is the same as that of IFFO, except that when R_k is greater than R, packets that cannot be accommodated in the current frame ($R_k - R + 1$ of them) are delayed until frame $k + 2$, at which point they are again subject to further delays if there is again a large backlog. We refer to these packets as "excess packets." Clearly, these protocols violate the flush-out condition, and do not belong to the class of IFFO protocols. We may consider PR-IFFL and F-IFFL versions of these schemes, whose definitions follow from those given earlier for the IFFO schemes. System evolution is again characterized by the first-order Markov chain R_k. However, note that a slight reinterpretation of R_k is needed. It was previously defined as the number of reserved slots in frame k. Since the excess packets are delayed until frame $k + 2$, it must now be interpreted as the number of packets for which reservations are needed at the beginning of frame k (including the excess packets, which have to be delayed). For PR-IFFL we have

$$R_{k+2} = A_k + \max(R_k - R + 1, 0)$$

By straightforward modification of the expressions for PR-IFFO, the elements of the transition probability matrix for PR-IFFL may now be written as

$$p_{ij} \triangleq \Pr(R_{k+2} = j \mid R_k = i)$$
$$= \begin{cases} \binom{MR}{j} \lambda^j (1-\lambda)^{MR-j} & 0 \le i \le R-1 \\ \binom{MR}{j-i+R-1} \lambda^{j-i+R-1} (1-\lambda)^{MR-(j-i+R-1)} & i \ge R-1 \end{cases}$$

For F-IFFL we observe that for $R_k < R - 1$ the transition probabilities derived for F-IFFO apply, whereas for $R_k \ge R - 1$ those for PR-IFFL must be used; hence the transition probabilities for F-IFFL are

$$p_{ij} \triangleq \Pr(R_{k+2} = j \mid R_k = i)$$
$$= \begin{cases} \binom{M(i+1)}{j} \lambda^j (1-\lambda)^{M(i+1)-j} * c_{N_k}(j) & 0 \le i \le R-1 \\ \binom{MR}{j-i+R-1} \lambda^{j-i+R-1} (1-\lambda)^{MR-(j-i+R-1)} & i \ge R-1 \end{cases}$$

4.1 Performance Results

Performance evaluation for the IFFL schemes proceeds in the same manner as that described earlier for the IFFO schemes. Figure 5 shows the expected value of R_k under PR-IFFL as a function of input rate. Note that, since the frame length is fixed at R slots, the maximum throughput that can be achieved is $(R-1)/R$ because one slot in each frame (the status slot) is needed for overhead. We may define the "efficiency" achieved under the IFFL protocols to be the throughput multiplied by the ratio $R/(R-1)$. For example, for $R =$ 12, a throughput rate of 0.9 packets/slot (the maximum shown in Figure 5)

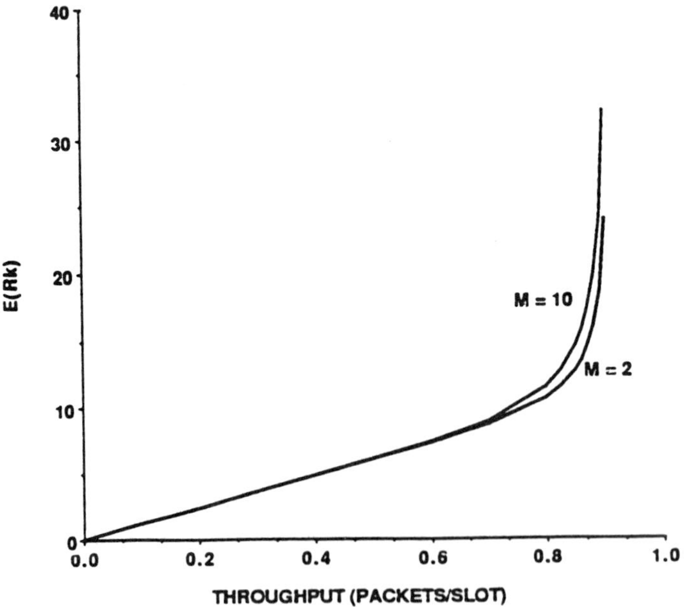

Figure 5 Expected value of R_k under PR-IFFL ($R = 12$, $M = 2, 10$).

corresponds to an efficiency of 0.9818. In contrast, under the IFFO protocols, the input rate (and thus the throughput) can be arbitrarily close to 1 because the frame length expands to accommodate all reservations.

5. NON-INTERLEAVED-FRAME FIXED-LENGTH (NIFFL) SCHEMES

Under the IFFO and IFFL schemes, the system state in even-numbered frames is independent of that in odd-numbered frames. Thus it is possible, for example, for the even-numbered slots to build up large backlogs (high values of R_k) while the odd-numbered slots are lightly loaded. Since the IFFO schemes flush out all k-packets by the end of frame $k + 2$, no inefficiency arises from this behavior. However, under IFFL, whenever $R_k > R - 1$ the excess packets ($R_k - R + 1$ of them, as discussed earlier) will be postponed to frame $k + 2$. If there are some unreserved slots in frame $k + 1$, it would be advantageous to transmit some or all of these excess packets in those slots.

To address this situation we consider a variation of the IFFL protocols that we call the non-interleaved-frame fixed-length (NIFFL) protocols. We show that the PR-NIFFL version is again characterized by an underlying first-order Markov chain, and thus can be evaluated using techniques similar to those used for the IFFO schemes. However, the description of F-NIFFL requires a second-order Markov chain, which makes performance evaluation considerably more difficult. The system evolution for the NIFFL protocols may be described as follows:

$$R_{k+2} = R_{k+2}^{(k)} + R_{k+2}^{(k+1)}$$

where $R_{k+2}^{(k)}$ is the number of k-packets that are included in R_{k+2}, that is, all arrivals in frame k, except those that were transmitted successfully in the contention slots of frame k. Thus we have

$$R_{k+2}^{(k)} = A_k - S_k$$

The term $R_{k+2}^{(k+1)}$ is the number of excess packets that are carried over from frame $k + 1$. We have

$$R_{k+2}^{(k+1)} = \max\{[R_{k+1} - (R - 1)], 0\}$$

The Markov chain now is the pair (R_{k+1}, R_k), and thus we need

$$p_{ij\to jm} \triangleq \Pr(R_{k+2} = m, R_{k+1} = j \mid R_{k+1} = j, R_k = i)$$

A brute-force system description would require a state probability vector $\Pr(R_{k+1}, R_k)$ of dimension N^2, where N is the truncation value as discussed earlier. We can write

$$\Pr(R_{k+1}, R_k) \triangleq [\mathbf{q}_0 \quad \mathbf{q}_1 \quad \cdots \quad \mathbf{q}_{N-1}]$$

where $\mathbf{q}_j$ is the row vector whose entries are

$$q_{ji} \triangleq \Pr(R_{k+1} = j, R_k = i)$$

Thus a transition probability matrix of size $N^2 \times N^2$ would be needed. The dimensions of the problem can be reduced somewhat by making the observation that R_{k+1} includes all of the excess packets contained in R_k. Thus, given R_{k+1}, the exact value of R_k is needed only if it is less than $R-1$. We define an aggregate state $R_k = R-1$ that actually contains all states for which $R_k \geq R-1$. Now the state (R_{k+1}, R_k) can be described by a vector of dimension NR rather than N^2. The state (R_{k+2}, R_{k+1}) still requires a vector of dimension N^2, however. The resulting transition probability matrix is of dimension $NR \times N^2$. This is still of unmanageable size. Before discussing a method to decompose the problem into one that requires N matrices of dimension $R \times N$ (still a very large size, but considerably smaller than before), we demonstrate how the PR-NIFFL protocol can be evaluated using a first-order Markov chain.

Under PR-NIFFL, $R_{k+2}^{(k)} = A_k$ because $S_k = 0$ (since there are no contention transmissions in a pure reservation system). Thus

$$R_{k+2} = A_k + \max\{[R_{k+1} - (R-1)], 0\}$$

Since the frame length is constant, A_k does not depend on R_k. It is binomially distributed with parameter λ over MR trials. Thus the system description for PR-NIFFL can be reduced to a first-order Markov chain as follows:

$$p_{ij} \triangleq \Pr(R_{k+2} = j \mid R_{k+1} = i)$$
$$= \begin{cases} \binom{MR}{j} \lambda^j (1-\lambda)^{MR-j} & 0 \leq i \leq R-1 \\ \binom{MR}{j-i+R-1} \lambda^{j-i+R-1} (1-\lambda)^{MR-(j-i+R-1)} & i \geq R-1 \end{cases}$$

Note that this expression is identical to that for PR-IFFL, except that R_k is replaced here by R_{k+1}.

The system evolution of F-NIFFL cannot be described by a first-order Markov chain. However, the second-order Markov chain can be decomposed to generate a collection of smaller problems, as we now discuss. We start from the observation made earlier that R_{k+1} is common to both states (i.e., origin and destination) of each transition. This led to the state probability description in terms of the vectors of the form $\mathbf{q}_j$ with elements q_{ji}; the index j takes on all possible values of R_{k+1} (i.e., from 0 to $N-1$) and i takes on all possible values of R_k (i.e., from 0 to $R-1$).* For each value of R_{k+1} we consider the transitions from R_k to R_{k+2}. The corresponding transition probabilities are denoted as

$$p_{i \xrightarrow{j} m} \triangleq p_{ij \to jm} \triangleq \Pr(R_{k+2} = m, R_{k+1} = j \mid R_{k+1} = j, R_k = i)$$

Whenever $j \leq R-1$, $R_{k+2}^{(k+1)} = 0$, in which case $R_{k+2} = R_{k+2}^{(k)}$. Whenever $j > R-1$, $R_{k+2}^{(k+1)} = j-(R-1)$ packets must be added to $R_{k+2}^{(k)}$. By an appropriate modification of the expressions previously derived for F-IFFL, we have the following transition probabilities for F-NIFFL:

$$p_{i \xrightarrow{j} m} = \begin{cases} \binom{M(i+1)}{m} \lambda^m (1-\lambda)^{M(i+1)-m} * c_{N_k}(m) & 0 \leq j \leq R-1 \\ \binom{M(i+1)}{m-j+R-1} \lambda^{m-j+R-1} & \\ \quad \times (1-\lambda)^{M(i+1)-(m-j+R-1)} * c_{N_k}(m) & j > R-1 \end{cases}$$

Recall that we must have $i \leq R-1$ since $R-1$ is an aggregate state (the excess packets are incorporated into $j = R_{k+1}$). Again, the convolution vanishes whenever $i = R-1$. Note that the two expressions given here correspond to different ranges of j, rather than i. There are N transition probability matrices of this type (one for each value of j, denoted $\mathbf{P}_{(j)}$), each of dimension $R \times N$. Each $\mathbf{q}_j$ vector is multiplied by the corresponding transition probability matrix $\mathbf{P}_{(j)}$. The result is again a collection of N probability vectors for R_{k+2}, each of dimension N, one for each value of R_{k+1}. The vector corresponding to $R_{k+1} = j$, which we denote $\mathbf{r}_j$, consists of the elements†

$$r_{jm} = \Pr(R_{k+1} = j, R_{k+2} = m)$$

In preparation for the next iteration, the state probabilities are rearranged in the form of the $\mathbf{q}_j$ vectors, where all states for which $R_{k+1} \geq R-1$ are combined in the aggregate state $R-1$. This procedure is repeated until convergence is achieved.

In our discussion of the IFFO protocols we made the observation that the performance of PR-IFFO bounds closely that of F-IFFO in the limit of

---

*As noted earlier, the state $R_k = R-1$ is an aggregate state that actually contains all values of $R_k \geq R-1$.

†Note that $\mathbf{q}_j$ refers to a probability vector in which the state in the later of two slots is held constant (i.e., $R_{k+1} = j$, while R_k varies), whereas $\mathbf{r}_j$ refers to a probability vector in which the state in the earlier of two slots is held constant (i.e., $R_{k+1} = j$, while R_{k+2} varies).

high throughput rates. Similarly, at high throughputs the performance of PR-NIFFL provides a tight upper bound on that of F-NIFFL. This is true because most frames have no unreserved slots at high throughput rates, in which case $R_k = R - 1$ (the aggregate state). Thus no k-packets can be transmitted in frame k, in which case F-NIFFL functions in the same manner as PR-NIFFL. Therefore, at high throughput rates (for which large transition probability matrices are needed), the performance of F-NIFFL can be bounded and closely approximated by that of PR-NIFFL, which is characterized by a much simpler description (i.e., a first-order Markov chain).

6. AN IFFL PROTOCOL FOR INTEGRATED VOICE/DATA SYSTEMS

The communication systems that gave rise to the models discussed thus far in this paper can be modified slightly to handle voice traffic in addition to data packets. This is an important extension in communication system design, and represents our main objective here. A customary model for voice assumes that voice calls are generated at idle terminals according to a Bernoulli process, and that they are geometrically distributed in length; thus the probability that a call is completed in any particular frame is also a Bernoulli process. There are M_V voice users in the system and M data users. The time constants associated with voice traffic are considerably larger than those associated with data traffic; voice calls will typically last from tens to hundreds of frames. This difference in time durations plays a key role in the development of an approximate system model for this protocol.

To accommodate the needs of both voice and data traffic, we consider a channel-access protocol under which a reservation scheme is used for voice traffic and IFFL (which combines reservation and contention) is used for data.* We call these the voice/data IFFL (VD-IFFL) protocols. Under these schemes, once a voice call is accepted by the system, it is guaranteed access to one slot each frame until its completion.† The standard idea of a "movable-boundary" mechanism is used to partition each frame between voice and data

*Although better performance can be expected under the NIFFL schemes, we consider IFFL schemes here because they are easier to model.

†Here we implicitly assume that the slot length has been selected in conjunction with the frame length (which is equal to the propagation delay) so that the voice burst rate results in the required symbol rate needed for real-time voice transmission. We could actually use any frame length greater than or equal to the propagation delay such that the voice transmission rate is correct.

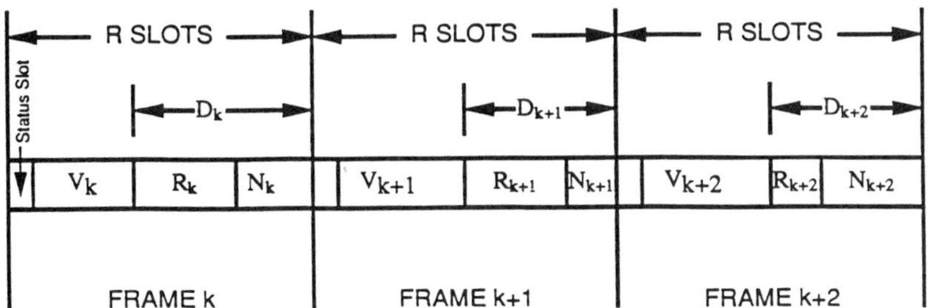

Figure 6 Frame structure for the VD-IFFL protocols.

operation, as shown in Figure 6. A fixed-length frame structure is necessary to accommodate the real-time requirements of voice traffic.

Voice calls are accepted by the system as long as the number of calls does not exceed some maximum value $V_{\max}$, which must be less than R. If a slot is not available for a new call, the call is assumed to be lost; there is no buffering of voice calls. The slots not used by voice calls (including empty slots in the voice portion of the frame) are used for data traffic, which is transmitted by using one of the IFFL protocols. Since the decision to accept new voice calls depends only on whether or not the threshold $V_{\max}$ is exceeded, voice traffic is unaffected by data traffic; however, the operation of the data protocol is dependent on voice traffic because data traffic is permitted to use unneeded slots in the voice portion. Thus this problem is similar to variable service rate queuing systems in which the service rate depends on another process. Our model can be extended to consider systems in which the decision to accept new voice calls also depends on the system backlog (i.e., the value of R_k). However, the analysis of such systems is considerably more difficult and is not addressed here.

As shown in Figure 6, the first slot of every frame is once again the status slot, during which each terminal transmits its reservations for packets that arrived in the previous slot. The next V_k slots are reserved for voice traffic, where V_k is the number of voice calls in progress at the beginning of slot k. The remainder of the frame consists of D_k data slots, where

$$D_k = R - 1 - V_k$$

As with the protocols designed purely for data, R_k is the number of data packets for which reservations are needed at the beginning of frame k. Whenever $R_k < D_k$, N_k^v slots are available for contention transmission, where

$$N_k^v = \max(D_k - R_k, 0) = \max(R - 1 - v - R_k, 0)$$

for each particular value of $V_k = v$. Whenever $R_k > D_k$, the excess packets are delayed until frame $k+2$. Operation of the data portion of VD-IFFL can thus be viewed as that of IFFL with a variable number of slots (D_k) available for data traffic, where D_k depends on V_k. In contrast, under IFFL exactly $R-1$ slots are available for data in each frame.

The VD-IFFL protocols can be characterized by a two-dimensional first-order Markov chain (R_k, V_k) with transition probabilities $\Pr(R_{k+2}, V_{k+2} \mid R_k, V_k)$.

6.1 An Exact Model for VD-IFFL

A brute-force approach would consider a probability vector containing all possible pairs of V_k and R_k. The maximum value of V_k would be the threshold value $V_{\max}$; R_k would have a maximum value of $N-1$ as in the evaluation of the data-only protocols. Thus a transition probability matrix of dimension $(V_{\max}+1)N \times (V_{\max}+1)N$ would be needed. However, not all transitions are possible, and significant reductions can be made by decomposing the problem into separate voice and data portions. We do this by recognizing that the voice-call process does not depend on the data-message process.* Thus

$$\Pr(R_{k+2}, V_{k+2} \mid R_k, V_k) = \Pr(R_{k+2} \mid R_k, V_k) \Pr(V_{k+2} \mid V_k)$$

The transition from R_k to R_{k+2} depends on V_k (because V_k determines D_k), but not on V_{k+2}. The transition from V_k to V_{k+2} does not depend on R_k or R_{k+2}. Therefore, we can consider the transitions corresponding to the data process separately for each value of V_k.

Corresponding to each value of V_k there is an $N \times N$ transition probability matrix for the data message process with elements

$$p_{ij}^{v} \triangleq \Pr(R_{k+2} = j \mid R_k = i, V_k = v)$$

These transition probabilities are easily obtained from those for the IFFL protocols. Under IFFL, $R-1$ slots are available for packet transmission in each frame. Under VD-IFFL, this number is reduced to $D_k = R - 1 - V_k$. Thus for each value of $V_k = v$ we replace $R-1$ by $R-1-v$. This yields

*This is not true for systems in which the decision on whether or not to accept a voice call is permitted to depend on R_k, however.

for PR-VD-IFFL:

$$p_{ij}^{v} = \begin{cases} \binom{MR}{j} \lambda^{j}(1-\lambda)^{MR-j} & 0 \le i \le R-1-v \\ \binom{MR}{j-i+R-1-v} \lambda^{j-i+R-1-v} \\ \quad \times (1-\lambda)^{MR-(j-i+R-1-v)} & i \ge R-1-v \end{cases}$$

Similarly, for F-VD-IFFL we obtain

$$p_{ij}^{v} = \begin{cases} \binom{M(i+1)}{j} \lambda^{j}(1-\lambda)^{M(i+1)-j} * c_{N_k^v}(j) & 0 \le i \le R-1-v \\ \binom{MR}{j-i+R-1-v} \lambda^{j-i+R-1-v} \\ \quad \times (1-\lambda)^{MR-(j-i+R-1-v)} & i \ge R-1-v \end{cases}$$

Thus at each iteration $(V_{\max} + 1)$ matrix multiplications (each of size $N \times N$, one for each value of v) must be carried out to determine the data transitions.

Next, we consider the voice transitions. The probability that a new call is generated at an idle terminal during any particular frame is denoted as λ_V. The probability that an ongoing call completes service during any particular frame is denoted as μ_V. When a call is blocked (because the threshold has been reached), it is dropped from the system and the terminal reenters the idle state. It is important to note that there is no frame interleaving of the voice call process, since a call occupies a time slot in every frame from its start to its completion. Thus the transition from frame k to frame $k+2$ represents a two-phase transition. We define $\mathbf{S}^{(2)}$ to be the $(V_{\max} + 1) \times (V_{\max} + 1)$ transition probability matrix with elements

$$s_{ij}^{(2)} \triangleq \Pr(V_{k+2} = j \mid V_k = i)$$

An exact evaluation of the equilibrium distribution of the voice-call process requires one matrix multiplication at each iteration. No truncation is required because the number of voice calls is characterized by a finite Markov chain with a maximum state value of $V_{\max} \le R - 1$. Thus this calculation is easily performed.

To evaluate $\mathbf{S}^{(2)}$ we start by considering the single-frame transition, which is characterized by the transition probability matrix $\mathbf{S}$ with elements

$$s_{ij} \triangleq \Pr(V_{k+1} = j \mid V_k = i)$$

Clearly, $\mathbf{S}^{(2)} = \mathbf{S}^2$. We may write

$$V_{k+1} = V_k + \Delta V_k^{+} - \Delta V_k^{-}$$

where ΔV_k^+ is the number of new voice calls accepted by the system (all new arrivals are accepted unless the threshold $V_{\max}$ would be exceeded) and ΔV_k^- is the number of completed voice calls (departures) in the frame.*

First, consider the currently idle users ($M_V - V_k$ of them), each of which will generate a new call with probability λ_V. If the number of voice arrivals is denoted as A_k^V, we have

$$\Pr(A_k^V = j \mid V_k = i) = \binom{M_V - V_k}{j} \lambda_V^j (1 - \lambda_V)^{M_V - V_k - j}$$

and

$$\Delta V_k^+ = \begin{cases} A_k^V & A_k \le V_{\max} - V_k \\ V_{\max} - V_k & A_k > V_{\max} - V_k \end{cases}$$

Now consider the V_k active voice users (not including the new additions), at each of which a call will be completed with probability μ_V. We have

$$\Pr(\Delta V_k^- = j \mid V_k = i) = \binom{i}{j} \mu_V^j (1 - \mu_V)^{i-j}$$

The ΔV_k^+ and ΔV_k^- are conditionally independent, given V_k. Thus it is straightforward to obtain the transition probability matrix $\mathbf{S}$, from which $\mathbf{S}^{(2)}$ follows directly.

6.2 An Approximate Model for the VD-IFFL Protocols

We now consider an approximate model that may simplify performance evaluation. Since the interarrival times and holding times of voice calls are considerably longer than a slot duration, V_k is a slowly changing quantity. The system may be described by a number of "super-states," each of which is characterized by a different value of V_k, and which contain all possible values of R_k. First, we assume that V_k is constant for all time. For each possible value of V_k we determine the equilibrium distribution of R_k by using the transition probabilities p_{ij}^v discussed earlier. Then we average the distribution of R_k over all possible values of V_k, whose distribution can be determined exactly following the procedure we just discussed. Courtois's concept of near-complete decomposability [6] appears to be applicable here.

This approach appears to be reasonable because the voice-call transitions are totally decoupled from data-message transitions, although the converse is not true in the sense that the data-message transition probabilities depend on

*Note that a call arriving in frame k cannot be completed in frame k; it actually begins service in frame $k + 1$. Also note that a call completing service in frame k does not open up space for another call until frame $k + 1$.

V_k. It is expected that agreement between the approximate and exact models will be close in the limit of very long interarrival and holding times. However, computational results are not available yet to confirm this expectation.

It is difficult to predict with certainty whether the approximate model will result in significantly reduced computation time. If the average number (over all values of V_k) of iterations needed using the approximate model is less than the number of iterations under the exact model, improvement will be realized. We anticipate that the average number of iterations will be reduced when V_k is held fixed because transitions for different values of V_k are completely decoupled from each other. Again, this plausible argument needs to be confirmed via computational results in the future. Another benefit of the approximate model is a reduced storage requirement because only one $N \times N$ matrix is used at any time (instead of $V_{\max} + 1$ of them).

7. CONCLUSIONS

In this paper we have considered the Markov chains that model a variety of multiple-access protocols in communication systems applications. We described the underlying physical model to clarify how the protocol dynamics reflect into the structure of the chains. The objective of performance evaluation of these protocols requires the computation of the first moment of the packet delay. To do so we wish to compute the stationary vector itself. The computational techniques considered range from standard truncation to dimensionality reduction techniques that exploit some of the properties of the protocols.

We believe that there is room for improvement in this computation. We are interested in techniques that can exploit further the special structure of these chains. Improved numerical methods for such computations will have significant impact on the application area of multiple-access protocols.

REFERENCES

[1] J. E. Wieselthier and A. Ephremides, A new class of protocols for multiple access in satellite networks, *IEEE Trans. Automatic Control* **AC-25** pp. 865–879 (October 1980).

[2] J. E. Wieselthier and A. Ephremides, Protocols of multiple access (a survey) and the IFFO protocols (a case study)—Invited, *Proceedings of the NATO Advanced Study Institute on New Concepts in Multi-User Communication, Norwich, U. K. 1980*, Sijthoff and Noordhoff, Alphen aan den Rijn, The Netherlands (1981).

[3] F. A. Tobagi, Multiaccess protocols in packet communication systems, *IEEE Trans. Commun.* **COM-28** 468–488 (April 1980).

[4] J. E. Wieselthier, A New Class of Multi-Access Protocols for Packet Communication Over a Satellite Channel—The Interleaved Frame Flush-Out Protocols, Ph.D. Dissertation, University of Maryland (March 1979).

[5] F. G. Foster, On the stochastic matrices associated with certain queueing processes, *Ann. Math. Statist.* **24** pp. 355–360 (1953).

[6] P. J. Courtois, *Decomposability: Queueing and Computer System Applications*, Academic Press, New York (1977).

14

A Stochastic Model for a Computer Communication Network Node with Phase Type Timeout Periods

S. CHAKRAVARTHY Department of Science and Mathematics, GMI Engineering and Management Institute, Flint, Michigan

K. V. RAVI Department of Electrical and Computer Engineering, GMI Engineering and Management Institute, Flint, Michigan

ABSTRACT

A finite capacity queuing system is considered to model the low and error control mechanisms at the higher layers, such as the transport layer of the open systems interconnection (OSI). Under the assumptions of transmission times and acknowledgment periods, and with phase type timeout periods, numerically stable algorithms are presented to compute various steady-state system performance measures, such as the throughput of the system, and means and variances of the number of messages in various buffers. A number of numerical examples are discussed.

1. INTRODUCTION

One of the prime objectives of the protocols in a computer communication network is to provide reliable communication. Typical problems that occur in such a network are noise on the transmission line, corruption of data within intermediate switching nodes, and loss of an acknowledgment. Many protocols implement error recovery by retransmitting a message for which no acknowledgment was received during a (constant or random) timeout period. Flow control is used to regulate the flow information between a pair of communication nodes, to prevent one node from sending more information than the receiving node can handle.

In this paper, a finite capacity queuing system is considered to model the flow and error control mechanisms at the higher layers of the OSI, such as the transport layer. We assume that messages (packets) arrive at the source node according to a Poisson process with rate λ. At any time at most $N + 1$ new messages may be present: one in the transmitting unit (service) and at most N waiting in the primary buffer. Any new arrival when the primary buffer is full is considered lost. The transmission times of the messages are independent and identically distributed random variables having a common *exponential* distribution with parameter μ_1. A copy of the transmitted message is kept in the secondary buffer and is removed once an acknowledgment for it is received from the destination mode. The secondary buffer can store at most S (window size) messages.

It is assumed that the arrival of acknowledgments at the source node follows a *Poisson* process with rate μ_2. A transmitted message may require a subsequent transmission if no acknowledgment for it is received within a specific *timeout* period. Messages that are to be retransmitted are stored in an auxiliary buffer and have a *nonpreemptive* priority over the messages waiting in the primary buffer.

Since the timeout period depends on a number of quantities such as the total time for a message to reach the destination, the processing of a message by the receiver, variable size of the messages, different paths used by the messages, congestion in the network, and availability of the protocol resources, it is reasonable to assume that the timeout period is random. For more details on the timeout period and on the higher layers of the OSI, we refer the reader to any introductory book on computer networks, such as Stallings [8], Tanenbaum [9], or Sloman and Kramer [7].

Hence we assume that the timeout period is random and follows a *phase type* distribution (PH-distribution) of order m with representation $(\underline{\alpha}, T)$. That is, the common probability distribution $F(\cdot)$ of the timeout period is given by

$$F(x) = 1 - \underline{\alpha} \exp(Tx)e, \qquad \text{for} \quad x \geq 0 \tag{1}$$

where $e' = (1, 1, \ldots, 1) \in \mathcal{R}^m$, $\underline{\alpha} \geq 0$, $\underline{\alpha}e \leq 1$, and where T is a *stable* matrix with negative diagonal entries, nonnegative off-diagonal entries and nonpositive row sums. To avoid uninteresting complications, we assume that $\underline{\alpha}e = 1$, which entails that $F(0+) = 0$. It is trivial to verify that the mean of $F(\cdot)$ is

$$\nu^{-1} = -\underline{\alpha} T^{-1} e \tag{2}$$

For later use, let $-Te = T^{\circ}$.

PH-distributions and PH-renewal processes were introduced by Neuts [4]. The class of PH-distributions include many well-known distributions such as generalized Erlang, hyperexponential, etc., as special cases. For example, if $F(\cdot)$ is generalized Erlang of order m with parameters $\lambda_1, \lambda_2, \ldots, \lambda_m$, then $(\underline{\alpha}, T)$ is given by

$$\underline{\alpha} = [1, 0, \ldots, 0] \qquad T = \begin{bmatrix} -\lambda_1 & \lambda_1 & 0 & \cdots & 0 & 0 \\ 0 & -\lambda_2 & \lambda_2 & \cdots & 0 & 0 \\ \cdot & \cdot & \cdot & \cdots & \cdot & \cdot \\ 0 & 0 & 0 & \cdots & -\lambda_{m-1} & \lambda_{m-1} \\ 0 & 0 & 0 & \cdots & 0 & -\lambda_m \end{bmatrix}$$

The class of PH-distributions is dense in the class of all distributions on $[0, \infty)$ and has a number of interesting closure properties. A detailed discussion of the properties of PH-distributions and their uses in stochastic modeling may be found in Neuts [5].

Using Erlangian timeout periods, Sethi and Ghosal [6] considered a finite queuing model to study the transmission process at a layer, such as the data link layer, in which the timeout periods are assumed to be almost constants. While it is possible to have more than one acknowledgment during a timeout period in their model, we assume in our model that each message has its own timeout clock, which is triggered only when it is at the head of the queue in the secondary buffer. However, with suitable modifications our model can be used to study various schemes of the acknowledgment process. Our analysis of the model is completely different from the one discussed [6].

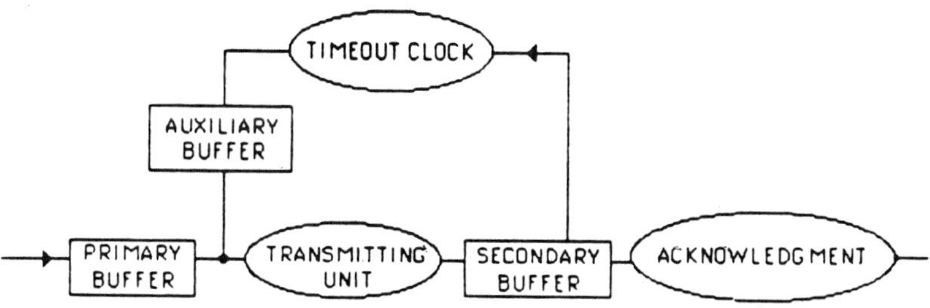

Figure 1

Thus, the system under study consists of a transmitting unit, a primary buffer to store the incoming new messages, a secondary buffer to store the transmitted messages waiting for an acknowledgment, and an auxiliary buffer to store the messages that need to be retransmitted. The messages are transmitted on a first-come, first-served basis. The timeout period is active as long as there is at least one message in the secondary buffer. This system is described pictorially in Figure 1.

2. THE MARKOV PROCESS

The system outlined in section 1 can now be described by a Markov process with $1 + 2W(N + 1) + mNW(W - 1) + m(W^2 + N)$ states. The state space may be partitioned into $N + 4$ sets of states and one state. These are denoted by $\Delta_1, \Delta_2, \mathbf{0}, \mathbf{1}, \ldots, \mathbf{N}, \mathbf{N+1}$, and $*$ and their descriptions are given in Table 1.

We have found it somewhat convenient to order the sets of state lexicographically.

In the sequel the notation $\mathbf{e}$ stands for a column vector of 1's, e_i a unit column vector with 1 in the ith position and 0 elsewhere, and I an identity matrix of appropriate dimensions. In places where the order of the matrices is to be mentioned explicitly, we use $e_i(r)$ and I_r to denote respectively a unit column vector of order r and an identity matrix of order r. The prime symbol ($'$) will denote the transpose of a matrix and the symbol $\otimes$ will denote the Kronecker product of two matrices. Specifically, $L \otimes M$ stands for the matrix made up of the block $L_{ij}M$. For more details on the Kronecker products, we refer the reader to Bellman [1] or Marcus and Minc [3].

Before we display the generator of the Markov process governing the system described above, we define a number of notations that are needed for later use.

$$B = \begin{bmatrix} -(\lambda+\mu_1)I & \lambda I & 0 & \cdots & 0 & 0 \\ 0 & -(\lambda+\mu_1)I & \lambda I & \cdots & 0 & 0 \\ 0 & 0 & -(\lambda+\mu_1)I & \cdots & 0 & 0 \\ \cdot & \cdot & \cdot & \cdots & -(\lambda+\mu_1)I & \lambda I \\ 0 & 0 & 0 & \cdots & 0 & -\mu_1 I \end{bmatrix}$$

$$c_1 = \mathbf{e}_1(N+1) \otimes [\mathbf{e}_1'(W) \otimes M_1, \mathbf{e}_1'(W-1) \otimes M_2, \ldots, \mathbf{e}_1'(2) \otimes M_{W-1}, M_W]$$

$$c_i = \mathbf{e}_1(N+1) \otimes [\mathbf{e}_1'(W_1) \otimes M_1, \mathbf{e}_1'(W-1) \otimes M_2, \ldots, \mathbf{e}_1'(2) \otimes M_{W-1}, M_W] \quad \text{for} \quad 2 \le i \le N+1$$

with $M_i = \mu_1(e_i \otimes \underline{\alpha})$ for $1 \le i \le W$, and $W_1 = 1 + [W(W-1)/2]$.

Table 1

State	Description
$*$	The system is idle.
$\Delta_1 = \{(i',j) : 1 \le i \le N+1, 0 \le j \le W-1\}$	There are i new messages in the system including the one that is currently being transmitted, and j messages are waiting for retransmission.
$\Delta_2 = \{(i,j') : 0 \le i \le N, 1 \le j \le W\}$	There are i new messages in the system, and j messages including the one in service are to be retransmitted.
$\mathbf{0} = \{0,j',r,k) : 0 \le j \le W-1, 1 \le r \le W-j, 1 \le k \le m\}$	There is no new message in the system, j messages including the one in service are to be retransmitted, r messages are waiting for an acknowledgment, and the phase of the timeout period is in k.
$\mathbf{i} = \{(i',j,r,k) : 0 \le j \le W-2, 1 \le r \le W-1-j, 1 \le k \le m\} U \{(i,0,W,k) : 1 \le k \le m\} U \{(i,j',r,k) : 1 \le j \le W-1, 1 \le r \le W-j, 1 \le k \le m\}$	There are i new messages in the system, j messages are waiting for retransmissions, r messages are waiting for an acknowledgment, and the phase of the timeout period is in k. If the transmitter is idle (because the secondary buffer is full) or if server is busy, then either a new message or the one that needs a retransmission is in service, for $1 \le i \le N$.
$\mathbf{N+1} = \{(N+1',j,r,k) : 0 \le j \le W-2, 1 \le r \le W-1-j, 1 \le k \le m\}$	There are $N+1$ new messages, j messages are waiting for retransmission, r messages are waiting for an acknowledgment, and the phase of the timeout period is in k.

NOTE: The prime denotes which message (new or one that requires retransmission) is currently being transmitted.

For $1 \le i \le N$,

$$D_i = \mathbf{e}_i'(N+1) \otimes \begin{bmatrix} \mathbf{e}_1(W) \otimes S_1 \\ \mathbf{e}_1(W-2) \otimes S_2 \\ \vdots \\ \mathbf{e}_1(2) \otimes S_{W-2} \\ \mathbf{e}_1(W_1) \otimes S_{W-1} \end{bmatrix}$$

and

$$D_{N+1} = \mathbf{e}_{N+1}'(N+1) \otimes \begin{bmatrix} \mathbf{e}_1(W) \otimes S_1 \\ \mathbf{e}_1(W-2) \otimes S_2 \\ \vdots \\ \mathbf{e}_1(2) \otimes S_{W-2} \\ S_{W-1} \end{bmatrix}$$

with $S_i = [\mu_2(\mathbf{e}_i'(W) \otimes \mathbf{e}) + \mathbf{e}_{i+1}'(W) \otimes T^\circ)]$ for $1 \le i \le W-1$.

$$E_1 = \mathbf{e}_1'(N+1) \otimes \begin{bmatrix} \mathbf{e}_1(W) \otimes \mathbf{e}_1'(W) \otimes T^\circ \\ \mathbf{e}_1(W-2) \otimes S_1 \\ \vdots \\ \mathbf{e}_1(2) \otimes S_{W-2} \\ S_{W-1} \end{bmatrix}$$

and

$$E_i = \mathbf{e}_i'(N+1) \otimes \begin{bmatrix} 0 \\ \mathbf{e}_1(W-1) \otimes S_1 \\ \mathbf{e}_1(W-2) \otimes S_2 \\ \vdots \\ \mathbf{e}_1(2) \otimes S_{W-2} \\ S_{W-1} \end{bmatrix}$$

for $2 \leq i \leq N + 1$.

$$F = \begin{bmatrix} F_0 & \hat{F}_0 & 0 & 0 & \cdots & 0 & 0 & 0 \\ \tilde{F}_1 & F_1 & \hat{F}_1 & 0 & \cdots & 0 & 0 & 0 \\ 0 & \tilde{F}_2 & F_2 & \hat{F}_2 & \cdots & 0 & 0 & 0 \\ \cdot & \cdot & \cdot & \cdot & \cdots & \cdot & \cdot & \cdot \\ 0 & 0 & 0 & 0 & \cdots & \tilde{F}_{W-2} & F_{W-2} & \hat{F}_{W-2} \\ 0 & 0 & 0 & 0 & \cdots & 0 & \tilde{F}_{W-1} & F_{W-1} \end{bmatrix}$$

with

$$F_0 = (I_W \otimes [T - (\lambda + \mu_2)I]) + F^*_W$$
$$F_j = (I_{W-j} \otimes [T - (\lambda + \mu_1 + \mu_2)I]) + F^*_{W-j}, \qquad \text{for} \quad 1 \leq j \leq W - 1$$

The square matrices (of order rm) F^*_r are such that the $(i, i-1)$th (block of order m) entry, for $2 \leq i \leq r, 2 \leq r \leq W$, is given by $\mu_2 \mathbf{e}\underline{\alpha}$ and all other entries are zeros. The square matrix F^*_1 of order m has all its entries equal to 0.

The matrix $\tilde{F}_0$ is a square matrix of order Wm and its $(i, i+1)$th (block) entry is $\mu_1 I$, for $1 \leq i \leq W - 1$, and all other entries are zeros. The matrices $\tilde{F}_r$ of order $(W-r)m \times (W+1-r)m$ are such that their $(i, i+1)$th entries are given by $\mu_1 I$, for $1 \leq i \leq W-r, 1 \leq r \leq W-1$, and all other entries are zeros.

The matrices $\hat{F}_r$, for $0 \leq r \leq W-2$, are of order $(W-r)m \times (W-1-r)m$ and are such that their $(i, i-1)$th entry is given by $T^\circ \underline{\alpha}$ for $2 \leq i \leq W-r$ and all other entries are zero.

The matrix G_1 of order $[W(W+1)/2]m \times [W(W-1)+1]m$ is such that its (i,i)th entries, for $1 \leq i \leq W$, and its $(W+i, [W(W-1)/2] + 1 + i)$th entries, for $1 \leq i \leq (W(W-1)/2)$, are λI_m. All other entries are zeros.

The matrix G_2 of order $[W(W-1)+1]m \times [W(W+1)/2]m$ is such that its (1,1)th entry is given by $\tilde{F}_0$ and its (i,i)th entries, for $2 \leq i \leq W-1$, are given by $\tilde{F}_i$, and all other entries are zeros.

The matrices H, J, U, and V are defined by

$$H = \begin{bmatrix} H_0 & \hat{H}_0 & 0 & \cdots & 0 & 0 & \tilde{H}_0 & 0 & 0 & \cdots & 0 & 0 \\ 0 & F_2 & \hat{F}_2 & \cdots & 0 & 0 & 0 & 0 & 0 & \cdots & 0 & 0 \\ 0 & 0 & F_3 & \cdots & 0 & 0 & 0 & 0 & 0 & \cdots & 0 & 0 \\ \cdot & \cdot & \cdot & \cdots & \cdot & \cdot & \cdot & \cdot & \cdot & \cdots & \cdot & \cdot \\ 0 & 0 & 0 & \cdots & F_{W-2} & \hat{F}_{W-2} & 0 & 0 & 0 & \cdots & 0 & 0 \\ 0 & 0 & 0 & \cdots & 0 & F_{W-1} & 0 & 0 & 0 & \cdots & 0 & 0 \\ \tilde{F}_1 & 0 & 0 & \cdots & 0 & 0 & F_1 & \hat{F}_1 & 0 & \cdots & 0 & 0 \\ 0 & 0 & 0 & \cdots & 0 & 0 & \tilde{F}_2 & F_2 & \hat{F}_2 & \cdots & 0 & 0 \\ \cdot & \cdot & \cdot & \cdots & \cdot & \cdot & \cdot & \cdot & \cdot & \cdots & \cdot & \cdot \\ 0 & 0 & 0 & \cdots & 0 & 0 & 0 & 0 & 0 & \cdots & \tilde{F}_{W-1} & F_{W-1} \end{bmatrix}$$

where H_0 is a square matrix of order Wm such that its (i,i)th entries, for $1 \leq i \leq W-1$, are given by $[T-(\lambda+\mu_1+\mu_2)I]$; its $(i,i-1)$th entries, for $2 \leq i \leq W$, are given by $\mu_2 e \underline{\alpha}$; and its (W,W)th entry is given by $[T-(\lambda+\mu_2)I]$. All other elements are zeros. The matrix $\hat{H}_0$ of order $Wm \times (W-2)m$ is such that its $(i,i-1)$th entries, for $2 \leq i \leq W-1$, are given by $T^\circ \underline{\alpha}$ and all other elements are zeros. The matrix $\tilde{H}_0$ of order $Wm \times (W-1)m$ is such that its $(W, W-1)$th entry is $T^\circ \underline{\alpha}$ and all other entries are zeros.

$$J = \begin{bmatrix} \tilde{F}_0 & 0 & 0 & \cdots & 0 & 0 & 0 & 0 & 0 & \cdots & 0 & 0 & 0 \\ 0 & 0 & 0 & \cdots & 0 & 0 & \tilde{F}_2 & 0 & 0 & \cdots & 0 & 0 & 0 \\ 0 & 0 & 0 & \cdots & 0 & 0 & 0 & \tilde{F}_3 & 0 & \cdots & 0 & 0 & 0 \\ \cdot & \cdot & \cdot & \cdots & \cdot & \cdot & \cdot & \cdot & \cdot & \cdots & \cdot & \cdot & \cdot \\ 0 & 0 & 0 & \cdots & 0 & 0 & 0 & 0 & 0 & \cdots & \tilde{F}_{W-2} & 0 & 0 \\ 0 & 0 & 0 & \cdots & 0 & 0 & 0 & 0 & 0 & \cdots & 0 & \tilde{F}_{W-1} & 0 \\ 0 & 0 & 0 & \cdots & 0 & 0 & 0 & 0 & 0 & \cdots & 0 & 0 & 0 \\ 0 & 0 & 0 & \cdots & 0 & 0 & 0 & 0 & 0 & \cdots & 0 & 0 & 0 \\ \cdot & \cdot & \cdot & \cdots & \cdot & \cdot & \cdot & \cdot & \cdot & \cdots & \cdot & \cdot & \cdot \\ 0 & 0 & 0 & \cdots & 0 & 0 & 0 & 0 & 0 & \cdots & 0 & 0 & 0 \end{bmatrix}$$

$$U = \begin{bmatrix} \tilde{F}_1 & 0 & 0 & \cdots & 0 & 0 & 0 & 0 & 0 & \cdots & 0 & 0 & 0 \\ 0 & 0 & 0 & \cdots & 0 & 0 & \tilde{F}_2 & 0 & 0 & \cdots & 0 & 0 & 0 \\ 0 & 0 & 0 & \cdots & 0 & 0 & 0 & \tilde{F}_3 & 0 & \cdots & 0 & 0 & 0 \\ \cdot & \cdot & \cdot & \cdots & \cdot & \cdot & \cdot & \cdot & \cdot & \cdots & \cdot & \cdot & \cdot \\ 0 & 0 & 0 & \cdots & 0 & 0 & 0 & 0 & 0 & \cdots & \tilde{F}_{W-2} & 0 & 0 \\ 0 & 0 & 0 & \cdots & 0 & 0 & 0 & 0 & 0 & \cdots & 0 & \tilde{F}_{W-1} & 0 \end{bmatrix}$$

$$V = \begin{bmatrix} V_1 & \hat{F}_1 & 0 & \cdots & 0 & 0 & 0 \\ 0 & V_2 & \hat{F}_2 & \cdots & 0 & 0 & 0 \\ \cdot & \cdot & \cdot & \cdots & \cdot & \cdot & \cdot \\ 0 & 0 & 0 & \cdots & 0 & V_{W-2} & \hat{F}_{W-2} \\ 0 & 0 & 0 & \cdots & 0 & 0 & V_{W-1} \end{bmatrix}$$

with

$$V_j = (I_{W-j} \otimes [T - (\mu_1 + \mu_2)I]) + F^*_{W-j}, \qquad \text{for} \quad 1 \leq j \leq W - 1$$

The matrix K of order $[W(W-1)+1]m$ is such that its (i,i)th entries, for $1 \leq i \leq W-1$ and for $W+1 \leq i \leq 1 + [W(W-1)/2]$, are given by λI_m and all other elements are zeros.

The square matrix L of order $[1 + W(W-1)]m$ is such that its (W,W)th entry and its (i,i)th entries, for $2 + [W(W-1)/2] \leq i \leq 1 + W(W-1)$, are given by λI_m. All other entries are zeros.

The Markov process has the infinitesimal generator Q given by

$$Q = \begin{bmatrix} * & \Delta_1 & \Delta_2 & \mathbf{0} & \mathbf{1} & \mathbf{2} & \mathbf{3} & \mathbf{4} & \cdots & \mathbf{N-1} & \mathbf{N} & \mathbf{N+1} \\ -\lambda & \lambda \mathbf{e}_1' & \mathbf{0} & \mathbf{0} & \mathbf{0} & \mathbf{0} & \mathbf{0} & \mathbf{0} & \cdots & \mathbf{0} & \mathbf{0} & \mathbf{0} \\ \mathbf{0} & B & 0 & C_1 & C_2 & C_3 & C_4 & C_5 & \cdots & C_N & C_{N+1} & 0 \\ \mathbf{0} & 0 & B & C_1 & C_2 & C_3 & C_4 & C_5 & \cdots & C_N & C_{N+1} & 0 \\ \mu_2 \mathbf{e}_1 \otimes \mathbf{e} & 0 & E_1 & F & G_1 & 0 & 0 & 0 & \cdots & 0 & 0 & 0 \\ \mathbf{0} & D_1 & E_2 & G_2 & H & \lambda I & 0 & 0 & \cdots & 0 & 0 & 0 \\ \mathbf{0} & D_2 & E_3 & 0 & J & H & \lambda I & 0 & \cdots & 0 & 0 & 0 \\ \mathbf{0} & D_3 & E_4 & 0 & 0 & J & H & \lambda I & \cdots & 0 & 0 & 0 \\ \cdot & \cdot & \cdot & \cdot & \cdot & \cdot & \cdot & \cdot & \cdots & \cdot & \cdot & \cdot \\ \mathbf{0} & D_{N-1} & E_N & 0 & 0 & 0 & 0 & 0 & \cdots & H & \lambda I & 0 \\ \mathbf{0} & D_N & E_{N+1} & 0 & 0 & 0 & 0 & 0 & \cdots & J & H + L & K \\ \mathbf{0} & D_{N+1} & 0 & 0 & 0 & 0 & 0 & 0 & \cdots & 0 & U & V \end{bmatrix}$$

3. STEADY-STATE EQUATIONS

In this section, we discuss the computation of steady-state probabilities by exploiting the special structure of the generator Q. The row vector x partitioned in the form $\mathbf{x} = [x^*, \underline{\xi}_1, \underline{\xi}_2, \mathbf{x}(0), \mathbf{x}(1), \ldots, \mathbf{x}(N+1)]$ according to the states $*$, Δ_1, Δ_2, **0**, **1**, ..., **N + 1** is the unique solution of the system

$$\mathbf{x}Q = 0, \qquad \mathbf{xe} = 1 \tag{3}$$

In view of the high order of the matrix Q, it is essential to exploit its special structure to evaluate the components of $\mathbf{x}$. The equations in (3) may now be written as

$$-\lambda x^* + \mu_2 x(0)(e_1 \otimes e) = 0 \tag{4}$$

$$\lambda x^* e_i' + \underline{\xi}_1 B + \sum_{\nu=1}^{N+1} x(\nu) D_\nu = \mathbf{0} \tag{5}$$

$$\underline{\xi}_2 B + \sum_{\nu=0}^{N} x(\nu) E_{V+1} = \mathbf{0} \tag{6}$$

$$(\underline{\xi}_1 + \underline{\xi}_2)C_1 + x(0)F + x(1)G_2 = \mathbf{0} \tag{7}$$

$$(\underline{\xi}_1 + \underline{\xi}_2)C_2 + x(0)G_1 + x(1)H + x(2)J = \mathbf{0} \tag{8}$$

$$(\underline{\xi}_1 + \underline{\xi}_2)C_{i+1} + \lambda x(i-1) + x(i)H + x(i+1)J = \mathbf{0}, \quad 2 \le i \le N-1 \tag{9}$$

$$(\underline{\xi}_1 + \underline{\xi}_2)C_{N+1} + \lambda x(N-1) + x(n)(H+L) + x(N+1)U = \mathbf{0} \tag{10}$$

$$x(N)K + x(N+1)V = \mathbf{0} \tag{11}$$

with the normalizing equation

$$x^* + \underline{\xi}_1 e + \underline{\xi}_2 e + \sum_{\nu=0}^{N+1} x(\nu)e = 1 \tag{12}$$

We further partition the components of x as $\underline{\xi}_1 = (\underline{\xi}_{11}, \underline{\xi}_{12}, \ldots, \underline{\xi}_{1,N+1})$, $\underline{\xi}_{1j}$, $1 \le j \le N+1$, are of order W; $\underline{\xi}_2 = (\underline{\xi}_{20}, \underline{\xi}_{21}, \ldots, \underline{\xi}_{2,N})$, $\underline{\xi}_{2j}$, $0 \le j \le N$, are of order W; and $x(0) = (a_0, a_1, \ldots, a_{W-1})$, where $a_j = (a_{j1}, a_{j2}, \ldots, a_{j,W-j})$, for $0 \le j \le W-1$, and all the components of a_j are of order m.

$$x(i) = \left(y_0^{(i)}, y_1^{(i)}, \ldots, y_{W-2}^{(i)}, z_1^{(i)}, z_2^{(i)}, \ldots, z_{W-1}^{(i)}\right)$$

for $1 \le i \le N$, where $y_0^{(i)} = (y_{01}^{(i)}, y_{02}^{(i)}, \ldots, y_{0W}^{(i)})$.

$$y_j^{(i)} = \left(y_{j1}^{(i)}, y_{j2}^{(i)}, \ldots, y_{W-1-j}^{(i)}\right), \qquad \text{for} \quad 1 \le j \le W-2$$

$$z_j^{(i)} = \left(z_{j1}^{(i)}, z_{j2}^{(i)}, \ldots z_{j,W-j}^{(i)}\right), \qquad \text{for} \quad 1 \le j \le W-1$$

and all the components of $y_j^{(i)}$, $0 \leq j \leq W-2$, and all the components of $z_j^{(i)}$, $1 \leq j \leq W-1$, are of order m.

$$x(N+1) = (b_0, b_1, \ldots, b_{W-2}) \qquad \text{where} \quad \mathbf{b}_j = (b_{j1}, b_{j2}, \ldots, b_{j,W-1-j})$$

for $0 \leq j \leq W-2$, and the components of b_j are of order m.

By using the special structure of the Kronecker products, which arise as coefficient matrices in equations (4)–(6), we can express x^*, $\underline{\xi}_1$, and $\underline{\xi}_2$ in terms of quantities of smaller dimensions. The final expressions are given as:

THEOREM 1 We have

$$\begin{aligned}
x^* &= (\mu_2/\lambda) a_{01} e \\
\underline{\xi}_{11} &= (\lambda+\mu_1)^{-1} \Big(\lambda x^* + \mu_2 y_{01}^{(1)} e, y_{01}^{(1)} T^\circ + \mu_2 y_{11}^{(1)} e, \ldots, \\
&\qquad y_{W-3,1}^{(1)} T^\circ + \mu_2 y_{W-2,1}^{(1)} e, y_{W-2,1}^{(1)} T^\circ \Big) \\
\underline{\xi}_{1j} &= (\lambda+\mu_1)^{-1} \Big(\lambda \underline{\xi}_{1j-1} + \Big(\mu_2 y_{01}^{(j)} e, y_{01}^{(j)} T^\circ + \mu_2 y_{11}^{(j)} e, \ldots, \\
&\qquad y_{W-3,1}^{(1)} T^\circ + \mu_2 y_{w-2,1}^{(j)} e, y_{w-2,1}^{(j)} T^\circ \Big) \Big), \qquad 2 \leq j \leq N \\
\underline{\xi}_{1,N+1} &= (\lambda/\mu_1) \underline{\xi}_{1,N} + 1/\mu_1 (\mu_2 b_{01} e, b_{01} T^\circ + \mu_2 b_{11} e, \ldots \\
&\qquad b_{w-3,1} T^\circ + \mu_2 b_{w-2,1} e, b_{w-2,1} T^\circ) \qquad\qquad (13) \\
\underline{\xi}_{20} &= (\lambda+\mu_1)^{-1} (a_{01} T^\circ + \mu_2 a_{11} e, a_{11} T^\circ + \mu_2 a_{21} e, \ldots, \\
&\qquad a_{w-2,1} T^\circ + \mu_2 a_{w-1,1} e, a_{w-1,1} T^\circ) \\
\underline{\xi}_{2j} &= (\lambda+\mu_1)^{-1} \Big(\lambda \underline{\xi}_{2j-1} + \Big(\mu_2 z_{11}^{(j)} e, z_{11}^{(j)} T^\circ + \mu_2 z_{21}^{(j)} e, \ldots, \\
&\qquad z_{w-2,1}^{(j)} T^\circ + \mu_2 z_{w-1,1}^{(j)} e, z_{w-1,1}^{(j)} T^\circ \Big) \Big), \quad 1 \leq j \leq N-1 \\
\underline{\xi}_{2N} &= (\lambda/\mu_1) \underline{\xi}_{2,N-1} + 1/\mu_1 \Big(\mu_2 z_{11}^{(N)} e, z_{11}^{(N)} T^\circ + \mu_2 z_{21}^{(N)} e, \ldots, \\
&\qquad z_{w-2,1}^{(N)} T^\circ + \mu_2 z_{w-1,1}^{(N)} e, z_{w-1,1}^{(N)} T^\circ \Big)
\end{aligned}$$

Once again, by exploiting the special structure of the coefficient matrices appearing in equations (7)–(11), we will now recast these equations into a form that makes them appropriate for solution by block Gauss–Seidel iteration. We evaluate the four inverses

$$A_1 = [(\lambda+\mu_2)I - T]^{-1} \qquad A_2 = [(\lambda+\mu_1+\mu_2)I - T\}^{-1}$$
$$A_3 = [\mu_2 I - T]^{-1} \qquad A_4 = [(\mu_1+\mu_2)I - T\}^{-1}$$

that are needed for the iterative procedure. In Theorem 3.2.1 of [5], it is shown that the above inverses do indeed exist and are nonnegative. The final equations are given in [2]. These equations are solved by Gauss–Seidel iteration. The successive solution vectors are kept within a compact polytope by forcing them to satisfy the normalizing equation (12). The successive vectors $\underline{\xi}_1$ and $\underline{\xi}_2$ and the successive scalar x^* are easily computed from equation (13).

3.1 Accuracy Checks

Algorithms for general PH-distributions have a powerful accuracy check, which is based on the following property. If T is an irreducible, stable matrix with eigenvalue of maximum real part $-c < 0$, and if $\underline{\alpha}$ is chosen to be the corresponding left eigenvector, normalized by $\underline{\alpha}e = 1$, then $\underline{\alpha}$ is a positive vector and the PH-distribution with the representation $(\underline{\alpha}, T)$ is the exponential distribution with parameter c. First we obtained the numerical solution for the exponential distributions in their simple form. Next, we implemented the general algorithm, but chose the representation of the PH-distribution so that *it was in fact exponential.* The general algorithm does not utilize this fact in any manner, but the two sets of numerical results agreed very much.) A number of other accuracy checks are also available. For example, the result of Theorem 2 below may be used to check whether the computed quantities satisfy it. We conclude this section with the following theorem.

THEOREM 2 We have, for $1 \leq j \leq N$,

$$\lambda\left[x(j-1)e + \underline{\xi}_{1j}e + \underline{\xi}_{2,j-1}e\right] = \mu_1\left[y^{(j)}e - y^{(j)}_{0W}e\right] - \mu_2\sum_{r=0}^{W-2} y^{(j)}_{r1}e + \sum_{r=0}^{W-2} y^{(j)}_{r1}T^{\circ}$$

and

$$\lambda\left[y^{(N)}e - y^{(N)}_{0W}e\right] = \mu_1 x(N+1)e + \mu_2\sum_{r=0}^{W-2} b_{r1}e + \sum_{r=0}^{W-2} b_{r1}T^{\circ}$$

Proof. We shall outline the proof for $j = 1$ and $j = 2$ as the others are similar. Postmultiplying each one of the equations corresponding to state 0 by e and adding the resulting equations and using equations (13), we get the equation for $j = 1$. Now postmultiplying each one of the equations corresponding to state 1 by e and by adding the resulting equations and by using the equation obtained for $j = 1$ and the equations (13), we get the equation for $j = 2$.

4. MEASURES OF SYSTEM PERFORMANCE

Once the steady-state probability vector x has been computed, several interesting system performance measures, useful in the interpretation of the physical behavior of the model, may be derived. Below, we list a number of such measures along with the corresponding formulas, needed in their derivation.

4.1 Throughput of the System

The throughput γ of the system is defined as the rate at which the messages are successfully transmitted and is given by

$$\gamma = \lambda\left[1 - \underline{\xi}_{1,N+1}e - \underline{\xi}_{2n}e - \sum_{j=1}^{W-1} z_j^{(N)}e - y_{0W}^{(N)}e - x(N+1)e\right]$$
$$= \mu_2[1 - x^* - \underline{\xi}_1 e - \underline{\xi}_2 e] \tag{14}$$

4.2 Overflow Probability

The overflow probability η is the probability that an arriving message finds the primary buffer full. For Poisson arrivals, this is also the probability that at an arbitrary time the primary buffer is full. It is given by

$$\eta = \left[\underline{\xi}_{1,N+1}e + \underline{\xi}_{2N}e + \sum_{j=1}^{W-1} z_j^{(N)}e + y_{0W}^{(N)}e + x(N+1)e\right] \tag{15}$$

4.3 Fraction of Time the Transmitter Is Busy

We denote by P_{new} and P_{old}, respectively, the proportion of time the transmitter is transmitting a new or old message. These quantities are given by

$$P_{\text{new}} = \underline{\xi}_1 e + x(N+1)e + \sum_{i=1}^{N}\left[\sum_{j=1}^{W-1} y_{0j}^{(i)}e + \sum_{r=1}^{W-2} y_r^{(i)}e\right]$$

$$P_{\text{old}} = \underline{\xi}_2 e + \sum_{j=1}^{W-1} a_j e + \sum_{i=1}^{N}\sum_{j=1}^{W-1} z_j^{(i)}e \tag{16}$$

4.4 Fraction of Time the Transmitter Is Idle

The fraction of time the transmitter is idle is given by

$$P_{\text{idle}} = x^* + \sum_{i=1}^{N} y_{0W}^{(i)} e + a_0 e \tag{17}$$

4.5 Fraction of Time a Primary Message Is Blocked

The fraction of time a primary (new) message is blocked by the transmission of a secondary (old) message is given by

$$P_{\text{block}} = \sum_{j=1}^{N} \underline{\xi}_{2j} e + \sum_{i=1}^{N} \sum_{j=1}^{W-1} z_j^{(i)} e \tag{18}$$

4.6 Density of Number of Transmissions Required

The probability of density $\{\chi_n\}$ of the number of transmissions required to successfully transmit a message is given by

$$\chi_n = (1-p)p^{n-1}, \qquad n \geq 1 \tag{19}$$

where

$$p = 1 - \underline{\alpha}\mu_2[\mu_2 I - T]^{-1} e \tag{20}$$

is the probability that a timeout period expires before an acknowledgment arrives.

4.7 Density of Number in the Primary Buffer at an Arbitrary Time

The marginal stationary probability density (φ_i) of the number in the primary buffer at an arbitrary time is given by

$$\varphi_0 = x^* + \underline{\xi}_{11} e + \underline{\xi}_{20} e + x(0)e + \sum_{j=1}^{W-2} y_j^{(1)} e + \sum_{j=1}^{W-1} y_{0j}^{(1)} e$$

$$\varphi_i = \underline{\xi}_{1,i+1} e + \underline{\xi}_{2i} e + y_{0W}^{(i)} e + \sum_{j=1}^{W-1} z_j^{(i)} e$$

$$+ \sum_{j=1}^{W-2} y_j^{i+1} e + \sum_{j=1}^{W-1} y_{0j}^{(i+1)} e \qquad 1 \leq i \leq N-1$$

$$\varphi_N = \underline{\xi}_{1,N+1}e + \underline{\xi}_{2N}e + y_{0W}^{(N)}e + \sum_{j=1}^{W-1} z_j^{(N)}e + x(N+1)e$$

4.8 Density of Number in the Secondary Buffer at an Arbitrary Time

The marginal stationary probability density $\{\vartheta_r\}$ of the number in the secondary buffer at an arbitrary time is given by

$$\vartheta_0 = x^* + \underline{\xi}_1 e + \underline{\xi}_2 e$$

$$\vartheta_r = \sum_{j=0}^{W-r} a_{j,r}e + \sum_{j=0}^{W-1-r} \sum_{i=1}^{N} y_{j,r}^{(i)}e + \sum_{j=1}^{W-r} \sum_{i=1}^{N} z_{j,r}^{(i)}e$$
$$+ \sum_{j=0}^{W-1-r} b_{j,r}e \qquad 1 \le r \le W-1$$

$$\vartheta_W = a_{0W}e + \sum_{i-1}^{N} y_{0W}^{(i)}e$$

4.9 Density of Number in the Auxiliary Buffer at an Arbitrary Time

The marginal stationary probability density of $\{\psi_j\}$ of the number in the auxiliary buffer at an arbitrary time is given by

$$\psi_0 = x^* + \sum_{i=1}^{N+1} \xi_{1,i,0} + \sum_{i=0}^{N} \xi_{2,i,1} + a_1 e + b_0 e + \sum_{i=1}^{N} y_0^{(i)}e + \sum_{i=1}^{N} z_1^{(i)}e$$

$$\psi_j = \sum_{i=1}^{N+1} \xi_{1,i,j} + \sum_{i=0}^{N} \xi_{2,i,j+1} + a_{j+1}e + b_j e + \sum_{i=1}^{N} y_j^{(i)}e$$
$$+ \sum_{i=1}^{N} z_{j+1}^{(i)}e \qquad 1 \le j \le W-2$$

$$\psi_{W-1} = \sum_{i=1}^{N+1} \xi_{1,i,W-1} + \sum_{i=0}^{N} \xi_{2,i,W}$$

5. NUMERICAL EXAMPLES

In order to test the algorithm proposed in this paper, a FORTRAN code was developed and tested on a large number of examples. Below we discuss a few interesting examples.

We consider the following four PH-distributions for the timeout periods.

I. E(5, 50) (21)

II. Exponential(10) (22)

III. 0.5 E(2, 20) + 0.495 Exponential(100)
+ 0.005 Exponential(1/9.01) (23)

IV. 0.9 Exponential(100) + 0.099 Exponential(10)
+ 0.001 Exponential(1/81.1) (24)

where $E(k,r)$ denotes the Erlang distribution with density

$$f(x) = (r^k/\lceil k)\, e^{-rx} x^{k-1} \qquad x \geq 0$$

While each of these PH-distributions has the same mean, 0.1, the distributions are qualitatively very different. The variances of these distributions are respectively $2 \times 10^{-3}, 10^{-2}, 809.4 \times 10^{-3}$ and 13146.716×10^{-3}.

EXAMPLE 1 Here we consider four different timeout periods, which have their respective PH-distributions given by (21)–(24). For each of these N was varied by fixing $W = 5$, $\lambda = 50$, and $\mu_1 = \mu_2 = 100$. Figures 2–7 give the graphs of the six system performance measures, the throughput, the fraction of time the transmitter is busy retransmitting the messages, the fraction of time the transmitter is idle, the fraction of time the new messages are blocked, and the mean and the standard deviation of the number of messages in the primary buffer, plotted against N, for these four PH-distributions.

Examining Figures 2–7, the following observations are in order.

1. As N increases the throughput increases, as is to be expected, to a saturation point. Any further increases in N have no effect on the throughput. For the timeout period that has a small variance, such as the one in I or II, the saturation point is attained for small values of N. However, when the timeout period has a large variance, such as the one in IV, large buffer size is needed to attain a saturation point.
2. For all values of N, the fraction of time the transmitter is busy with retransmitting messages and the fraction of time the new messages are blocked appear to increase as the variance of the timeout periods increases. On the other hand, the probability that the transmitter is idle appears to decrease as the variance of the timeout periods increases,

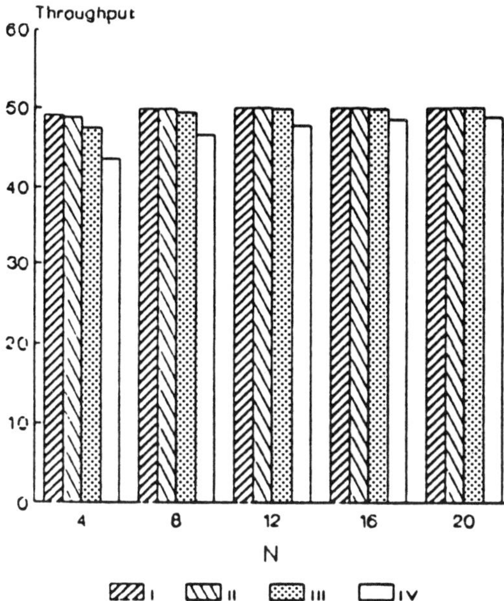

Figure 2

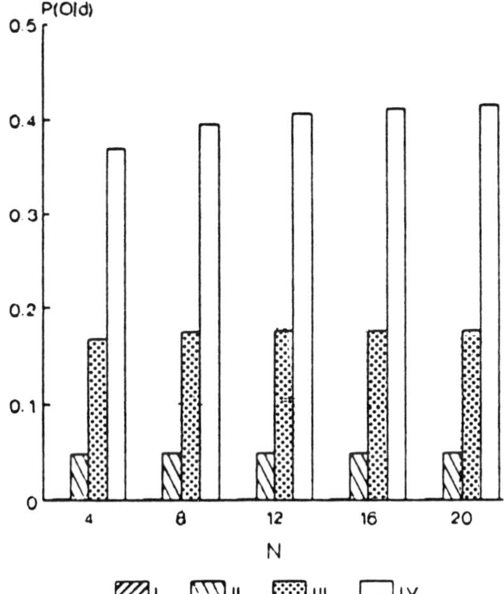

Figure 3

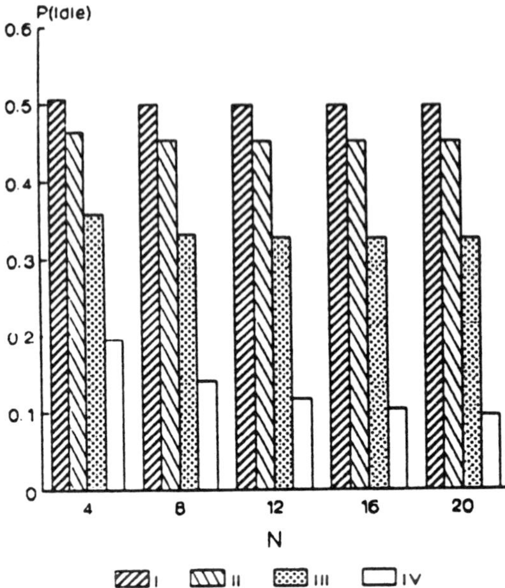

Figure 4

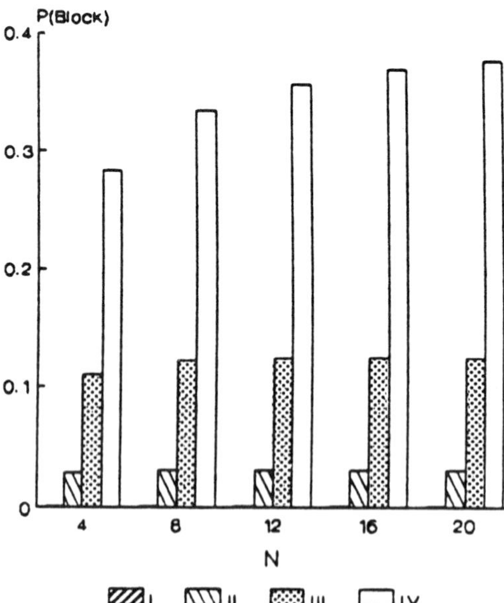

Figure 5

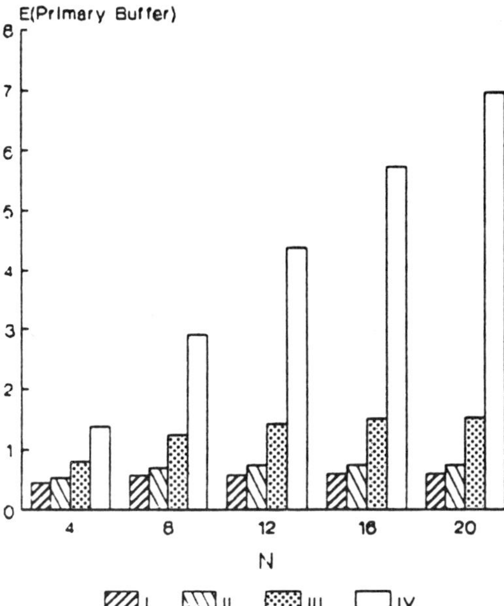

Figure 6

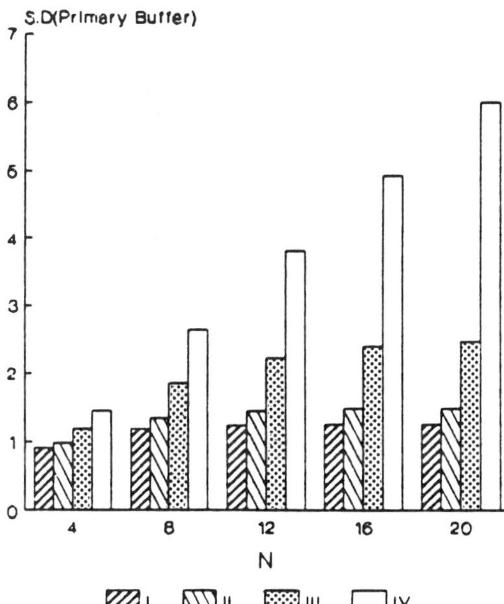

Figure 7

for all values of N. Also, this probability decreases asymptotically to a saturation point as N increases. These may be intuitively explained as follows. For a timeout period that has a large variance, there will be shorter timeout periods separated by longer ones. During longer timeout periods, there is a high probability of receiving an acknowledgment. However, during shorter timeout periods the probability that a message in the secondary buffer leaves the system is small.

3. The expected number of messages in the primary buffer appears to increase significantly as the variance of the timeout periods increases. Also, this expected number appears to increase very significantly as N increases only when the timeout periods have a large variance such as the one in IV. A similar observation also holds good for the standard deviation of the number of messages in the primary buffer.

The next example is to illustrate the expected behavior of the performance measures as functions of the arrival rate.

EXAMPLE 2 Here we again consider four different timeout periods, which have their respective PH-distributions given by (21)–(24). For each of these, the arrival rate was increased from 10 to 100 by fixing $N = 15$, $W = 5$, $\mu_1 = \mu_2 = 50$. Figures 8–13 give the graphs of the six system performance measures considered in example 1, plotted against λ, for these four PH-distributions.

An examination of the graphs in Figures 8–13 shows that:

1. As λ increases, five performance measures—the throughput, the fraction of time the transmitter is busy with retransmissions, the fraction of time the transmitter is idle, the fraction of time the new messages are blocked, and the expected number of messages in the primary buffer—asymptotically approach the corresponding saturation points.
2. The standard deviation of the number of messages in the primary buffer, for all the four timeout period distributions considered, increases initially and then decreases asymptotically to corresponding saturation points as λ increases.

REMARK Our computational experience indicates the following relationships between $P(\text{new})$, $P(\text{old})$, μ_1, γ, and the average rate of retransmissions required to transmit a message successfully:

$$P(\text{new}) = p^{-1}(1-p)P(\text{old}) \tag{25}$$

$$\gamma = \mu_1 P(\text{new}) \tag{26}$$

where γ, $P(\text{new})$, $P(\text{old})$ and p are, respectively, as given in equations (14), (16), and (20). Note that the quantity $p(1-p)^{-1}$ is the expected number

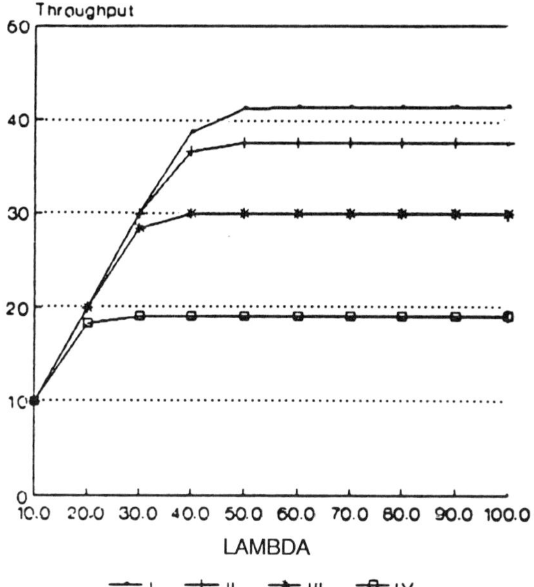

Figure 8

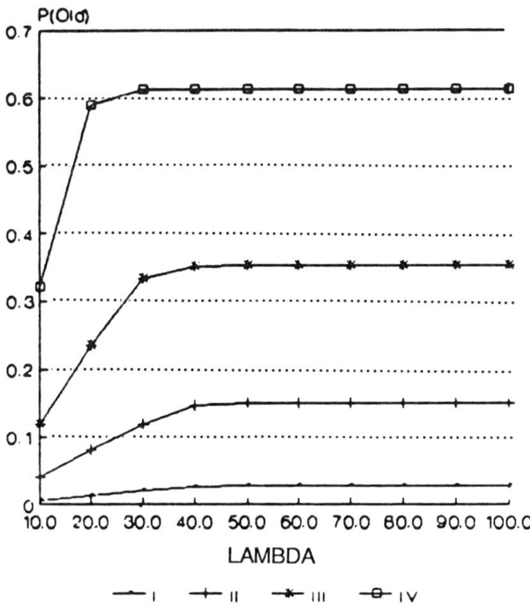

Figure 9

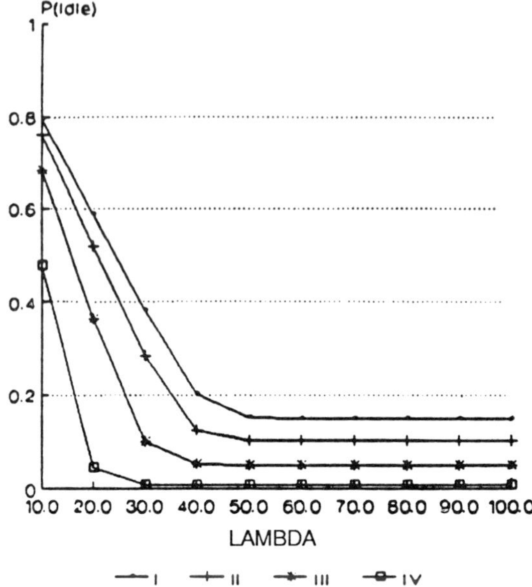

Figure 10

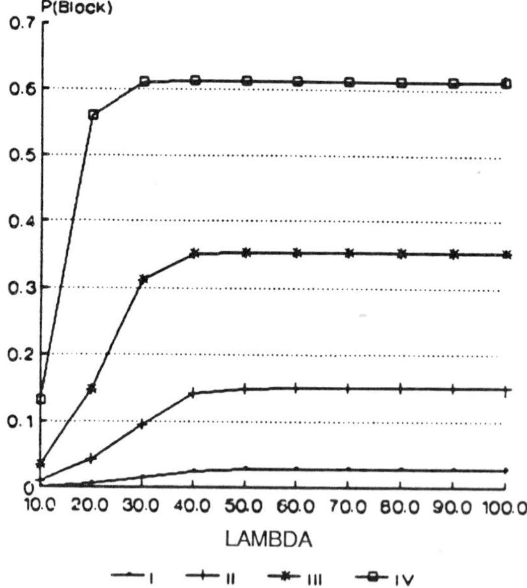

Figure 11

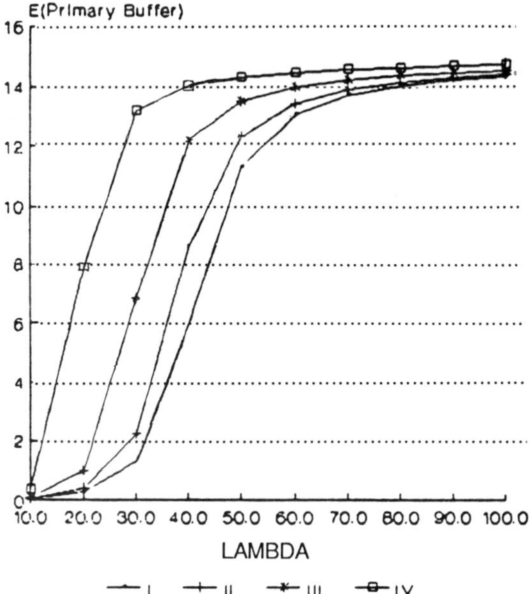

Figure 12

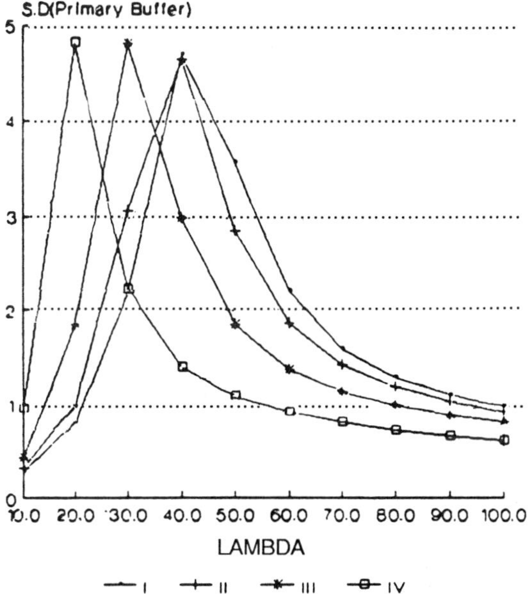

Figure 13

of retransmissions, since the density of number of transmissions required to transmit a message successfully has a geometric density with parameter p. Equation (26) is very intuitive and gives yet another expression for the throughput of the system. Combining equations (25) and (26) we get another intuitive way of looking at the throughput of the system in terms of the fraction of time the transmitter is busy with retransmitting the messages.

6. CONCLUDING REMARKS

In this paper, we considered a finite capacity queuing system to model the flow and error control mechanism at the higher layers of the OSI. We assumed the timeout periods to be random. For our model, our numerical experience supports the intuitive belief that modeling timeout periods to be almost constant yields comparatively better performance measures. However, it would be of interest to see how the performance measures behave if various other acknowledgment schemes are considered. One such scheme is as follows. The timeout period starts as soon as the secondary buffer gets a message and runs until there is an absorption in the underlying Markov process. This scheme allows the possibility of having more than one acknowledgment during a timeout period. Our present methodology can easily be carried out for this scheme. A more realistic scheme would be to have separate timeout periods for each of the messages waiting in the secondary buffer. But this scheme is considerably more complex to analyze mathematically.

REFERENCES

[1] Bellman, R. (1960). *Introduction to Matrix Analysis*, McGraw-Hill, New York.

[2] Chakravarthy, S. and Ravi, K. V. (1989). A Stochastic Model for a Computer Communication Network Node with Phase Type Timeout Periods, Technical Report, GMI Engineering and Management Institute, Flint, Mich. 48504.

[3] Marcus, M. and Minc, H. (1964). *A Survey of Matrix Theory and Matrix Inequalities*, Allyn and Bacon, Boston.

[4] Neuts, M. F. (1975). *Probability Distributions of Phase Type*, Liber Amicorum Prof. Emeritus H. Florin, Department of Mathematics, University of Louvain, Belgium, 173–206.

[5] Sethi, A. S., and Ghosal, R. (1986). *An Analytical Model for Window Mechanisms with Error Control*, Proc. INFOCOM86, 162–171.

[6] Sloman, M., and Kramer, J. (1987). *Distributed Systems and Computer Networks*, Prentice-Hall, Englewood Cliffs, N.J.

[7] Stallings, W. (1985). *Data and Computer Communications*, Macmillan, New York.

[8] Tanenbaum, A. S. (1981). *Computer Networks*, Prentice-Hall, Englewood Cliffs, N.J.

15

Numerical Comparison of the Replacement Process Approach with the Aggregation–Disaggregation Algorithm for Row-Continuous Markov Chains*

USHIO SUMITA and MARIA RIEDERS William E. Simon Graduate School of Business Administration, University of Rochester, Rochester, New York

ABSTRACT

In a recent paper by Sumita and Rieders (1990), a new algorithm has been developed for computing the ergodic probability vector for large Markov chains. Decomposing the state space into M lumps, the algorithm generates a sequence of replacement processes on individual lumps in such a way that in the limit the ergodic probability vector for a replacement process on one lump will be proportional to the ergodic probability vector of the original Markov chain restricted to that lump. In a sequel to the original paper by the same authors, this idea is applied to row-continuous Markov chains, where the underlying skip-free structure enables one to construct the replacement distributions explicitly without involving any inverse matrices. This paper discusses the numerical study of the replacement process approach for such

*This paper has been partially supported by the IBM Program of Support for Education in the Management of Information Systems and the Nippon Telegram and Telephone Research Fund.

chains in comparison with Takahashi's modified aggregation–disaggregation algorithm and a modification of Van der Heyden's algorithm. When successive substitution is employed as a means to solve systems of linear equations involved in either algorithm, the extensive numerical experiments suggest that the replacement process approach is more efficient than Takahashi's modified algorithm by a factor of three to five. Furthermore, the efficiency gap increases as the size of the Markov chain increases.

1. INTRODUCTION

Row-continuous Markov chains or quasi-birth–death processes play an important role in the performance analysis of computer and production systems. Accordingly, many algorithms have been developed for computing the ergodic probability vector (see Hajek [1], Keilson, Sumita and Zachmann [3], Keilson and Zachmann [4], Latouche, Jacobs, and Gaver [5], Neuts [6], and Ramaswami and Lucantoni [7], among others).

One of the most prevalent algorithmic methods for computing the ergodic probability vector of large-scale Markov chains with general structure is the iterative aggregation–disaggregation algorithm developed by Takahashi [12] and Takahashi and Takami [13]. In this approach, states are grouped to form a set of lumps. The aggregation procedure produces lumped probabilities, while the disaggregation procedure allocates a lumped probability to individual states in that lump. By alternately solving an aggregated system and a disaggregated system, the algorithm generates the ergodic probability vector. By exploiting the skip-free structure, the aggregation–disaggregation algorithm for row-continuous Markov chains can be simplified where the aggregation part can be totally eliminated [12, 13]. We call this algorithm Takahashi's modified aggregation–disaggregation algorithm. The aggregation–disaggregation approach has been studied subsequently by many researchers and substantial literature exists. The reader is referred to two excellent survey papers by Haviv [2] and Schweitzer [8], and references therein.

In a recent paper by Sumita and Rieders [10], a new algorithm has been developed for computing the ergodic probability vector for large Markov chains. This algorithm is based on a decomposition of the state space S into lumps:

$$\mathcal{L} = \{L(1),\ldots,L(M)\} \qquad S = \bigcup_{i=1}^{M} L(i). \tag{1}$$

Using positive probability vectors (typically a uniform distribution) on each lump in $\mathcal{L} \setminus L(1)$ as input, all states of individual lumps except $L(1)$ are aggregated. For this modified process, the replacement process on $L(1)$ is generated involving matrices of the sizes $|L(1)|$ and an inverse matrix of size $(M - 1)$, where $|L(m)|$ denotes the number of states in $L(m)$. Using the resulting ergodic probability vector of the replacement process together with the input probability vector on $L(i)$, $i = 3, \ldots, M$, the replacement process for $L(2)$ can be constructed. The algorithm proceeds in a similar manner until it converges. It is shown that in the limit the ergodic probability vectors of the replacement processes are proportional to the ergodic probability vector of the original Markov chain restricted to individual lumps. The weights for individual lumps are then easily found by manipulating a square matrix of size M.

For general Markov chains, finding the replacement distributions requires the computation of inverse matrices of size $M - 1$. For row-continuous Markov chains, however, the underlying skip-free property enables one to construct the replacement distributions explicitly without computing any inverse matrices at all. Based on this observation, a subsequent paper by the same authors [11] has developed a new algorithm for row-continuous Markov chains. In this paper, we discuss the numerical performance of the new algorithm through extensive experiments. In particular, a numerical comparison of the new algorithm with Takahashi's modified aggregation–disaggregation algorithm [12] and a modification of Van der Heyden's algorithm is discussed.

In Section 2, we briefly summarize the new algorithm from [11]. Section 3 describes the modified aggregation–disaggregation method based on Takahashi [12]. In Section 4, the numerical results are presented. Concluding remarks are given in Section 5.

2. THE REPLACEMENT PROCESS ALGORITHM FOR ROW-CONTINUOUS MARKOV CHAINS

Let $J(k)$ be an ergodic Markov chain on $S = \{1,\ldots,N\}$ governed by $\mathbf{P} = [P_{ij}]_{ij \in S}$, where $\mathbf{P}$ is of the form

$$
\mathbf{P} = \begin{bmatrix}
\mathbf{P}_1^{(0)} & \mathbf{P}_1^{+} & & & & & & & & \\
\mathbf{P}_2^{-} & \mathbf{P}_2^{(0)} & \mathbf{P}_2^{+} & & & & & & & \\
 & \mathbf{P}_3^{-} & \mathbf{P}_3^{(0)} & \ddots & & & & 0 & & \\
 & & \ddots & \ddots & \ddots & & & & & \\
 & & & \ddots & \mathbf{P}_{m-1}^{(0)} & \mathbf{P}_{m-1}^{+} & & & & \\
 & & & & \mathbf{P}_m^{-} & \mathbf{P}_m^{(0)} & \mathbf{P}_m^{+} & & & \\
 & & & & & \mathbf{P}_{m+1}^{-} & \mathbf{P}_{m+1}^{(0)} & \ddots & & \\
 & 0 & & & & & \ddots & \ddots & \ddots & \\
 & & & & & & & \ddots & \mathbf{P}_{M-1}^{(0)} & \mathbf{P}_{M-1}^{+} \\
 & & & & & & & & \mathbf{P}_M^{-} & \mathbf{P}_M^{(0)}
\end{bmatrix} \tag{2}
$$

Throughout the paper, we assume that $J(k)$ is ergodic having the ergodic probability vector $\pi = [\pi_i]_{i\in S}$. Let $\mathcal{L} = \{L(1)\ldots,L(M)\}$ be a decomposition of S, where $L(m)$ corresponds to the mth block in (2) for $m \in \mathcal{M} = \{1,\ldots,M\}$.

Given two probability vectors $\mathbf{y}_{L(m-1)}^T$ and $\mathbf{y}_{L(m+1)}^T$ on $L(m-1)$ and $L(m+1)$, respectively, a stochastic process $\tilde{J}_m^{(m)}(k)$ on $L(m)$ can be constructed by specifying its one step transition probability matrix $\mathbf{P}_m^{**}[\mathbf{y}_{L(m-1)}^T, \mathbf{y}_{L(m+1)}^T]$, where

$$
\mathbf{P}_m^{**}[\mathbf{y}_{L(m-1)}^T, \mathbf{y}_{L(m+1)}^T] = \mathbf{P}_m^{(0)} + \frac{\alpha_m^- \xi_{m-1}^{+T}}{\xi_{m-1}^{+T}\mathbf{e}_{L(m)}} + \frac{\alpha_m^+ \xi_{m+1}^{-T}}{\xi_{m+1}^{-T}\mathbf{e}_{L(m)}} \tag{3}
$$

Here,

$$
\alpha_m^- = \mathbf{P}_m^- \mathbf{e}_{L(m-1)}; \qquad \alpha_m^+ = \mathbf{P}_m^+ \mathbf{e}_{L(m+1)}; \tag{4}
$$

$$
\xi_{m-1}^{+T} = \mathbf{y}_{L(m-1)}^T \mathbf{P}_{m-1}^+; \qquad \xi_{m+1}^{-T} = \mathbf{y}_{L(m+1)}^T \mathbf{P}_{m+1}^-; \tag{5}
$$

and $\mathbf{e}_{L(m)}$ is the column vector of size $|L(m)|$ having all components equal to one.

The Markov chain $\tilde{J}_m^{(m)}(k)$ can be interpreted as the replacement process on $L(m)$ generated from a Markov chain $J_m(k)$ on $L(m) \cup \{(m-1)^*, (m+1)^*\}$ governed by

$$\hat{\mathbf{P}}_m = \begin{bmatrix} 1 - \xi_{m-1}^{+T}\mathbf{e}_{L(m)} & \xi_{m-1}^{+T} & 0 \\ \alpha_m^- & \mathbf{P}_m^{(0)} & \alpha_m^+ \\ 0 & \xi_{m+1}^{-T} & 1 - \xi_{m+1}^{-T}\mathbf{e}_{L(m)} \end{bmatrix} \tag{6}$$

Here, the state $(m-1)^*$ is the aggregated state corresponding to $L(m-1)$ and $(m+1)^*$ is the aggregated state corresponding to $L(m+1)$. It has been shown that if $\mathbf{y}_{L(m-1)}^T$ and $\mathbf{y}_{L(m+1)}^T$ are proportional to the ergodic subprobability vectors $\pi_{L(m-1)}^T$ and $\pi_{L(m+1)}^T$ of the original Markov chain, then $\mathbf{y}_{L(m)}^T$ obtained as the ergodic probability vector of $\mathbf{P}_m^{**}(\mathbf{y}_{L(m-1)}^T, \mathbf{y}_{L(m+1)}^T)$ is also proportional to $\pi_{L(m)}^T$. This motivates the following algorithm developed in [11]. The matrix $\hat{\mathbf{P}}$ below should be understood symbolically when the lump sizes are different.

ALGORITHM 2.1: REPLACEMENT PROCESS ALGORITHM FOR ROW-CONTINUOUS MARKOV CHAINS

Input: a transition probability matrix $P(N \times N)$ in the form

$$\hat{\mathbf{P}} = \begin{bmatrix} 0 & | & \mathbf{P}_1^{(0)} & | & \mathbf{P}_1^+ \\ P_2^- & | & P_2^{(0} & | & P_2^+ \\ \vdots & | & \vdots & | & \vdots \\ \mathbf{P}_{M-1}^- & | & \mathbf{P}_{\mathbf{M}-1}^{(0)} & | & \mathbf{P}_{\mathbf{M}-1}^+ \\ \mathbf{P}_{\mathbf{M}}^- & | & \mathbf{P}_{\mathbf{M}}^{(0)} & | & 0 \end{bmatrix}$$

lumping: $\mathcal{L} = \{L(1), \ldots, L(\mathcal{M})\}$

vector: $\mathbf{y}_0^T = [\mathbf{y}_{0:L(1)}^T, \ldots, \mathbf{y}_{0:L(\mathcal{M})}^T] > 0^T \qquad \mathbf{y}_{0:L(r)}^T\mathbf{e}_{L(r)} = 1 \qquad r \in \mathcal{M}$

level of accuracy: ϵ

Output: a probability vector $\mathbf{q}^T = [\mathbf{q}_{L(1)}^T, \ldots, \mathbf{q}_{L(\mathcal{M})}^T]$ upon convergence.

1. Compute $\mathcal{A} = [\alpha_m^-, \alpha_m^{(0)}, \alpha_m^+]_{m \in \mathcal{M}}$, which is $(N \times 3)$, where
$$\alpha_m^- = \mathbf{P}_m^-\mathbf{e}_{L(m-1)}, \qquad \alpha_m^{(0)} = \mathbf{P}_m^{(0)}\mathbf{e}_{L(m)} \qquad \text{and} \qquad \alpha_m^+ = \mathbf{P}_m^+\mathbf{e}_{L(m+1)}.$$
2. Set $\mathbf{y}^T = \mathbf{y}_1^T = \mathbf{y}_0^T$.
3. LOOP1: Set $m = 1$.
4. LOOP2: Compute $\xi_{m-1}^{+T} = \mathbf{y}_{L(m-1)}^T\mathbf{P}_{m-1}^+$ and $\xi_{m+1}^{-T} = \mathbf{y}_{L(m+1)}^T\mathbf{P}_{m+1}^-$.

5. Set $\zeta_{m-1}^{+T} = \xi_{m-1}^{+T}/\xi_{m-1}^{+T}\mathbf{e}_{L(m)}$ and $\zeta_{m+1}^{-T} = \xi_{m+1}^{-T}/\xi_{m+1}^{-T}\mathbf{e}_{L(m)}$.
6. Set $\mathbf{P}_m^{**} = \mathbf{P}_m^{(0)} + \alpha_{L(m)}^{-}\zeta_{m-1}^{+T} + \alpha_{L(m)}^{+}\zeta_{m+1}^{-T}$, which is $(|L(m)| \times |L(m)|)$.
7. Find $\mathbf{x}_{L(m)}^T > 0^T$ such that $\mathbf{x}_{L(m)}^T = \mathbf{x}_{L(m)}^T\mathbf{P}_m^{**}$ and $\mathbf{x}_{L(m)}^T\mathbf{e}_{L(m)} = 1$.
8. Set $\mathbf{y}_{L(m)}^T = \mathbf{x}_{L(m)}^T$.
9. If $m < M$, set $m = m + 1$ and go to LOOP2.
10. If $\|\mathbf{y} - \mathbf{y}_1\| \geq \alpha$ set $\mathbf{y}_1 = \mathbf{y}$ and go to LOOP1.
11. Compute $\mathcal{B} = [\mathbf{y}_{L(m)}^T\alpha_{L(m)}^{-}, \mathbf{y}_{L(m)}^{-}\alpha_{L(m)}^{(0)}\mathbf{y}_{L(m)}^T\alpha_m^{+}]_{m\in\mathcal{M}}$, which is $(M \times 3)$.
12. Find $\gamma^T > 0^T$ on $\mathcal{M}$ as the probability vector of the birth–death process with transition probabilities given in $\mathcal{B}$.
13. Construct $\mathbf{q}^T = [\mathbf{q}_{L(1)}^T, \ldots, \mathbf{q}_{L(M)}^T]$ by setting $\mathbf{q}_{L(m)}^T = \gamma_m\mathbf{y}_{L(m)}^T$, $m \in \mathcal{M}$.

REMARK 2.2 In Algorithm 2.1, $\mathbf{P}_m^{**}(\mathbf{y}_{L(m-1)}^T, \mathbf{y}_{L(m+1)}^T)$ is constructed first and then $\mathbf{x}_{L(m)}^T$ is obtained. In actual computation, however, it may be more efficient to implement the power method by computing

$$\mathbf{v}_2^T = \mathbf{v}_1^T\mathbf{P}_m^{(0)} + \frac{\mathbf{v}_1^T\alpha_m^{-}}{\xi_{m-1}^{+T}\mathbf{e}_{L(m)}}\xi_{m-1}^{+T} + \frac{\mathbf{v}_1^T\alpha_m^{+}}{\xi_{m+1}^{-T}\mathbf{e}_{L(m)}}\xi_{m+1}^{-T} \tag{7}$$

repeatedly until $\|\mathbf{v}_2 - \mathbf{v}_1\| < \epsilon$, and then setting $\mathbf{x}_{L(m)}^T = \mathbf{v}_2^T$. The number of divisions required for computing $\mathbf{x}_{L(m)}^T$ via equation (7) is $2|L(m)|$ if ξ_{m-1}^{+T} and ξ_{m+1}^{-T} are normalized outside the iteration. Alternatively, the two divisions may be performed within each iteration where the inner products $\mathbf{v}_1^T\alpha_m^{-}$ and $\mathbf{v}_1^T\alpha_m^{+}$ are divided by the row sums of ξ_{m-1}^{+T} and ξ_{m+1}^{-T}. For most cases, the latter seems to be more efficient.

3. TAKAHASHI'S MODIFIED AGGREGATION–DISAGGREGATION ALGORITHM

A modified aggregation–disaggregation algorithm has been proposed by Takahashi [12] for computing the ergodic probability vector for large-scale row-continuous Markov chains. The idea behind the modification is to eliminate the aggregation procedure and to simplify the disaggregation procedure by taking advantage of the skip-free structure. This algorithm for one inner loop corresponding to lump $L(m)$ is described below.

ALGORITHM 3.1: MODIFIED AGGREGATION–DISAGGREGATION ALGORITHM (ONE INNER LOOP FOR LUMP $L(M)$)
Input: two probability vectors $\mathbf{y}_{L(m-1)}^T$ on $L(m-1)$ and $\mathbf{y}_{L(m+1)}^T$ on $L(m+1)$.
Output: a probability vector $\mathbf{y}_{L(m)}^T$ on $L(m)$.

1. Find $\pi_{L(m)}^{-T}$ by solving

$$\pi_{L(m)}^{-T} = \pi_{L(m)}^{-T}\mathbf{P}_m^{(0)} + \xi_{m-1}^{+T} \tag{8}$$

2. Find $\pi_{L(m)}^{+T}$ by solving

$$\pi_{L(m)}^{+T} = \pi_{L(m)}^{+T}\mathbf{P}_m^{(0)} + \xi_{m+1}^{-T} \tag{9}$$

3. Compute $\hat{\pi}_{L(m)}^{T} = A\pi_{L(m)}^{-T} + B\pi_{L(m)}^{+T}$, where

$$A = \pi_{L(m)}^{+T}\alpha_m^{-}; B = \xi_{m-1}^{+T}\mathbf{e}_{L(m)} - \pi_{L(m)}^{-T}\alpha_m^{-}$$

4. Produce $\mathbf{y}_{L(m)}^{T}$ by $\mathbf{y}_{l(m)}^{T} = \hat{\pi}_{L(m)}^{T}/\hat{\pi}_{L(m)}^{T}\mathbf{e}_{L(m)}$.

In Algorithm 3.1, for $m = 1$, Equation (8) is ignored and $\pi_{L(1)}^{-T} = \mathbf{0}^T$, while for $m = M$, Equation (9) is ignored and $\pi_{L(M)}^{+T} = \mathbf{0}^T$. Starting with initial probability vectors on individual lumps, Algorithm 3.1 can be applied repeatedly until convergence. While Equations (8) and (9) preserve the form of the disaggregation procedure in a simplified form, the aggregation procedure is totally eliminated.

We note that $\mathbf{y}_{L(m)}^{T}$ produced by the replacement process approach is exactly the same as $\mathbf{y}_{L(m)}^{T}$ obtained from Algorithm 3.1 of Takahashi. Furthermore, $\mathbf{y}_{L(m)}^{T}$ is proportional to the subergodic vector on $L(m)$ of $J_m(k)$ governed by $\hat{\mathbf{P}}_m$ of equation (6). Consequently, the third alternative for computing $\mathbf{y}_{L(m)}^{T}$ is to apply the power method directly to $\hat{\mathbf{P}}_m$ and then to normalize the resulting subergodic vector on $L(m)$. This approach can be interpreted as a modification of Van der Heyden's algorithm reported in Haviv [2] and is implicit in Takahashi [12]. Although all three algorithms generate exactly the same probability vector on $L(m)$, the ideas behind the algorithms and their performance characteristics are quite different. For convenience, we name the three algorithms in the following manner.

RP-Algo: the replacement process algorithm.
AD-Algo: Takahashi's modified aggregation–disaggregation algorithm.
VH-Algo: the modification of Van der Heyden's algorithm.

For a complexity analysis of the three algorithms, the reader is referred to [11].

4. NUMERICAL RESULTS

For examining the efficiency and accuracy of RP-Algo, extensive numerical experiments have been conducted, comparing RP-Algo with AD-Algo and

VH-Algo. As a means to solve systems of linear equations, successive substitution is employed in all three algorithms. The purpose of this section is to report these numerical results. We begin this by describing the frame of the numerical experiments.

4.1 State Space Size and Lumping

State space size N	No. of rows M	Row size L
100	10	10
200	10	20
300	10	30
400	10	40
500	10	50
600	10	60

4.2 Sparsity Level ρ

In order to assure ergodicity, the transition structure of an almost periodic Markov chain is used as a basis for each $P_m^{(0)}$; see Figure 1. Additional transitions among the states are then generated randomly. The levels of sparsity for those submatrices are 0, 20, 40, 60, and 80%, where sparsity level 0% corresponds to completely dense matrices $\mathbf{P}_m^-$, $\mathbf{P}_m^{(0)}$, and $\mathbf{P}_m^+$, while 80% implies the matrices have only 20% nonzero entries.

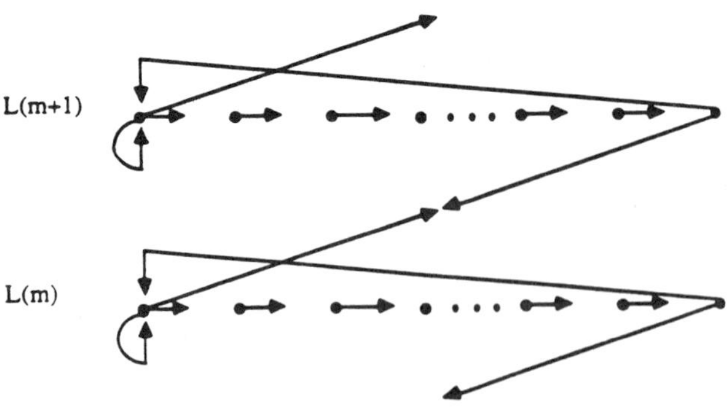

Figure 1 Basic Markov chain.

4.3 Number of Experiments

For each pair of N and sparsity level ρ, five random experiments have been conducted.

4.4 Level of Accuracy ϵ

In all experiments, the level of accuracy ϵ is $10^{-5}/N$, where N is the state space size.

4.5 Performance Measures

As performance measures for comparing the three algorithms we use the number of nonzero operations for additions, multiplications and divisions, separately. Given a common accuracy level, the three algorithms always have the same number of outer loops. However, the speed of convergence of the successive substitution for individual lump computations can be quite different, which is also of our interest.

In Table 1 the average number of outer loops is reported for each pair of sparsity level ρ and state space size N. In general, the number of outer loops increases as the sparsity level increases.

In Figures 2–4 the ratios of the number of nonzero operations for RP-Algo versus that for AD-Algo are plotted as a function of N for $\rho = 0$ through $\rho = 0.8$, for additions, multiplications and divisions, respectively. Figure 5 illustrates the corresponding ratios for the average number of iterations required for the successive substitution within lumps.

For additions and multiplications, RP-Algo dominates AD-Algo uniformly in such a way that this efficiency gap increases as the state space size increases or the sparsity level decreases. For $N = 600$, RP-Algo supersedes AD-Algo

Table 1 Average Number of Outer Loops

Sparsity level ρ	State space size N					
	100	200	300	400	500	600
0	6.0	5.2	5.0	5.0	5.0	5.0
20	7.0	6.0	6.0	6.0	5.0	5.0
40	7.6	7.0	6.0	6.0	6.0	6.0
60	8.8	8.0	7.0	7.0	6.0	6.0
80	11.0	10.2	9.0	9.0	8.0	7.0

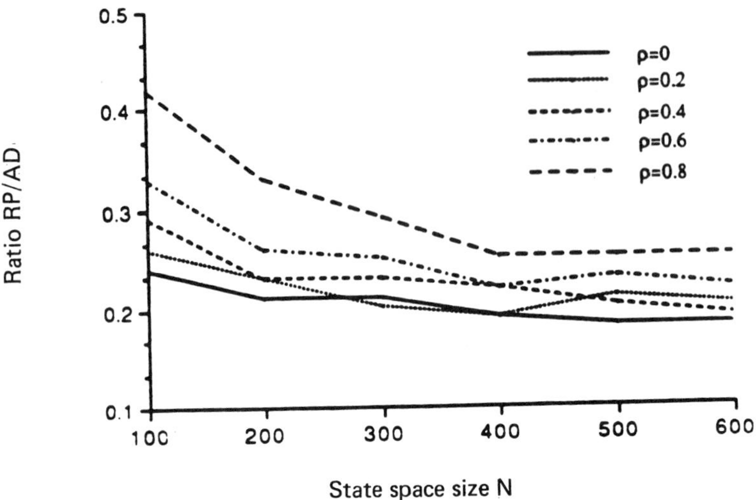

Figure 2 Number of additions: ratio RP-Algo/AD-Algo.

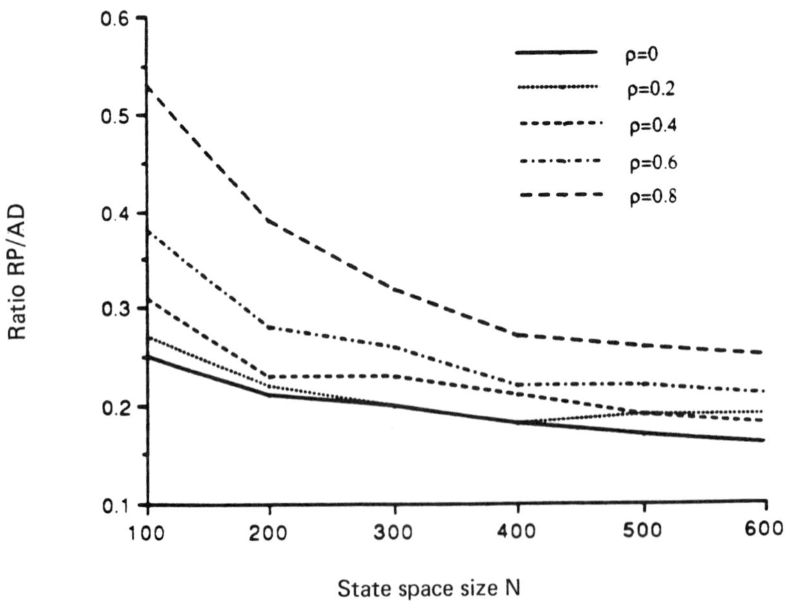

Figure 3 Number of multiplications: ratio RP-Algo/AD-Algo.

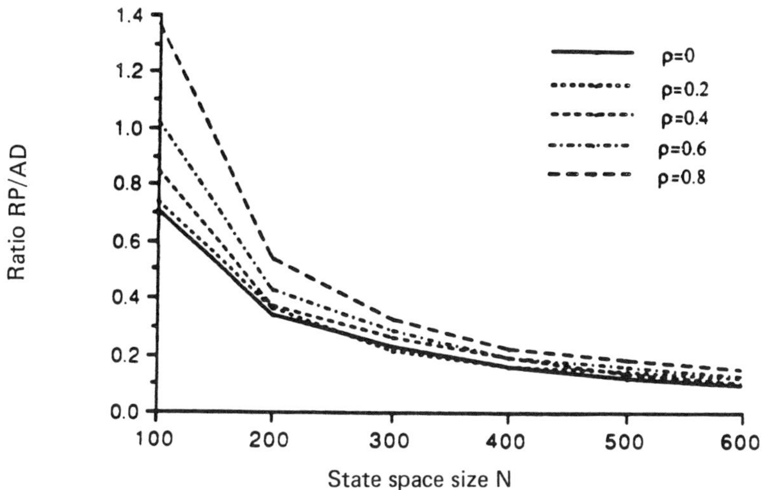

Figure 4 Number of divisions: ratio RP-Algo/AD-Algo.

by a factor of at least four. For divisions, the second approach stated in Remark 2.3 is employed for RP-Algo. When $N = 100$, the number of divisions required for RP-Algo is more or less comparable with AD-Algo. However, as N increases, RP-Algo performs much better than AD-Algo. In particular,

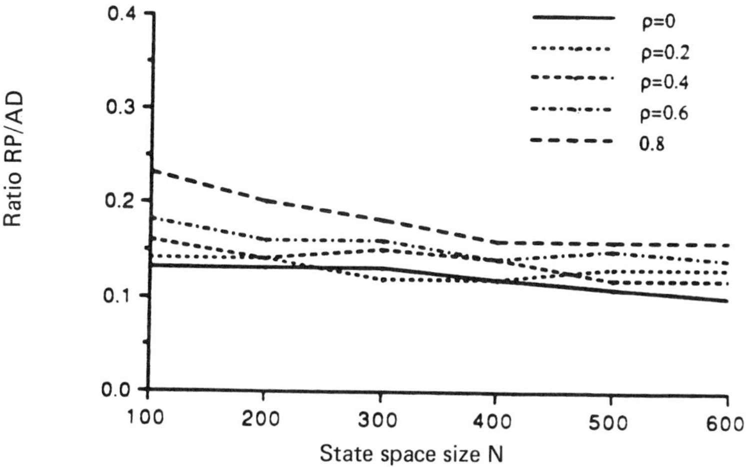

Figure 5 Average number of iterations within lumps: ratio RP-Algo/AD-Algo.

Table 2 Absolute Numbers of Nonzero Operations (given in thousands)

	Sparsity level ρ		
	0	0.4	0.8
$N = 100$			
Additions			
RP-Algo	51.9	54.6	64.8
AD-Algo	215.5	190.9	153.4
Ratio RP/AD	0.24	0.29	0.42
Multiplications			
RP-Algo	48.7	52.3	63.8
AD-Algo	195.3	166.7	119.6
Ratio RP/AD	0.25	0.31	0.53
Divisions			
RP-Algo	431.2	653.0	1514.8
AD-Algo	610.0	770.0	1110.0
Ratio RP/AD	0.71	0.85	1.37
$N = 600$			
Additions			
RP-Algo	1000.4	814.0	464.1
AD-Algo	5667.2	4278.6	1893.5
Ratio RP/AD	0.18	0.19	0.25
Multiplications			
RP-Algo	896.1	748.2	437.8
AD-Algo	5482.4	4114.2	1749.0
Ratio RP/AD	0.16	0.18	0.25
Divisions			
RP-Algo	280.0	377.2	614.8
AD-Algo	3010.0	3610.0	4210.0
Ratio RP/AD	0.09	0.10	0.15

the ratio is dominated by 0.15 for $N = 600$. The number of iterations required for the successive substitution in RP-Algo is much less than that for AD-Algo. In particular, for N greater than 400, the ratio is dominated by 0.16.

We have seen that the performance ratios of RP-Algo/AD-Algo decrease as N increases. This implies that the actual savings in absolute numbers of

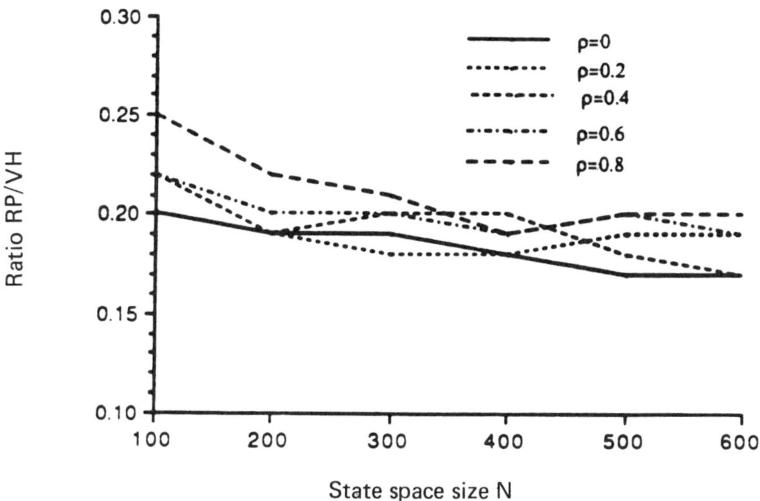

Figure 6 Number of additions: ratio RP-Algo/VH-Algo.

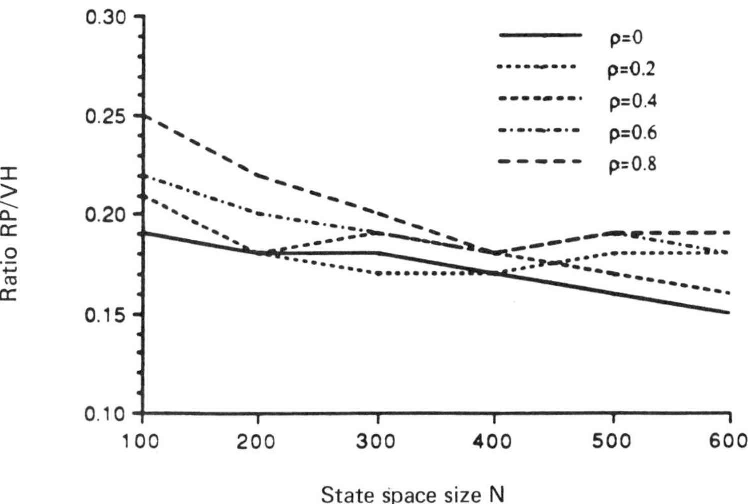

Figure 7 Number of multiplications: ratio RP-Algo/VH-Algo.

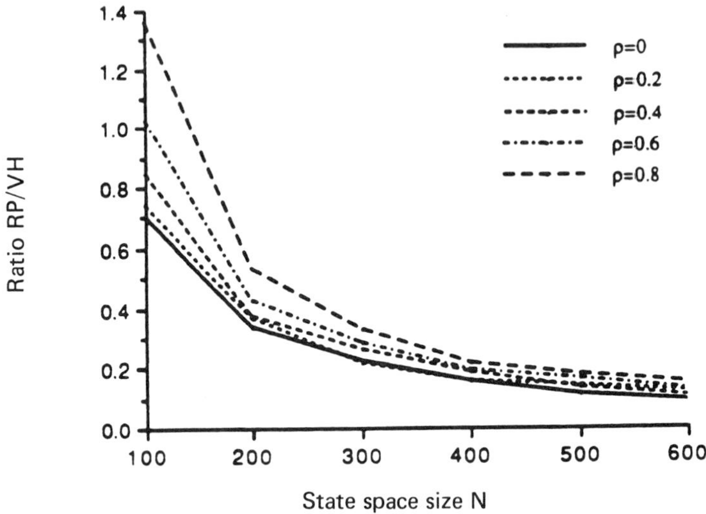

Figure 8 Number of divisions: ratio RP-Algo/VH-Algo.

nonzero operations could increase substantially. To illustrate this point, these absolute numbers are summarized for $N = 100$ and $N = 600$ in Table 2.

Figures 6–9 are similar to Figures 2–5 but for the performance comparison of RP-Algo versus VH-Algo. We observe that the numerical results here are more or less similar to the previous results. It is worth noting that AD-Algo uniformly dominates VH-Algo.

5. CONCLUDING REMARKS

The following observations can be summarized through the numerical experiments presented in this paper.

1. Among the three algorithms, RP-Algo dominates both AD-Algo and VH-Algo uniformly. AD-Algo dominates VH-Algo uniformly.
2. The performance efficiency of RP-Algo in comparison with AD-Algo and VH-Algo increases as the size of the Markov chain increases or as the sparsity level decreases, that is, as matrices become more dense.
3. For N greater than 400, RP-Algo supersedes the other two algorithms by a factor of four or more.

These numerical experiments are still far from being complete and further numerical study is needed to understand the performance characteristics of the three algorithms. For this purpose, testing the algorithms for substantially

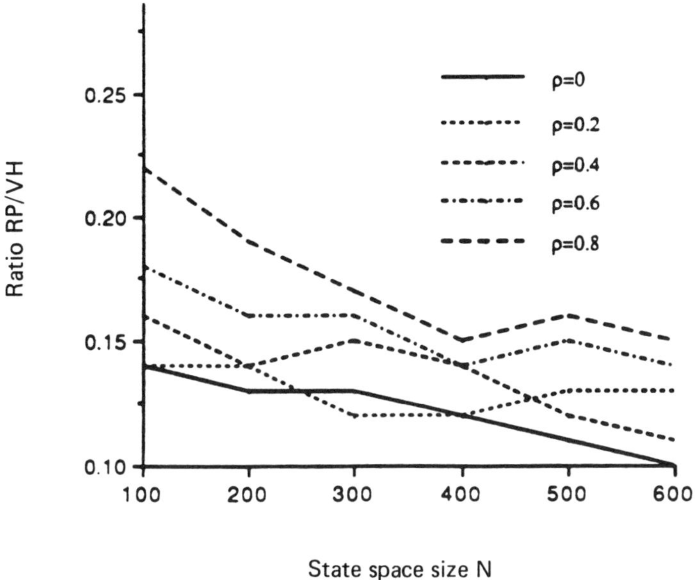

Figure 9 Average number of iterations within lump: ratio RP-Algo/VH-Algo.

larger Markov chains with specific structure (e.g., Ph/Ph/c queueing systems and tandem queueing systems) is in progress. A performance comparison of RP-Algo with other numerical algorithms (e.g., the algorithm of Seelen [9]) is underway.

REFERENCES

[1] Hajek, B. (1982), Birth and death processes on the integers with phases and general boundaries, J. Appl. Probability, 19, 488–499.

[2] Haviv, M. (1987), Aggregation/disaggregation methods for computing the stationary distribution of a Markov chain, SIAM J. Numer. Anal., 24(4), 952–966.

[3] Keilson, J., U. Sumita, and M. Zachmann (1987), Row-continuous finite Markov chains—structure and algorithms, J. Oper. Res. Society of Japan, Vol. 30, No. 3, 291–314.

[4] Keilson, J. and M. Zachmann (1988), Homogeneous row-continuous bivariate Markov chains with boundaries, J. Appl. Prob., Special Vol. No. 25a, 237–256.

[5] Latouche, D. M., P. A. Jacobs, and D. P. Gaver (1984), Finite Markov chain models skip-free in one direction, Naval Res. Logistic Q., 31, 571–588.

[6] Neuts, M. F. (1981), *Matrix-Geometric Solutions in Stochastic Models—An Algorithmic Approach*, Johns Hopkins University Press, Baltimore.

[7] Ramaswami, V., and D. M. Lucantoni (1984), Algorithms for the multiserver queue with phase type service, Technical Report, Bell Comm. Res., Inc., and AT&T Bell Labs.

[8] Schweitzer, P. (1984), Aggregation methods for large Markov chains, in *Mathematical Computer Performance and Reliability*, G. Iazeolla, P. J. Courtois, A. Hordijk, eds., Elsevier North Holland, Amsterdam, 275–286.

[9] Seelen, L. P. (1986), An algorithm for Ph/Ph/c queues, Eur. J. Operations Res., 23, 118-127.

[10] Sumita, U., and M. Rieders (1990), A new algorithm for computing the ergodic probability vector for large Markov chains: replacement process approach, Probability Eng. Info. Sci., Vol. 4, 89–116.

[11] Sumita, U., and M. Rieders (1988), Application of the replacement process approach for computing the ergodic probability vector of large-scale row-continuous Markov chains, William E. Simon School of Business Administration, University of Rochester, Working Paper Series No. QM88-10.

[12] Takahashi, Y. (1975), A lumping method for numerical calculations of stationary distributions of Markov chains, Res. Rep. No. B-18, Dept. of Information Sciences, Tokyo Institute of Technology.

[13] Takahashi, Y., and Takami, Y. (1976), A numerical method for the steady-state probabilities of a GI/G/c queueing system in a general case, J. Operations Res. Soc. Jpn., 19(2), 147–157.

16

Analysis of a Loss System with Mutual Overflow in a Markovian Environment

UDO R. KRIEGER Research Institute of the Deutsche Bundespost, Darmstadt, Germany

ABSTRACT

We investigate a loss system modeling mutual overflow between two fully available trunk groups. Assuming exponentially distributed call holding times and two independent Markov modulated Poisson processes to be offered as arrival streams, a Markovian model is derived and its steady-state distribution is computed by numerical methods. We extend Schweitzer's iterative A/D method to a larger class of regular splittings and improve Haviv's error analysis of this procedure. Moreover, formulas for the call-congestion rates of the arrival streams are derived.

1. INTRODUCTION

Recently, considerable attention has been devoted to the analysis of advanced routing schemes in nonhierarchical, circuit-switched digital networks based on efficient modern signalling systems such as CCITT CCS No. 7. Especially adaptive routing procedures that balance the effects of random traffic fluctuations within the network have been studied intensively.

In this context, a new loss system modeling mutual overflow between two fully available trunk groups has been introduced and investigated (cf. [27], [22], [23]). It describes a telecommunication network that consists of an exchange A being connected to two exchanges B and C of the long-distance network by two distinct trunk groups (see Figure 1). Each route carries peaked traffic originating in A. If one route is blocked the corresponding stream is allowed to use the other route. Hence, the traffic follows a mutual overflow routing scheme, called symmetric grading in the case of one line per group (cf. [22], [36, p. 39]).

In this paper we study the traffic behavior in this telecommunication model assuming exponentially distributed call holding times and two independent Markov modulated Poisson processes to be offered as arrival streams. Thus we extend the assumptions of previous studies (cf. [27], [22]).

We describe simple algorithms for the calculation of the relevant steady-state performance characteristics of the model. Considering such procedures for the computation of the stationary distribution of a homogeneous continuous-time Markov chain (CTMC), our analysis offers new insights into the behavior of iterative methods accelerated by aggregation–disaggregation (A/D) steps. We extend Haviv's error analysis for iterative A/D methods (cf. [16], [5]). Our results are a natural extension of Chatelin and Miranker's work [5] to the case of semiconvergent matrices.

This paper is structured as follows. In Section 2 we give a detailed mathematical description of the system by means of a Markovian model. Section 3

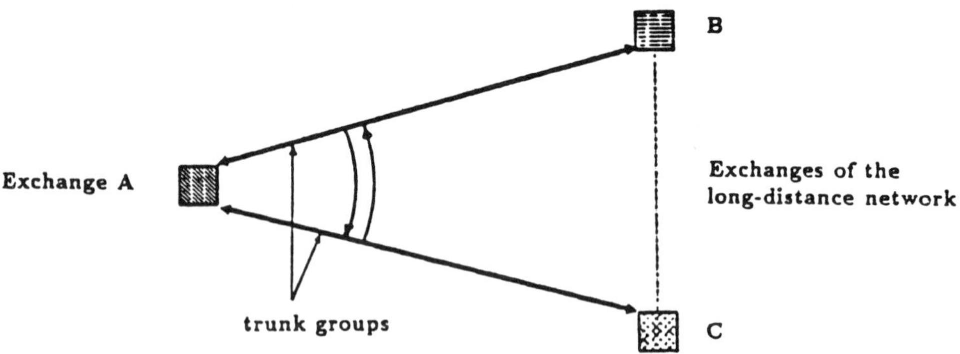

Figure 1 Network with mutual overflow routing.

provides the mathematical background for the calculation of the steady-state probability vector of a Markov chain. In Section 4 an extension of Schweitzer's A/D algorithm (cf. [40], [41]) to a larger class of iterative procedures is presented and Haviv's error analysis is improved. Furthermore, a stochastic interpretation of the block Gauss–Seidel method is provided. In Section 5 the proposed procedures are applied to the computation of the steady-state vector of the mutual overflow model. In Section 6 formulas are derived for the overall and individual call-congestion rates of the arrival streams.

2. A MATHEMATICAL MODEL OF THE OVERFLOW SYSTEM

The investigated network can be described by a loss system composed of two fully available trunk groups, called systems 1 and 2, with N_1 and N_2 lines. Following a standard approach in teletraffic theory (cf. [29], [30]), the offered traffic streams are modeled by two IPP-renewal processes resulting from a two-moment approximation of the peaked traffic streams 1 and 2. They will be represented by two mutually independent Markov modulated Poisson processes (MMPP) (see Appendix) with generator matrices

$$Q_1 = \begin{bmatrix} -\gamma_1 & \gamma_1 \\ \omega_1 & -\omega_1 \end{bmatrix} \qquad Q_2 = \begin{bmatrix} -\gamma_2 & \gamma_2 \\ \omega_2 & -\omega_2 \end{bmatrix}$$

and rate vectors $\hat{\lambda}_1 = \left[\begin{smallmatrix}\lambda_1 \\ 0\end{smallmatrix}\right]$ and $\hat{\lambda}_2 = \left[\begin{smallmatrix}\lambda_2 \\ 0\end{smallmatrix}\right]$, respectively (cf. [29], [30], [28], [26]). Here, λ_i is the intensity of the Poisson process associated with the IPP stream i, $1/\gamma_i$ its mean on-time, and $1/\omega_i$ its mean off-time, $i \in \{1, 2\}$ (cf. [26, p. 438]).

The arrival streams 1 and 2 follow a mutual overflow routing scheme. This means that upon arrival at system 1 a call of flow 1, for instance, is searching for a free line. If possible, a free trunk is selected in a random manner and occupied. If system 1 is busy and there are free lines in system 2, the incoming call from flow 1 will immediately overflow to system 2 upon arrival and occupy a line selected at random. If both systems are busy, the call will be blocked and lost without further impact on the system (lost calls cleared) (see Figure 2).

Call holding times are supposed to be mutually independent, exponentially distributed random variables with a common finite mean $1/\mu$. They are also assumed to be independent of the arrival processes.

Let us denote the phase of the controlling CTMC of the MMPP $i \in \{1, 2\}$ at time $t \geq 0$ by $Y_i(t)$. Its associated irreducible generator is Q_i. The Markovian environment resulting from the composition of both arrival streams is given

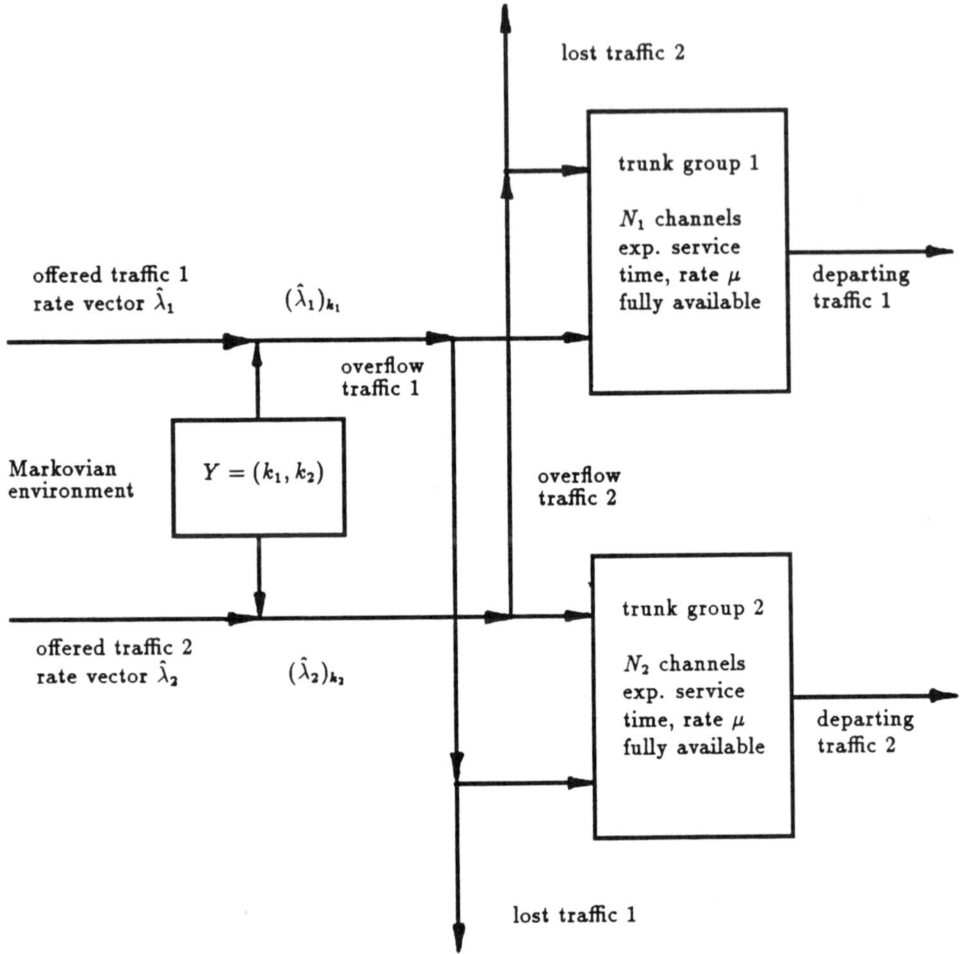

Figure 2 Queuing model of a loss system with mutual overflow in a Markovian environment.

by $Y(t) = (Y_1(t), Y_2(t))$. It possesses the irreducible generator

$$Q = Q_1 \oplus Q_2 = \begin{bmatrix} -\gamma_1 - \gamma_2 & \gamma_2 & \gamma_1 & 0 \\ \omega_2 & -\gamma_1 - \omega_2 & 0 & \gamma_1 \\ \omega_1 & 0 & -\omega_1 - \gamma_2 & \gamma_2 \\ 0 & \omega_1 & \omega_2 & -\omega_1 - \omega_2 \end{bmatrix} \in \mathbb{R}^{m \times m}$$

with $m = 4$. Its states $Y(t) = (k_1, k_2) \equiv k$ will be ordered lexicographically and enumerated by integers $k \in \{1,2,3,4\}$. The occupation of groups will be modeled by $X(t) = (X_1(t), X_2(t))$, $t \geq 0$, where the state variables $X_i(t)$ denote the number of busy trunks in the groups $i \in \{1,2\}$ at time t.

This overflow system in Markovian environment can be described by an irreducible CTMC $Z(t) = (X(t), Y(t))$, $t \geq 0$, with a finite state space $S = \{(i,j,k) \mid 1 \leq k \leq 4; 0 \leq i \leq N_1; 0 \leq j \leq N_2\}$. Its limiting distribution $P = (P_{ijk})_{i=0,\ldots,N_1; j=0,\ldots,N_2; k=1,\ldots,4}$, $P_{ijk} = \lim_{t\to\infty} P(Z(t) = (i,j,k))$, is the unique solution of the normalization condition $\sum_{i=0}^{N_1} \sum_{j=0}^{N_2} \sum_{k=1}^{4} P_{ijk} = 1$ and the balance equations

$$\begin{aligned} &[\Lambda(1-\delta_{i,N_1}\delta_{j,N_2}) + i\mu I + j\mu I - Q^t] \cdot P_{ij} \\ &= (\Lambda_1 + \Lambda_2 \delta_{j,N_2})(1-\delta_{i,0}) \cdot P_{i-1j} + (i+1)\mu(1-\delta_{i,N_1}) \cdot P_{i+1j} \\ &\quad + (\Lambda_2 + \Lambda_1 \delta_{i,N_1})(1-\delta_{j,0}) \cdot P_{ij-1} + (j+1)\mu(1-\delta_{j,N_2}) \cdot P_{ij+1} \end{aligned} \tag{1}$$

$0 \leq i \leq N_1$, $0 \leq j \leq N_2$, with $\delta_{k,l} = 1$ for $k = l$ and 0 otherwise. Here, $P_{ij}^t = (P_{ij1}, \ldots, P_{ij4})$ is the portion of the steady-state vector on the aggregate (i,j), $\Lambda_1 = \text{Diag}(\hat{\lambda}_1) \otimes I_2 = \text{Diag}(\lambda_1, \lambda_1, 0, 0)$, $\Lambda_2 = I_2 \otimes \text{Diag}(\hat{\lambda}_2) = \text{Diag}(\lambda_2, 0, \lambda_2, 0)$ and $\Lambda = \Lambda_1 + \Lambda_2$ are the arrival rate matrices. Assuming a lexicographical ordering of states, the steady-state probability vector $P^t = (P_{00}^t, P_{01}^t, \ldots, P_{0N_2}^t, \ldots, P_{N_10}^t, \ldots, P_{N_1N_2}^t)$ is the unique positive, normalized solution of the homogeneous linear system $A \cdot P = 0$. $A = -\tilde{Q}^t \in \mathbb{R}^{N \times N}$, $N = (N_1+1)\cdot(N_2+1)\cdot m$, is the negative transpose of the generator matrix $\tilde{Q}$ associated with $\{Z(t), t \geq 0\}$. It is an irreducible Q-matrix (cf. [37], [24]) having a block tridiagonal structure

$$A = \begin{bmatrix} E_0 - B & D_0 & 0 & \cdots & & \cdots & \cdots & 0 \\ B & E_1 - D_0 - B & D_1 & \ddots & & \cdots & \cdots & \vdots \\ 0 & B & \ddots & \ddots & & \ddots & \cdots & \vdots \\ \vdots & \ddots & B & E_k - D_{k-1} - B & D_k & & \ddots & \vdots \\ \vdots & \cdots & \ddots & \ddots & & \ddots & \ddots & 0 \\ \vdots & \cdots & \cdots & \ddots & & \ddots & \ddots & D_{N_1-1} \\ 0 & \cdots & \cdots & \cdots & & 0 & B & E_{N_1} - D_{N_1-1} \end{bmatrix} \tag{2}$$

with

$$\begin{aligned} B &= -(I_{N_2+1} \otimes \Lambda_1 + e_{N_2+1} \cdot e_{N_2+1}^t \otimes \Lambda_2) \\ D_i &= -(i+1)\cdot\mu\cdot I_L \qquad i = 0, \ldots N_1 - 1 \end{aligned}$$

where $L = (N_2 + 1) \cdot m$, and with irreducible block tridiagonal Q-matrices

$$E_i = \begin{bmatrix} -Q^t - G_i & T_0 & 0 & \cdots & \cdots & \cdots & 0 \\ G_i & -Q^t - T_0 - G_i & T_1 & \ddots & \cdots & \cdots & \vdots \\ 0 & G_i & \ddots & \ddots & \ddots & \cdots & \vdots \\ \vdots & \ddots & G_i & -Q^t - T_{k-1} - G_i & T_k & \ddots & \vdots \\ \vdots & \cdots & \ddots & \ddots & \ddots & \ddots & 0 \\ \vdots & \cdots & \cdots & \ddots & \ddots & \ddots & T_{N_2-1} \\ 0 & \cdots & \cdots & \cdots & 0 & G_i & -Q^t - T_{N_2-1} \end{bmatrix}$$

for $i = 0, \ldots N_1$ of order L along its diagonal, where

$$G_i = -(\Lambda_2 + \Lambda_1 \delta_{i,N_1})$$
$$T_j = -(j + 1) \cdot \mu \cdot I_m \qquad j = 0, \ldots N_2 - 1.$$

Here I_k is the identity matrix of order k and e_k the kth unit vector. Thus, A is a 2-cyclic consistently ordered Q-matrix with respect to this block partition (cf. [24], [45], [37], [3]). Taking advantage of the block structure of A, the steady-state vector P may be computed by a block iterative scheme derived from an R-regular splitting of A or an accelerated point iterative scheme such as JOR or SOR (cf. [24], [37], [30]).

3. MATHEMATICAL FOUNDATION OF COMPUTATIONAL METHODS FOR MARKOVIAN MODELS

In this section we provide the mathematical background for the calculation of the steady-state distribution p of a CTMC with finite state space $S = \{1, \ldots, n\}$ and irreducible generator matrix $Q \in \mathbb{R}^{n \times n}$. We know that p is the unique positive solution of the linear system

$$A \cdot x = 0 \tag{3}$$

which satisfies additionally the normalization condition $e^t \cdot x = 1$. Here e denotes the vector with all ones and $A = -Q^t$ is the irreducible Q-matrix associated with Q (cf. [37], [24]).

In the following we adopt the notation of Berman, Plemmons [3, Chap. 2, p. 26] w.r.t. vector and matrix orderings: Let $x \in \mathbb{R}^n$, then $x \gg 0 \Leftrightarrow x_i > 0$ for each $i \in \{1, \ldots, n\}$, $x > 0 \Leftrightarrow x_i \geq 0$ for each $i \in \{1, \ldots, n\}$ and $x_j > 0$ for some $j \in \{1, \ldots, n\}$, $x \geq 0 \Leftrightarrow x_i \geq 0$ for each $i \in \{1, \ldots, n\}$.

Solving the system (3), we may employ either direct or iterative methods. The latter are usually based on a regular matrix splitting of the form $A = M - N$, $M \in \mathbb{R}^{n \times n}$, $N \in \mathbb{R}^{n \times n}$ (cf. [39, Def. 2.3, p. 410]). The associated nonnegative iteration matrix is $J = M^{-1} \cdot N$.

It is known that every regular splitting of an irreducible singular M-matrix is also graph compatible, weak regular (cf. [39, Def. 2.3, p. 410]). Thus we conclude from [39, Theorem 4.4, p. 420] and [3, Theorem 6.4.16, p. 146] that the spectral radius $\rho(J) = 1$ is a simple eigenvalue of J, especially $\text{index}_1(J) = 1$. Given a positive initial vector $x^{(0)}$, the convergence of the sequence

$$x^{(k+1)} = J \cdot x^{(k)} \qquad k = 0, 1, \ldots$$

to a solution of the eigenvalue problem

$$x = J \cdot x \tag{4}$$

is guaranteed if J is semiconvergent (cf. [3, p. 197]), that is, if J has no further eigenvalues on the unit circle apart from $\rho(J) = 1$. As this condition may be violated, it is necessary to enforce semiconvergence by proceeding to the extrapolated iteration matrix

$$J_\omega = (1 - \omega)I + \omega J \qquad 0 < \omega < 1$$

It is well known that the sequence $\{x^{(k)} : k \in \mathbb{N}_0\}$ defined by the iterative scheme

$$x^{(k+1)} = J_\omega \cdot x^{(k)} \qquad k = 0, 1, \ldots, \tag{5}$$

called stationary first-order Richardson extrapolation, converges to a positive solution of equations (4) and (3), respectively, provided the initial vector $x^{(0)}$ is positive (cf. [2, p. 173]). The unique normalized solution coincides with p.

The convergence of the procedure (5) may be accelerated by inserting some aggregation–disaggregation (A/D) steps during the iteration (cf. [40], [41], [16], [4], [5], [34]). In order to apply Schweitzer's convergence result [41, Theorem 4, p. 328] in this context, we have to define a fallback procedure $x^{(k+1)} = T \cdot x^{(k)}$. It is based on a semiconvergent stochastic matrix T that converges to the normalized eigenvector x^* corresponding to the eigenvalue $\rho(T) = 1$ which is related to the stationary distribution p by some transformation (see equation (6); cf. [33, p. 126]).

Usually, neither J nor J_ω is stochastic. In order to construct a stochastic iteration matrix, we proceed to a nonnegative matrix T, called dual iteration matrix, by a similarity transformation:

$$T = M \cdot J \cdot M^{-1} = N \cdot M^{-1}$$

$$T_\omega = M \cdot J_\omega \cdot M^{-1} = (1 - \omega)I + \omega T \qquad 0 < \omega < 1$$

Then T_ω, $0 < \omega \leq 1$, is column stochastic and $\|T_\omega\|_1 = \rho(T_\omega) = 1$ as well as $\text{index}_1(T_\omega) = 1$ hold. Furthermore, T_ω, $0 < \omega < 1$, is semiconvergent and $\rho(T_\omega) = 1$ is a simple eigenvalue. For all vectors $x > 0$ we conclude $T_\omega \cdot x > 0$.

PROPOSITION 1 Let $Q \in \mathbb{R}^{n \times n}$ be an irreducible generator matrix of a CTMC with steady-state distribution $p \gg 0$. Set $A = -Q^t$. Given a nontrivial regular splitting $A = M - N$, we denote the dual iteration matrix by $T = N \cdot M^{-1} \geq 0$ and set $T_\omega = (1 - \omega)I + \omega T$ for some $0 < \omega < 1$. Then the sequence

$$x^{(k+1)} = T_\omega \cdot x^{(k)} \qquad k = 0, 1, \ldots$$

converges to a vector $x^* > 0$ with $e^t \cdot x^* = 1$, provided the initial vector $x^{(0)}$ satisfies $x^{(0)} > 0$, $e^t \cdot x^{(0)} = 1$. In this case the relation

$$p = \frac{M^{-1} \cdot x^*}{e^t \cdot M^{-1} \cdot x^*} \tag{6}$$

holds.

4. THE A/D ACCELERATED DUAL ITERATION

Let $T_\omega \geq 0$, $0 < \omega < 1$, be a stochastic, extrapolated dual iteration matrix corresponding to a regular splitting of the irreducible Q-matrix $A = M - N \in \mathbb{R}^{n \times n}$. Subsequently the subscript ω will be omitted. Let $x^* > 0$ denote the unique normalized eigenvector corresponding to the spectral radius $\rho(T) = 1$.

In this section we show that Schweitzer's iterative A/D procedure (IAD) and the corresponding convergence result [41, Theorem 4, p. 328] may be applied beyond point or block Jacobi and Gauss–Seidel splittings to a larger class of regular splittings. First we extend Haviv's error analysis of the IAD method (cf. [16]).

4.1 Error Analysis

We choose a partition $\Gamma = \{J_1, \ldots, J_m\}$ of the state space $S = \{1, \ldots, n\}$ into $m \geq 2$ disjoint sets J_i with $n_i \geq 1$ elements each. Without loss of generality we assume the elements of these sets to be enumerated in a consecutive order such that $i < j$ holds if $i \in J_l$, $j \in J_k$, $l < k$. Furthermore, w.l.g. let T and $x^{*t} = (x_1^{*t}, \ldots, x_m^{*t})$ be arranged according to this state space partition and ordering.

Following the approach of Chatelin and Miranker (cf. [5], [16], [4]), we define an aggregation matrix $R \in \mathbb{R}^{m \times n}$ by

$$R_{ij} = \begin{cases} 1 & \text{if } j \in J_i \\ 0 & \text{otherwise} \end{cases} \qquad 1 \le i \le m, 1 \le j \le n \tag{7}$$

For a fixed vector $x = \begin{bmatrix} x_1 \\ \vdots \\ x_m \end{bmatrix} > 0$ with $e^t \cdot x = 1$ the prolongation matrix $P_{(x)} \in \mathbb{R}^{n \times m}$ is given by

$$P_{(x)_{ij}} = \begin{cases} (y_j)_i & \text{if } i \in J_j \\ 0 & \text{otherwise} \end{cases} \qquad 1 \le i \le n, 1 \le j \le m \tag{8}$$

where the vector $y = y_{(x)} = \begin{bmatrix} y_1 \\ \vdots \\ y_m \end{bmatrix} > 0$ is defined for $j \in \{1, \ldots, m\}$ as follows:

$$\mathbb{R}^{n_j} \ni y_{(x)_j} = \begin{cases} x_j / \alpha_{(x)_j} & \text{if } x_j > 0 \\ 1/n_j \cdot e & \text{if } x_j = 0 \end{cases}$$

$$\alpha_{(x)_j} = e^t \cdot x_j$$

According to the construction, the relations $e^t \cdot y(x)_j = 1$ and $x_j = \alpha_{(x)_j} \cdot y_{(x)_j}$ hold for each $j \in \{1, \ldots, m\}$. Also, $x > 0$, $e^t \cdot x = 1$ implies $e^t \cdot \alpha = 1$ for $\alpha = \alpha_{(x)} = (\alpha_{(x)_1}, \ldots, \alpha_{(x)_m})^t > 0$. Hence, the matrices

$$R = \begin{bmatrix} e^t & 0 & \ldots & 0 \\ 0 & e^t & \ddots & \vdots \\ \vdots & \ddots & \ddots & 0 \\ 0 & \ldots & 0 & e^t \end{bmatrix} > 0 \qquad P_{(x)} = \begin{bmatrix} y_1 & 0 & \ldots & 0 \\ 0 & y_2 & \ddots & \vdots \\ \vdots & \ddots & \ddots & 0 \\ 0 & \ldots & 0 & y_m \end{bmatrix} > 0$$

satisfy $e^t \cdot P_{(x)} = e^t$, $e^t \cdot R = e^t$, and $R \cdot P_{(x)} = I$. For fixed $x > 0$ we define the nonnegative projection matrix by

$$\Pi = \Pi_{(x)} = P_{(x)} \cdot R = \begin{bmatrix} y_1 e^t & 0 & \ldots & 0 \\ 0 & y_2 e^t & \ddots & \vdots \\ \vdots & \ddots & \ddots & 0 \\ 0 & \ldots & 0 & y_m e^t \end{bmatrix} \in \mathbb{R}^{n \times n} \tag{9}$$

Then Π is column stochastic, that is, $e^t \cdot \Pi = e^t$. Furthermore, $R \cdot x = \alpha_{(x)}$ and $P_{(x)} \cdot \alpha_{(x)} = x$ induce $\Pi_{(x)} \cdot x = x$. $\Pi_{(x)} \cdot \Pi_{(x)} = \Pi_{(x)}$ implies that the matrix $\Pi_{(x)}$ is a projection onto the subspace $\Pi_{(x)}(\mathbb{R}^n)$ along $\mathrm{Kern}(\Pi_{(x)})$ (cf. [5, p. 20]).

Defining the projection of the dual iteration matrix T by

$$G = G_{(x)} = \Pi_{(x)} \cdot T$$

we see that $G \geq 0$ is a stochastic matrix satisfying $\rho(G) = 1$ and $\mathrm{index}_1(G) = 1$. Hence, the group inverse $(I - G)^{\#} = (I - \Pi_{(x)} \cdot T)^{\#}$ corresponding to $I - G$ exists (cf. [31, p. 445], [3, p. 119]). Obviously, the related projection matrix

$$F = F_{(x)} = T \cdot \Pi_{(x)}$$

is a stochastic matrix, too. Denoting the aggregated iteration matrix by

$$B = B_{(x)} = R \cdot T \cdot P_{(x)} \in \mathbb{R}^{m \times m} \tag{10}$$

it follows that $B \geq 0$ is stochastic. Hence, there exists a vector $\alpha_{(x)} > 0$ in $\mathbb{R}^m$ satisfying

$$B_{(x)} \cdot \alpha_{(x)} = \alpha_{(x)} \tag{11}$$

and $e^t \cdot \alpha_{(x)} = 1$. The disaggregated vector $\tilde{x}_{(x)} \in \mathbb{R}^n$ defined by

$$\tilde{x}_{(x)} = P_{(x)} \cdot \alpha_{(x)} \geq 0 \tag{12}$$

fulfills $e^t \cdot \tilde{x}_{(x)} = 1$ and $\Pi_{(x)} \cdot \tilde{x}_{(x)} = P_{(x)} \cdot \alpha_{(x)} = \tilde{x}_{(x)}$. Thus, there exists a nonnegative vector $\tilde{x}^* \in \Pi_{(x)}(\mathbb{R}^n)$ with $\Pi_{(x)} \cdot (I - T) \cdot \Pi_{(x)} \cdot \tilde{x}^* = 0$ and $e^t \cdot \tilde{x}^* = 1$. This solution approach may be interpreted as a Galerkin approximation in $\Pi_{(x)}(\mathbb{R}^n)$ (cf. [5, p. 20ff]). Obviously, the solution $\tilde{x}_{(x)}$ fulfills

$$(I - \Pi_{(x)} \cdot T) \cdot \tilde{x}_{(x)} = 0$$

As x^* is the unique nonnegative, normalized eigenvector of T corresponding to the spectral radius $\rho(T) = 1$, we conclude that

$$(I - \Pi_{(x)} \cdot T) \cdot (x^* - \tilde{x}_{(x)}) = (I - \Pi_{(x)}) \cdot (x^* - x) \tag{13}$$

holds (cf. [16, p. 954], [5, (4.3), p. 31]).

An A/D step consists of an aggregation step (11) followed by a disaggregation step (12). Equation (13) provides a simple relationship between the errors before an A/D step, $\epsilon = x^* - x$, and after an A/D step, $\tilde{\epsilon} = x^* - \tilde{x}_{(x)}$.

The following proposition reveals a relation between the steady-state vectors of the stochastic matrices G, F, and B. Here, Kern denotes the nullspace of a linear operator.

PROPOSITION 2 Let $T \in \mathbb{R}^{n \times n}$ be a stochastic matrix. Suppose $x \in \mathbb{R}^n$ satisfies $x > 0$ and $e^t \cdot x = 1$. Define the prolongation matrix $P_{(x)} \in \mathbb{R}^{n \times m}$, the restriction matrix $R \in \mathbb{R}^{m \times n}$, and the projection matrix $\Pi_{(x)} \in \mathbb{R}^{n \times n}$ according to equations (8), (7), and (9).

Then the stochastic matrices $\Pi_{(x)} \cdot T$, $T \cdot \Pi_{(x)}$, and $R \cdot T \cdot P_{(x)}$ fulfill the following relations:

$$
\begin{aligned}
\mathrm{Kern}(I - \Pi_{(x)} \cdot T) &= P_{(x)}(\mathrm{Kern}(I - R \cdot T \cdot P_{(x)})) \\
R(\mathrm{Kern}(I - \Pi_{(x)} \cdot T)) &= \mathrm{Kern}(I - R \cdot T \cdot P_{(x)}) \\
T(\mathrm{Kern}(I - \Pi_{(x)} \cdot T)) &= \mathrm{Kern}(I - T \cdot \Pi_{(x)}) \\
\Pi_{(x)}(\mathrm{Kern}(I - T \cdot \Pi_{(x)})) &= \mathrm{Kern}(I - \Pi_{(x)} \cdot T) \\
\mathrm{Kern}(I - \Pi_{(x)} \cdot T) &\subseteq \mathrm{Kern}(I - \Pi_{(x)})
\end{aligned}
$$

Standard results from linear algebra yield the following relations.

REMARK 1

$$
\begin{aligned}
\mathrm{Dim}(T(\mathrm{Kern}(I - \Pi_{(x)} \cdot T))) &= \mathrm{Dim}(\mathrm{Kern}(I - \Pi_{(x)} \cdot T)) \\
\mathrm{Dim}(R(\mathrm{Kern}(I - \Pi_{(x)} \cdot T))) &= \mathrm{Dim}(\mathrm{Kern}(I - \Pi_{(x)} \cdot T)) \\
\mathrm{Dim}(P_{(x)}(\mathrm{Kern}(I - R \cdot T \cdot P_{(x)}))) &= \mathrm{Dim}(\mathrm{Kern}(I - R \cdot T \cdot P_{(x)})) \\
\mathrm{Dim}(\Pi_{(x)}(\mathrm{Kern}(I - T \cdot \Pi_{(x)}))) &= \mathrm{Dim}(\mathrm{Kern}(I - T \cdot \Pi_{(x)}))
\end{aligned}
$$

PROPOSITION 3 Suppose the assumptions of Proposition 2 hold. Then the following conditions are equivalent:

$$\mathrm{Dim}(\mathrm{Kern}(I - \Pi_{(x)} \cdot T)) = 1 \tag{14}$$

$$\mathrm{Dim}(\mathrm{Kern}(I - R \cdot T \cdot P_{(x)})) = 1 \tag{15}$$

$$\mathrm{Dim}(\mathrm{Kern}(I - T \cdot \Pi_{(x)})) = 1 \tag{16}$$

In this case, there exist vectors $\tilde{x}^* \in \mathbb{R}^n$, $z^* \in \mathbb{R}^n$, $\alpha^* \in \mathbb{R}^m$ satisfying

$$
\begin{aligned}
&\tilde{x}^* > 0 \qquad e^t \cdot \tilde{x}^* = 1 \qquad z^* > 0 \qquad e^t \cdot z^* = 1 \\
&\alpha^* > 0 \qquad e^t \cdot \alpha^* = 1
\end{aligned} \tag{17}
$$

and

$$
\begin{aligned}
\mathrm{Kern}(I - \Pi_{(x)} \cdot T) &= \mathrm{span}(\tilde{x}^*) \\
\mathrm{Kern}(I - R \cdot T \cdot P_{(x)}) &= \mathrm{span}(\alpha^*) \\
\mathrm{Kern}(I - T \cdot \Pi_{(x)}) &= \mathrm{span}(z^*)
\end{aligned} \tag{18}
$$

These vectors fulfill the relations:

$$\tilde{x}^* = P_{(x)} \cdot \alpha^* \qquad \alpha^* = R \cdot \tilde{x}^* \qquad \tilde{x}^* = \Pi_{(x)} \cdot z^* \qquad z^* = T \cdot \tilde{x}^* \tag{19}$$

As the group inverse of the stochastic matrix $G = \Pi_{(x)} \cdot T$ exists (cf. [31, Theorem 2.1, p. 445]), the matrix

$$W = W_{(x)} = I - (I - G_{(x)}) \cdot (I - G_{(x)})^{\#}$$

is well defined and a projection onto Range(W) = Kern($I - G_{(x)}$) along Kern(W) = Range($I - G_{(x)}$). Furthermore, $e^t \cdot W = e^t$ and $W \cdot \tilde{x}^* = \tilde{x}^*$ hold for the steady-state vector $\tilde{x}^*$ of $G_{(x)}$.

PROPOSITION 4 Suppose the assumptions of Proposition 2 and one of the conditions (14), (15), or (16) in Proposition 3 are fulfilled. Then

$$W = I - (I - \Pi_{(x)} T) \cdot (I - \Pi_{(x)} T)^{\#} = \tilde{x}^* \cdot e^t = P_{(x)} \cdot \alpha^*_{(x)} \cdot e^t > 0 \qquad (20)$$

holds, where $\tilde{x}^*$ and $\alpha^*_{(x)}$ satisfy the relations (17), (18), and (19) in Proposition 3. Furthermore, W is stochastic.

Proof. Cf. [32, Lemma 1, p. 142], [31, Theorem 2.3, p. 449], [15, Lemma 2.5, p. 37].

As the vectors x^* and $\tilde{x}_{(x)}$ from (12) are normalized, the error $\tilde{\epsilon}$ after an A/D step satisfies $e^t \cdot \tilde{\epsilon} = 0$. Taking into account equation (13), we are now able to prove an extension of the error result [5, (4.3), p. 31] derived by Chatelin and Miranker.

THEOREM 1 Suppose the assumptions of Proposition 2 and one of the conditions (14), (15), or (16) in Proposition 3 hold. Let x^* be the stationary distribution of T. Denote the error before and after an A/D step by $\epsilon = x^* - x$ and $\tilde{\epsilon} = x^* - \tilde{x}_{(x)}$, respectively.

Then

$$\tilde{\epsilon} = (I - \Pi_{(x)} \cdot T + W)^{-1} \cdot (I - \Pi_{(x)}) \cdot \epsilon \qquad (21)$$

$$= (I - \Pi_{(x)} \cdot T)^{\#} \cdot (I - \Pi_{(x)}) \cdot \epsilon \qquad (22)$$

hold with W as in equation (20). $Z_{(\tilde{x}^*)} = (I - \Pi_{(x)} \cdot T + W)^{-1}$ is the fundamental matrix associated with the stochastic matrix $\Pi_{(x)} \cdot T$.

Proof. Equation (21) immediately follows from the assumptions, equation (13), and Proposition 4. In this case we conclude from [15, Corollary 3.3, p. 83] that the generalized fundamental matrix $(I - \Pi_{(x)} \cdot T + u \cdot e^t)^{-1}$ exists (cf. [19]). It has the representation

$$(I - \Pi_{(x)} \cdot T + u \cdot e^t)^{-1} = (I - \Pi_{(x)} \cdot T)^{\#} \cdot (I - u \cdot e^t) + \tilde{x}^* \cdot e^t \qquad (23)$$

for any vector u satisfying $e^t \cdot u = 1$. Hence, equations (20) and (21) yield equation (22).

Suppose the assumptions of Theorem 1 hold. Then the existence of the generalized fundamental matrix $(I - \Pi_{(x)} \cdot T + u \cdot e^t)^{-1}$ is guaranteed for any vector u provided $e^t \cdot u \neq 0$ holds (cf. [19]). Hence, equation (23) yields an explicit representation of the error $\tilde{\epsilon}$ in terms of the initial vector x.

COROLLARY 1 Suppose the assumptions of Theorem 1 hold. Let $x > 0$, $e^t \cdot x = 1$ be the initial vector and let u be an arbitrary vector satisfying $e^t \cdot u = 1$. Then

$$\tilde{\epsilon} = (I - \Pi_{(x)} \cdot T + u \cdot e^t)^{-1} \cdot (I - \Pi_{(x)}) \cdot \epsilon$$

especially

$$\tilde{\epsilon} = (I - \Pi_{(x)} \cdot T + x \cdot e^t)^{-1} \cdot (I - \Pi_{(x)}) \cdot \epsilon \tag{24}$$

hold.

The error formula (22) has been derived by Haviv (cf. [16, Theorem, p. 954]). The equivalent representations (21) and (24) have the advantage of being easily computable. They reveal the influence of the initial vector x, the prolongation matrix $P_{(x)}$, and the projection matrix $\Pi_{(x)}$ onto the error $\tilde{\epsilon}$ after an A/D step. Haviv [16] has imposed the condition that $\Pi_{(x)}$ is irreducible in order to guarantee equation (22). The next result shows that this is the strongest condition, which is of little practical interest.

LEMMA 1 Suppose the assumptions of Proposition 2 hold. Furthermore, assume $\mathrm{Dim}(\mathrm{Kern}(I - \Pi_{(x)})) = 1$ is satisfied. A sufficient condition is the irreducibility of $\Pi_{(x)}$.

Then the conditions (14), (15), and (16) are satisfied. Moreover,

$$\mathrm{Kern}(I - \Pi_{(x)} \cdot T) = \mathrm{Kern}(I - \Pi_{(x)}) = \mathrm{span}(x)$$

and $\tilde{x}^* = x$ hold where $\tilde{x}^*$ is the unique stationary distribution of $\Pi_{(x)} \cdot T$.

Proof. The result immediately follows from Proposition 2 and $\Pi_{(x)} \cdot x = x$.

An important feature of the iterative A/D procedure is the error reduction that is achieved by inserting an A/D step (11) and (12) during the iteration $x^{(k+1)} = T \cdot x^{(k)}$. This improvement has been computed by Chatelin and Miranker for the case of a convergent iteration matrix T (cf. [5, section 4.2.4, p. 36f]). The next theorem extends their result to semiconvergent iteration matrices.

THEOREM 2 Suppose the assumptions of Proposition 2 and one of the conditions (14), (15), or (16) in Proposition 3 are satisfied.

Then $\text{Kern}(I - T \cdot \Pi_{(x)}) = \text{span}(z^*)$ holds where z^* is the unique stationary distribution of $T \cdot \Pi_{(x)}$.

Let $\epsilon = x^* - x$ be the error before an A/D step and $\hat{\epsilon} = x^* - \hat{x}_{(x)}$ the error after an A/D step (11) and (12) followed by an iteration step $\hat{x} = \hat{x}_{(x)} = T \cdot \tilde{x}_{(x)}$. Then

$$\hat{\epsilon} = (I - T \cdot \Pi_{(x)} + H)^{-1} \cdot T \cdot (I - \Pi_{(x)}) \cdot \epsilon \tag{25}$$

$$= (I - T \cdot \Pi_{(x)})^{\#} \cdot T \cdot (I - \Pi_{(x)}) \cdot \epsilon \tag{26}$$

with

$$H = I - (I - T \cdot \Pi_{(x)}) \cdot (I - T \cdot \Pi_{(x)})^{\#} = z^* \cdot e^t \tag{27}$$

follows. Moreover, $z^* = \hat{x}_{(x)}$ holds.

Proof. The assumptions imply $\text{Dim}(\text{Kern}(I - \Pi_{(x)} \cdot T)) = 1$. Proposition 2 yields $\Pi_{(x)}(\text{Kern}(I - T \cdot \Pi_{(x)})) = \text{Kern}(I - \Pi_{(x)} \cdot T) = \text{span}(\tilde{x}^*)$ where $\tilde{x}^*$ is the unique stationary distribution of $\Pi_{(x)} \cdot T$. According to Proposition 3 $\text{Kern}(I - T \cdot \Pi_{(x)})$ is generated by the unique stationary distribution z^* of $T \cdot \Pi_{(x)}$. Furthermore, the projection matrix H exists and satisfies equation (27).

From $\tilde{x}^* = \Pi_{(x)} \cdot \tilde{x}^*$ we conclude $\Pi_{(x)} \cdot \hat{x} = \Pi_{(x)} \cdot T \cdot \Pi_{(x)} \cdot \tilde{x}^* = \tilde{x}^*$. Hence, $\hat{x} = T \cdot \Pi_{(x)} \cdot \hat{x}$ holds, implying $z^* = \hat{x}_{(x)} > 0$, as $e^t \cdot \hat{x}_{(x)} = 1$, and

$$(I - T \cdot \Pi_{(x)}) \cdot (x^* - \hat{x}_{(x)}) = T \cdot (I - \Pi_{(x)}) \cdot (x^* - x)$$

This equation yields equation (25) if we take into account $H \cdot \hat{\epsilon} = z^* \cdot e^t \cdot (x^* - \hat{x}_{(x)}) = 0$. The proof of equation (26) follows along the lines of equation (23).

COROLLARY 2 Suppose the assumptions of Theorem 2 hold. Let $x > 0$, $e^t \cdot x = 1$ be the initial vector and let u be an arbitrary vector satisfying $e^t \cdot u = 1$. Then

$$\hat{\epsilon} = (I - T \cdot \Pi_{(x)} + u \cdot e^t)^{-1} \cdot T \cdot (I - \Pi_{(x)}) \cdot \epsilon$$

especially

$$\hat{\epsilon} = (I - T \cdot \Pi_{(x)} + x \cdot e^t)^{-1} \cdot T \cdot (I - \Pi_{(x)}) \cdot \epsilon \tag{28}$$

hold.

The gain of an iteration step following the A/D step may be computed by the ratio $\hat{\beta}$ of magnitudes of the errors $\hat{\epsilon} = x^* - \hat{x}_{(x)}$ and $\tilde{\epsilon} = x^* - \tilde{x}_{(x)}$. Taking into account $(I - \Pi_{(x)})^2 = I - \Pi_{(x)}$, we conclude from equations (28) and (13):

$$\hat{\beta} = \frac{\|x^* - \hat{x}_{(x)}\|}{\|x^* - \tilde{x}_{(x)}\|} = \frac{\|\hat{\epsilon}\|}{\|(I - \Pi_{(x)})\epsilon\|} \frac{\|(I - \Pi_{(x)})\epsilon\|}{\|\tilde{\epsilon}\|}$$

$$\leq \|(I - T \cdot \Pi_{(x)} + x \cdot e^t)^{-1}\| \cdot \|T \cdot (I - \Pi_{(x)})\| \cdot \|I - \Pi_{(x)} \cdot T\| \tag{29}$$

This upper bound (29) extends the corresponding result [5, (4.18), p. 37] of Chatelin and Miranker.

Analyzing the proofs, we recognize that only the assumptions $P_{(x)} \geq 0$, $e^t \cdot P_{(x)} = e^t, R \geq 0, e^t \cdot R = e^t, R \cdot P_{(x)} = I, \Pi_{(x)} = P_{(x)} \cdot R \geq 0, e^t \cdot \Pi_{(x)} = e^t$, and $\Pi_{(x)} \cdot x = x$ have to be imposed on the prolongation, aggregation, and projection matrix to guarantee our results. Thus, the error representations (21), (22), (24), (25), (26), and (28) also hold for partial aggregation methods such as the methods e and f in Haviv [16, p. 957f] (see also [44]).

4.2 Convergence Results

First we define a continuous semi-norm in $\mathbb{R}^n$ by

$$r(x) = \|(I - T) \cdot x\|_1 \qquad x \in \mathbb{R}^n$$

and a compact set $K = \{x \in \mathbb{R}^n : x \geq 0, e^t \cdot x = 1\}$. Let $L = \{x \in K : r(x) = 0\} = \{x \in K : x = T \cdot x\} = \{x^*\}$. We define a family of compact sets by $K_\varepsilon = \{x \in K : r(x) \leq \varepsilon\}$, $\varepsilon > 0$. Obviously, $\lim_{n\to\infty} K_{1/n} = \bigcap_{n>0} K_{1/n} = L$. According to Proposition 1 the scheme

$$x^{(k+1)} = T \cdot x^{(k)} \qquad k = 0, 1, \ldots$$

is a convergent fallback procedure for any $x^{(0)} > 0$, and $r(Tx) \leq r(x)$, $x \in \mathbb{R}^n$, holds due to $\|T\|_1 = \rho(T) = 1$. Now we may construct an iterative A/D algorithm according to the scheme of Schweitzer and Kindle for the generator Q of a CTMC (cf. [41, p. 326f]).

Dual IAD Algorithm

Assumption: Let $A = M - N$ be a regular splitting of the irreducible Q-matrix $A = -Q^t \in \mathbb{R}^{n \times n}$ with extrapolated dual iteration matrix $T = I - \omega A M^{-1}$ for some $\omega \in (0, 1)$. Select a partition $\Gamma = \{J_1, \ldots, J_m\}$ of $\{1, \ldots, n\}$ into $m \geq 2$ disjoint sets.

1. Initialization:
   Select an initial vector $x^{(0)} \gg 0$, $e^t x^{(0)} = 1$, and three real numbers $0 < \varepsilon, c_1, c_2 < 1$. Construct the matrices $P_{(x)} \in \mathbb{R}^{n \times m}$, $R \in \mathbb{R}^{m \times n}$, $T \in \mathbb{R}^{n \times n}$, and $B_{(x)} \in \mathbb{R}^{m \times m}$ according to equations (8), (7), and (10). Set $k = 0$.
2. A/D step:

Solve	$B_{(x^{(k)})} \cdot \alpha_{(x^{(k)})} = \alpha_{(x^{(k)})}$
subject to	$e^t \cdot \alpha_{(x^{(k)})} = 1$, $\alpha_{(x^{(k)})} > 0$
and compute	$\tilde{x} = P_{(x^{(k)})} \cdot \alpha_{(x^{(k)})}$

3. Iteration step:
   Compute $\qquad x^{(k+1)} = T \cdot \tilde{x}$

4. Convergence test:

   If $\quad r(\tilde{x}) \leq c_1 \cdot r(x^{(k)})$

   then $\quad$ go to step 5

   else $\quad$ compute $x^{(k+1)} = T^m \cdot x^{(k)}$

   $\quad$ with $m = m(x^{(k)}) \in \mathbb{N}$ such that

   $$r(x^{(k+1)}) \leq c_2 \cdot r(x^{(k)})$$

   endif

5. Termination test:

   If $\quad \|x^{(k+1)} - x^{(k)}\|_1 / \|x^{(k)}\|_1 = \|x^{(k+1)} - x^{(k)}\|_1 < \varepsilon$

   then $\quad$ go to step 6

   else $\quad$ k=k+1

   $\quad$ go to step 2

   endif

6. Normalization:

$$p = \frac{M^{-1} \cdot x^{(k+1)}}{e^t \cdot M^{-1} \cdot x^{(k+1)}}$$

From the mathematical point of view this algorithm is reasonable as seen by the following lemma.

LEMMA 2 Let Q be the irreducible generator matrix of a CTMC and choose a regular splitting $A = M - N$ of the Q-matrix $A = -Q^t$, such as an M-splitting like the block Gauss–Seidel splitting $A = D - L - U$. $T_1 = NM^{-1}$ is the corresponding dual stochastic iteration matrix and $T = (1-\omega)I + \omega T_1$, $0 < \omega < 1$, the extrapolated variant.

Then the dual IAD algorithm converges to the steady-state distribution p of Q for any initial vector $x^{(0)} \gg 0$ with $e^t x^{(0)} = 1$.

Moreover, $r(x^{(k+1)}) \leq \max(c_1, c_2) \cdot r(x^{(k)})$, $k \geq 0$ holds.

Proof. The proof follows the lines of Schweitzer and Kindle [41, Theorem 4].

4.3 Stochastic Interpretation of the Block Gauss–Seidel Method

The IAD procedure may be applied to any regular splitting of $A = -Q^t$, especially to a point or block Gauss–Seidel splitting $A = (D-L)-U$ that generates an M-splitting. Hence, the results are still valid if we apply the procedure to a variant of Rose's R-regular block splitting (cf. [37], [24, Def. 14, p. 67]).

Mitra and Tsoucas [33] have pointed out that, in analogy to the point Jacobi method, the point Gauss–Seidel procedure has a probabilistic interpretation (cf. [24, p. 56]). They have shown that the iterative scheme evolves identically in law with a related DTMC, called high-stepping random walk. It is obtained by observing the embedded jump chain associated with the original CTMC at selected instants only. As the derivation of this interpretation is, to our point of view, somewhat cumbersome, we shall describe an equivalent but rather simple approach that is also applicable to the block Gauss–Seidel procedure.

Let $Q = (L + U) - D$ be the irreducible generator matrix of a CTMC that is partitioned into $m \geq 2$ blocks. Let $D = \text{Diag}(D_{11}, \ldots, D_{mm})$ be the block diagonal, and $L \geq 0$ the strictly lower and $U \geq 0$ the strictly upper block triangular part of Q. As the diagonal blocks D_{ii}, $i = 1, \ldots, m$, are regular, D is a regular M-matrix. We construct a generalized jump chain with t.p.m. $P = D^{-1}(L+U)$. If a point partition is chosen, P is the t.p.m. of the associated jump chain embedded at successive transition epochs.

Now we construct a new DTMC by associating with each state i of the jump chain a new state i'. The original states, called up-states, are collected in a set S. The new states, called down-states, constitute a set S'. The block partition induces a natural aggregation of the states. The transition behavior of the new chain is specified as follows: Considering the natural ordering of numbers, only increasing transitions between up-state aggregates are possible. The transition probability for a transition from i to j is determined by the corresponding probability $(D^{-1}U)_{ij}$ of the jump chain. Each down-state i' can reach its corresponding up-state i with probability 1, but no other states. Hence, transitions are not allowed between down-states. Each down-state i' can be reached from an up-state j with probability $(D^{-1}L)_{ji}$. Thus the t.p.m. P^{udc} of the new chain has the following form:

$$P^{udc} = \begin{array}{c} \\ \end{array}\begin{array}{cc} S' & S \end{array} \\ P^{udc} = \begin{bmatrix} 0 & I \\ D^{-1}L & D^{-1}U \end{bmatrix} \begin{array}{c} S' \\ S \end{array}$$

Now we construct the reduced Markov chain on the set S' of down-states according to Grassmann's state reduction approach (cf. [11], [24, p. 28ff]). This means that we observe the new chain only during its visits in the set of down-states. Hence, the t.p.m. of this reduced DTMC is given by (cf. [11], [38, p. 261])

$$P_{S'} = (I - D^{-1}U)^{-1} \cdot D^{-1}L$$

Regarding the original generalized jump chain, this procedure has a simple stochastic interpretation. First, we divide each sample path of the jump chain into consecutive sections of a (perhaps empty) monotone increasing sequence of aggregates followed by a single nonincreasing step with respect to the natural ordering of aggregates. The length of the increasing path may be zero if two nonincreasing transitions between aggregates follow each other. Let us assume that i and j are two successive states reached after such an excursion. Then the probability for such a transition from i to j is given by $(P_{S'})_{ij}$.

If we associate the M-matrix $A = -Q^t = (D-(U+L))^t$ with the generator matrix Q, the iteration matrix of the forward Gauss–Seidel procedure, based on the splitting $A = M - N, M = D^t - U^t, N = L^t$, is defined by $J = (D^t - U^t)^{-1} \cdot L^t$. A similarity transformation with the matrix M yields

$$T_1 = M \cdot J \cdot M^{-1} = L^t \cdot (D^t - U^t)^{-1} = L^t \cdot D^{-t} \cdot (I - U^t \cdot D^{-t})^{-1}$$
$$= [(I - D^{-1}U)^{-1} \cdot D^{-1} \cdot L]^t = (P_{S'})^t$$

Hence, the dual iteration matrix T_1 corresponding to the Gauss–Seidel method has the same spectral properties as the stochastic matrix $P_{S'}$ associated with the constructed DTMC.

The extrapolated version $T_\omega = (1-\omega)I + \omega T_1, 0 < \omega < 1$, called stationary first-order Richardson extrapolation, is related to the t.p.m. of a modified generalized jump chain. This modification stems from damping the transition probabilities of the jump chain by a term ω and adding new transitions from each state into itself with probabilities $1-\omega$ (cf. [20, Chap. V: Ergodic Markov Chains, Theorem 5.1.1, p. 99], [24, section 6.4]).

5. CALCULATION OF THE STATIONARY DISTRIBUTION OF THE OVERFLOW MODEL

As no simple analytical expression for the stationary distribution P of the Markov chain $\{Z(t), t \geq 0\}$ is available, the steady-state vector has to be calculated as solution of the balance equations (1) by appropriate direct or iterative numerical methods (cf. [24], [18], [9], [11]). All procedures discussed subsequently are based on the Q-matrix representation (2) according to Section 3.

Considering the sparsity of the generator matrix $\tilde{Q}$, we restrict our attention to iterative solution techniques. Due to the ease and efficiency of implementation, especially the use of vectorizable and parallel executable algorithms, standard procedures based on point splittings $A = D-L-U$ such as the point Jacobi and Gauss–Seidel methods have again attracted considerable attention. Obviously, the latter are nontrivial M-splittings.

It can be shown that $A = -\tilde{Q}^t$ is an irreducible, 2-cyclic matrix with respect to the point partition and has property A. Thus, the point Jacobi matrix $J = D^{-1}(L + U)$ is cyclic of index 2, hence not semiconvergent (cf. [3, Theorem 2.2.30, p. 35], [24, Theorem 29, p. 56]), whereas the corresponding JOR and SOR procedures with the iteration matrices $J_\omega = (1 - \omega)I + \omega J$ and $L_\omega = (D - \omega L)^{-1}((1 - \omega)D + \omega U)$ are convergent for each relaxation parameter

$\omega \in (0, 1)$ (cf. [24 p. 73], [2, p. 174], [1, Cor. 3, p. 395], [17, Theorem 3.4, p. 191]).

As $M_{21} = (D - L)_{21} = -\gamma_2 \neq 0$ and $N_{12} = U_{12} = \omega_2 \neq 0$ hold, we conclude from [39, Cor. 3.8, p. 417] that the point Gauss–Seidel procedure is convergent (see also [37, Cor. 2, p. 139], [24, Theorem 35, p. 66]). This point iteration may be accelerated by applying the standard relaxation technique (cf. [18]) or by employing a semi-iterative technique such as the stationary or nonstationary Chebyshev method or Eiermann's stationary fourth-order scheme (cf. [45, section 5], [6], [35], [7, Lemma 8.4, p. 28], [2], [12], [13]). The main difficulty with respect to these procedures is the determination of "optimal" relaxation parameters. As there is no a priori information about the location of the eigenvalues of the iteration matrices, heuristic procedures estimating approximately optimal parameters seem to be the only practicable approach (cf. [14, section 9.5, p. 223ff], [42], [12], [13]).

An alternative is given by block iterative schemes such as the block Gauss–Seidel procedure or its modified versions based on Rose's R-regular splitting (cf. [37], [30], [25]). Due to [37, Cor. 2, p. 139] the block Gauss–Seidel scheme derived from the given block tridiagonal structure (2) of $-\tilde{Q}^t$ is convergent as the diagonal blocks are irreducible regular M-matrices. All methods may be combined with A/D steps if Schweitzer's IAD procedure is used (see Section 4.2; cf. [34], [21], [44]).

6. COMPUTATION OF THE CONGESTION RATES

In teletraffic theory the most important steady-state performance characteristics of a loss system are the congestion rates. Recall that we distinguish between time-congestion and call-congestion rates with respect to the arrival streams of a model. In the subsequent section we assume the reader to be familiar with the notions of time-stationary and customer-stationary distributions and time congestion as well as customer-dependent and average call congestion of a $G_n/M/N/0$ model (cf. [8, p. 58ff], [10, section 11, p. 517ff]).

Furthermore, we assume the vector process $Z(t) = (X_1(t), X_2(t), Y(t))$ to be in steady state. Its steady-state distribution is denoted by P.

6.1 Overall Congestion Rates

The steady-state distribution Π of the occupation process $X(t) = (X_1(t), X_2(t))$ is given by the marginal distribution

$$\Pi_{ij} = \lim_{t \to \infty} P(X(t) = (i,j)) = e^t \cdot P_{ij}$$

Thus, the overall time-congestion of the overflow model is $\Pi_{N_1N_2} = e^t P_{N_1N_2}$. The time-congestion rates of groups 1 and 2 are $\Pi_{N_1} = \sum_{j=0}^{N_2} e^t P_{N_1j}$ and $\Pi_{N_2} = \sum_{i=0}^{N_1} e^t P_{iN_2}$.

Let $R_i, i \in \{1,2\}$, denote the overall call-congestion rate of MMPP i, that is, the steady-state probability that a call of the arrival stream i selected at random finds no free trunk in the overflow system upon its arrival. Following the lines of Meier-Hellstern [29], we conclude that these overall call-congestion rates have the form

$$R_1 = \frac{P^t \cdot I_{(N_1+1)(N_2+1)} \otimes \Lambda_1 \cdot (e_{(N_1+1)(N_2+1)} \otimes e)}{e^t \cdot I_{(N_1+1)(N_2+1)} \otimes \Lambda_1 \cdot P}$$

$$= \frac{P_{N_1N_21} + P_{N_1N_22}}{\sum_{i=0}^{N_1} \sum_{j=0}^{N_2} (P_{ij1} + P_{ij2})}$$

$$R_2 = \frac{P^t \cdot I_{(N_1+1)(N_2+1)} \otimes \Lambda_2 \cdot (e_{(N_1+1)(N_2+1)} \otimes e)}{e^t \cdot I_{(N_1+1)(N_2+1)} \otimes \Lambda_2 \cdot P}$$

$$= \frac{P_{N_1N_21} + P_{N_1N_23}}{\sum_{i=0}^{N_1} \sum_{j=0}^{N_2} (P_{ij1} + P_{ij3})}$$

6.2 Flow-Dependent Call-Congestion Rates

In this section we analyze an isolated trunk group of the mutual overflow model and determine its corresponding call-congestion rates with respect to the offered traffic streams.

Considering group 1, for instance, there are two flows offered to this subsystem: the MMPP 1 and the stream of calls overflowing from group 2 to 1. The overflow stream is created by selecting the arrival instants $T_n^{(2)}$ corresponding to those customers of the second MMPP that find the system in one of the states (i, N_2, k), $i \in \{0, \ldots, N_1\}$, $k \in \{1,3\}$. If we mark incoming calls of the offered MMPPs by two different colors representing distinct customer types, the isolated primary trunk group may be described by a special $G_2/M/N_1/0$ model (cf. Franken [8, p. 61]).

Let us denote the individual call-congestion rates of the customer stream j offered to group i by B_{ij}, $i,j \in \{1,2\}$. Then these flow-dependent call-congestion rates with respect to groups 1 and 2 are given by

$$B_{11} = \frac{e^t \cdot (e_{N_1+1} e^t_{N_1+1}) \otimes I_{N_2+1} \otimes \Lambda_1 \cdot P}{e^t \cdot I_{(N_1+1)(N_2+1)} \otimes \Lambda_1 \cdot P} = \frac{\sum_{j=0}^{N_2} (P_{N_1j1} + P_{N_1j2})}{\sum_{i=0}^{N_1} \sum_{j=0}^{N_2} (P_{ij1} + P_{ij2})}$$

$$B_{12} = \frac{e^t \cdot (e_{(N_1+1)(N_2+1)} e^t_{(N_1+1)(N_2+1)}) \otimes \Lambda_2 \cdot P}{e^t \cdot I_{N_1+1} \otimes (e_{N_2+1} e^t_{N_2+1}) \otimes \Lambda_2 \cdot P} = \frac{P_{N_1N_21} + P_{N_1N_23}}{\sum_{i=0}^{N_1} (P_{iN_21} + P_{iN_23})}$$

$$B_{21} = \frac{e^t \cdot (e_{(N_1+1)(N_2+1)} e^t_{(N_1+1)(N_2+1)}) \otimes \Lambda_1 \cdot P}{e^t \cdot (e_{N_1+1} e^t_{N_1+1}) \otimes I_{N_2+1} \otimes \Lambda_1 \cdot P} = \frac{P_{N_1 N_2 1} + P_{N_1 N_2 2}}{\sum_{j=0}^{N_2} (P_{N_1 j 1} + P_{N_1 j 2})}$$

$$B_{22} = \frac{e^t \cdot I_{N_1+1} \otimes (e_{N_2+1} e^t_{N_2+1}) \otimes \Lambda_2 \cdot P}{e^t \cdot I_{(N_1+1)(N_2+1)} \otimes \Lambda_2 \cdot P} = \frac{\sum_{i=0}^{N_1} (P_{i N_2 1} + P_{i N_2 3})}{\sum_{i=0}^{N_1} \sum_{j=0}^{N_2} (P_{ij1} + P_{ij3})}$$

7. CONCLUSIONS AND PERSPECTIVES

We have investigated a loss system that describes the mutual overflow of two Markov modulated Poisson processes between two fully available trunk groups. In order to compute the steady-state distribution of the associated CTMC modeling the behavior of the system, we have presented convergent point and block iterative methods based on regular splittings of the corresponding generator matrix.

In this context, we extended Schweitzer's iterative A/D algorithm to a larger class of splittings and improved Haviv's error analysis. Furthermore, a stochastic interpretation of the block Gauss–Seidel method was discussed.

The proposed procedures may be employed to calculate the steady-state performance characteristics of the model, which have been specified in terms of the overall and individual call-congestion rates of the arrival streams.

At present we are developing a software package that provides the user with a Markovian model world describing circuit-switched networks with adaptive routing schemes and advanced flow-control mechanisms. The proposed accelerated point iteration methods and some direct methods for the computation of the steady-state vector of a CTMC (cf. [11], [9], [18], [2], [24]) constitute the numerical solver of the system. The package is endowed with a user-friendly graphical interface and will be running on a workstation.

Further investigations concerning the efficient implementation of direct and iterative methods on a vector computer (IBM 3090 Vf) and on a distributed-memory multiprocessor architecture (Transputer system), the comparison of these algorithms, and their application to the performance analysis of adaptive routing schemes in circuit-switched networks are a subject of current research and will be reported elsewhere.

APPENDIX: DEFINITION OF THE MARKOV MODULATED POISSON PROCESS

Informally speaking, a Markov modulated Poisson process (MMPP) $\{N(t), t \geq 0\}$ with $m \in \mathbb{N}$ phases is a variant of a nonhomogeneous Poisson process whose intensity function $\lambda(t)$ is a stochastic process $\lambda(t) = \sum_{i=1}^{m} \lambda_i \, 1_{\{i\}}(Y(t))$ governed by an irreducible, homogeneous, continuous-time Markov chain $\{Y(t), t \geq 0\}$ with finite state space $\{1,\dots,m\}$. This means that $\{\lambda(t), t \geq 0\}$ is a CTMC with state space $\{\lambda_1,\dots,\lambda_m\}$. The resulting point process is a special case of a double stochastic Poisson process that is also known as conditional Poisson process ([43], [28]). Therefore, it is convenient to call an MMPP a Poisson process in a Markovian environment.

An MMPP is parametrized by the specification of the initial probability vector $p = (p_i) \in \mathbb{R}^m$ and the generator matrix $Q \in \mathbb{R}^{m \times m}$ of the controlling CTMC and the rate vector $0 < \lambda = [\lambda_1,\dots,\lambda_m]^t \in \mathbb{R}^m$. Notice that some intensities may be zero. It is convenient to associate a diagonal matrix $\Lambda = \mathrm{Diag}(\lambda)$ with the rate vector. Henceforth, $Y(t)$ is called the phase of the controlling CTMC of the MMPP at time $t \geq 0$.

Serfozo [43] has proved that any Markov modulated Poisson process may be represented by an equivalent two-dimensional pure Markov jump process. Regarding this equivalent characterization, we introduce an MMPP by the following definition.

DEFINITION 1 A nonnegative integer-valued stochastic process $\{N(t), t \geq 0\}$ is called Markov modulated Poisson process with $m \in \mathbb{N}$ phases and representation (Q, Λ, p) if there exists a two-dimensional regular Markov chain $\{Z(t) = (N(t), Y(t)), t \geq 0\}$ with values in $\mathbb{N}_0 \times \{1,\dots,m\}$ such that the following conditions are satisfied:

1. $0 < \lambda = [\lambda_1,\dots,\lambda_m]^t \in \mathbb{R}^m$, $\Lambda = \mathrm{Diag}(\lambda)$, and $-Q \in \mathbb{R}^{m \times m}$ is an irreducible M-matrix with $Q \cdot e = 0$; $p = (p_i) \in \mathbb{R}^m$ is a probability vector.
2. $Z(t)$ can be represented by its embedded jump chain, that is, there exists a Markov process $\{(N_n, Y_n, T_n), n \in \mathbb{N}_0\}$ with values in $\mathbb{N}_0 \times \{1,\dots,m\} \times [0,\infty)$ and $P(N_0 = 0, Y_n = i, T_0 = 0) = p_i$, $1 \leq i \leq m$, satisfying

$$P(N_{n+1} - N_n = k, Y_{n+1} = j, T_{n+1} - T_n > t \mid N_n, Y_n = i, T_n)$$

$$= \begin{cases} \dfrac{Q_{ij}}{\lambda_i - Q_{ii}} e^{-(\lambda_i - Q_{ii})t} & k = 0, i \neq j \\[2ex] \dfrac{\lambda_i}{\lambda_i - Q_{ii}} e^{-(\lambda_i - Q_{ii})t} & k = 1, i = j \\[2ex] 0 & \text{otherwise} \end{cases}$$

such that

$$(N(t), Y(t)) = (N_n, Y_n) \qquad \text{if} \quad T_n \leq t < T_{n+1}$$

holds.

Obviously, the MMPP $\{N(t), t \geq 0\}$ may be characterized as a special semi-Markovian point process with embedded Markov chain $R = (\Lambda - Q)^{-1}\Lambda$ and semi-Markov matrix $F(x) = (P(\tilde{Y}_{n+1} = j, \tilde{T}_{n+1} - \tilde{Y}_n \leq x \mid \tilde{Y}_n = i))_{i,j=1,\dots,m} = (I - e^{(Q-\Lambda)x}) \cdot R, x \geq 0$ (cf. [28, p. 9]). Here $\tilde{Y}_n$ denotes the phase of the controlling Markov chain $Y(t)$ immediately after the nth event of the MMPP $N(t)$, and $\tilde{T}_n$ is its corresponding arrival instant. Notice that we assume that $t = 0$ is an arrival instant. This means that we determine the synchronous version of the point process $N(t)$ whose distribution is given by the Palm distribution.

The MMPP has been studied by Serfozo [43] and Meier [28], among others. For details the reader is referred to their articles and the references therein.

REFERENCES

[1] G. P. Barker and R. J. Plemmons. Convergent iterations for computing stationary distributions of Markov chains. SIAM J. Alg. Disc. Methods, 7(3):390–398, 1986.

[2] G. P. Barker and S. J. Yang. Semi-iterative and iterative methods for singular M-matrices. SIAM J. Matrix Anal. Appl., 9(2):168–180, 1988.

[3] A. Berman and R. J. Plemmons. *Nonnegative Matrices in the Mathematical Sciences*. Academic Press, New York, 1979.

[4] F. Chatelin. Iterative aggregation/disaggregation methods. in P. J. Courtois, G. Iazeolla and A. Hordijk, editors, *Mathematical Computer Performance and Reliability*, pp. 199–207, North-Holland, Amsterdam, 1984.

[5] F. Chatelin and W. Miranker. Acceleration by aggregation of successive approximations methods. Linear Algebra Appl., 43:17–47, 1982.

[6] M. Eiermann, I. Marek, and W. Niethammer. On the solution of singular linear systems of algebraic equations by semiiterative methods. Numer. Math., 53:265–283, 1988.

[7] M. Eiermann, R. Varga, and W. Niethammer. Iterationsverfahren für nichtsymmetrische Gleichungssysteme und Approximationsmethoden im Komplexen. Jahresbericht Math.-Verein., Teubner, 89:1–32, 1987.

[8] P. Franken, D. König, U. Arndt, and V. Schmidt. *Queues and Point Processes*. John Wiley, New York, 1982.

[9] R. Funderlic and R. J. Plemmons. A combined direct-iterative method for certain M-matrix linear systems. SIAM J. Alg. Disc. Methods, 5:33–42, 1984.

[10] B. W. Gnedenko and D. König. *Handbuch der Bedienungstheorie II*. Akademie-Verlag, Berlin, 1984.

[11] W. Grassmann, M. Taksar, and D. Heymann. Regenerative analysis and steady-state distributions for Markov chains. Operations Res., 33(5):1107–1116, 1985.

[12] A. Hadjidimos. On the optimization of the classical iterative schemes for the solution of complex singular linear systems. SIAM J. Alg. Disc. Methods, 6(4):555–566, 1985.

[13] A. Hadjidimos. Optimum stationary and nonstationary iterative methods for the solution of singular linear systems. Numer. Math., 51:517–530, 1987.

[14] L. A. Hageman and D. M. Young. *Applied Iterative Methods*. Academic Press, New York, 1981.

[15] W. J. Harrod. Rank Modification Method for Certain Singular Systems of Linear Equations. Ph.D. thesis, University of Tennessee, Knoxville, 1982.

[16] M. Haviv. Aggregation/disaggregation methods for computing the stationary distribution of a Markov chain. SIAM J. Numer. Anal., 42:183–198, 1982.

[17] M. Neumann, J. Buoni, and R. Varga. Theorems of Stein–Rosenberg type: III, The singular case. Linear Algebra Appl., 41:183–198, 1982.

[18] L. Kaufman, Matrix methods for queueing problems. SIAM J. Sci. Stat. Comput., 4:525–552, 1983.

[19] J. G. Kemeny, Generalization of a fundamental matrix. Linear Algebra Appl., 38:193–206, 1981.

[20] J. G. Kemeny and L. J. Snell. *Finite Markov Chains*. Springer, New York, 1976.

[21] J. R. Koury, D. F. McAllister, and W. J. Stewart. Iterative methods for computing stationary distributions of nearly completely decomposable Markov chains. SIAM J. Alg. Disc. Methods, 5(2):164–186, 1984.

[22] U. Krieger. Analysis of a loss system with mutual overflow. In *International Teletraffic Seminar on Teletraffic and Network, Beijing, September 12–16, 1988, Proceedings*, pp. 331–340, 1988.

[23] U. Krieger. Analysis of a loss system with mutual overflow and external traffic. In *Stochastiche Modelle und Methoden in der Informationstechnik, Nürnberg, April 12–14, 1989, Proceedings, ITG-Fachbericht 107*, pp. 145–153, VDE-Verlag, Berlin, 1989.

[24] U. Krieger. *Computational Methods for Markovian Queueing Models*. Technical Report 4302 TB 10E, Research Institute of the DBP, Darmstadt, 1989.

[25] U. Krieger. Untersuchung eines Nachrichtenverkehrsmodelles mit wechselseitigem Überlauf und Externverkehr. Preprint, V 88–12–01, Research Institute of the DBP, Darmstadt, 1988.

[26] A. Kuczura. The interrupted Poisson process as an overflow process. Bell System Tech. J. 52(3):437–448, 1973.

[27] F. LeGall and J. Bernussou. An analytical formulation for grade of service determination in telephone networks. IEEE Trans. Commun., 31(3):420–424, 1983.

[28] K. Meier. A Statistical Procedure for Fitting Markov-Modulated Poisson Processes. Ph.D. thesis, University of Delaware, 1984.

[29] K. Meier-Hellstern. The analysis of a queue arising in overflow models. IEEE Trans. Commun., 37(4):367–372, 1989.

[30] K. Meier-Hellstern. Parcel overflows in queues with multiple inputs. In *Proceedings, ITC 12, Turino, Italy*, pp. 5.1B.3.1–5.1B.3.8, 1988.

[31] C. D. Meyer. The role of the group generalised inverse in the theory of finite Markov chains. SIAM Rev., 17:443–464, 1975.

[32] C. D. Meyer and M. W. Stadelmaier. Singular M-matrices and inverse positivity. Linear Algebra Appl., 22:139–156, 1978.

[33] D. Mitra and P. Tsoucas. Convergence of relaxations for numerical solutions of stochastic problems. In O. J. Boxma, G. Iazeolla, and P. J. Courtois, editors, *Applied Mathematics and Performance/Reliability Models of Computer/Communication Systems*, pp. 119–133, North-Holland, Amsterdam, 1987.

[34] B. Müller. Zerlegungsorientierte, numerische Verfahren für Markovsche Rechensystemmodelle. Ph.D. thesis, Universität Dortmund, Abteilung Informatik, 1980.

[35] W. Niethammer and R. Varga. The analysis of k-step iterative methods for linear systems from summability theory. Numer. Math.,41:177–206, 1983.

[36] A. Ridder. Stochastic Inequalities for Queues. Ph.D. thesis, Rijksuniversiteit te Leiden, 1987.

[37] D. Rose. Convergent regular splittings for singular M-matrices. SIAM J. Alg. Disc. Methods, 5(1):133–144, 1984.

[38] R. Schassberger. An aggregation principle for computing invariant probability vectors in Semi-Markovian models. In P. J. Courtois, G. Iazeolla and A. Hordijk, editors, *Mathematical Computer Performance and Reliability*, pp. 259–272, North Holland, Amsterdam, 1984.

[39] H. Schneider. Theorems on M-splittings of a singular M-matrix which depend on graph structure. Linear Algebra Appl., 58:407–424, 1984.

[40] P. J. Schweitzer. Aggregation methods for large Markov chains. In P. J. Courtois, G. Iazeolla and A. Hordijk, editors, *Mathematical Computer Performance and Reliability*, pp. 275–285, North Holland, Amsterdam, 1984.

[41] P. J. Schweitzer and K. W. Kindle. An iterative aggregation-disaggregation algorithm for solving linear equations. Appl. Math. Comput.,18: 313–353, 1986.

[42] L. P. Seelen. An Algorithm for $PH/PH/c$ Queues. Research Report 131, December 1984, Vrije Universiteit, Dept. of Actuarial Sciences and Economics, Amsterdam, 1984.

[43] R. F. Serfozo. Conditional Poisson processes. J. Appl. Prob., 9:288–302, 1972.

[44] U. Sumita and M. Rieders. A New Algorithm for Computing the Probability Vector for Large Markov Chains: Replacement Process Approach. Working Paper Series No. QM88–08, William E. Simon Graduate School of Business Administration, University of Rochester, 1988.

[45] R. Varga. *Matrix Iterative Analysis*. Prentice-Hall, Englewood Cliffs, N. J., 1962.

17

Computing Conditional Distributions of Queuing Network Matrices

PIERRE SEMAL and PIERRE–JACQUES COURTOIS Philips Research Laboratories, Louvain-la-Neuve, Belgium

ABSTRACT

This paper addresses the problem of computing conditional steady-state distributions in large Markovian queuing networks.

An algebraic framework is introduced to describe and analyze decomposition techniques which yield these distributions. First, this yields a new proof that the so-called short-cut or Norton's method [1][5] gives exact results under conditions of local balance. As opposed to previous work, this proof is constructive in the sense that the Norton's method is directly derived from the global and the local balance equations. A by-product of the proof is to generalize the class of network routing schemes to which the theorem applies. Besides, it is shown that the exact conditional distributions can also be obtained by decomposition under conditions that are weaker than local balance. These conditions are called here "subsystem balance" and the corresponding exact decomposition "bounded population range" (BPR).

In the second part of the paper, properties of polyhedra of Perron–Frobenius eigenvectors are used to analyze the accuracy of these methods when they are applied to more general queuing networks that do not satisfy any type of balance. We show, then, how the accuracy depends on the structure of the entire network and can thereby be arbitrary poor in the absence of any special network property. On the other hand, the methods are guaranteed to be accurate when the network is "nearly completely decomposable." Finally, for networks without any specific property, the BPR method provides better guarantees than the Norton's method; another advantage of the BPR method is that it always converges as the amount of computation involved is increased.

1. INTRODUCTION

Steady-state probability distributions that are conditioned on some parts of the network being in a certain state play an important role in the analysis of large queuing network models. These distributions are essential when, in the case of large state spaces, decomposition techniques are needed to compute the complete state probability distribution. They are also useful when, for instance, one is interested in certain network critical behaviors that can occur only when certain conditions are fulfilled.

This paper addresses the problem of computing conditional steady-state distributions in large Markovian queuing networks. Two important types of conditional distributions, that is, two ways of restricting the state space, are considered: either the state of a part σ of the queuing network or the population in a part of the network is assumed to remain constant. In Section 2, an algebraic characterization of the two corresponding conditional distributions shows that, without any additional property, their exact determination is as expensive as the computation of the complete distribution π.

Section 3 mainly focuses on networks satisfying *local balance* [2]. It is well known that the *short-cut method* [1][5][16], also known as Norton's theorem, is guaranteed to yield the exact conditional distributions under this property. A new derivation of this result is given. As opposed to previous work, this proof is constructive in the sense that the Norton's method is directly derived from the global and the local balance equations. A by-product of the proof is to generalize the class of network routing schemes to which the theorem applies.

A new method called the *bounded population range* (BPR) method is then proposed. This method is proved to yield the exact distribution when the network satisfies the subsystem balance, a condition weaker than the local balance.

Section 4 analyzes, by means of polyhedra of Perron–Frobenius eigenvectors [10][11], the accuracy of these methods when applied to nonproduct form networks. In such networks, both methods provide approximations of the conditional distributions only. It is then shown that the approximation yielded by the short-cut method is guaranteed to be accurate for nearly completely decomposable [8][14] networks only.

The same analysis shows that the BPR method provides better guarantees than the short-cut method. Finally, a criterion is given to ensure that the approximation yielded by the BPR method is accurate. However, for the BPR method to meet this criterion could result, in some cases, in large computational efforts.

The discussion is limited in this paper to closed queuing networks, although all the used principles remain valid for open queuing networks. The results presented here could therefore be extended, without too much difficulty, to open queuing networks.

2. CONDITIONAL DISTRIBUTIONS IN QUEUING NETWORKS

The state of a network composed of b queues is generally represented by a b-tuple x,

$$x = [x_1 \ldots x_b]$$

with

$$x_i = [x_{i1} \ldots x_{is_i}] \qquad 1 \leq i \leq b$$

where x_{ij} denotes the number of jobs in jth stage of service at queue i and s_i is the number of service stages at queue i. We will use $x - 1_j^i + 1_l^k$ to denote the state derived from x by transferring one job from stage j of queue i to stage l of queue k. $\mathcal{S}$ will be used to denote the state space, that is, the set of all feasible states. The behavior of the system can then be modeled by a discrete time Markov chain with state space $\mathcal{S}$. The transition probabilities between all the states of $\mathcal{S}$ are given by the stochastic matrix P with Perron–Frobenius eigenvector π:

$$\pi P = \pi \tag{1}$$

The component of π corresponding to the state x is the steady-state probability that the system is in state x:

$$\pi_x = \lim_{t \to \infty} \text{Prob}[x(t) = x]$$

Assume now that the queuing network is partitioned into two subsystems, denoted σ and β respectively, and that the network state descriptor is partitioned accordingly,

$$x = (x_\sigma, x_\beta)$$

Then two important types of conditional probability distributions can be of interest. One can be interested in the steady-state probability distribution of the system when the subsystem σ is in a precise state x_σ. This conditional distribution is denoted $\pi_{|x_\sigma}$ and is given by

$$\forall x_\beta : \pi_{|x_\sigma}[x_\beta] \stackrel{\text{def}}{=} \lim_{t\to\infty} \text{Prob}[x_\beta(t) = x_\beta | x_\sigma(t) = x_\sigma]$$

where $\pi_{|x_\sigma}[x_\beta]$ denotes the x_β component of the vector $\pi_{|x_\sigma}$. Instead of fixing the precise state of the subsystem σ, only its population, that is, the number of jobs in it, can be fixed. The steady-state probability distribution of the system when the population in σ is equal to n is the conditional probability distribution $\pi_{|n}$ given by

$$\forall x_\sigma, x_\beta : \pi_{|n}[x_\sigma, x_\beta] \stackrel{\text{def}}{=} \lim_{t\to\infty} \text{Prob}[x_\sigma(t) = x_\sigma, x_\beta(t) = x_\beta \mid \sharp x_\sigma(t) = n]$$

where $\sharp x_\sigma$ denotes the population of σ when σ is in state x_σ. (For conditional probability distributions based on job classes, [7] constitutes a good reference.)

The conditional distribution $\pi_{|x_\sigma}$ is, at a normalization constant, equal to π_{x_σ}, the part of the complete probability distribution π corresponding to the states with σ in state x_σ:

$$\pi_{|x_\sigma} = \pi_{x_\sigma} \left[\sum_{x_\beta} \pi_{x_\sigma}[x_\beta] \right]^{-1} = \pi_{x_\sigma} \left[\sum_{x_\beta} \pi[x_\sigma, x_\beta] \right]^{-1} \tag{2}$$

and since the distribution π satisfies the Perron–Frobenius equation (1), this equation (1) can be transformed [4] to yield

$$\pi_{|x_\sigma} = \pi_{|x_\sigma} \left[P_{x_\sigma}^{x_\sigma} + P_{\overline{x_\sigma}}^{x_\sigma} \left(I - P_{\overline{x_\sigma}}^{\overline{x_\sigma}}\right)^{-1} P_{x_\sigma}^{\overline{x_\sigma}} \right] \stackrel{\text{def}}{=} \pi_{|x_\sigma} [P_{x_\sigma}^{x_\sigma} + C_{x_\sigma}^{x_\sigma}] \tag{3}$$

where $P_{y_\sigma}^{x_\sigma}$ denotes the matrix of transition probabilities from states of the form $(x_\sigma \cdot)$ to states of the form $(y_\sigma \cdot)$ and where $\overline{x_\sigma}$ denotes any state $(y_\sigma \cdot)$ with $y_\sigma \neq x_\sigma$. This shows that the conditional distribution $\pi_{|x_\sigma}$ is given by the Perron–Frobenius eigenvector of the stochastic matrix $[P_{x_\sigma}^{x_\sigma} + C_{x_\sigma}^{x_\sigma}]$. The $(x_\beta; y_\beta)$ component of the matrix $P_{x_\sigma}^{x_\sigma}$ is the probability of going from a state (x_σ, x_β) to a state (x_σ, y_β) in one single transition. The same component of the matrix $C_{x_\sigma}^{x_\sigma}$ gives the probability of going from a state (x_σ, x_β) to a state (x_σ, y_β) through any nonempty sequence of intermediate states $\{(z_\sigma^1, z_\beta^1), (z_\sigma^2, z_\beta^2), \ldots\}$ with $z_\sigma^r \neq x_\sigma, r \geq 1$. The expression (3) can also be obtained using probabilistic arguments only (see e.g. [13], pp. 114–115).

Similarly, the conditional distribution $\pi_{|n}$ is equal, at a normalization constant, to π_n, the part of the complete distribution π corresponding to states with a population of n jobs in σ:

$$\pi_{|n} = \pi_n \left[\sum_{x_\beta} \sum_{x_\sigma} \pi_n[x_\sigma, x_\beta] \right]^{-1} = \pi_n \left[\sum_{x_\beta} \sum_{x_\sigma | \sharp x_\sigma = n} \pi[x_\sigma, x_\beta] \right]^{-1} \tag{4}$$

and can be similarly shown to satisfy the equation:

$$\pi_{|n} = \pi_{|n} \left[P_n^n + P_{\bar{n}}^n (I - P_{\bar{n}}^{\bar{n}})^{-1} P_n^{\bar{n}} \right] \stackrel{\text{def}}{=} \pi_{|n}[P_n^n + C_n^n] \tag{5}$$

where P_m^n denotes the matrix of transition probabilities from all the states with a population of n jobs in σ to all the states with a population of m jobs in σ and where $\bar{n}$ denotes σ populations of size different from n. There are several ways of going from a state (x_σ, x_β) to a state (y_σ, y_β) with $\sharp x_\sigma = \sharp y_\sigma = n$: either in a single step (the probabilities of such transitions are given by the matrix P_n^n), or through a sequence of intermediate states $\{(z_\sigma^1, z_\beta^1), (z_\sigma^2, z_\beta^2), \cdots\}$ with $\sharp z_\sigma^r \neq n, r \geq 1$ (the probabilities of such transitions are given by the matrix C_n^n) As for equation (3), equation (5) can be obtained using probabilistic arguments only (see ([13]), pp. 114–115).

The conditional distributions $\pi_{|x_\sigma}$ and $\pi_{|n}$ can thus be determined by the equations (3) and (5). However, the computation of the matrices $C_{x_\sigma}^{x_\sigma}$ and C_n^n requires large inverses to be determined, a computational effort almost as heavy as the resolution of the initial system (1). One is therefore interested in *more easily computable* matrices, say $\hat{C}_{x_\sigma}^{x_\sigma}$ and $\hat{C}_n^n$, such that the solutions of the equations

$$\begin{aligned} \hat{p}_{|x_\sigma} &= \hat{p}_{|x_\sigma}[P_{x_\sigma}^{x_\sigma} + \hat{C}_{x_\sigma}^{x_\sigma}] \\ \hat{p}_{|_n} &= \hat{p}_{|n}[P_n^n + \hat{C}_n^n] \end{aligned} \tag{6}$$

are identical or close to $\pi_{|x_\sigma}$ and $\pi_{|n}$, respectively. The use of matrices $\hat{C}_{x_\sigma}^{x_\sigma}$ and $\hat{C}_n^n$ which produce exact solutions is called *exact aggregation*. Exact aggregation will of course be obtained with any matrices $\hat{C}_{x_\sigma}^{x_\sigma}$ and $\hat{C}_n^n$ that satisfy

$$\begin{aligned} \pi_{|x_\sigma} \hat{C}_{x_\sigma}^{x_\sigma} &= \pi_{|x_\sigma} C_{x_\sigma}^{x_\sigma} \\ \pi_{|n} \hat{C}_n^n &= \pi_{|n} C_n^n \end{aligned} \tag{7}$$

Each of the equation systems (7) is composed of k (the size of the vector $\pi_{|x_\sigma}$ or $\pi_{|n}$) equations in k^2 variables (the elements of the matrix $\hat{C}_{x_\sigma}^{x_\sigma}$ or $\hat{C}_n^n$); thus many such matrices exist. In the following subsections we propose certain matrices that satisfy (7) when the queuing system satisfies the local balance equations or the subsystem balance equations.

3. NETWORKS WITH LOCAL OR SUBSYSTEM BALANCE

In any network, the steady-state probability vector π is defined by the *global balance* equations

$$\forall x : \sum_{(i,j)} \sum_{(k,l)} \pi[x + 1_l^k - 1_j^i] \operatorname{serv}_{(k,l)}(x_k + 1_l^k) R[k,l;i,j] = \pi[x] \sum_{(i,j)} \operatorname{serv}_{(i,j)}(x_i) \quad (8)$$

where the probability of a service in stage l of queue k, $\operatorname{serv}_{(k,l)}(x_k)$, may depend on the state x_k of queue k and where $R[k,l;i,j]$ denotes the probability that after the completion of a service in stage l of queue k, this job is routed to stage j of queue i. In this section, we will focus on queuing networks whose steady-state probability vector π satisfies either the *local balance* equations

$$\forall x : \sum_{(k,l)} \pi[x + 1_l^k - 1_j^i] \operatorname{serv}_{(k,l)}(x_k + 1_l^k) R[k,l;i,j] = \pi_x \operatorname{serv}_{(i,j)}(x_i) \quad (9)$$

or the *subsystem balance* equations

$$\forall x : \sum_{(i,j)\in\sigma} \sum_{(k,l)} \pi[x + 1_l^k - 1_j^i] \operatorname{serv}_{(k,l)}(x_k + 1_l^k) R[k,l;i,j] = \pi[x] \sum_{(i,j)\in\sigma} \operatorname{serv}_{(i,j)}(x_i) \quad (10)$$

These sets of equations express balance for three sets of ergodic flow rates when the system is in steady-state:

1. The global balance equations (which by definition are always verified) state that the probability of entering a state x by means of a job entering any stage of service of the whole system is equal to the probability of leaving that same state x by means of a service completion at any stage of the system.
2. The local balance equations state [5] that the probability of entering a state x by means of a job entering some stage (i,j) is equal to the probability of leaving that same state x by means of a service completion at that same stage (i,j).

3. The subsystem balance equations state that the probability of entering a state x by means of a job entering any stage of the subsystem σ is equal to the probability of leaving that same state x by means of a service completion at any stage of the subsystem σ.

When the local or the subsystem balance equations are satisfied, the global balance equations can be transformed in order to obtain a simpler definition for the steady-state probability vector π. Such transformations also yield matrices $\hat{C}_{x_\sigma}^{x_\sigma}$ and $\hat{C}_n^n$, which, when introduced in equation (6), allow the conditional distributions $\pi_{|x_\sigma}$ and $\pi_{|n}$ to be exactly computed. (Another kind of balance, the *station balance*, has been considered [6] in the literature. This balance is used to explain why some service disciplines yield product form solutions.)

Let us first rewrite the global balance equations (8) by decomposing the double sum into four terms

$$
\begin{aligned}
\forall x = (x_\sigma, x_\beta): \\
&\sum_{(i,j)\in\sigma}\sum_{(k,l)\in\sigma} \pi[x_\sigma + 1_l^k - 1_j^i, x_\beta]\,\mathrm{serv}_{(k,l)}(x_k + 1_l^k) R[k,l;i,j] \\
&+ \sum_{(i,j)\in\sigma}\sum_{(k,l)\in\beta} \pi[x_\sigma - 1_j^i, x_\beta + 1_l^k]\,\mathrm{serv}_{(k,l)}(x_k + 1_l^k) R[k,l;i,j] \\
&+ \sum_{(i,j)\in\beta}\sum_{(k,l)\in\sigma} \pi[x_\sigma + 1_l^k, x_\beta - 1_j^i]\,\mathrm{serv}_{(k,l)}(x_k + 1_l^k) R[k,l;i,j] \\
&+ \sum_{(i,j)\in\beta}\sum_{(k,l)\in\beta} \pi[x_\sigma, x_\beta + 1_l^k - 1_j^i]\,\mathrm{serv}_{(k,l)}(x_k + 1_l^k) R[k,l;i,j] \\
&= \pi[x_\sigma, x_\beta] \sum_{(i,j)} \mathrm{serv}_{(i,j)}(x_i)
\end{aligned}
\tag{11}
$$

Using π_{x_σ} to denote the part of π that corresponds to states of the form $(x_\sigma \cdot)$ and using the structure of the stochastic matrix P as depicted in Figure 1, this last system of equations reduces to the following vector form:

$$
\begin{aligned}
\forall x_\sigma: \\
&\sum_{(i,j)\in\sigma}\sum_{(k,l)\in\sigma} \pi_{x_\sigma + 1_l^k - 1_j^i} P_{x_\sigma}^{x_\sigma + 1_l^k - 1_j^i} + \sum_{(i,j)\in\sigma} \pi_{x_\sigma - 1_j^i} P_{x_\sigma}^{x_\sigma - 1_j^i} \\
&+ \sum_{(k,l)\in\sigma} \pi_{x_\sigma + 1_l^k} P_{x_\sigma}^{x_\sigma + 1_l^k} + \pi_{x_\sigma} P_{x_\sigma}^{x_\sigma} = \pi_{x_\sigma}
\end{aligned}
\tag{12}
$$

Note that by construction, the diagonal element of $P_{x_\sigma}^{x_\sigma}$ are the complement to 1 of the row sums of P. When the local balance equations (9) are satisfied, this last equation can then be transformed to obtain the following result.

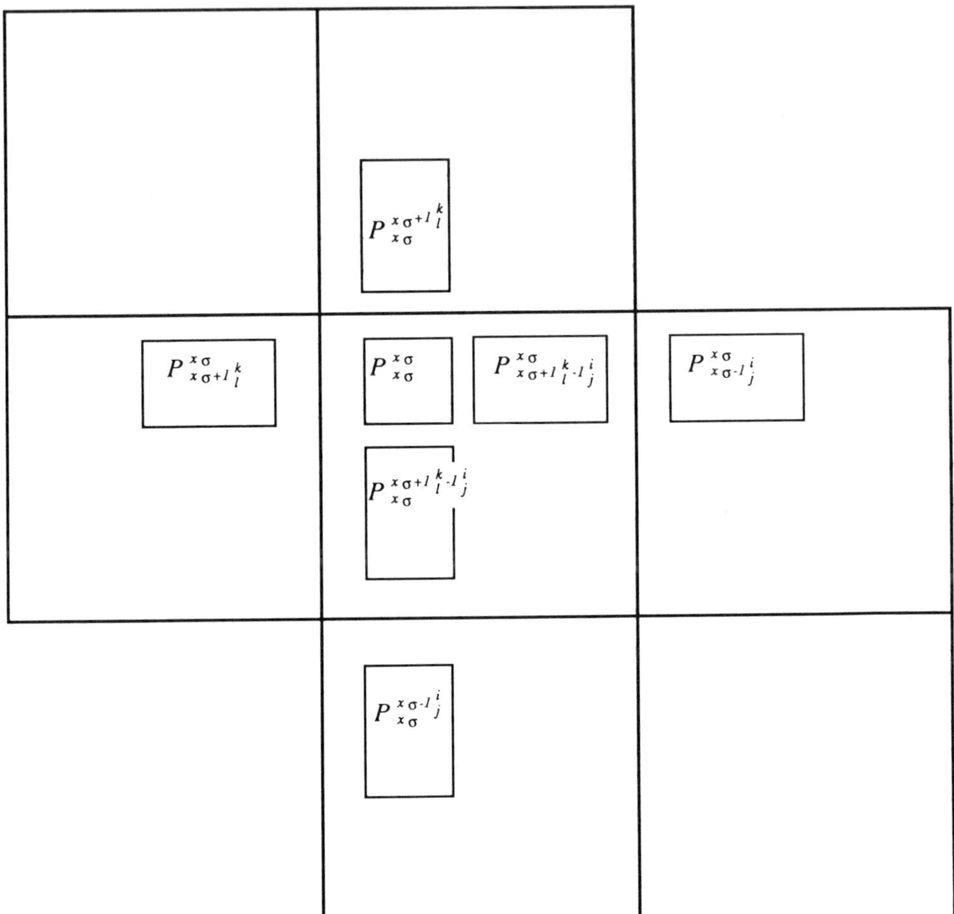

Figure 1 Matrix P structure.

THEOREM 1 In queuing networks that satisfy the local balance equations (9), the conditional probability distribution $\pi_{|x_\sigma}$ is exactly given by the following equation

$$\pi_{|x_\sigma} = \pi_{|x_\sigma}[P^{x_\sigma}_{x_\sigma} + \hat{C}^{x_\sigma}_{x_\sigma}] \tag{13}$$

with

$$\hat{C}^{x_\sigma}_{x_\sigma} \stackrel{\text{def}}{=} \hat{S}^{x_\sigma}_{x_\sigma} + \hat{L}^{x_\sigma}_{x_\sigma}$$

The two matrices $\hat{S}_{x_\sigma}^{x_\sigma}$ and $\hat{L}_{x_\sigma}^{x_\sigma}$ are defined as follows:

$$\hat{S}_{x_\sigma}^{x_\sigma} = \text{diag}\left[\sum_{(i,j)\in\sigma}\sum_{(k,l)\in\sigma} P_{x_\sigma+1_l^k-1_j^i}^{x_\sigma}\mathbf{1}^T + \sum_{(i,j)\in\sigma} P_{x_\sigma-1_j^i}^{x_\sigma}\mathbf{1}^T\right]$$

$$[\hat{L}_{x_\sigma}^{x_\sigma}]_{x_\beta+1_j^i-1_l^k}^{x_\beta} = \text{serv}_{(k,l)}(x_k)R[k,l;\sigma](I - R[\sigma;\sigma])^{-1}R[\sigma;i,j],$$

$$\begin{cases}\forall(k,l)\in\beta\\ \forall(i,j)\in\beta\end{cases} \tag{14}$$

where $\mathbf{1}^T$ denotes a column vector of all ones of appropriate size and the operator diag[z] builds a square matrix with the vector z on the diagonal. The elements of the matrix R_γ^δ, ($\delta,\gamma \in \{\sigma,\beta\}$), are the routing probabilities from (queue, stage) pairs of δ to (queue, stage) pairs of γ. Thus, the quantity

$$R[k,l;\sigma](I - R[\sigma;\sigma])^{-1}R[\sigma;i,j]$$

is the probability that the first pair (queue, stage) of β which is visited by a job which leaves β from the pair (k,l) is the pair (i,j), assuming zero service time in σ.

Proof. This theorem results from the following two equalities:

$$\sum_{(i,j)\in\sigma}\sum_{(k,l)\in\sigma} \pi_{x_\sigma+1_l^k-1_j^i}P_{x_\sigma}^{x_\sigma+1_l^k-1_j^i} + \sum_{(i,j)\in\sigma}\pi_{x_\sigma-1_j^i}P_{x_\sigma}^{x_\sigma-1_j^i} = \pi_{x_\sigma}\hat{S}_{x_\sigma}^{x_\sigma} \tag{15}$$

$$\sum_{(k,l)\in\sigma}\pi_{x_\sigma+1_l^k}P_{x_\sigma}^{x_\sigma+1_l^k} = \pi_{x_\sigma}\hat{L}_{x_\sigma}^{x_\sigma} \tag{16}$$

Summing the local balance equations (9) over all pairs (i,j) of σ leads to the subsystem balance equations (10), which, when rewritten in vector form, yield the first equality (15). The second equality is obtained by applying recursively the local balance equations. A formal proof of this equality is given in Appendix A. Introducing equalities (15) and (16) in (12) proves that the subvector π_{x_σ} satisfies equation (13). Theorem 1 results then from the parallelism (defined in equation (2)) between π_{x_σ} and $\pi_{|x_\sigma}$.

Theorem 1 shows that exact aggregation is obtained with the matrix $P_{x_\sigma}^{x_\sigma} + \hat{S}_{x_\sigma}^{x_\sigma} + \hat{L}_{x_\sigma}^{x_\sigma}$. Such an exact aggregation has already been proposed in [5] where it was called the *short-cut method* or *Norton's theorem*. In that paper [5], it was indeed proved that the conditional distribution $\pi_{|x_\sigma}$ can be obtained by considering the subsystem β in isolation and by forcing the jobs that leave β to immediately return to β, assuming zero service times in σ. An identical interpretation can be derived from Theorem 1. Indeed, the elements of $\hat{S}_{x_\sigma}^{x_\sigma}$

correspond to the service completions in σ. These elements are added to their complements on the diagonal of $P_{x_\sigma}^{x_\sigma}$. This indicates that all the transitions corresponding to service completions in σ can be ignored. The subsystem β can therefore be studied in isolation. The definition (14) of the matrix $\hat{L}_{x_\sigma}^{x_\sigma}$ and the equality

$$\sum_{(k,l)\in\sigma} P_{x_\sigma+1_l^k}^{x_\sigma}\mathbf{1}^T = \hat{L}_{x_\sigma}^{x_\sigma}\mathbf{1}^T$$

show that all the jobs which leave β to σ are forced to immediately return to β, assuming zero service times in σ.

Unlike the results in [5], Theorem 1 does not require any assumption on the routing probability matrix. This suggests that the name Norton's theorem, which assumes that the subsystem σ is connected to β through a single bridge and reciprocally, can be misleading. An extension to more general routing matrices was already proposed in [1] for product-form networks. The proofs in [5] and [1] are based on the fact that the steady-state probability vector π has a product-form expression. The originality of our proof comes from the fact that the method is directly derived from the global and the local balance equations.

Equations (13) and (14) also show that only the size but not the components of $P_{x_\sigma}^{x_\sigma}$ and $\hat{C}_{x_\sigma}^{x_\sigma}$ in equation (13) depend on the precise state x_σ. This means that the conditional distribution $\pi_{|x_\sigma}$ is identical for all the states x_σ with same σ population.

An approach similar to the one used for $\pi_{|x_\sigma}$ leads to the following theorem for the conditional distribution $\pi_{|n}$.

THEOREM 2 In queuing networks that satisfy the local balance equations, the conditional probability distribution $\pi_{|n}$ is exactly given by the equation

$$\pi_{|n} = \pi_{|n}(P_n^n + \hat{C}_n^n)$$

with

$$\hat{C}_n^n \overset{\text{def}}{=} \hat{S}_n^n + \hat{L}_n^n$$

and

$$[\hat{S}_n^n]_{(x_\sigma+1_j^i-1_l^k,x_\beta)}^{(x_\sigma,x_\beta)} \overset{\text{def}}{=} \text{serv}_{(k,l)}(x_k)R[k,l;\beta](I-R[\beta;\beta])^{-1}R[\beta;i,j],$$

$$\begin{cases} \forall(i,j)\in\sigma \\ \forall(k,l)\in\sigma \end{cases}$$

$$[\hat{L}_n^n]^{(x_\sigma, x_\beta)}_{(x_\sigma, x_\beta + 1^i_j - 1^k_l)} \stackrel{\text{def}}{=} \text{serv}_{(k,l)}(x_k) R[k,l;\sigma](I - R[\sigma;\sigma])^{-1} R[\sigma;i,j],$$

$$\begin{cases} \forall (i,j) \in \beta \\ \forall (k,l) \in \beta \end{cases}$$

Proof. The details of the proof of this theorem are omitted since it is quite similar to that of Theorem 1. Conceptually, using the block structure of the matrix P depicted in Figure 2, the global balance equations (8) can be

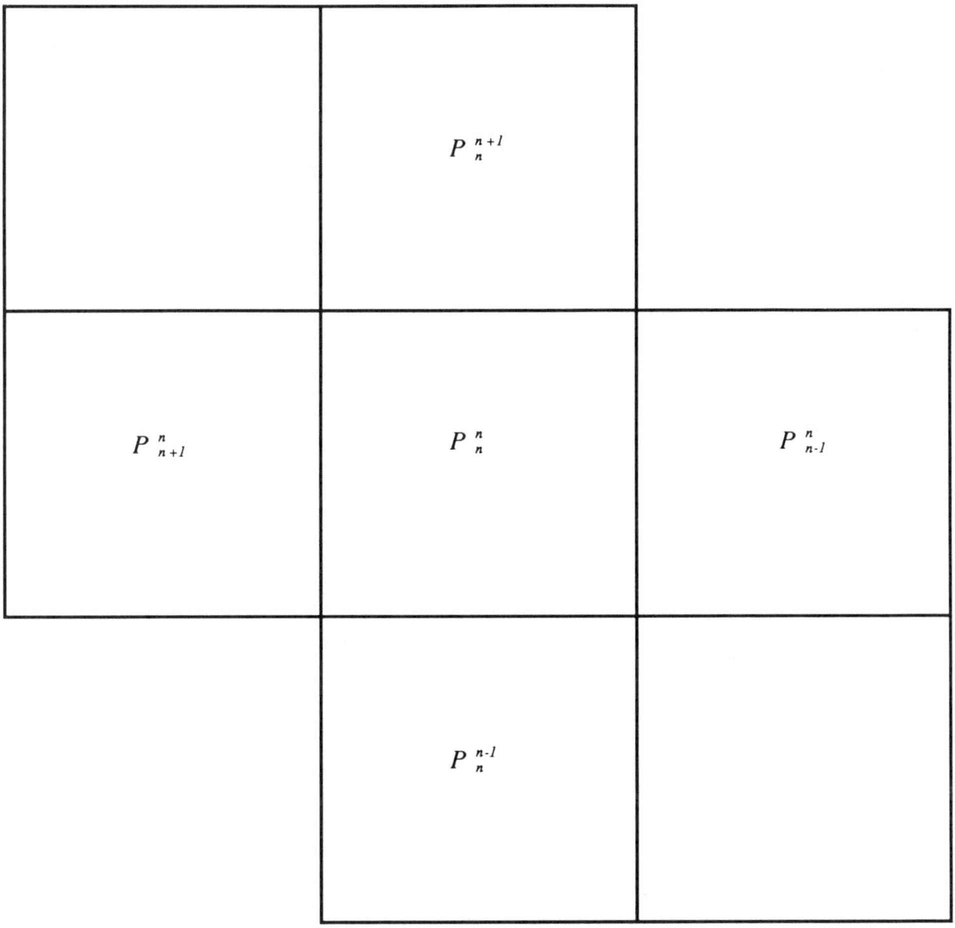

Figure 2 Matrix P structure.

rewritten in the vector form

$$\pi_{n-1}P_n^{n-1} + \pi_n P_n^n + \pi_{n+1}P_n^{n+1} = \pi_n \tag{17}$$

where π_n denotes the part of the steady-state probability vector π corresponding to states with a population of n jobs in σ. As for the proof of Theorem 1, the two equations

$$\pi_{n-1}P_n^{n-1} = \pi_n \hat{S}_n^n$$

$$\pi_{n+1}P_n^{n+1} = \pi_n \hat{L}_n^n$$

which are only vector transcriptions of the equations

$$\begin{aligned}
&\forall(x_\sigma,x_\beta) \mid \sharp x_\sigma = n : \\
&\qquad \sum_{(i,j)\in\sigma}\sum_{(k,l)\in\beta} \pi[x_\sigma - 1_j^i, x_\beta + 1_l^k]\,\mathrm{serv}_{(k,l)}(x_k + 1_l^k)R[k,l;i,j] \\
&\qquad = \sum_{(i,j)\in\sigma}\sum_{(r,s)\in\sigma} \pi[x_\sigma + 1_s^r - 1_j^i, x_\beta]\,\mathrm{serv}_{(r,s)}(x_r + 1_s^r)R[r,s;\beta] \\
&\qquad\qquad \times [I - R[\beta;\beta]]^{-1}R[\beta;i,j]
\end{aligned} \tag{18}$$

and

$$\begin{aligned}
&\forall(x_\sigma,x_\beta) \mid \sharp x_\sigma = n : \\
&\qquad \sum_{(i,j)\in\beta}\sum_{(k,l)\in\sigma} \pi[x_\sigma + 1_l^k, x_\beta - 1_j^i]\,\mathrm{serv}_{(k,l)}(x_k + 1_l^k)R[k,l;i,j] \\
&\qquad = \sum_{(i,j)\in\beta}\sum_{(r,s)\in\beta} \pi[x_\sigma, x_\beta + 1_s^r - 1_j^i]\,\mathrm{serv}_{(r,s)}(x_r + 1_s^r)R[r,s;\sigma] \\
&\qquad\qquad \times [I - R[\sigma;\sigma]]^{-1}R[\sigma;i,j]
\end{aligned} \tag{19}$$

can be shown to hold. Equation (19) was already proved in Appendix A and equation (18) can be established in the same way.

A word of comparison between Theorems 1 and 2 might be useful. In Theorem 1, $\pi_{|x_\sigma}$ was determined by forcing the jobs that leave β to immediately return to β, assuming zero service times in σ (the matrix $\hat{L}_{x_\sigma}^{x_\sigma}$), and by neglecting the transitions corresponding to service completions in σ (the matrix $\hat{S}_{x_\sigma}^{x_\sigma}$ vanishes with its complement on the diagonal of $P_{x_\sigma}^{x_\sigma}$). In Theorem 2, the matrix $\hat{L}_n^n$ is similar to $\hat{L}_{x_\sigma}^{x_\sigma}$ and means that the jobs that leave β are again forced to immediately return to β. On the other hand, the transitions corresponding to service completions in σ are no longer neglected. The transitions by which jobs move from σ to σ keep the populations constant in σ; they correspond to off-diagonal blocks of P_n^n (see Figure 1 and 2) and are not added to their complements on the diagonal of P_n^n. The remaining transitions, which correspond to jobs leaving σ for β, must therefore be taken into account, assuming zero service time in β (the matrix $\hat{S}_n^n$).

Both Theorem 1 and 2 require the local balance equations to be satisfied. Next we give exact aggregations that require the subsystem balance equations (10) only to be satisfied.

THEOREM 3 In queuing networks that satisfy the subsystem balance equations (10), the conditional probability distribution $\pi_{|n}$ is exactly given by the equation

$$\pi_{|n} = \pi_{|n}(P_n^n + \hat{C}_n^n) \tag{20}$$

with $\hat{C}_n^n$ chosen among one of the three following possibilities:

$$\begin{aligned}\hat{C}_n^n = &(i) \quad I - P(\sigma)_n^n + P_{n+1}^n[I - P(\sigma)_{n+1}^{n+1}]^{-1}P_n^{n+1}\\ &(ii) \quad P_{n-1}^n[I - P(\beta)_{n-1}^{n-1}]^{-1}P_n^{n-1} + I - P(\beta)_n^n\\ &(iii) \quad P_{n-1}^n[I - P(\beta)_{n-1}^{n-1}]^{-1}P_n^{n-1} + P_{n+1}^n[I - P(\sigma)_{n+1}^{n+1}]^{-1}P_n^{n+1}\end{aligned}$$

where $P(\sigma)$ and $P(\beta)$ are stochastic matrices identical to P except that in $P(\sigma)$ (respectively $P(\beta)$), all the transitions corresponding to services in β (respectively σ) are discarded.

Proof. Details of the proof are given in Appendix B. Basically, one assumes that the subsystem balance equations (10) are satisfied for σ populations of n jobs, that is, for all the states $x = (x_\sigma, x_\beta)$ such that $\sharp x_\sigma = n$. Rewriting these equations in vector form gives, after some manipulations:

$$\pi_{n-1}P_n^{n-1} = \pi_n[I - P(\sigma)_n^n] \tag{21}$$

and

$$\pi_{n+1}P_n^{n+1} = \pi_n[I - P(\beta)_n^n]$$

If the subsystem balance equations are also satisfied for σ populations of $n+1$ jobs, equation (21) can be rewritten as

$$\pi_n P_{n+1}^n[I - P(\sigma)_{n+1}^{n+1}]^{-1} = \pi_{n+1} \tag{22}$$

Using now equations (21) and (22) in the global balance equations (17) shows that π_n satisfies equation (20) with $\hat{C}_n^n$ defined by (i). Theorem 3 (i) then results from the parallelism between $\pi_{|n}$ and π_n (see equation (4)). Similarly, the aggregation matrix (ii) and (iii) require the subsystem balance equations to be satisfied for σ populations of n and $n-1$ jobs, and of $n-1$ and $n+1$ jobs, respectively.

The proof shows that Theorem 3 holds when the subsystem balance equations are satisfied for σ population of $n-1$, n, and $n+1$ jobs only. When the subsystem balance equations are satisfied for other σ populations, other exact aggregation matrices can be obtained. For example, let us assume that these equations are satisfied for σ populations of $n+j$ and $n-k$ jobs, respectively. One therefore has

$$\pi_{n+j-1}P_{n+j}^{n+j-1} = \pi_{n+j}[I - P(\sigma)_{n+j}^{n+j}] \tag{23}$$

and

$$\pi_{n-k+1}P_{n-k}^{n-k+1} = \pi_{n-k}[I - P(\beta)_{n-k}^{n-k}] \tag{24}$$

Then, the following recurrent definitions [12] of the matrices $Y(n+j)_{n+1}$ and $X(n-k)_{n-1}$

$$Y(n_j)_{n+j-0} \stackrel{\text{def}}{=} I - P(\sigma)_{n+j}^{n+j}$$

$$Y(n+j)_{n+j-i} \stackrel{\text{def}}{=} I - P_{n+j-i}^{n+j-i} - P_{n+j-(i-1)}^{n+j-i}[Y(n+j)_{n+j-(i-1)}]^{-1}P_{n+j-i}^{n+j-(i-1)}$$

$$1 \leq i \leq j-1 \tag{25}$$

$$X(n-k)_{n-k+0} \stackrel{\text{def}}{=} I - P(\beta)_{n-k}^{n-k}$$

$$X(n-k)_{n-k+i} \stackrel{\text{def}}{=} I - P_{n-k+i}^{n-k+i} - P_{n-k+i-1}^{n-k+i}[X(n-k)_{n-k+i-1}]^{-1}P_{n-k+i}^{n-k+i-1}$$

$$1 \leq i \leq k-1$$

allow us to state the following theorem:

THEOREM 4 In queuing networks that satisfy the subsystem balance equations for two σ populations only, $n-k$ and $n+j$ jobs respectively, the conditional probability distribution $\pi_{|n}$ is exactly given by the following equation

$$\pi_{|n} = \pi_{|n}(P_n^n + \hat{C}_n^n) \tag{26}$$

with

$$\hat{C}_n^n = P_{n-1}^n[X(n-k)_{n-1}]^{-1}P_n^{n-1} + P_{n+1}^n[Y(n+j)_{n+1}]^{-1}P_n^{n+1}$$

Proof. see Appendix C.

Theorems 3 and 4 introduce new exact aggregation methods that have natural physical interpretations. For example, the term

$$P_{n+1}^n[I - P(\sigma)_{n+1}^{n+1}]^{-1}P_n^{n+1}$$

in Theorem 3 indicates that the transitions that correspond to jobs leaving β for σ (the elements of P_{n+1}^n) are recovered by forcing the subsystem σ, brought by this arrival in some state with a σ population of $n + 1$ jobs, to send one job back to β, after any sequence of job transitions inside σ (i.e., which keeps the σ populations at $n + 1$). Equivalently, the term

$$P^n_{n+1}[Y(n+j)_{n+1}]^{-1}P^{n+1}_n$$

and the equation (33) established in Appendix C

$$\pi_{n+1}P^{n+1}_n = \pi_n P^n_{n+1}[Y(n+j)_{n+1}]^{-1}P^{n+1}_n$$

show that the transitions by which jobs leave β for σ (those of P^n_{n+1}) are recovered by forcing the system brought by this arrival in some state with a σ population of $n + 1$ jobs to go to a state with a σ population of n jobs, after any sequence of job transitions which maintains the σ population at values between $n + 1$ and $n + j$. Identical interpretations can be given for jobs leaving σ for β. This physical interpretation suggests that one name the methods based on the matrices defined in Theorem 3 or 4, the bounded population range method.

The purpose of Theorems 3 and 4 was to show that other exact aggregation methods can be used when the network satisfies certain flow rate balance properties. The particular interest of these theorems is to show that these properties must be satisfied at subsystem level only, a condition weaker than the local balance equations. This enlarges the class of networks that can be exactly aggregated. So far, unfortunately, we have failed to characterize such networks. But, when dealing with networks that do not satisfy any balance property, the approximation obtained by assuming subsystem balance will be proved to be more accurate than the approximation obtained by assuming the local balance. This might be the major interest of subsystem balance.

4. NON-PRODUCT FORM NETWORKS

The aggregation methods proposed in the previous section yield the exact solutions when the network satisfies local or subsystem balance. Many queuing networks, however, do not satisfy any of these balances. Nevertheless, these aggregation methods can still be used in order to obtain approximations. In order to evaluate the quality of these approximations, we first need the following two lemmas.

LEMMA 1 If $a = bZ$ with $a, b, Z > 0, Z \neq 0, a\mathbf{1}^T = 1$, then $a \in \mathcal{P}[Z]$, where $\mathcal{P}[Z]$ denotes the polyhedron whose vertices are given by the normalized nonnull rows of the matrix Z.

Proof. Let $d \stackrel{\text{def}}{=} Z\mathbf{1}^T$. The inverse of the following nonnegative diagonal matrix Λ_Z

$$[\Lambda_Z]_{ij} = \begin{cases} 0 & \text{if} \quad i \neq j \\ d_i & \text{if} \quad i = j \quad \text{and} \quad d_i \neq 0 \\ 1 & \text{if} \quad i = j \quad \text{and} \quad d_i = 0 \end{cases}$$

normalizes the nonnull rows of Z:

$$\Lambda_Z^{-1} Z \mathbf{1}^T = \mathbf{1}_Z^T$$

where 1_Z^T denotes a column vector of ones and zeros:

$$[\mathbf{1}_Z^T]_i = \begin{cases} 1 & \text{if} \quad d_i \neq 0 \\ 0 & \text{if} \quad d_i \neq 0 \end{cases}$$

One can now easily derive the equations

$$a = (b\Lambda_Z)\Lambda_Z^{-1} Z$$
$$(b\Lambda_Z) \geq 0$$
$$(b\Lambda_Z)\mathbf{1}_Z^T = 1$$

which proves that the vector a is given by the convex combination $(b\Lambda_Z)$ of the normalized *nonnull* rows of the matrix Z.

Lemma 1 gives the conditions under which the normalized nonnegative vector a is guaranteed to belong to the polyhedron $\mathcal{P}[Z]$, whatever the nonnegative vector b is. Any point of the polyhedron could therefore be chosen as an approximation of the vector a. The smaller the polyhedron could therefore be chosen as an approximation will be. The discussion given in Appendix D shows that the size of this polyhedron is mainly determined by the difference between the modulus of the two largest eigenvalues of Z. Typically, if

$$\text{sp}[Z] \stackrel{\text{def}}{=} \{\lambda_1, \ldots, \lambda_m\} = \{\lambda_1, 0, \ldots, 0\}$$

where sp$[Z]$ denotes the spectrum of Z, then Z is of rank one and the polyhedron reduces to a single point. In this case, the error made by approximating a by a point of $\mathcal{P}[Z]$ is of course null. If, on the other hand,

$$|\lambda_i| = |\lambda_1| \qquad 1 \leq i \leq m$$

then the matrix $\Lambda_Z^{-1} Z$ reduces to a permutation matrix and the polyhedron $\mathcal{P}[Z]$ is the whole $[0, 1]^m$ hypercube. The error made on a is in this case infinite.

LEMMA 2

$$\forall W, Z \geq 0, \quad W \neq 0 \neq Z, \quad \mathcal{P}[WZ] \subset \mathcal{P}[Z]$$

Proof. $\forall a \in \mathcal{P}[WZ], a\mathbf{1}^T = 1$. Furthermore, $\exists b \geq 0$, with

$$a = b\Lambda_{WZ}^{-1}WZ = b\Lambda_{WZ}^{-1}W\Lambda_Z\Lambda_Z^{-1}Z \stackrel{\text{def}}{=} c\Lambda_Z^{-1}Z$$

which shows with

$$c \geq \mathbf{0}$$

$$c\mathbf{1}_Z^T = c\Lambda_Z^{-1}Z\mathbf{1}^T = a\mathbf{1}^T = 1$$

that a belongs to $\mathcal{P}[Z]$.

Lemma 2 proves that the polyhedron $\mathcal{P}[WZ]$ is not greater than the polyhedron $\mathcal{P}[Z]$. Typically, $\mathcal{P}[WZ]$ varies from a single point when W is of rank one to $\mathcal{P}[Z]$ when W is equal, at a row normalization matrix, to a permutation matrix.

Let us now proceed to the analysis of the accuracy of the aggregation methods proposed in Section 3 if one applies these methods to non-product form networks. The conditional distribution $\pi_{|x_\sigma}$ is exactly given by the expression (3), which can be transformed to yield

$$\pi_{|x_\sigma} = \pi_{|x_\sigma} C_{x_\sigma}^{x_\sigma}(I - P_{x_\sigma}^{x_\sigma})^{-1}$$

Since all the quantities satisfy the conditions of Lemma 1, one can write

$$\pi_{|x_\sigma} \in \mathcal{P}[(I - P_{x_\sigma}^{x_\sigma})^{-1}] \stackrel{\text{def}}{=} \mathcal{P}[Z_{x_\sigma}^{x_\sigma}]$$

which shows that the exact solution $\pi_{|x_\sigma}$ is given by a precise point of the polyhedron $\mathcal{P}[Z_{x_\sigma}^{x_\sigma}]$. If this polyhedron is small, any of its points provides a good approximation of the exact solution. The polyhedron will be small when the modulus of the two largest eigenvalues of $Z_{x_\sigma}^{x_\sigma}$ are sufficiently apart. Since the eigenvalues o $Z_{x_\sigma}^{x_\sigma}$ and of $P_{x_\sigma}^{x_\sigma}$ are related by

$$\forall\lambda \in \text{sp}[P_{x_\sigma}^{x_\sigma}] : (1 - \lambda)^{-1} \in \text{sp}[Z_{x_\sigma}^{x_\sigma}]$$

and since the matrix $P_{x_\sigma}^{x_\sigma}$ is substochastic, the polyhedron $\mathcal{P}[Z_{x_\sigma}^{x_\sigma}]$ will be small whenever the largest eigenvalue of $P_{x_\sigma}^{x_\sigma}$ is close to 1 and its second largest eigenvalue has a modulus much smaller than one. These conditions are met in *nearly completely decomposable* (ncd) networks [8][9]. Indeed, if the transition rates inside the subsystem β (i.e., those of $P_{x_\sigma}^{x_\sigma}$) are much larger than the transitions inside σ (i.e., those of $P^{x_\sigma}_{x_\sigma+1_l^k-1_j^i}$) and than the transitions between the subsystems (i.e., those of $P^{x_\sigma}_{x_\sigma+1_l^k}$ and of $P^{x_\sigma}_{x_\sigma-1_j^i}$), then the matrix $P_{x_\sigma}^{x_\sigma}$ is almost stochastic (its spectral radius is then close to unity) and its second largest eigenvalue has a modulus much smaller than 1. That means that is such systems, any aggregation method that approximates the conditional distribution $\pi_{|x_\sigma}$ by a point of the polyhedron $\mathcal{P}[Z_{x_\sigma}^{x_\sigma}]$ provides an accurate solution.

In non-product form networks, the short-cut method determines the approximated solution $\hat{p}_{|x_\sigma}$ by the system

$$\hat{p}_{|x_\sigma} = \hat{p}_{|x_\sigma}(P^{x_\sigma}_{x_\sigma} + \hat{C}^{x_\sigma}_{x_\sigma})$$

which can be rewritten as

$$\hat{p}_{|x_\sigma} = \hat{p}_{|x_\sigma}\hat{C}^{x_\sigma}_{x_\sigma}(I - P^{x_\sigma}_{x_\sigma})^{-1} = \hat{p}_{|x_\sigma}\hat{C}^{x_\sigma}_{x_\sigma}Z^{x_\sigma}_{x_\sigma}$$

which shows, by Lemma 1, that $\hat{p}_{|x_\sigma} \in \mathcal{P}[Z^{x_\sigma}_{x_\sigma}]$. This method, therefore, necessarily yields accurate solutions for ncd systems. However, for non-ncd systems, the polyhedron can be arbitrary large and the approximations quite poor.

The same development can be done for the exact conditional distribution $\pi|_n$. Rewriting equation (5) gives

$$\pi|_n = \pi|_n C^n_n(I - P^n_n)^{-1} = \pi|_n P^n_{\bar{n}}(I - P^{\bar{n}}_{\bar{n}})^{-1}P^{\bar{n}}_n(I - P^n_n)^{-1} \tag{27}$$

and

$$\pi|_n \in \mathcal{P}[P^{\bar{n}}_n(I - P^n_n)^{-1}] \stackrel{\text{def}}{=} \mathcal{P}[P^{\bar{n}}_n Z^n_n] \subset \mathcal{P}[Z^n_n]$$

where the inclusion results from Lemma 2. Similarly, the polyhedron $\mathcal{P}[Z^n_n]$ is small if P^n_n is almost stochastic, that is, in systems with transition rates between the two subsystems much smaller than those inside the subsystems. Again, ncd systems meet this condition. Both aggregation methods introduced by Theorems 2 and 3 approximate the exact solution by a point of the polyhedron $\mathcal{P}[Z^n_n]$ and provide therefore accurate solutions for ncd systems. Furthermore, the matrix $\hat{C}^n_n$ introduced in (iii) of Theorem 3 can be rewritten in the form $BP^{\bar{n}}_n$, which shows that the corresponding aggregation method provides a solution that belongs to the polyhedron $\mathcal{P}[P^{\bar{n}}_n Z^n_n]$ and presents therefore more guarantees. Again, for non-ncd systems, the approximation can be arbitrarily poor. Note also that the aggregation method introduced by Courtois [8] can also be proved to yield an approximation that belongs to $\mathcal{P}[Z^n_n]$.

In order to analyze the accuracy of the bounded population range method, let us first develop equation (27), which defines the exact conditional distribution $\pi|_n$. Let us therefore define the matrices

$$Z^n_n(n+j, n-k) \stackrel{\text{def}}{=}$$
$$(I - P^n_n - P^n_{n+1}[\bar{Y}(n+j)_{n+1}]^{-1}P^{n+1}_n - P^n_{n-1}[\bar{X}(n-k)_{n-1}]^{-1}P^{n-1}_n)^{-1}$$

where $\bar{X}$ and $\bar{Y}$ have definitions similar to those of X and Y (see equation (25)):

$$\bar{Y}(n+j)_{n+j-0} \stackrel{\text{def}}{=} I - P^{n+j}_{n+j}$$

$$\bar{Y}(n+j)_{n+j-i} \stackrel{\text{def}}{=} I - P_{n+j-i}^{n+j-i} - P_{n+j-(i-1)}^{n+j-i}[\bar{Y}(n+j)_{n+j-(i-1)}]^{-1}P_{n+j-i}^{n+j-(i-1)}$$
$$1 \le i \le j-1$$
$$\bar{X}(n-k)_{n-k+0} \stackrel{\text{def}}{=} I - P_{n-k}^{n-k}$$
$$\bar{X}(n-k)_{n-k+i} \stackrel{\text{def}}{=} I - P_{n-k+i}^{n-k+i} - P_{n-k+i-1}^{n-k+i}[\bar{X}(n-k)_{n-k+i-1}]^{-1}P_{n-k+i}^{n-k+i-1}$$
$$1 \le i \le k-1$$

Using the block tridiagonal structure of the matrix P, the following relations can then be established.

LEMMA 3

(*i*) $\pi_{|n} = \pi_{|n}(P_n^n + P_{n+1}^n[\bar{Y}(N)_{n+1}]^{-1}P_n^{n+1} + P_{n-1}^n[\bar{X}(0)_{n-1}]^{-1}P_n^{n-1})$

$\forall k,j,: 1 \le k \le n, 1 \le j \le N-n$

(*ii*) $\pi_{|n} \in \mathcal{P}[Z_n^n(n+j,n-k)]$ (28)

(*iii*) $\mathcal{P}[Z_n^n(n+j+1,n-(k+1))] \subseteq \cdot\mathcal{P}[Z_n^n(n+j,n-k)]$

Proof. See Appendix E.

Lemma 3 shows that the exact solution $\pi_{|n}$ is guaranteed to belong to any of the polyhedra $\mathcal{P}[Z_n^n(n+j,n-k)]$ and that the larger j and k are, the smaller the polyhedron is. Note that statement (iii) remains valid if j and k are increased independently. Eventually, one can state that the polyhedron $\mathcal{P}[Z_n^n(n+(N-n),n-n)] = \mathcal{P}[Z_n^n(N,0)]$ reduces to a single point. Indeed, the matrix

$$P_n^n + P_{n+1}^n[\bar{Y}(N)_{n+1}]^{-1}P_n^{n+1} + P_{n-1}^n[\bar{X}(0)_{n-1}]^{-1}P_n^{n-1}$$

is stochastic and the inverse $[Z_n^n(N,0)]$ does not exists. However, the row-normalized inverse exists and is equal [10] to the rank one matrix

$$\mathbf{1}^T\pi_{|n}$$

Any point of a polyhedron $\mathcal{P}[Z_n^n(n+j,n-k)]$ can therefore be used as an approximation of $\pi_{|n}$; the larger j and k are, the smaller the polyhedron and the better the approximation will be. The size of a polyhedron $\mathcal{P}[Z_n^n(n+j,n-k)]$ depends on the difference between the modulus of the two largest eigenvalues of $Z_n^n(n+j,n-k)$. Again, one has

$$\forall\lambda \in \text{sp}[P_n^n + P_{n+1}^n[\bar{Y}(n+j)_{n+1}]^{-1}P_n^{n+1} + P_{n-1}^n[\bar{X}(n-k)_{n-1}]^{-1}P_n^{n-1}]:$$
$$(1-\lambda)^{-1} \in \text{sp}[Z_n^n(n_{+j},n-k)]$$

which shows that the polyhedron is small when the matrix

$$P_n^n + P_{n+1}^n[\bar{Y}(n+j)_{n+1}]^{-1}P_n^{n+1} + P_{n-1}^n + P_{n-1}^n[\bar{X}(n-k)_{n-1}]^{-1}P_n^{n-1}$$

is almost stochastic and presents large couplings between its states. This is the case when the system corresponding to states $(x_\sigma x_\beta)$ with $(n-k \leq \sharp x_\sigma \leq n+j)$ is nearly completely decomposable. In such systems, any point of $\mathcal{P}[Z_n^n(n+j,n-k)]$ constitutes a good approximation of $\pi_{|n}$.

Let us now analyze the bounded population range method. Let $\hat{p}_{|n}(n+j,n-k)$ denote the solution of the system

$$\hat{p}_{|n}(n+j,n-k) = \hat{p}_{|n}(n+j,n-k)$$
$$(P_n^n + P_{n-1}^n[X(n-k)_{n-1}]^{-1}P_n^{n-1} + P_{n+1}^n[Y(n+j)_{n+1}]^{-1}P_n^{n+1}) \quad (29)$$

Then, using the same mathematical tools as in Appendix E, the following relation can be established

$$\hat{p}_{|n}(N,n-k) \in \mathcal{P}[Z_n^n(N,n-(k-1))] \qquad 1<k<n-1$$
$$\hat{p}_{|n}(n+j,0) \in \mathcal{P}[Z_n^n(n+(j-1),0] \qquad 1<j<N-n$$
$$\hat{p}_{|n}(n+j,n-k) \in \mathcal{P}[Z_n^n(n+j-1,n-(k-1))] \qquad 1<k<n-1, \quad 1<j<N-n$$

This shows that $\hat{p}_{|n}(n+j,n-k)$ is guaranteed to belong to a polyhedron which also contains the exact solution. Furthermore, the larger j and k are, the smaller this polyhedron is. If the values j and k are such that the system composed of the states $(x_\sigma x_\beta)$ with $(n-k \leq \sharp x_\sigma \leq n+j)$ is ncd, the approximation $\hat{p}_{|n}(n+j,n-k)$ will be accurate. The amount of computation required by this method comes from the determination of the inverses $[X(n-k)_{n-1}]^{-1}$ and $[Y(n+j)_{n+1}]^{-1}$ and from the resolution of the system (29); it is approximately equal to $(k+j+1)*c$, where c is the time needed to determine the inverse of a Y or X matrix.

5. CONCLUSIONS

Different methods for the determination of conditional distributions in queuing networks have been analyzed. If the network satisfies the local balance property, the short-cut method yields the exact solution and appears adequate if explicit closed-form expressions are not available. If the network satisfies the subsystem balance property, the bounded population range method, with minimum range $(n-1,n)$ or $(n,n+1)$, gives the exact solution. For the networks that do not present any of these properties, the short-cut method and the bounded population range methods yield only approximations. These approximations will be accurate if the network is nearly completely decomposable. Finally, for non-ncd networks that do not present any specific property, the bounded population range method still yields accurate approximations if a sufficiently large range $(n - k, n + j)$ of population values is considered.

APPENDIX A

Proof. In order to prove the equality (16), we recursively use the local balance equations. This leads to the following development:

$$
\begin{aligned}
&\pi[x_\sigma + 1_l^k, x_\beta - 1_j^i]\,\mathrm{serv}_{(k,l)}(x_k + 1_l^k) \\
&= \sum_{(r_0,s_0)\in\beta} \pi[x_\sigma, x_\beta + 1_{s_0}^{r_0} - 1_j^i]\,\mathrm{serv}_{(r_0,s_0)}(x_{r_0} + 1_{s_0}^{r_0}) R[r_0,s_0;k,l] \\
&\quad + \sum_{(r_0,s_0)\in\sigma} \pi[x_\sigma + 1_{s_0}^{r_0}, x_\beta - 1_j^i]\,\mathrm{serv}_{(r_0,s_0)}(x_{r_0} + 1_{s_0}^{r_0}) R[r_0,s_0;k,l] \\
&= \sum_{(r_0,s_0)\in\beta} \pi[x_\sigma, x_\beta + 1_{s_0}^{r_0} - 1_j^i]\,\mathrm{serv}_{(r_0,s_0)}(x_{r_0} + 1_{s_0}^{r_0}) R[r_0,s_0;k,l] \\
&\quad + \sum_{(r_1,s_1)\in\beta} \pi[x_\sigma, x_\beta + 1_{s_1}^{r_1} - 1_j^i]\,\mathrm{serv}_{(r_1,s_1)}(x_{r_1} + 1_{s_1}^{r_1}) \\
&\quad \times \sum_{(r_0,s_0)\in\sigma} R[r_1,s_1;r_0,s_0]R[r_0,s_0;k,l] \\
&\quad + \sum_{(r_1,s_1)\in\sigma} \pi[x_\sigma + 1_{s_1}^{r_1}, x_\beta - 1_j^i]\,\mathrm{serv}_{(r_1,s_1)}(x_{r_1} + 1_{s_1}^{r_1}) \\
&\quad \times \sum_{(r_0,s_0)\in\sigma} R[r_1,s_1;r_0,s_0]R[r_0,s_0;k,l] \\
&= \sum_{(r_0,s_0)\in\beta} \pi[x_\sigma, x_\beta + 1_{s_0}^{r_0} - 1_j^i]\,\mathrm{serv}_{(r_0,s_0)}(x_{r_0} + 1_{s_0}^{r_0}) R[r_0,s_0;k,l] \\
&\quad + \sum_{(r_1,s_1)\in\beta} \pi[x_\sigma, x_\beta + 1_{s_1}^{r_1} - 1_j^i]\,\mathrm{serv}_{(r_1,s_1)}(x_{r_1} + 1_{s_1}^{r_1}) \\
&\quad \times \sum_{(r_0,s_0)\in\sigma} R[r_1,s_1;r_0,s_0]R[r_0,s_0;k,l] \\
&\quad + \sum_{(r_2,s_2)\in\beta} \pi[x_\sigma + 1_{s_2}^{r_2}, x_\beta - 1_j^i]\,\mathrm{serv}_{(r_2,s_2)}(x_{r_2} + 1_{s_2}^{r_2}) \\
&\quad \times \sum_{(r_0,s_0)\in\sigma} \sum_{(r_1,s_1)\in\sigma} R[r_2,s_2;r_1,s_1]R[r_1,s_1;r_0,s_0]R[r_0,s_0;k,l] \\
&\quad + \sum_{(r_2,s_2)\in\sigma} \pi[x_\sigma, x_\beta + 1_{s_2}^{r_2} - 1_j^i]\,\mathrm{serv}_{(r_2,s_2)}(x_{r_2} + 1_{s_2}^{r_2})
\end{aligned}
$$

$$\times \sum_{(r_0,s_0)\in\sigma} \sum_{(r_1,s_1)\in\sigma} R[r_2,s_2;r_1,s_1]R[r_1,s_1;r_0;s_0]R[r_0,s_0;k,l]$$

which can be indefinitely iterated for the elements of the form

$$\pi[x_\sigma + 1_{s_k}^{r_k}, x_\beta - 1_j^i]\,\mathrm{serv}_{(r_k,s_k)}(x_{r_k} + 1_{s_k}^{r_k})$$

By renaming the summation indices, one obtains

$$\begin{aligned}
&\pi[x_\sigma + 1_l^k, x_\beta - 1_j^i]\,\mathrm{serv}_{(k,l)}(x_k + 1_l^k)\\
&= \sum_{(r_0,s_0)\in\beta} \pi[x_\sigma, x_\beta + 1_{s_0}^{r_0} - 1_j^i]\,\mathrm{serv}_{(r_0,s_0)}(x_{r_0} + 1_{s_0}^{r_0})R[r_0,s_0;k,l]\\
&\quad + \sum_{(r_0,s_0)\in\beta} \pi[x_\sigma, x_\beta + 1_{s_0}^{r_0} - 1_j^i]\,\mathrm{serv}_{(r_0,s_0)}(x_{r_0} + 1_{s_0}^{r_0})\\
&\quad \times \sum_{(r_1,s_1)\in\sigma} R[r_0,s_0;r_1,s_1]R[r_1,s_1;k,l]\\
&\quad + \sum_{(r_0,s_0)\in\beta} \pi[x_\sigma, x_\beta + 1_{s_0}^{r_0} - 1_j^i]\,\mathrm{serv}_{(r_0,s_0)}(x_{r_0} + 1_{s_0}^{r_0})\\
&\quad \times \sum_{(r_1,s_1)\in\sigma} \sum_{(r_2,s_2)\in\sigma} R[r_0,s_0;r_1,s_1]R[r_1,s_1;r_2,s_2]R[r_2,s_2;k,l] + \cdots\\
&= \sum_{(r_0,s_0)\in\beta} \pi[x_\sigma, x_\beta + 1_{s_0}^{r_0} - 1_j^i]\,\mathrm{serv}_{(r_0,s_0)}(x_{r_0} + 1_{s_0}^{r_0})\\
&\quad \times \Bigg[R[r_0,s_0;k,l] + \sum_{(r_1,s_1)\in\sigma} R[r_0,s_0;r_1,s_1]R[r_1,s_1;k,l]\\
&\quad + \sum_{(r_1,s_1)\in\sigma} \sum_{(r_2,s_2)\in\sigma} R[r_0,s_0;r_1,s_1]R[r_1,s_1;r_2,s_2]R[r_2,s_2;k,l] + \cdots\Bigg]\\
&= \sum_{(r_0,s_0)\in\beta} \pi[x_\sigma, x_\beta + 1_{s_0}^{r_0} - 1_j^i]\,\mathrm{serv}_{(r_0,s_0)}(x_{r_0} + 1_{s_0}^{r_0})\\
&\quad \times e_{r_0,s_0}[R[\beta;\sigma] + R[\beta;\sigma]R[\sigma;\sigma] + R[\beta;\sigma]R[\sigma;\sigma]R[\sigma;\sigma] + \cdots]e_{k,l}^T\\
&= \sum_{(r_0,s_0)\in\beta} \pi[x_\sigma, x_\beta + 1_{s_0}^{r_0} - 1_j^i]\,\mathrm{serv}_{(r_0,s_0)}(x_{r_0} + 1_{s_0}^{r_0})e_{r_0,s_0}R[\beta;\sigma]\\
&\quad \times [I - R[\sigma;\sigma]]^{-1}e_{k,l}^T\\
&= \sum_{(r_0,s_0)\in\beta} \pi[x_\sigma, x_\beta + 1_{s_0}^{r_0} - 1_j^i]\,\mathrm{serv}_{(r_0,s_0)}(x_{r_0} + 1_{s_0}^{r_0})R[r_0,s_0;\sigma]
\end{aligned}$$

$$\times [I - R[\sigma;\sigma]]^{-1} e_{k,l}^T$$

where $e_{i,j}$ is a row vector of all zeros except the (i,j) component, which is equal to one. Substituting this last expression in the third sum of equation (11) leads to the equation

$$\sum_{(i,j)\in\beta}\sum_{(k,l)\in\sigma} \pi[x_\sigma + 1_l^k, x_\beta - 1_j^i]\,\text{serv}_{(k,l)}(x_k + 1_l^k)R[k,l;i,j]$$

$$= \sum_{(i,j)\in\beta}\sum_{(k,l)\in\sigma}\sum_{(r_0,s_0)\in\beta} \pi[x_\sigma, x_\beta + 1_{s_0}^{r_0} - 1_j^i]\,\text{serv}_{(r_0,s_0)}(x_{r_0} + 1_{s_0}^{r_0})$$

$$\times R[r_0,s_0;\sigma][I - R[\sigma;\sigma]]^{-1} e_{k,l}^T R[k,l;i,j]$$

$$= \sum_{(i,j)\in\beta}\sum_{(r_0,s_0)\in\beta} \pi[x_\sigma, x_\beta + 1_{s_0}^{r_0} - 1_j^i]\,\text{serv}_{(r_0,s_0)}(x_{r_0} + 1_{s_0}^{r_0})R[r_0,s_0;\sigma]$$

$$\times [I - R[\sigma;\sigma]]^{-1} R[\sigma;\beta] e_{i,j}^T$$

$$= \sum_{(i,j)\in\beta}\sum_{(r_0,s_0)\in\beta} \pi[x_\sigma, x_\beta + 1_{s_0}^{r_0} - 1_j^i]\,\text{serv}_{(r_0,s_0)}(x_{r_0} + 1_{s_0}^{r_0})R[r_0,s_0;\sigma]$$

$$\times [I - R[\sigma;\sigma]]^{-1} R[\sigma;i,j]$$

which in vector form is equation (16).

APPENDIX B

Note that by construction of the matrices $P(\sigma)$ and $P(\beta)$, one has

$$P(\sigma)_{n-1}^n = P_{n-1}^n, \qquad P(\beta)_{n-1}^n = 0$$
$$P(\sigma)_{n+1}^n = 0, \qquad P(\beta)_{n+1}^n = P_{n+1}^n$$
$$P(\sigma)_n^n + P(\beta)_n^n = I + P_n^n$$

The proof of Theorem 3 can be derived as follows. Decomposing the subsystems balance equation as follows

$$\sum_{(i,j)\in\sigma}\sum_{(k,l)\in\sigma} \pi[x_\sigma + 1_l^k - 1_j^i, x_\beta]\,\text{serv}_{(k,l)}(x_k + 1_l^k)R[k,l;i,j]$$

$$+ \sum_{(i,j)\in\sigma}\sum_{(k,l)\in\beta} \pi[x_\sigma - 1_j^i, x_\beta + 1_l^k]\,\text{serv}_{(k,l)}(x_k + 1_l^k)R[k,l;i,j]$$

$$= \pi[x_\sigma, x_\beta] \sum_{(i,j)\in\sigma} \text{serv}_{(i,j)}(x_i)$$

and rewriting it in vector form yields

$$\pi_{n-1}P_n^{n-1} = \pi_n[i - P(\sigma)_n^n] \tag{30}$$

By subtracting the subsystem balance equation (10) from the global balance equation, one obtains

$$\sum_{(i,j)\in\beta}\sum_{(k,l)\in\sigma} \pi[x_\sigma + 1_l^k, x_\beta - 1_j^i]\,\mathrm{serv}_{(k,l)}(x_k + 1_l^k)R[k,l;i,j]$$
$$+ \sum_{(i,j)\in\beta}\sum_{(k,l)\in\beta} \pi[x_\sigma, x_\beta + 1_l^k - 1_j^i]\,\mathrm{serv}_{(k,l)}(x_k + 1_l^k)R[k,l;i,j]$$
$$= \pi[x_\sigma, x_\beta] \sum_{(i,j)\in\beta} \mathrm{serv}_{(i,j)}(x_i)$$

which can similarly be transformed to give

$$\pi_{n+1}P_n^{n+1} = \pi_n[I - P(\beta)_n^n] \tag{31}$$

The equations (30) and (31) can now be rewritten for the next indices to yield

$$\pi_{n+1} = \pi_n P_{n+1}^n [I - P(\sigma)_{n+1}^{n+1}]^{-1}$$
$$\pi_{n-1} = \pi_n P_{n-1}^n [I - P(\beta)_{n-1}^{n-1}]^{-1} \tag{32}$$

Introducing equations (30), (31), or (32) in the global balance equations

$$\pi_{n-1}P_n^{n-1} + \pi_n P_n^n + \pi_{n+1}P_n^{n+1} = \pi_n$$

shows that the vector π_n satisfies equation (20). Theorem 3 results then from the parallelism between π_n and $\pi_{|n}$.

APPENDIX C

From equation (23) and the global balance equation

$$\pi_{n+j-2}P_{n+j-1}^{n+j-2} + \pi_{n+j}P_{n+j-1}^{n+j} = \pi_{n+j-1}[I - P_{n+j-1}^{n+j-1}]$$

one obtains

$$\pi_{n+j-2}P_{n+j-1}^{n+j-2} = \pi_{n+j-1}[I - P_{n+j-1}^{n+j-1} - P_{n+j}^{n+j-1}[I - P(\sigma)_{n+j}^{n+j}]^{-1}P_{n+j-1}^{n+j}]$$
$$= \pi_{n+j-1}[I - P_{n+j-1}^{n+j-1} - P_{n+j}^{n+j-1}[Y(n+j)_{n+j}]^{-1}P_{n+j-1}^{n+j}]$$

which is similar to equation (23), except that the indices have all been decremented. Repeating these substitutions leads to

$$\pi_n P_{n+1}^n = \pi_{n+1}[I - P_{n+1}^{n+1} - P_{n+2}^{n+1}[Y(n+j)_{n+2}]^{-1}P_{n+1}^{n+2}]$$

which allows the expression

$$\pi_{n+1}P_n^{n+1} = \pi_n P_{n+1}^n[Y(n+j)_{n+1}]^{-1}P_n^{n+1} \tag{33}$$

to be written. The same development made with equation (24) leads to

$$\pi_{n-1}P_n^{n-1} = \pi_n P_{n-1}^n[X(n-k)_{n-1}]^{-1}P_n^{n-1}$$

which shows, by substituting in the global balance equation (17), that π_n satisfies (26). The parallelism between π_n and $\pi_{|n}$ ends the proof.

APPENDIX D

By the classical theory of nonnegative matrices (see [3]), the relations

$$Z \geq 0 \qquad Z \neq 0$$

guarantee that

$$\rho(Z) = \lambda_1 \in \text{sp}[Z] \qquad \exists \begin{cases} x \geq 0, \\ y^T \geq 0^T \end{cases} \qquad \text{with} \quad \begin{cases} xZ = \lambda_1 x \\ Zy^T = \lambda_1 y^T \end{cases}$$

Using now Stewart's theory of invariant subspaces [15], there exists an $(n-1,n)$ matrix X and an $(n,n-1)$ matrix Y with bounded norms that satisfy

$$\begin{bmatrix} x \\ X \end{bmatrix}^{-1} = \begin{bmatrix} y \\ Y \end{bmatrix}^T$$

and that allow the following development to be done:

$$\begin{aligned} Z &= \begin{bmatrix} y \\ Y \end{bmatrix}^T \begin{bmatrix} x \\ X \end{bmatrix} Z \begin{bmatrix} y \\ Y \end{bmatrix}^T \begin{bmatrix} x \\ X \end{bmatrix} \\ &= \begin{bmatrix} y \\ Y \end{bmatrix}^T \begin{bmatrix} \lambda_1 & o \\ 0 & XZY^T \end{bmatrix} \begin{bmatrix} x \\ X \end{bmatrix}^{-1} \\ &= \lambda_1 y^T x + Y^T XZY^T X \end{aligned}$$

where the matrix XZY^T contains all the eigenvalues of Z except λ_1. This last equation shows that the vertices of the polyhedron $\mathcal{P}(Z)$ are determined by two terms that have to be normalized. The first one is identical for all the vertices and is given by x weighted by λ_1; the second one can vary from one vertex to the other but is bounded and is mainly ruled by the largest eigenvalue of XZY^T, that is, the second largest eigenvalue of Z. This shows that the size of the polyhedron is mainly defined by the difference (or the ratio) between the modulus of the two largest eigenvalues of Z.

APPENDIX E: PROOF OF LEMMA 3

(i) The conditional distribution $\pi_{|n}$ is defined by

$$\pi_{|n} = \pi_{|n}(P_n^n + P_{\bar{n}}^n(I - P_{\bar{n}}^{\bar{n}})^{-1}P_n^{\bar{n}})$$

where

$$P_{\bar{n}}^n(I - P_{\bar{n}}^{\bar{n}})^{-1}P_n^{\bar{n}}$$

$$= [0\cdots 0 P_{n-1}^n P_{n+1}^n 0\cdots 0]\begin{bmatrix} I - P_{0,n-1}^{0,n-1} & 0 \\ 0 & I - P_{n+1,N}^{n+1,N} \end{bmatrix}^{-1}\begin{bmatrix} 0 \\ \vdots \\ 0 \\ P_n^{n-1} \\ P_n^{n+1} \\ 0 \\ \vdots \\ 0 \end{bmatrix}$$

$$= P_{n-1}^n[(I - P_{0,n-1}^{0,n-1})^{-1}]_{n-1,n-1)}P_n^{n-1} + P_{n+1}^n[(I - P_{n+1,N}^{n+1,N})^{-1}]_{(n+1,n+1)}P_n^{n+1}$$

The inverse of the first term of the right-hand side can be developed as

$$[(I - P_{0,n-1}^{0,n-1})^{-1}]_{(n-1,n-1)}$$

$$= \left[\begin{pmatrix} I - P_0^0 & -P_1^0 & \cdots & 0 \\ -P_0^1 & I - P_1^1 & \cdots & 0 \\ \vdots & \vdots & & \vdots \\ 0 & 0 & \cdots & I - P_{n-1}^{n-1} \end{pmatrix}^{-1}\right]_{(n-1,n-1)}$$

$$= \left[\begin{pmatrix} I - P_1^1 - P_0^1(I - P_0^0)^{-1}P_1^0 & -P_2^1 & \cdots & 0 \\ -P_1^2 & I - P_2^2 & \cdots & 0 \\ \vdots & \vdots & & \vdots \\ 0 & 0 & \cdots & I - P_{n-1}^{n-1} \end{pmatrix}^{-1}\right]_{(n-1,n-1)}$$

$$= \cdots = [\bar{X}(0)_{n-1}]^{-1}$$

A similar development for the second term ends the proof of (i).

(ii) The following relations

$$[\bar{X}(n - (k + 1))_{n-1}]^{-1} \geq [\bar{X}(n - k)_{n-1}]^{-1}$$

$$[\bar{Y}(n+(j+1))_{n+1}]^{-1} \geq [\bar{Y}(n+j)_{n+1}]^{-1} \tag{34}$$

allow us to make the development

$$\begin{aligned}\pi_{|n} &= \pi_{|n}(P_n^n + P_{n+1}^n[\bar{Y}(N)_{n+1}]^{-1}P_n^{n+1} + P_{n-1}^n[\bar{X}(0)_{n-1}]^{-1}P_n^{n-1}) \\ &= \pi_{|n}(P_n^n + P_{n+1}^n[\bar{Y}(n+j)_{n+1}]^{-1}P_n^{n+1} + P_{n-1}^n[\bar{X}(n-k)_{n-1}]^{-1}P_n^{n-1}) \\ &\quad + \pi_{|n}T \\ &= \pi_{|n}TZ_n^n(n+j,n-k)\end{aligned}$$

By Lemma 1, the nonnegativity of T that results from equation (34) ends the proof of (ii).

(iii) The relations (34) allow us to write

$$[Z_n^n(n+j+1,n-(k+1))]^{-1} = [Z_n^n(n+j,n-k)]^{-1}T$$

where T is again a nonnegative matrix. Transforming this last equation gives

$$Z_n^n(n+j+1,n-(k+1)) = (I + Z_n^n(n+j+1,n-(k+1))T)Z_n^n(n+j,n-k)$$

which, by the nonnegativity of all terms, leads to (iii).

REFERENCES

[1] Balsamo, S., and Iazeolla, G., An extension of Norton's Theorem for queueing networks, *IEEE Trans. Soft. Eng.*, se-8, No. 4, July 1982, pp. 298–305.

[2] Baskett, F., Chandy, K. M., Muntz, R. R. and Palacios, F. G., Open, closed and mixed networks of queues with different classes of customers, *J. Assoc. Comput. Mach.*, Vol. 22, No. 2, 1975, pp. 248–260.

[3] Berman, A., and Plemmons, R. J., *Nonnegative Matrices in the Mathematical Sciences*, Academic Press, New York, 1979.

[4] Bodewig, E., *Matrix Calculus*, North Holland, Amsterdam, 1956.

[5] Chandy, K. M., Herzog, U., and Woo, L., Parametric analysis of queueing networks, *IBM J. Res. Develop.*, Vol. 19, Jan. 1975, pp. 43–49.

[6] Chandy, K. M., Howard, J. H., and Towsley, D. F., Product form and local balance in queueing networks, *J. Assoc. Comput. Mach.*, 24, 1977, pp. 250–263.

[7] Conway, A. E., and Georganas, N. D., Decomposition and aggregation by class in closed queueing networks, *IEEE Trans. Soft Eng.*, se-12, No. 10, 1986, pp. 1025–1040.

[8] Courtois, P.-J., Error analysis in nearly-completely decomposable stochastic systems, *Econometrica*, 43, 1975.

[9] Courtois, P.-J., *Decomposability: Queueing and Computer System Application*, Academic Press, New York, 1977.

[10] Courtois, P.-J, and Semal, P., Bounds for the positive eigenvectors of nonnegative matrices and for their approximations by decomposition, *J. Assoc. Comput. Mach.*, 31, 1984, pp. 804–825.

[11] Courtois, P.-J., and Semal, P., On polyhedra of Perron–Frobenius eigenvectors, *Linear Algebra Appl.*, 65, 1985, pp. 157–170.

[12] Courtois, P.-J., and Semal, P., Computable bounds for conditional steady-state probabilities in large Markov chains and queueing models, *IEEE Selected Areas Commun.*, sac-4, No. 6, September 1986, pp. 926–937.

[13] Kemeny, J. G., and Snell, J. L., *Finite Markov Chain*, van Nostrand, New York, 1960.

[14] Simon, H. A., and Ando, A., Aggregation of variables in dynamic systems, *Econometrica* 29, 1961, pp. 111-138.

[15] Stewart, G. W., Computable error bounds for aggregated Markov chains, *J. Assoc. Comput. Mach.*, 30, 1983, pp. 271–285.

[16] Vantilborgh, H., Exact aggregation in exponential queueing networks, *J. Assoc. Comput. Mach.*, 25, 1978, pp. 620–629.

18

Finding Transient Solutions in Markovian Event Systems Through Randomization

WINFRIED K. GRASSMANN Department of Computational Science, University of Saskatchewan, Saskatoon, Saskatchewan, Canada

ABSTRACT

In 1953, Jensen proposed a very efficient method to find transient solutions of continuous-time Markov chains. This method, known by the names "randomization" or "uniformization," has become very popular. In this paper, we describe Jensen's method in the setting of Markovian event systems, and we show its potentials and limitations. We also point out the importance of transient solutions in general, and we show that in many cases of interest, transient solutions are more easily obtained than equilibrium solutions.

1. INTRODUCTION

This paper gives a review of the randomization method and its use for Markov modeling. The randomization method, as it is now commonly referred to, was originally proposed by Jensen [1953] as a method for finding transient solutions for continuous-time Markovian systems, and we will therefore call it

Jensen's method. The term "randomization" is really a generic term, and should not be used for a particular instance. Thus, even though Jensen's method is a randomization method, it is not the only one. Jensen's method is also called uniformization, but this term has the same shortcomings as the term randomization: it is generic and lacks specificity. We, therefore, will not use it either.

Basic to Jensen's method is the following idea: in a discrete Markov chain, changes to the system can only occur at times 0, 1, 2, ... If one replaces these unit time intervals by exponential random variables, one "randomizes" the process, and it can be shown that in this way, one can construct almost any continuous-time Markov chain of interest. This construct allows one to convert continuous-time Markov chains into their discrete-time analogues and determine the transient solutions in discrete time, which can be done without difficulty.

Jensen's method has proven to be a very efficient method for finding transient solutions in a great variety of continuous-time Markov chains. It has been used to analyze queuing problems (Grassmann [1977a], Gross and Miller [1984], Melamed and Yadin [1984a, b]), reliability problems (Kohlas [1982], Reibman and Trivedi [1988]), and maintenance problems (Love, [1977]). According to Reibman and Trivedi, it is still one of the most efficient methods available to find transient solutions of Markovian systems. Since it does not contain any subtractions, it is very resistant against rounding errors. A further advantage is its probabilistic interpretation, and the fact that it can easily be combined with integration as will be shown.

Jensen's method is particularly well suited to do Markov modeling. In Markov modeling, one converts queuing and reliability problems into continuous-time Markov chains. Since most problems of interest have several dimensions, the resulting Markov chains tend to be very large, and it is in such large Markov chains where Jensen's method shines. To make full use of its potential, one needs efficient and user-friendly methods to formulate Markovian systems. Such methods have been introduced by Irani and Wallace [1971], and have been further developed by Gross and Miller [1984] and Grassmann [1979, 1983]. Even though these methods at first glance seen to be very simple, even trivial, they are extremely powerful, and I cannot perceive how to do Markovian modeling without them. They are therefore described in some detail. This also allows us to demonstrate the application of Jensen's method in a natural setting.

2. MARKOV MODELING

Why is discrete event simulation so much more successful than Markov modeling? One reason is obviously the great flexibility of simulation, which allows the user to formulate and solve a great variety of different models. One of the main causes of this flexibility is the notion of events. An event is, of course, anything that changes the state of the system. Events have been, and should be, used in Markov modeling as well. The RQA analyzer, a program written by Wallace and Rosenberg (Wallace and Rosenberg [1966]), makes heavy use of events. This idea of using events for Markov modeling was later adopted by myself (Grassmann [1979]), and it is described in detail by Gross and Miller [1984], who coined the name SERT for their approach. SERT is an abbreviation for state-variables, event, rate, and transitions, and it means that one should first identify the state variables, then the events acting on the state variables, and finally, their rates. This information can then be used to obtain the transition matrix of the resulting Markov chain. A system similar to a Markovian event system has also been described by Stewart [1988].

Recently, we introduced the term Markovian event system (Grassmann [1989]). Markovian event systems are essentially discrete event systems like the ones used in discrete event simulation, except that the events are not scheduled. Instead, events are Poisson, and if the system is in state s, event k will occur at a rate λ_{sk}, where λ_{sk} depends only on s and k. We now demonstrate by means of an example how a Markovian event system is formulated.

EXAMPLE A service system consists of three servers. Each server has a separate waiting line. The first two servers work in parallel. They are followed by the third server. The line of each server includes the customer in service. If we want to exclude the customer in service, we use the word queue. Servers 1 and 2 both work at a rate of μ, and their service times are exponential. Server 3 works at a rate of μ_3, and his service time is Erlang-2. Arrivals are Poisson with a rate of λ. An arriving customer either joins line 1 or 2, depending on which of the two lines is shorter. Once served by server 1 or 2, he (she) joins the queue in front of server 3 to wait for his or her turn. After the service with server 3 is complete, he or she leaves the system.

The state of the system can be described by four state variables, namely X_1, X_2, X_3, and X_4. X_1 and X_2 represent the length of line 1 and 2, respectively. The fact that server 3 has an Erlang service time requires one to treat the customer in service separately from the other customers in line 3. To reflect this, we define X_3 as the queue length in front of server 3, and X_4 as the number of phases that still have to be completed before the customer presently in service can leave. If server 3 is idle, $X_4 = 0$. Note that $X_i \geq 0$, $i = 1, 2, 3, 4$. If one wants to keep the state-space finite, one has to bound X_i, $i = 1, 2, 3$. Since the service time of server 3 is Erlang-2, $X_4 \leq 2$. An event is not allowed to occur if the resulting state contains state variables that are outside their admissible range. For instance, if $X_1 = 0$, no departure from line 1 can take

Table 1 Example of a Markovian Event System

Event	Effect on X_1	X_2	X_3	X_4	Rate	Condition
Arrival 1	+1				λ	$X_1 \leq X_2$
Dept. 1	−1		+1		μ	
Arrival 2		+1			λ	$X_1 > X_2$
Dept. 2		−1	+1		μ	
Start 3			−1	+2	∞	
Next phase				−1	$2\mu_3$	

place. The system can now be represented as in Table 1, which is essentially the input for our queueing programs.

The information in this table should be self-explanatory. Arrival 1, for instance, increases X_1, which is the length of line 1, leaving the other lines unchanged. It occurs at a rate of λ, but only if $X_1 \leq X_2$. The other events should be interpreted in a similar fashion. Note that Start 3 has a rate of infinity, that is, this event will take place as soon as $X_3 > 0$ and $X_4 = 0$.

Formally, Markovian event systems are characterized by d state variables and e events. Every state variable is an integer, that is, the state space S is a subset of I^d, where I is the set of integers. The events consist of an effect and a rate. The effect of event k is a mapping $t_k : S \rightarrow S$, which converts the present state $i \in S$ into a target state $j \in S$. The rate of event k, $k = 1, 2, \ldots, e$ is λ_{ik}, $i \in S$. Frequently, it is convenient to give conditions, and unless the conditions are met, the event cannot take place. As mentioned, an event k is not allowed to happen if its target state $j \notin S$. The number of states is $n = |S|$. The term n can be finite or infinite.

The conversion of a Markovian event system into a Markov chain is relatively simple, and a solution was already provided in 1966 (Wallace and Rosenberg [1966]). We will discuss possible solution methods in Section 4, and we will show how these methods can be tailored to fit Jensen's method.

3. JENSEN'S METHOD

We assume in this section that the states are numbered from 1 to n. The problem is to find $\pi_j(t)$, the probabilities of being in state j ($j = 1, 2, \ldots, n$) at time $t > 0$. The initial probabilities $\pi_i(0)$, $i = 1, 2, \ldots, n$, are assumed to be known. Once the $\pi_j(t)$ are known, expectations and other measures of interest can be obtained without difficulty. For the purpose of discussion, let q_{ij} be the rate of going from state i to state j, $i \neq j$, and let

$$q_i = \Sigma q_{ij} \tag{1}$$

The parameters q_i, $i = 1, 2, \ldots, n$ will be called the leaving rates of state i. If we set $q_{ii} = -q_i$, we can define $Q = [q_{ij}]$ as the infinitesimal generator.

Transient solutions are relatively easy to obtain if all q_i are equal. The idea of Jensen was to make all q_i equal by padding the system with null-events, that is, with events that have no effect. The leaving rates of all states are thus unified, hence the name unification. In detail, one proceeds as follows: one chooses a value q that equals or exceeds all q_i, $i = 1, 2, \ldots, n$. For every state i with $q_i < q$, one introduces null-events at a rate of $q_{ii}^* = q - q_i > 0$. Normally, one will use the smallest possible value for q, that is, one chooses

$$q = \max q_i$$

Obviously, this approach is not applicable if any rate is infinite, because in this case, q also becomes infinite.

Events and null-events result in what we call "jumps." According to our construction, the rate of jumps is always q, and it is independent of the state. In Markovian systems, the number of jumps in the interval from 0 to t is Poisson with parameter qt, that is,

$$P(k \text{ jumps until } t) = p(k; qt) = e^{-qt}(qt)^k / k!$$

Let X_k, $k > 0$, be the state of the system immediately following jump k, and let X_0 be the starting state. It is easy to see that the process $\{X_k; k \geq 0\}$ is a discrete-time Markov chain. The transition probabilities p_{ij} of this chain are

$$p_{ij} = P(X_k = j \mid X_{k-1} = i) = q_{ij}/q \qquad i \neq j$$
$$p_{ii} = q_{ii}^*/q = 1 - q_i/q$$

π_j^k, the probability of being in state j after k jumps, can now be obtained in the normal way. Clearly, $\pi_i^0 = \pi_i(0)$, and

$$\pi_j^k = \sum_{i=1}^{n} \pi_i^{k-1} p_{ij} \qquad k > 0 \tag{2}$$

Equation (2) allows one to find the π_j^k recursively. Since the probability of making k jumps in the interval $(0, t)$ is $p(k; qt)$, one has

$$\pi_j(t) = \sum_{k=0}^{\infty} \pi_j^k p(k; qt) \tag{3}$$

Equations (2) and (3) can also be obtained through a Taylor expansion. If $\underline{\pi}(t) = [\pi_j(t)]$ is the row vector consisting of the $\pi_j(t)$, one has

$$\begin{aligned}\underline{\pi}(t)\exp(qt) &= \underline{\pi}(0)\exp(Qt)\exp(qt)\\ &= \underline{\pi}(0)\exp[(Q/q+I)qt]\\ &= \underline{\pi}(0)\exp(Pqt)\\ &= \sum \underline{\pi}(0)P^k(qt)^k/k!\end{aligned}$$

Since $P = (Q/q + I) = [p_{ij}]$, and since $p(k;qt) = \exp(-qt)(qt)^k k!$, equations (2) and (3) follow readily.

When using equation (3) for the numerical evaluation of $\pi_j(t)$, one must truncate the sum at the right-hand side at some finite number m. To determine the size of the resulting truncation error, one can use the following bounds for $\pi_j(t)$ (Grassmann [1977b]).

$$\sum_{k=0}^{m} \pi_j^k p(k;qt) \leq \pi_j(t) \leq \sum_{k=0}^{m} \pi_j^k p(k;qt) + \sum_{k=m+1}^{\infty} p(k;qt)$$

It follows that the absolute truncation error is bounded by the Poisson distribution. When $\pi_j(t)$ is small, the relative truncation error is more meaningful than the absolute error. In this case, one can approximate $\pi_j^k, k > m$, by $\pi_j(t)$. This leads to the following approximation:

$$\pi_j(t) \approx \sum_{k=0}^{m} \pi_j^k p(k;qt)/\left(1 - \sum_{k>m} p(k;qt)\right)$$

The reader may verify that this approximation is also valid if P is periodic, as long as m is large compared to the period of the chain. Hence, the Poisson distribution also approximates the relative error of $\pi_j(t)$. Table 2 gives the values for m corresponding to different tail probabilities of the Poisson distribution.

For large qt, the Poisson distribution can be approximated by the normal distribution. By using the normal approximation, one can find m as

$$m = qt + z_\alpha(qt)^{1/2} \tag{4}$$

When dealing with transient solutions, one frequently has integrals of the form

$$I = \int_0^T c_i \pi_i(t)\, dt \tag{5}$$

For instance, if c_i is the cost for being in state i, and if the system runs from 0 to T, I is the total cost during this time period. If $\pi_i(t)$ in equation (5) is

Table 2 m for selected values of qt and selected tail probabilities

	qt							
Tail	0.1	0.2	0.5	1.0	2.0	4.0	8.0	16.0
0.0005	2	3	4	6	8	12	19	31
0.00005	3	3	5	7	10	14	21	34
0.000005	3	4	6	8	11	16	23	37

(Note: m is always rounded to next higher value)

replaced by equation (3), one can easily integrate, using

$$q(k;qT) = \int_0^T p(k;qt)\,dt = \sum_{\nu=k+1}^{\infty} p(\nu;qT)/q$$

After some minor calculation, one finds (see Grassmann [1987])

$$I = \sum_{k=0}^{\infty} E_k q(k;fT) \tag{6}$$

with

$$E_k = \sum_{i \in S} c_i \pi_i^k$$

In the past, we sometimes also used the following formula for calculating I (see also Gross and Miller [1984]):

$$I = \sum_{k=0}^{\infty} E_k/(k+1)p(k;qT) \tag{7}$$

We could show, however, that equation (7) always converges more slowly than equation (6) (Grassmann [1987]), and we therefore no longer recommend equation (7). The method to find the time integral can also be used in many other connections, including the evaluation of the variance of time averages (Grassmann [1987]), an entity that is of great importance in the context of simulation.

4. THE ALGORITHMIC IMPLEMENTATION OF JENSEN'S METHOD

Jensen's method can be implemented in a dynamic or in a static fashion. In the static implementation, the entire transition matrix is generated before the algorithm starts. This is only possible if the number of states is finite. In dynamic implementations, on the other hand, one creates the transition matrix as needed, which means that problems with n infinite can be analyzed. The description here may give the impression that the problem is a very simple one. This is only partially true. The real difficulties lie in the efficient computer implementation, which lies outside the scope of this paper.

In static implementation, one converts the Markovian event system, such as the one given in Table 1, into a transition matrix as follows. To find the rows of the transition matrix, one enumerates all states, which can be done, using recursive function calls. If i is a state enumerated in this fashion, the row corresponding to i is given by $q_{ij} = \lambda_{ik}$ if $j = t_\gamma(i)$ for some event γ, and $q_{ij} = 0$ otherwise. This means that there are at most e nonzero entries per row, that is, the matrix is sparse, and sparse matrix techniques are appropriate. For instance, one could store each entry by a triple, namely, its row, its column, and its rate.

Jensen's method is not applicable when the model contains rates that are infinite, because in this case q would also become infinite. For this reason, one must remove all infinite rate events, and possibly even some events with very high rates in order to reduce q. Wallace and Rosenberg [1966] and Grassmann [1979] used the following method to do this. Every immediate event was combined with the event which preceded it. For instance, the event Start 3 in Table 1 can only occur after either the event Dept. 1 or Dept. 2 (if the server is idle) or after the event next phase (if X_4 reduces to 0). If one amalgamates Dept. 1 with Start 3, one obtains a finite rate event, which can only occur if the server is idle ($X_4 = 0$). The other triggering events can incorporate Start 3 in a similar fashion. If the triggering event also has an infinite rate, a further amalgamation has to take place. This process continues until a finite rate event is encountered. Unfortunately, this process can lead to an infinite loop. This happens if a finite rate event triggers the immediate event A, and if A triggers A itself, either directly or indirectly. This may happen even if the problem is well defined. We are presently investigating possible solutions to overcome this difficulty.

After all immediate events are eliminated, one can calculate q_i for all $i \in S$ according to equation (1), and the p_{ij} can be obtained. Obviously, there are at most $e + 1$ nonzero p_{ij} per row. According to equation (2), one has to take the sum of all $\pi_i^{k-1} p_{ij}$ to find the π_i^k. This addition is best done as follows. One fixes i, and multiplies π_i^{k-1} by all possible p_{ij}, and adds the products to the respective π_j^k. In other words, the evaluation of the π_j^k can be done by rows. The number of flops to do this is clearly $O(e)$ per row or $O(en)$ for the entire matrix. As soon as all π_j^k are calculated, one multiplies them by $p(k;qt)$ and adds these products to obtain the $\pi_i(t)$ according to equation (3). This requires $O(n)$ flops. Since $k = 1, 2, \ldots, m$, one can thus obtain all $\pi_i(t)$ in $O(nem)$ flops, or, in view of equation (4), in $O(neqt)$ flops. The array space needed for all this is obviously $3n$.

A major problem hampering the application of Markov modeling in general is the fact that n increases exponentially with the dimension d. Moreover, e, the number of events, also increases as the number of state variables increases. The reason is that with each state variable added, one has to add at least one event that can change this new state variable in some way. Consequently, ne increases significantly as the number of state variables increases, and the static implementation of Jensen's method is thus only applicable if the model in question has only few variables. For large models, all numerical methods break down, and one has to use discrete event simulation, a technique that is not affected by the curse of dimensionality.

In dynamic methods, the number of active states is dynamically adjusted. At time zero, only a finite number of states have a nonzero initial probability. The set of these states shall be denoted by S_0. When the first jump occurs, new states may arise, and these new states are combined with the states of S_0 to form the new set of states S_1. S_1, in turn, gives rise to a set S_2, the set of active states after two jumps. In this way, one continues until S_m is obtained. While doing this, one can readily calculate the π_i^k and accumulate the $\pi_i(t)$ Obviously, the sum of all $\pi_i(t)$, $i \notin S_k$, is negligible. Moreover, since the number of events is finite, the number of states added in m steps must still be finite.

In detail, one proceeds as follows. At the start of iteration k, one sets $S_k = S_{k-1}$. For all $i \in S_k$, and all events γ, $\gamma = 1, 2, \ldots, e$, one now generates the effects $j = t_\gamma(i)$. If $j \notin S_k$, a new state is created and added to S_k. With the creation of a new state, the appropriate accumulator for the calculation of the π_j^{k+1} is initialized to $\pi_i^k \lambda_{i\gamma}/q$, where $\lambda_{i\gamma}$ is the rate of event γ in state i. If the state j already exists, then $\pi_i^k \lambda_{i\gamma}/q$ is merely added to the appropriate counter. At the end of each iteration, the π_j^{k+1} can be multiplied by $p(k;qt)$ and added in accordance to equation (3). To implement this algorithm efficiently, one has to have a method that can quickly determine if state j is already part of S_k, and if so, find the necessary accumulators for π_j^k. In our implementations, we used hashing functions to accomplish this. In this fashion, one has a computation effort per iteration that is proportional to $|S_k|e$. In the case of immediate events, the above algorithm can be modified as follows. For each new state created, a check is made to see if there is an immediate event associated with this state. If an immediate event, say g, is possible, then the new state is $t_g(t_\gamma(i))$ rather than the unstable state $t_\gamma(i)$. Of course, $t_g(t_\gamma(i))$ may again be unstable, in which case one has to do another mapping. This continues until a stable state is reached. As before, this might result in infinite loops.

If we ignore infinite rate events, the number of flops per iteration is $O(|S_k|e)$. Unfortunately, it is not possible to find a generally applicable formula for $|S_k|$. To get a feel for how $|S_k|$ behaves as k increases, we restrict the problem as follows. We assume that the system starts empty and idle, and that any event can increase $X_1 + X_2 + \cdots + X_d$ by at most 1. Both conditions are met, for instance, in open queuing networks that start empty and idle. Moreover, only trivial expansions of the model are needed to cover many additional cases of interest. Under the conditions stated above, one has, using a result from the occupancy problem (Feller [1957], p. 54),

$$|S_k| = \binom{d+k}{k}$$

the number of flops to do all the m iterations of Jensen's method is proportional to $\sum |S_k|e$ operations, which is (see Knuth [1969], p. 54)

$$\sum_{k=0}^{m} |S_k|e = s\sum_{k=0}^{m}\binom{d+k}{k} = e\binom{d+m+1}{m}$$

Again, the number of states increases very fast as the number of dimensions increases, which is what one would expect.

5. PROBLEMS AND EXTENSIONS

As mentioned earlier, the time complexity of Jensen's method increases in direct proportion with q, or, if q is set to $\max q_i$, in direct proportion to the largest leaving rate. Hence, a few very large q_i can degrade the performance of Jensen's method considerably, and the question is whether this can somehow be avoided. A mixture of high and low leaving rates indicates that the problem is stiff. Several methods have been suggested to overcome stiffness, but none seems to be completely satisfactory. One method is selective randomization (Melamed and Yadin [1984,a,b], Miller [1983]). In this method, one divides the states into two subsets, namely D and $S - D$. In set D, the largest $q_i, i \in D$, is denoted by q^1, and in set $S - D$, the largest $q_i, i \in S - D$, is denoted by q^2. We define $\pi_i^{n,m}$ to be the probability that the system is in state i after n jumps, and that m of the n jumps occur while the starting state is in D. One has

$$\pi_{i(t)} = \sum_{n,m} \pi_i^{n,m} p(m; q^1 t) p(n-m; q^2 t)$$

In some cases, in particular if D and $S-D$ are two periodicity classes after uniformization within each partition, the $\pi_i^{n,m}$ are obtainable without problems. In general, however, m would be an additional state variable, and introducing this extra state variable is an expensive proposition. Hence, selective randomization is only applicable in special circumstances.

A more promising approach for reducing q was given by Bobbio and Trivedi [1986] (see also Reibman et al. [1989]). In this approach, the state space is partitioned into two sets, namely the set of fast states (S^f), and the set of slow states ($S - S^f$). The states S^f are characterized by high leaving rates, and the states of ($S - S^f$) by low ones. Generally, high leaving rates mean that equilibrium is reached quickly. Hence, if the partitions would form ergodic subchains of their own, S^f would tend to reach equilibrium much more quickly than $S - S^f$. Suppose now that the time between changes in S^f is very small compared to the time between two changes in the remaining chain. In this case, S^f may have reached a relative equilibrium while the rest of the chain stays in a particular state. This fact allows one to determine an approximate transient solution as follows. One restricts the calculation of transient solution to $S - S^f$, and assumes that S^f is in a relative equilibrium. For details of this method, see Reibman et al. [1989].

Jensen's method can also be used for transition matrices that vary with time, provided the variation occurs in a stepwise fashion. Problems arise in the case where the q_{ij} change in a continuous way. What can be done in this case? As mentioned earlier, Jensen's method essentially amounts to a Taylor expansion of $\underline{\pi}(t)\exp(qt)$. In the case where Q is independent of t, this expansion can be obtained easily. The Taylor series expansion is considerably more difficult to find if Q depends on t. Moreover, the terms of the expansion can no longer be guaranteed to be nonnegative, and this seems to preclude a probabilistic interpretation of the expansion. It is therefore unlikely that Jensen's method generalizes for cases where Q varies continuously with time. The only thing one can do is to use the standard solvers for differential equations to find $\underline{\pi}(t)$, or, preferably, to find $\underline{\pi}(t)\exp(qt)$.

6. TRANSIENT SOLUTIONS VERSUS EQUILIBRIUM SOLUTIONS

In this world, everything is in constant change, and equilibrium solutions are therefore nonexistent. The question is thus not why one should calculate transient solutions. The real question is why one should be interested in equilibrium solutions. One of the main reason for this is, of course, that equilibrium solutions are easier to obtain than transient solutions, and they may be good enough, provided the convergence to equilibrium is reasonably fast. This argument gives rise to two questions.

1. What is the computational complexity of equilibrium methods, when compared to Jensen's method?
2. How fast do transient systems approach their equilibrium?

We resolve the computational complexity issue first. As mentioned, Jensen's method requires order $O(neqt)$ flops to find a solution. In the case of equilibrium, one has different methods with different complexities. Simple elimination methods are $O(n^3)$. Thus, as long as eqt is small when compared to n^2, direct methods are not competitive. Even using a bandsolver does not help as much as one would hope. One needs numerically stable bandsolvers, and we could show (Grassmann [1989]) that within the context of Markovian event systems, numerically stable bandsolvers still have a complexity of $O(n^3)$. For large systems, it is generally agreed that iterative methods, such as the method of Gauss–Seidel, are the methods of choice. These methods typically have a complexity per iteration of $O(ne)$ flops per iteration, the same complexity per iteration as Jensen's method. Indeed, the number of flops for one iteration in Gauss–Seidel is almost exactly the same as the number of operations required to find the π_j^k from π_i^{k-1}. The question is thus how many iterations are needed in the respective methods. In the case of Jensen's method with $qt \leq 16$, the number of iterations is given in Table 2, and it is at most 37. To assess the number of iterations for the case of iterative methods, we rely on published figures. Kaufman [1983] solved a number of problems, using different versions of Gauss–Seidel. The number of iterations she needed to solve these problems varied between 37 and 94. Koury et al. [1984] solved a great number of problems, using a variety of iterative techniques to determine equilibrium solutions. For the problems with more than 100 states, the number of iterations needed to find the equilibrium solution varied between 6 and 200. Mitra and Tsoucas [1987] used Gauss–Seidel to find equilibrium solutions in Markov chains, and they provided a number of graphs to show the convergence of this method under different ordering strategies. The graphs clearly indicate that for large problems, one regularly needs more the 40 iterations to find the result. Thus if we concentrate on large problems, one regularly needs more than 40 iterations to find the result. Thus, if we concentrate on large problems, and on $ft \leq 16$, Jensen's method is at the very least competitive, and most likely superior to Gauss–Seidel. For small problems, the computation times are minimal and not normally of concern. Moreover, in Jensen's method, one can predict the number of iterations before solving the problem, something that is clearly impossible in Gauss-Seidel and related methods. We conclude that equilibrium solutions do not have any significant computational advantage for $qt \leq 16$. Thus, unless either q or t is high, one should seriously consider calculating transient solutions.

According to my knowledge, there are no dynamic methods available to find equilibrium solutions. This is in marked contrast to Jensen's method. The reason is that in Jensen's method, the total mass of the probabilities not yet calculated can be estimated, something that seems to be impossible when calculating equilibrium probabilities. Hence, there are not, and maybe there never will be, generally applicable methods to find equilibrium probabilities for the infinite state Markov chains.

How fast do real world Markovian systems approach equilibrium? In the book *Applications of Queueing Theory* (Newell [1982]), about half of the examples deal with transient queues. Newell analyzes, in particular, queues in which the traffic intensity is above 1 for a short time, and for such queues, equilibrium approximations would be inappropriate. Gordon Newell would not necessarily endorse Jensen's method as a tool for finding transient solutions, since his pets are fluid and diffusion approximations. I am sure, however, that he would agree with me that transient solutions are of great importance for solving practical problems. The need for transient solutions is not restricted to queuing problems in which the traffic intensity ρ occasionally exceeds unity. In the M/M/1 queue, for instance, the relaxation time increases with the third power of $1 - \rho$, which means that for high ρ, it will take exceedingly long until equilibrium is reached.

Jensen's method can also be used to make a number of indirect contributions. In particular, it has been used to find waiting-time and sojourn-time distributions (Grassmann [1977b], Melamed and Yadin [1984b]). To do this, one merely complements the state space S by a state 0, which is used to indicate the fact that the wait (or the sojourn) is over. If the time is set such that the waiting time starts at time 0, then $\pi_0(t)$ is the probability that the wait is shorter than t. This construct allows one to find waiting times for very complex systems, and for very complex queuing disciplines. Jensen's method can also be used in connection with embedded Markov chains. For details, see Grassmann [1982].

Jensen's method is no panacea. In particular, if the leaving rates q_i vary greatly with i, Jensen's method is slow. Large differences between the q_i are an indication of widely differing time scales, a problem that also affects other methods for solving ODEs. Jensen's method cannot be used if the $\lambda_{i\gamma}$ change with time, unless the changes are discrete. For small values of n, spectral analysis methods may be superior. However, Jensen's method is a good and stable method. Theoretical considerations (Grassmann [1982]) and experimental results (Reibman and Trivedi [1988]) show that Jensen's method is numerically very stable. Its performance is very predictable, and it is easy to program. For all these reason, Reibman and Trevedi conclude that Jensen's method is the method of choice for a typical problem, a conclusion that I fully support.

REFERENCES

Bobbio, A., and K. S. Trevedi. 1986. An aggregation method for the transient analysis of stiff Markov chains. *IEEE Trans. Comput.* C-35: 9, pp. 803– 814.

Feller, W. 1957. *An Introduction to Probability Theory and Its Applications*, vol. 1, John Wiley, New York.

Grassmann, W. K. 1977a. Transient solutions in Markovian queueing systems. *Comput. Operations Res.* 4, pp. 47–53.

Grassmann, W. K. 1977b. Transient solutions in Markovian queues. *Eur. J. Operations Res.* 1, pp. 396–402.

Grassmann, W. K. 1979. Modelbuilding without simulation. *Management Science Proceedings, ASAC 1979 Conference, University of Saskatchewan*, pp. 37–46.

Grassmann, W. K. 1982. The GI/PH/1 queue: A method to find the transition matrix. *INFOR.*, vol. 20, pp. 144–156.

Grassmann, W. K. 1983. Markov modelling. *Proceedings of the 1983 Winter Simulation Conference*, pp. 613–619. Arlington, Virginia.

Grassmann, W. K. 1987. Means and variances of time averages in Markovian environments, *Eur. J. Operations Res.*, vol. 31, pp. 132–139.

Grassmann, W. K. 1989. Numerical solutions for Markovian event systems. In *Quantitative Methoden in den Wirtschafswissenschaften.* P. Kall et al. (editors), Springer, Berlin, pp. 73–87.

Gross, D., and D. R. Miller. 1984. The randomization technique as a modelling tool and solution procedure for transient Markov processes, *Operations Res.* 32, pp. 343–361.

Irani, K. B. and V. L. Wallace. 1971. On the Network Linguistics and the Conversational Design of Queueing Networks. *J. Assoc. Comput. Mach.* 18, pp. 616–629.

Jensen, A. 1953. Markoff chains as an aid in the study of Markoff processes. *Skand. Akuarietidskrift.* 36, pp. 87–91.

Kaufman, L. 1983. Matrix methods for queueing problems. *SIAM J. Sci. Stat. Comput.* 4, pp. 525–552.

Kohlas, J. 1982. *Stochastic Methods of Operations Research.* Cambridge University Press, Cambridge.

Knuth, D. E. 1969. *The Art of Computer Programming*, vol. 1. Addison-Wesley, Reading, Mass.

Koury, J. R., D. F. McAllister, and W. J. Stewart. 1984. Iterative Methods for computing stationary distribution of nearly decomposable Markov chains. *SIAM J. Alg. Disc. Methods*, vol. 5, pp. 164–186.

Love, C. E. 1977. Purchase/replacement rules for decaying service facilities. *Comput. Operations Res.*, vol. 4, pp. 111–118.

Melamed, B., and M. Yadin. 1984a. Randomization procedures in the computation of cumulative-time distributions over discrete state Markov processes. *Operations Res.* 32, pp. 926–943.

Melamed, B. and M. Yadin. 1984b. Numerical computation of sojourn-time distributions in queueing networks. *J. ACM.* 31, pp. 839–854.

Miller, D. R. 1983. Reliability Calculation Using Randomization for Markovian Fault-Tolerant Computing Systems. 13th Annual International Symposium Fault-Tolerant Computing, IEEE Computer Society, Silver Spring, MD.

Mitra, D. and P. Tsoucas. 1987. Relaxation for the numerical solutions of some stochastic problems. Preprint.

Newell, G. F. 1982. *Applications of Queueing Theory*, 2nd ed., Chapman and Hall, London.

Reibman, A. and K. Trivedi, 1988. Numerical transient analysis of Markov models. *Comput. Operations Res.* 15, pp. 19–36.

Reibman, A., K. Trivedi, S. Kumar, and G. Ciardo. 1989. Analysis of stiff Markov chains. *ORSA J. Comput.* 1, pp. 126–133.

Stewart, W. J. 1988, MARCA: Markov Chain Analyser. Computer Science Technical Report. North Carolina State University.

Wallace, V. L., and R. S. Rosenberg. 1966. RQA-1. The Recursive Queue Analyser. System Eng. Lab, University of Michigan, Ann Arbor, Technical Report.

19

A Splitting Technique for Markov Chain Transient Solution

ANDREW REIBMAN AT&T Bell Laboratories, Holmdel, New Jersey

ABSTRACT

Many application require transient solutions of Markov chains. Several authors have considered efficient algorithms for the transient solution of acyclic Markov chains. In this paper, we investigate an approximate numerical technique for solving cyclic Markov chains. The basic approach is to break the Markov chain into two acyclic subchains, and iteratively combine their solution. An easily computable error bound allows the accuracy to be bounded before the computation begins.

1. INTRODUCTION

Markov chains are widely used in performance and reliability modeling. Many applications require transient solutions of Markov chains. Several authors have considered efficient algorithms for the transient solutions of acyclic

Markov chains. In this paper, we consider an approximate numerical technique for solving general (cyclic) Markov chains. The basic approach is to split a cyclic Markov chain into two more easily solved acyclic subchains, solve the subchains separately, and iteratively combine the solution.

A continuous-time Markov chain (CTMC) is a stochastic process $\{\mathcal{X}(t), t \geq 0\}$. We assume that the state is finite and of size n. We let $q_{ij}, i \neq j$, by the rate of transition from state i to state j. Define $q_{ii} = -\sum_{j \neq i} q_{ij}$. The matrix $Q = [q_{ij}]$ is the *infinitesimal generator* of the CTMC. If $p_i(t)$ is the probability the CTMC is in a state i at the time t, the row vector $\underline{p}(t)$ is called the *state probability vector* of the CTMC. The system of Kolmogorov differential equations describes the behavior of the state probability vector as a function of t:

$$\dot{\underline{p}}(t) = \underline{p}(t)Q, \qquad \underline{p}(0) = \underline{p}_0 \tag{1}$$

Here, $\underline{p}_0$ is the *initial condition* or *initial state vector.*

Often, CTMC's are analyzed in steady-state. In equation (1), this corresponds to the condition $\dot{\underline{p}}(t) = \underline{0}$, where $\underline{0}$ is a row vector of size n with all zero entries. So to solve for the steady-state probability vector, $\underline{\pi}$, we need only solve the linear algebraic system

$$\underline{\pi}Q = \underline{0}, \qquad \underline{\pi}\underline{e}^T = 1 \tag{2}$$

where $\underline{e}$ is a row vector of ones and the superscript T denotes transpose. The problem of steady-state solution has been examined in detail in the literature [Ste78][SG86].

Although steady-state analysis is more widely considered, many applications require CTMC transient analysis. In contrast to steady-state solution, transient analysis requires the solution of the Kolmogorov equations for a set of time points; linear differential equations must be solved instead of linear algebraic equations. Exact numerical transient solution is discussed in detail in [RT88] and [RTS89]. In the rest of this section, we review the pertinent results on exact and approximate CTMC transient solution. In Section 2, we introduce an approximate technique based on the Trotter product formula for splitting the matrix exponential. In Section 3, we give a small numerical example.

1.1 Exact Numerical Solution

The general solution for the entire system (1) is given by

$$\underline{p}(t) = \underline{p_0}e^{Qt} \tag{3}$$

where e^{Qt}, the *matrix exponential* [GVL83][MV78], is defined by the Taylor series

$$e^{Qt} = \sum_{i=0}^{\infty} \frac{(Qt)^i}{i!}. \tag{4}$$

For $t \gg 0$, the direct computation of the matrix exponential Taylor series is subject to severe roundoff error [MV78]. So a Taylor series is not suitable as a general purpose solution method. Similarly, classical methods from control theory, based on eigenanalysis [GVL83][MV78], can be slow and numerically unstable, particularly for large problems. Two alternatives to techniques based on eigenanalysis are differential equation solution and uniformization [RT88][RTS89].

Differential equation solution methods are often readily available. Many of these solvers use *explicit* differential equation solution formulas. *Explicit* differential equation solvers use only function evaluation—for Markov chains they use only matrix–vector multiplications and not linear algebraic system solution. Such explicit methods are not suitable for solving *stiff* systems. The presence of fast and slow rates in the same model is called *stiffness*. *Fast* transition rates are large relative to $1/t$, where t is the solution interval. *Slow* rates are small relative to $1/t$. Stiffness is common in Markov reliability models, where component failure rates are slow, and repairs and reconfigurations are fast. Stiffness often causes difficulties with conventional numerical solution methods.

Stable *implicit* methods, like the trapezoid rule, require the solution of a linear algebraic system at each time step. These methods suffer little degradation in the face of increasing stiffness. However, they are less accurate and incur substantial performance penalties on non-stiff problems [RT88].

Though the direct evaluation of the matrix exponential series (4) is not a practical method for solving CTMC's, another approach based on the evaluation of infinite series is more promising. Uniformization is a series technique derived from the Taylor series expansion of the matrix exponential [Gra77][MY84][RT88]. It has a natural stochastic interpretation based on the reduction of a CTMC to a discrete-time Markov chain (DTMC) subordinated to a Poisson process.

From a CTMC we obtain a DTMC $\mathcal{Z} = \{\mathcal{Z}_i, i = 0, 1, \ldots\}$ and a Poisson process $\{\mathcal{N}(t), t \geq 0\}$. $\mathcal{N}(t)$ is independent of $\mathcal{Z}$ and has a rate q. We form $\mathcal{X}(t) = \mathcal{Z}_{\mathcal{N}(t)}$ in the following way: Let $q \geq \max_{1 \leq i \leq n} |q_{ii}|$. Consider an equivalent process for which the transition rate from state i is q and a $1 - |q_{ii}|/q$ fraction of these transitions returns immediately to state i. The transition probability matrix of the DTMC $\mathcal{Z}$ is $Q^* = Q/q + I$. This process is the *uniformization* (or *randomization*) of the original CTMC. Once a Markov process is uniformized, the transient solution is computed by conditioning on $\mathcal{N}(t)$, the number of transitions in $[0,t]$.

$$p_j(t) = \Pr\{\mathcal{X}(t) = j\} = \sum_{i=0}^{\infty} \Pr\{\mathcal{X}(t) = j \mid \mathcal{N}(t) = i\} \Pr\{\mathcal{N}(t) = i\} \qquad (5)$$

Thus

$$\underline{p}(t) = \sum_{i=0}^{\infty} \underline{\Pi}(i) e^{-qt} \frac{(qt)^i}{i!} \qquad (6)$$

where $\underline{\Pi}(i) = \{\Pi_1(i), \Pi_2(i), \ldots, \Pi_n(i)\}$ is the state probability vector of the subordinated DTMC $\mathcal{Z}$ at step i, with $\underline{\Pi}(0) = \underline{p}(0)$. This equality can also be derived algebraically. Note that the DTMC transient state probability vector can be computed by successive matrix–vector multiplications;

$$\underline{\Pi}(i) = \underline{\Pi}(i-1)Q^* = \underline{p}(0)(Q^*)^i \qquad (7)$$

Because the matrix Q^* is stochastic and therefore nonnegative, only terms of the same sign (positive) will be added. This avoids potentially severe roundoff error encountered when directly evaluating the matrix exponential series.

The simple error bound for uniformization is based on a simple right-side truncation of the Poisson distribution. If the series is truncated after the term k, the truncation error in each element of $\underline{p}(t)$ is bounded by ϵ, where ϵ is easily computed using

$$\epsilon = e^{-qt} \sum_{i=k+1}^{\infty} \frac{(-qt)^i}{i!} = 1 - e^{-qt} \sum_{i=0}^{k} \frac{(qt)^i}{i!} \qquad (8)$$

This inequality can be used before the actual computation to determine the number of terms needed to obtain a desired absolute error tolerance.

On non-stiff problems, uniformization provides an accurate and economical solution technique. As the largest transition rate (q) and solution interval (t) grow, uniformization's performance degrades.

1.2 Special Case and Approximate Numerical Solution

In addition to "exact" numerical techniques that are applicable to all models, both "approximate" and "special case" solution methods can be used to efficiently solve many CTMC's. Birth–death processes are one widely-considered set of special case models that can be easily solved. Another set of examples, which we will use later, is acyclic Markov chains. Several solution algorithms have been proposed for acyclic Markov chains [Sev69] [MRT87][Cra88].

A second possible approach to more efficient solution is approximation. Several techniques for approximate transient solution of CTMC's have been proposed. Many are based on time scale decomposition. In time-scale decomposition, the fast parts of the model are treated as if they were in steady-state with respect to the slow parts of the model. One example of this approach is found in [BT86].

A second approach to approximation is to replace the original model by a series of more easily solved special case models. In [Cra88], cyclic chains are approximated by a sequence of more easily solved acyclic chains. Basically, additional states are added to represent the number of times a given cycle has been passed through. Cycles are "unrolled" into infinite, but acyclic, sequences of states. The infinite sequences are truncated to yield an acyclic, finite-state chain. More states can be added to the approximate chain, until the desired level of accuracy is reached. A disadvantage of this approach is that it can greatly increase the size of the state space.

In the approach we will discuss here, no additional states will be added. We will break the chain into two parts (either structurally or by time scales), solve the parts separately, and recombine the solution.

2. MATRIX SPLITTING TECHNIQUE

In this section we will discuss an approximate CTMC solution approach based on splitting the generator matrix. Given Q, the generator matrix for a CTMC, we define a *splitting* of Q as two matrices A and B, with $Q = A + B$.

We will find it convenient to choose A and B so that they are Markov generators of *subchains* of Q. We define a *subchain* of Q as a Markov chain that has the same state space as Q, but that includes only a subset of the original transitions. If A and B were chosen so that they were subchains of Q, then all the off-diagonal entries in A and B would have corresponding entries in Q, and each off-diagonal entry of Q is also found in either A or B, but not both. More formally, for all $0 < i, j \leq n$ and $i \neq j$, either $q_{ij} = a_{ij}$ and $b_{ij} = 0$, or $q_{ij} = b_{ij}$ and $a_{ij} = 0$. The diagonal of Q is apportioned so that for all $0 < i \leq n$, $q_{ii} = a_{ii} + b_{ii}$, $a_{ii} = -\sum_{j\, j \neq i} a_{ij}$, and $b_{ii} = -\sum_{j\, j \neq i} b_{ij}$.

For example, consider a three-state reliability model for a two-component fault-tolerant system with repair. Suppose the system fails if both components are down at the same time. The components fail independently at rate λ and are repaired at rate μ. The generator for the model is given by

$$Q = \begin{bmatrix} -2\lambda & 2\lambda & 0 \\ \mu & -(\lambda+\mu) & \lambda \\ 0 & 0 & 0 \end{bmatrix} \tag{9}$$

A natural splitting would be

$$Q = A + B = \begin{bmatrix} -2\lambda & 2\lambda & 0 \\ 0 & -\lambda & \lambda \\ 0 & 0 & 0 \end{bmatrix} + \begin{bmatrix} 0 & 0 & 0 \\ \mu & -\mu & 0 \\ 0 & 0 & 0 \end{bmatrix} \tag{10}$$

Ideally, we would like to solve the two halves of the splitting separately and combine the result.

An obstacle to combining the splitting results is the fact that the exponential of the sum of two matrices is the product of their exponentials if and only if the matrices commute [GVL83]:

$$e^{A+B} = e^{A}e^{B} \Longleftrightarrow AB = BA \tag{11}$$

In theory, the actual value of $e^{A}e^{B}$ can be computed using Cambell's theorem [MKS66], although this is not a practical approach for computation. A more useful result is the Trotter product formula [Tro59][RS80]

$$e^{A+B} = \lim_{r\to\infty} (e^{A/r}e^{B/r})^{r} \tag{12}$$

Moler and Van Loan [MV78] suggest the truncation of this sequence as an approximate solution technique for the matrix exponential:

$$e^{A+B} \approx (e^{A/r}e^{B/r})^{r} \tag{13}$$

In the next three subsections, we discuss an error bound for this truncated sequence, possible splittings that are appropriate for Markov chain solution, and approaches for solving the chain once it has been split.

2.1 Error Bound

An error bound from [MV78] helps gauge the usefulness of the splitting technique and helps determine how many terms need to be computed in order to achieve a desired level of accuracy. Define

$$S_r = e^{(A+B)/r} \tag{14}$$

and

$$T_r = e^{A/r}e^{B/r} \tag{15}$$

Using the identity

$$S_r^r - T_r^r = \sum_{m=0}^{r-1} S_r^m (S_r - T_r) T_r^{r-1-m} \tag{16}$$

Reed and Simon [RS80] show that $\|S_r^r - T_r^r\| \le C/r^2$ where C depends on A and B. Using this identity, Moler and Van Loan [MV78] bound the L_2 norm [GVL83] of the error introduced by truncating equation (12) at term r:

$$\|e^{A+B} - (e^{A/r} e^{B/r})^r\| \le \frac{\|[A,B]\|}{2r} e^{\|A\|+\|B\|} \tag{17}$$

Here, $[A,B] = AB - BA$, the commutator of A and B. Moler and Van Loan refine this bound for the case where A and B are self-adjoint and the case where $A+B$ is a matrix in companion form. We are currently investigating the possibility of improving this error bound by exploiting the stochastic structure of Markov chain generator matrices.

For general matrices, Moler and Van Loan suggest this approach is probably not as efficient as other techniques. But, for Markov chains (and particularly the stiff Markov chains that arise in reliability modeling), splittings arise naturally and lead to computationally efficient techniques.

2.2 Splittings for Markov Chains

Splittings for Markov chains can be based on the stochastic structure of the problem, transition rate time-scales, or the zero/nonzero structure of the generator matrix. The goal is to split the matrix so that each half is much easier to solve than the original CTMC.

A first approach to splitting is to choose two subchains that have a simpler zero/nonzero structure than the original chain, or that have simple closed-form solutions. For example, a generator matrix Q can be divided into upper and lower triangular generators, U and L, so that $Q = U + L$. We assume that the diagonal is apportioned between L and U, so that U and L are the generators of two acyclic Markov chains. These chains can then be solved using efficient algorithms [MRT87]. Depending on the structure of the original Markov chain, other splittings based on the physical structure of the chain may be feasible.

A second approach to splitting, based on the magnitude of the elements, may be useful for solving stiff Markov models that are difficult to solve with conventional numerical solution methods. If a stiff CTMC is split so that all the fast entries go in one subchain and all the slow entries are in the other, neither subchain will be stiff. The models can be solved separately, using a conventional solver, and the solution combined, avoiding the need for special stiff CTMC solution techniques.

2.3 Solving the Split Chain

Once the chain is split, we can cast equation (3) in the form of equation (13) and "unroll" it to get

$$\underline{p}(t) \approx \underline{p_0} e^{Ut/r} e^{Lt/r} \cdots e^{Ut/r} e^{Lt/r} e^{Ut/r} e^{Lt/r} \tag{18}$$

with r instances of the factor $e^{Ut/r}e^{Lt/r}$. There are two approaches to solving this equation.

The first approach is to compute the matrix exponentials $e^{Ut/r}$ and $e^{Lt/r}$. Computing matrix exponentials may not be desirable for the general Markov chains, but for acyclic chains it can be done efficiently [MRT87]. The result, $(e^{Ut/r}e^{Lt/r})^r$, is computed using successive multiplications or *binary powering*. In binary powering, $(e^{Ut/r}e^{Lt/r})^2$, $(e^{Ut/r}e^{Lt/r})^4$, etc. are computed by squaring, until the solution is obtained. If r is not a power of 2, the appropriate terms are multiplied to get the final solution. This approach is a good choice if the cost of computing the matrix exponential is small, and the value of r needed is large. For an n-state chain, solving for the matrix exponential at a single time point is never worse than solving for the solution vector n times (and it could be much better than this.) So when $r \gg n$, the binary powering approach is probably preferred. When n is large, one problem with this approach is that the matrices $e^{Ut/r}$ and $e^{Lt/r}$ are generally dense. In this case sparse matrix methods may be more appropriate.

A second approach is to treat the problem like a nonhomogeneous Markov chain. The chain alternates between the U and L transition rate matrices at intervals of length t/r. Effectively, the solution vector computed for each time point kt/r becomes the initial state vector for a new chain that runs from kt/r to $(k + 1)t/r$. The solution can be obtained using any conventional CTMC solver (e.g. an ODE solution technique or a stepwise application of uniformization). The effort needed is similar to solving the original chain over the interval $[0, 2t]$, but with a matrix that has been greatly simplified. If r is very large, one potential drawback is that many extra grid points may be introduced in order to match the solution at the points where the generator changes from U to L or vice versa.

Depending on what sort of software is already available, both of the approaches described could be used. Which approach is more efficient depends on the values of n, r, and t that are being considered.

3. NUMERICAL RESULTS

To illustrate the use of the splitting approximation technique, we consider a small example. We consider an *M/M/1/k* queue with the arrival rate λ and service rate μ. This is a convenient example, because it has a closed-form solution and frequently has been used elsewhere.

The generator matrix for the *M/M/1/k* queue,

$$Q = \begin{bmatrix} -\lambda & \lambda & 0 & \cdots & 0 & 0 & 0 \\ \mu & -(\lambda+\mu) & \lambda & \cdots & 0 & 0 & 0 \\ & & \ddots & & & & \\ & & & \ddots & & & \\ & & & & \ddots & & \\ 0 & 0 & 0 & \cdots & \mu & -(\lambda+\mu) & \lambda \\ 0 & 0 & 0 & \cdots & 0 & \mu & -\mu \end{bmatrix} \tag{19}$$

can be split into generators for a simple birth process and a simple death process,

$$Q = U + L = \begin{bmatrix} -\lambda & \lambda & 0 & \cdots & 0 & 0 & 0 \\ 0 & -\lambda & \lambda & \cdots & 0 & 0 & 0 \\ & & \ddots & & & & \\ & & & \ddots & & & \\ & & & & \ddots & & \\ 0 & 0 & 0 & \cdots & 0 & -\lambda & \lambda \\ 0 & 0 & 0 & \cdots & 0 & 0 & 0 \end{bmatrix} + \begin{bmatrix} 0 & 0 & 0 & \cdots & 0 & 0 & 0 \\ \mu & -\mu & 0 & \cdots & 0 & 0 & 0 \\ & & \ddots & & & & \\ & & & \ddots & & & \\ & & & & \ddots & & \\ 0 & 0 & 0 & \cdots & \mu & -\mu & 0 \\ 0 & 0 & 0 & \cdots & 0 & \mu & -\mu \end{bmatrix} \tag{20}$$

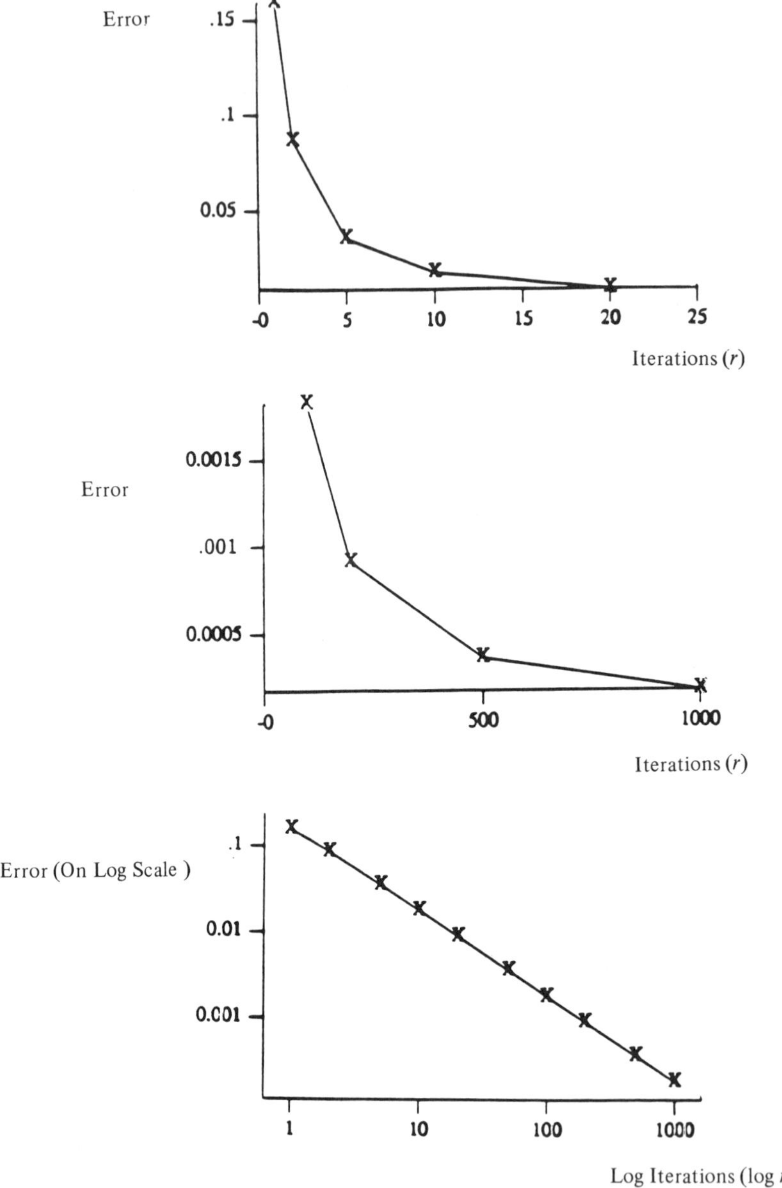

Figure 1 (a) Error as a function of the number of iterations ($r = 1$ to $r = 20$). (b) Error as a function of the number of iterations ($r = 50$ to $r = 1000$). (c) Error as a function of the number of iterations ($r = 1$ to $r = 1000$ on log–log scale).

For illustration, we numerically analyze the $M/M/1/k$ queue for $t = 1.0$, $k = 10$, $\lambda = 0.9$, and $\mu = 1.0$. We consider a range of r values: 1, 2, 5, 10, 20, 50, 100, 500, and 1000. In Figure 1, we plot the L_2 norm (the square root of the sum of the squares of the elements) of the error vector (the difference between the exact and approximate solution vectors) as a function of the value of the r. Figure 1a shows the error from 1 to 25 iterations. Figure 1b shows the error from 50 to 1000 iterations. Figure 1c shows the whole range, 1–1000, on a log–log scale. The error decreases as the number of terms used increases.

4. CONCLUSION

We have discussed an approximation technique for the numerical solution of Markov chains based on splitting the generator matrix. The method has an error bound that converges linearly in the number of terms used.

The method's main advantage is that it allows a CTMC to be split into two subchains that can be chosen so that they are more easily solved than the original problem. One possible strategy is to break the chain into acyclic subchains. A second strategy, suitable for stiff problems, is to break the chain so that all fast transitions are in one subchain and all slow transitions are in the other. The splitting method is easily integrated with existing numerical solution methods based on matrix exponential solution, uniformization, or differential equation solution.

We are implementing a general-purpose code for splitting that is compatible with existing numerical Markov chain solvers. We plan to compare the performance of the splitting technique presented here with other numerical methods on a range of example problems. Our current theoretical work focuses on trying to exploit the stochastic interpretation of the approximation technique to improve the convergence rate and find a better error bound.

REFERENCES

[BT86] A. Bobbio and K. S. Trivedi. An aggregation technique for the transient analysis of stiff Markov chains. *IEEE Trans. Comput.* C-35(9):803–814, September 1986.

[Cra88] K. Crank. A method for approximating the probability functions of a Markov chain. *J. Appl. Probability*, 25:808–814, 1988.

[Gra77] W. K. Grassmann. Transient solution in Markovian queueing systems. *Comput. Operations Res.*, 4:47–56, 1977.

[GVL83] G. H. Golub and C. F. Van Loan. *Matrix Computations*. John Hopkins University Press, Baltimore, 1983.

[MKS66] W. Magnus, A. Karrass, and D. Solitar. *Combinatorial Group Theory*. John Wiley & Sons, New York, 1966.

[MRT87] R. A. Marie, A. L. Reibman, and K. S. Trivedi. Transient solution of acyclic Markov chains. *Perform. Evaluation*, 7(3):175–194, 1987.

[MV78] C. Moler and C. F. Van Loan. Nineteen dubious ways to compute the exponential of a matrix. *SIAM Rev.* 20(4):801–835, October 1978.

[MY84] B. Melamed and M. Yadin. Randomization procedures in the computation of cumulative-time distributions over discrete state Markov processes. *Operations Res.*, 32(4):926–944, 1984.

[RS80] M. Reed and B. Simon. *Functional Analysis*. Academic Press, New York, 1980.

[RT88] A. L. Reibman and K. S. Trivedi. Numerical transient analysis of Markov models. *Comput. Operations Res.* 15(1):19–36, 1988.

[RTS89] A. Reibman, K. Trivedi, and R. Smith. Markov and Markov reward model transient analysis; an overview of numerical approaches. *Eur. J. Operations Res.*, 40(2):257–267, 1989.

[Sev69] N. C. Severo. A recursion theorem on solving differential-difference equations and applications to some stochastic process. *J. Appl. Probability*, 6:673–681, 1969.

[SG86] W. Stewart and A. Goyal. Matrix Methods in Large Dependability Models. Research Report RC-11485, IBM T. J. Watson Research Center, 1986.

[Ste78] W. Stewart. A comparison of numerical techniques in Markov modeling. *Commun. ACM*, 21(2):144–152, February 1978.

[Tro59] H. F. Trotter. On the product of semi-groups of operations. *Proc. AMS*, 10:545–552, 1959.

20

Transient Solutions of Time-Inhomogeneous Markov Reward Models with Discontinuous Rates

MARK K. SMOTHERMAN Department of Computer Science, Clemson University, Clemson, South Carolina

ABSTRACT

A time-inhomogeneous Markov reward model allows globally time-dependent transitions and incorporates cumulative reward measures to provide figures of merit for work performed. Transient solutions of such models are useful in combined reliability and performance evaluation. When this approach is generalized to the modeling of phased-mission systems, phase changes appear in the model as transitions with discontinuous rates. This paper discusses the solution of such models using an ordinary differential equation solver in combination with event queue and recalculation list control, adaptive step size techniques, and explicit reward measure integration for calculating the expected value of accumulated reward. A solution tool, called PUMA, has been implemented to demonstrate these concepts.

1. INTRODUCTION

Markov models are valuable tools in the analysis of the performance and reliability of systems. Performance evaluation uses both transient solutions and equilibrium solutions, while reliability evaluation and combined performance-reliability evaluation (called performability) require transient solutions. This paper is concerned with the latter case.

If $\{X(t) \mid t \geq 0\}$ is a finite-state stochastic process with state probabilities $p_i(t) = \Pr[X(t) = i]$, then from the Markov assumption we can derive the following differential equations [15]

$$p_i'(t) = \sum_j p_j(t) a_{ij}(t) \tag{1}$$

where $a_{ij}(t)$ is the transition rate from state i into state j. The system of equations can be rewritten as $p'(t) = p(t)A(t)$, where $p(t) = (p_0(t), p_1(t), \ldots, p_{n-1}(t))$ is the row vector of state probabilities and $A(t) = [a_{ij}(t)]_{n \times n}$ is the transition rate matrix.

Markov models can be classified as time-homogeneous and time-inhomogeneous. Time-homogeneous is the special case in which all transition rates are independent of time, that is, $A(t) = A$ above. It is probably the most popular form of Markov model due to its ease of solution and the variety of solution techniques available [10, 11, 16]. Using this approach in reliability and performability modeling means that component failure and repair behavior must be described by exponential distributions, that is, the transition rates must be constant. This is too restrictive for some systems [6, 14].

The use of globally time-dependent transitions provides additional flexibility in reliability modeling by allowing the use of time-dependent failure processes, such as described by the Weibull distribution [6, 15]. The transition rate functions, $a_{ij}(t)$, are typically continuous; however, an analytic solution to the general time-inhomogeneous model is usually difficult to obtain. Instead, reliability modeling using this approach typically makes use of a numerical solution from an ordinary differential equation solver [3, 6].

An extension of the Markov model that provides instantaneous and cumulative measures of weighted state occupancy is called a reward model. Each state has an associated weight, called a reward rate or yield, which represents the relative value of being in the state. Examples are productivity rates, such as jobs/hour or transactions/second, and economic rates, such as profit/day. Negative rates are allowed and are called costs. Rates may also be time dependent. Howard has explored the use of reward rates with semi-Markov processes and allows weights to be associated with transitions as well as states (called bonuses, or penalties when negative)[7].

Formally, let $r(t) = (r_0(t), r_1(t), \ldots, r_{n-1}(t))^T$ be the column vector of reward rates at time t and $p(t)$ be defined as above. Then $p(t)r(t)$ is the instantaneous reward of the system at time t. If $Y(t)$ is the accumulated reward until t, then the expected value of $Y(t)$ is defined by

$$E[Y(t)] = \int_0^t p(u)r(u)\,du = \sum_i \int_0^t p_i(u)r_i(u)\,du \tag{2}$$

With the proper rates, this measure can give information on expected work performed or value received, or on expected time spent in a certain subset of states (i.e., a reward rate of 1). In the latter case, the accumulated reward measures provide life cycle measures, such as expected duty time and expected time under repair.

Solutions to time-homogeneous Markov reward models can use the same solution methods as the underlying Markov model, since the defining equations are a straightforward extension to the Markov case [10, 12, 16]. That is, let

$$m(t) = \int_0^t p(u)\,du$$

Then a direct integration of (1) yields

$$m'(t) = m(t)A + p(0), \qquad m(0) = 0 \tag{3}$$

and then $E[Y(t)] = m(t)r$.

The rate matrix and forcing function in equation (3) are time-independent, so the solution presents little difficulty and can be done in parallel with the underlying Markov model. Unfortunately, such an extension is not available in the time-inhomogeneous case, that is, whenever $A(t)$ and $r(t)$ are time dependent. Thus the integration of equation (2) must be performed explicitly.

2. PERFORMABILITY MODELS OF PHASED-MISSION SYSTEMS

The reliability and performability modeling of phased-mission systems requires either a series of submodels to be solved [1, 5], or the extended interpretation of a Markov model in which the concept of a state transition is generalized to include phase changes, as well as failures and repairs [13]. In this approach, different phases are represented as different subsets of states in a single model, and phase changes are represented by time-varying transitions among these subsets. These phase-change transitions are typically discontinuous.

The formulation of a single model representation of a phased-mission system can, of course, include the impulse function as a holding-time density; therefore, the traditional approach of a series of submodels becomes a special case. Advantages of the single model framework include the provision of state-dependent phase changes, since phase-change transitions out of the different states in a given phase can have different rates or impulse functions.

Repairs are not generally modeled in time-inhomogeneous Markov models, since a "good as new" repair is assumed to return the failure process of a component to time $t = 0$. With general time-varying failure rates the global time dependence cannot be set back to zero. The time at repair, τ, is a random variable, and no simple offset calculation will suffice (unless τ is deterministic, such as a scheduled maintenance action). For time-homogeneous Markov models, the repair assumption does not present a difficulty since each transition erases all influence of the past.

In the phased-mission model, the repair assumptions can be handled as "continuous wear"; or, for those submodels in which there are no time-varying rates other than the phase changes, a "good as entry" repair policy can be represented. In the latter case, the failure processes for components can be logically set back to entry time into that phase.

3. A SIMPLE EXAMPLE

Figure 1 depicts a simple one-component system that experiences a phase change according to the hazard rate $h_1(t)$. The state numbers (i,j) represent the number of active components, i, and the phase number, j. If we consider the uniform distribution of phase-change time $F_U(x;a,b)$, with corresponding hazard rate $h_U(x;a,b)$, and equivalent failure rates before and after the phase change, the model demonstrates the effect of the uniform distribution of phase change time on the state probabilities. This is shown in Figure 2, for which the parameters are $\lambda = \lambda_1 = \lambda_2 = 0.001$/unit and $F_U(x;3,7)$ is the uniform distribution between time units 3 and 7. Note that the failure transitions introduce the negative exponential effect on the probability transfer from state 1, 1 to state 1, 2; that is, the sum $p_{1,1}(t) + p_{1,2}(t) = e^{(-\lambda t)}$, as expected. The uniform hazard rate $h_U(x;3,7)$ is also superimposed onto Figure 2. The discontinuities and steep ascent of the hazard rate curve are evident.

If we consider the one-component model with parameters $\lambda_1 = 0.01$/unit and $\lambda_2 = 0.03$/unit, we can investigate the effect on the mission reliability at 10 time units (i.e., $p_{1,2}(10)$) of uncertainty in the phase-change time. Given an impulse function $u(t-x)$, with corresponding hazard rate $h(x-t)$, and choosing the fixed phase-change time of 5 time units, the reliability at 10 time units is 0.819. However, if the phase change could occur up to one time unit

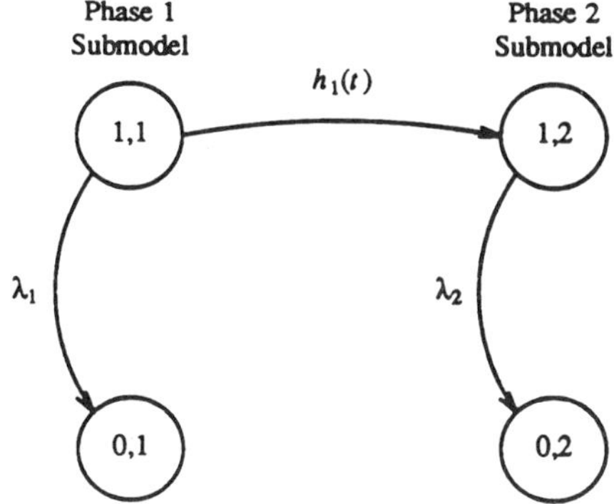

Figure 1 Single-component system.

before the nominal time, that is, according to $F_U(x;4,5)$ and $h_U(x;4,5)$, then the reliability decreases slightly to 0.811.

Because of the simplicity of this model, it is straightforward to show that the solution of this time-inhomogeneous Markov model using $F_U(x;a,b)$ distribution to describe the phase-change transition yields the same value of expected reliability as does the doubly stochastic approach of Iyer [8]. The system reliability of the model is equal to $p_{1,1}(t) + p_{1,2}(t)$, and as a function of total time t_m and phase change time t_{pc} can be written

$$R_{sys}(t_m, t_{pc}) = \begin{cases} e^{-\lambda_1 t_m} & t_m \le t_{pc} \\ e^{-\lambda_1 t_{pc}} e^{-\lambda_2 (t_m - t_{pc})} & t_m > t_{pc} \end{cases}$$

The expected value of the system reliability at time t_m, given that the distribution function of the phase-change time t_{pc} is $F_U(x;a,b)$ and that $t_m > b$, is

$$E[R_{sys}(t_m)] = \int_{-\infty}^{\infty} R_{sys}(t_m, t_{pc})\, dCdf(t_{pc}) = \int_a^b \frac{R_{sys}(t_m, t_{pc})}{b-a}\, dt_{pc}$$

Solving, we have

$$E[R_{sys}(t_m)] = \frac{e^{-\lambda_2 t_m}}{(b-a)(\lambda_1 - \lambda_2)} [e^{-(\lambda_1 - \lambda_2)a} - e^{-(\lambda_1 - \lambda_2)b}] \tag{4}$$

The solution of the time-inhomogeneous model can be found by using the convolution integrals [6]

$$p_0(t) = e^{\int_0^{t_m} a_{00}(x)\, dx}$$

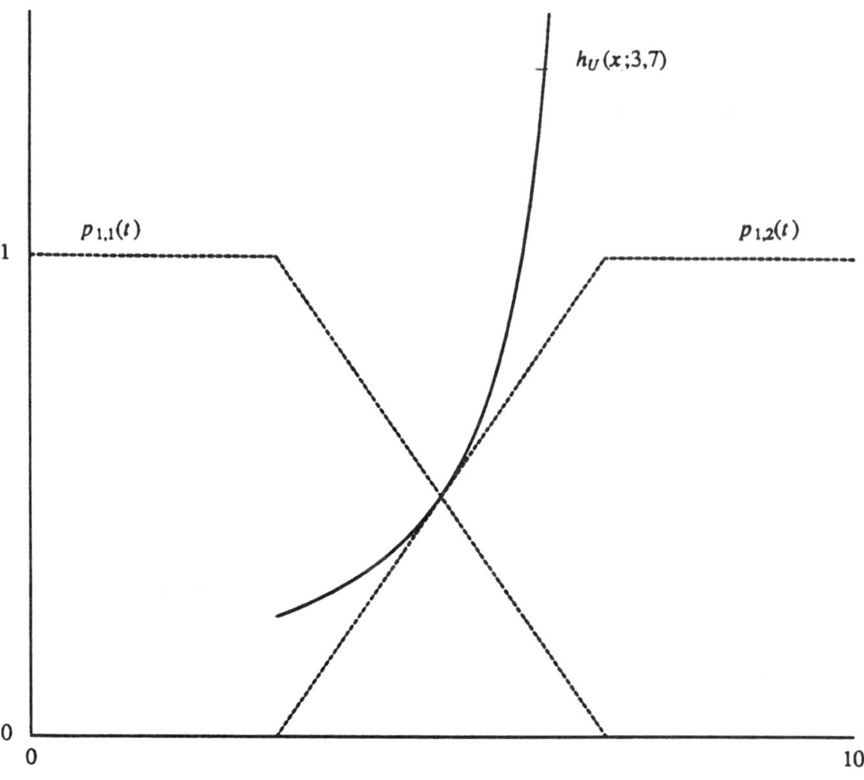

Figure 2 Single-component model solution.

$$p_1(t) = \int_0^{t_m} p_0(x) a_{10}(x) e^{\int_x^{t_m} a_{11}(u)\,du}\, dx$$

where

$$a_{00}(x) = \begin{cases} -\lambda_1 & x < a \\ -\lambda_1 - \dfrac{1}{b-x} & a \le x < b \\ \text{undefined} & x \ge b \end{cases}$$

$$a_{10}(x) = \begin{cases} 0 & x < a \\ \dfrac{1}{b-x} & a \le x < b \\ \text{undefined} & x \ge b \end{cases}$$

$$a_{11}(x) = -\lambda_2$$

and

$$p_0(x) = \begin{cases} e^{-\lambda_1 x} & x < a \\ e^{-\lambda_1 a + \int_a^x (-\lambda_1 - \frac{1}{b-u})\, du} & a \le x < b \\ 0 & x \ge b \end{cases}$$

For the case $a \le x < b$, $p_0(t)$ can be simplified:

$$p_0(x) = \frac{b-x}{b-a} e^{-\lambda_1 x}$$

Note that for $x \ge a$

$$p_1(x) = e^{-\lambda_2 x} \int_a^{\min(x,b)} \frac{p_0(u)}{b-u} e^{-\lambda_2 u}\, du$$

From these definitions, $p_1(t_m)$ given $t_m > b$ can be shown to equal equation (4). Thus, the use of a phase-change distribution obviates the second integration required by the doubly stochastic model.

4. NUMERICAL SOLUTION

A prototype tool for solving time-inhomogeneous Markov reward models has been developed at Clemson University. The tool, called PUMA for phase model unification for mission analysis, currently provides for fixed-time phase changes and for phase-change times described according to uniform distributions. Within those phase submodels with constant failure rates, an assumption of "good as entry" repair is made; other phases can use Weibull failure rates and "continuous wear" repair. PUMA currently exists as approximately 1200 lines of C code.

Because of the presence of discontinuous rates, the solution of the underlying system of differential equations is driven by an event-queue technique similar to those found in many simulation programs. All phase changes and result display times are placed as events on the event queue. A step size control adjusts the next step in the solution so as to not overstep the next event.

A phase change event includes information on the type of change, the exiting state, the entry state(s), and the branching probabilities for multiple entry states. There is one event notice for each fixed-time phase change, and there are two notices for each transition governed by a uniform distribution (i.e., begin and end). Of course, multiple events may happen at the same time.

A fixed-time phase-change event causes a transfer of state probability from one state to another. That is, for exit state i and entry state j with branching probability of one, the fixed-time phase-change event is interpreted as $p_j(t) =$

$p_j(t) + p_i(t), p_i(t) = 0$. When the branching probability q_{ij} is less than one, several fixed-time phase-change events exist to describe the complete phase change out of state i, and each must perform $p_j(t) = p_j(t) + q_{ij} \times p_i(t)$. Only the last fixed-time phase-change event performs $p_i(t) = 0$.

A phase change that is described by a uniform distribution is represented by a starting event and an ending event. The starting event causes a new term to be inserted into the appropriate matrix entry list. The term contains the time-dependent rate that must be recomputed as $q_{ij} \times h_U(x; a, b)$ at each step along the solution. The ending event will remove the term.

The system of differential equation is solved by the fifth-order Runge–Kutta method, RKQC, given in Press et al. [9]. This method was chosen for its robust nature and its ease of implementation. Before each derivative evaluation, the values of the matrix entries representing time-dependent failure transitions and phase changes are recalculated.

To bound the local error of each step in the solution, a standard approach is used for adaptive step size control for Runge–Kutta methods [2,9]. This is in addition to the event step size control mentioned above. Minimum and maximum step sizes are also specified, and a provision is made in which a fixed-time phase-change transition is performed near the ending time of a uniform distribution if the value of the associated transition rate grows too large. This is an instantaneous transfer of the residual probability of the exiting state into the entry state(s), and it is used by the adaptive step size control whenever the step size required to meet the local error tolerance is smaller than the minimum specified step size. Using the error tolerance and minimum step size parameters, a user can adjust the accuracy of the solution at the expense of efficiency.

The state probabilities at each step are collected by a Simpson's rule numerical integration, yielding the accumulated reward measures as given in equation (2). There may be multiple reward measures desired from the model, and at each time step, each reward measure counter is updated according to its list of associated states and reward rates. Simpson's rule was chosen since the Runge–Kutta method provides state probability values at the endpoints and the midpoint of a time step. This relationship demonstrates that the Markov model solution technique must be chosen in coordination with the reward measure integration.

5. EFFICIENT PROCESSING OF LARGE STATE SPACES

Since the single model framework requires the representation of all states of all phases, models of complex systems have potentially large state spaces.

Sparse matrix techniques are used in PUMA to store the transition rate matrix. The diagonals are stored in a separate vector for ease of direct access, and the off-diagonal elements are stored in doubly linked lists, one per row. The doubly linked lists provide for ease of introducing and eliminating the phase-change hazard rates.

Before each derivative evaluation, the values of the matrix entries representing time-dependent failure transitions and phase changes must be recalculated. To speed this processing, a recalculation list control is implemented. A separate, doubly linked list is maintained that contains pointers to those terms in the off-diagonal matrix row lists that must be recalculated. The starting event of a phase change inserts a term pointer into the recalculation list along with the term placed into a matrix row list. The ending event removes both the term and the recalculation list entry.

The recalculation of the diagonal matrix entries is also affected by the recalculation list. Before each derivative evaluation, the vector of diagonals is loaded with precalculated off-diagonal row sums of the constant terms of the row entries. As the recalculation list is processed, the appropriate diagonal entries are updated.

In addition to the recalculation list, a potential savings in solution time is available based on the phase structure. Since only a few phases are active at a given time (e.g., a single current phase, only the phases involved in a multiple-objective mission, or only the phases that overlap during a phase a phase change), state subset calculations can reduce solution time. In this approach all states are represented in the matrix, but the main loop of the differential equation solver is restructured to only calculate the states and transitions corresponding to active phases. This technique has not yet been implemented in PUMA.

Other techniques may also offer savings in the storage of the transition rate matrix. Dugan has proposed the use of indicator variables [4], in which those phases that cannot be active at the same time are mapped onto the same state space. All the transition rates must be represented, but each is multiplied by a phase indicator. The indicator takes on the value 1 when that phase is active and 0 otherwise. Thus, for the traditional fixed-time phase-change approach (i.e., a single active phase at a time), the representation of the states is reduced from the sum of the states in each phase to just the maximum phase state size. With multiple active phases, the maximum sum of active states over all possible concurrently active phases bounds the storage requirements.

Another possible storage optimization parallels the state subset calculation technique above. Here the transition rate matrix is reformulated after the end of each phase change. This has the same effect as indicator variables, but avoids the extra multiplications and additions at each derivative evaluation at the price of extra overhead at the end of each phase change.

However, even with these techniques, which are made available because of the active phase nature of phased-mission models, standard state aggregation techniques such as behavioral decomposition [3,6] and cyclic models may be necessary to represent the more complex systems.

6. A LARGER EXAMPLE

A common example in the phased-mission analysis literature is a three-phase, three component system [1,5]. In Figure 3 the configurations of the phases are shown. Assuming identical component failure processes within each phase, the composite time-inhomogeneous Markov model is given in Figure 4. The differential equations of the model are

$$p'(t) = p(t)\left[\begin{array}{cccc|ccc|cc}
a_{00}(t) & 3\lambda_1 & 0 & 0 & h_1(t) & 0 & 0 & 0 & 0 \\
\mu_1 & a_{11}(t) & 2\lambda_1 & 0 & 0 & (2/3)h_1(t) & (1/3)h_1(t) & 0 & 0 \\
0 & 2\mu_1 & a_{22}(t) & \lambda_1 & 0 & 0 & h_1(t) & 0 & 0 \\
0 & 0 & 0 & 0 & 0 & 0 & 0 & 0 & 0 \\
\hline
0 & 0 & 0 & 0 & a_{44}(t) & 2\lambda_2 & \lambda_2 & h_2(t) & 0 \\
0 & 0 & 0 & 0 & \mu_2 & a_{55}(t) & 2\lambda_2 & 0 & h_2(t) \\
0 & 0 & 0 & 0 & 0 & 0 & 0 & 0 & 0 \\
\hline
0 & 0 & 0 & 0 & 0 & 0 & 0 & a_{77}(t) & 3\lambda_3(t) \\
0 & 0 & 0 & 0 & 0 & 0 & 0 & 0 & 0
\end{array}\right]$$

The subsets representing the three phases are delimited in the matrix above, and it can be seen that the different subsets are connected by the phase-change transitions.

In this model repair with multiple repairmen is allowed as long as the configuration remains operational. The state numbers represent the number of operational components and the phase number. Branching probabilities are associated with the phase change out of state (2,1); these represent the prob-

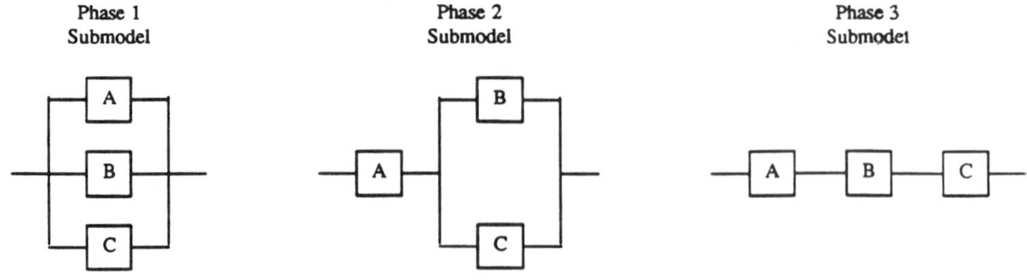

Figure 3 Phase configurations of three-component system.

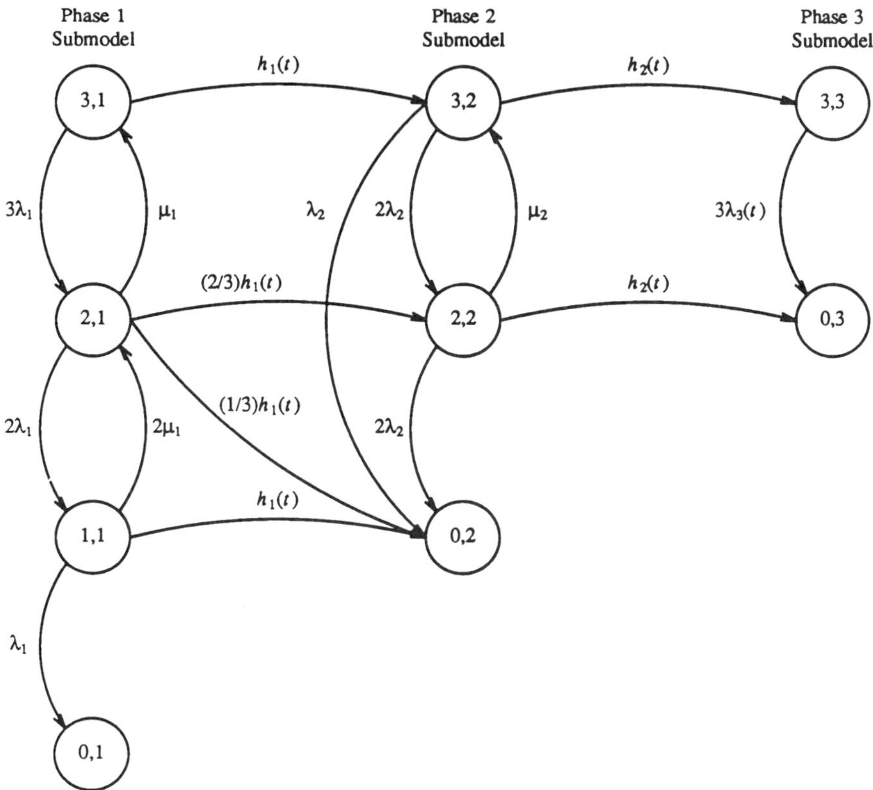

Figure 4 Three-component system.

abilities of either component A or one of the other components having failed. The use of such probabilities allows smaller submodels to be used when the components have identical failure rates. This model also demonstrates the use of time-dependent failure rates, that is, the component failure behavior in phase 3.

Model parameters for the three-component system are given in Table 1. Note that the uniform distributions used for phase-change times in the second column of values are centered about the fixed times in the first column of values and thus have the equivalent mean phase-change times. The model solution using the first column of parameter values yields a reliability at 100 time units (i.e., $p_{3,3}(100)$) of 0.796, while use of the second column of values

Table 1 Three-Component Model Parameters

Symbol	Interpretation	Values	
λ_1	component failure rate during phase 1	10^{-4}/unit	(same)
λ_2	component failure rate during phase 2	5×10^{-4}/unit	(same)
$\lambda_3(t)$	component failure rate during phase 3	$\lambda + \alpha^{-1}\beta(t-\gamma)^{\beta-1}$	(same)
λ	constant part of $\lambda_3(t)$	10^{-4}/unit	(same)
α	scale parameter of Weibull distribution	10^4 unit	(same)
β	shape parameter of Weibull distribution	1.5	(same)
γ	location parameter of Weibull distribution	30 units	25 units
μ_1	repair rate during phase 1	0.1/unit	(same)
μ_2	repair rate during phase 2	0.05/unit	(same)
$h_1(t)$	phase change to phase 2	$h(x-20)$	$h_U(x;15,25)$
$h_2(t)$	phase change to phase 3	$h(x-30)$	$h_U(x;25,35)$
$r_i(t)$	reward rate (for acc. time)	1/unit	(same)
ϵ	local error tolerance	0.00001	(same)
step_{max}	normal step size	0.25 unit	(same)
step_{min}	minimum step size	0.0001 unit	(same)

yields 0.784. Other results, including reward measures, are contained in Table 2. Note that the maximum possible mission value, E[mv], would be with no failures and is equal to 70.

7. CONCLUSIONS

Solutions to time-homogeneous Markov models are well understood, and the extension to a time-homogeneous Markov reward model is simplified by the two observations, that the defining equations can be directly integrated and that the resulting set of time-homogeneous differential equations can be

Table 2 Three-Component Model Results

Symbol	Interpretation	Values	
$p_{3,3}(100)$	system reliability at 100 time units	0.796	0.784
E[rt1]	E[time under repair, phase 1]	0.337 units	0.338 units
E[rt2]	E[time under repair, phase 2]	0.174 units	0.172 units
E[mv]	E[mission value; time in state 3, 3]	62.65 units	62.08 units
	derivative evaluations	4800	6024
	trial steps	400	502
	taken steps	400	455
	execution time on Concurrent 3230	12.2 seconds	15.9 seconds

solved with the same techniques as the underlying Markov model. The extension to a time-inhomogeneous Markov reward model is more problematic, since a numerical ODE solution becomes the solution method of choice and the direct integration of the defining equations is no longer available as a shortcut to the reward measures. Thus an explicit integration for the reward measures is necessary, and the Markov model solution technique must be chosen with the reward integration in mind. Finally, the extension to discontinuous rates for the time-inhomogeneous Markov reward model, as seen in models of phased-mission systems, requires event queue control to manage the rate changes.

The composite nature of phased-mission models engenders large state spaces. While sparse matrix techniques are helpful, they are likely not enough. Techniques such as recalculation list control and subset calculations can reduce solution time, and techniques such as indicator variables and matrix reformation can reduce storage requirements.

The PUMA analysis tool at Clemson University has been developed to meet the requirements of large, time-inhomogeneous Markov reward models with discontinuous rates. By providing reliability and expected accumulated reward measures, this tool supports performability analysis in the phased-mission environment. Future work on PUMA includes refinement of the numerical routines, continued experimentation with time and storage optimizations, and the provision of higher moments and distributions of accumulated reward.

REFERENCES

[1] C. A. Clarotti, S. Contini, and R. Somma, Repairable multiphase systems—Markov and fault-tree approaches for reliability evaluation, in G. Apostolakis, S. Garribba, and G. Volta (eds.), *Synthesis and Analysis Methods for Safety and Reliability Studies*. New York: Plenum Press, 1980, pp. 45–58.

[2] G. Dahlquist and A. Bjorck, *Numerical Methods*. Englewood Cliffs, N.J.: Prentice-Hall, 1974.

[3] J. B. Dugan, K. S. Trivedi, M. K. Smotherman, and R. M. Geist, The hybrid automated reliability predictor, *AIAA J. Guidance, Control, Dynamics*, vol. 9, 1986, pp. 319–331.

[4] J. B. Dugan, Automated analysis of phased mission reliability, *IEEE Trans. Reliability*, in press.

[5] J. D. Essary and H. Ziehms, Reliability analysis of phased missions, in R. E. Barlow, J. B. Fussell, and N. D. Singpurwalla (eds.), *Reliability and Fault Tree Analysis*. Society of Industrial and Applied Mathematics, 1975, pp. 213–236.

[6] R. M. Geist, M. K. Smotherman, K. S. Trivedi, and J. B. Dugan, The reliability of life-critical computer systems, *Acta Informatica*, vol. 23, Nov. 1986, pp. 621–642.

[7] R. A. Howard, *Dynamic Probabilistic Systems, Vol. II: Semi-Markov and Decision Processes*. New York: Wiley, 1971.

[8] R. K. Iyer, Reliability evaluation of fault-tolerant systems—Effect of variability in failure rates, *IEEE Trans. Computers*, vol. C-33, Feb. 1984, pp. 197–200.

[9] W. H. Press, B. P. Flannery, S. A. Teukolsky, and W. T. Vetterling, *Numerical Recipes: The Art of Scientific Computing*. Cambridge: Cambridge University Press, 1986.

[10] A. L. Reibman, R. M. Smith, and K. S. Trivedi, Markov and Markov reward model transient analysis: An overview of numerical approaches, *Eur. J. Operations Res.* vol. 40, May 1989, pp. 257–267.

[11] A. L. Reibman and K. S. Trivedi, Numerical transient analysis of Markov models, *Comput. Operations Res.* vol. 15, 1988, pp. 19–36.

[12] R. M. Smith, K. S. Trivedi, and A. V. Ramesh, Performability analysis: Measures, an algorithm and a case study, *IEEE Trans. Computers*, vol. C-37, April 1988, pp. 406–417.

[13] M. K. Smotherman and K. Zemoudeh, A nonhomogeneous Markov model for phased mission reliability analysis, *IEEE Trans. Reliability*, vol. 38, December 1989, pp. 585–590.

[14] F. A. Tillman, C. H. Lie, and C. L. Hwang, Simulation model of mission effectiveness for military systems, *IEEE Trans. Reliability*, vol. 27, 1978, pp. 191–194.

[15] K. S. Trivedi, *Probability and Statistics with Reliability, Queueing, and Computer Science Applications*. Englewood Cliffs, N. J. Prentice-Hall, 1982.

[16] K. S. Trivedi, A. L. Reibman, and R. M. Smith, Transient analysis of Markov and Markov reward models, in G. Iazeolla, P. J. Courtois, and O. J. Boxma (eds.), *Computer Performance and Reliability*. New York: Elsevier Science Publishers, 1988, pp. 535–545.

21

First Passage Times in Nearly Decomposable Markov Chains

GUY LATOUCHE Free University of Brussels, Brussels, Belgium

ABSTRACT

Consider a finite, irreducible, aperiodic Markov chain, for which the states are grouped into disjoint subsets. Assume that transition probabilities within a subset are large compared to the transition probabilities between different subsets. We analyze the first passage times from a state to a subset.

1. INTRODUCTION

We consider an irreducible, aperiodic Markov chain on the state space $E = \{1, 2, \ldots, N\}$ that is partitioned into m subsets E_1, E_2, ... E_m called aggregates. The stochastic matrix P is block partitioned as follows:

$$P = \begin{bmatrix} P_{11} & P_{12} & \dots & P_{1m} \\ P_{21} & P_{22} & \dots & P_{2m} \\ \vdots & & & \\ P_{m1} & P_{m2} & \dots & P_{mm} \end{bmatrix} \tag{1}$$

for $1 \le j, j' \le m$, $P_{jj'}$ is a submatrix of dimensions $n_j \times n_{j'}$, where n_j is the number of states in the aggregate E_j.

Furthermore, we assume that the Markov chain is nearly completely decomposable: the elements in the off-diagonal blocks of P are significantly smaller than the nonzero elements in the diagonal blocks. More formally, we write

$$P_{jj} = P^*_{jj} + \epsilon G_{jj}, \qquad P_{jj'} = \epsilon G_{jj'}, \qquad 1 \le j \ne j' \le m \tag{2}$$

where ϵ is a small positive real number, the submatrices P^*_{jj} are irreducible, aperiodic, and stochastic, and the submatrices $G_{jj'}$ are such that P is irreducible for $\epsilon > 0$.

Let Ω be a subset of one of the aggregates, such that $\Omega \subset E_1$. We define τ_i as the number of transitions, starting from state i, before reaching any state in Ω. The moment generating function of τ_i has been analyzed in [3]. In particular, we have shown that the moments of τ_i have a Laurent series expansion around $\epsilon = 0$, with a pole at $\epsilon = 0$. Our purpose here is to further examine the first moment of τ_i when Ω is one of the aggregates. Our motivation for considering this special case is described below.

Some queueing systems may be described as a Markov chain on a bidimensional state space $E = \{(j,s), 0 \le j \le m, l \le s \le n_j\}$, where j represents the number of active users of the system, and s is a descriptor of the systems state. (For instance, in a model for a multiprogrammed computer system, j is the number of active jobs, s may be the number of jobs that request the CPU, the remaining $(j - s)$ jobs being engaged in I/O activities.) If the rate of change within the system is much higher than the rate of change in the number of users, then the Markov chain has the structure given in equations (1) and (2); each aggregate is characterized by the number of active users. In such a case, one might be more interested in the first passage time to a given level of activity than to a specific state of the system.

We shall assume throughout this paper that $\Omega = E_1$, and we shall analyze $h_i = E[\tau_i]$, that is, the expected first passage time to the aggregate E_1, starting from state i at time 0.

We define the vectors $\mathbf{h}_1, \mathbf{h}_2, \dots, \mathbf{h}_m$, where $\mathbf{h}_j$ is a column vector with n_j components, one for each state in E_j. We also define the vector $\mathbf{h}$ with

$(N - n_1)$ components, and the matrix $\tilde{P}$ of order $(N - n_1)$ as

$$\mathbf{h} = \begin{bmatrix} \mathbf{h}_2 \\ \vdots \\ \mathbf{h}_m \end{bmatrix} \tag{3}$$

$$\tilde{P} = \begin{bmatrix} P_{22} & \ldots\ldots & P_{2m} \\ \vdots & & \\ P_{m2} & \ldots\ldots & P_{mm} \end{bmatrix} \tag{4}$$

Since the matrix P is irreducible, we have the standard results that $\mathbf{h} = (I - \tilde{P})^{-1}\mathbf{1}$ where $\mathbf{1}$ is a column vector of ones, and $\mathbf{h}_1 = \mathbf{1} + \sum_{2 \le j \le m} P_{1j}\mathbf{h}_j$. If N is large and if P has the structure (2), then the matrix $(I - \tilde{P})$ is a large, nearly singular matrix.

In the next section, we use techniques from perturbation theory, and construct a converging sequence of approximations to $\mathbf{h}$. In Section 3, we determine bounds for the error at each approximation step, and show that the error is geometrically decreasing.

1.1 Notational Conventions

In general, we use capital letters to denote matrices and lower-case boldface letters to denote vectors; the context indicates whether they are row or column vectors. In order to facilitate the reading of the formulae, the expression $\mathbf{uv}$ always represents the inner product of a row vector $\mathbf{u}$ by a column vector $\mathbf{v}$, while $\mathbf{v} \times \mathbf{u}$ represents the product of a column vector $\mathbf{v}$ by a row vector $\mathbf{u}$.

We shall restrict the use of indices such as i to represent individual states of the Markov chain, while using indices such as A or B to represent aggregates. Thus, in the sequel, $P_{ii'}$ will represent one entry in the matrix P, while $P_{A,B}$ will represent a block of elements $P_{ii'}$, where $i \in$ aggregate E_A and $i' \in$ aggregate E_B.

We represent by $\mathbf{0}$ any vector with all elements equal to zero, by $\mathbf{1}$ a column vector with all elements equal to one.

2. PERTURBATION ANALYSIS

It is proved in [3] that for ϵ sufficiently small,

$$\mathbf{h}_A = \sum_{k \ge -1} \epsilon^k \mathbf{h}_A^{(k)}, \qquad \text{for } A \ne 1$$

$$= \sum_{k \geq 0} \epsilon^k \mathbf{h}_A^{(k)}, \qquad \text{for } A = 1 \tag{5}$$

In order to evaluate the vectors $\mathbf{h}_A$, $1 \leq A \leq m$, one may successively determine the vectors $\mathbf{h}_A^{(-1)}$, $\mathbf{h}_A^{(0)}$, $\mathbf{h}_A^{(1)}$, etc. and truncate the series in the right-hand side of equation (5): $\mathbf{h}_A^{[n]} = \sum_{-1 \leq k \leq n} \epsilon^k \mathbf{h}_A^{(k)}$. The sequence $\left\{\mathbf{h}_A^{[n]}, n \geq -1\right\}$ automatically converges to $\mathbf{h}_A$, provided that ϵ is within the radius of convergence of equation (5). We shall first analyze $\mathbf{h}_A$ for $A \neq 1$, and examine $\mathbf{h}_1$ at the end of this section.

By an elementary probabilistic argument, we show that

$$\mathbf{h}_A = \mathbf{1} + \sum_{2 \leq X \leq m} P_{AX} \mathbf{h}_X$$

Using equation (2), this may be written as

$$(I - P_{AA}^*)\mathbf{h}_A = \mathbf{1} + \epsilon \sum_{2 \leq X \leq m} G_{AX} \mathbf{h}_X, \qquad 2 \leq A \leq M$$

By equating the coefficients of equal powers of ϵ in the equation above, we obtain the systems

$$(I - P_{AA}^*)\mathbf{h}_A^{(-1)} = \mathbf{0} \tag{6a}$$

$$(I - P_{AA}^*)\mathbf{h}_A^{(0)} = \mathbf{1} + \sum_{2 \leq X \leq m} G_{AX} \mathbf{h}_X^{(-1)} \tag{6b}$$

$$(I - P_{AA}^*)\mathbf{h}_A^{(k)} = \sum_{2 \leq X \leq m} G_{AX} \mathbf{h}_X^{(k-1)}, \qquad k \geq 1, \quad 2 \leq A \leq m \tag{6c}$$

We denote by π_A^* the stationary probability vector associated to P_{AA}^* ($\pi_A^* P_{AA}^* = \pi_A^*$, $\pi_A^* \mathbf{1} = 1$), for all A. If we premultiply equations (6b) and (6c) by π_A^*, we further obtain that

$$\sum_{2 \leq X \leq m} \pi_A^* G_{AX} \mathbf{h}_X^{(-1)} = -1 \tag{7a}$$

$$\sum_{2 \leq X \leq m} \pi_A^* G_{AX} \mathbf{h}_X^{(k)} = 0, \qquad k \geq 0, \quad 2 \leq A \leq m \tag{7b}$$

It is shown in [3] that the systems (6) and (7) completely determine the vectors $\mathbf{h}_A^{(k)}$, $k \geq -1$, $2 \leq A \leq m$; an explicit expression for $\mathbf{h}_A^{(-1)}$ is given there. We determine here simple and explicit expressions for the coefficients $\mathbf{h}_A^{(k)}$, for all $k \geq 0$.

We define the matrix Q of order m by $Q(A,B) = \pi_A^* G_{AB}\mathbf{1}$, $1 \leq A, B \leq m$, and the matrix T of order $m-1$ by

$$T = \begin{bmatrix} Q(2,2) & \dots & Q(2,m) \\ \vdots & & \vdots \\ Q(m,2) & \dots & Q(m,m) \end{bmatrix} \tag{8}$$

The matrix T is nonsingular, and we have that (see [3], Theorem 1) $\mathbf{h}_B^{(-1)} = t_B\mathbf{1}$, where the $(m-1)$-column vector $\mathbf{t}$ is given by

$$\mathbf{t} = (-T^{-1})\mathbf{1} \tag{9}$$

This may be written in matrix notations as follows. For any matrix H, we use the notation $(H)_{i,.}$ to represent its ith row. We then have that

$$\mathbf{h}^{(-1)} = S\mathbf{1} \tag{10}$$

where the matrix S has dimensions $(N - n_1) \times (m-1)$ and is given by

$$S = \begin{bmatrix} \mathbf{1} \times (-T^{-1})_{2,.} \\ \vdots \\ \mathbf{1} \times (-T^{-1})_{m,.} \end{bmatrix} \tag{11}$$

In order to determine $\mathbf{h}^{(k)}$ for $k \geq 0$, we need a preliminary lemma and some further notation.

LEMMA 2.1 Consider a matrix H that has eigenvalue zero of multiplicity one. Let $\mathbf{u}$ and $\mathbf{v}$ respectively represent the corresponding left and right eigenvectors, normalized by $\mathbf{uv} = 1$.

The linear system $H\mathbf{x} = \mathbf{b}$, $\mathbf{ux} = \alpha$, where $\mathbf{ub} = 0$, has a unique solution, given by $\mathbf{x} = H^{\#}\mathbf{b} + \alpha\mathbf{v}$, where $H^{\#}$ denotes the Drazin inverse of H.

Proof. The Jordan normal form of H may be written as

$$H = U \begin{bmatrix} 0 & \mathbf{0} \\ \mathbf{0} & H' \end{bmatrix} U^{-1}$$

where H' is nonsingular. By Campbell and Meyer [1], Theorem 7.2.1, we have that

$$H^{\#} = U \begin{bmatrix} 0 & \mathbf{0} \\ \mathbf{0} & (H')^{-1} \end{bmatrix} U^{-1}$$

Clearly, for any $\lambda \neq 0$, the matrix $H + \lambda\mathbf{v} \times \mathbf{u}$ is nonsingular, and its inverse is equal to $H^{\#} + \lambda^{-1}\mathbf{v} \times \mathbf{u}$.

Any solution of the system $H\mathbf{x} = \mathbf{b}$, $\mathbf{u}\mathbf{x} = \alpha$, is also a solution of the system $(H + \lambda\mathbf{v} \times \mathbf{u})\mathbf{x} = \mathbf{b} + \lambda\alpha\mathbf{v}$, where $\lambda \neq 0$. The second system has a unique solution, which proves the first claim. Furthermore,

$$\begin{aligned}\mathbf{x} &= (H + \lambda\mathbf{v} \times \mathbf{u})^{-1}(\mathbf{b} + \lambda\alpha\mathbf{v}) \\ &= (H^{\#} + \lambda^{-1}\mathbf{v} \times \mathbf{u})(\mathbf{b} + \lambda\alpha\mathbf{v})\end{aligned}$$

which proves the second claim, since $\mathbf{u}\mathbf{v} = 1$, $\mathbf{u}\mathbf{b} = 0$ and $H^{\#}\mathbf{v} = \mathbf{0}$.

Finally, we define the square matrices M_B of order n_B, $1 \leq B \leq m$, as $M_B = (I - P^*_{BB})^{\#}$.

THEOREM 2.1 For ϵ sufficiently small, we have that $\mathbf{h} = \sum_{k\geq -1} \epsilon^k \mathbf{h}^{(k)}$. The vector $\mathbf{h}^{(-1)}$ is given by equation (10). Moreover,

$$\mathbf{h}^{(k)} = R\mathbf{h}^{(k-1)}, \qquad \text{for all} \quad k \geq 0$$

where the matrix R of order $N - n_1$ is defined by

$$R_{AB} = M_A G_{AB} + \sum_{2\leq X,Y\leq m} (-T^{-1})(A,Y)\mathbf{1} \times \pi^*_Y G_{YX} M_X G_{XB}, \qquad \text{for } 2 \leq B, B' \leq m \tag{12}$$

Proof. If we apply Lemma 2.1 to the equations (6b) and (6c), we readily obtain that

$$\mathbf{h}^{(k)}_A = M_A \sum_{2\leq B\leq m} G_{AB}\mathbf{h}^{(k-1)}_B + (\pi^*_A\mathbf{h}^{(k)}_A)\mathbf{1}, \qquad k \geq 0 \tag{13}$$

Thus, we only have to show that

$$\pi^*_A\mathbf{h}^{(k)}_A = \sum_{2\leq B,X,Y\leq m} (-T^{-1})(A,Y)\pi^*_Y G_{YX} M_X G_{XB}\mathbf{h}^{(k-1)}_B \tag{14}$$

If we replace in equation (7b) the vectors $\mathbf{h}^{(k)}_X$ by the right-hand side of equation (13), we obtain that

$$\sum_{2\leq X\leq m} Q(A,X)(\pi^*_X\mathbf{h}^{(k)}_X) = -\sum_{2\leq B,X\leq m} \pi^*_A G_{AX} M_X G_{XB}\mathbf{h}^{(k-1)}_B$$

from which equation (14) follows, since the matrix T defined by equation (8) is nonsingular.

For the case when the initial state i belongs to E_1, we use $\mathbf{h}_1 = \mathbf{1} + \sum_{2 \le A \le m} P_{1A}\mathbf{h}_A$, and readily obtain that, for ϵ sufficiently small,

$$\mathbf{h}_1 = \sum_{k \ge 0} \epsilon^k \mathbf{h}_1^{(k)}$$

with

$$\mathbf{h}_1^{(0)} = \mathbf{1} + \sum_{2 \le A \le m} G_{1A}\mathbf{1}t_A \tag{15}$$

$$\mathbf{h}_1^{(k)} = \sum_{2 \le A \le m} G_{1A}\mathbf{h}_A^{(k-1)}, \qquad k \ge 1$$

where $\mathbf{t}$ is given by equation (9).

REMARKS (A). If the initial state i belongs to E_1, there is a probability $1 - O(\epsilon)$ that $\tau_i = 1$, that is, that the next transition leaves the Markov chain in E_1. The density of τ_i, however, has a very heavy tail, so that $E[\tau_i] \ne 1$ even in first approximation, as shown by equation (14).

(B). Consider a Markov chain on the state space $\{1, 2, \ldots, m\}$ and transition matrix $\hat{P}$ with $\hat{p}(A,B) = \pi_A^* P_{AB}\mathbf{1}$, $1 \le A, B \le m$. One easily verifies that $\frac{1}{\epsilon}t_A$, where $\mathbf{t}$ is given by equation (9), is the expected first passage time from state A to state 1 in that Markov chain. This observation, together with equation (10), leads to the following conclusion. If one is only interested in first passage times across different aggregates, one may in first approximation work with the matrix $\hat{P}$, where each aggregate has been collapsed into a single state.

3. ERROR BOUNDS

In this section, we determine bounds for the $(N - n_1)$-vectors $\mathbf{r}^{(n)}$, $n \ge 0$, defined by

$$\mathbf{r}^{(n)} = \sum_{k \ge n} \epsilon^k \mathbf{h}^{(k)} = \mathbf{h} - \sum_{-1 \le k \le n-1} \epsilon^k \mathbf{h}^{(k)} \tag{16}$$

From Theorem 2.1, one obtains that

$$\mathbf{r}^{(n+1)} = \epsilon R \mathbf{r}^{(n)}, \qquad n \ge 0 \tag{17}$$

and also that

$$\mathbf{r}^{(n)} = \epsilon^n \mathbf{h}^{(n)} + \mathbf{r}^{(n+1)} = \epsilon^n R \mathbf{h}^{(n-1)} + \epsilon R \mathbf{r}^{(n)}, \qquad n \ge 0 \tag{18}$$

From equation (17), we readily conclude that if ϵ is sufficiently small, then the convergence of $\mathbf{r}^{(n)}$ to $\mathbf{0}$ is *asymptotically* geometric, with rate $\epsilon\,\mathrm{sp}(R)$. Instead of considering the whole matrix R, we shall use its detailed structure (12) and proceed as in [4]. We shall then prove that convergence is at least geometric for all n, with a rate that is easily determined.

LEMMA 3.1 Let $\mathbf{u}$ and $\boldsymbol{\pi}$ be two row vectors, and $\mathbf{v}$ a column vector such that $\boldsymbol{\pi}\mathbf{v} = 0$. Then

$$|\mathbf{u}\mathbf{v}| \leq \mathcal{L}(\mathbf{u}, \boldsymbol{\pi})\|\mathbf{v}\|_1 \tag{19}$$

where

$$\mathcal{L}(\mathbf{u}, \boldsymbol{\pi}) = \min_{\alpha \epsilon R} \|\mathbf{u} - \alpha\boldsymbol{\pi}\|_\infty$$

Proof. One has that

$$\begin{aligned} |\mathbf{u}\mathbf{v}| &= |(\mathbf{u} - \alpha\boldsymbol{\pi})\mathbf{v}|, \qquad \text{for all} \quad \alpha \\ &\leq \|\mathbf{u} - \alpha\boldsymbol{\pi}\|_p \|\mathbf{v}\|_q \end{aligned}$$

for any integers p and q such that $1/p + 1/q = 1$, by Hölder's inequality. The lemma follows from the particular choice $q = 1$.

This lemma is inspired from [2]. A straightforward application of Hölder's inequality would yield $|\mathbf{u}\mathbf{v}| \leq \|\mathbf{u}\|_\infty \|\mathbf{v}\|_1$. Here, we use the knowledge that $\mathbf{v}$ is orthogonal to a vector $\boldsymbol{\pi}$ to get the tighter bound (19).

LEMMA 3.2 Let $\mathbf{w}$ be any $(N - n_1)$-column vector such that

$$\sum_{2 \leq B \leq m} \boldsymbol{\pi}_A^* G_{AB} \mathbf{w}_B = 0.$$

Then

$$\left| \sum_{2 \leq B \leq m} M_A G_{AB} \mathbf{w}_B \right| \leq \mathbf{b}_A \|\mathbf{w}\|_\infty$$

where

$$\begin{aligned} b_i &= 2\mathcal{L}((M_A)_{i,\cdot}, \boldsymbol{\pi}_A^*)\mathbf{e}|G_{A,A}|\mathbf{1}, \qquad i \epsilon E_A, \\ \mathbf{e} &= (\mathbf{1} \cdots \mathbf{1}) \end{aligned} \tag{20}$$

Proof. In order to simplify the notations, we shall use $\mathbf{u}_i = (M_A)_{i,\cdot}$. Then,

$$\begin{aligned}\left|\sum_{2\le B\le m} \mathbf{u}_i G_{AB}\mathbf{w}_B\right| &= \left|\mathbf{u}_i\left(\sum_{2\le B\le m} G_{AB}\mathbf{w}_B\right)\right| \\ &\le \mathcal{L}(\mathbf{u}_i, \boldsymbol{\pi}_A^*)\left\|\sum_{2\le B\le m} G_{AB}\mathbf{w}_B\right\|_1, \qquad \text{by Lemma 3.1} \\ &\le \mathcal{L}(\mathbf{u}_i, \boldsymbol{\pi}_A^*)\left\|\sum_{2\le B\le m} |G_{AB}||\mathbf{w}_B|\right\|_1 \\ &\le \mathcal{L}(\mathbf{u}_i, \boldsymbol{\pi}_A^*)\left\|\sum_{2\le B\le m} |G_{AB}|\mathbf{1}\right\|_1 \|\mathbf{w}\|_\infty\end{aligned}$$

Since the matrices P and P_{AA}^* are stochastic, we have that $G_{AB} > 0$, for $A \neq B$, and $\sum_{1\le B\le m} G_{AB}\mathbf{1} = 0$. Hence,

$$\begin{aligned}\sum_{2\le B\le m} |G_{AB}|\mathbf{1} &= \sum_{\substack{2\le B\le m\\ B\neq A}} G_{AB}\mathbf{1} + |G_{AA}|\mathbf{1} \\ &\le -G_{AA}\mathbf{1} + |G_{AA}|\mathbf{1} \le 2|G_{AA}|\mathbf{1}\end{aligned}$$

Since $\||G_{AA}|\mathbf{1}\|_1 = \mathbf{e}|G_{AA}|\mathbf{1}$, the lemma is proved.

We now define the square matrix K of order $(m-1)$ by

$$K(A,B) = \mathcal{L}(\boldsymbol{\pi}_A^* G_{AB}, \boldsymbol{\pi}_B^*)(\mathbf{e}\mathbf{b}_B)$$

and the $(m-1)$-column vector $\boldsymbol{\mu}$ by

$$\boldsymbol{\mu} = (-T^{-1})K\mathbf{1}$$

LEMMA 3.3 Let $\mathbf{w}$ be any $(N-n_1)$-column vector such that

$$\sum_{2\le B\le m} \boldsymbol{\pi}_A^* G_{AB}\mathbf{w}_B = 0, \qquad \text{for all} \quad A.$$

Then $|R\mathbf{w}| \le \|\mathbf{w}\|_\infty \boldsymbol{\rho}$, where

$$\boldsymbol{\rho}_A = \mathbf{b}_A + \mu_A\mathbf{1}, \qquad 1\le A\le m.$$

Proof. By equation (12), we have that

$$|(R\mathbf{w})_A| \le \left| \sum_{2\le B\le m} M_A G_{AB} \mathbf{w}_B \right| + \left| \sum_{2\le B,X,Y\le m} (-T^{-1})(A,Y)\mathbf{1} \times \pi_Y^* G_{YX} M_X G_{XB} \mathbf{w}_B \right|$$

and we directly apply Lemma 3.2 to the first term in the right-hand side. To determine a bound for the second term, we observe that

$$\begin{aligned}
&\left| (\pi_Y^* G_{YX}) \left(\sum_{2\le B\le m} M_X G_{XB} \mathbf{w}_B \right) \right| \\
&\le \mathcal{L}(\pi_Y^* G_{YX}, \pi_X^*) \left\| \sum_{2\le B\le m} M_X G_{XB} \mathbf{w}_B \right\|_1 \\
&\le \mathcal{L}(\pi_Y^* G_{YX}, \pi_X^*)(\mathbf{e}\mathbf{b}_X)\|\mathbf{w}\|_\infty = K(Y,X)\|\mathbf{w}\|_\infty
\end{aligned}$$

the first inequality results from Lemma 3.1, since $\pi_X^* M_X = \mathbf{0}$; the second inequality results from Lemma 3.2. The remainder of the proof is routine.

We are now in a position to prove the main result of this section. We firstly consider bounds on $|\mathbf{r}^{(n)}|$ for $n \ge 1$. The case of $n = 0$ is more involved and is treated separately.

THEOREM 3.1

$$|\mathbf{r}^{(n)}| \le \epsilon^n (1 - \epsilon\|\rho\|_\infty)^{-1} \|\mathbf{h}^{(n-1)}\|_\infty \rho, \qquad n \ge 1 \tag{21}$$

$$\|\mathbf{r}^{(n)}\|_\infty \le \|\mathbf{r}^{(0)}\|_\infty (\epsilon\|\rho\|_\infty)^n, \qquad n \ge 1 \tag{22}$$

Proof. From equations (7b) and (16), we conclude that $\mathbf{h}^{(n)}$ and $\mathbf{r}^{(n)}$ satisfy the assumption of Lemma 3.3, for all $n \ge 0$. We then apply Lemma 3.3 to equation (17), and obtain that $|\mathbf{r}^{(n+1)}| \le \epsilon\|\mathbf{r}^{(n)}\|_\infty \rho$, from which equation (22) directly results. When we apply Lemma 3.3 to equation (18), we obtain that

$$\begin{aligned}
\|\mathbf{r}^{(n)}\|_\infty &\le \|\epsilon^n R\mathbf{h}^{(n-1)}\|_\infty + \|\epsilon R\mathbf{r}^{(n)}\|_\infty \\
&\le \left\{ \epsilon^n \|\mathbf{h}^{(n-1)}\|_\infty + \epsilon\|\mathbf{r}^{(n)}\|_\infty \right\} \|\rho\|_\infty
\end{aligned}$$

which proves (21).

For $n = 0$, we must use another argument, since $\mathbf{r}^{(0)} = R\mathbf{h}^{(-1)} + \epsilon R\mathbf{r}^{(0)}$ by equation (18), but we may not apply Lemma 3.3 to the product $R\mathbf{h}^{(-1)}$

[see equation (7a)]. By repeating the argument of Lemmas 3.2 and 3.3, we eventually obtain that

$$|R\mathbf{h}^{(-1)}| \le \|\mathbf{h}^{(-1)}\|_\infty \tilde{\rho}$$

where

$$\begin{aligned}
\tilde{\rho}_A &= \tilde{b}_A + \tilde{\mu}_A \mathbf{1}, \qquad \text{all} \quad A \\
\tilde{b}_i &= 2\|(M_A)_{i..}\|_\infty \mathbf{e} G_{AA} \mathbf{1}, \qquad \text{all} \quad i \epsilon E_A, \ \text{all}\, A \\
\tilde{K}(A,B) &= \mathcal{L}(\pi_A^* G_{AB}, \pi_B^*)(\mathbf{e}\tilde{\mathbf{b}}_B), \qquad \text{all} \quad A \text{ and } B \\
\tilde{\mu} &= (-T^{-1})\tilde{K}\mathbf{1}
\end{aligned}$$

We finally obtain that

$$|\mathbf{r}^{(0)}| \le \|\mathbf{h}^{(-1)}\|_\infty \tilde{\rho} + \epsilon \|\mathbf{r}^{(0)}\|_\infty \rho \tag{23}$$

$$\|\mathbf{r}^{(0)}\|_\infty \le (1 - \epsilon\|\rho\|_\infty)^{-1} \|\tilde{\rho}\|_\infty \|\mathbf{h}^{(-1)}\|_\infty \tag{24}$$

We observe from Theorem 3.1 that $1/\|\rho\|_\infty$ is a lower bound on the radius of convergence for the series in equation (5).

4. CONCLUDING REMARK

The same approach may be used to determine tighter error bounds for $|\mathbf{r}^{(n)}|$, approximations and error bounds for higher moments of the first passage time to an aggregate, and also in the case when Ω is a proper subset of one of the aggregates.

REFERENCES

[1] Campbell, S. L., and Meyer, C.D. *Generalized Inverses of Linear Transformations*. Pitman Publishing Limited, London, 1979.

[2] Haviv, M., and Van der Heyden, L. Perturbation bounds for the stationary probabilities of a finite Markov chain. *Adv. Appl. Prob. 16*, 1984, 804–818.

[3] Latouche, G., and Louchard, G. Return times in nearly completely decomposable stochastic processes. *J. Appl. Prob. 15*, 1978, 251–267.

[4] Louchard, G., and Latouche, G. Geometric bounds on iterative approximations for nearly completely decomposable Markov chains. *J. Appl. Prob. 27*, 1990, to appear.

22

Bounds for Transient Characteristics of Large or Infinite Markov Chains

PIERRE-JACQUES COURTOIS and PIERRE SEMAL Philips Research Laboratory, Louvain-la-Neuve, Belgium

ABSTRACT

A technique that combines state aggregation and bounds on visit rates is proposed to compute bounds on transient characteristics of Markov chains that have large or infinite states spaces.

1. INTRODUCTION

The purpose of this paper is to suggest an approach to evaluate transient behavior characteristics of (ergodic) time homogeneous Markov chains that have large or infinite state spaces, and to provide bounds for these characteristics.

To study the transient behavior of a Markov chain, it is often convenient to modify the process by making certain states absorbing. We take here a somewhat different approach. We exploit the fact that the transient characteristics of a time-homogeneous Markov chain, which is ergodic or not, can often be computed via a transformed process that is ergodic. Aggregation can then be used to analyze this process if its state space is too large and is beyond the capabilities of classical numerical methods.

In the context of large absorbing chains, the state aggregation technique was already used by Bobbio and Trivedi [2], who proposed a method to approximate the transient behavior of *stiff* Markov systems. The technique proposed here provides bounds on transient characteristics, such as the drift and the moments of the time before absorption; no specific structure is assumed for the Markov chain.

We shall assume that the stochastic transient matrix of a large Markov chain can be organized into blocks:

$$\begin{bmatrix} P_{00} & \cdots & P_{0I} & \cdots & P_{0N} & P_{0A} \\ \vdots & & \vdots & & \vdots & \vdots \\ P_{I0} & \cdots & P_{II} & \cdots & P_{IN} & P_{IA} \\ \vdots & & \vdots & & \vdots & \vdots \\ P_{N0} & \cdots & P_{NI} & \cdots & P_{NN} & P_{NA} \\ R_0 & \cdots & R_I & \cdots & R_N & P_A \end{bmatrix} \tag{1}$$

Each principal submatrix $P_{00}, \ldots, P_{NN}, P_A$ corresponds to a particular subset of states. This partitioning is determined by the system characteristic that is under investigation. In models of resource usage, for instance, the subsets I, $I = 0, \ldots, N$, may contain all the states that correspond to a population of I users. In the context of reliability models, the subset I generally contains all the states with I failed components (see e.g. [9]).

The subset A can be viewed as containing those states that correspond to large user populations, that is, larger than N, or to system states with more than N failed components. The system characteristics in which we will be interested are the time spent by the system before reaching A when started from some initial state, and the tendency (drift) of the system toward states of A when it is in a subset I. Subset A may be of infinite size in certain models (e.g., in queuing models with infinite populations). The drift gives then information on the stability of the system.

When the matrix P, defined by equation (1), is ergodic, we shall use π to denote its limiting probability distribution: $\pi P = \pi$. And we shall define on π the same partition as on P:

$$\pi = (X_0 \pi_0 \cdots X_I \pi_I \cdots X_N \pi_N X_A \pi_A)$$

where X_I is the limiting (marginal) distribution of being in subset I and π_I the limiting (conditional) probability vector of the states of subset I. The vector $(X_0 \cdots X_N X_A)$ is also limiting probability vector of the aggregation matrix P^{agg}:

$$(X_0 \cdots X_I \cdots X_N X_A) P^{\text{agg}} = (X_0 \cdots X_I \cdots X_N X_A)$$

where

$$P^{\text{agg}} = \begin{bmatrix} P_{00}^{\text{agg}} & \cdots & P_{0I}^{\text{agg}} & \cdots & P_{0N}^{\text{agg}} & P_{0A}^{\text{agg}} \\ \vdots & & \vdots & & \vdots & \vdots \\ P_{I0}^{\text{agg}} & \cdots & P_{II}^{\text{agg}} & \cdots & P_{IN}^{\text{agg}} & P_{IA}^{\text{agg}} \\ \vdots & & \vdots & & \vdots & \vdots \\ P_{N0}^{\text{agg}} & \cdots & P_{NI}^{\text{agg}} & \cdots & P_{NN}^{\text{agg}} & P_{NA}^{\text{agg}} \\ R_0^{\text{agg}} & \cdots & R_I^{\text{agg}} & \cdots & R_N^{\text{agg}} & R_A^{\text{agg}} \end{bmatrix}$$

where

$$P_{IJ}^{\text{agg}} \stackrel{\text{def}}{=} \pi_I P_{IJ} \mathbf{1}^T, \qquad I = 0,\ldots,N; \qquad J = 0,\ldots,N,A$$

$$R_J^{\text{agg}} \stackrel{\text{def}}{=} \pi_A R_J \mathbf{1}^T \qquad J = 0,\ldots,N,A$$

and $\mathbf{1}^T$ denotes a column vector of ones. This aggregation matrix will turn out to be of importance in the determination of the transient characteristics of P that are introduced in Sections 2 and 3. But all the conditional distributions π_I $(I = 0,\ldots,N)$ and π_A have to be known to determine P^{agg}. If the state space is large (more than 10^4 as in the example of Section 7), this can be computationally too heavy. The technique proposed in Section 4 is intended to alleviate this problem.

2. EXPECTED DRIFT FUNCTION

A first indication of the direction in which the system can be expected to move when it is in subset I is given by the function

$$d(I) \stackrel{\text{def}}{=} \sum_J (J - I)(\pi_I P_{IJ} \mathbf{1}_J^T) = \sum_J (J - I) P_{IJ}^{\text{agg}}$$

Typically, $d(I) > 0$ indicates that the system is more likely to move toward subsets of larger indices. This can be an unequivocal sign of unstability in systems with infinite state spaces. The example of Section 5 illustrates this point.

The function $d(I)$ is expressed in terms of the ergodic flow rate out of subset I only. Unfortunately, the conditional distribution π_I of this subset cannot be exactly obtained in isolation from the other $\pi_J, J \neq I$. But we will see that it is possible to bound this distribution independently from the others, a technique that is useful when the state space is large.

3. TIME BEFORE ABSORPTION

Another quantity of interest in many applications is the time needed by a stochastic process to reach for the first time a critical or a faulty state or set of such states. In this respect, it is useful to observe that the Markov process defined by the matrix

$$P = \begin{bmatrix} P_{00} & \cdots & P_{0I} & \cdots & P_{0N} & P_{0A}\mathbf{1}^T \\ \vdots & & \vdots & & \vdots & \vdots \\ P_{I0} & \cdots & P_{II} & \cdots & P_{IN} & P_{IA}\mathbf{1}^T \\ \vdots & & \vdots & & \vdots & \vdots \\ P_{N0} & \cdots & P_{NI} & \cdots & P_{NN} & P_{NA}\mathbf{1}^T \\ r_0 & \cdots & r_I & \cdots & r_N & 0 \end{bmatrix} \tag{2}$$

where r is a row vector, has the same behavior as the chain with one absorbing state

$$\begin{bmatrix} P_{00} & \cdots & P_{0I} & \cdots & P_{0N} & P_{0A}\mathbf{1}^T \\ \vdots & & \vdots & & \vdots & \vdots \\ P_{I0} & \cdots & P_{II} & \cdots & P_{IN} & P_{IA}\mathbf{1}^T \\ \vdots & & \vdots & & \vdots & \vdots \\ P_{N0} & \cdots & P_{NI} & \cdots & P_{NN} & P_{NA}\mathbf{1}^T \\ 0 & \cdots & 0 & \cdots & 0 & 1 \end{bmatrix} \tag{3}$$

if one assumes that whenever this chain reaches the absorbing state (say A), it is restated with the initial distribution vector $r = (r_0 \cdots r_N)$. The ergodic behavior of equation (2) describes the behavior of the absorbing chain system (3) over an infinite number of runs, each started with the initial distribution r. In these conditions, the mean of the time before absorption of the absorbing chain (3) is given by

$$\tau \stackrel{\text{def}}{=} (r_0 \cdots r_N) \left[I - \begin{bmatrix} P_{00} & \cdots & P_{0I} & \cdots & P_{0N} \\ \vdots & & \vdots & & \vdots \\ P_{I0} & \cdots & P_{II} & \cdots & P_{IN} \\ \vdots & & \vdots & & \vdots \\ P_{N0} & \cdots & P_{NI} & \cdots & P_{NN} \end{bmatrix} \right]^{-1} \mathbf{1}^T \tag{4}$$

which can be rewritten as

$$\tau \stackrel{\text{def}}{=} (r_0 \cdots r_N)[I - G_A]^{-1}\mathbf{1}^T \tag{5}$$

where G_A is the matrix of the transition probabilities between all the transient states (G_A is substochastic and has no absorbing class itself). This time τ can be derived from the steady-state distribution π of the stochastic matrix P given in equation (2). Indeed, we have

THEOREM 1 If r is the mean time before absorption of the chain

$$\begin{bmatrix} G_A & (I - G_A)\mathbf{1}^T \\ \mathbf{0} & 1 \end{bmatrix}$$

when started with the initial distribution r, then

$$\tau = \frac{1}{X_A} - 1 \tag{6}$$

where X_A is the last component of the steady-state distribution of the matrix

$$\begin{bmatrix} G_A & (I - G_A)\mathbf{1}^T \\ r & 0 \end{bmatrix} \tag{7}$$

Proof. The proof is given in the Appendix. Intuitively, $1/X_A$ is the mean time between two visits to the last state of matrix (7) and each time the system is in this last state, one further step is needed to restart the absorbing chain with the initial distribution r. The time τ is thus given by equation (6).

Since τ is defined by the steady-state probability of state A only, the states of G_A can be aggregated. For instance, τ can be derived from the steady-state vector $X = (X_0 \cdots X_A)$ of the aggregated matrix

$$P^{\text{agg}} = \begin{bmatrix} \pi_0 P_{00} \mathbf{1}_0^T & \cdots & \pi_0 P_{0N} \mathbf{1}_N^T & \pi_0 P_{0A} \mathbf{1}^T \\ \vdots & & \vdots & \vdots \\ \pi_N P_{N0} \mathbf{1}_0^T & \cdots & \pi_N P_{NN} \mathbf{1}_N^T & \pi_N P_{NA} \mathbf{1}^T \\ r_0 \mathbf{1}_0^T & \cdots & r_N \mathbf{1}_N^T & 0 \end{bmatrix}$$

In Section 4, it is shown that bounds on the distributions π_I can be obtained separately for each block I, $(I = 0, \ldots, N)$; and from these bounds, a lower and an upper bound matrix on P^{agg} can be constructed, from which bounds on X_A, and thus on τ, can be derived. Obtaining bounds on τ in this way requires less computational effort than obtaining τ via equation (4).

Moreover, the technique continues to work, albeit with less accuracy, if the initial distribution is not known. As a consequence, it continues to work also if, instead of a single absorbing state A, an arbitrary class A has been considered, not all the states of the class being not necessarily associated with a same initial distribution r.

The same approach can be followed to bound the variance of the time before absorption.

THEOREM 2 If τ_2 is the variance of the time before absorption of the chain

$$\begin{bmatrix} G_A & (I - G_A)\mathbf{1}^T \\ 0 & 1 \end{bmatrix}$$

when started with the initial distribution r, then

$$\tau_2 = 2\tau\tau' - \tau - \tau^2$$

where τ is the mean time before absorption, that is

$$\tau = \frac{1}{X_A} - 1, \qquad X \begin{bmatrix} G_A & (I - G_A)\mathbf{1}^T \\ r & 0 \end{bmatrix} = X$$

and where τ' is given by

$$\tau' = \frac{1}{X'_A} - 1$$

where X'_A is the last component of the steady-state distribution of the matrix

$$\begin{bmatrix} G_A & (I - G_A)\mathbf{1}^T \\ r' & 0 \end{bmatrix}$$

with

$$r' = \frac{1}{1-X_A}(X_0\pi_0 \cdots X_N\pi_N)$$

Proof. The proof is given in Appendix. Note that τ' is the *ergodic exit time* (see e.g. [8]) and can be seen as the mean time before absorption of the chain G_A with an initial distribution r' equal to the normalized part of the steady-state vector π corresponding to the transient states. As for X_A, the quantity X'_A can be derived from the steady-state analysis of the aggregated matrix P'^{agg}:

$$P'^{agg} = \begin{bmatrix} P'^{agg}_{00} & \cdots & P'^{agg}_{0I} & \cdots & P'^{agg}_{0N} & P'^{agg}_{0A} \\ \vdots & \vdots & \vdots & \vdots & & \\ P'^{agg}_{I0} & \cdots & P'^{agg}_{II} & \cdots & P'^{agg}_{IN} & P'^{agg}_{IA} \\ \vdots & \vdots & \vdots & \vdots & & \\ P'^{agg}_{N0} & \cdots & P'^{agg}_{NI} & \cdots & P'^{agg}_{NN} & P'^{agg}_{NA} \\ \dfrac{X_0}{1-X_A} & \cdots & \dfrac{X_1}{1-X_A} & \cdots & \dfrac{X_N}{1-X_A} & 0 \end{bmatrix}$$

where

$$P'^{agg}_{IJ} = \pi'_I P_{IJ} \mathbf{1}^T$$

Thus, bounds on τ_2 can be obtained from bounds on τ and τ'. Bounds on τ', like those on τ, can be obtained blockwise from bounds on π'_I $(I = 0, \ldots, N)$; and from these bounds, a lower and an upper bound matrix on P'^{agg} can be constructed from which bounds on X'_A can derived.

Therefore, bounds on the drift, on the mean, and the variance of the time before absorption require bounds on conditional distributions, and lower and upper bounds on aggregate matrices. Let us now see how such bounds can be obtained.

4. BOUNDS ON CONDITIONAL DISTRIBUTIONS

We will assume that the transient characteristics of the model are adequately captured by the ergodic matrix P define by equation (1). If we define for each block I the following subvectors and submatrices:

$$\pi_I^0 \stackrel{\text{def}}{=} [X_0\pi_0 \cdots X_{I-1}\pi_{I-1} \quad X_{I+1}\pi_{I+1} \cdots X_N\pi_N]$$

$$R_I^0 \stackrel{\text{def}}{=} [R_0 \cdots R_{(I-1)} \quad R_{(I+1)} \cdots R_N]$$

$$E_I \stackrel{\text{def}}{=} [P_{I0} \cdots P_{I(I-1)} \quad P_{I(I+1)} \cdots P_{IN}]$$

$$F_I \stackrel{\text{def}}{=} \begin{bmatrix} P_{0I} \\ \vdots \\ P_{(I-1)I} \\ P_{(I+1)I} \\ \vdots \\ P_{NI} \end{bmatrix}$$

$$G_I \stackrel{\text{def}}{=} \begin{bmatrix} P_{00} & \cdots & P_{O(I-1)} & P_{O(I+1)} & \cdots & P_{0N} \\ \vdots & & \vdots & \vdots & & \vdots \\ P_{(I-1)0} & \cdots & P_{(I-1)(I-1)} & P_{(I-1)(I+1)} & \cdots & P_{(I-1)N} \\ P_{(I+1)0} & \cdots & P_{(I+1)(I-1)} & P_{(I+1)(I+1)} & \cdots & P_{(I+1)N} \\ \vdots & & \vdots & \vdots & & \vdots \\ P_{N0} & \cdots & P_{N(I-1)} & P_{N(I+1)} & \cdots & P_{NN} \end{bmatrix}$$

then the global steady-state equations $\pi P = \pi$, where P is defined by matrix (1), yield the following system:

$$\begin{cases} X_1\pi_I E_I + \pi_I^0 G_I + X_A\pi_A R_I^0 = \pi_I^0 \\ X_1\pi_I P_{II} + \pi_I^0 F_I + X_A\pi_A R_I = X_I\pi_I \end{cases}$$

Solving this system in $X_I\pi_I$ gives

$$X_I\pi_I = X_A\pi_A(R_I + R_I^0[I - G_I]^{-1}F_I)[I - P_{II} - E_I[I - G_I]^{-1}F_I]^{-1}$$

and since π_I is a normalized vector, we write

$$\pi_I = \frac{X_A\pi_A}{X_I} B_I[I - P_{II}^*]^{-1} = \frac{\pi_A B_I[I - P_{II}^*]^{-1}}{\pi_A B_I[I - P_{II}^*]^{-1}\mathbf{1}^T} \tag{8}$$

with

$$B_I \stackrel{\text{def}}{=} (R_I + R_I^0[I - G_I]^{-1}F_I), \qquad P_{II} \stackrel{\text{def}}{=} P_{II}^* + E_I[I - G_I]^{-1}F_I$$

Expression (8) shows that π_I is given by the non-negative combination $\pi_A B_I$ of the rows of the inverse $[I - P_{II}^*]^{-1}$. The matrix B_I describes how the block I is entered for the first time from the set of states A: the term R_I is the probability of entering directly some precise state of the block I, while the second term is the probability of entering in the block I after any nonempty sequence of transitions in other blocks. This expression of π_I can be further developed. Let us define the following diagonal matrix Δ_Q

$$[\Delta_Q]_{i,j} = \begin{cases} 0, & \text{if } j \neq i \\ 1, & \text{if } j = i, \quad \text{and} \quad e_i Q \mathbf{1}^T = 0 \\ e_i Q \mathbf{1}^T & \text{if } j = i, \quad \text{and} \quad e_i Q \mathbf{1}^T \neq 0 \end{cases}$$

the matrix Δ_Q^{-1} is a matrix operator that normalizes to one the nonnull rows of Q. Introducing this operator in the expression (8) of π_I yields

$$\pi_I = \frac{\pi_A B_I \Delta_{[I-P_{II}^*]^{-1}}}{\pi_A B_I \Delta_{[I-P_{II}^*]^{-1}} \mathbf{1}^T} \qquad \Delta^{-1}_{[I-P_{II}^*]^{-1}} [I - P_{II}^*]^{-1} = \zeta_I Z_I$$

with

$$\zeta_I \stackrel{\text{def}}{=} \frac{\pi_A B_I \Delta_{[I-P_{II}^*]^{-1}}}{\pi_A B_I \Delta_{[I-P_{II}^*]^{-1}} \mathbf{1}^T}, \qquad Z_I \stackrel{\text{def}}{=} \Delta^{-1}_{[I-P_{II}^*]^{-1}} [I - P_{II}^*]^{-1}$$

The vector π_I is thus given by the convex combination, ζ_I, of the rows of Z_I.

When the state set A reduces to one single state, $\pi_A = 1$. The determination of the quantities in the expression (9) requires the computation of the inverses $[I - G_I]^{-1}$ and $[I - P_{II}^*]^{-1}$. Since A corresponds to one single state only, the dimension of G_I is $n - n_I - 1$, where n_I denotes the size of the block I. The exact determination of π_I would require therefore the computation of inverses of sizes n_I and $n - n_I - 1$, respectively, an amount of work similar to that required to solve the whole system $\pi P = \pi$ of size n. For larger sets A, the size of G_I reduces to $n - n_I - n_A$, where n_A is the size of A. If n_A is sufficiently large, the inverse of $[I - G_I]$ can be obtained at an acceptable cost. However, the exact determination of π_I requires now the knowledge of π_A.

Fortunately, information on π_I can be obtained without requiring the knowledge of π_A or of the exact inverses $[I - G_I]^{-1}$ and $[I - P_{II}^*]^{-1}$. Since π_I is given [equation (9)] by a convex combination ζ_I of the rows of Z_I, π_I belongs to the polyhedron whose vertices are the rows of Z_I. When the convex combination ζ_I is not known, that is, when the vector π_A is not available, bounds on the polyhedron can provide bounds on π_I. These bounds will be all the most tight as the polyhedron is small. If the computation of the inverse $[I - P_{II}^*]^{-1}$, from which the polyhedron Z_I is derived, is itself too expensive, any other polyhedron that includes Z_I and that is easier to obtain will also provide bounds on π_I. Section 6 discusses the tightness of the bounds derived from such polyhedra and also the computational effort required to obtain them.

One important consequence of equation (9) should not be overlooked at this point. *The matrix B_I, and the submatrix R, are not necessary for the computation of the bounds of π_I. Therefore, there is no need to assume a specific matrix*

R when an associated ergodic matrix P is constructed. The bounds on τ and τ', which are obtained in this case, are then valid for every possible return matrix R.

Note that if information on B_I is available, other polyhedra that guarantee to contain π_I can be obtained by introducing in expression (8) a matrix that normalizes the nonnull rows of the product

$$B_I[I - P^*_{II}]^{-1}$$

instead of one that normalizes the rows of the inverse only. One then obtains

$$\pi_I = \frac{\pi_A \Delta_{B_I[I-P^*_{II}]^{-1}}}{\pi_A \Delta_{B_I[I-P^*_{II}]^{-1}} \mathbf{1}^T_{B_I}} \Delta^{-1}_{B_I[I-P^*_{II}]^{-1}} B_I[I - P^*_{II}]^{-1} = \omega_I W_I \tag{10}$$

with

$$\omega_I \stackrel{\text{def}}{=} \frac{\pi_A \Delta^{-1}_{B_I[I-P^*_{II}]^{-1}}}{\pi_A \Delta^{-1}_{B_I[I-P^*_{II}]^{-1}} \mathbf{1}^T_{B_I}}, \qquad W_I \stackrel{\text{def}}{=} \Delta^{-1}_{B_I[I-P^*_{II}]^{-1}} B_I[I - P^*_{II}]^{-1}$$

where by construction the column vector $\mathbf{1}^T_{B_I}$ is given by

$$\mathbf{1}^T_{B_I} = \Delta^{-1}_{B_I[I-P^*_{II}]^{-1}} B_I[I - P^*_{II}]^{-1} \mathbf{1}^T$$

The ith component of $\mathbf{1}^T_{B_I}$ is thus equal to one if the corresponding rows of B_I is non-null and is equal to zero if the corresponding row of B_I is null. The equation (10) shows that π_I is given by the convex combination ω_I of only the non-null rows of W_I. Thus the polyhedron bases on the non-null rows of the matrix W_I contains the vector π_I. If ω_I is not known, bounds on π_I can be derived from this polyhedron. The tightness of these bounds and the computational effort needed to obtain them are discussed in Section 6.

5. BOUNDS ON MARGINAL DISTRIBUTIONS

We have seen that in order to obtain bounds on τ (and τ'), we also need to bound the limiting distribution vector $(X_0 \ldots X_N X_A)$ of the matrix P^{agg} or at least its last component X_A. The matrix P^{agg} is not exactly known since its elements are given by

$$P^{\text{agg}}_{IJ} \stackrel{\text{def}}{=} \pi_I P_{IJ} \mathbf{1}^T, \qquad I = 0, \ldots, N; \qquad J = 0 \ldots N, A$$

and only bounds on the conditional distributions distribution π_I are available. However, the following technique can be used.

Let us first remark that, from the bounds on the conditional distributions, bounds on the elements of the aggregated matrix can be derived:

$$P^{\min}_{IJ} \leq P^{\text{agg}}_{IJ} \leq P^{\max}_{IJ}$$

Furthermore, it has been shown in [4] the if $L \leq P$, elementwise, where P is a stochastic matrix, then the Perron–Frobenius eigenvector X of the matrix P, that is, the vector X satisfying $XP = X$, is a convex combination of the normalized rows of $[I - L]^{-1}$. The vector X belongs thus to the polyhedron constructed on the normalized rows of the inverse $[I - L]^{-1}$. In our problem, the polyhedron based on the normalized rows of the matrix $[I - P^{\min}]^{-1}$ is therefore guaranteed to contain the marginal distribution vector X and bounds on this vector can be derived from bounds on this polyhedron.

When the aggregated matrix has a special structure, tighter bounds can be obtained. If, for example, P^{agg} is tridiagonal, then the marginal probability X_A will be maximized by maximizing the probability of the transitions that move the system close to A and by minimizing those that move the system away from A. One can then construct the matrix

$$P^{sup} = \begin{bmatrix} \times & P_{01}^{\max} & \mathbf{0} & \cdots & \mathbf{0} & \mathbf{0} \\ P_{10}^{\min} & \times & P_{12}^{\max} & \cdots & \mathbf{0} & \mathbf{0} \\ \mathbf{0} & P_{21}^{\min} & \times & \cdots & \mathbf{0} & \mathbf{0} \\ \vdots & \vdots & \vdots & & \vdots & \vdots \\ \mathbf{0} & \mathbf{0} & \mathbf{0} & \cdots & \times & P_{NA}^{\max} \\ r_0\mathbf{1}_0^T & r_1\mathbf{1}_1^T & r_2\mathbf{1}_2^T & \cdots & r_N\mathbf{1}_N^T & 0 \end{bmatrix}$$

where the diagonal components $\times$ are the complement to one of the row sums. The Perron–Frobenius eigenvector X^{sup} of P^{sup} then satisfies

$$X_A \leq X_A^{sup}$$

Similarly, a lower bound X_A^{inf} on X_A will be given by the last component of the Perron–Frobenius eigenvector X^{inf} of the matrix P^{inf}

$$P^{inf} = \begin{bmatrix} \times & P_{01}^{\min} & \mathbf{0} & \cdots & \mathbf{0} & \mathbf{0} \\ P_{10}^{\max} & \times & P_{12}^{\min} & \cdots & \mathbf{0} & \mathbf{0} \\ \mathbf{0} & P_{21}^{\max} & \times & \cdots & \mathbf{0} & \mathbf{0} \\ \vdots & \vdots & \vdots & & \vdots & \vdots \\ \mathbf{0} & \mathbf{0} & \mathbf{0} & \cdots & \times & P_{NA}^{\min} \\ r_0\mathbf{1}_0^T & r_1\mathbf{1}_1^T & r_2\mathbf{1}_2^T & \cdots & r_N\mathbf{1}_N^T & 0 \end{bmatrix}$$

This technique was already used in [11].

The same procedure can be used when the aggregation matrix has the Hessenberg structure (see e.g. [9]; see also [6], which uses the particular structure of some Markov chains to evaluate moments of the time before absorption).

6. TIGHTNESS AND COMPUTATIONAL COMPLEXITY

Let us first discuss the tightness of the bounds derived from the polyhedron based on the normalized rows of $[I - P^*_{II}]^{-1}$. The size of this polyhedron is mainly [5] determined by the substochasticity of the matrix P^*_{II}: the more stochastic the matrix P^*_{II} is, the more parallel the rows of the inverse are, and the smaller the polyhedron is. The matrix P^*_{II} is given by

$$P^*_{II} \stackrel{\text{def}}{=} P_{II} + E_I[I - G_I]^{-1}F_I$$

where G_I is a matrix of transition probabilities between states that do not belong to the sets I or A. When A is empty, the matrix G_I gives the transition probabilities between all the states that are not in the Ith block. In this case, it can be shown that the matrix P^*_{II} is stochastic so that the polyhedron reduces to a single point. The other extreme case corresponds to an empty matrix G_I, that is, a situation where all the states that are not in I have been reassembled in A. In this case, P^*_{II} reduces to P_{II} and the polyhedron size depends on the substochasticity of P_{II}. The size of the polyhedron based on $[I - P^*_{II}]^{-1}$ varies thus between that of the polyhedron based on $[I - P^*_{II}]^{-1}$ and a single point. The largest the product $E_I[I - G_I]^{-1}F_I$ is, the closer to a stochastic matrix P^*_{II} is and the smaller the polyhedron based on $[I - P^*_{II}]^{-1}$ will be. The states corresponding to G_I should therefore be chosen as those that maximize the quantity

$$E_I[I - G_I]^{-1}F_I$$

Indeed, this is the choice that brings the matrix P^*_{II} as close as possible to a stochastic matrix and that therefore leads to the smallest polyhedron. This choice consists in reassembling in G_I all the states that have strong connections with the states of I (see [3], for example).

From the point of view of the computational complexity, the determination of the polyhedron based on $[I - P^*_{II}]^{-1}$ requires the computation of P^*_{II} and of the inverse $[I - P^*_{II}]^{-1}$. The computation of an inverse of size n_I can be done in $(2/3)n_I^3$ operations if one uses the LU decomposition [$(1/3)n_I^3$ operation for the decomposition of $[I - P^*_{II}]$ into the product LU and $(1/3)n_I^3$ operations for the computation of the product $U^{-1}L^{-1}$]; or in $(7/6)n_I^3$ operations with the QR decomposition [$(2/3)n_I^3$ operation for the decomposition of $[I - P^*_{II}]$ into the product QR and $(1/2)n_I^3$ operations for the product $R^{-1}Q^H$]. Note that the LU decomposition can introduce numerical rounding errors while the QR decomposition is guaranteed to be numerically stable.

The determination of P^*_{II} mainly consists in the computation of the inverse $[I - G_I]^{-1}$ and in the product $E_I[I - G_I]^{-1}F_I$. The product requires at most $(n_I + n_{G_I})n_I n_{G_I}$ operations where n_{G_I} is the size of G_I. As detailed in the previous paragraph, the inverse of $[I - G_I]$ requires $kn_{G_I}^3$ operations, with k close to one ($2/3 \leq k \leq 7/6$). The complexity of the computation of the polyhedron based on Z_I is thus

$$C = kn_I^3 + kn_{G_I}^3 + (n_I + n_{G_I})n_I n_{G_I} \tag{11}$$

or

$$C = O([\max(n_I, n_{G_I})]^3)$$

This clearly shows that the cheapest polyhedron (obtained with $n_{G_I} = 0$) is the largest (the one based on $[I - P_{II}]^{-1}$), and that the most expensive polyhedron ($n_{G_I} = n - n_I$) is also the smallest (i.e., a single point). This enlightens the compromise that has to be found between the tightness of the bounds and an admissible computational effort.

The polyhedron based on W_I requires the determination of the product

$$B_I[I - P_{II}^*]^{-1} = (r_I + r_I^0[I - G_I]^{-1}F_I)[I - P_{II} - E_I[I - G_I]^{-1}F_I]^{-1}$$

The computational effort needed to compute the inverse $[I - P_{II}^*]^{-1}$ is given by equation (11). The determination of B_I and the multiplication $B_I[I - P_{II}^*]^{-1}$ can be done in

$$kn_{G_I}^3 + n_{G_I}n_I(n_A + n_{G_I}) + n_A n_I^2$$

operations. However, this constitutes a theoretical bound only since, in practice, most of the rows of the matrix B_I are null and n_A should be replaced by a quantity close to $\max(n_I, n_{G_I})$. This shows that the computation of B_I and the premultiplication of the inverse by B_I requires an amount of work similar to that needed to compute the inverse $[I - P_{II}^*]^{-1}$. The determination of W_I is therefore approximately two times as expensive as the determination of Z_I.

From the point of view of the tightness, it can easily be shown [12] that the premultiplication of $[I - P_{II}^*]^{-1}$ by B_I can only lead to a smaller polyhedron. Intuitively, each normalized row of the product is a convex combination of the normalized rows of $[I - P_{II}^*]^{-1}$. Therefore, the polyhedron based on the normalization rows of the product can only be smaller than the polyhedron based on the normalized rows of $[I - P_{II}^*]^{-1}$.

7. STABILITY ANALYSIS OF A SLOTTED ALOHA SYSTEM

Consider a system consisting of M stations each provided with B buffers. A station with one packet ready for transmission transmits the packet in the next slot with probability $p(i)$ if i ($1 \leq i \leq B$) is the number of packets waiting for transmission in the queue. The packet will be correctly received if and only if all the other stations do not transmit in that slot. If it is not correctly received, the packet remain at the head of the queue, ready for transmission in the next slot. The generation process of new packets is identical for all stations and is Bernoulli of parameter σ: more precisely, in each station, and in each slot, a packet is generated with the probability σ. Since each station is provided with a finite number of buffers, arrivals occurring when these buffers are occupied are lost.

ALOHA systems have been extensively studied (see [1] for a general overview or [10] for a more recent bibliography). Most of the studies (see for example [13], [14], or [10]) address stability questions that are directly related to the steady-state distribution of the system. Here we focus on transient characteristics.

The system can be modeled by a Markov chain that describes the state of the system at the end of each slot. This state is given by the M-tuples

$$(s_1 \cdots s_k \cdots s_M)$$

where s_k denotes the number of packets awaiting transmission in station k. The state space can be very large [$(B+1)^M$ states]. However, since individual stations need not be identified, one can use the state descriptor

$$(i_B \cdots i_j \cdots i_1) \tag{12}$$

where $i_j (1 \leq j \leq B)$ gives the number of stations that have j awaiting packets. This leads to a state space of much smaller size:

$$\binom{M+B}{M}$$

Since the number of packets in the system is the variable of interest, the following state descriptor is more appropriate:

$$(m i_B \cdots i_j \cdots i_2)$$

where m is the total number of awaiting packets. This state descriptor is similar to descriptor (12) since

$$i_1 = m - \sum_{j=2}^{B} j i_j$$

Note that the number of distinct states with an identical population m remains independent of the number of stations M in the system, for all values $m < M$. The Markov chain is therefore characterized by the stochastic matrix P that has the following Hessenberg structure:

$$P = \begin{bmatrix} P_{00} & P_{01} & P_{02} & \cdots & P_{O(N-1)} & P_{0N} & P_{0A} \\ P_{10} & P_{11} & P_{12} & \cdots & P_{1(N-1)} & P_{1N} & P_{1A} \\ 0 & P_{21} & P_{22} & \cdots & P_{2(N-1)} & P_{2N} & P_{0A} \\ \vdots & \vdots & \vdots & & \vdots & \vdots & \vdots \\ 0 & 0 & 0 & \cdots & P_{N(N-1)} & P_{NN} & P_{NA} \\ 0 & 0 & 0 & \cdots & 0 & R_N & P_A \end{bmatrix}$$

where P_{IJ} is the block of transition probabilities from all the states corresponding to a populations of I packets in the system to the states corresponding to a population of J packets. The block P_A contains all the system states corresponding to a population larger that N packets. The zero blocks result from the fact that at most one packet can be successfully transmitted in a slot.

Let us first study the macroscopic drift

$$d(I) \stackrel{\text{def}}{=} \sum_J (J - I)(\pi_I P_{IJ} \mathbf{1}_J^T)$$

This function gives information on the behavior of the system when exactly I packets are waiting. Indeed, each block P_{IJ} can be decomposed into two terms:

$$P_{IJ} = P_{IJ}^{suc} + P_{IJ}^{nos}$$

where the elements of P_{IJ}^{suc} are the transition probabilities that correspond to the arrival of $(J - I + 1)$ packets and to a successful transmission, and those of P_{IJ}^{nos} are the transition probabilities which correspond to the arrival of $(J - I)$ packets with no successful transmission. (Note that $P_{I(I-1)} = P_{I(I-1)}^{suc}$.) With this decomposition, the growth rate $\text{gr}(I)$ of the packet population is given by

$$\text{gr}(I) = \sum_{J \geq I} (J - I)\pi_I \left(P_{IJ}^{nos} \mathbf{1}_J^T + P_{I(J-I)}^{suc} \mathbf{1}_J^T\right)$$

and the drop rate $\text{dr}(I)$ is given by

$$\text{dr}(I) = \sum_{J \geq I} \pi_I P_{I(J-1)}^{suc} \mathbf{1}_J^T$$

Then the following results hold:

$$\begin{aligned} &\text{gr}(I) - \text{dr}(I) \\ &= \sum_{J \geq I} (J - I)\pi_I P_{IJ}^{nos} \mathbf{1}_J^T + \sum_{J \geq I} (J - I - 1)\pi_I P_{I(J-1)}^{suc} \mathbf{1}_{(J-1)}^T \end{aligned}$$

$$= -\pi_I P^{suc}_{I(I-1)} \mathbf{1}^T_{(I-1)} + \sum_{J \geq I} (J - I) \pi_I P^{nos}_{IJ} \mathbf{1}^T_J$$

$$+ \sum_{J \geq I+1} (J - I - 1) \pi_I P^{suc}_{I(J-1)} \mathbf{1}^T_{(J-1)}$$

$$= -\pi_I P^{suc}_{I(I-1)} \mathbf{1}^T_{(} I - 1) + \sum_{J \geq I} (J - I) \pi_I P^{nos}_{IJ} \mathbf{1}^T_J + \sum_{J \geq I} (J - I) \pi_I P^{suc}_{IJ} \mathbf{1}^T_J$$

$$= \sum_{J \geq I-1} (J - I) \pi_I P_{IJ} \mathbf{1}^T_J = d(I)$$

This shows that the drift $d(I)$ is equal to the difference between the growth rate and the drop rate of the system when I packets are awaiting transmission.

Bounds on the drift can be derived as follows. First, introducing equation (10) in the expression of the drift gives

$$d(I) \stackrel{\text{def}}{=} \sum_J (J - I)(\omega_I W_I P_{IJ} \mathbf{1}^T_J) = \omega_I \sum_J (J - I)(W_I P_{IJ} \mathbf{1}^T_J)$$

where all the components of ω_I corresponding to nonnull rows of W_I are nonnegative and sum up to one. This allows us to write

$$\min_{i \in \mathcal{W}_I} \left[\sum_J (J - I)(W_I P_{IJ} \mathbf{1}^T_J) \right]_i \leq d(I)$$

$$\leq \max_{i \in \mathcal{W}_I} \left[\sum_J (J - I)(W_I P_{IJ} \mathbf{1}^T_J) \right]_i \qquad (13)$$

where $\mathcal{W}_I$ denotes the set of indices corresponding to the nonnull rows of W_I. These bounds can be shown to be much tighter than those obtained by first bounding the vector π_I,

$$\pi_I^{\min} \leq \pi_I \leq \pi_I^{\max}$$

with

$$[\pi_I^{\min}]_k \stackrel{\text{def}}{=} \min_{i \in \mathcal{W}_I} [W_I]_{ik}, \qquad [\pi_I^{\max}]_k \stackrel{\text{def}}{=} \max_{i \in \mathcal{W}_I} [W_I]_{ik}$$

and by then bounding $d(I)$:

$$\sum_J (J - I)(\pi_I^{\min} P_{IJ} \mathbf{1}^T_J) \leq d(I) \leq \sum_J (J - I)(\pi_I^{\max} P_{IJ} \mathbf{1}^T_J)$$

Tables 1–4 give the bounds obtained with (13) on the drift $d(10)$, for various value of the load. These results call for the following comments.

Table 1 Drift: $M = 20, B = 4, \sigma = 0.005, P(i) = (0.3, 0.2, 0.1, 0.05)$

N	$N = 10$	$N = 11$	$N = 12$	$N = 13$	$N = 14$	$N = 15$
$u(d(10))$	−0.0595	−0.0860	−0.1003	−0.1073	−0.1107	−0.1124
$l(d(10))$	−0.1550	−0.1274	−0.1193	−0.1164	−0.1152	−0.1147

Table 2 Drift: $M = 20, B = 4, \sigma = 0.010, P(i) = (0.3, 0.2, 0.1, 0.05)$

N	$N = 10$	$N = 11$	$N = 12$	$N = 13$	$N = 14$	$N = 15$
$u(d(10))$	+0.0537	+0.0248	−0.0034	−0.0267	−0.0436	−0.0547
$l(d(10))$	−0.1337	−0.1011	−0.0880	−0.0817	−0.0783	−0.0762

As explained earlier, the tightness of the bounds is essentially determined by the substochasticity of the matrix P^*_{II}

$$P^*_{II} \stackrel{\text{def}}{=} P_{II} + E_I[I - G_I]^{-1}F_I$$

In this example, G_I corresponds to the transition probabilities between the states corresponding to a population of J packets ($J = 0, \ldots, I-1, I+1, \ldots, N$). The larger N is, the larger the quantity $E_I[I - G_I]^{-1}F_I$ is, the closer to a stochastic matrix P^*_{II} is, and the tighter the bounds will be. This is clearly illustrated by each table. On the other hand, for a same value of

Table 3 Drift: $M = 20, B = 4, \sigma = 0.015, P(i) = (0.3, 0.2, 0.1, 0.05)$

N	$N = 10$	$N = 11$	$N = 12$	$N = 13$	$N = 14$	$N = 15$
$u(d(10))$	+0.1624	+0.1396	+0.1103	+0.0764	+0.0430	+0.0151
$l(d(10))$	−0.0965	−0.0714	−0.0592	−0.0525	−0.0483	−0.0454

Table 4 Drift:$M = 20, B = 4, \sigma = 0.020, P(i) = (0.3, 0.2, 0.1, 0.05)$

N	$N = 10$	$N = 11$	$N = 12$	$N = 13$	$N = 14$	$N = 15$
$u(d(10))$	+0.2678	+0.2520	+0.2296	+0.1990	+0.1608	+0.1198
$l(d(10))$	−0.0341	−0.0172	−0.0077	−0.0018	+0.0023	+0.0055

N, the matrix P^*_{II} becomes more deficient with respect to stochasticity as the load increases. This explains why the tightness of the bounds decreases as the load σ increases.

Table 5 gives the number of millions of floating point operations (MFlop) that were required to obtained the bounds. It also gives, for each value of N, the size n of the state space which has been taken into consideration only, a fraction varying between 1×10^{-2} and 3×10^{-2} of the total state space, which, in this example, is of size 10,626. This table shows that the determination of the bounds has a computational complexity of order n^3, as predicted by the analysis of Section 4.

To obtain the mean time before absorption into states corresponding to packet population exceeding the value N, one must compute the conditional distributions π_I for $I = 0, \ldots, N$. Rough bounds have been used in order to maintain the computational complexity below that of the determination of the whole inverse (4). More precisely, for the determination of π_I, the matrix G_I has been chosen as the matrix of transition probabilities between the states with a population smaller that I only

$$G_I = \begin{bmatrix} P_{00} & P_{00} & \cdots & P_{O(I-2)} & P_{O(I-1)} \\ P_{10} & P_{10} & \cdots & P_{1(I-2)} & P_{1(I-1)} \\ \vdots & \vdots & & \vdots & \vdots \\ \mathbf{0} & \mathbf{0} & \cdots & P_{(I-1)(I-2)} & P_{(I-1)(I-1)} \end{bmatrix}$$

This choice is advantageous; because of the Hessenberg structure of the G_I matrices, the matrices P^*_{II} can be determined recursively. Bounds on the elements of the matrix P^{agg} haven then been computed and the technique described in Section 5 has been used to obtain bounds on X_A, where A contains all the states with a population larger than 10 packets. From the bounds on X_A, the bounds $l(\tau)$ and $u(\tau)$ on the average time τ spent in states with a total population not larger than $N = 10$ packets were obtained. The results are given in Table 6.

Table 5 State space size and computational complexity: $M = 20$, $B = 4$.

	$N = 10$	$N = 11$	$N = 12$	$N = 13$	$N = 14$	$N = 15$
n	94	121	155	194	241	295
MFlop	0.13	0.27	0.55	1.20	2.61	5.44

Table 6 Mean Time before Absorption: $N = 10$, $M = 20, B = 4, P(i) = (0.3, 0.2, 0.1, 0.05)$

	$\sigma = 0.005$	$\sigma = 0.010$	$\sigma = 0.015$	$\sigma = 0.020$
$u(\tau)$	2.54×10^7	4.28×10^4	1.60×10^3	2.55×10^2
τ	1.23×10^7	1.54×10^4	6.45×10^2	1.41×10^2
$l(\tau)$	8.10×10^5	7.47×10^2	1.10×10^2	5.42×10^1
Mflop	0.159	0.159	0.159	0.159

Although rather loose, these bounds can be useful for many practical purposes. Consider, for instance, the bounds for the load $\sigma = 0.005$. They show that more than 8×10^5 slots are needed on the average before the number of packets awaiting transmission exceeds 10. During this time, the population never exceeds 10 packets; and since each station has $B = 4$ buffers, at most 2 stations are ever full at the same time. The total arrival rate during this time period is therefore at least equal to $(M - 2)\sigma = 0.09$ packets/slot. Thus a total average of at least 7.2×10^4 packets have arrived in the stations and have been successfully transmitted, since only 10 are left in the system. This type of observation is a guarantee of enough stability for many practical purposes.

Acknowledgments: We would like to thank Albert Greenberg, AT &T, Bell Labs., Murray Hill, N.J., and Guy Latouche, ULB, Brussels, Belgium, for useful comments on an earlier version of this paper.

APPENDIX

To prove equation (6), we simply rewrite the Perron–Frobenius equation

$$[X_0\pi_0 \cdots X_N\pi_N X_A] \begin{bmatrix} G_A & (I - G_A)\mathbf{1}^T \\ r & 0 \end{bmatrix} = [X_0\pi_0 \cdots X_N\pi_N X_A]$$

for all except the last column of the matrix. This yields

$$[X_0\pi_0 \cdots X_N\pi_N]G_A + X_A r = [X_0\pi_0 \cdots X_N\pi_N]$$

which can be transformed to give

$$X_A r[I - G_A]^{-1} = [X_0\pi_0 \cdots X_N\pi_N] \tag{14}$$

and by equation (5),

$$\tau = r[I - G_A]^{-1}\mathbf{1}^T = \frac{1}{X_A}[X_0\pi_0 \cdots X_N\pi_N]\mathbf{1}^T = \frac{1 - X_A}{X_A} \tag{15}$$

which is the desired result.

The *variance of the time before absorption* is given (see [7]) by

$$\tau_2 \stackrel{\text{def}}{=} r(2[I - G_A]^{-1} - I)[I - G_A]^{-1}\mathbf{1}^T - \tau^2$$

which can be rewritten as

$$\tau^2 = 2\frac{X_A}{1-X_A}\tau' - \tau - \tau^2 = 2\tau\tau' - \tau - \tau^2$$

with

$$\tau' \stackrel{\text{def}}{=} \frac{1-X_A}{X_A} r[I - G_A]^{-1}[I - G_A]^{-1}\mathbf{1}^T$$

The quantity τ' can be seen as the mean time before absorption of the chain G_A with the initial distribution $r' = (r'_0 \cdots r'_N)$:

$$\tau' = r'[I - G_A]^{-1}\mathbf{1}^T \tag{16}$$

with

$$r' = \frac{X_A}{1-X_A}[r_0 \cdots r_N][I - G_A]^{-1}$$

which, by equation (14), is equal to

$$r' = \frac{1}{1-X_A}[X_0\pi_0 \cdots X_N\pi_N]$$

that is, the normalized part of the steady-state distribution π corresponding to the transient states. As for τ, instead of exactly computing τ' through equation (16), this time can be derived from the steady-state analysis of the matrix P':

$$P' = \begin{bmatrix} p_{00} & \cdots & p_{0I} & \cdots & p_{0N} & p_{0A} \\ \vdots & & \vdots & & \vdots & \\ p_{I0} & \cdots & p_{II} & \cdots & p_{IN} & p_{IA} \\ \vdots & & \vdots & & \vdots & \\ p_{N0} & \cdots & p_{NI} & \cdots & p_{NN} & p_{NA} \\ \frac{X_0}{1-X_A}\pi_0 & \cdots & \frac{X_I}{1-X_A}\pi_I & \cdots & \frac{X_N}{1-X_A}\pi_N & \cdots & 0 \end{bmatrix}$$

Indeed, if $(X'_0\pi'_0 \cdots X'_N\pi'_N X'_A)$ is the steady-state eigenvector of P', τ' is given by

$$\tau' = \frac{1-X'_A}{X'_A}$$

which ends the proof of Theorem 2.

REFERENCES

[1] Bertsekas, D., and Gallager, R. *Data Networks*, Prentice-Hall Englewood Cliffs, N.J., 1987.

[2] Bobbio, A., and Trivedi, K. S., An aggregation technique for the transient analysis of stiff Markov chains, *IEEE Trans. Computers*, Vol. C-35, NO. 9, 1986, 803–814.

[3] Courtois, P.-J., *Decomposability: Queueing and Computer System Application*, Academic Press, New York, 1977.

[4] Courtois, P.-J., and Semal, P., Bounds for the positive eigenvectors of nonnegative matrices and for their approximations by decomposition, *J. Assoc. Comput. Mach.*, 31. 1984, 804–825.

[5] Courtois, P.-J., and Semal, P., On polyhedra of Perron-Frobenius eigenvectors, *Linear Algebra Appl.*, 65, 1985, 157–170.

[6] Gaver, D. P., Jacobs, P. A., and Latouche, G., Finite birth-and-death models in randomly changing environments, *Adv. Appl. Prob.* 16, 1984, 715–731.

[7] Kemeny, J. G., and Snell, J. L., *Finite Markov Chain*, D. van Nostrand Co., New York, 1960.

[8] Keilson, J. *Markov Chain Models—Rarity and Exponentiality*, Springer-Verlag, New York, 1979.

[9] Muntz, R. R., de Souza e Silva, E., and Goyal, A., Bounding availability of repairable computer systems, *Performance Eval. Rev.*, Vol. 17, N. 1, May 1989, 29–38.

[10] Rao, R. R., and Ephremides A., On the stability of interacting queues in a multiple-access system, *IEEE Trans. Info. Theory*, Vol. 34, N. 5, 1988, 918–930.

[11] Semal, P., and Courtois, P.-J., Stability analysis of large Markovian models, in *Performance '87*, P.-J. Courtois and G. Latouche, eds., 363–382, North-Holland, Amsterdam, 1987.

[12] Semal, P., and Courtois, P.-J., Computing conditional distributions of queueing network matrices, this volume, Chapter 17.

[13] Szpankowski, W., Stability Conditions for Multidimensional Queueing Systems and Applications to Analysis of Computer Systems, Purdue University Report CSD-TR-601, 1986.

[14] Tsybakov, B., and Mikhailov W., Ergodicity of slotted ALOHA systems, Problems Info. Transmission, Vol. 15, N. 4, 1979, 73–87.

23

Evaluating Bounds on Steady-State Availability of Repairable Systems from Markov Models*

JOHN C. S. LUI AND RICHARD R. MUNTZ Department of Computer Science, University of California at Los Angeles, Los Angeles, California

ABSTRACT

System availability is an important reliability measure for computer system designers. Most often Markov models are used in representing systems for reliability analysis. Due to the complex interactions between components, in general no closed-form solution can be obtained. Also, the model often has an unmanageable state space and it quickly becomes impractical to even generate all the states in the system model. In this paper, we will present a method for bounding steady-state availability and at the same time, drastically reduce the size of the Markov chains that must be solved. In [7], a method of analysis is developed in which part of the state space is represented in detail but the vast majority is approximated with a few states. The methodology provides bounds on steady-state availability. Here we extend the work in [7], which requires an *a priori* decision to be made concerning the amount of

*This work was supported by IBM through a University of California MICRO grant.

detail to represent (and therefore the tightness on the bounds). This paper extends those results to allow piecewise generation of the transition matrix such that at each step the bounds can be incrementally improved. We believe the approach may be applicable more generally, but at present we require certain assumptions that are valid (and reasonable) for reliability models.

1. INTRODUCTION

System availability and reliability are crucial measures for system designers, particularly life-critical situations or when large financial loss is possible. In order to evaluate reliability measures for those complex systems, Markov models are often used [4]. Unfortunately, Markov models of real systems generally have a very large state space [10] and it becomes impractical even to generate the entire transition rate matrix. Various methods have been proposed to calculate various performance and reliability measures from large Markov models. Among these are the aggregation/disaggregation method for chains with *nearly completely decomposable structure* [1], the iterative aggregation/disaggregation method [11] [12], and matrix-geometric based methods [8].

A Markov chain is nearly completely decomposable if the interactions between groups of states are not comparable with the interactions within the groups. This is usually not the case for reliability models of computer systems and the aggregation/disaggregation method is unlikely to yield a sufficiently accurate approximation. The iterative aggregation/disaggregation method uses a method similar to the Gauss–Seidel iteration technique. One problem is that the convergence rate may be slow. Also, existence and uniqueness are difficult to show. The matrix-geometric techniques require certain regularity of structure that is often missing from reliability models. In this paper, we propose a method that allows for efficient calculation of bounds on steady-state availability by explicitly taking advantage of the properties of reliability models.

Exact reliability measures are generally not required but rather sufficiently accurate approximate results. If the approximate results can be given in terms of upper and lower bounds then this is ideal. Recently, the results reported in [7] provided a way to compute bounds on steady-state availability for models of repairable computer systems. Rather than generating all the states in the system model, the approach calls for generating those states that account for most of the probability mass while all other states are grouped into a small number of aggregate states. It is shown that bounds can be obtained by suitably replacing some transition rates out of aggregate states by lower bounds on those rates and replacing others by upper bounds. This approach is "one step" in that a decision is made a priori as to how much of the transition rate matrix to expand in detail. In this paper, we extend the method to a multistep procedure in which we successively generate more of the transition rate matrix. At each step we obtain tighter bounds on steady-state availability. The following section will provide a more detailed description of the problem and the work reported in [7]. We present some background in Section 2 and Section 3, the procedure itself in Section 4, and a discussion of the results in Section 5.

2. BACKGROUND

We are interested in availability analysis for computer/communication systems. The behavior of such a system is assumed to be specified by a continuous-time, discrete-state homogeneous Markov process. Unfortunately, the characteristics of availability models (complex interactions between components, scheduling and maintenance policies, complex criteria for a system to be operational, etc.) preclude the possibility of closed-form solutions in general. Thus numerical solution methods are most widely used. Yet one of the major limiting factors of numerical methods is the large state space cardinality of realistic models. A real-life availability model can have tens of millions of states and thereby outstrip memory and processor capacities. In this paper, we present an algorithm that can provide enormous state space reduction for numerical solution and perhaps more importantly, also provide error bounds.

Many performance and reliability measures can be expressed in terms of a reward function. If $r(i)$ is the reward rate for state i, the expected reward rate $\mathcal{M}$ can be expressed as

$$\mathcal{M} = \sum_{i \in S} r(i)\pi(i) \tag{1}$$

with S being the state space of the Markov model and $\pi(i)$ being the steady-state probability of state i. Availability is a special case in which the "operational" states have reward of 1 and the "nonoperational" states have reward of 0.

Systems are generally designed to have a high level of availability. It is reasonable therefore to expect that during the life time of the system, most of the components are operational. With this in mind, we partition the state space of the model into $\mathcal{F}_i$, $0 \leq i \leq n$, where n is the number of system components and where $\mathcal{F}_i$ contains all the states that have *exactly* i failed components. The idea is to represent the detailed behavior of the model for $\mathcal{F}_i$, $i \leq K$

$$\begin{bmatrix} \mathcal{Q}_{00} & \mathcal{Q}_{01} & \mathcal{Q}_{02} & & \cdots & & \mathcal{Q}_{0n} \\ \mathcal{Q}_{10} & \mathcal{Q}_{11} & \mathcal{Q}_{12} & & \cdots & & \mathcal{Q}_{1n} \\ 0 & \mathcal{Q}_{21} & \mathcal{Q}_{22} & & \cdots & & \\ 0 & 0 & \mathcal{Q}_{32} & \mathcal{Q}_{33} & & & \cdot \\ \cdot & & 0 & \mathcal{Q}_{43} & \mathcal{Q}_{44} & & \cdot \\ \cdot & \cdots & & \ddots & & \ddots & \cdot \\ \cdot & & & & & & \\ 0 & \cdots & & & 0 & \mathcal{Q}_{n-1,n} & \mathcal{Q}_{nn} \end{bmatrix}$$

Figure 1 Transition matrix.

for some small value of K and approximate the remainder of the model via aggregation.

The transition rate matrix can then be viewed as shown in Figure 1, in which submatrix Q_{ii} corresponds to $\mathcal{F}_i$. In this figure the submatrices denoted by 0 contain all zero elements. This is a consequence of an assumption that there is zero probability of two or more components becoming operational at exactly the same time. Note that this does not preclude multiple repair facilities or any other common feature of reliability models. (In particular note that the case in which a dormant component that becomes active due to the repair of a second component does not violate the assumption. We simply do not consider such a dormant component as failed in the state partitions.)

We now summarize one of the results from [7]. Consider three sets of states:

$$\mathcal{G}_0 = \mathcal{F}_0$$

$$\mathcal{G}_1 = \left\{ \bigcup_{i=1}^{K} \mathcal{F}_i \right\}$$

$$\mathcal{G}_2 = \left\{ \bigcup_{i=K+1}^{n} \mathcal{F}_i \right\}$$

Figure 2 illustrates the transition matrix G in which G_{ii} is the principal submatrix corresponding to states in $\mathcal{G}_i$. (The submatrix shown as 0 is a con-

$$\begin{bmatrix} G_{00} & G_{01} & G_{02} \\ G_{10} & G_{11} & G_{12} \\ 0 & G_{21} & G_{22} \end{bmatrix}$$

Figure 2 Matrix G

$$\left[\begin{array}{cc|cc} G_{00} & G_{01} & 0 & G_{02} \\ G_{10} & G_{11} & 0 & G_{12} \\ \hline G_{10} & 0 & G_{11} & G_{12} \\ 0 & 0 & G_{21} & G_{22} \end{array}\right]$$

Figure 3 Duplication of states. Matrix G'.

sequence of the "nearest neighbor" property discussed previously.) Now construct a new transition matrix as shown in Figure 3. The relationship between the process defined by G and that defined by G' is illustrated in Figure 4. Basically, in the new process there are two sets of states corresponding to the states $\mathcal{G}_1$ of the original model. Let us call them $\mathcal{G}'_{1u}$ and $\mathcal{G}'_{1d}$ as shown in Figure 4. The idea behind this transformation can be explained as follows. Assume the system starts in the "all components up" state, that is $\mathcal{F}_0$. As components fail and are repaired the system will stay in states $\mathcal{G}_0$ and $\mathcal{G}'_{1u}$ until the first time that there are $K + 1$ or more failed components. At this point the system is in a state of $\mathcal{G}_2$. However when the number of failed components falls to K, the system now enters a state of $\mathcal{G}'_{1d}$. (Now the notation is explainable; "u" stands for "going Up" and "d" stands for "going Down." As the number of failed components goes up, the system visits states in $\mathcal{G}'_{1u}$ and after $K + 1$ failures have been reached, it visits the states in $\mathcal{G}'_{1d}$ as the number of failed components goes down.)

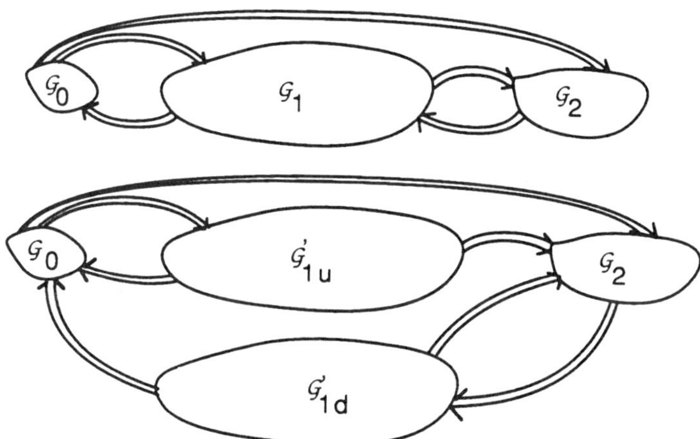

Figure 4 Relationship of G and G'.

From the construction it is easy to show that the two transition matrices are such that the steady-state probabilities of the original process (G) can be calculated from the steady-state probabilities of the second process (G'). Specifically, if

$$[\pi'_0, \pi'_1, \pi'_{1_2}, \pi'_2] \text{ is the solution of } \pi' G' = 0 \text{ and } \sum \pi'(i) = 1$$

then

$$[\pi'_0, \pi'_{1_2} + \pi_{1_2}, \pi'_2] \text{ is the solution of } \pi G = 0 \text{ and } \sum \pi(i) = 1$$

There is a natural mapping of the states in G' to states in G. In terms of rewards, the reward function for each state in G' is the same reward as the associated state in G. It is clear then that the mean availabilities of the two systems are identical.

Now consider aggregation of the states $\mathcal{F}_i, i \geq K+1$, and $F'_i, 1 \leq i \leq K$, as illustrated in Figure 5. The corresponding transition rate matrix is shown in Figure 6. The results from [7] show that if the rates indicated by a "+" are replaced by upper bounds (on the actual values) and the rates indicated by "−" are replaced by lower bounds then bounds on the steady-state availability, $\mathcal{A}$, can be expressed as

$$\sum_{i \in \mathcal{D}} r(i)\pi'(i) \leq \mathcal{A} \leq \sum_{i \in \mathcal{D}} r(i)\pi'(i) + \left(1 - \sum_{i \in \mathcal{D}} \pi'(i)\right) \tag{2}$$

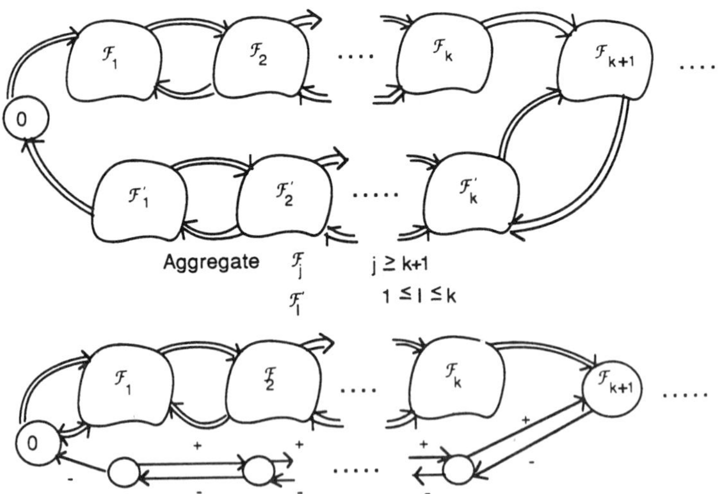

Figure 5 Aggregation of states.

$$\left[\begin{array}{cc|cccccc}
G_{00} & G_{01} & G_{0K+1'} & & \cdots & & & G_{0n} \\
G_{10} & G_{11} & G_{1K+1'} & & \cdots & & & G_{1n} \\
\hline
- & 0\ldots0 & \bullet & + & \cdots & & + & + \\
0 & & - & \bullet & & & + & + \\
\vdots & \vdots & 0 & - & \ddots & & + & + \\
 & & & 0 & - & & \bullet & + \\
0 & 0\ldots0 & \cdots & & 0 & & - & \bullet
\end{array}\right]$$

Figure 6 Form of transition matrix after aggregation and putting bounds in rates.

where $\mathcal{D} = \mathcal{G} \cup \mathcal{G}'_{1u}$ and $\pi'(i)$ is the steady-state probability for state i in the model with the indicated replacement of transition rates by bounds.

Since the above result is valid regardless of the set of operational states, it follows trivially that

$$\pi'(i) \leq \pi(i)\, \forall \text{ states } i$$

To show this for a particular state i, simply define the reward function as:

$$r(j) = \begin{cases} 1 & j = i \\ 0 & j \neq i \end{cases} \tag{3}$$

3. PRELIMINARIES

In the "single step" procedure described in the previous section, the decision as to the dimension of the matrix G_{11} is made *a priori*. This implies that once the bounds have been calculated there is no means provided for utilizing the work that has been already done in solving the first portion of the matrix to further tighten the bounds. The extension that we present in this section alleviates these difficulties. It allows an incremental generation of the transition matrix. At each step a new portion of the matrix is generated. Further, at each step the results from the previous steps are used to form a transition rate matrix whose solution allows us to bound the stationary-state probabilities for an additional set of states. This allows us to incrementally improve the bounds on availability.

At each step (after the first) there are three sets of states of the original model that we will need to distinguish.

$\mathcal{D}'$ = set of states for which lower bounds on the stationary state probabilities were obtained in previous steps

$$\begin{bmatrix} Q_{D'D'} & Q_{D'D} & Q_{D'A} \\ Q_{DD'} & Q_{DD} & Q_{DA} \\ 0 & Q_{AD} & Q_{AA} \end{bmatrix}$$

Figure 7 Initial matrix, G.

$\mathcal{D}$ = set of states that are the center of attention for this step and for which a lower bound on the stationary state probabilities will be calculated in this step

$\mathcal{A}$ = the complement of $\mathcal{D}' \cup \mathcal{D}$

Figure 7 illustrates this partitioning of the state space in terms of the transition matrix G. In the following we describe a short sequence of transformations to G. Each transformation is such that in the resulting model, the stationary state probabilities for the states in $\mathcal{D}$ are always (individually) bounded from below. We start by constructing the matrix G_1 from G as illustrated in Figure 8. G_1 corresponds to a model in which the states in $\mathcal{D}$ have been replicated. The set of replicate states (referred to as "clones") will be denoted by $\mathcal{C}$. Note that in G_1 the submatrix Q_{CC} is equal to Q_{DD}. We use the notation Q_{CC} because it enhances readability.

We use the notation π_{D/G_i} to denote the vector of stationary state probabilities for states in the set of states $\mathcal{D}$ when the transition rate matrix is G_i.

In the previous section while reviewing the results from [7] a similar construction was described. From that discussion it is clear that if

$$[\pi_{D'/G_1}, \pi_{D/G_1}, \pi_{C/G_1}, \pi_{A/G_1}] \text{ is the solution of } \pi_1 G_1 = 0 \text{ and } \sum \pi_{G_1}(i) = 1$$

then

$$[\pi_{D'/G_1}, \pi_{D/G_1} + \pi_{C/G_1}, \pi_{A/G_1}] \text{ is the solution of } \pi G = 0 \text{ and } \sum \pi_G(i) = 1$$

It follows immediately that $\pi_{D/G_1} \leq \pi_{D/G}$, since $\pi_{C/G_1} \geq 0$.

In the next several transformations we make use of the fact that it is possible to perform exact aggregation of a transition rate matrix. We assume that

$$\begin{bmatrix} Q_{D'D'} & Q_{D'D} & 0 & Q_{D'A} \\ Q_{DD'} & Q_{DD} & 0 & Q_{DA} \\ Q_{DD'} & 0 & Q_{CC} & Q_{DA} \\ 0 & 0 & Q_{AD} & Q_{AA} \end{bmatrix}$$

Figure 8 Introduction of "clone" states. Matrix G_1.

$$\begin{bmatrix} \bullet & R_{d'D} & 0 & R_{d'A} \\ Q_{Dd'} & Q_{DD} & 0 & Q_{DA} \\ Q_{Dd'} & 0 & Q_{CC} & Q_{DA} \\ 0 & 0 & Q_{AD} & Q_{AA} \end{bmatrix}$$

Figure 9 After exact aggregation of the states in $\mathcal{D}'$. Matrix G_2.

the reader is familiar with the basic aggregation/disaggregation approximation procedure as described in [1]. Later we show that exact aggregation is not actually required in the computation of the bounds. We merely use exact aggregation in the intermediate steps of the development.

G_2 (see Figure 9) is formed from G_1 by exact aggregation of the states in $\mathcal{D}'$. We will refer to the single state that replaces $\mathcal{D}'$ as d'. Since exact aggregation is assumed we have that $\pi_{D/G_2} = \pi_{D/G_1}$.

G_3 (see Figure 10) is exactly the same as G_2 except that the transition from d' to states in $\mathcal{D}$ and $\mathcal{C}$ are modified. In G_3 the submatrices $R'_{d'D}$ and $R'_{d'C}$ are required to have nonnegative elements and to be such that

$$R'_{d'D} + R'_{d'C} = R_{d'D} \tag{4}$$

A probabilistic interpretation is that the original transitions from d' to states in $\mathcal{D}$ are each "split" so that part remains to the state in D and part goes to the corresponding "clone" state in C.

From the construction of G_3 it is easy to show that if

$[\pi_{d'/G_2}, \pi_{D/G_2}, \pi_{C/G_2}, \pi_{A/G_2}]$ is the solution of $\pi G_2 = 0, \sum \pi_{G_2}(i) = 1$

and

$[\pi_{d'/G_3}, \pi_{D/G_3}, \pi_{C/G_3}, \pi_{A.G_3}]$ is the solution of $\pi G_3 = 0, \sum \pi_{G_3}(i) = 1$

then

$$\pi_{d'/G_3} = \pi_{d'/G_2}$$

$$\pi_{D/G_3} + \pi_{C/G_3} = \pi_{D/G_2} + \pi_{C/G_2}$$

$$\pi_{A/G_3} = \pi_{A/G_2}$$

$$\begin{bmatrix} \bullet & R'_{d'D} & R'_{d'C} & R_{d'A} \\ Q_{Dd'} & Q_{DD} & 0 & Q_{DA} \\ Q_{Dd'} & 0 & Q_{CC} & Q_{DA} \\ 0 & 0 & Q_{AD} & Q_{AA} \end{bmatrix}$$

Figure 10 Modified rates from state d'. Matrix G_3.

The result that we really want is expressed in the following theorem.

THEOREM 1 $\pi_{D/G_3} \le \pi_{D/G_2}$.

Proof. The proof is given in the Appendix.

Now we consider aggregation of the subsets of states in $\mathcal{C}$ and $\mathcal{A}$. By definition,

$$\mathcal{D} = \bigcup_{j=L_i}^{H_i} \mathcal{F}_j$$

$$\mathcal{C} = \bigcup_{j=L_i}^{H_i} \mathcal{F}'_j$$

and

$$\mathcal{A} = \bigcup_{j=H_i+1}^{N} \mathcal{F}_j$$

where L_i and H_i are integers associated with the ith step and denote the minimum and maximum number of failed components for states in $\mathcal{D}$.

We form one aggregate state for each subset $\mathcal{F}'_j$ in $\mathcal{C}$ and $\mathcal{F}_j$ in $\mathcal{A}$. The resulting matrix G_4 is shown in Figure 11. In this figure we have also interchanged the ordering of the state d' and the set of states $\mathcal{D}$. Since we have assumed exact aggregation in forming G_4 it is clear that

$$\pi_{D/G_4} = \pi_{D/G_3}$$

At this point we note that π_{D/G_4} provides a lower bound on $\pi_{D/G}$, that is $\pi_{D/G_4} \le \pi_{D/G}$. However, there are some practical problems in applying the procedure exactly as described: values for the transition rates out of aggregate states are required. To actually calculate these would require solving the original model in detail, and this is what we are trying to avoid.

Comparing the matrix in Figure 6 with the matrix G_4 in Figure 11, we note that they have the same form. Therefore the result quoted in the last section applies. Specifically, if the elements shown in Figure 12 as "+" are replaced by upper bounds on those rates and the elements shown as "−" are replaced by lower bounds then the solution for the stationary state probabilities will yield a lower bound for the state probabilities for states in D, that is, $\pi_{D/G_5} \le \pi_{D/G_4}$.

In summary, $\pi_{D/G_5} \le \pi_{D/G}$, and clearly G_5 has a much reduced states space than G. A remaining issue is the calculation of the upper and lower bounds

$$\left[\begin{array}{cc|ccccc|ccccc}
Q_{D,D} & Q_{D,d'} & 0 & 0 & \cdots & \cdots & 0 & Q_{D,A_1} & Q_{D,A_2} & \cdots & \cdots & Q_{D,A_n} \\
R'_{d',D} & \bullet & r_{d',C_1} & r_{d',C_2} & \cdots & \cdots & r_{d',C_J} & r_{d',A_1} & r_{d',A_2} & \cdots & \cdots & r_{d',A_n} \\
\hline
0\ldots0 & r_{C_1,D'} & \bullet & r_{C_1,C_2} & \cdots & \cdots & r_{C_1,C_j} & r_{C_1,A_1} & r_{C_1,A_2} & \cdots & \cdots & r_{C_1,A_n} \\
0\ldots0 & 0 & r_{C_2,C_1} & \bullet & \cdots & \cdots & r_{C_2,C_j} & r_{C_2,A_1} & r_{C_2,A_2} & \cdots & \cdots & r_{C_2,A_n} \\
\vdots & \vdots & 0 & r_{C_3,C_2} & \bullet & \cdots & r_{C_3,C_4} & r_{C_3,A_1} & \cdots & \cdots & \cdots & r_{C_2,A_n} \\
\vdots & \vdots & \vdots & \vdots & \vdots & \ddots & \vdots & \vdots & \vdots & \vdots & \vdots & \vdots \\
0\ldots0 & \vdots & 0 & \cdots & \cdots0 & r_{C_J,C_{J-1}} & \bullet & r_{C_3,A_1} & \cdots & \cdots & \cdots & r_{C_2,A_n} \\
\hline
0\ldots0 & 0 & 0 & 0 & 0 & \cdots & r_{A_1,C_J} & \bullet & r_{A_1,A_2} & \cdots & \cdots & r_{A_1,A_n} \\
\vdots & 0 & 0 & 0 & 0 & \cdots & 0 & r_{A_2,A_1} & \bullet & \cdots & \cdots & r_{A_2,A_n} \\
\vdots & 0 & 0 & 0 & 0 & \cdots & 0 & 0 & r_{A_3,A_2} & \bullet & \cdots & r_{A_3,A_n} \\
\vdots & \vdots & \vdots & \vdots & \vdots & \cdots & \vdots & \vdots & \vdots & \cdots & \ddots & \vdots \\
0\ldots0 & 0 & 0 & 0 & 0 & \cdots & 0 & 0 & 0 & \cdots & r_{A_n,A_{n-1}} & \bullet
\end{array}\right]$$

Figure 11 Aggregation of states in $\mathcal{C}$ and $\mathcal{A}$. Matrix G_4.

on the transition rates required for G_5. This issue is addressed and a detailed specification of the multistep algorithm is given in the following section.

4. DESCRIPTION OF THE ALGORITHM

The bounds on the transition rates in G_5 are of three types. The rates denoted by "+" in Figure 12 are to be replaced by upper bounds. A simple upper

$$\left[\begin{array}{cc|ccccc|ccccc}
Q_{D,D} & Q_{D,d'} & 0 & 0 & \cdots & \cdots & 0 & Q_{D,A_1} & Q_{D,A_2} & \cdots & \cdots & Q_{D,A_n} \\
R'_{d',D} & \bullet & + & + & \cdots & \cdots & + & + & + & \cdots & \cdots & + \\
\hline
0\ldots0 & - & \bullet & + & \cdots & \cdots & + & + & + & \cdots & \cdots & + \\
0\ldots0 & 0 & - & \bullet & \cdots & \cdots & + & + & + & \cdots & \cdots & + \\
\vdots & \vdots & 0 & - & \bullet & \cdots & + & + & \cdots & \cdots & \cdots & + \\
\vdots & \vdots & \vdots & \vdots & \vdots & \ddots & \vdots & \vdots & \vdots & \vdots & \vdots & \vdots \\
0\ldots0 & \vdots & 0 & \cdots & \cdots0 & - & \bullet & + & \cdots & \cdots & \cdots & + \\
\hline
0\ldots0 & 0 & 0 & 0 & 0 & \cdots & - & \bullet & + & \cdots & \cdots & + \\
\vdots & 0 & 0 & 0 & 0 & \cdots & 0 & - & \bullet & \cdots & \cdots & + \\
\vdots & 0 & 0 & 0 & 0 & \cdots & 0 & 0 & - & \bullet & \cdots & + \\
\vdots & \vdots & \vdots & \vdots & \vdots & \cdots & \vdots & \vdots & \vdots & \cdots & \ddots & \vdots \\
0\ldots0 & 0 & 0 & 0 & 0 & \cdots & 0 & 0 & 0 & \cdots & - & \bullet
\end{array}\right]$$

Figure 12 Replacement of transition rates with bounds. Matrix G_5.

bound is easily seen to be the sum of the failure rates of all components. Similarly, the rates denoted by "$-$" are to be replaced by lower bounds. A simple lower bound that suffices is the minimum repair rate of all components.

Based on the construction of the previous section, submatrix $R'_{d'D}$ in G_5 is required to contain lower bounds on the conditional rates from d' to states in $\mathcal{D}$. If $\pi_{D'/G}$ is the vector of stationary state probabilities for states in $\mathcal{D}$ then $(\pi_{D'/G} \sum \pi_{D'/G}(i))\, Q_{D'D}$ is the vector of conditional transition rates to states in $\mathcal{D}$. It follows immediately that if

$$\pi'_{D'} \leq \pi_{D'/G}$$

then

$$\pi'_{D'} Q_{D'D} \leq (\pi_{D'/G} / \sum \pi_{D'/G}(i)) Q_{D'D}$$

This provides the needed lower bound provided only that a lower bound on $\pi_{D'/G}$ is available.

In the multistep procedure we describe next, at each step the set of states $\mathcal{D}'$ corresponds exactly to the set of states for which a lower bound has been found on previous steps. Therefore these lower bound stationary state probabilities provide the needed data for the following step.

4.1 Multistep Procedure

The multistep bounding procedure is as follows:

1. (Step 1) Generate lower bounds on the stationary state probabilities for states $\mathcal{F}_0$ through $\mathcal{F}_{H_1}$ using the results from [7]. This provides lower bounds on the state probabilities of these states as well as an initial set of bounds on steady-state availability. If the bounds are tight enough then terminate.
2. (Step $i \geq 2$) Generate the portion of G corresponding to $\mathcal{F}_{L_i}$ through $\mathcal{F}_{H_i}$. Construct the matrix corresponding to G_5 described in the previous section. The submatrices Q_{DD} and $Q_{Dd'}$ are generated from the model definition. Let $\pi'_{D'}$ be the vector of lower bounds on the state probabilities computed from previous steps. Now set the submatrix $R'_{d'D}$ equal to $\pi'_{D'} Q_{D'D}$ and solve for π_{D/G_5}, which provides lower bounds on the state probabilities for states in $\mathcal{D}$. Compute upper and lower bounds on steady-state availability using equation (1) from Section 2. If the bounds are tight enough then terminate, else repeat Step 2.

5. EXAMPLE

In this section, we present a simple example to illustrate the multistep bounding procedure. We use a model of a fault-tolerant heterogeneous database system as depicted in Figure 13. The components of this system are a front end, two databases, and three processing subsystems consisting of a switch, a memory and two processors. Components may fail and be repaired according to the rates given in Table 1. If the processor fails, it has a 0.05 probability of contaminating a database. Components are repaired by a single repair facility that gives preemptive priority to components in the order: front end, databases, switches, processors, and lastly the memories. (Ties are broken by random selection.) The database system is considered operational if the front end is operational, at least one database is operational, and at least one processing subsystem is operational. A processing subsystem is operation if the switch, the memory, and at least on processor are operational. Also, this system is in *active breakdown mode*, meaning that components fail even when the system is nonoperational.

In Table 2, we present the bounds on steady-state availability. We note that for each step, the bounds of the availability are significantly tightened. In step one we set $\mathcal{D}_1 = \left\{\bigcup_{i=0}^{2} \mathcal{F}_i\right\}$. In step two $\mathcal{D}_2 = \left\{\bigcup_{i=3}^{5} \mathcal{F}_i\right\}$, and in step three $\mathcal{D}_3 = \left\{\bigcup_{i=6}^{7} \mathcal{F}_i\right\}$.

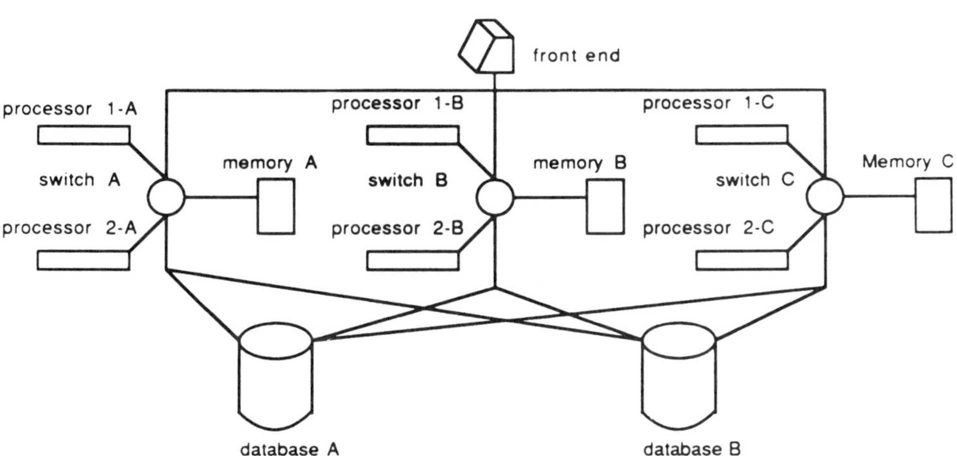

Figure 13 A fault-tolerant distributed database system.

Table 1 Failure and Repair Rates (per Hour)

Component	Mean Failure Rates	Mean Repair Rates
Front-end	1/8000	2.0
Processor 1-A	1/300	1.3
Processor 2-A	1/100	1.3
Switch A	1/550	2.0
Memory A	1/500	1.0
Processor 1-B	1/400	1.1
Processor 2-B	1/300	1.4
Switch B	1/600	2.1
Memory B	1/550	1.5
Processor 1-C	1/350	1.2
Processor 2-C	1/275	1.4
Switch C	1/650	2.5
Memory C	1/600	2.0
Database A	1/4000	2.0
Database B	1/5000	2.1

Table 2 Upper and Lower Bounds on Steady-State Availability of the Database System

Step number	Lower bound	Upper bound	No. of states in D
1	0.999878304	0.999939472	120
2	0.999930212	0.999938779	4823
3	0.999937012	0.999937590	11440

6. CONCLUSION

We have presented a method of determining bounds on the steady-state availability of a repairable computer system from a Markov model. This work is an extension of [7] that extends the one-step procedure into a multistep procedure for evaluating the bounds of availability. At each step a tighter bound on availability can be obtained by generating more of the transition matrix of the original model.

The method not only provides the flexibility to tradeoff computational resources and tighter bounds but it allows the tradeoff to be made dynamically; that is, if at some step the bounds are not tight enough then more of the matrix can be generated and the bounds tightened without loss of the previous calculation.

The procedure described relies on several properties of reliability models. First, to obtain reasonably tight bounds and yet generate only a small portion of the transition rate matrix requires that the stationary state probability distribution be highly skewed. Second, we utilized the assumption that the aggregated Markov process has the "skip-free to the left" property, that is, is upper Hessenberg. In the availability modeling domain this corresponds to an assumption that there is zero probability of more than two components being repaired at the same instant. In addition, we used the nature of the application domain in choosing the definition of the aggregate states and in placing upper and lower bounds on the transitions between aggregates.

There is ongoing work investigating certain properties of the algorithm presented and comparing with alternatives. For example, rather than exploring more of the state space, it may sometimes be advantageous to refine the estimates of state probabilities already calculated, for example, using the scheme in [2]. Also, with the procedure presented in this paper, there is some "unrecoverable" error at each step, which cannot be eliminated by further steps. (This is the difference between the lower bounds and the actual stationary state probabilities.) This error can be eliminated by the iterative approach. The tradeoff between the "forward only" and iterative approach remains to be investigated. It is also interesting to consider whether it is possible to provide efficient guidelines (either *a priori* or dynamic) to choose between the two.

It appears that there are a number of applications for which the assumption of a highly skewed stationary state probability distribution is reasonable. See for example [3, 6] applications in probabilistic communication protocol verification. It remains to be investigated whether the other properties (e.g., "skip free to the left") will be applicable in other domains or whether the method can be generalized to relax these conditions.

APPENDIX: PROOF FOR THEOREM 1

THEOREM 1 $\pi_{D/G_3} \leq \pi_{D/G_2}$

Proof. Let G be the transition rate matrix for a finite, irreducible continuous time Markov chain $\mathcal{G}$. Then $\mathcal{G}$ is uniformizable, which means it can be transformed to a discrete time Markov chain with transition probability matrix P, which has the same stationary probability vector π [9]. This transformation is achieved by:

$$P = I + \lambda^{-1}G$$

where λ is greater than or equal to the largest absolute diagonal element of G. With this in mind, we can transform G_2 and G_3 into P_2 and P_3 with $\lambda = \max(\lambda_2, \lambda_3)$ where $\lambda_2(\lambda_3)$ is the largest of the absolute values of the diagonal elements of $G_2(G_3)$.

Let us define the following notation:

- p_{ij} — the transition probability from state i to state j in P_2
- p'_{ij} — the transition probability from state i to state j in P_3
- d_{ij} — $= p_{ij} - p'_{ij}$
- $r(i)$ — reward (either 0 or 1) for state i
- $T_2[r(i)]$ — one-step expected reward of P_2 given the present state is i, that is, $T_2[r(i)] = \sum_j p_{ij} r(j)$
- $T_3[r(i)]$ — one-step expected reward of P_3 given the present state is i, that is, $T_3[r(i)] = \sum_j p'_{ij} r(j)$
- $R_2^k(i)$ — k-step ($k-1$ transitions plus initial position) accumulative reward for P_2 given the initial state is i, that is,

$$R_2^k(i) = \sum_{l=0}^{k-1} T^l[r(i)] \qquad \text{for} \quad k = 1, 2, 3, \ldots$$

  and $T_2^0[r(i)]$ is an identity function
- $R_3^k(i)$ — k-step ($k-1$ transitions plus initial position) accumulative reward for P_3 given the initial state is i.
- $1\{c\}$ — an indicator function equal to 1 if the condition c is true, else 0

A special case of the results in [13, 14] is that the expected reward for $\mathcal{G}_2$ is greater than or equal to the expected reward for $\mathcal{G}_3$ if and only if

$$(T_2 - T_3)R_2^k(i) \geq 0 \qquad \forall i \text{ and } \forall k$$

Let ϕ be a one-to-one mapping from $\mathcal{D}$ to $\mathcal{C}$ that maps each state $s_D \in \mathcal{D}$ to the corresponding state $s_C \in \mathcal{C}$.

Let f be any nonnegative function applied to state i of the Markov chain. Since the only difference between G_2 and G_3 is in the rates out of d', then for any nonnegative function f:

$$(T_2 - T_3)f(i) = 1\{i = d'\} \left\{ \sum_{s_D \in \mathcal{D}} p_{d',s_D} f(s_D) + P_{d',d'} f(d') \right.$$

$$+ \sum_{a_i \in \mathcal{A}} p_{d',a_i} f(a_i) - \sum_{s_D \in \mathcal{D}} p'_{d',s_D} f(s_D) - p'_{d',d'} f(d')$$

$$\left. - \sum_{s_c \in \mathcal{C}} p_{d',s_C} f(s_C) - \sum_{a_i \in \mathcal{A}} p'_{d',a_i} f(a_i) \right\}$$

Since

$$p_{d',d'} = p'_{d',d'}$$

$$p_{d',a_i} = p'_{d',a_i} \qquad \forall a_i \in \mathcal{A}$$

$$p_{d',s_D} \geq p'_{d',s_D} \qquad \forall s_D \in \mathcal{D}$$

it follows that

$$(T_2 - T_3) f(i) = 1\{i = d'\} \left\{ \sum_{s_D \in \mathcal{D}} d_{d',s_D} f(s_D) + \sum_{s_C \in \mathcal{C}} d_{d',s_C} f(s_C) \right\}$$

Since by construction of G_3 from G_2,

$$d_{d',s_D} = -d_{d',\phi(s_D)} \qquad \forall s_D \in \mathcal{D}$$

we have:

$$(T_2 - T_3) f(i) = 1\{i = d'\} \left\{ \sum_{s_D \in \mathcal{D}} d_{d',s_D} [f(s_D) - f(\phi(s_D))] \right\}$$

A sufficient condition for the above expression to be greater than or equal to 0 is

$$[f(s_D) - f(s_C)] \geq 0 \qquad \forall s_D \in \mathcal{D} \qquad \text{and} \quad s_C = \phi(s_D)$$

Letting the function f be $R^k(i)$, then the condition:

$$(T_2 - T_3) R_2^k(i) \geq 0 \qquad \forall i, k$$

is satisfied if

$$R_2^k(s_D) - R_2^k(s_C) \geq 0 \qquad \forall s_D \in \mathcal{D} \qquad \text{and} \quad \forall k$$

where $s_C = \phi(s_D)$

The above sufficient conditions can be easily proved by induction, for each choice of $s_D \in \mathcal{D}$ and $s_C = \phi(s_D)$.

For $k = 0$, since $R_2^0(i) = 0$ for any state i, the inequalities hold.

Assume $R_2^k(s_D) - R_2^k(s_C) \geq 0$ for $k \leq n$. For $k = n + 1$ we have:

$$R_2^{n+1}(s_D) - R_2^{n+1}(s_C)$$

$$= \left\{ r(s_D) + \sum_{i_D \in \mathcal{D}} p_{s_D,i_D} R_2^n(i_D) + p_{s_D,d'} R_2^n(d') + \sum_{a_i \in \mathcal{A}} p_{s_D,a_i} R_2^n(a_i) \right.$$

$$\left. - r(s_C) - \sum_{i_C \in \mathcal{C}} p_{s_C,i_C} R_2^n(i_C) - p_{s_C,d'} R_2^n(d') - \sum_{a_i \in \mathcal{A}} p_{s_C,a_i} R_2^n(a_i) \right\}$$

Since we want to bound the stationary state probabilities for states in $\mathcal{D}$, we assign the following reward:

$$r(x) = \begin{cases} 1 & x = s_D \\ 0 & \text{otherwise} \end{cases}$$

and based on the construction of G_3, we have:

$$\begin{aligned} p_{s_D,d'} &= p_{s_C,d'} & \\ p_{s_D,a_i} &= p_{s_C,a_i} & \forall s_C = \phi(s_D) \text{ and } \forall a_i \in \mathcal{A} \\ p_{s_D,i_D} &= p_{s_C,i_C} & \forall s_C = \phi(s_D) \text{ and } \forall i_C = \phi(i_D) \end{aligned}$$

It follows that

$$\begin{aligned} R_2^{n+1}(s_D) - R_2^{n+1}(s_C) = r(s_D) + \left\{ \sum_{i_D \in \mathcal{D}} p_{s_D,i_D} [R_2^n(i_D) - R_2^n(\phi(i_D))] \right\} \\ \geq 0 \end{aligned}$$

REFERENCES

[1] P. J. Courtois. *Decomposability—Queueing and Computer System Approximation.* Academic Press, New York, 1977.

[2] P. J. Courtois and P. Semal. Bounds for the positive eigenvectors of non-negative matrices and for their approximations, *JACM*, Oct. 1984, pp. 804–825.

[3] D. D. Dimitrijevic and M. Chen. An Integrated Algorithm for Probabilistic Protocol Verification and Evaluation, IBM Tech. Report RC 13901, 1988.

[4] A. Goyal, S. S. Lavenberg and K. S. Trivedi. Probabilistic modeling of computer system availability. *Ann. of Operations Res.*, vol. 8, pp. 285–306, 1986.

[5] J. G. Kemeny, and J. L. Snell. *Finite Markov Chains*. Van Nostrand Company, New York, 1960.

[6] N. F. Maxemchuk and K. Sabnani. Probabilistic verification of communication protocols, in *Protocol Specification, Testing and Verification*, VII, eds. H. Rubin and C. H. West, Elsevier, New York, 1987, pp. 307–320.

[7] R. R. Muntz, E. De Souza Silva, and A. Goyal. Bounding availability of repairable computer systems, *SIGMETRICS 1989*, pp. 29–38, also to appear in a special issue of *IEEE-TC* on performance evaluation, Dec. 1989.

[8] M. F. Neuts. *Matrix-Geometric Solutions in Stochastic Models—An Algorithmic Approach*. Johns Hopkins University Press, Baltimore, 1981.

[9] S. M. Ross. *Stochastic Processes*. Wiley, New York, 1983.

[10] W. J. Stewart, and A. Goyal. Matrix Methods in Large Dependability Models. IBM Research Report, 11485, Nov. 4, 1985.

[11] Y. Takahashi. Some Problems for Applications of Markov Chains, Ph.D. Thesis, Tokyo Institute of Technology, March, 1972.

[12] Y. Takahashi. A lumping method for numerical calculations of stationary distributions of Markov chains, Research Reports on Information Science, Tokyo Institute of Technology, No. B-18, June 1975.

[13] N. M. van Dijk. Simple bounds for queueing systems with breakdowns. *Performance Evaluation* 8, 1988, pp. 117–128.

[14] N. M. van Dijk. The Importance of Bias-terms for Error Bounds and Comparison Results, Tech. Report, 1989-36, 1989. Free University, The Netherlands.

24

Projection Methods for the Numerical Solution of Markov Chain Models

YOUCEF SAAD* NASA Ames Research Center, Moffett Field, California

ABSTRACT

This paper gives an overview of projection methods for computing stationary probability distributions for Markov chain models. A general projection method is a method that seeks an approximation from a subspace of small dimension to the original problem. Thus, the original matrix problem of size N is approximated by one of dimension m, typically much smaller than N. A particularly successful class of methods based on this principle is that of Krylov subspace methods that utilize subspaces of the form span $\{v, Av, \ldots, A^{m-1}v\}$. These methods are effective in solving linear systems (i.e., conjugate gradients, GMRES) and eigenvalue problems (i.e., Lanczos, Arnoldi) as well as nonlinear equations. They can be combined with more traditional iterative methods such as SOR or SSOR, or with incomplete factorization methods to enhance convergence.

---

*Current affiliation: University of Minnesota, Minneapolis, Minnesota

1. INTRODUCTION

In Markov chain modeling, one often seeks the stationary probability distributions of a system that can occupy a number of N different states. If we number these states from 1 to N, and call π_i the probability of the system to be in state i at equilibrium, that is, in the long run, then the basic equation to compute these probabilities is

$$\pi P = \pi \tag{1}$$

where $\pi = [\pi_1, \pi_2, \ldots, \pi_N]$ is the row vector comprised of the stationary probabilities, and P is the matrix of transition probabilities: p_{ij} is the probability that the system switches from state i to state j for $i \neq j$ and $p_{ii} = 1 - \sum_{j \neq i} p_{ij}$. The system (1) can be viewed as an eigenvalue problem, or more precisely an eigenvector problem, since the eigenvalue is known to be unity. In fact the right eigenvector is known and we seek the left eigenvector. Alternatively, defining

$$Q = I - P \tag{2}$$

we may rewrite equation (1) as

$$\pi Q = 0 \qquad \text{or} \qquad Q^T \pi^T = 0 \tag{3}$$

which is a homogeneous linear system to solve. The reason why we distinguish between these two viewpoints is that there are methods that are well known for linear systems but have no equivalent for eigenvalue problems and vice versa. We can choose effective algorithms from both camps. For example, subspace iteration, which is one of the methods used in Markov chain modeling, is not directly applicable for solving linear systems.

When the number of states is small then there are a number of reliable techniques that can be used to solve equation (1), for example, an inverse iteration approach based on Gaussian elimination. The difficulty is that in practice the number of states for realistic systems can be enormous and the cost and storage of the standard methods become prohibitive. The main philosophy of projection methods is to avoid costly manipulations on the original matrix. Rather, the main operations performed with the matrix are matrix by vector multiplications. The matrix P is usually very sparse so that storage is not a big burden and matrix by vector multiplications are inexpensive.

The organization of the paper is as follows. We start by describing general projection processes for both eigenvalue problems and linear systems in Section 2. In Section 3 we take on the eigenvalue point of view and describe two techniques based on this approach. The linear systems point of view will then be described in Section 4 with some specific approaches. Some numerical tests will then be reported.

2. PROJECTION METHODS

2.1 Projection Techniques for Solving Linear Systems

We start with the linear system of linear equations

$$Ax = b \tag{4}$$

where A is a large sparse nonsymmetric matrix. A projection process to solve equation (4) is a technique that computes an approximation to equation (4) for a subspace of dimension m by enforcing some orthogonality condition on the residual vector $r = b - Ax$. More precisely, let K be a subspace of dimension m from which we seek the approximation to x and let L be another subspace of dimension m which will define the orthogonality conditions. We define the approximate solution by writing that

$$\tilde{x} \in K \tag{5}$$

$$b - A\tilde{x} \perp L \tag{6}$$

Since the approximation x lies in a subspace of dimension m and equation (6) imposes m conditions, there will in general, but not always, exist a unique solution to the above problem, but we will come back to this problem later.

The approach just described is known as the Petrov–Galerkin projection method. A particular case of importance is when $L = K$. Then the method is called an orthogonal projection method or a Galerkin method.

Let P be the orthogonal projector onto K: for any x, Px is uniquely defined by $Px \in K, (I - P)x \perp K$. Similarly, let Q be the (oblique) projector onto K and orthogonal to L: for any x, Qx is uniquely defined by $Qx \in K, (I - Q)x \perp L$. For this projector to be defined it is assumed that no vector of K is orthogonal to L, or equivalently, that no vector of L is orthogonal to K. Then the Petrov–Galerkin method consists of replacing the original problem by the problem of finding $\tilde{x}$ satisfying the equations $P\tilde{x} = \tilde{x}$, and $Q(b - A\tilde{x}) = 0$. In other words, $\tilde{x}$ is a nontrivial solution of

$$Q(b - AP\tilde{x}) = 0 \tag{7}$$

which takes the form

$$\tilde{A}\tilde{x} = Qb \tag{8}$$

where we have set $\tilde{A} = QAP$.

The question that arises next is: given the two subspaces K and L, how accurate should we expect the approximate solution to be? It is in general difficult to answer this question. The element of K that is the closest to the exact solution x^* in the 2-norm sense is clearly Px^*. So the error in the 2-norm sense must be larger than $\|(I - P)x^*\|_2$. This is an upper bound but we wish to find a lower bound. Typically one would like to establish an upper bound on the residual norm $b - A\tilde{x}$ in terms of, for example, $(I - P)x^*$. We know of no simple error bounds of this type for general projection methods in the nonhermitian case. However, we can easily establish an *a priori* residual bound for the approximate problem. More precisely:

THEOREM 2.1 Let $\gamma = \|QA(I - P)\|_2$. The exact solution x^* satisfies the following residual condition with respect to the approximate problem (8):

$$\|Qb - \tilde{A}x^*\|_2 \leq \gamma\|(I - P)x^*\|_2 \tag{9}$$

Proof. We have

$$Qb - \tilde{A}x^* = QAx^* - QAPx^* = QA(I - P)x^* \tag{10}$$

Since $I - P$ is a projector this gives

$$Qb - \tilde{A}x^* = QA(I - P)(I - P)x^* \tag{11}$$

which immediately yields the result.

In many of the projection methods, b is a vector of K so that we have $Qb = b$. This theorem shows that when the distance between the exact solution x^* and the subspace K is small then a good approximation can be obtained provided that the norm of Q is not to large and that the projected problem is not too poorly conditioned. In the orthogonal projection case, $Q = P$ and we get $\gamma \leq \|A\|_2$.

In practice, one needs to work with some convenient basis of the subspace K. Given a basis $V = [v_1, v_2, \ldots, v_m]$ of K and a basis $W = [w_1, w_2, \ldots, w_m]$ of L, and writing $\tilde{x} = V_m y$ where y is in $\mathbb{R}^m$, we see immediately that y must satisfy

$$W_m^T(AV_m y - b) = 0 \tag{12}$$

or

$$Hy = W_m^T b \tag{13}$$

where we have set $H = W_m^T A V_m$.

There are two important particular cases for L. The case when $L = K$, we have already mentioned. The second case of importance is when $L = AK$ because it is equivalent to minimizing the residual norm on the subspace K. More precisely, we have the following theorem: see for example [16].

THEOREM 2.2 When $L = AK$ then the approximate solution $\tilde{x}$ minimizes the residual norm over the subspace K.

We refer to this class of projection methods as the class of minimal residual methods. An example of such a technique will be described later in Section 3.

It is important to consider the effect of using a nonzero initial guess for approximating the solution. If we had a guess x_0 to the solution of equation (4) we would attempt to find a vector δ in K so that the update $x_0 + \delta$ satisfies the usual Petrov–Galerkin conditions. In other words, we would write

$$b - A(x_0 + \delta) \perp L \qquad \text{or} \qquad r_0 - A\delta \perp L \tag{14}$$

with the usual notation $r_0 = b - Ax_0$.

In Markov chain modeling, the matrix A is singular and $b = 0$ and we would like to show how to adapt the methods just described to this situation. A direct application of the previous principles would lead to a projected problem of the form

$$Hy = 0 \tag{15}$$

Unfortunately, there is no reason for H to be a singular matrix and as a result the only solution to equation (15) would be $y = 0$ in general. The difficulty does not arise if we take the nonzero initial guess formulation discussed above. Then the projected problem becomes

$$Hy = Q^T r_0 \tag{16}$$

in which $r_0 = b - Ax_0 = -Ax_0$. Notice that if the matrix H happens to be singular, then one can take the solution of equation (16) to be an eigenvector of the matrix H associated with the eigenvalue zero. With this modification one can say that the projected problem always has a solution. The only possible concern might be the practicality of the decision to declare a matrix singular. The real question is whether or not a nearly singular matrix could lead to a substantially different result than would be obtained by considering it to be singular. However, if the matrix is nearly singular, then the solution of equation (16) is equivalent to a form of inverse iteration and the result, after normalization, should be an accurate approximation to the exact eigenvector associated with the eigenvalue zero.

For the minimal residual methods referred to earlier, there is no difficulty in defining the approximate solution in the formulation where $x_0 \neq 0$. This is because the approximate solution minimizes the 2-norm of $r_0 - A\delta$ over $\delta \in K$ and this optimization problem always has a unique solution.

2.2 Projection Techniques for Solving Eigenvalue Problems

We now consider the standard eigenvalue problem

$$Ax = \lambda x \tag{17}$$

Similarly to the previous section, we assume that we are given two subspaces K and L, and we seek an approximate eigenvalue $\tilde{\lambda} \in \mathbb{C}$ and an approximate eigenvector $\tilde{x}$ from the subspace K, by imposing the Petrov–Galerkin condition

$$A\tilde{x} - \tilde{\lambda}\tilde{x} \perp L \tag{18}$$

Similarly to the previous section, if P is the orthogonal projector onto K and Q is the oblique projector onto K orthogonally to L, then the original problem is replaced by the projected problem

$$\tilde{A}\tilde{x} = \tilde{\lambda}\tilde{x} \tag{19}$$

and we can show a theorem similar to the main theorem of the previous section, namely:

THEOREM 2.3 Let $\gamma = \|QA(I-P)\|_2$. The exact eigenvector x associated with the exact eigenvalue λ satisfies the following residual condition with respect to the approximate problem (16),

$$\|(\tilde{A} - \delta I)x\|_2 \leq \sqrt{\gamma^2 + |\delta|^2}\,\|(I-P)x\|_2 \tag{20}$$

The proof of this result is similar to the one shown in the previous section for linear systems and can be found in [13].

3. METHODS FOR EIGENVALUE PROBLEMS

We consider two sample eigenvalue methods used in Markov chain modeling. The first is a technique based on the subspace iteration method of Bauer [3]. It has been used with good success for Markov chain modeling [7, 18]. The second is a method due to Arnoldi [1] in 1951.

3.1 Subspace Iteration

One of the simplest methods for computing invariant subspaces is the so-called subspace iteration [7, 18] methods well known to the structural engineers. In its simplest form, the method is equivalent to a projection method onto the subspace $K = \text{span}\{A^m V_0\}$ where V_0 is an initial set of p columns, that is, an $N \times p$ matrix. Note that the dimension of the subspace K is constant. One of the forms of the subspace iteration algorithm can be described as follows.

1. Choose an initial orthonormal system $V_0 \equiv [v_1, v_2, \ldots, v_m]$ and an integer k.

2. Compute $X = A^k V_0$ and orthonormalize X to get V.
3. Perform a projection process with V, i.e., compute the eigenvalues and the matrix U of eigenvectors of the matrix $C = V^T AV$.
4. Test for convergence. If satisfied then exit, else continue.
5. Take $V_0 = VU$, the set of approximate eigenvectors choose, a new k and go to 2.

A well-known alternative consists of replacing the set of eigenvectors U in steps 3 and 5 by the Schur vectors of the matrix C, that is, the column vectors of the matrix that transforms C in upper quasi-triangular form [9, 15].

The above algorithm utilizes the matrix A only to compute successive matrix by vector products $w = Av$, so sparsity can be exploited. However, it faces the drawback that it is generally a slow method.

Often, Chebyshev iteration is used to accelerate convergence: step 2 is replaced by $X = t_k(A)V_0$, where t_k is obtained from the Chebyshev polynomial of the first kind, of degree k, by a linear change of variables. The three-term recurrence of Chebyshev polynomials allows us to compute a vector $w = t_k(A)v$ at almost the same cost as $A^k v$. Moreover, it is then possible to compute the rightmost (or leftmost) eigenvalues of A. Also, performance is improved as this is the usual primary reason for using Chebyshev iteration. Details on implementation can be found in [14].

Subspace iteration as described above is seldom competitive with methods that use a combination of preconditioning and Krylov subspace methods described later. If we use a block size of p, then the convergence rate for the eigenvalue $\lambda_1 = 1$ is of the order $|\lambda_{p+1}|$, if we order the eigenvalues by decreasing order of magnitude.

3.2 Arnoldi's Method

A second method used in the literature is the Arnoldi process [1, 11], which is a projection process onto the so-called Krylov subspace

$$K_m = \text{span}\left\{v_1, Av_1, \ldots, A^{m-1}v_1\right\} \tag{21}$$

The algorithm starts with some nonzero vector v_1 and generates the sequence of vectors v_i from the following algorithm.

Algorithm: Arnoldi

1. *Initialize*: Choose an initial vector v_1 of norm unity.
2. *Iterate*: Do $j = 1, 2, \ldots, m$.

    1. Compute $w := Av_j$.

2. Compute a set of j coefficients h_{ij} so that

$$w := w - \sum_{i=1}^{j} h_{ij} v_i \tag{22}$$

is orthogonal to all previous v_i's.

3. Compute $h_{j+1j} = \|w\|_2$ and $v_{j+1} = w/h_{j+1j}$.

By construction, the above algorithm produces an orthonormal basis of the Krylov subspace $K_m = \text{span}\,\{v_1, Av_1, \ldots, A^{m-1}v_1\}$. The $m \times m$ upper Hessenberg matrix H_m, consisting of the coefficients h_{ij} computed by the algorithm, represents the restriction of the linear transformation A to the subspace K_m, with respect to this basis, that is, we have

$$H_m = V_m^T A V_m \tag{23}$$

where $V_m = [v_1, v_2, \ldots, v_m]$. Approximations to some of the eigenvalues of A can be obtained from the eigenvalues of H_m. This is Arnoldi's method in its simplest form.

Note the useful relation for later use,

$$AV_m = V_{m+1}\bar{H}_m \tag{24}$$

$$= V_m H_m + h_{m+1,m} v_{m+1} e_m^T \tag{25}$$

where $\bar{H}_m$ is the $(m+1) \times m$ upper Hessenberg matrix whose nonzero elements are the h_{ij} defined in the above algorithm. In other words, $\bar{H}_m$ is obtained from H_m by appending the row $[0, 0, \ldots, 0, h_{m+1,m}]$ to it.

As m increases, the eigenvalues of H_m that are located in the outermost part of the spectrum start converging toward corresponding eigenvalues of A. In practice, however, one difficulty with the above algorithm is that as m increases cost and storage increase rapidly. One solution is to use the method iteratively: m is fixed and the initial vector v_1 is taken at each new iteration as a linear combination of some of the approximate eigenvectors. Moreover, there are several ways of accelerating convergence by preprocessing v_1 by a Chebyshev iteration before restarting, that is, by taking $v_1 = t_k(A)z$ where z is again a linear combination of eigenvectors.

A technique related to Arnoldi's method is the nonsymmetric Lanczos algorithm [9, 4], which delivers a nonsymmetric tridiagonal matrix instead of a Hessenberg matrix. Unlike Arnoldi's process, this method requires multiplications by both A and A^T at every step. On the other hand, it has the big advantage of requiring little storage (five vectors). Although no comparisons of the performances of the Lanczos and the Arnoldi type algorithms have been made, the Lanczos methods are usually recommended whenever the number of eigenvalues to be computed is large.

4. METHODS FOR LINEAR SYSTEMS

4.1 FOM and GMRES

In this section we take the point of view that we want to solve the linear system $Ax = b$. We start by assuming that the system is nonsingular. The methods to be described here are based on Krylov subspaces. Assume that we have an initial guess x_0 with residual vector $r_0 = b - Ax_0$. We will take

$$v_1 = r_0/\beta \qquad \text{where} \quad \beta \equiv \|r_0\|_2 \tag{26}$$

and run m steps of Arnoldi's methods starting with the vector v_1 thus defined. To apply a Galerkin projection process onto the subspace K_m, we need to seek a vector $\delta_m \in K_m$ and the orthogonality condition

$$V_m^T(r_0 - Av_m y_m) = 0$$

yields immediately

$$y_m = H_m^{-1}\beta e_1 \tag{27}$$

where we have set $\beta \equiv \|r_0\|_2$ and used the orthogonality of V_m and the relation (23). Thus the approximate solution takes the form

$$x_m = x_0 + V_m H_m^{-1}\beta e_1 \tag{28}$$

The method described above, which consists of generating the Arnoldi basis V_m with v_1 defined by (26) and the approximate solution via (28), is referred to as the Arnoldi process for linear systems or the "Full Orthogonalization Method" [12]. A detail that is important for the implementation is that the residual norm of the approximate solution can be determined, without explicitly computing the solution, via the formula

$$\|b - Ax_m\|_2 = |h_{m+1,m} e_m^T y_m| \tag{29}$$

which is a direct consequence of equation (25).

Although not proved rigorously, it is usually observed that as m increases, the approximate solution x_m rapidly approaches the exact solution. Ideally, one would like to use a large enough m that x_m is as close as desired to the solution. However, this is not feasible in practice because of the rapid increase in the storage requirement as the dimension m of the subspace increases. Therefore, the basic idea described above is usually implemented with restarting. The dimension m is fixed, and the method is after each outer loop consisting of the process just described the initial guess x_0 is reset to be equal to x_m and the process is restarted until convergence is achieved.

One drawback of the above algorithm is that from the theoretical point of view there is little known concerning convergence. For this reason, several authors have instead turned to methods with optimal properties: for example one may seek a method that minimizes the residual norm over the whole subspace K. As was seen before, this is realized by taking $L = AK$. One such version is the GMRES algorithm [17]. In GMRES, one seeks to minimize the residual norm of the approximate solution in the affine subspace $x_0 + K_m$. This means that the approximate solution $x_0 + V_m y$ must minimize $J(y) = \|b - AV_m y\|_2$ over $y \in \mathbb{R}^m$. Utilizing the relation (24) we get

$$\begin{aligned} J(y) &= \|b - A(x_0 + V_m y)\|_2 = \|r_0 - AV_m y\|_2 \\ &= \|\beta v_1 - V_{m+1}\bar{H}_m y\|_2 = \|V_{m+1}[\beta e_1 - \bar{H}_m y]\|_2 \\ &= \|\beta e_1 - \bar{H}_m y\|_2 \end{aligned}$$

The last equality is a consequence of the orthogonality of the vectors v_i's. Therefore the only difference between this approach and the FOM approach is the way in which the vector y is obtained. In one case it is obtained by solving $m \times m$ linear system (27) and in the second by solving the least squares problem

$$\text{Find } y_m \text{ solution of: } \min_{y \in \mathbb{R}^m} \|\beta e_1 - \bar{H}_m y\|_2 \tag{30}$$

Similarly to the FOM method, there exists a formula that allows us to compute the residual norm of x_m without computing x_m. This is based on the trivial equality

$$\|b - Ax_m\|_2 = \min_{y \in \mathbb{R}^m} \|\beta e_1 - \bar{H}_m y\|_2 \tag{31}$$

In a practical implementation of the GMRES algorithm, the Hessenberg matrix is progressively reduced to upper triangular form by using Givens rotations. The same rotations are applied to the right-hand-side βe_1. Because the last row of the resulting matrix is zero row, the above minimum can be seen to be equal to the bottom element of the right-hand-side after these rotations. This provides the residual norm of the current iterate for free without having to compute the iterate itself.

In summary we can put the two methods within the same framework as follows.

Algorithm: FOM/GMRES

1. *Start*: Choose x_0.
2. *Arnoldi process*:

   - Compute $r_0 := b - Ax_0$
   - Compute $\beta = \|r_0\|_2$; and $v_1 = r_0/\beta$.
   - Arnoldi Process. For $j = 1, 2, \ldots,$ do:

a. Form Av_j and orthogonalize it against the previous $v_1, \ldots, v_j$ via

$$h_{ij} = (Av_j, v_i) \qquad i = 1, 2, \cdots, j$$

$$\hat{v}_{j+1} = Av_j - \sum_{i=1}^{j} h_{ij} v_i$$

$$h_{j+1j} = \|\hat{v}_{j+1}\|_2$$

and

$$v_{j+1} = \hat{v}_{j+1}/h_{j+1j} \tag{32}$$

b. Compute the residual norm $\rho_j = \|b - Ax_j\|_2$, of the solution x_j via the formulas (29) and (31).

c. If $\rho_j \leq \epsilon$ set $m = j$ and go to 3.

3. *Form the approximate solution*:
   Define $\bar{H}_m$ to be the $(m+1) \times m$ (Hessenberg) matrix whose nonzero entries are the coefficients $h_{ij}, 1 \leq i \leq j+1,\ 1 \leq j \leq m$, and H_m the $m \times m$ submatrix obtained from $\bar{H}_m$ by removing its last row. Let $V_m \equiv [v_1, v_2, \cdots, v_m]$, and where $e_1 = [1, 0, \ldots, 0]^T$. Then:
   FOM:

   - Compute $x_m = x_0 + \beta V_m H_m^{-1} e_1$.

   GMRES:

   - Find the vector y_m that minimizes $\|\beta e_1 - \bar{H}_m y\|_2$, over all vectors y in $\mathbb{R}^m$.
   - Compute $x_m = x_0 + V_m y_m$.

4. *Stopping test*: If x_m is determined to be a good enough approximate solution to equation (4), then stop, else set $x_0 := x_m$ and go to 2.

4.2 Preconditionings

Often the projection methods themselves are not sufficient to achieve good performance and preconditioning is necessary. Typical preconditioning techniques consist of approximating the original matrix by a "close by" matrix M and then solving a preconditioned system such as

$$M^{-1}Ax = M^{-1}b \tag{33}$$

by some suitable iterative method. For this approach to be practical it is necessary that the linear system solution with the matrix M be inexpensive and easy to implement. In the core of the Krylov subspace algorithm that is used, the matrix by vector product is replaced by a matrix vector product followed by a linear system solve with the matrix M. Thus the modifications to the algorithm FOM/GMRES displayed above is simply to replace all occurrences of A by $M^{-1}A$, and the right-hand side b by $M^{-1}b$. One can also precondition by solving the system

$$AM^{-1}z = b \tag{34}$$

whose solution z is related to the solution of the original system by $x = M^{-1}z$. There are no *a priori* reasons for using this right preconditioning approach rather than the left preconditioning approach, except that an approximate solution $\tilde{z}$ for the system(34) has the same residual vector as for the original system.

The simplest preconditioning technique is the incomplete LU factorization that consists of factoring A as

$$A + LU + E \tag{35}$$

where the matrix LU matches A everywhere where there are nonzero elements and E is a remainder. The matrix $M = LU$ is then used as a preconditioning matrix. Note that the exact LU factorization of A would require far more computation and storage than the incomplete LU factorization whose cost is typically of the order of NZ, the number of nonzero elements of A. The simplest form of incomplete factorization is one in which the L matrix has the same structure as the lower part of A and the U has the same structure as the upper triangular part of A. This is referred to as the ILU(0) preconditioning to account for the fact that no fill-in is allowed.

The incomplete LU factorization outlined above exists for M matrices and can be computed by a very simple procedure, which consists of performing the LU factorization and replacing by zero any nonzero element that is introduced outside of the nonzero structure of A during the process.

We are interested in the incomplete factorization of the matrix $A + Q^T$, but since this matrix is singular the classical results on the existence of the ILU factorization [8] do not readily apply to this case. However, these classical results can be trivially extended to such matrices by, for example, adapting the results in [2], page 42.

The quality of the ILU(0) preconditioning can be improved in several ways by allowing more fill-in. A notion that is used in this regard is that of level of fill-in: initially all elements have level of fill-in equal to zero. Thereafter, at each step the level of fill-in of an element is updated by adding one to the sum of the levels of fill-ins of its parents in L and U. Here the parent-child relation corresponds to the creation of a fill-in element by the basic operation in Gaussian elimination. ILU(k) will then correspond to dropping all elements whose level of fill-in exceeds k. Thus ILU(0) is the usual incomplete LU factorization with no fill-in.

As can be seen, this incomplete factorization technique relies entirely on the structure of the matrix and not at all on the actual values of its elements. For a large class of problems that come from PDE's this is usually sufficient because of the fact that these matrices are M-matrices. On the other hand, the situation may be different for other classes of problems.

An alternative used is to drop the elements during the ILU factorization according to their magnitude rather than their position. Several such techniques exist [6, 20]. Some are implemented in the context of direct solvers such as in MA28 [5] and in Y12M [20]. Here a rather accurate incomplete factorization is usually performed and a simple technique such as iterative refinement is used as an iterative procedure. In the context of iterative solvers, one can simplify the factorization process enormously [6, 10]. One drawback of this approach is that it is rather difficult to predict the storage that will be necessary during the factorization. A second alternative would be to only keep a given number of elements per row during the incomplete factorization. Two techniques based on the two approaches outlined above have been implemented and tested on realistic Markov chain problems in [10].

In addition to incomplete factorization preconditionings one can also use the more traditional relaxation methods such as the SOR, or SSOR iteration, as preconditioners. Our experience in [10] with these techniques on real Markov chain problems is that they are not as efficient as the ILU type preconditioners. For further details see [10].

5. NUMERICAL TESTS

The following illustration is taken from [10]. It compares a few of the methods described in this paper and includes the power method referred to here as the fixed-point iteration and a direct solver (GE).

The example deals with the system architecture of a time-shared multiprogrammed, paged, virtual memory computer. The system is composed of a set of N terminals, a central processing unit, a secondary memory, and a filing device. This real-life example models requests to each of the devices of the system that are queued and scheduled on a first-come, first-served basis. For further details see [10]. The matrix Q obtained here is of dimension 1771 with 11,011 nonzero elements.

Table 1 shows various statistics associated with the performance of several methods for solving the matrix equation $Q^T x = 0$. The direct solver called GE is without pivoting and without any reordering. All runs have been made on an Ardent Titan computer in double precision. The compiler option used was -O3. but no additional optimization of the code was performed.

Table 1 Performance Results for test Example $N = 1,771$, $NZ = 11,011$; NCD Case

Method	Method parameters	Total time	Setup time	Iter. time	Flops	Additional memory	Iters	Residual norm
GE		3.3			4.7	111,989		0.219E−16
ARNOLDI	$m = 10$	51.4			32.9	17,971	*1010	0.121E−04
	$m = 20$	46.5			51.5	36,341	*1020	0.131E−03
	$m = 25$	83.1			96.8	45,676	*1625	0.184E−03
PCARN/								
+ILU0	$m = 5$	19.8	0.3	19.3	4.8	23,408	160	0.324E−10
	$m = 10$	15.7	0.3	15.2	5.6	32,263	150	0.811E−10
+ILUK	$m = 10, K = 5$	9.1	3.3	5.6	2.5	30,050	70	0.291E−10
	$m = 10, K = 10,$	5.9	4.3	1.4	0.5	38,735	10	0.409E−11
+ILUTH	$m = 10, \tau = .01$	15.1	1.6	13.4	7.1	26,104	230	0.543E−10
	$m = 10, \tau = .001$	11.6	1.8	9.7	3.0	32,142	80	0.205E−10
GMRES/								
+ILU0	$m = 5$	149.1	0.3	148.6		23,408	*1600	0.210E−06
	$m = 10$	16.4	0.3	15.9	5.2	32,263	140	0.632E−10
	$m = 20$	16.5	0.3	16.0	8.7	49,973	160	0.715E−10
+ILUK	$m = 10, K = 5$	7.5	3.3	4.1	1.8	30,050	50	0.922E−10
	$m = 10, K = 10$	5.8	4.2	1.4	0.5	38,735	10	0.438E−11
+ILUTH	$m = 10, \tau = .01$	13.0	1.6	11.2	5.5	26,104	180	0.579E−10
	$m = 5, \tau = .001$	92.4	1.7	90.5		23,287	*1000	0.298E−06
	$m = 10, \tau = .001$	97.2	1.8	95.2		32,142	*1000	0.204E−06
	$m = 20, \tau = .001$	9.9	1.8	8.0	4.3	49,852	80	0.746E−10
GMRES/								
+SOR	$m = 10, \omega = 1.0$	94.5			37.3	17,710	*1000	0.529E−06
	$m = 10, \omega = 1.95$	49.8			18.7	17,710	*500	0.416E−03

In the figure ILUTH refers to an ILU factoriztion with threshold τ, which consists of dropping all elements during the factorization that are smaller than τ in magnitude. ILUK refers to a strategy whereby only the K largest elements in L and U are kept.

As can be seen from this example, Gaussian elimination is faster than the iterative solvers, although it requires far more storage. However, for the larger problems treated in [10] Gaussian elimination was too slow or required too much memory. In some cases it also failed to produce an answer. Among the iterative solvers tested and not shown in the table were the SOR method and the preconditioned power method. The SOR method was generally not reliable because of the demand for an optimal parameter. As for the power method, it converged more slowly than the methods shown here. One surprising observation made in the experiments in [10] is the fact that the FOM method was often more reliable than its GMRES counterpart. Another point that is not too well understood is the effect of improving the incomplete factorization by taking, for example, a smaller threshold in *ILUTH*, or a larger K in ILUK. Whereas the quality of the factorization improves, since the error matrix E is smaller, this does not always mean that the number of iterations in the preconditioned Krylov Subspace method will be smaller. This is clearly illustrated in the table. The phenomenon may be due to the fact that the L and U factors produced by a more accurate factorization have in fact a worse condition number than with the less accurate factorization. The relation between the accuracy in the preconditioning technique and the overall performance is not clear. Note however, that the best results are obtained with the more accurate preconditioners.

REFERENCES

[1] W. E. Arnoldi. The principle of minimized iteration in the solution of the matrix eigenvalue problem. *Q. Appl. Math.*, 9:17–29, 1951.

[2] O. Axelsson and V. A. Barker. *Finite Element Solution of Boundary Value Problems*. Academic Press, Orlando, Fla., 1984.

[3] F. L. Bauer. Das verfahren der treppeniteration und verwandte verfahren zur losung algebraischer eigenwertprobleme. *ZAMP*, 8:214–235, 1957.

[4] J. Cullum and R. Willoughby. A Lanczos procedure for the modal analysis of very large nonsymmetric matrices. In *Proceedings of the 23rd Conference on Decision and Control, Las Vegas*, 1984.

[5] I. S. Duff, A. M. Erisman, and J. K. Reid. *Direct Methods for Sparse Matrices*. Clarendon Press, Oxford, 1986.

[6] A. Jennings and G. M. Malik. Partial elimination. *J. Inst. Math. Appl.*, 20:307–316, 1977.

[7] A. Jennings and W. J. Stewart. A simultaneous iteration algorithm for real matrices. *ACM Trans. Math. Software*, 7:184–198, 1981.

[8] J. A. Meijerink and H. A. van der Vorst. An iterative solution method for linear systems of which the coefficient matrix is a symmetric M-matrix. *Math. Comp.*, 31(137):148–162, 1977.

[9] B. N. Parlett, D. R. Taylor, and Z. S. Liu. A look-ahead Lanczos algorithm for nonsymmetric matrices. *Math. Comput.*, 44:105–124, 1985.

[10] B. Philippe, Y. Saad, and W. J. Stewart. Numerical methods in Markov chain modeling. Technical Report 89.39, RIACS, NASA Ames, Mofett Field, Calif., 1989.

[11] Y. Saad. Variations on Arnoldi's method for computing eigenelements of large unsymmetric matrices. *Linear Algebra Appl.*, 34:269–295, 1980.

[12] Y. Saad. Krylov subspace methods for solving large unsymmetric linear systems. *Math. Comput.*, 37:105–126, 1981.

[13] Y. Saad. Projection methods for solving large sparse eigenvalue problems. In B. Kagstrom and A. Ruhe, editors, *Matrix Pencils, proceedings, Pitea Havsbad*, pp. 121–144, Berlin, 1982. University of Umea, Sweden, Springer Verlag. Lecture notes in Math. Series, Number=973.

[14] Y. Saad. Chebyshev acceleration techniques for solving nonsymmetric eigenvalue problems. *Math. Comput.*, 42:567–588, 1984.

[15] Y. Saad. Numerical solution of large nonsymmetric eigenvalue problems. *Comput. Phys. Commun.*, 53, 1989.

[16] Y. Saad and M. H. Schultz. Conjugate gradient-like algorithms for solving nonsymmetric linear systems. *Math. Comput.*, 44(170)417–424, 1985.

[17] Y. Saad and M. H. Schultz. GMRES: A generalized minimal residual algorithm for solving nonsymmetric linear systems. *SIAM J. Sci. Statist. Comput.*, 7:856–869, 1986.

[18] G. W. Stewart. Simultaneous iteration for computing invariant subspaces of nonhermitian matrices. *Numer. Math.*, 25:123–136, 1976.

[19] G. W. Stewart. SRRIT—A fortran subroutine to calculate the dominant invariant subspaces of a real matrix. Technical Report TR-514, University of Maryland, College Park, Md., 1978.

[20] Z. Zlatev. Use of iterative refinement in the solution of sparse linear systems. *SIAM J. Numer. Anal.*, 19:381–399, 1982.

25

The Biconjugate Gradient Method for Obtaining the Steady-State Probability Distributions of Markovian Multiechelon Repairable Item Inventory Systems

DONALD GROSS, BINGCHANG GU†, AND RICHARD M. SOLAND
Department of Operations Research, The George Washington University, Washington, D.C.

ABSTRACT

The biconjugate gradient method is used to compute the steady-state probability distribution for finite state-space continuous-time Markov processes that arise in the modeling of two-echelon repairable item inventory systems. Alternative systems of linear equations that express the steady-state conditions are examined, generation and storage of the transition rate matrix are discussed briefly, and various initializations and stopping criteria are tested. Numerical results are given for problems with up to 200,000 states. Good results are obtained with a two-phase algorithm that uses the Gauss–Seidel method first and the biconjugate gradient method subsequently.

Research supported by NSF grant ECS-8519284.

†Current affiliation: Chinese Aeronautical Radio Electronics Research Institute, Shanghai, China

1. INTRODUCTION

The steady-state probability distribution for a finite state-space continuous-time Markov process (CTMP) with n states is determined by the solution of the set of linear equations

$$Q^T \pi^T = \mathbf{0} \tag{1}$$

$$\mathbf{e}^T \pi^T = 1 \tag{2}$$

where Q is the $n \times n$ rate matrix ($q_{ij}, i \neq j$, is the transition rate from state i to state j, and $q_{ii} = -\sum_{j \neq i} q_{ij}$), $\mathbf{0}$ is a column vector with n zero elements, $\mathbf{e}$ is a column vector with n elements equal to one, and π is the n-component row vector of steady state probabilities. (We adopt the convention that all vectors are column vectors unless otherwise indicated. T denotes transpose.)

Since one equation in (1) is redundant, the system shown in equations (1) and (2) can be modified to the nonsingular system

$$\bar{Q}\pi^T = \mathbf{e}_i \tag{3a}$$

where $\mathbf{e}_i$ is an n-component column vector with $n - 1$ zero elements and the ith element equal to one, and $\bar{Q}$ is the Q^T matrix with the ith row replaced by elements all equal to one.

Another modification that changes the singular system of (1) and (2) to a nonsingular system is to arbitrarily set one $\pi_i = 1$ and reduce the set of equations (1) to an $n - 1$ by $n - 1$ nonsingular set

$$\hat{Q}\hat{\pi}^T = \mathbf{f} \tag{3b}$$

where $\hat{\pi}$ is an $(n - 1)$-element row vector, $\hat{Q}$ is the result of the "changed" Q^T matrix, and $\mathbf{f}$ is the resultant $(n - 1)$-element right-hand side (no longer $\mathbf{0}$ or $\mathbf{e}_i$). After solving for the $n - 1$ $\hat{\pi}_j$, these, together with $\pi_i = 1$, can be renormalized.

The difficulty in solving this problem is that the Q matrix may be very large when it is generated from certain classes of CTMPs, for example, those arising from multiechelon repairable item inventory systems (MERIIS). Such systems may yield CTMPs with tens or even hundreds of thousands of states.

The class of MERIIS studied here consists of two field locations (say bases) with a central repair facility (say a depot). Each base has a "local" repair capability and keeps spares on hand. The depot also has spares.

When an item fails at a base, there is a known probability that it can be repaired at the base. If this is the case, the item goes into base repair and a spare, if available, is immediately put into service. If no spare is available, the base must wait until a unit comes out of the base repair shop.

If the failed unit is not base repairable, it is dispatched to the depot. If a depot spare is available it is immediately dispatched (zero travel time assumed). If the depot has no spares, a backorder is established. When an item comes out of depot repair, if backorders exist, the item is dispatched to the appropriate base according to a backorder filling rule that, when both bases are in backorder, sends the item to the base that has the largest percent deficit below its allocated number of items. If no backorders are on the books, the unit goes into depot spares inventory.

The system is shown schematically in Figure 1. Failure times and repair times are assumed to follow exponential probability distributions.

The remainder of the paper is organized as follows. Section 2 presents the biconjugate gradient method for solving the steady-state balance equations of this non-Jacksonian Markov network (it is non-Jacksonian since routing probabilities associated with filling depot backorders are state dependent). Section 3 deals with stopping criteria for the iterative solution procedure, while Section 4 deals with initial solutions required to start the iterative process. Section 5 treats the generation and storage of the infinitesimal generator (or rate) matrix Q of the CTMP. Section 6 presents results, including some for a two-phase algorithm that uses both the Gauss–Seidel and biconjugate gradient methods.

The work reported here is part of a larger project in which several different iterative methods for the solution of (1) and (2), (3a), or (3b) have been tested on several different computers. A significant part of the effort has been

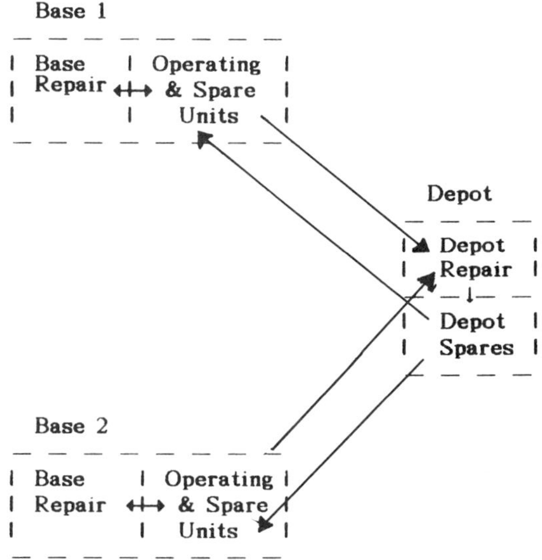

Figure 1 Schematic diagram of a two-base, two-echelon system.

devoted to the generation and storage of the matrix Q, but here we just briefly treat these aspects of the overall problem. A full description of the project is given by Gross et al. (1990).

2. THE BICONJUGATE GRADIENT METHOD

Consider the general set of linear equations

$$A\mathbf{x} = \mathbf{b} \tag{4}$$

where $A = \{a_{ij}\}$ is of order N, and $\mathbf{x}$ and $\mathbf{b}$ are N-component column vectors. Iterative methods for the solution of equation (4) include the Jacobi (J) and Gauss–Seidel (GS) procedures [see Young (1971)] and the conjugate gradient (CG) method [see Jacobs (1981)].

The biconjugate gradient (BCG) algorithm was used by Lanczos (1950) as a means of computing the eigenvalues of an asymmetric matrix A. Fletcher (1976) used it for solving indefinite systems. The BCG procedure described below is a generalization of the CG method. It can be used for asymmetric and/or singular matrices A. When A is symmetric and positive definite, BCG is the same as CG.

The BCG algorithm defines two residual vectors $\mathbf{r}$ and $\bar{\mathbf{r}}$ ($\bar{\mathbf{r}}$ is called a biresidual vector) and two search direction vectors $\mathbf{d}$ and $\bar{\mathbf{d}}$ ($\bar{\mathbf{d}}$ is called a bidirection vector), and it proceeds as follows [see Jacobs (1981)]. Given $\mathbf{x}^0$ as an initial solution to $A\mathbf{x} = \mathbf{b}$, set

$$\mathbf{r}^0 = \mathbf{b} - A\mathbf{x}^0$$
$$\bar{\mathbf{r}}^0 = \mathbf{r}^0$$
$$\mathbf{d}^0 = \mathbf{r}^0$$
$$\bar{\mathbf{d}}^0 = \bar{\mathbf{r}}^0$$

1. Then for $k = 0, 1, 2, \ldots,$ calculate:

   (a) $\gamma_k = (\bar{\mathbf{r}}^k)^T \mathbf{r}^k / (\bar{\mathbf{d}}^k)^T A \mathbf{d}^k$

   (b) $\mathbf{x}^{k+1} = \mathbf{x}^k + \gamma_k \mathbf{d}^k$

   (c) $\mathbf{r}^{k+1} = \mathbf{r}^k - \gamma_k A \mathbf{d}^k$

   (d) $\bar{\mathbf{r}}^{k+1} = \bar{\mathbf{r}}^k - \gamma_k A^T \bar{\mathbf{d}}^k$

2. Check convergence. If a stopping criterion is met, then stop. Otherwise:
3. Calculate:

   (a) $\beta_k = (\bar{\mathbf{r}}^{k+1})^T \mathbf{r}^{k+1} / (\bar{\mathbf{r}}^k)^T \mathbf{r}^k$

   (b) $\mathbf{d}^{k+1} = \mathbf{r}^{k+1} + \beta_k \mathbf{d}^k$

(c) $\bar{\mathbf{d}}^{k+1} = \bar{\mathbf{r}}^{k+1} + \beta_k \bar{\mathbf{d}}^k$

(d) Go to step 1(a).

It can be shown [see Jacobs (1981)] that the expressions for γ_k and β_k given above ensure the biconjugacy conditions

$$(\bar{\mathbf{d}}^i, A\mathbf{d}^j) = 0 \qquad \text{for} \quad i \neq j$$

$$(\mathbf{d}^i, A^T\bar{\mathbf{d}}^j) = 0 \qquad \text{for} \quad i \neq j$$

and the biorthogonality conditions

$$(\bar{\mathbf{r}}^i, \mathbf{r}^j) = 0 = (\mathbf{r}^i, \bar{\mathbf{r}}^j) \qquad \text{for} \quad i > j$$

$$(\bar{\mathbf{r}}^i, \mathbf{d}^j) = 0 = (\mathbf{r}^i, \bar{\mathbf{d}}^j) \qquad \text{for} \quad i > j$$

where the inner product $(\mathbf{x}, \mathbf{y}) = \mathbf{x}^T\mathbf{y}$.

The biconjugacy conditions and the biorthogonality conditions together imply that the BCG algorithm is a theoretically exact procedure, that is, with exact arithmetic, and if the procedure does not break down because of division by zero, then $\mathbf{x}^k$ satisfies $A\mathbf{x}^k = \mathbf{b}$ for some $k \leq N$. In practice, however, roundoff errors on any computer preclude the exact result $A\mathbf{x}^k = \mathbf{b}$. More importantly, though, since N is so large as to make N iterations of the BCG algorithm prohibitively long, BCG behaves like an iterative procedure in that $\mathbf{x}^k$ usually converges to a solution of $A\mathbf{x} = \mathbf{b}$ and closely approximates a solution for relatively small values of k.

3. STOPPING CRITERIA

There are a variety of stopping criteria that can be used for iterative procedures. Generally, they fall into two major classes, one that stops when a check on the solution vector $\mathbf{x}^k$ is satisfied, and one that stops when a check on the residual vector $\mathbf{r}^k = \mathbf{b} - A\mathbf{x}^k$ is satisfied. After preliminary experimentation, we considered two in each class.

3.1 Stopping Criteria Involving the Solution Vector

We preset an "error" tolerance ϵ. The first solution vector stopping criterion we consider is the very commonly used Cauchy criterion:

$$\max_i |x_i^k - x_i^{k-1}| \leq \epsilon/N \tag{5}$$

Thus, in order to stop, all differences between the respective elements of the last two solution vectors must be less than or equal to ϵ/N. The reason for dividing ϵ by N is that our x_i are probabilities and the larger the state space N, the smaller on average each x_i will be since they must sum to one.

The second stopping criterion is suggested by Mitra and Tsoucas (1987). They found it worked well for Markovian tandem queuing network problems:

$$\max_i \left\{ \frac{x_i^k}{x_i^{k-1}} \right\} - \min_i \left\{ \frac{x_i^k}{x_i^{k-1}} \right\} \leq \epsilon \tag{6}$$

Here the absolute value of the largest negative fractional change in a solution element is added to the largest positive fractional change in a solution element and the sum must not exceed ϵ.

3.2 Stopping Criteria Involving Residuals

Again letting $\mathbf{r}^k$ denote the residual vector after the kth iteration, that is,

$$\mathbf{r}^k = \mathbf{b} - A\mathbf{x}^k$$

we consider the following two stopping criteria, which we call residual 1 and residual 2, respectively:

$$\max_i |r_i^k| / \max_{i,l} |a_{ij}| \cdot \max_i |x_i^k| \leq \epsilon/N \tag{7}$$

$$\max_i |r_i^k| / \max_i |r_i^{(0)}| \leq \epsilon \tag{8}$$

4. STARTING SOLUTIONS

It is well known that for iterative procedures the choice of a starting solution affects the accuracy of the results and the number of iterations until convergence is achieved. We tested several starting solutions in order to study their influence.

For ease of notation we denote by l the index of a state with largest steady-state probability and by s the index of a state with smallest steady-state probability. Prior to solving a new problem we do not know l and s, of course, but subsequently we do know their values. Also, let $m = \lfloor \sqrt{n} \rfloor$, where $\lfloor y \rfloor$ denotes the integer part of y; m is the index of an arbitrarily chosen "middle" state.

4.1 $n \times n$ Nonsingular Case [Equation (3a)]

For the nonsingular case (3a) where a row of Q^T must be replaced by elements each equal to one, we tried three different possibilities: row l, row m, and row

s. For each of these, we considered the following starting solution vectors:

(a) set $\pi_l = 1$ and $\pi_i = 0\ (i \neq l)$
(b) set $\pi_m = 1$ and $\pi_i = 0\ (i \neq m)$
(c) set $\pi_s = 1$ and $\pi_i = 0\ (i \neq s)$
(d) set $\pi_i = 1/n\ \forall i$
(e) set $\pi_i = 0\ \forall i$

4.2 $n - 1 \times n - 1$ Nonsingular Case [Equation (3b)]

For this case, it is necessary to delete a row and column from the Q matrix. We considered here the three possibilities of deleting row and column l, m, and s, respectively. For each of these, we used the following starting solutions:

(a) set $\pi_i = 1/n\ \forall_i$
(b) set $\pi_i = 0\ \forall_i$

Further, when row and column l were deleted, we also considered

(c) set $\pi_m = 1$ and $\pi_i = 0\ (i \neq m)$
(d) set $\pi_s = 1$ and $\pi_i = 0\ (i \neq s)$

When row and column m were deleted, we further considered

(e) set $\pi_l = 1$ and $\pi_i = 0\ (i \neq l)$
(f) set $\pi_s = 1$ and $\pi_i = 0\ (i \neq s)$

Finally, when row and column s were deleted we further considered

(g) set $\pi_l = 1$ and $\pi_i = 0\ (i \neq l)$
(h) set $\pi_m = 1$ and $\pi_i = 0\ (i \neq m)$

4.3 Singular Case [Equation (1)]

For the $n \times n$ singular Q matrix, we considered the following starting solutions:

(a) $\pi_l = 1, \pi_i = 0\ (i \neq l)$
(b) $\pi_m = 1, \pi_i = 0\ (i \neq m)$
(c) $\pi_s = 1, \pi_i = 0\ (i \neq s)$
(d) $\pi_i = 1/n\ \forall i$

5. GENERATION AND STORAGE OF THE Q MATRIX

The first step for any solution procedure is the generation of the rate matrix Q. The off-diagonal elements of Q are the transition rates between states and the diagonal elements are defined so the row sums equal zero. Even when there are only two bases and one depot and only a few items at each base, the number of states may be very large. For example, when there are five items at one base, four items at the other base, and two spare items at the depot, we have 375 states; also, 18 items at one base, 13 at the other base, and three at the depot results in 20,748 states and a Q matrix of size 20,748 by 20,748. In general, if there are N_1 items at one base, N_2 items at the other base, and N_d items at the depot, the number of states is

$$(N_1 + 1) \times (N_1 + 2) \times (N_2 + 1) \times (N_2 + 2)/4 + (N_1 + 1) \times (N_2 + 1) \times N_d$$

5.1 Ordering of the States

The number of events that account for all the state transitions in this kind of system is 4×(number of bases)+2, or 10 for a two-base system. Table 1 lists the possible system events. Note that events of both types FBsD and FDd cannot occur from the same state, as FBsD can only occur when no spares are available at the depot and FDd can occur only when spares are available at the depot. Similarly, events of both types RDs and RDd cannot occur from the same state. Therefore, each state can interact with at most 4×(number of bases) states. For the two-base model, each state can interact with at most eight other states, so each row and column of the Q matrix contains at most eight nonzero off-diagonal elements; therefore the Q matrix is sparse and we can greatly reduce storage by storing only the nonzero elements of Q and their coordinates.

We can achieve still greater efficiency for some of the iterative techniques [see Kaufman (1983)] by reducing the half bandwidth of Q, where half bandwidth is defined to be the distance from the diagonal element of a row to the furthest nonzero element in the same row, maximized over all rows [see Strang (1980)]. We change the half bandwidth of Q by changing the order of the states, thereby rearranging the rows and columns of Q. In order to best describe the state space, we have associated with each state j a six-dimensional vector, defined as follows:

$$j = [\#\text{U1}, \#\text{R1}, \#\text{U2}, \#\text{R2}, \#\text{Ud}, \#\text{Rd}],$$

Table 1 Possible System Events

Symbol	Description of event
FBs	Failure at base s ($s = 1, 2$) that is base repairable.
RBs	Repair completed at base s.
FBsD	Failure at base s that is depot repairable, and no spares are available at the depot.
RDs	Repair completed at the depot and sent to base s.
FDd	Failure at any base that is depot repairable and spares are available at the depot.
RDd	Repair completed at depot when there are no backorders from any base, so that the item becomes a depot spare.

where

$$\#\text{Us} = \text{number of items "up" at base } s,\ s = 1, 2$$

$$\#\text{Rs} = \text{number of items in local repair at base } s$$

$$\#\text{Ud} = \text{number of spare items available at the depot}$$

$$\#\text{Rd} = \text{number of items in repair at the depot}$$

We try to minimize the half bandwidth of Q by placing the states that communicate with each other as close together as possible. In Gross et al. (1990) we describe a heuristic algorithm that we found to produce good results.

It is interesting, and important, to note, however, that the BCG algorithm is invariant to the order of the states.

5.2 Storing the Q Matrix

The next step, after the states have been generated and simultaneously ordered and numbered, is to determine which states communicate, that is, where the nonzero entries of the Q matrix are located. We have programmed two methods to accomplish this.

One method first transforms the six-dimensional state descriptor vectors into scalars so that they can be sorted lexicographically and searched efficiently. This is accomplished by treating the six components of each state descriptor vector as if they were six binary numbers and concatenating them. Five bits are allocated to each of #U1, #R1, #U2, and #R2, four bits are allocated to #Ud, and seven bits are allocated to #Rd. This procedure limits the number of items per base to a maximum of 31, which should be sufficient for many applications. For example, the state (0, 2, 1, 0, 0, 1) is stored as the binary scalar 00000 00010 00001 00000 0000 0000001. These binary scalars are sorted into numerical order from largest to smallest. A binary search is then performed on them to find the states that communicate with a particular state.

Another method for determining which states communicate, without the necessity of employing a search routine, divides all the state vectors into several blocks according to their first two components. A block contains all vectors whose first two components are the same. The communication of states in the system is then divided into two categories: communication between blocks and communication within blocks. Details are provided by Gross et al. (1990). This method was used in the computational testing reported in Section 6.

Once the determination is made as to which states communicate, the entries for the Q matrix must be stored. Storing the entire Q matrix in the computer would be very inefficient, and not feasible for large models, so we employ a technique called SERT to generate the information contained in Q and store it as two sets of t (number of events; see Table 1) vectors each [see Gross and Miller (1984)]. There are t"rate" vectors and t "source" vectors, each with n (size of the state space) elements. These can be regarded as $t \times n$ matrices R and S, where corresponding elements of R and S are "paired." As mentioned previously, the value of t is 10 for a two-base system; more generally it is $4b + 2$, where b is the number of bases.

Thus there is a rate vector and a source vector associated with each of the t events (for event k, the associated vectors are the kth rows of S and R). The value of the jth element of the kth source vector (s_{kj}) gives the state in which the system was (state i, the "coming from" or "source" state) before event k transited the system to state j. Suppose, for example, $s_{kj} = i$ (i.e., when the system is in state i, an event of type k takes the system to state j). The corresponding element in the jth column and kth row of R, r_{kj}, gives the transition rate q_{ij}. A zero entry as the jth element of a source vector means there is no state from which its associated event can take the system into state j, and the corresponding element in the rate vector is also zero. Summarizing, we can write $q_{ij} = r_{kj}$, where k is such that $s_{kj} = i$. Note that the events are defined so that for each pair (i,j) there is at most one event that can take the system from state i to state j.

We note that SERT as described here is a slight modification of the original procedure as developed in Gross and Miller (1984) in that as originally developed the s_{ki} gave the state j to which event k takes the system when it is in state i (as opposed to our s_{kj} being the state i in which the system *was* before occurrence of event k). Thus instead of a source vector, original SERT used a target vector (hence the acronym SERT—state, event, rate, target). We continue the use of the SERT acronym, since the philosophy is the same.

Thus, using SERT, we need to store only two $t \times n$ matrices (R and S), rather than the $n \times n$ matrix Q. In large (realistic size) problems, this is a significant storage savings since $t << n$. Further savings in storage are achieved through a packing procedure that stores the pair of rate and source elements (r_{kj}, s_{kj}) in the first empty row of the proper column (column j) in both matrices.

6. RESULTS

Seven different two-base, two-echelon problems were used for numerical testing. Their parameter values are given in Table 2, and Table 3 provides definitions of the parameters. The seven problems range in size from 24 states (for a total of four items) to 200,175 states (for a total of 65 items).

We used the singular form of the A matrix and six different nonsingular forms; these are presented in Table 4. Combining each of these seven versions of A with the possible starting solutions described in Section 4, we obtained a total of 31 combinations. Each of these was used together with each of the four stopping criteria of Section 3 to yield a total of 124 cases. For problems 1 and 2, each of the 124 cases was run on four different computers: (1) Vax 11/780, (2) Alliant FX/8, (3) Cyber 205, and (4) Cray XMP/48. For problem 3, each of the 124 cases was run on the Alliant and Cyber computers. Problems 4 through 7 were run on the Alliant computer, but only for selected combinations.

In making comparisons, we examined both CPU time until convergence and accuracy of the solution obtained. We judged accuracy in terms of the percentage errors in three availability measures. AVs, $s = 1, 2$, is the availability at base s, defined as the steady-state probability that the number of operational items at base s is at least equal to the number desired (i.e., $\Pr\{\#\text{Us} \geq \text{Os}\}$); it is computed by summing the steady-state probabilities of the appropriate states. The other availability measure is AV12, the joint availability at both bases, defined as $\Pr\{\#\text{U1} \geq \text{O1}, \#\text{U2} \geq \text{O2}\}$. We set an upper limit of 1% relative error in all three availability measures and looked for the combinations with smallest CPU times meeting this constraint.

Space limitations preclude the presentation here of detailed results, so we shall simply summarize them. The best stopping criterion for use with the BCG algorithm was residual 2 [equation (8)]. For problems of modest size (i.e., problems 1, 2, and 3), the best combination was the singular matrix Q and starting solution $\pi_l = 1$. Given that l is initially unknown, it is important to note that comparable results were obtained by starting with $\pi_m = 1$.

Table 2 Problems Run: 2 Bases, 2 Echelons (see Figure 1)

Problem no.	N_1	O_1	C_1	α_1	λ_1	μ_1	N_2	O_2	C_2	α_2	λ_2	μ_2	N_d	C_d	μ_d	No. of states n
1	2	1	1	.5	.6	.8	1	1	1	.6	.6	.8	1	1	.8	24
2	5	3	2	.5	.4	.6	4	2	2	.7	.4	.5	2	2	.3	375
3	10	6	4	.5	.4	.6	8	4	4	.7	.4	.5	4	4	.3	3366
4	18	14	2	.67	.2	1.0	13	10	2	.67	.143	1.0	3	4	.5	20,748
5	25	19	3	.65	.2	1.0	15	12	2	.65	.13	1.0	6	4	.67	50,232
6	32	25	4	.65	.2	1.0	17	13	2	.65	.12	1.0	7	5	.64	100,089
7	33	26	4	.65	.19	1.0	24	19	3	.65	.12	1.0	8	6	.60	200,175

Table 3 Parameter Definitions

Parameter	Description
N_s	Total number of items at base s, $s = 1, 2$.
O_s	Number of operating items desired at base s ($N_s - O_s$ = number of spares at base s).
C_s	Number of repair channels in the base s repair facility.
N_d	Number of depot spares.
C_d	Number of depot repair channels.
α_s	Probability that a failed item at base s is locally repairable.
λ_s	Failure rate of an operating item at base s.
μ_s	Base s repair rate for an item.
μ_d	Depot repair rate.

For larger problems (i.e., problems 4 through 7), the best BCG combination was $\bar{Q}_l$ (see Table 4) with starting solution $\boldsymbol{\pi} = \mathbf{0}$. On these larger problems, results were sensitive to a knowledge of l, so a two-phase algorithm was devised and implemented. It initially uses the Gauss–Seidel (GS) method, with singular matrix Q and starting solution $\pi_i = 1/n\ \forall i$, in order to estimate l. When l is determined with sufficient confidence (our operational definition was the same apparent value of l for 15 consecutive iterations), the algorithm switches to BCG and uses $\boldsymbol{\pi} = \mathbf{0}$ as its starting solution. Two different versions of the two-phase algorithm use the $\bar{Q}_l$ and $\hat{Q}_l$ versions of BCG, respectively. Table 5 presents comparative results for pure GS (Q_{gs}), pure BCG ($\bar{Q}_b$), assuming l is known, and the two-phase methods ($\bar{Q}$ and

Table 4 A Matrix Combinations

Symbol	Singularity	Size	Row and column manipulated (see Section 4)
Q	Singular	$n \times n$	
$\bar{Q}_l$	Nonsingular	$n \times n$	l
$\bar{Q}_m$	Nonsingular	$n \times n$	m
$\bar{Q}_s$	Nonsingular	$n \times n$	s
$\hat{Q}_l$	Nonsingular	$(n-1) \times (n-1)$	l
$\hat{Q}_m$	Nonsingular	$(n-1) \times (n-1)$	m
$\hat{Q}_s$	Nonsingular	$(n-1) \times (n-1)$	s

Table 5 Comparison of Gauss–Seidel, Biconjugate Gradient, and Two-Phase Methods: Alliant Computer, $\epsilon = 10^{-3}$

Problem	Q	l	Iterations		% Error			Setup time (s)	Iteration time (s)		Total iteration time (s)
			GS	BCG	AV1	AV2	AV12		GS	BCG	
1	Q_{gs}	—	11	—	1.1E−2	6.7E−3	1.4E−2	1.1E−2	1.0E−2	—	1.0E−2
(n = 24)	$\underline{Q}_b$	9	—	11	8.5E−3	1.4E−1	1.3E−1	1.1E−2	—	2.9E−2	2.9E−2
	$\underline{Q}$	9	16	11	8.5E−3	1.4E−1	1.3E−1	1.1E−2	1.5E−2	2.9E−2	4.4E−2
	$\hat{\underline{Q}}$	9	16	13	1.5E−3	1.4E−2	1.6E−2	1.1E−2	1.5E−2	3.2E−2	4.7E−2
2	Q_{gs}	—	99	—	4.7E−2	2.8E−2	5.5E−2	1.9E−1	1.6	—	1.6
(n = 375)	$\underline{Q}_b$	37	—	25	1.6E−2	3.8E−2	3.9E−2	1.9E−1	—	9.8E−1	9.8E−1
	$\underline{Q}$	37	21	25	1.6E−2	3.8E−2	3.9E−2	1.9E−1	3.3E−1	9.8E−1	1.3
	$\hat{\underline{Q}}$	37	21	34	8.5E−3	1.4E−3	8.4E−3	1.9E−1	3.3E−1	1.3	1.6
3	Q_{gs}	—	252	—	5.6E−2	3.7E−2	6.9E−2	2.0	3.8E1	—	3.8E1
(n = 3366)	$\underline{Q}_b$	679	—	28	4.8E−1	1.4E−2	6.0E−1	2.0	—	1.1E1	1.1E1
	$\underline{Q}$	777	24	34	6.4E−1	1.6E−1	7.5E−1	2.0	3.6	1.3E1	1.7E1
	$\hat{\underline{Q}}$	777	24	50	3.5E−4	2.2E−4	3.2E−4	2.0	3.6	1.9E1	2.3E1

4	Q_{gs}	—	235	—	5.6E−4	3.2E−3	2.5E−4	1.4E1	2.3E2	—	2.3E2
(n = 20,748)	$\bar{Q}_b$	3837	—	22	6.8E−1	2.2E−1	5.7E−1	1.4E1	—	5.7E1	5.7E1
	$\bar{Q}$	3836	46	22	3.4E−1	4.4E−1	1.1E−1	1.4E1	4.5E1	5.7E1	1.0E2
	$\hat{Q}$	3836	46	32	5.5E−2	6.9E−2	4.1E−2	1.4E1	4.5E1	8.0E1	1.2E2
5	Q_{gs}	—	361	—	1.1E−3	1.3E−3	1.6E−3	3.4E1	8.7E2	—	8.7E2
(n = 50,232)	$\bar{Q}_b$	10297	—	22	4.9E−1	1.5E−1	8.3E−2	3.4E1	—	1.4E2	1.4E2
	$\bar{Q}$	10296	58	23	3.3E−1	2.8E−1	9.6E−2	3.4E1	1.4E2	1.5E2	2.9E2
	$\hat{Q}$	10296	58	30	1.7E−1	9.3E−3	1.2E−1	3.4E1	1.4E2	1.8E2	3.2E2
6	Q_{gs}	—	456	—	1.1E−3	5.3E−4	1.3E−3	6.9E1	2.2E3	—	2.2E3
(n = 100,089)	$\bar{Q}_b$	21799	—	24	2.6E−1	7.7E−2	1.6E−2	6.9E1	—	3.1E2	3.1E2
	$\bar{Q}$	21799	78	24	2.6E−1	7.7E−2	1.6E−2	6.9E1	3.8E2	3.1E2	6.8E2
	$\hat{Q}$	21799	78	33	9.8E−2	1.6E−2	4.9E−2	6.9E1	3.8E2	4.1E2	7.8E2
7	Q_{gs}	—	509	—	4.3E−4	3.1E−4	5.3E−4	1.4E2	4.9E3	—	4.9E3
(n = 200,175)	$\bar{Q}_b$	42586	—	24	7.7E−1	2.3E−1	3.7E−1	1.4E2	—	6.2E2	6.2E2
	$\bar{Q}$	42586	79	24	7.7E−1	2.3E−1	3.7E−1	1.4E2	7.7E2	6.2E2	1.4E3
	$\hat{Q}$	42586	79	34	8.6E−2	4.1E−2	8.2E−2	1.4E2	7.7E2	8.5E2	1.6E3

$\hat{Q}$); for the largest problems the two-phase methods took about twice as long as pure BCG, but still only one-third as long as pure GS.

It is of interest to note the relative speeds of the various computers we used with the BCG and two-phase algorithms. For problem 2, of size $n = 375$, the relative CPU times for the same BCG algorithm are as follows: Cray 1.0, Cyber 1.9, Alliant 13, and Vax 40. This yields a ratio of Alliant time to Cyber time of 6.7. On the larger problems 3 through 6, this ratio increased slightly to 7.0. It should be noted that the larger problems could not be run on the Vax because of time limitations, and some of them could not be run on the Cray and Cyber because of storage limitations.

In summary, the two-phase algorithm appears to be a reasonable solution approach for moderately large finite state-space CTMPs emanating from multiechelon repairable item inventory systems.

REFERENCES

[1] Fletcher, R. (1976). Conjugate gradient methods for indefinite systems. In *Proceedings of the Dundee Conference on Numerical Analysis*, edited by G. A. Watson. Springer Verlag, Berlin.

[2] Gross, D., B. Gu, R. M. Soland, J. W. Stern, and E. J. Zaldivar (1990). Iterative solution methods for Markovian multi-echelon repairable item inventory systems in steady state. GWU/IMSE/Serial T-523/90, Institute for Management Science and Engineering, George Washington University, Washington, DC.

[3] Gross, D., and D. R. Miller (1984). The randomization technique as a modeling tool and solution procedure for transient Markov processes. *Operations Res.*, *Vol. 32*, pp. 343–361.

[4] Jacobs, D. A. H. (1981). Preconditioned conjugate gradient methods for solving systems of algebraic equations. Note No. RD/L/N193/80, Central Electricity Research Labs, Leatherhead, Surrey.

[5] Kaufman, L. (1983). Matrix methods for queueing problems. *SIAM J. Sci. Stat. Comput.*, *Vol. 4*, pp. 525–552.

[6] Lanczos, C. (1950). An iterative method for the solution of the eigenvalue problem of linear differential and integral operators. *J. Res. Nat. Bur. Standards*, *Vol. 45*, pp. 255-282.

[7] Mitra, D., and P. Tsoucas (1987). Relaxations for the numerical solutions of some stochastic problems. Technical Report, AT &T Bell Laboratories, Murray Hill, NJ.

[8] Strang, G. (1980). *Linear Algebra and Its Applications*, 2nd ed. Academic Press, New York.

[9] Young, D. M. (1971). *Iterative Solutions of Large Linear Systems*. Academic Press, New York.

26

Computing the Stationary Distribution Vector of an Irreducible Markov Chain on a Shared-Memory Multiprocessor

R. B. MATTINGLY Department of Mathematical and Computer Science, Youngstown State University, Youngstown, Ohio

C. D. MEYER* Department of Mathematics, North Carolina State University, Raleigh, North Carolina

ABSTRACT

A direct method for computing the stationary distribution vector of an irreducible Markov chain using a concept known as stochastic complementation is presented. Parallel algorithms based on this method are applied to nearly completely reducible systems and tested on a shared-memory multiprocessor. This method is compared to parallel implementations of other standard direct methods.

1. INTRODUCTION

For an n-state irreducible Markov chain with transition matrix P, there exists a unique stationary distribution vector π. This vector may be computed using

*The work of this author was supported in part by the National Science Foundation grant DMS-8521154.

a quantity that was first defined in [8] and that was given the name stochastic complement in [9]. Although the theory of stochastic complementation applies to any irreducible Markov chain, the problems under consideration in this paper are all loosely coupled systems that give rise to transition matrices that are almost completely reducible.

Algorithms based on stochastic complementation are well suited to shared-memory parallel computers, such as the Sequent Balance. In this paper, the results of numerical experiments comparing stochastic complementation algorithms to standard techniques of solution are presented.

In Section 2, some background on the Sequent Balance architecture is presented. Section 3 discusses the parallel implementation of some standard techniques for computing the stationary distribution vector. In Section 4, the concept of stochastic complementation is introduced, along with the implementation of several algorithms based on this concept. Experimental results are reported in Section 5.

2. SEQUENT BALANCE 21000 ARCHITECTURE

All the parallel algorithms under consideration were implemented on the Sequent Balance 21000 located at the Advanced Computing Research Facility at Argonne National Laboratory. The configuration of this machine at the time of these experiments included 24 processors sharing 24 megabytes of memory. The codes were written in Balance FORTRAN, which includes the standard FORTRAN 77 language plus a set of compiler directives for parallelizing sections of code.

The most common areas for parallelization are FORTRAN DO loops. If the loop iterations are independent of one another, the work in the DO loop may be spread out among the available processors, with each processor being responsible for executing certain iterations of the loop. The variables within the loop must be classified as either shared, local, or reduction. Shared variables are any variables that are read-only in the loop, or arrays in which no element is referenced by more than one loop iteration. Only one copy of each of these variables is required in memory. Local variables are those variables that are reinitialized with every new loop iteration. Each processor needs to have its own copy of every local variable. An example of a reduction variable would be a variable used to accumulate the result of a dot product. Only one copy of a reduction variable should exist; however, all processors must be able to reference it. There are other classifications possible for DO loop variables, but they did not occur in any of the codes developed for this work. For further details on parallelizing code for the Sequent Balance, see [1].

3. IMPLEMENTATION OF STANDARD METHODS

3.1 Power Method

For regular Markov chains, it is well known [8] that the iteration

$$\pi^{(i+1)} = \pi^{(i)}P \tag{1}$$

will converge to the stationary distribution vector π independent of the starting guess. However, the convergence rate of this method is dependent on the size of the second-largest eigenvalue of the matrix P. If this eigenvalue is sufficiently close to 1 in magnitude, the convergence will be extremely slow. For problems in which the Markov chain is almost completely reducible, the associated matrix can be permuted to a nearly block diagonal form, where each diagonal block is nearly stochastic. Since all stochastic matrices have 1 as an eigenvalue, and since the eigenvalues of a matrix are continuous functions of the entries in the matrix, matrices with the structure given above will always have several eigenvalues close to 1 in magnitude. Thus, the power method will be unsuitable for the problems under consideration. However, some of the methods that are to be considered divide the original problem into smaller subproblems. Often these subproblems can be solved by using the power method, so this algorithm will be of interest as a means of obtaining intermediate results.

The implementation of the power method on a shared-memory multiprocessor requires only one copy of the transition matrix, and two vectors: the current distribution vector $\pi^{(i)}$, and the new distribution vector $\pi^{(i+1)}$, which is computed according to equation (1). Clearly, different columns of the matrix can be assigned to different processors. Each processor multiplies its particular columns of P by $\pi^{(i)}$, which is shared by all processors, and stores the results in the corresponding locations in $\pi^{(i+1)}$. Then, if

$$\|\pi^{(i+1)} - \pi^{(i)}\|_\infty < \epsilon \tag{2}$$

where ϵ is some specified tolerance, the algorithm terminates. Otherwise, another iteration is performed.

3.2 LU Factorization

Instead of using an iterative technique, the stationary distribution vector can be found directly by solving the equations below:

$$(P^T - I)\pi^T = 0 \qquad e^T\pi^T = 1 \tag{3}$$

where I represents the $n \times n$ identity matrix and e represents a column vector of all ones whose length is determined from the context.

If P is an irreducible stochastic matrix, it will have one eigenvalue equal to 1. Therefore, the matrix $P^T - I$ will have one zero eigenvalue, and will thus be rank $n - 1$. It is known that any of the equations can be replaced by the normalization condition $\pi e = 1$, resulting in a nonsingular $n \times n$ system. In particular, consider the matrix

$$A = \begin{bmatrix} B \\ e^T \end{bmatrix} \tag{4}$$

where B is the $(n - 1) \times n$ matrix formed by deleting row n of the matrix $P^T - I$. The LU factorization of A can be found using the LINPACK routine DGEFA [4], which yields

$$LU\pi^T = e_n \tag{5}$$

where e_n is a vector containing zeros everywhere except for a 1 in the nth place. This system can be solved using DGESL, another LINPACK routine that uses forward substitution to solve $Ly = e_n$ and then uses backward substitution to solve $U\pi^T = y$.

This algorithm does not parallelize well. The factorization of the matrix must proceed one column at a time. The forward and backward substitution operations are also ill-suited for parallelism because, for example, the computation of π_k in the backward substitution algorithm requires that π_{k+1} through π_n already be known.

In this work, a flop is defined to be any of the four basic arithmetic operations. With this definition, it is easy to see that the LU factorization takes $(2/3)n^3 + O(n^2)$ flops. This is the major computation in the algorithm because the forward and backward substitution routines are each $O(n^2)$ operations.

Finally, it should be noted that other variations using the LU factorization are possible [7]. Observe that since L is unit lower-triangular, so is L^{-1}, which implies that $L^{-1}e_n = e_n$. Therefore, equation (5) may be multiplied by L^{-1} and the system may be solved without using forward substitution. This is only slightly more efficient than our implementation, because the forward substitution adds only a few operations to the total operation count. It is also possible to find the LU factorization of the rank-deficient matrix $I - P^T$. However, additional work is required to scale the computed solution so that the normalization condition is satisfied.

3.3 QR Factorization

In [5], an algorithm that uses the QR factorization to compute the stationary distribution vector is presented. If the QR factorization of the matrix

$I - P$ is formed, the stationary distribution vector can be computed merely by taking the transpose of the last column of Q and then normalizing the elements so that their sum is equal to 1. Thus, the QR algorithm for computing the stationary distribution vector of a Markov chain involves nothing more than computing the QR factorization of the matrix $I - P$, and then constructing the last column of Q, which is given by $M_1 M_2 \cdots M_{n-1} e_n$, where each M_i is a Householder transformation matrix. This product should be formed from right to left, that is, first forming $M_{n-1} e_n$, then multiplying on the left by M_{n-2}, and so on. The advantages of performing the computations in this order are discussed in [6]. Finding the QR factorization of the matrix requires $(4/3)n^3 + O(n^2)$ flops. Constructing the last column of Q requires only $O(n^2)$ flops.

4. STOCHASTIC COMPLEMENTATION

4.1 Introduction

None of the methods discussed previously take advantage of the special structure the matrix P has in the case where the Markov chain consists of several loosely coupled subsystems. In fact, some of the methods can perform very poorly on these problems. On the other hand, aggregation techniques [3][10], which are designed with these loosely coupled problems in mind, generally cannot be expected to compute the exact stationary distribution vector. The algorithm presented in this section uncouples the original Markov chain into two or more smaller chains, which have the following properties:

If the original chain is irreducible, the smaller chains are also irreducible and hence, each one has a unique stationary distribution vector v_i.

The v_i's can be computed independently of one another.

The v_i's can be easily coupled together in order to produce the stationary distribution vector π for the original problem.

In the discussion that follows, the matrix P is divided into blocks as

$$P = \begin{bmatrix} P_{11} & P_{12} & \cdots & P_{1s} \\ P_{21} & P_{22} & \cdots & P_{2s} \\ \vdots & \vdots & \ddots & \vdots \\ P_{s1} & P_{s2} & \cdots & P_{ss} \end{bmatrix} \tag{6}$$

where the diagonal blocks are square. Furthermore, if A represents any vector or matrix, the notation $A \geq 0$ means that each element of A is nonnegative, and $A > 0$ means that each element is strictly positive.

It is well known [2] that if P is a stochastic matrix, then $I - P$ is a singular M-matrix of rank $n-1$, but every principal submatrix of $I - P$ of order less than n is a nonsingular M-matrix. In particular, if P_i represents the principal submatrix of P obtained by deleting the ith row and column of blocks from P, then for $i = 1, \ldots, s$, $I - P_i$ is a nonsingular M-matrix, which implies

$$(I - P_i)^{-1} > 0 \qquad i = 1, \ldots, s \tag{7}$$

DEFINITION 4.1 For an irreducible stochastic matrix

$$P = \begin{bmatrix} P_{11} & P_{12} \\ P_{21} & P_{22} \end{bmatrix}$$

in which P_{11} and P_{22} are square, the stochastic complement of P_{11} is defined in [9] to be the matrix

$$S_{11} = P_{11} + P_{12}(I - P_{22})^{-1}P_{21} \tag{8}$$

and the stochastic complement of P_{22} is defined to be the matrix

$$S_{22} = P_{22} + P_{21}(I - P_{11})^{-1}P_{12} \tag{9}$$

This terminology stems from the facts that S_{11} and S_{22} are stochastic matrices, and that the matrix S_{11} is [I-Schur complement($I - P_{22}$)] and S_{22} is [I-Schur complement($I - P_{11}$)] in the matrix $I - P$.

The definition of stochastic complementation can be extended to higher-level partitions. For example, suppose the matrix P is given by

$$\begin{bmatrix} P_{11} & P_{12} & P_{13} \\ P_{21} & P_{22} & P_{23} \\ P_{31} & P_{32} & P_{33} \end{bmatrix}$$

where the diagonal blocks are again square. Then, for example, the stochastic complement of P_{22} is given by

$$S_{22} = P_{22} + [P_{21} \quad P_{23}] \begin{bmatrix} I - P_{11} & -P_{13} \\ -P_{31} & I - P_{33} \end{bmatrix}^{-1} \begin{bmatrix} P_{12} \\ P_{32} \end{bmatrix} \tag{10}$$

The general definition given in [9] is stated below:

DEFINITION 4.2 For an $n \times n$ irreducible stochastic matrix P partitioned as in equation (6), the stochastic complement of P_{ii} is given by the matrix

$$S_{ii} = P_{ii} + P_{i*}(I - P_i)^{-1}P_{*i} \qquad i = 1, \ldots, s \tag{11}$$

where P_{i*} refers to the ith row of blocks with the diagonal block removed, and P_{*i} refers to the ith column of blocks with the diagonal block removed.

For example, in equation (10), P_{2*} is given by

$$[P_{21} \quad P_{23}]$$

P_{*2} is given by

$$\begin{bmatrix} P_{12} \\ P_{32} \end{bmatrix}$$

and P_2 is given by

$$\begin{bmatrix} P_{11} & P_{13} \\ P_{31} & P_{33} \end{bmatrix}$$

It is shown in [9] that the stochastic complements defined above are always irreducible stochastic matrices. As such, there exists for each stochastic complement a unique stationary distribution vector v_i, whose relationship to the stationary distribution vector π of the original transition matrix P is described in the following theorem.

THEOREM 4.1 Let P be defined and partitioned as in Definition 4.2, and let S_{ii} be the stochastic complement of P_{ii}, for $i = 1, \ldots, s$. Let the stationary distribution vector of P be denoted by $\pi = [\pi_1 \quad \pi_2 \quad \cdots \quad \pi_s]$, where the dimensions of each π_i correspond to the dimensions of P_{ii}. Define

$$v_i = \frac{\pi_i}{\pi_i e} \qquad i = 1, \ldots, s \tag{12}$$

Then for each i, v_i is the unique stationary distribution vector of the stochastic complement S_{ii}.

Proof. See [8] or [9].

Using standard techniques, the vectors v_1 through v_s can be computed independently from one another. Once they have been found, it is possible to recouple them to produce π. In order to show how this is done, two other definitions are needed.

DEFINITION 4.3 Let P be defined and partitioned as in Definition 4.2, and let S_{ii} be the stochastic complement of P_{ii} for each i. Furthermore, let v_i be the unique stationary distribution vector for S_{ii}. Define the coupling matrix to be an $s \times s$ matrix C whose entries are given by

$$c_{ij} = v_i P_{ij} e \qquad i = 1,\ldots,s, \quad j = 1,\ldots,s \tag{13}$$

Clearly, since $v_i > 0$, $e > 0$, and $P_{ij} \geq 0$, C must be nonnegative. Furthermore,

$$\sum_{j=1}^{s} c_{ij} = \sum_{j=1}^{s} v_i P_{ij} e = v_i \sum_{j=1}^{s} P_{ij} e = v_i e = 1 \qquad i = 1,\ldots,s \tag{14}$$

which implies that C must be a stochastic matrix. Moreover, it can be shown that C is irreducible and as such, it possesses a unique stationary distribution vector ξ. Define the coupling coefficients to be the elements $\xi_1, \xi_2, \ldots, \xi_s$ of the stationary distribution vector.

The relationship between the stationary distribution vectors π and ξ of the matrices P and C is given in the following theorem.

THEOREM 4.2 Let P be defined and partitioned as in Definition 4.2, and let $\pi = [\,\pi_1 \quad \pi_2 \quad \cdots \quad \pi_s\,]$ be the stationary distribution vector for P partitioned as in Theorem 4.1. If C is the coupling matrix given in Definition 4.3 with unique stationary distribution vector ξ, then the following relationship holds:

$$\xi_i = \pi_i e \qquad i = 1,\ldots,s \tag{15}$$

Proof. Since ξ is unique, the proof is immediate by observing that if the elements of ξ are defined by equation (15), then the equations $\xi C = \xi$ and $\xi e = 1$ are satisfied.

The preceding discussion justifies the following procedure for computing the stationary distribution vector π of an irreducible stochastic matrix P partitioned as in equation (6):

1. Determine the stochastic complements S_{ii} for $i = 1, 2, \ldots, s$.
2. Determine the stationary distribution v_i for each S_{ii}.
3. Determine the coupling matrix C.
4. Determine the stationary distribution ξ for C.
5. Form the stationary distribution π for P as

$$\pi = [\,\xi_1 v_1 \quad \xi_2 v_2 \quad \cdots \quad \xi_s v_s\,] \tag{16}$$

For Markov chains that consist of a large number of loosely coupled subsystems, a natural way to approach the computation of π on a multiprocessor would be to partition the transition matrix so that all of the states making up a particular subsystem are grouped into one block. Then, a stochastic complement corresponding to each diagonal block would be formed. Since each can be computed independently, the computation of each stochastic complement and its associated stationary distribution vector could be assigned to a different processor. This paper, however, focuses on Markov chains that consist of at most four loosely coupled systems. Dividing the work among processors as described above would automatically limit the maximum possible speedup to four. Thus, a better way to approach the parallel implementation of these problems would be to spread out the work in computing each stochastic complement and its associated stationary distribution vector among all available processors. Specific programs for partitioning P and forming two, three, or four stochastic complements have been written. Each of these implementations is discussed in greater detail below.

4.2 Implementation of the 2 × 2 Block SC Algorithm (SC2)

For Markov chains consisting of two loosely coupled blocks, a natural way to use stochastic complementation would be to partition P into a 2×2 block system and then form two stochastic complements S_{11} and S_{22} according to equations (8) and (9).

The formation of S_{11} can be broken down into two major steps:

1. Solve $(I - P_{22})X_1 = P_{21}$ for X_1.
2. Form $S_{11} = P_{11} + P_{12}X_1$.

To carry out the first step, the LU factorization of $(I - P_{22})$ is found using the LINPACK routine DGEFA. This subroutine was modified in order to run it in parallel on the Sequent Balance. To solve for X_1, the routine DGESL (also from LINPACK) is called j times, where j is the number of columns in P_{21}. Here, different processors can call DGESL with different columns of P_{21}, solving for the corresponding columns in X_1. Since $I - P_{22}$ is read-only within this part of the code, the processors may share it. The remainder of the computation of S_{11} only involves matrix multiplication and matrix addition, both of which are parallelizable. S_{22} can be computed in similar fashion.

If the Markov chain consists of two loosely coupled systems, then the transition matrix is nearly block diagonal (or could be permuted to such a form), and the second-largest eigenvalue of P is close to 1. In this case, the power method should not be used to compute π. However, experience has shown that the power method can be used to solve for v_1 and v_2, the stationary distribution vectors of S_{11} and S_{22}.

The stationary distribution vector for the coupling matrix C is needed. In order to find this, only the off-diagonal elements of C, c_{12} and c_{21} need to be computed because ξ is given by

$$\xi = \frac{1}{c_{12} + c_{21}} [c_{21} \quad c_{12}] \tag{17}$$

The computation of c_{12} and c_{21} is performed according to equation (13) and involves only dot products of vectors and is easily parallelized. Forming the stationary distribution vector of the original Markov chain using equation (16) with $s = 2$ involves only scalar-vector multiplication.

4.3 Implementation of the 3 × 3 Block SC Algorithm (SC3)

If a Markov chain consists of three loosely coupled subsystems, then the corresponding transition matrix can be permuted to a form that will be close in norm to a 3×3 block diagonal matrix. Thus, a natural implementation of the SC algorithm would be one that created three stochastic complements, S_{11}, S_{22}, and S_{33}, whose dimensions correspond to the dimensions of the diagonal blocks of P. If the blocks are roughly the same size, then the effect of this algorithm is to find the stationary distribution vector of the original matrix by finding the stationary distribution vectors v_1, v_2, and v_3 of three related matrices, each of which is only one-third the size of the original problem. It turns out, however, that this is not an advantage, because the majority of the work is not done in finding v_1, v_2, and v_3; most of the work is done in forming S_{11}, S_{22}, and S_{33}. As an example, S_{11} is formed according to

$$S_{11} = P_{11} + [P_{12} \quad P_{13}] \begin{bmatrix} I - P_{22} & -P_{23} \\ -P_{32} & I - P_{33} \end{bmatrix}^{-1} \begin{bmatrix} P_{21} \\ P_{31} \end{bmatrix} \tag{18}$$

Note that during an intermediate step, a matrix whose size is two-thirds the size of the original problem must be formed. Thus, the SC3 algorithm is not expected to be as efficient as SC2.

4.4 Implementation of the 4 × 4 Block SC Algorithm (SC4)

The situation described in the 3×3 block case becomes even worse in cases where the Markov chain consists of four loosely coupled subsystems. In this case, it is necessary to form four stochastic complements, each of which is roughly one-fourth the size of the original problem. Unfortunately, in order to compute each of these complements, a matrix that is three-fourths the size of the original matrix must be formed. The savings found in solving the four smaller Markov chains is not enough to warrant the additional computation necessary to form these smaller problems.

4.5 Implementation of a Successive SC Algorithm (SSC)

An alternative to the SC4 algorithm discussed above would be to form four stochastic complements in the following manner. First compute two stochastic complements, say S and T, by applying SC2 to P. Each of these matrices will be roughly half the size of the original problem. Then SC2 can be applied to S, producing stochastic complements SS and ST. In similar fashion, stochastic complements TS and TT can be produced. This two-step process will produce four stochastic complements whose sizes correspond to the diagonal blocks of the original loosely coupled system. Forming the complements in this manner insures that no intermediate computation will involve a matrix whose size is more than half the size of the original matrix, avoiding the excessive overhead of the SC4 algorithm. Another advantage to this approach is that all of the coupling matrices needed are 2×2. Thus, at each step, the coupling coefficients can be found simply by computing the two off-diagonal entries of the appropriate coupling matrix.

For problems that naturally decompose into larger numbers of subproblems, the generalization of this approach can be easily seen.

4.6 Operation Counts

The major computation in the stochastic complementation algorithm is in forming the stochastic complements. Often, the stationary distribution vector of each of the stochastic complements can be found with only a few iterations of the power method. The cost of these iterations will not be included in the following analysis.

Assume that from an $n \times n$ transition matrix P, k stochastic complements are to be formed. Assume for simplicity that k divides n and that each stochastic complement is of order n/k. Forming the stochastic complement S_{ii} can be broken down into three major steps:

1. Find the LU factorization of $I - P_i$. This matrix is of order $(k-1)n/k$ and thus, the LU factorization requires $(2/3)(k-1)^3n^3/k^3 + O(n^2)$ flops.
2. Use forward and backward substitution to solve $(I - P_i)X = P_{*i}$ for X. Doing forward and backward substitution to find one column of X

requires $2(k-1)^2n^2/k^2$ flops. The matrix X has n/k columns. Thus, the total number of flops required is $2(k-1)^2n^3/k^3$.

3. Form $P_{ii} + P_{i*}X$. The matrix multiplication is accomplished in $2(k-1)n^3/k^3$ flops. The matrix addition is $O(n^2)$.

The total number of flops required for forming k stochastic complements (neglecting lower-order terms) is therefore found to be

$$k\left(\frac{2(k-1)^3}{3k^3} + \frac{2(k-1)^2}{k^3} + \frac{2(k-1)}{k^3}\right)n^3 = \frac{2k^3-2}{3k^2}n^3 \tag{19}$$

After each stochastic complement is formed, the stationary distribution vectors must be computed. Experience has shown that this can be accomplished with a few iterations of the power method and can be assumed to be $O(n^2)$. Once the stationary distribution vectors have been found, the coupling matrix C must be computed. The formation of the entries of C is easily seen to be $O(n^2)$ also. Finally, the stationary distribution vector of C must be found, but since C is of order k, and since $k \ll n$ in the cases under consideration, this computation is negligible. Finally, the stationary distribution vectors must be scaled and combined to form the stationary distribution vector of the original Markov chain. This is an $O(n)$ operation. Thus, the previously computed flop count can be expected to give a reasonable estimate for the number of flops required by each algorithm. The number of flops required by $k \times k$ block SC algorithms for the cases $k = 2$, 3, and 4, excluding the lower-order terms, are:

$k = 2$ $\quad (7/6)n^3$

$k = 3$ $\quad (52/27)n^3$

$k = 4$ $\quad (21/8)n^3$

The SSC algorithm consists of the SC2 algorithm applied to the matrix P, followed by the SC2 algorithm applied to two matrices of order $n/2$. Thus the flop count for the SSC algorithm (again neglecting lower-order terms) can be computed as

$$(7/6)n^3 + 2(7/6)(n/2)^3 = (35/24)n^3 \tag{20}$$

All of the above algorithms were tested along with the LU and QR algorithms on a simple test problem of size 60×60. The tests were run on a single processor of the Sequent Balance 21000. The timings indicate that these estimates for the number of flops seem to predict rather well the relative performance of each of these algorithms, as shown in Table 1.

Depending on how many iterations of the power method are needed, these algorithms can require a lot more work in order to find the stationary distri-

Table 1 Relative Performance of SC Algorithms on a Single Processor

Algorithm	Number of flops	Flop rate relative to LU	Time (s)	Ratio of algorithm time to LU time
LU	$(2/3)n^3$	1	2.60	1
SC2	$(7/6)n^3$	1.75	5.45	2.1
QR	$(4/3)n^3$	2	5.70	2.2
SSC	$(35/24)n^3$	2.1875	6.83	2.6
SC3	$(52/27)n^3$	2.8889	8.57	3.3
SC4	$(21/8)n^3$	3.9375	11.45	4.4

bution vectors than either the LU or QR factorization methods. Since these $O(n^2)$ operations were neglected, the theoretical flop counts are lower than what would actually be required in practice.

Suppose the SSC algorithm is extended. For example, after forming two stochastic complements of order $n/2$ and dividing them into four stochastic complements of order $n/4$, these could be further divided into eight complements each of order $n/8$. At each stage, subdividing k stochastic complements each of order n/k requires

$$k\,\frac{7n^3}{6k^3} = \frac{7n^3}{6k^2} = \frac{7n^3}{6(4^p)} \tag{21}$$

flops, assuming that $k = 2^p$ for some positive integer p.

An upper bound for the number of flops required by the extended SSC algorithm is given by

$$\frac{7n^3}{6}\sum_{p=0}^{\infty}\frac{1}{4^p} = \frac{14}{9}n^3 \tag{22}$$

Comparing this with equation (19), it is clear that the SSC algorithm will require fewer flops than any $k \times k$ block SC algorithm with $k > 2$.

4.7 Comparison to a Simple Aggregation Technique

The stochastic complementation algorithm can be classified as an exact aggregation technique [11]. That is to say, any error in computing the stationary distribution vector is due to machine roundoff error and is not a function of the size of the off-diagonal blocks. There are cases where inexact aggregation techniques can introduce large errors, even when the off-diagonal blocks are relatively small. For instance, consider the following example:

Define

$$P = \begin{bmatrix} 1-3\epsilon & 2\epsilon & \epsilon & 0 \\ \epsilon & 1-\epsilon & 0 & 0 \\ 0 & 0 & 1-2\epsilon & 2\epsilon \\ 0 & \epsilon & \epsilon & 1-2\epsilon \end{bmatrix} \tag{23}$$

The SC2 algorithm computes S_{11} and S_{22} as follows:

$$S_{11} = \begin{bmatrix} 1-3\epsilon & 3\epsilon \\ \epsilon & 1-\epsilon \end{bmatrix} \tag{24}$$

$$S_{22} = \begin{bmatrix} 1-2\epsilon & 2\epsilon \\ 2\epsilon & 1-2\epsilon \end{bmatrix} \tag{25}$$

The corresponding stationary distribution vectors are $[1/4 \quad 3/4]$ and $[1/2 \quad 1/2]$.

The coupling matrix C is computed as

$$C = \begin{bmatrix} 1-\epsilon/4 & \epsilon/4 \\ \epsilon/2 & 1-\epsilon/2 \end{bmatrix} \tag{26}$$

which has stationary distribution vector $[2/3 \quad 1/3]$. Thus,

$$\pi = [(2/3)[1/4\ 3/4] \quad (1/3)[1/2\ 1/2]] = [1/6 \quad 1/2 \quad 1/6 \quad 1/6] \tag{27}$$

It is easily verified that this is the exact stationary distribution vector of P, independent of ϵ.

A simple aggregation technique can be defined as follows:

1. Compute $w_1 = P_{12}e$ and $w_2 = P_{21}e$.
2. Form $P'_{11} = P_{11} + \text{diag}(w_1)$ and $P'_{22} = P_{22} + \text{diag}(w_2)$.
3. Find stationary distribution vectors for P'_{11} and P'_{22}.
4. Recouple the vectors found in step 3 to form an estimate for π.

Thus, in each row of the diagonal blocks, the diagonal element is perturbed so that the row sum equals 1. For many cases where P_{11} and P_{22} are nearly stochastic anyway, this method works very well. Once the vectors v_1 and v_2 are found, the coupling matrix is computed in the same fashion as in the SC algorithm. For the problem under consideration, this algorithm produces the following computations:

$$P'_{11} = \begin{bmatrix} 1-2\epsilon & 2\epsilon \\ \epsilon & 1-\epsilon \end{bmatrix} \tag{28}$$

$$P'_{22} = \begin{bmatrix} 1-2\epsilon & 2\epsilon \\ \epsilon & 1-\epsilon \end{bmatrix} \tag{29}$$

The stationary distribution vectors are found to be [1/3 2/3] and [1/3 2/3].

The coupling matrix is found to be

$$C = \begin{bmatrix} 1-\epsilon/3 & \epsilon/3 \\ 2\epsilon/3 & 1-2\epsilon/3 \end{bmatrix} \tag{30}$$

with stationary distribution vector [2/3 1/3].

This produces for P a stationary distribution vector

$$[\,(2/3)[1/3\ 2/3] \quad (1/3)[1/3\ 2/3]\,] = [\,2/9 \quad 4/9 \quad 1/9 \quad 2/9\,] \tag{31}$$

Again, these computations are independent of ϵ.

The point of this example is to show that even if a problem is very loosely coupled, aggregation techniques might be susceptible to errors not due to machine roundoff. This is a difficulty avoided by the stochastic complementation algorithm. A comparison of the stochastic complementation algorithm to more sophisticated aggregation/disaggregation techniques could be the subject of another investigation.

5. EXPERIMENTAL RESULTS

5.1 Preliminaries

The algorithms to be tested on the Balance were coded in standard FORTRAN. The timings reported below were obtained using a timing routine supplied by Argonne National Laboratory. This routine had a resolution of 1/60 of a second. The times reported are averages of all runs made for each algorithm on a particular problem.

In addition to the CPU time required by each algorithm, two other quantities are reported: speedup and efficiency. Speedup is defined to be the ratio of the CPU time required by an algorithm to execute on a single processor and the time required to execute the same problem on more than one processor. Efficiency is the speedup divided by the number of processors.

5.2 Problem 1

This problem was a 60×60 matrix that was almost completely reducible into two 30×30 blocks. The matrix entries were random numbers between 0 and 1. The diagonal entries were scaled by a factor of 200. Finally, each row of the matrix was rescaled to make the matrix stochastic. Three algorithms were tested: LU, SC2, and QR. The timings, speedups and efficiencies are reported in Table 2. As expected, the LU algorithm was the fastest serial method. It, however, did not parallelize nearly as well as either the SC2 algorithm or even the QR algorithm, so the differences in the timings decreased as the number of processors increased. If the off-diagonal blocks are very small in norm relative to the diagonal blocks, the LU factorization might fail because the matrix it factors will be nearly rank-deficient. The following problem gives an example of this.

5.3 Problem 2

This test problem consisted of a 50×50 matrix that was almost completely reducible into two blocks, each of size 25×25. Thus, the block structure of the matrix was

$$P = \begin{bmatrix} P_{11} & P_{12} \\ P_{21} & P_{22} \end{bmatrix} \tag{32}$$

where $\|P_{12}\|_\infty = \|P_{21}\|_\infty = \epsilon$.

This problem was run several times, with ϵ varying between 10^{-6} and 10^{-15}. The matrix P was chosen to be doubly stochastic so that the computed solutions could be compared to e/n, the exact solution. None of the algorithms made use of this special structure.

Table 2 Test Results for Problem 1

	Time (s)			Speedup			Efficiency		
nprocs	*LU*	SC2	*QR*	*LU*	SC2	*QR*	*LU*	SC2	*QR*
1	2.57	5.30	5.67	1	1	1	1	1	1
4	1.10	1.63	1.80	2.3	3.3	3.2	0.58	0.81	0.79
8	0.67	0.97	1.08	3.8	5.5	5.3	0.48	0.68	0.66
12	0.55	0.77	0.90	4.7	6.9	6.3	0.39	0.57	0.52
16	0.50	0.62	0.83	5.1	8.5	6.8	0.32	0.53	0.43

The LU algorithm is unsuitable for this problem when $\epsilon \leq 10^{-10}$. The matrix A defined in equation (4) is ordinarily full rank. In this case, however, the second-largest eigenvalue of the matrix P is close to one, so that $P^T - I$ is close to a rank $n-2$ matrix, and A is nearly rank-deficient. Test results verify that the LU algorithm indeed fails to compute the stationary distribution vector with any degree of precision.

Experiments also showed that the QR algorithm also fails in that too many significant digits are lost. The QR algorithm operates on the matrix $I - P$, and does not incorporate the normalization condition into the matrix. Thus, the matrix processed by this algorithm is almost rank $n - 2$.

The SC2 algorithm avoids the problems that plagued the methods just discussed. When the stochastic complements S_{11} and S_{22} are formed, the spectrum of the matrix P splits into the spectra of the two complements [9]. That is to say, if P were completely reducible ($P_{12} = P_{21} = 0$), then the spectrum of P would simply be the union of the spectra of the diagonal blocks. In the case where P is loosely coupled, the eigenvalues of the diagonal blocks (and hence the eigenvalues of the stochastic complements) are each close to an eigenvalue of P. Thus, S_{11} and S_{22} each have one eigenvalue equal to 1, but no others near 1.

Recall that in the formation of the stochastic complements, the LU factorization is used on the matrices $I - P_{11}$ and $I - P_{22}$. Since P_{11} and P_{22} are nearly stochastic, the matrices factored with the LU algorithm have eigenvalues close to zero. One might expect the LU factorization to produce some inaccurate computations in this case, and this in fact might occur. However, since $\|P_{21}\|$ is small, and since the results of this computation are multiplied by P_{12}, whose norm is also very small, the update made to P_{11} to produce S_{11} is very small, and the effect of any inaccuracies in the computations should be minimal. The experimental results seem to support this claim because the stochastic complementation algorithm computed the stationary distribution vector correctly to machine precision, which is about 1.1×10^{-16} on the Sequent Balance.

For this problem, the stochastic complements also turn out to be doubly stochastic. When using the power method to find the stationary distribution vectors of the complements, an initial guess of e^T/n is often used. To insure that this problem did not give an unfair advantage to the SC2 algorithm, the problem was run using a different initial guess for the stationary distribution vector of each stochastic complement. The algorithm still computed the correct stationary distribution vector. It merely required additional iterations.

The experimental results for this problem are given in Table 3. The SC2 algorithm in all cases produced the exact answer to machine precision. The

Table 3 Relative Error in Computed Solution to Problem 2

ϵ	LU	QR	$SC2$
10^{-10}	2.3×10^{-7}	3.2×10^{-7}	0
10^{-15}	1.4×10^{-3}	6.0×10^{-3}	0

relative error was computed as the ∞-norm of the error vector divided by the ∞-norm of the answer.

5.4 Problem 3

This problem was a 60×60 matrix that was almost decomposable into three 20×20 blocks. The matrix was completely dense, but had no other special structure. In particular, the matrix was not doubly stochastic. The norm of the off-diagonal blocks was on the order of 10^{-2}. Table 4 gives the results of using the QR algorithm and the SC3 algorithm on this problem. Recall that the SC2 algorithm was faster than the QR algorithm. As expected, however, the additional overhead incurred by the SC3 algorithm caused it to be slower than QR.

5.5 Problem 4

This problem was again a 60×60 matrix. Three algorithms were tested on this problem: the QR algorithm, the SSC algorithm, and the SC4 algo-

Table 4 Test Results for Problem 3

	time (s)		speedup		efficiency	
nprocs	QR	$SC3$	QR	$SC3$	QR	$SC3$
1	5.68	8.62	1	1	1	1
4	1.82	2.67	3.1	3.2	0.78	0.81
8	1.12	1.70	5.1	5.1	0.63	0.63
12	0.88	1.28	6.5	6.7	0.53	0.56
16	0.80	1.22	7.1	7.1	0.44	0.44

Table 5 Test Results for Problem 4

	time (s)			speedup			efficiency		
nprocs	*QR*	*SSC*	*SC*4	*QR*	*SSC*	*SC*4	*QR*	*SSC*	*SC*4
1	5.70	6.83	11.45	1	1	1	1	1	1
4	1.82	2.30	3.93	3.1	3.0	2.9	0.78	0.74	0.73
8	1.10	1.40	2.28	5.2	4.9	5.0	0.65	0.61	0.63
12	0.92	1.12	2.07	6.1	6.1	5.5	0.52	0.51	0.46
16	0.83	0.85	1.48	8.0	8.0	7.7	0.43	0.50	0.48

rithm. As expected, the SC4 algorithm performed very poorly. The results are summarized in Table 5.

6. CONCLUSIONS

Stochastic complementation appears to be a viable technique for uncoupling loosely coupled Markov chains. The best implementations of stochastic complementation studied thus far are the SC2 algorithm for uncoupling two loosely coupled blocks, and the SSC algorithm, which is a generalization of the SC2 algorithm, for uncoupling several loosely coupled blocks. Moderate speedups on a shared-memory multiprocessor have been obtained. The SC2 algorithm runs faster than the *QR* algorithm, but not as fast as the *LU* algorithm. However, for loosely coupled chains, the solutions computed by the standard *LU* or *QR* techniques were significantly less accurate than the solution computed by the SC2 algorithm. Future investigations could examine the issue of accuracy in greater detail.

REFERENCES

[1] *Balance 8000 Guide to Parallel Programming*, Sequent Computer Systems, Inc., 1985.

[2] Berman, A., and Plemmons, R. J., *Nonnegative Matrices in the Mathematical Sciences*, Academic Press, New York, 1979.

[3] Cao, W. L., and Stewart, W. J., Iterative aggregation/disaggregation techniques for nearly uncoupled Markov chains, *J. ACM* 32 (1985), pp. 702–719.

[4] Dongarra, J. J., Moler, C. B., Bunch, J. R., and Stewart, G. W., *LINPACK Users' Guide*, SIAM, Philadelphia, 1979.

[5] Golub, G. H., and Meyer, C. D., Using the QR factorization and group inversion to compute, differentiate, and estimate the sensitivity of stationary probabilities for Markov chains, *SIAM J. Alg. Disc. Methods* 7 (1984), pp. 273–281.

[6] Golub, G. H., and Van Loan, C. F., *Matrix Computations*, Johns Hopkins University Press, Baltimore, 1983.

[7] Harrod, W. J., and Plemmons, R. J., Comparison of some direct methods for computing stationary distributions of Markov chains, *SIAM J. Sci. Stat. Comput.* 5 (1984), pp. 453–469.

[8] Kemeny, J. G., and Snell, J. L., *Finite Markov Chains*, Van Nostrand, Princeton, N.J., 1960.

[9] Meyer, C. D., Stochastic complementation, uncoupling Markov chains and the theory of nearly reducible systems, *SIAM Rev.* 31 (1989), pp. 240–272.

[10] Simon, H. A., and Ando, A., Aggregation of variables in dynamic systems, *Econometrica* 29 (1961), pp. 111–138.

[11] Zarling, R. L., Numerical solutions of nearly completely decomposable queueing networks, Ph. D. dissertation, University of North Carolina (1976).

27

Steady-State Behavior of Interacting Queues: A Numerical Approach

A. NAKASIS National Institute for Standards and Technology, Gaithersburg, Maryland

ANTHONY EPHREMIDES Electrical Engineering Department, University of Maryland, College Park, Maryland

ABSTRACT

In this paper, we consider the case of M queues that interact according to the model of multiple access transmissions over a collision channel; that is, each queue receives messages that it attempts to transmit in a classical slotted ALOHA fashion.

The steady-state behavior of such systems is unknown except for certain relatively simple cases. Here, we propose a numerical approach that permits the calculation, within any desired accuracy, of the joint queue size distribution as well as of the moments. Our approach uses a technique that involves auxiliary systems of queues that dominate, in a well-defined sense, the given ones. Consideration of such dominating systems permits sometimes the study of the ergodic region of the original systems. The methodology developed here is applicable to contexts more general than that of multiple access transmissions.

1. INTRODUCTION

Consider M discrete-time queuing systems that interact in the following way: Each system accepts messages that arrive according to a Bernoulli process. Let λ_i be the probability that a message arrives at queue i, $i = 1, 2, \ldots, M$, at any given time slot. The length of the message may be considered fixed, and the message arrivals at different queues are statistically independent. All queues are assumed to have infinite capacity (this assumption is not strictly necessary, but it corresponds to the most interesting case in our approach). The ith queue server attempts to transmit the head-of-the-line message in a given slot with constant probability q_i, and the transmission attempts by different queue servers are statistically independent. Service is completed when a message is successfully transmitted. If two or more servers attempt transmission in the same time slot, then none are successful. Therefore, the probability of successful transmissions by the ith user in a given time slot is equal to

$$q_i \prod_j (1 - q_j)$$

with j ranging over all nonempty queues, other than i, in that time slot. Thus, we see that the queues are coupled and interact.

This problem has been considered by several researchers in recent years. Fayolle and Iasnogorodski [1] considered two symmetric queues in continuous time in the context of coupled processors. Ephremides and Saadawi [2] have considered an approximate model for the case of M symmetric queues. Sidi and Segall [3,4] have considered the case of two queues (nonsymmetric) and obtained precise solutions in certain special cases, while proposing another approximate model for other cases. Szpankowski [5,6] has looked at the determination of the ergodic (stable) region of such systems and has obtained upper and lower bounds. He has also proposed approximate models for the steady-state analysis. There have been several other approximate techniques that have been proposed [7–11].

Interestingly enough, information theorists have looked at the same problem from another point of view. Their concern has been the determination of the maximum stable throughput (capacity) of the collision model just described. Massey and Mathys [12] have obtained the capacity for the symmetric, no-feedback case of M users. Tsybakov et al. [13] have obtained the capacity region for the asymmetric case ($M = 2$). Hui [14] has obtained the capacity region for the asymmetric, no-feedback case (arbitrary M). In [15], this region is derived for $M = 2$ in a new and much simpler way and shown to be identical to that obtained by Tsybakov and Hui without using the complicated mathematical tools from [16] that were needed in [13]. Also, bounds for the ergodicity region are obtained for $M > 2$ and the whole question of ergodicity is coupled to the idea of dominance as used here.

Aside from the question of determining the ergodic region, however, it is of interest to calculate the queue-size distribution (as well as its moments). In this paper, we show how, by defining a suitable "dominant" system of queues, we can perform this calculation to any desired level of accuracy. We use a numerical approach that is sufficiently general to apply to some other systems that can be described by similar Markov chains. Of course there are numerous studies of general methods for solving equilibrium equations and for establishing sensitivity of solutions [17–20]. Our approach is somewhat problem specific (ALOHA system framework or similar modes of queue interactions) and uses a novel idea of dominance.

2. DESCRIPTION OF THE NUMERICAL APPROACH

The purpose of this section is to outline a method that will allow us to transform the problem of approximating the steady-state vector of an infinite Markovian matrix to the problem of approximating the steady-state vector of a finite Markovian matrix. We will first try to appeal to the intuition of the reader, and we will then supply the mathematical details.

Assume that S is a partially ordered countable set and that $X(t)$ and $Y(t)$ are irreducible aperiodic Markov chains over S, such that all realizations satisfy the following relationship:

$$\text{if } X(t) \leq Y(t), \text{ then } X(t') \leq Y(t') \text{ for all } t' \geq t, w.p.1$$

In this instance we shall say that $X(t)$ is dominated by $Y(t)$. In the event that $X(t)$ and $Y(t)$ are identical–that is, if they have the same transition behavior but differ only in their initial conditions–we shall say that $X(t)$ is self dominated.

In what follows we shall assume that $Y(t)$ dominates $X(t)$ and that $X(t)$ is self dominated. The Markov chain $X(t)$ will correspond to a system A with transition probability matrix $P(A)$ and $Y(t)$ will correspond to a B with transition matrix $P(B)$. System A corresponds to the discrete-time ALOHA system of M users described in the introduction, while system B corresponds to a copy of A in which a user with an empty queue continues to attempt transmission of a "dummy" packet with the same probability as if his queue were not empty (see [15] for details).

Therefore, each state can be seen as a vector of M entries each of which is a nonnegative integer representing the queue size of each user. We can easily prove that the Markov process that corresponds to a system of these

M interacting queues is self dominated, by considering a system of $M + M$ queues such that:

1. The arrival streams at queues i and $i + M$ are identical, $i = 1, 2, \ldots, M$, and
2. If both queues i and $i + M$ are not empty, the ith server will attempt transmission if and only if the $(i + M)$th will $(i = 1, 2, \ldots, M)$.

Thus, the group of the first M queues corresponds to $X(t)$ and the group of the next M queues corresponds to another version of $X(t)$ with a possibly different initial condition.

Let now a set of states, U, be called *well-ordered* if and only if for every Y in U and for every $X, X \leq Y$ implies that X is in U (in other words, if a state Y is in U, then every state that can be reached from Y by one or more departures is also in U).

Let, then U be a well-ordered set, and let U^c be its complement. Let A and B be self-dominated chains such that B dominates A, and let $p_n(A)$ and $p_n(b)$ be defined as follows:

$$p_0(A) = p_0(B) = (1, 0, 0, \ldots)$$

and

$$p_n(A) = p_{n-1}(A)P(A)$$

and

$$p_n(B) = p_{n-1}(B)P(B) \qquad n = 1, 2, 3, 4, \ldots$$

[In other words, $p_n(A)$ and $p_n(B)$ are the n-step distributions of the two chains if the corresponding systems start with their queues empty at time 0.]

Because A is dominated by B, we have that for every n, $n = 0, 1, 2, 3, \ldots,$

$$p_n(A, U) \geq p_n(B, U)$$

where $p_n(\cdot, U)$ denotes the mass of $p_n(\cdot)$ that is concentrated on U, and that

$$p(A, U) \geq p(B, U) \tag{1}$$

That is, the n-step distribution for A has more probability mass over U than the n-step distribution for B and, conversely, over U^c, which is U's complement. Similarly, in the steady state, the probability that chain A is in U exceeds the probability that chain B is in U while the converse is true over U^c.

Because A dominates itself and because the null state (all queues empty) is less than every other state in our state space, for every n, $n = 1, 2, 3, 4, \ldots,$ we have

$$p_n(A, U) \geq p(A, U)$$

Similarly,

$$p_n(B,U) \geq p(B,U)$$

if B is self dominating.

Suppose now that system B is so constructed that it is easier to analyze and that we can find a *finite* well-ordered set U whose complement U^c satisfies

$$p(B,U^c) < h$$

for some $h > 0$, arbitrarily small. Then we know that $p(A,U^c)$ is also less than h and that if we succeed in approximating $p(A)$ over U, and if h is small enough, then we will have approximated $p(A)$ over the entire state space. One way of achieving this is to find a value of n such that the U-rows of the nth power of $P(A)$ are very similar to each other. Indeed, we know that the row entries of $P^n(A)$ represent the n-step transition probabilities and that for every U and n sufficiently large, the U-rows of $P^n(A)$ will be nearly identical. Since $p(A)$ itself is the weighted average of the rows $P^n(A)$ and since the weights outside U sum to less than h, and U-row of $P_n(A)$ will be a good approximation for $p(A)$. Of course, the problem is not trivial because in order to find the entries $P^n(A)$, one needs to perform many computations and store intermediate results over sets that are much bigger than U. For example, in order to approximate the steady-state vector of a two-queue system within 0.01,* one might need to take $U = \{0,1,2,\ldots,27\} \times \{0,1,2,3,\ldots,27\}$ and n of the order of 1500. In 1500 steps, one may visit, starting from a state in U, any state (i,j) with $0 \leq i \leq 1527$ and $0 \leq j \leq 1527$. Thus, an exact computation of the U-rows of $P^{1500}(A)$ may involve computations over very big sets (2.25 million states). Thus, one should try to approximate rather than compute the entries of the U-rows of $P^n(A)$.

Another potential problem area is that the region of stability (ergodicity) of the dominant chain B may be smaller than the region of A. If the parameters of the problem are such that A is ergodic but B is not, then the above approach will be useless. This problem has been addressed [15] where, in fact, interesting stability results follow from that consideration. These, however, are not addressed here. In a future paper we shall present further extensions of these ideas by which the solution of a system of $(M-1)$ queues can be used to study the ergodicity of a system of M queues.

We will now provide the mathematical details for the statements and claims made in this section in order to supply the necessary base of rigor and accuracy. Let A and B be self dominating irreducible aperiodic countable chains such that B dominates A and let $p(A)/p(B)$, $p_n(A)/p_n(B)$, and $P(A)A/P(B)$

*In the usual L_1 norm sense.

stand for their steady-state distributions, their n-step distributions, and their transition matrices respectively.

In what follows, we use the dominance concept to derive inequalities that allow us to approximate $p(A)$ by the steady-state vector of a related finite chain. Similarly, we can approximate the moments of $p(A)$ by the moments of the steady-state vector of such a finite chain. These results will follow easily from general dominance concepts as developed in [21]. For example, if g is a monotonic function that maps S onto the real numbers, then we have

1. If $g(x)$ is nondecreasing, then the expected value of $g(x)$ under $p(A)$ is less than or equal to the expected value of $g(x)$ under $p(B)$.
2. If $g(x)$ is nonincreasing, then the expected value of $g(x)$ under $p(A)$ is greater than or equal to the expected value of $g(x)$ under $p(B)$.

Also, if g is a nondecreasing function, and U is a well-ordered set (with U^c its complement), then if the total contribution of all states in U^c to the computation of $E(g(x) \mid B)$ is h or less, so is that contribution to the computation of $E(g(x) \mid A)$.

It is also quite clear that if U is a well-ordered set, then

$$p(A,U) \geq p(B,U)$$

Finally, assume that S has a minimum, called the null state and denoted by 0, that $p_0(A)$ is a probability vector that assigns probability one to the null state, and that $p_n(A)$ is the nth iterate of $p_0(A)$. Then for each well-ordered set U we have

$$p_n(A,U) \geq p(A,U)$$

since $p(A)$ is the weighted average of all possible nth iterates. However, owing to self dominance, the nth iterate that starts at a state $s \neq 0$ assigns less probability to U and more probability to U^c than the nth iterate that starts at the null state.

For some of the results of the following section we shall need the following definitions: A *probability transfer* is any vector over S whose L_1 norm is finite and whose entries sum to 0. A probability transfer is called *simple* if and only if all the entries are 0 except for two. These two entries correspond by convention to states s_a and s_b, and it is assumed that the s_b entry is positive. A simple probability transfer will be called *upward* (downward) if and only if $s_a < s_b (s_b < s_a)$. A sum of simple probability transfers is called *canonical* if and only if the L_1 norm of the sum is finite and equal to the sum of the L_1 norms of the summands. (This very simply means that if the s_x entry of one summand is positive (negative) all other s_x entries are positive (negative) or zero). A probability transfer vector is called *upward* (downward) if it can be expressed as a canonical sum of simple upward (downward) probability transfers.

It is clear that if δ is an upward (downward) probability transfer, then so is $\delta P(A)$, provided A is self-dominating. Also, if D and E are operators on probability vectors $p \epsilon S$ such that for all p, $pD - p$ is a downward probability transfer and $pE - p$ is an upward probability transfer, then for all n, $n > 0$, $p(P(A)D)^n$ is a downward probability transfer and $p(P(A)E)^n - pP(A)^n$ is an upward probability transfer. As a result, if U is any well-ordered set, the following holds:

$$p((P(A)D)^n, U) \geq p((P(A))^n, U) \geq p((P(A)E)^n, U)$$

In the following section, we will show how a simple numerical calculation can approximate a finite chain's steady-state vector (and by extension the expected value of any function f), and then we will couple the calculation to the above-described approach so as to handle the infinite case.

3. FINITE MARKOV CHAIN CASE

If the chain A is finite, in principle we can solve the balance equations numerically. Sometimes, relaxation is numerically more attractive than solution of equations. Thus, even in the finite case, we may be interested in approximating the steady-state solution $p(A)$ by $p_n(A)$ (for a suitably large value of n). Of course, we know that many birth–death chains are geometrically ergodic (in fact all finite ergodic chains are geometrically ergodic), and the problem then becomes to determine the coefficient of geometric ergodicity in order to determine what value of n will achieve the desired accuracy. Instead of looking at the eigenvalue spectrum of the transition matrix $P(A)$ (an approach that is burdensome and, of course, unless A is infinite), we will use a different rate of convergence that is based on a much simpler computation (see also [22] on a related contraction).

The mathematical preliminaries of the preceding section imply quite straightforwardly the following:

1. If δ is a simple probability transfer with norm $2p$ and nonzero entries i (negative) and j (positive) then

   $$\delta P(A) = p\,(j\text{th row of} \quad P(A) - i\text{th row of } P(A))$$

2. If $P(A)$ is such that there is a positive f such that the L_1 norm of the difference of any two rows of $P(A)$ is $2(1-f)$ or less, then we have that if δ is a probability transfer, then

   $$|\delta P(A)| \leq |\delta|(1-f)$$

(It suffices to consider a canonical decomposition of δ into a sum of simple probability transfers.)

We notice that if A is a finite irreducible aperiodic chain, there is always a value m and a positive number f such that the L_1 norm of any difference of rows of $(P(A))^m$ is $2(1-f)$ or less. In this case, remark 2 above implies that for any p,

$$|(p-p(A))P(A)^{mn}| \leq |p-p(A)|(1-f)^n \leq 2(1-f)^n$$

Therefore, if we wish to approximate $p(A)$ within ϵ it is sufficient to choose an arbitrary p and multiply it mn times by $p(A)$ where

$$n > \ln(\epsilon/2)/\ln(1-f)$$

Clearly, this result assumes that each multiplication introduces no error. If arithmetic errors are possible, then we may assume that each time we multiply a vector by $(P(A))^m$, we make an error, which is a probability transfer with norm equal to some value r or less (since the computed and the theoretical result are probability vectors, their difference is a probability transfer). It ensues that mn multiplications later the cumulative norm of the errors will be at most:

$$r(1-f)^{n-1} + r(1-f)^{n-2} + \cdots + r < r/f$$

In this case, we can always approximate $p(A)$ within $r/f + \epsilon$, provided that n is chosen sufficiently large, that is, $n > \ln(\epsilon/2)/\ln(1-f)$.

We would like now to use this derivation for the case where the matrix A is infinite. There are two problems in that case:

1. There is no guarantee that we can find an m such that the L_1 norm of the difference of any two rows of $(P(A))^m$ is less than two, and
2. Even if such an m exited, there is no guarantee that the supremum of the L_1 norms of the row differences of $(P(A))^m$ is less than two; thus there may be no positive f such that all the L_1 norms are $2(1-f)$ or less.

Thus, we must modify the technique in order to apply it to the infinite case.

4. INFINITE MARKOV CHAIN

The purpose of this section is to show how the results of the previous section can be combined with the properties of dominated and self dominated Markov chains so that relaxation over a finite space can be used in the approximation of the steady-state vector of a Markov chain A such as the one described in the previous sections. Unless we otherwise specify, we shall assume for the moment that all operations can be carried out without any arithmetic error and that all norms are L_1 norms. Again we want to explore the capability of the application of successive relaxation iterations on the probability vector to approximate the steady-state probability distribution of the chain. In all cases our algorithm will consist of iterating the multiplication of an arbitrary initial probability vector with a truncated transition matrix. The object of our approach is to develop a methodology for performing this truncation and for controlling the errors that result from the truncation, the "spillage" of probability mass outside the truncated region as the iterations are performed, and the arithmetic errors. Each of the next three propositions contributes towards the control of the errors and the definition of our method. Proposition 1 is especially useful when the dominant system B is fully known. Proposition 2 defines an alternative when B is not fully known, and Proposition 3 refines Proposition 2 for the case of arithmetic errors.

PROPOSITION 1 Let V be a finite subset of S and let T be a transition matrix over V such that any two rows of T differ in norm by $2(1-f)$ at most, for some $f > 0$. Assume that there is a subset U of V such that $p(A, U) > 1 - h_1$, for some small $h_1 > 0$, and that for some integer n any U-row of R, where $R = (P(A))^n$, and the corresponding U-row of T differ by at most $2h_2$. Let t and r, where $r = p(A)$, be the steady-state vectors T and R, respectively, Then the norm of $t - r$ does not exceed $(4h_1 + 2h_2)/f$.

(Note that T can be thought of as a transition matrix over S with all states outside V absorbing and t can be considered as a vector over S with all entries outside V zero.)

Proof. We notice that

$$t - r = (t - r)T + r(T - R) \tag{2}$$

We observe that there is a canonical decomposition of $t - r$ such that all summands having nonzero entries outside V sum in norm to $2h_1$ or less. Therefore, if the norm of $t - r$ is x, the norm of $(t - r)T$ is at most $x(1-f) + 2h_1$. We also observe that $r(T - R)$ is the weighted average of the rows of $T - R$. The U-rows have norms of at most $2h_2$ and the weights outside U sum to h_1 or less. Therefore, the norm of this term will be $2h_2 + 2h_1$ or less. Thus, from equation (2) we obtain

$$x \leq (1 - f)x + 2h_1 + 2h_2 + 2h_1 \tag{3}$$

which in turn proves that $x \leq (4h_1 + 2h_2)/f$.

This proposition can be used to derive a basic relaxation algorithm as follows:

Given h_1, one can use the dominant chain B, which is presumably easy to handle analytically, in order to obtain a finite well-ordered set U such that $p(B, U) > 1 - h_1$.

Let U be of the form $U_a = \{s \mid s \leq s_a\}$ and let n be such that the probability that a B-random walk that starts at s_a will visit the null state in n steps or less exceeds f. Then the same will hold for an A-random walk that starts in any state s in U_a. It follows that any two U-vectors of R will differ in norm by at most $2(1 - f)$.

Once we have determined U and n, we can use h_2 in order to find V. Indeed, assume that a well-ordered set is found in such a way that

$$\text{Prob}\,(Y(n) \text{ is in } V^c \mid Y(0) = s_a) < h_2$$

where Y is a B-random walk. Dominance implies that the same will hold for an A-random walk and for any initial state s, provided that $s \leq s_a$. Thus the U-entries of R sum outside V to h_2 or less.

Finally, let T be a transition matrix over V such that (a) the U-rows of T and R are identical over V with the possible exception of the column for the null state, and (b) every row outside U is identical to some U-row. We can show then that T satisfies our earlier lemmas and that its steady-state vector t can be a satisfactory approximation of $p(a)$.

This procedure can be modified so that it may be used in the presence of errors of arithmetic. Indeed, we do not need to compute R; any "good" approximation of its U-rows would do. Thus, we could start with a target error ϵ and approximate in norm the U-rows of R within ϵ. In this case the U-rows of T would differ by $2(1 - f + \epsilon)$ and the U-rows of T and R would differ by $2h_2 + \epsilon$. This is a minor difficulty insofar as ϵ could be absorbed in the other constants.

A greater difficulty would be to program our software in such a way that the total error per row would be ϵ or less. However, practice has shown that this is still not the biggest problem. Even when we can compute without roundoff and truncation errors, an exact computation of the U-rows of R may involve sets that are much bigger than V. Practice has shown that a typical n may be of the order of a few hundred to a few thousand iterations. An exact computation of the U-entries over V is not feasible unless it is run over a set that has some $(n/2)^M$ states, where M is the number of queues. This is so because during n steps, a queue can register up to $n/2$ net arrivals and still return to its original state. Thus, if $n = 2000$ and $M = 2$, approximately 10^6 states must be considered in the computation of the U-rows. Clearly, such a computation is prohibitive and we must find a way of computing R's rows that involves small finite sets and allows us to keep control of the probability spillages.* Finally, the dominating system B may not be completely amenable to analytic study, and thus we may not be able to extract from B all the information that we need. Indeed, we may know $p(B)$ but not necessarily the n-step transition probabilities of B, and in this case it would be useful to develop an algorithm that would use B as little as possible. This leads us to the following proposition:

PROPOSITION 2 Let $U_a = \{s \mid s \leq s_a\}$ be such that $p(B, U_a) > 1-h_1$ and let $U_b = \{s \mid s \leq s_b\}$ and n be such that:

1. An n-step random walk starting at s_a will not cross outside U_b with probability $1 - h_2$, and
2. An n-step random walk starting at s_a will visit the null state with probability at least f.

Then, the nth iterate of any vector having its mass over U and $p(A)$ will differ in norm by at most $2(1-f+h_1)$. Furthermore, if we introduce a matrix L that agrees with $P(A)$ over each U_b row and redirects all transitions that end outside U_b to $s_b+(1, 1, 1, \ldots, 1)$, then the nth iterate according to L of a vector that has all its mass over U_a and $p(A)$ differ in norm by $2((1-f)+h_1+h_2)$ or less.

Proof. Under these assumptions the U-rows of R, where $R = (P(A))^n$, differ in norm by no more than $2(1-f)$, and $p(A)$ is the weighted sum of the rows of R with the weights outside U summing to h_1 or less. Therefore, any U_a row of R differs from $p(A)$ by at most $2((1-f)+h_1)$. This result will remain true for any combination of U_a-rows, provided the weights are positive and sum to 1; therefore, it will be true for the nth iterate of any vector with its mass over U_a.

If instead of multiplying by $P(A)$ we multiply by L, the disturbed paths will sum to h_2 or less (probability of going from s_a to U_b's complement in n steps or less). Since A is self dominated, the same result holds for every other U_a-row and the total error introduced by using L rather than A will be $2h_2$ or less; hence the limit of $2((1-f)+h_1+h_2)$.

The thrust of Proposition 2 is to introduce a simpler computation that can be performed over a small finite set. Indeed, from Proposition 2 we derive the following generic algorithm.

*By the term "spillage," we mean the phenomenon in which the iteration computation starts yielding nonzero values for elements of the n-step probability vector that lie outside the truncated state space set.

ALGORITHM 1

Step 1: Given a target error value ϵ, choose target values for h_1, h_2. and f in such a way that

$$h_1 + h_2 + (1 - f) \leq \epsilon/2$$

(It is assumed that ϵ_1, h_1, h_2, f, and $1 - f$ are all positive numbers.)

Step 2: Given h_1, use B to derive a state $s(a)$ such that if $U(a) = \{s \mid s \leq s(a)\}$, then $p(B, U(a)) > 1 - h_1$.

Step 3: Given $s(a)$ and f, use B's iterates or whatever analytical forms are available to derive a n such that an n-step random walk starting at $s(a)$ will visit the origin (all queues empty) with probability f or more.

Step 4: Given $s(a)$, n, and h_2, use B's iterates or whatever analytical means are available to derive a $s(b)$ such that an n-step random walk starting at $s(a)$ will stay within $U(b)$, $U(b) = \{s \mid s \leq s(b)\}$, with probability $1 - h_2$ or more.

Step 5: Let p be a probability vector with all its mass at $s(a)$ and let L be a probability matrix that agrees with $P(A)$ over $U(b)$ and treats all other states as absorbing. Compute pL^n. The result is within ϵ (L_1-norm) of $p(A)$.

The weakness of this algorithm is that as a rule we do not know enough about B to carry out steps 3 and 4, and that a great degree of *ad hoc* heuristics may be needed for the derivation of analytical results. A way to bypass steps 3 and 4 is to use the following modified algorithm.

ALGORITHM 2

Step 1: Given a target value ϵ, decompose ϵ into h_1, h_2, and $(1 - f)$ as in step 1 of Algorithm 1.

Step 2: Given h_2, find $s(a)$ as in step 2 above.

Step 3: Guess a value $s(b)$, $s(a) \leq s(b)$, and let $L = L(b)$ be defined as in step 5 above (agrees with $P(A)$ over $U(b)$ and treats all other states as absorbing).

Step 4: Let $p(0)$ and $q(0)$ be probability vectors with all of their mass over $s(a)$. Define

$$p(i + 1) = p(i)L,$$
$$r = (q(i, 0), 0, 0, 0, 0, \ldots)$$
$$q(i + 1) = (q(i) - r)L + r$$

and compute the iterates of p and q until any of the following two statements is true:

a. The mass of $p(n)$ outside $U(b)$ exceeds h_2, or
b. $q(n, 0$ exceeds f.

If (b) occurs while (a) is false, n has been computed and the choice of $s(b)$ is justified. Go to step 6.

Step 5: Guess a better (bigger) $s(b)$, define $U(b)$ and L accordingly, and return to step 4.

Step 6: Done; the L_1 norm of $p(n) - p(A)$ is ϵ or less.

We note the following:

1. The theory guarantees that step 4 will terminate, whatever the choice of $s(b)$, because eventually (a) will become and remain true; similarly, theory guarantees that if all entries of $s(b)$ are sufficiently large, (b) will come to be true, while (a) is still false.
2. The value of $q(m, 0)$ is a lower estimate of the probability that an m-step random walk starting at $s(a)$ will visit the origin, while $q(m, 0)$ plus the mass of $q(m)$ outside $U(b)$ is an upper estimate for that probability.
3. The results of step 4 may guide the guess of $s(b)$ in step 5.

While the above algorithm is not entirely satisfactory, it may be the best one can do. Even when analytical results are feasible, it may be better to use the second algorithm if there are reasons to believe that the analytical estimates of $s(b)$ are extremely conservative and large to the point of being worthless. Our experience so far and our numerical results indicate that, as a rule, it is quite easy to guess an adequate $s(b)$, provided that we are not too near the instability region. We note that both of these algorithms, as stated, rely on B for the derivation of $s(b)$ (step 2). This presupposes that both A and B are ergodic. If the case is that A is ergodic but B is not, these algorithms cannot be used unless there are other analytical means such that for each positive h we can derive an $s(a)$ that depends on h and satisfies $p(A, U)(a)) > 1 - h$ (i.e., B equals A).

The preceding algorithm assumed that the L-iterates of p and q can be computed without any arithmetic error. What if arithmetic errors are present? One way to proceed might be to perform the computations only with truncations and without roundoff. In this way we can obtain an upper estimate for the total error, 1.0-(mass of p). This upper estimate can be rolled into the computation of h_2, thereby treating all truncated probability as paths that crossed outside U_b. Clearly, our ability to keep the total error small will depend on the number of bits, K_1, used to represent the entries of L and of our vectors, the number of bits, K_2, available for intermediate computations, the number of states in U_b, and the number of iterations n. It is easy, albeit not very efficient, to create an algorithm capable of allocating any desired number of bits to our vectors and to obtain extended precision. The same technique may also be used for the entries of A, or, alternatively, the following lemma can be used.

LEMMA Let $X(t)$ be an A-random walk such that $X(0) = 0$ and let $t(n)$ be the greatest integer such that $X(t(n)) = 0$ and $t(n) \leq n$. Assume that there is a constant K such that $E(n - t(n)) < K$. Furthermore, assume that there is a self dominating matrix A^- that is dominated by W and is such that if A is an (A^-)-random walk, then

$$\text{Prob}\,(W(t+1) = X(t+1) \mid W(t) = X(t)) > 1 - u$$

In this case the steady-state vectors of A^- and A differ in L_1 norm by Ku or less.

Proof. Using the earlier lemmas, we observe that since A dominates A^-, we have $W(0) = X(0) = 0$ and $W(t) \leq X(t)$ for all t. Thus if $X(t(n)) = 0$, we also have $W(t(n)) = 0$. Let $q(j) = \text{Prob}(t(n) = j)$. If $t(n) = j$, it follows that

$$\text{Prob}(X(n) = W(n)) > (1-u)^{n-j}$$

and that

$$\text{Prob}(X(n) \neq W(n)) < 1 - (1-u)^{n-j} < (n-j)u$$

Therefore, $\text{Prob}(X(n) \neq W(n) \mid X(0) = W(0) = 0)$ is less than

$$\sum_{j=0}^{n} q(j)(n-j)u \leq Ku$$

But, this implies that the nth iterates of A and A^- differ by at most Ku and that, therefore, the same is true for the steady-state vectors.

This lemma shows that if such a K can be found, then we may approximate A's steady-state vector by considering an A^- such as the one in the lemma above. An A^- can be easily constructed if one remembers how the self dominance of A was proven. One way, not necessarily the best, of constructing A^- is to truncate all the probabilities of the possible events at each transition and to construct an additional event whose probability is equal to what we truncated in total from the original events. When this event takes place, we go from s to the infimum of the states that are directly reachable from s.

The thrust of all this discussion is to demonstrate that it is possible, if enough information can be gleaned from B_1, to approximate $p(A)$ through $p(A^-)$, where the entries of A^- are exactly known. But any scheme that simply sums the truncation errors cannot be very efficient because it underestimates the fact that errors tend to offset each other. Therefore, we shall digress a little to describe a variation of that scheme that tries to take advantage of the fact that errors may cancel each other out. This will lead to the third proposition, which can be used to modify our second method.

Assume that pL cannot be computed accurately but that there are upward and downward probability transfer operators E and D such that, for all p, pLE and pLD are exactly representable in the machine. Let N be a matrix equal to L in all its entries except for the null state that is absorbing for N and assume that for each p, pNE_1 and pND_1 are exactly representable with E_1 and D_1 upward and downward probability transfers, respectively. Let then $p(0)$ be a probability vector with all its mass on s_a and let $p(n), p^+(n), p^-(n)$, $q(n)$, $q^+(n)$, and $q^-(n)$ be defined as follows:

$$p^+(0) = p^-(0) = q^+(0) = q^-(0) = q(0) = p(0)$$

and for $n > 0$

$$p^+(n) = p^+(n-1)LE \quad p(n) = p(n-1)L \quad p^-(n) = p^-(n-1)LD$$
$$q^+(n) = q^+(n-1)NE_1 \quad q(n) = q(n-1)N \quad q^-(n) = q^-(n-1)ND_1$$

We then prove the following.

PROPOSITION 3 For each $n > 0$, we have

$$p^+(n, U_b^c) \geq p(n, U_b^c) \geq p^-(n, U_b^c)$$
$$q^+(n, 0) \leq q(n, 0) \leq q^-(n, 0)$$

where $q(n, 0)$ represents the probability that an n-step path will visit 0 before crossing to U_b^c.

Proof. These inequalities follow directly from our results on upward/downward transfers described in the earlier lemmas. On the other hand, $q(n, 0)$ clearly represents all paths of length n or less which started at s_a and visited 0 before visiting U_b^c.

This proposition allows us to test whether a choice of U_b and n is good for our purpose. [It suffices that $p^+(n, U_b^c) < h_2$ and that $f < q^+(n, 0)$.] If this is not the case, we may use the estimates provided by p^- and q^- to decide whether our computation failed because of inadequate accuracy or because of a bad U_b choice.

We remark that when "good" U_b and n values are available, then for each well-ordered set U, we have $p^+(n, U) \leq p(n, U) \leq p^-(n, U)$. Therefore, we can have point estimates for all states in U_a as follows:

Let $U(s)$ be the set of all states that are strictly less than s and let $U_s = U(s) \bigcup \{s\}$. Then $p^-(n, U_s) - p^+(n, U(s)) \geq p(n, s) \geq p^+(n, U_s) -$

$p^-(n, U(s))$. Furthermore, we know that

$$p^+(n) = p(n) + d_1$$

and

$$p^-(n) = p(n) - d_2$$

where d_1 and d_2 are upward probability transfers. Therefore, $p^+(n) - p^-(n) = d_1 + d_2$ is known and is an upward probability transfer. As a rule, knowing $d_1 + d_2$ does not allow us to deduce d_1 and d_2 individually and the norm of $d_1 + d_2$ does not necessarily bound the norms of d_1 and d_2. But if q is equal to p plus an upward probability transfer, and if g is a nondecreasing function over the state space, then the expected value of g under p, that is, $E(g,p)$, is no greater than the expected value of $g(x)$ under q, that is, $E(g,q)$. This follows from the earlier lemmas. If g is chosen so that $g(s)$ is the sum of the coordinates of state s (number of customers in the system), then one can show that the norm of $q - p$ is bounded by $2(E(g,q) - E(g,p)))$. Therefore, knowledge of p^+ and p^- does allow us to derive bounds on the norms $(p^+) - p$ and of $p - (p^-)$, but these bounds will most likely not be tight.

To conclude: It is possible to address the issue of arithmetic errors during our computations with varying degrees of success. It is possible to use truncation, in which case the truncated mass represents the total error, and it is possible to use probability transfers and obtain upper and lower bounds on the probability of every well-ordered set in our space. Furthermore, we can derive bounds on the norms of the differences of the steady-state vector and the computed vectors.

It remains to be seen how these improvements can be incorporated in a more general theory. We have already obtained some tentative results in this area, which we will not include here so as not to diverge from this paper's main subject.

5. CONCLUSION

We have described how we can perform the necessary truncation for an infinite Markov chain and keep the error under control with computationally feasible means. Thus, a simple modified relaxation method can yield approximate values for the steady-state probability vector of the chain that can be, in principle, within any desired degree of accuracy from the true value.

The method utilized the concept of a dominant chain and was motivated by the well-known problem of delay analysis of a discrete-time slotted ALOHA system of multiple access that involves a finite number of buffered terminals.

We tried to emphasize that the numerical approach to which we were led is applicable to contexts more general than that of the ALOHA system. In fact, the whole question of determining the ergodicity region for the ALOHA system is connected to the approach developed here and is studied in more depth in [15], where the same concept of dominance as here is considered.

REFERENCES

[1] G. Fayolle and R. Iasnogorodski, Two coupled processors: The reduction to a Riemann–Hilbert problem, *Wahrscheinlichkeits theorie*, pp. 1–27, 1979.

[2] T. Saadawi and A. Ephremides, Analysis, stability, and Optimization of Slotted Aloha, *IEEE Trans. AC*, June 1981.

[3] M. Sidi, A. Segall, Two interfering queues in packet-radio networks, *IEEE Trans. Commun.*, Vol. 31, No. 1, pp. 123–129, January 1983.

[4] A. Segall and M. Sidi, Priority queueing systems with applications to packet-radio networks, *Proc. International Seminar on Modeling and Pre-Evaluation*, pp. 159–177, Paris, 1983.

[5] W. Szpankowski, A multiqueue problem: Bounds and approximations, in *Performance of Computer-Communication Systems* (H. Rubin and W. Bux, editors), IFIP, 1984.

[6] W. Szpankowski, Performance evaluation of a protocol for multiaccess systems, *Proc. Performance '83*, pp. 337–395, College Park, Md., 1983.

[7] S. S. Kamal and S. A. Mahmond, A study of users' buffer variations in random access satellite channels, *IEEE Trans. Comm.*, Vol. 27, pp. 857–868, June 1979.

[8] K. K. Kittal and A. N. Venetsanopoulos, On the dynamic control of the urn scheme for multiple access broadcast communication systems, *IEEE Trans. Commun.*, Vol 29, No. 7, pp. 962–970, July 1981.

[9] H. Takagi and L. Kleinrock, Queueing delays in contention packet broad-systems, *IEEE Trans. Commun.*, to be published.

[10] H. Kobayashi, Application of the diffusion approximation to queueing networks I: ks Equilibrium queue distributions, *J. ACM*, Vol. 21, No. 2, pp. 316–328, April 1974.

[11] Y. Ozonato and S. Noguchi, On the thrashing cusp in slotted Aloha systems, *IEEE Trans. Commun.*, Vol. 11, Nov. 1985.

[12] J. Massey and P. Mathys, The capacity of the collision channel without feedback, *IEEE Trans. Info. Theory*, March 1985.

[13] B. Tsybakov and W. Mikhailov, Ergodicity of slotted ALOHA systems, *Problems Info. Transmission*, Vol. 15, No. 4, pp. 73–87, 1979.

[14] J. Y. Hui, Fundamental Issues of Multiple Accessing, Ph.D. Thesis, MIT, Cambridge, Mass., 1983.

[15] R. Rao and A. Ephremides, On the ergodicity region of the slotted-ALOHA system, *IEEE Trans. Info. Theory*, September 1988.

[16] W. Malyschev, A classification of two dimensional Markov chains, and Piece-linear Martingales, *Doklady Akademy Nauk SSSR*, 3, pp. 526–528, 1972 (in Russian).

[17] D. Freedman, *Approximating Countable Markov Chains*, Academic Press, New York, 1972.

[18] Keilson, *Markov Chain Models—Rarity and Exponentiality*, Springer-Verlag, New York, 1979.

[19] L. Kaufman, Matrix methods for queuing problems, *SIAM J. Sci. Stat. Comput.*, Vol. 4, pp. 525–552, 1984.

[20] C. D. Meyer, The condition of a finite Markov chain and perturbation bounds for the limiting probabilities, *SIAM J. Alg. Disc. Methods*, Vol. 1, pp. 273–283, 1980.

[21] D. Stoyan, *Comparison of Queues and Other Stochastic Models*, John Wiley, New York, 1983.

[22] Seneta, *Non-negative Matrices and Markov Chains*, Springer-Verlag, New York, 1981.

[23] S. Halfin, The shortest queue problem, *J. Appl. Probability*, Vol. 22, pp. 865–878, 1985.

[24] W. Feller, *An Introduction to Probability Theory*, John Wiley, New York, 1950.

[25] A. Ephremides, P. Varaiya, and J. Walrand, A Simple dynamic routing problem, *IEEE Trans. Automatic Control*, Vol. AC-25, No. 4, pp. 690, 693, August 1980.

28

The Most Likely Steady State for Large Numbers of Stochastic Traveling Salesmen

ROBERT GEIST and ROBERT REYNOLDS Department of Computer Science, Clemson University, Clemson, South Carolina

ABSTRACT

A class of continuous time, ergodic Markov processes having extremely large state space is considered, and a technique for numerical determination of high-probability components of the steady-state solution vector is proposed. The technique first identifies the steady-state solution as a maximum entropy *Gibbs measure*, and then maps the associated energy function onto a *Hopfield neural network*. A numerical iteration scheme based on an unresolved conjecture due to Hillam is used to complete the computation. An application of this technique to computer graphics is offered in the form of a new algorithm for digital halftone resolution.

1. INTRODUCTION

The *stochastic traveling salesmen problem* may be stated as follows: n cities are connected by a network of roadways, but there need not be a direct route between any two cities. (The graph is not necessarily complete.) Associated with each city i is a customer need, N_i. A collection of $m \leq n$ salesmen move from city to city in an attempt to satisfy these customer needs. At most one salesman may occupy any city at any one time. If there is not a salesman present in their city, customers will drive to neighboring cities, but their desire to do so may depend upon the distance between the cities, as well as the need in each city. As a result, the attractiveness of city i to a salesman is given by

$$A(i) = N_i + \sum_{k \in nhbr(i)} (1 - 2\omega_k) f(i,k)$$

where $nhbr(i)$ is the set of cities that neighbor city i, $f(i,k)$ is an arbitrary nonnegative symmetric function that expresses the additional need in city k that would be satisfied by a salesman in city i, and ω_k is binary with value 1 if there is a salesman in k, which reduces the attractiveness of locating in i, and value 0 if not, which increases the attractiveness. The rate at which a salesman moves from city i to an unoccupied neighbor city j then depends upon the difference in attractiveness, $A(j) - A(i)$. A salesman is obliged to average at least one day in any city visited, and so, given that his next city is j, a salesman arriving to i will spend an exponentially distributed amount of time with mean $1 + e^{-(A(j)-A(i))}$. Note that if $f \approx 0$ and $N_j \gg N_i$, then the mean stay in i will be near the minimum, 1 day. The problem is to determine that assignment of salesmen to cities that is most likely to occur in the "long run" or steady state.

This problem is conveniently represented by a stochastic Petri net [11] with inhibitor arcs or, equivalently, a continuous-time Markov process with $\binom{n}{m}$ states representing the possible assignments of salesmen to cities. Nonzero transition rates exist between any two states with the property that one can be obtained from the other by moving a single salesman from a city i to some neighbor city j. In this case the transition rate is $1/(1 + e^{-(A(j)-A(i))})$.

Thus it would seem that the solution is numerically straightforward, but the complicating factor is the size of the problem. We want to consider applications with as many as 2^{16} cities and as many as 2^{15} salesmen, that is, Markov processes with $\binom{2^{16}}{2^{15}} \approx 10^{20,000}$ states. We will show that high-probability components of the steady-state solution vector of such processes can be obtained numerically by working in the "log space" that results from mapping the problem onto computational structures known as *Hopfield neural networks* [5].

We will construct this mapping in two steps. First, in Section 2, we discuss potentials and energy functions and show how our problem has a solution that is naturally represented as a maximum entropy Gibbs measure. Second, in Section 3, we show how to map such Gibbs measures onto Hopfield neural networks, which, in turn, allow us to numerically identify the high-probability components of the steady-state solution vector. In Section 4 we provide an application to computer graphics, specifically, a new algorithm for digital halftone resolution. Section 5 contains conclusions and current directions.

2. A GIBBS MEASURE REPRESENTATION

Let S be the set of cities, and let $T = \{0, 1, \ldots k-1\}$, the first k nonnegative integers, which we plan to use as labels on the elements of S. If we represent the power set of S by the cartesian product, $\{0, 1\}^{|S|}$, then a *potential* on S is a map

$$U : \{0, 1\}^{|S|} \times T^{|S|} \longrightarrow R$$

such that

1. $U(\emptyset, \omega) = 0$
2. $U(A, \omega) = U(A, \omega')$ whenever $A \otimes \omega = A \otimes \omega'$

where, if $A = (a_1, a_2, \ldots, a_n)$ and $\omega = (\omega_1, \omega_2, \ldots, \omega_n)$, then $A \otimes \omega = (a_1\omega_1, a_2\omega_2, \ldots, a_n\omega_n)$. The *energy* of the potential is then given by

$$U(\omega) = \sum_A U(A, \omega)$$

and the *Gibbs measure induced by U* [8] is the probability measure induced on the collection of possible labelings, $T^{|S|}$, given by

$$P(\omega) = \frac{e^{-U(\omega)}}{N}$$

where N is a normalizing constant.

The chief motivation for consideration of such measures is their relationship to entropy. If P is a probability measure on $T^{|S|}$, let

$$h(P) = \sum_\omega P(\omega)U(\omega) + \sum_\omega P(\omega)\log P(\omega)$$

If we fix some $\omega_l \in T^{|S|}$, then $P(\omega_l) = 1 - \sum_{\omega \neq \omega_l} P(\omega)$, and, for any $\omega_0 \neq \omega_l$ we have

$$\frac{\partial h}{\partial P(\omega_0)} = U(\omega_0) - U(\omega_l) + \log\left[\frac{P(\omega_0)}{P(\omega_l)}\right]$$

Setting $\frac{\partial h}{\partial P(\omega_0)} = 0$, we obtain

$$\frac{P(\omega_0)}{P(\omega_l)} = \frac{e^{-U(\omega_0)}}{e^{-U(\omega_l)}}$$

and, summing over $\omega_0 \neq \omega_l$, we find

$$P(\omega_l) = \frac{e^{-U(\omega_l)}}{\sum_\omega e^{-U(\omega)}}$$

so that

$$P(\omega_0) = \frac{e^{-U(\omega_0)}}{\sum_\omega e^{-U(\omega)}}$$

which is the Gibbs measure induced by U. The Gibbs measure is thus seen to minimize $h(P)$, and hence among all those measures that have the same expected energy, $E[U(\omega)] = \sum_\omega P(\omega)U(\omega)$, it is the one that maximizes the entropy, $-\sum_\omega P(\omega)\log P(\omega)$.

In our application, the label set T will be $\{0,1\}$ = {no salesman present, salesman present}. We define a potential U as follows: for each single-element set $\{i\} \subset S$,

$$U(\{i\},\omega) = (1-2\omega_i)N_i/2$$

and for each two-element set $\{i,j\} \subset S$ where i and j are neighbor cities,

$$U(\{i,j\},\omega) = (1-2\omega_i)(1-2\omega_j)f(i,j)/2$$

For all other sets $A \subset S$, $U(A,\omega) = 0$. If $U(\omega) = \sum_{A \subset S} U(A,\omega)$ is the energy of this potential and $P(\omega) = e^{-U(\omega)}/N$ is the Gibbs measure induced by U, then we use the same notation (with a different normalization),

$$P(\omega) = \frac{e^{-U(\omega)}}{M}$$

for the probability measure induced by P on the set of $\binom{n}{m}$ labelings containing exactly m 1's.

Now suppose that ω' is a labeling (assignment of salesmen to cities) obtained from a labeling ω by moving a single salesman from city i to city j. Then

$$\begin{aligned} U(\omega) - U(\omega') &= N_j - N_i + \sum_{k \in nhbr(j)} (1-2\omega_k)f(j,k) \\ &\quad - \sum_{k \in nhbr(i)} (1-2\omega_k)f(i,k) \\ &= A(j) - A(i) \end{aligned}$$

Thus, if Q is the transition rate matrix for our Markov process of Section 1, we have

$$\begin{aligned} Q(\omega,\omega') &= \frac{1}{1+e^{-(A(j)-A(i))}} \\ &= \frac{1}{1+e^{U(\omega')-U(\omega)}} \\ &= \frac{e^{-U(\omega')}}{e^{-U(\omega')}+e^{-U(\omega)}} \end{aligned}$$

and hence

$$\begin{aligned} P(\omega)Q(\omega,\omega') &= \frac{e^{-U(\omega)}e^{-U(\omega')}}{M\,\{e^{-U(\omega')}+e^{-U(\omega)}\}} \\ &= P(\omega')Q(\omega',\omega) \end{aligned}$$

from which we can conclude that our restricted Gibbs measure, P, is the steady-state solution of this process [9].

Finding that assignment ω with maximum probability is thus reduced to finding that ω with minimum energy, $U(\omega)$.

3. A NEURAL NETWORK REPRESENTATION

A neural network is a collection of simple analog processing elements designed to mimic biological neurons. The computational paradigm provided by such networks is a radical departure from that of the classical von Neumann architecture. The "input" to such a network is a matrix of interconnections among the processing elements, together with an initial voltage that is applied to each element. The networks are designed so that the stable output voltages of the analog elements are binary. The collection of all binary output levels is then interpreted as the "result" of the computation.

Of particular interest to us here is the class of neural networks proposed by Hopfield [5]. A four-element example, from [12], is shown in Figure 1. Neurons are represented by amplifiers, each providing both standard (voltage $\theta_i \in [-1, 1]$) and inverted outputs. Synapses are represented by the physical connections between input lines to the amplifiers and, in feedback, output lines from the amplifiers. Resistors are used to make these connections. If the input to amplifier i is connected to the output of amplifier j by a resistor with value R_{ij}, then the *conductance* of the connection is T_{ij}, whose magnitude is $1/R_{ij}$, and whose sign is determined by whether the connection to amplifier j is from the standard or inverted output. Hopfield showed that when the matrix T is symmetric with zero diagonal and the amplifiers are operated in

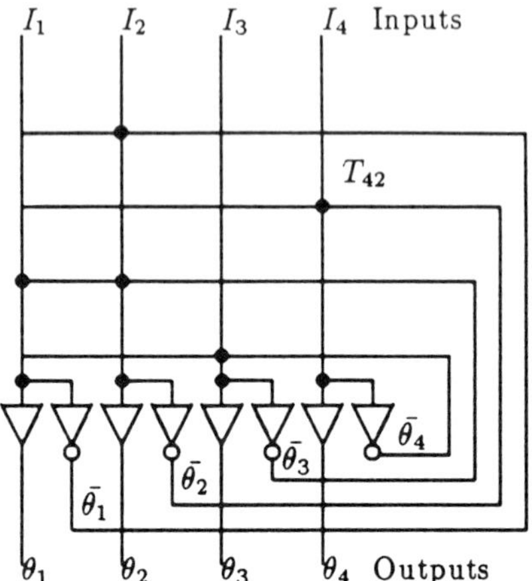

Figure 1 Four-element Hopfield network.

high-gain mode, the stable states of the network are binary ($\{-1, 1\}$) and are the local minima of the computational energy,

$$E(\theta) = (-1/2)\sum_{i=1}^{n}\sum_{j=1}^{n} T_{ij}\theta_i\theta_j - \sum_{i=1}^{n}\theta_i I_i \tag{1}$$

Here I_i is the external input to amplifier i.

We observe that our energy function $U(\omega)$ can be written in the form of the Hopfield computational energy. Specifically, if we assign

$$T_{ij} = T_{ji} = \begin{cases} -f(i,j)/2 & i,j \text{ neighbors} \\ 0 & \text{otherwise} \end{cases}$$
$$I_i = N_i/2$$
$$\theta_i = (2\omega_i - 1)$$

in the Hopfield computational energy expression (1), we obtain $U(\omega)$. We should note, however, that this specification alone will not suffice to ensure net convergence to a state with a specified number, m, of salesmen assigned to cities, since this net corresponds to the original Gibbs measure induced by U on the set of *all* labelings, not the measure induced on the restricted set of $\binom{n}{m}$ labelings of interest. To incorporate the additional constraint, we use a variation on a technique suggested in [12]: we add to $E(\theta)$ a summand of the form

$$C\left(\sum_{i=1}^{n}\frac{\theta_i+1}{2}-m\right)^2$$

where $C > 0$, and, to reinforce integral solutions while maintaining the Hopfield constraint of a zero-diagonal T matrix, we add another of the form

$$C\left(\sum_{i=1}^{n}\frac{1-\theta_i^2}{4}\right)$$

The net effect is to add $C(m - n/2)$ to each I_i and $-C/2$ to each nondiagonal T_{ij} in equation (1).

Thus a Hopfield network of 2^{16} amplifiers would enable us to locate the local minima of $U(\omega)$ and hence the high-probability components of the steady-state solution vector of our original Markov process having as many as $\binom{2^{16}}{2^{15}}$ states. Of course there is nothing magic here about 2^{16}, and much larger systems are feasible. Unfortunately, Hopfield networks of this size have not yet been built, and we are forced to resort to net simulation.

Net simulation is traditionally approached (e.g., [12]) as a numerical integration of the system of n differential equations describing the operation of the amplifiers [5]:

$$C_i\,du_i/dt = \sum_j T_{ij}g(u_j) - u_i/R_i + I_i \tag{2}$$

Here the u_i are internal input voltages to the amplifiers, and are related to the desired output voltages, the θ_i, by a sigmoidal *gain function*, $g(x)$. A reasonable choice for $g(x)$ is a scaled hyperbolic tangent, $g(x) = \tanh(\lambda x)$. Here λ is called the *gain*. The C_i are the input capacitances of the amplifiers, and $R_i = 1/(1/\rho + \sum_j |T_{i,j}|)$, where ρ is amplifier input resistance.

We have found numerical integration of large (2^{16} neuron) systems of the form equation (2) to be extremely time consuming, and therefore have developed an alternative approach. Any equilibrium of equation (2) is given by

$$0 = \sum_j T_{i,j}g(u_j) - u_i/R_i + I_i$$

that is,

$$u_i = R_i\left(\sum_j T_{i,j}g(u_j) + I_i\right)$$

or simply

$$u = G(u)$$

where $G(u) = \text{diag}(R)(Tg(u) + I)$, diag$(R)$ has R_i's on the diagonal and 0's elsewhere, and $g(u) = (g(u_1), g(u_2), \ldots)$. Thus we seek a fixed point of a certain n-dimensional function. If $|\ \ |$ denotes the max norm on Euclidean n-space and $\|\ \ \|$ its induced matrix norm, then since $R_i < 1/\sum_j |T_{ij}|$ we have

$$\begin{aligned} |G(u) - G(u')| &= |\,\text{diag}(R)T(g(u) - g(u'))| \\ &\leq \|\,\text{diag}(R)T\,\| \cdot |g(u) - g(u')| \\ &< |g(u) - g(u')| \\ &\leq \lambda |u - u'| \end{aligned}$$

Thus, for gain $\lambda < 1$, convergence of the simple iteration scheme, $u^{k+1} = G(u^k)$, is straightforward (see, e.g., [6]). Unfortunately, the Hopfield result speaks only of high-gain operation, and we must consider $\lambda > 1$, where the simple iteration is likely to diverge. Fortunately, there is an intriguing alternative.

In [4] Hillam established a remarkable result for functions on the real line: if $f : [a,b] \to [a,b]$ satisfies $|f(x) - f(y)| \leq M|x - y|$, then the iteration scheme

$$x_{n+1} = \frac{1}{M+1} f(x_n) + \frac{M}{M+1} x_n \tag{3}$$

converges to a fixed point of f. On the conjecture that this result might extend to higher dimensions, Hillam noted that a completely new approach would be needed, since his proof relied heavily upon the total ordering of the real line. To our knowledge, this conjecture remains unresolved.

Nevertheless, we have found substantial empirical evidence to support it, at least within the neural net environment. Using equation (3) with $M = \lambda$, we find that convergence to a fixed point of G (that is, average component error $|u_i - G(u_i)| < 10^{-10}$) usually requires fewer than 200 iterations, even for these large (2^{16} neuron) systems. We have not found a net for which this scheme fails to converge.

4. AN APPLICATION TO COMPUTER GRAPHICS

The digital halftone resolution problem may be stated as follows: given an $n \times n$ array V of real numbers, $V(i,j) \in [0, 1]$, produce an $n \times n$ array ω of binary integers, $\omega(i,j) \in \{0, 1\}$, such that ω, when displayed on a binary output device such as a computer monitor or laser printer, is a "good" representation of the real information, the intensities contained in V. Interpretations of "good" are many and involve metrics on R^{n^2}, object detection/recognition, aesthetics, and even stochastic power spectra [13]. The obvious resolution algorithm, round the values in V, fails to meet most of these goodness measures. If $V(i,j) = .4999999$ for all i, j, then $\omega(i,j) = 0$ for all i, j, and a desired gray image is displayed as white. Consideration of neighborhood intensities seems imperative.

Many halftone resolution algorithms have been proposed (see [7]), each successful on some measure, but none considered entirely satisfactory. The most commonly used resolution algorithm is probably the *ordered dither* [2], in which we tile the image matrix V with a smaller fixed array D of threshold values, and then turn on the pixel (set $\omega(i,j) = 1$) if and only if $V(i,j)$ exceeds the corresponding threshold value. A standard 4×4 tile is

$$D = \begin{bmatrix} 1/32 & 17/32 & 5/32 & 21/32 \\ 25/32 & 9/32 & 29/32 & 13/32 \\ 7/32 & 23/32 & 3/32 & 19/32 \\ 31/32 & 15/32 & 27/32 & 11/32 \end{bmatrix}$$

Note that a uniform intensity of 0.5 would cause 8 of every 16 pixels (every other one) to be turned on. Also note that the entries in the tile are carefully chosen to break horizontal and vertical lines, which are easily recognized by the eye.

In Figure 2 we show a 256×256 pixel image of a digitized photo resolved by this ordered dither. This image was printed on a conventional 300 pixel per inch laser printer at an expanded resolution of 75 pixels per inch.

Figure 2 Digitized photo resolved by ordered dither.

One of the standard complaints lodged against this technique is that it imparts an artificial texture to the image. This "computery" look is quite evident in our figure.

As an alternative, let us consider a 256×256 array of pixels as a lattice of $n = 65{,}536$ cities with both orthogonal and diagonal connecting roadways. If we let $m = \lfloor \sum V(i,j) + 0.5 \rfloor$, then any assignment of m salesmen to these n cities is a halftone resolution with exactly the correct average intensity per pixel.

Our neural net simulator (3) can be used to produce a minimum energy resolution with this correct average intensity. Note that the lattice structure for connections means that each of the n cities has at most 8 neighbors, so that storage required by equation (3) is linear in n. (The additional $-C/2$ term in each $T_{i,j}$ can be handled separately.)

The required potential, U, or attractiveness, A, can be selected in many ways. A choice that offers an interesting resolution is given as follows. For each $V(i,j) \in [0,1]$ let $v(i,j) = 2V(i,j) - 1$, that is, scale over $[-1,1]$. Then let

$$N_{(i,j)} = 2v(i,j) - 0.1 \sum_{(k,l) \in nhbr(i,j)} v(k,l)$$

$$f((i_1,j_1),(i_2,j_2)) = \begin{cases} h(i_1,j_1,i_2,j_2) \\ \quad \times (2 - |v(i_1,j_1) + v(i_2,j_2)|) & (i_1 - i_2)^2 + (j_1 - j_2)^2 = 1 \\ 0.75h(i_1,j_1,i_2,j_2) \\ \quad \times (2 - |v(i_1,j_1) + v(i_2,j_2)|) & (i_1 - i_2)^2 + (j_1 - j_2)^2 = 2 \\ 0 & \text{otherwise} \end{cases}$$

where $h(i_1,j_1,i_2,j_2) = |H(i_1 \bmod 4, j_1 \bmod 4) - H(i_2 \bmod 4, j_2 \bmod 4)|$ and

$$H = \begin{bmatrix} 0.4 & 2 & 0.6 & 2.2 \\ 1.8 & 0.2 & 1.6 & 0 \\ 0.8 & 2.4 & 1 & 2.6 \\ 3 & 1.4 & 2.8 & 1.2 \end{bmatrix}$$

With this assignment, we see that as $V(i,j)$ (or $v(i,j)$) becomes larger it becomes more important to turn that pixel on (set $\omega(i,j) = 1$), but that there are some attenuating factors. The magnitude of $N_{(i,j)}$ can be reduced by neighboring intensities of value comparable to $v(i,j)$. Further, $f((i_1,j_1),(i_2,j_2))$, which is the force with which we insist that adjacent pixels assume opposite parity, is at a maximum when desired intensities $v(i_1,j_1)$ and $v(i_2,j_2)$ are of equal magnitude and opposite sign, such as one black ($v = 1$) and one white ($v = -1$) or both gray ($v = 0$). Finally, the H matrix is used to ensure that in

Figure 3 Digitized photo resolved by neural net.

local regions of uniform intensity we still have significant differences in the f values in each of the eight neighboring directions. This tends to reduce local randomness in the resolution and speeds convergence.

In Figure 3 we show the image of Figure 2 resolved by our neural net simulation (3) using this assignment. Gain was set at 1.6. Convergence required 188 iterations from a uniform gray starting point.

Although we cannot make any claims for aesthetic improvement over ordered dither, we have succeeded in eliminating the patterning and in turning on (nearly) the correct number of pixels. Net specification includes many parameter values, and as yet we have only rough guidelines for their selection. Much empirical work remains before we can translate these parameter settings to aesthetic values. The reader is invited to experiment.

Another interesting approach to halftone resolution using neural nets can be found in [1], where a frequency-weighted mean squared error function is minimized. The weights are determined from psychophysical experiments reported in [10].

5. CONCLUSIONS

We have considered a class of continuous-time, ergodic Markov processes that represent the stochastic traveling salesmen problem. We have proposed a technique for numerical determination of high-probability components of the steady-state solution vector of such processes. The intended use of the technique is the case where the Markov state space becomes extremely large, such as $10^{20,000}$. The technique first identifies the steady-state solution as a maximum entropy Gibbs measure, and then maps the associated energy function onto a Hopfield neural network. An iteration scheme whose convergence is based upon an unresolved conjecture (Hillam) completes the computation. As an example of the use of this technique, we have provided an application to computer graphics, a new algorithm for digital halftone resolution.

The power of this technique lies in the fact that we can attack a class of problems whose state space grows exponentially with problem parameter n, while using space and time that grow only linearly with n. Although we do not obtain complete steady-state information, nor do we necessarily find *the* maximum probability component, identification of components with relatively high probability can be of value.

We are investigating several extensions. Our choice of potential for the halftone resolution algorithm was fairly straightforward and may be naive. We are conducting experiments to determine how potential definition affects image quality.

A more important extension is the mapping itself. We have established a path connecting stochastic Petri nets, reversible Markov processes, Gibbs measures, and neural networks. Other classes of neural networks have been proposed [3], and a widening of the path to target such structures remains a most intriguing possibility.

Finally, a resolution of Hillam's conjecture would allow us to remove convergence "warning labels" from our algorithm.

REFERENCES

[1] D. Anastassiou. Neural net based digital halftoning of images. Proc. IEEE Int. Symp. on Circuits and Systems, June 1988.

[2] J. Foley and A. Van Dam. *Fundamentals of Interactive Computer Graphics*. Addison-Wesley, Reading, Mass., 1984.

[3] R. Hecht-Nielsen. Neurocomputing: Picking the human brain. IEEE Spectrum, March 1988.

[4] B.P. Hillam. A generalization of Krasnoselski's theorem on the real line. Math. Mag., 48:167–168, 1975.

[5] J.J. Hopfield. Neurons with graded response have collective computational properties like those of two-state neurons. Proc. Natl. Acad. Sci. USA, 81:3088–3092, 1984.

[6] E. Isaacson and H.B. Keller. *Analysis of Numerical Methods*. John Wiley & Sons, New York, 1966.

[7] J. Jarvis, C. Judice, and W. Ninke. A survey of techniques for the display of continuous tone pictures on bilevel displays. Comp. Graphics Image Process., 5:13–40, 1976.

[8] J. Kemeny, J. Snell, and A. Knapp. *Denumerable Markov Chains*. Springer-Verlag, New York, second edition, 1976.

[9] R. Kindermann and J Snell. *Contemporary Mathematics: Vol. 1, Markov Random Fields and Their Applications*. AMS, Providence, R.I., 1980.

[10] J. Mannos and D. Sakrison. The effects of a visual fidelity criterion on the encoding of images. IEEE Trans. Info. Theory, IT-20, July 1974.

[11] M.K. Molloy. Performance analysis using stochastic petri nets. IEEE Trans. Comput., C-31, 1982.

[12] E. Page and G. Tagliarini. Algorithm development for neural networks. In Proc. SPIE Symp. on Innovative Science and Technology, Los Angeles, January 1988.

[13] R. Ulichney. Dithering with blue noise. Proc. IEEE, 76:56–79, 1988.

29

Combinatorial Optimization, Markov Chains, and Stochastic Automata

EUGENE SHRAGOWITZ and RUNG-BIN LIN Department of Computer Science, University of Minnesota Minneapolis, Minnesota

ABSTRACT

It has been known for some time [KIR83, MIT86] that Markov chain techniques can be applied to solving combinatorial optimization problems. Under certain conditions a nonstationary Markov chain of states in the solution space of the combinatorial problem asymptotically converges in probability to a global minimum. A similar set of ideas was introduced earlier in the domain of automata theory for adaptive optimization of the unknown function of parameters. Random search algorithms were interpreted there as stochastic automata and were investigated by means of discrete Markov chains. In this paper we prove the equivalence between some popular randomized algorithms for combinatorial optimization (simulated annealing, SA), certain types of stochastic automata (S-type GH-stochastic automata with variable structure), and nonstationary Markov chains. By using Markov chain techniques, it was demonstrated that many stochastic automata have properties similar to those of simulated annealing and are not inferior to SA in computations.

Large computational experiments were conducted on a network of Apollo computers.

1. INTRODUCTION

It is difficult to attribute introduction of random search algorithms for combinatorial optimization to one particular author. Some random search algorithms for combinatorial optimization were discussed for example in [REI65] and [REI66]. These algorithms are characterized by probabilistic procedures for generating probes and probabilistic procedures for making moves according to the results of such probes. It is possible to distinguish two kinds of random search algorithms: those with and those without adaptation of the parameters of the algorithm.

Simultaneously researchers in the United States and Soviet Union worked on a formal theory of learning systems and interpretation of learning algorithms as deterministic or probabilistic automata. Among the first papers we can name are [BUS58], [McM66], [TSE61], [VAR63], and [VOR65], a book [TSE69], and many others.

Random search algorithms were applied to one of the types of learning systems: adaptive optimization of an unknown function of the parameters. Interpretation of random search algorithms for adaptive control as stochastic automata with learning capabilities was proposed in [SHA69]. Low convergence speed was mentioned there as a major impediment for applications of these methods to functions of many variables. In this theory, a random search algorithm is considered as a stochastic automaton. It has a binary (or continuous) input and a set of actions coinciding with the set of possible combinations of values of the variables. Random actions of the automaton result in selection of variable values, which produce a value for the objective function. In response to the increase or decrease of the objective function, the automaton changes the transition matrix and its state in accordance with a new transition matrix. It then performs a new action again, etc. The majority of works in this area investigated the behavior of automata with the binary (0 or 1) penalty function. The behavior of this type of automata in a stationary random medium was described by means of discrete Markov chains, and final probabilities of automata states were determined as final probabilities of the states of the chain.

The distinction between different automata is made based on the nature of the input, definition and number of states, rules of updating the transition matrix, the set of actions associated with each state, etc.

It will be shown that many random search algorithms used so far for combinatorial optimization (including SA) can be interpreted in terms of stochastic automata with or without learning capabilities. We shall also describe new stochastic automata, which belong to the same class. We will provide an experimental comparison of the SA algorithm with other automata for a graph partitioning problem. A method of generating graphs for this comparison together with parameters for the SA algorithm are borrowed from the well-known work of D. Johnson et al. [JOH87].

2. STOCHASTIC AUTOMATA AND COMBINATORIAL OPTIMIZATION

An optimization problem can be interpreted as environment where a stochastic algorithm performs its functions (Fig. 1). The input to this environment is a vector of parameters x. In the case which interests us, these parameters are discrete and defined in the space S_M of combinatorial objects induced by the discrete finite set D. The output or response of the environment is defined as the bounded scalar function F where $F = F(x)$.

In the theory of learning automata function F is not known. The problem is defined as $F(x') \rightarrow$ extremum $(x' \in S_M)$. Such definition of the problem is similar to formulation of the combinatorial optimization problem with this difference: function F is defined only by its realizations in the first case and analytically defined in the second case. In both cases an optimal solution for the problem can be defined as $x' = \arg\text{extremum}\, F(x'), x' \in S_M$. When analytical formulation of the function F is not known, the problem can be solved by one of the following techniques: (1) by identification of the objective function followed by solution of the corresponding integer programming problem; (2) by search, that is, by a series of interactions with the problem wherein the algorithm will successively request values of the objective function for states $x_1, x_2, \ldots$ with the purpose of converging asymptotically to the state x^*; or (3) by combination of the previous two techniques, which results in extrapolation models. When the objective function is known, the set of alternatives is almost the same. The second of these alternatives, the search technique, will be considered here.

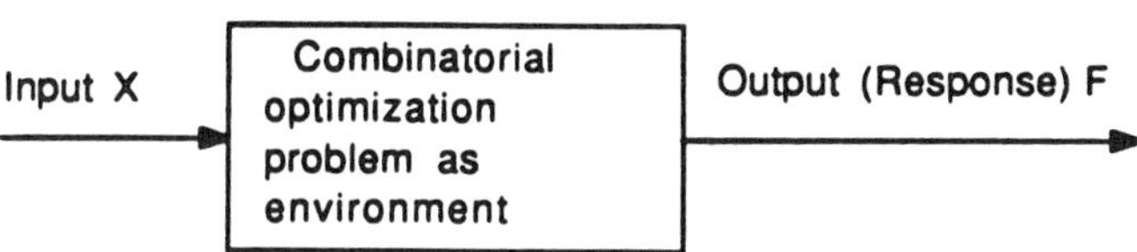

Figure 1 Environment.

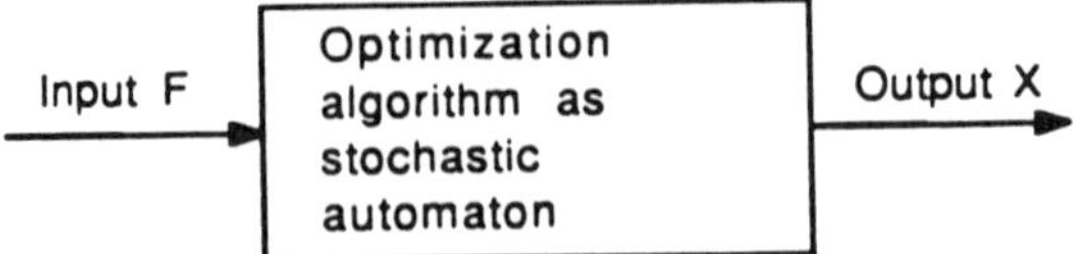

Figure 2 Stochastic automaton.

A stochastic algorithm for finding x^* can be interpreted as a stochastic automaton (Fig. 2). Behavior of the stochastic automaton on the ith step of the search process is usually characterized by k probes, that is, by obtaining the value of the objective function for states $x_i^1, x_i^2, \ldots, x_i^k$, selected randomly from the vicinity of the state x_{i-1} according to a certain law.

Then collected information is used to produce a new state of the system from the old one, that is, $x_{i+1} = f(m_i, x_i^1, x_1^2, \ldots, x_i^k)$, where m_i is a memory vector, which describes the history of the search. Vector m_i is adapted by the learning algorithm based on information obtained after $(i-1)$ steps, where $m_i = f(m_{i-1}, x_{i-1}, x_i^1, x_i^2, \ldots, x_i^k)$. If $m_i = 0$, for any i then the search algorithm is independent of the history (Fig. 3).

2.1 Classification of Stochastic Automata

A stochastic automaton [TZ77] is a sextuplet $\{Y, A, X, \psi, \eta, a_1\}$ where Y is the input set, $A = \{a_1, a_2, \ldots, a_n\}$ is the set of internal states, $X = \{x_1, x_2, \ldots, x_b\}$ is the output or action set, ψ is a mapping ψ: $A^*Y \rightarrow A$ giving the state associated with each input and previous state, $\eta : A^*Y \rightarrow X$ is a mapping giving the output associated with each state and input a_1 is the initial state of the automaton. If both mappings ψ and η are deterministic then the automaton is called deterministic. If one of these mappings or both of them are probabilistic then the automaton is called stochastic.

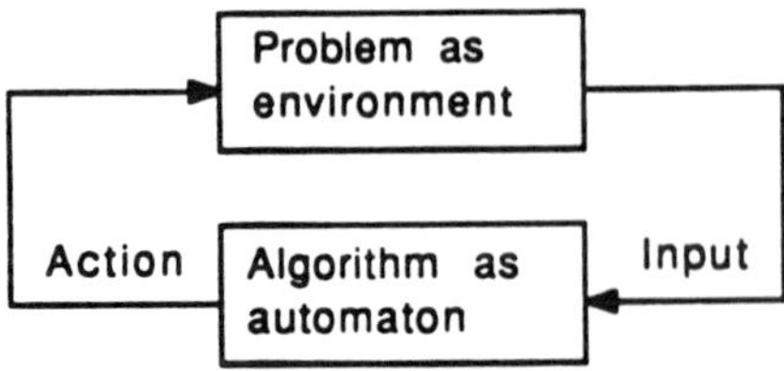

Figure 3 Learning automaton.

A stochastic automaton with the deterministic mapping η and the probabilistic mapping ψ is called a G-type stochastic automaton. A stochastic automaton with the probabilistic mapping η and deterministic mapping ψ is called an H-type stochastic automaton. When both ψ and η transformations are probabilistic then the automaton is called a GH-type, or Mealy, stochastic automaton. When stochastic mapping $\eta : A \rightarrow X$ is independent from output Y, then such automaton is called GH-type Moore automaton.

If matrices of transitions $\|\psi_n^{ij}(y)\|$ and actions $\|\eta_n^{ik}(y)\|$ are time-invariant, then we say that the automaton has a constant structure. Otherwise, it is called an automaton with variable structure.

An automaton with a binary input function is known as a P-model. An automaton with an input set that consists of a finite collection of distinct symbols is named a Q-model. If the input set is an interval [0, 1] then the automaton is named an S-model. The optimization of a multimodal function can be interpreted as an S-model problem. S-Model problems can be solved by the algorithms developed for the P-model problems [VAR73] if the objective function is bounded.

2.2 P- and S-Model Stochastic Automata for Unconstrained Optimization

For a P-model automaton, the input variable y may take only two values: 0 and 1. The value $y = 0$ is assigned when the objective function decreases as a result of step m, that is, $F_m - F_{m-1} < 0$, and it is called nonpenalty. The value $y = 1$, called penalty, is assigned when $F_m - F_{m-1} > 0$. Changes in the states of the automaton as responses to the input Y are described by the transition matrix. The transition matrix for the deterministic automaton is very simple. Each row contains only one element equal to 1 and all the other elements are equal to 0. An element of the transition matrix of the stochastic automaton is the probability of transition from the ith state to the jth state for a given value of input variable Y. The probability is defined from the interval $(0, 1)$. Obviously, the deterministic automaton is a special case of stochastic automata. The probability of generating a penalty as the result of action x_i is denoted c_i $(i = 1, \ldots, n)$ and called the penalty probability. Then $(1 - c_i)$ is the probability of a nonpenalty. The transition probability p_{ij} from the state a_i to the state a_j for the P-model automaton is given by the formula:

$$P_{ij} = c_i p_{ij}(1) + (1 - c_i) p_{ij}(0)$$

and the matrix $\mathbf{P}$ of interaction between the automaton and the environment has form $\mathbf{P} = \|p_{ij}\|$, $i,j = 1, \ldots, n$. The matrix $\mathbf{P}$ is stochastic and can be interpreted as a Markov chain. If the medium is stationary [TSE69], then this chain is ergodic, and final probabilities π_i of the automata states exist and are independent of its initial state. Probabilities of these states can be defined as a solution of the linear equation $\pi\mathbf{P} = \pi$, where $\mathbf{P}$ is the matrix of transition probabilities and π is the vector of final probabilities of states. An additional normalizing constraint is imposed on π:

$$\sum_{i=1}^{n} \pi_i = 1$$

If $g_k, k = 1, 2, \ldots, n$ is the probability of action x_k performed by all states of an automaton and g is a vector of action probabilities, then the mathematical expectation E of a penalty for the automaton is given by the formula

$$E(g,c) = \sum_{i=1}^{n} g_i c_i$$

If actions are chosen at random with equal probability, then the average penalty is denoted E_0 and can be written as follows:

$$E_0(g,c) = 1/n \sum_{i=1}^{n} c_i$$

If $E(g,c) < E_0(g,c)$, then the automaton is called *expedient.*

If an average loss is minimal then such an automaton is called *optimal*, that is,

$$E(g,c) = c^*$$
$$c^* = \min_i \{c_i\}$$

If an average loss is bounded by $c^* + \epsilon$, where $\epsilon > 0$, then an automaton is called *ϵ-optimal.*

$$E(g,c) < c^* + \epsilon$$

A stochastic automaton with probabilities of actions changing with the time is called *absolutely expedient* if the average loss satisfies the condition

$$E(g(m+1),c) < E(g(m),c) \qquad m = 1,2,\ldots$$

A stochastic automaton is called asymptotically optimal if

$$\lim_{m\to\infty} E(g(m),c) = c^*$$

In the class of stochastic asymptotically optimal automata, we will identify a subclass of automata with the function $g = g(m,c(m))$, where $c(m)$ is a monotonically increasing (decreasing) function of m. This is equivalent to the assumption that media interacting with the automaton is nonstationary. P-Model deterministic and stochastic automata in random media were studied by [TSE61], [VAR63], [NAR74], [LAK73], and many others. An objective function that can be optimized by such an automaton corresponds to the average number of successful steps in the process of search. Such a process does not necessarily converge to the optimal solution for the global optimization problem.

Historically automata algorithms were first used for adaptive control. The application of automata algorithms for adaptive control is based on the assumption that probabilistic properties of the media are not known beforehand. An *a priori* knowledge of the penalty probabilities $c_1, c_2, \ldots, c_n$ would make it possible to construct an automaton having only states and actions with the maximum probability of nonpenalty $c_{\max}$. This consideration is also relevant for combinatorial optimization problems with well-defined objective functions. The interpretation of the combinatorial optimization problem as a random media interacting with an automaton may be used for selection of states with the lower value of the objective function. An attempt can be made to use information acquired by random probes for adaptation of the algorithm for improved expediency.

The problem of minimizing function F is more adequately described by an S-type automaton. Let us assume that the objective function F is normalized and takes values from the interval [0, 1]. It was proposed in [VAR73] to transform F into a series of zeros and ones by a random number generator, which chooses one with the probability F and zero with the probability $(1-F)$. The result of this transformation can be considered as a penalty for the selection of a variant. Such a pattern assumes that not all "bad" moves will be penalized and, thus, assumes "hill climbing" capabilities for an automaton.

3. METRICS IN COMBINATORIAL SPACES

The SA algorithm, like many other algorithms for combinatorial optimization, generates a new point in the solution space by a "slight" perturbation of an old solution. In many descriptions of the SA algorithm given in the literature, the function *perturb* is not defined quantitatively [ROM84, MIT86]. To determine a notion of perturbation, a distance in the space of combinatorial objects should be defined. To be measurable a distance should obey the axioms of the metrics. Let us call a sphere with the center x_0 and a radius r, a set of points X in the space S_M, such that $p(x_0,x_i) \leq r$, where $x_0, x_i \in S_M$.

Let us call a neighborhood of $x_0 \in S_M$ any sphere with the center point x_0 and radius $\leq r$.

3.1 Metric in the Space of Permutations

If N is a set of combinatorial objects, then the set of all possible permutations S_M contains $|N|!$ elements. Let x_i and x_j be elements of the set of permutations. The distance between elements x_i and x_j $\rho(x_i,x_j)$, is defined as the minimum number of elementary permutations transforming x_i into x_j, where an elementary permutation is an exchange of positions of two elements of the ordered set N.

It is easy to show that ρ satisfies two axioms of metric spaces.

1. $\rho(x_i,x_j) = 0$ iff $x_i = x_j$.
2. $\rho(x_i,x_j) = r(x_j,x_i)$.
3. The third axiom is also satisfied. Let x_i, x_j, $x_k \in S_M$. Then $\rho(x_i,x_j) + \rho(x_j,x_k) \geq \rho(x_i,x_k)$, because x_i is transformed into x_k by $\rho(x_i,x_j)+\rho(x_j,x_k)$ elementary permutations, but this number is not necessarily minimal.

From this definition of metrics, it follows that a sphere of radius one includes all pairwise permutations from the initial point, a sphere of radius two includes all points reachable in not more than two permutations, etc.

3.2 Metrics in the Space of Samples

Let N be a set of objects from which samples are selected. Let B denote the set of all samples. Let x and y be two different samples from the set B. Let us introduce a metric defined by $\rho(x,y) = |(x \cup y) \setminus (x \cap y)|$. It is necessary to show that ρ satisfies the axioms of metric spaces. From the definitions it easily follows that the first two axioms are satisfied. The proof for the third axiom is provided.

For $x, y, z \in B$, $\rho(x,y) + \rho(x,z) = |(x \cup y) \setminus (x \cap y)| + |(x \cup z) \setminus (x \cap z)| \geq |[(x \cup y) \setminus (x \cap y)] \cup [(x \cup z) \setminus (x \cap z)]| = |(x \cup y \cup z) \setminus (x \cap y \cap z)| \geq |(y \cup z) \setminus (y \cap z)| = \rho(y,z)$. This result is based on the condition $x \cap y \cap z \neq \varnothing$ and the inclusion $x \cup y \cup z \supset (y \cup z) \setminus (y \cap z)$.

The metrics stated here certainly do not exhaust the set of all possible metrics in the space of combinatorial objects. There are several other metrics mentioned in the literature [SER81, COH80, CAM88]. The selection of a metric is problem-specific and reflects the internal structure of the problem. For example, for the graph bipartitioning problem, one convenient metric would be a metric of the sample space.

REMARK The selection of the metric for methods based on a neighborhood search is determined as a compromise between two factors: size of the neighborhood for a given metric, and the maximal numbers of steps necessary to reach any point in the solution space. For example, if any point in the solution space is reachable in one step, then the size of the neighborhood for the NP-hard combinatorial problem is not less than $O(2^N)$, where N is a problem size. A metric is reasonable, when the neighborhood size and the maximal number of steps are both $O(N^k)$. In light of these considerations, the exchange of position between two elements two elements of the ordered set N for combinatorial problems based on ordering generates a neighborhood of the size $O(N^2)$, and the minimum path length between any two states is no more than $N - 1$ permutations.

4. RANDOMIZED ALGORITHMS FOR COMBINATORIAL OPTIMIZATION AS STOCHASTIC AUTOMATA

4.1 The Hill-Climbing Algorithm as Stochastic Automaton

The probabilistic hill-climbing algorithms for combinatorial optimization [REI65, VAR73, ROM84], which we denote HC, randomly generate next points in the solution space and can "climb hills," that is, they accept solutions with a higher value of the objective function than the current solution. A decision to "climb the hill" is made when random numbers from the interval $(0, 1)$ become smaller than the value of the three-argument continuous function $g(S, \text{NewS}, T)$, defined on the interval $(0, 1)$. The general form of this algorithm is given in Figure 4.

Here, S is the current solution to the optimization problem. Its initial value, S_0, may be any feasible solution; *perturb* is the randomized function that generates a new feasible solution from the current solution S according to a given metric, and $g(S, \text{NewS}, T)$ is a continuous function of the current position in the solution space S, the new position NewS, and the parameter T, defined on the interval $(0, 1)$. The term $r = \mathit{random}$, where *random* is a pseudo-random number form the interval $(0, 1)$. The questions are whether the hill-climbing algorithm can be described as an automaton, and, if so, what kind of automaton this algorithm represents?

THEOREM 1 The HC algorithm is an S-model GH-stochastic automaton with variable structure.

```
procedure ProbabilisticHillClimbing ;
  S := S0;  {initial solution}
  T := T0; {  time-dependant parameter}
 repeat
    NewS := perturb (S);
      If (F(NewS) < F(S)) or (r < g (S,NewS,T))
        then   S:= NewS;
    UpdateT ;
  until "termination criterion for algorithm";
end; {of ProbabilisticHillClimbing }
```

Figure 4 Probabilistic hill-climbing algorithm.

Proof. Let all points in the solution space of the combinatorial problem be defined as the internal states of automaton $A = \{a_1, a_2, \ldots, a_n\}$. An action set X_i, which is contained in X, is determined by the set of neighbors for any point a_i in a given metric ρ. Mapping ψ produces the state associated with each input and previous state. This mapping is defined by two matrices. One matrix $P_0 = \|p_{ij}^0\|$ defines probabilities of transitions from the state i to the state j for $\Delta F = F(a_j) - F(a_i) \leq 0$. Another matrix $P_1 = \|p_{ij}^1\|$ defines probabilities of transitions for $\Delta F = F(a_j) - F(a_i) > 0$. It follows from the description of the HC algorithm given by Fig. 4 and our definitions of an automaton state that elements of the row i of matrix P_0 are $p_{ij}^0 = 1/|A_i|; j \in A_i$; $p_{ij}^0 = 0; j \notin A_i$. Elements of the row i of matrix P_1 are defined according to the formulas:

$$p_{ij}^1 = q(r < g(F(a_i), F(a_j), c))/|A_i| \qquad \text{for} \quad j \in A_i, i \neq j$$

$$p_{ij}^1 = 1 - \sum_{m=1 \text{ and } m \neq 1}^{|A_i|} q(r < g(F(a_i), F(a_m), c))/|A_i| \qquad \text{for} \quad i = j$$

$$p_{ij}^1 = 0 \qquad j \notin A_i$$

The first matrix P_0 for the HC algorithm is deterministic, and the second matrix P_1 is stochastic, where q is a probability that the condition defined by the *if* operator is satisfied for $DF = F(a_j) - F(a_i) > 0$. Thus, mapping ψ is stochastic for the HC algorithm. The action x_i consists of random selection of one state in the neighborhood of the current state i. Mapping η gives the action associated with each state and input. This mapping is also stochastic, since probes are generated randomly according to the uniform distribution of probabilities of neighborhood states. Parameter T in the HC algorithm changes with the time; thus, the HC algorithm has a variable structure.

From the onset of automata theory, Markov chains were applied to their analysis. We will also apply Markov chain techniques to study convergence properties of GH-stochastic automata.

THEOREM 2 A Markov chain induced by the HC algorithm with the function $T(n)$ = const in a finite metric space of combinatorial objects is ergodic.

Proof. To prove the state of the theorem it is necessary to show that Markov chain is irreducible, aperiodic, and positive persistent. Some additional constraints should also be imposed on the function *perturb* and $g(S, \text{NewS}, T)$. According the theorem II.1.1 from [ISA76], a Markov chain is irreducible if and only if all pairs of states intercommunicate. If the space of combinatorial objects is a finite metric space it implies that a distance between any pair of points is finite and positive according to the selected metric. If the HC algorithm satisfies the following additional conditions:

1. Procedure *perturb* is defined in such a way that it generates with the nonzero probability all points that belong to the sphere of radius $\rho = 1$ with the uniform probability.
2. Function $g(S, \text{NewS}, T)$ is such that the transition from the state i to any neighboring state j occurs with the probability $p_{ij} > 0$ for each point j in the neighborhood of the current point i.

(We will demonstrate later that these conditions are easy to satisfy).

Then a minimal number of transitions between any two points i and j in the solution space S_M is a finite positive integer $m = \rho(i,j) = \rho(j,i) \geq 0$, that is, all pairs of states intercommunicate and the Markov chain is irreducible. In a finite irreducible Markov chain all states are positive persistent. If states intercommunicate they are of the same type, so it is enough to demonstrate that one state has a period one to prove aperiodicity. Let us assume that j is a state where $F_j = \min F$. Thus $F_j < F_i$ for any i, and the probability $p_{jj}^{(1)} > 0$ according to the definition of function g (i.e., 1) is the G.C.D. of all those m's for which $p_{jj}(m) > 0$. Therefore Markov chain induced by HC is aperiodic.

It follows from the stationarity of Markov chain that the vector π probabilities of the states (points in the solution space) can be determined from the vector equation $\pi \mathbf{P} = \pi$, where $\mathbf{P}$ is a transition matrix. The resulting distribution $\pi = (\pi_1, \pi_2, \ldots)$ is called the stationary probability distribution for the chain.

In the following paragraph it will be demonstrated that some well-known randomized algorithms for the combinatorial optimization are in fact S-type GH-stochastic automata with variable structures. We will introduce some new algorithms that belong to this class and will study their asymptotic convergence properties.

5. SIMULATED ANNEALING ALGORITHM AS A STOCHASTIC ASYMPTOTICALLY OPTIMAL AUTOMATON

In the previous paragraph it was shown that randomized hill-climbing algorithms with the procedure *perturb*, and acceptance function $g(S, \text{NewS}, T)$, specified according to some mild constraints, are GH-type stochastic automata. The set of points in the solution space generated by such an automaton is a Markov chain with the stationary probabilities of states.

The next question is, what can be done to make such an automaton asymptotically optimal, that is, to obtain convergence with probability 1 to the set of global optima? The obvious candidate for manipulation is the function g and one of its arguments the function $T(m)$ in particular. We will demonstrate that one well-known algorithm for combinatorial optimization is in fact an asymptotically optimal GH-stochastic automaton.

The SA algorithm in its generic form can be described by Figure 5.

THEOREM 3 The SA algorithm is an S-model GH-stochastic automaton with a variable structure.

Proof. It is evident that the SA algorithm is a particular case of the HC algorithm.

Several proofs were published, [MIT86], [AAR85], [GID85], and others, stating that if certain conditions on the function *perturb*, which generates a specific neighbor, and the acceptance function $g(T_k)$ are satisfied, then the SA algorithm converges to a set of global optima with the probability 1. It was required that parameter T_k asymptotically decrease to 0 when k approaches infinity.

THEOREM 4 The SA algorithm is an asymptotically optimal S-type GH-stochastic automaton with variable structure.

```
 S := S0;  {initial solution}
 T := T0;  {initial parameter value}
 repeat
    NewS := perturb (S);
      If (F(NewS) < F(S)) or (random  <  e(F(S)-F(NewS))/T)
         then  S:= NewS;
   UpdateT;
  until "termination criteria are satisfied";
end; {of Simulated Annealing }
```

Figure 5 Simulated annealing algorithm.

Proof. This follows from the convergence proofs mentioned above and definition of the asymptotically optimal automaton in Section 2.2.

6. ANOTHER STOCHASTIC ASYMPTOTICALLY OPTIMAL AUTOMATON

In Section 5 we described a class of stochastic hill-climbing automata and demonstrated that members of this class have stationary probability distributions of states when parameters of the algorithm are fixed. At the end of Section 5, we stated that one member of the class of HC automata, known as the SA algorithm, has an additional property—convergence to the set of global optima, when the time-dependent parameter T_k monotonically decreases to zero at a very slow rate. The next question is whether asymptotic convergence is a feature of only one automaton from this class, and, if not, whether there are other automata with the same property. This question is answered in the affirmative in this section. Let us consider a function $g(S, \text{NewS}, T_k) = (F(S)/F(\text{NewS}))^{T_k}$, where $T_k \to \infty$, when $k \to \infty$. As one can see, this acceptance function is an exponent of a simple relation between values of the objective function in the probed point $j\,(F_j = F(\text{NewS}))$ and the initial point $i(F_i = F(S))$. If $F(\text{NewS}) > F(S)$, then $g < 1$, therefore this function satisfies constraints imposed by the HC algorithm. The parameter T_k is not decreasing to zero as in SA, but increasing to infinity. We would like to demonstrate that this automaton has an asymptotic convergence property of SA. This acceptance function is based on *direct division* of one value of the objective function by another. So we will call this algorithm DD automaton.

Let us establish some equivalences:

$$\log(g_{ij}(T)) = \log(F_i/F_j)^T = T\log(F_i/F_j)$$
$$= T(\log F_i - \log F_j) = -T(\log F_j - \log F_i)$$

Simultaneously,

$$\log(e^{-T(\log F_j - \log F_i)}) = -T(\log F_j - \log F_i)$$

Thus,

$$g_{ij}(T) = (F_i/F_j)^T = e^{-T(\log F_j - \log F_i)}$$

Let $N(i)$ be the neighborhood of the state i in the finite metric space S_M and $h(i,j)/h(i)$ be the probability of generating j from i where $h(i,j)$ is the "weight" of each of the neighbors of i and $h(i)$ is a normalizing function to ensure that $\frac{1}{h(i)}\sum_{j\in N(i)} h(i,j) = 1$. Therefore, the transition probabilities of the Markov chain can be represented as follows:

$$p_{ij}(T) = \frac{h(i,j)}{h(i)} \min\{1, e^{-T(\log F_j - \log F_i)}\} \qquad \text{if} \quad j \in N(i)$$
$$p_{ij}(T) = 0 \qquad \text{otherwise,}$$
$$p_{ii}(T) = 1 - \sum_{j\in N(i)} p_{ij}(T)$$

A nonstationary Markov chain, with the transition probability matrix $\mathbf{P}(\mathbf{T}_k)$, is formed when $T = T_k, k = 0, 1, 2, 3 \ldots$ is specified. Here, $T_k \to \infty$ as $k \to \infty$. Many properties of SA, which were described in [MIT86], also can be found in this Markov chain.

1. The quasi-stationary probability of a state i in the solution space can be defined as a function of F_i and T:

$$\pi_i(T) = \frac{h(i)e^{-T\log F_i}}{H(T)}$$

where $H(T)$ is a scaling factor, such that $\|\pi(T)\| = \|\sum_1^n \pi_i(T)\| = 1$, and $n = |S_M|$.
2. The stationarity of the Markov chain can be checked directly by proving the sufficient condition of the detailed balance. If $h(i,j) = h(j,i)$ and $\pi_i(T)$ is defined as above, then it is easy to see that the global balance holds, that is, $\pi(T)\mathbf{P}(T) = \pi(T)$ where $\mathbf{P}(T) = \{p_{ij}(T)\}$ is a one-step transition probability matrix of the Markov chain.
3. For each $i \in S^*$, where S^* is the set of global minima, the probability $\pi_i(T_k)$ is monotonically increasing with k, that is,

$$\pi_i(T_{k+1}) - \pi_i(T_k) > 0 \qquad \text{for all} \quad k \geq 0$$

and for $i \notin S^*$, there exists a unique integer $0 \leq m_i < \infty$ such that

$$\pi_i(T_{k+1}) - \pi_i(T_k) > 0 \quad \text{for} \quad 0 \leq k \leq m_i - 1$$
$$\pi_i(T_{k+1}) - \pi_i(T_k) < 0 \quad \text{for} \quad k \geq m_i$$

4. The Markov chain induced by the DD algorithm with the following *Update* function is weakly ergodic for $\Gamma \geq \gamma \log \bar{\omega}$:

$$T_k = \frac{\log(k + k_0 + 1)}{\Gamma}$$

   where $0 \leq k_0 < \infty$, γ is the radius of the underlying transition graph of the Markov chain, and $\bar{\omega}$ id defined as $\max_{i \in S_M} \max_{j \in N(i)} |F_j - F_i|$.
5. The Markov chain induced by the DD algorithm is strongly ergodic.

7. YET ANOTHER STOCHASTIC AUTOMATON

Let us consider another hill-climbing algorithm, which we will call LD, with the acceptance probability described by the following law $a_{ij} = 1$ for $F_j < F_i$ and $a_{ij} = T_k/(F_j - F_i)$ for $F_j > F_i$, where $T_k < 1$ and $T_k \to 0$ for $k \to \infty$.

This acceptance probability function was considered in [NAH85], where it was called a linear difference heuristic. The following properties of the LD algorithm can be stated:

1. The LD algorithm is an S-type GH-stochastic automaton with a variable structure. This property was established for the whole class of the HC automata and LD obviously belong to the HC class.
2. To keep LD from being trapped in the local minimum, some additional conditions must be imposed on the parameter T_k. These conditions are similar to those imposed on the parameter of the SA algorithm. Let us denote $c_{oi} = \min 1/(F_j - F_i)$, where i is a local minimum. Then the condition to prevent the LD algorithm from being trapped in the local minimum is $\sum_{k=1}^{\infty} T_k c_{0i} = \infty$ or just $\sum_{k=1}^{\infty} T_k = \infty$. It is easy to see that this condition is not satisfied for $T_k = a^{(k-1)} T_0$ for any $a < 1$ but is satisfied for $T_k = T_0 / \log k$.
3. The stationary probability distribution π of the finite homogeneous Markov chain induced by the LD algorithm with the parameter $T = \text{const}$ exists. This fact follows from Theorem 2.

THEOREM 5 The nonstationary Markov chain induced by the LD automaton is weakly ergodic.

Proof. The elements of one-step transition probability for LD are defined by the following formulas:

$$p_{ij}(T_k) = \frac{T_k}{\xi(F_j - F_i)} \quad \text{for} \quad F_j > F_i, j \in N(i), \qquad \xi = |N(i)|, i \neq j \quad (1)$$

$$p_{ij}(T_k) = 1/\xi \quad \text{for} \quad F_j \leq F_i, j \in N(i), \qquad \xi = |N(i)|, i \neq j \quad (2)$$

$$p_{ij}(T_k) = 0 \quad \text{for} \quad j \notin N(i) \quad (3)$$

Let $\bar{\omega} = \max_{i \in S_M} \max_{j \in N(i)} |F(j) - F(i)|$. Then for any nondiagonal entries in the matrix $P_{ij}(T_k) \geq T_k/(\xi\bar{\omega})$, assuming that $T_k/(\xi\bar{\omega}) < 1$. With T_k monotonically decreasing, entries of the type (1) are monotonically decreasing. The entries of the type (2) remain unchanged. Since the sum of all entries in each row is equal to 1, diagonal entries are monotonically increasing with k. Therefore, some k_0, $0 \leq k_0 \leq \infty$, exists such that $p_{ii}(T_m) \geq T_m/(\xi\bar{\omega})$, where $m \geq r(k_0 - 1)$; r is the radius of the transition graph.

For every i and $i' \in S_M$ and $m \geq k_0 r$,

$$P_{ii'}(m - r, m) \geq \prod_{i=m-r}^{m-1} \frac{T_i}{\xi\bar{\omega}} \geq (T_{m-1}/(\xi\bar{\omega}))^r$$

The coefficient of ergodicity τ of P is defined as $\tau(P) = 1 - \alpha(P) = 1 - \min_{i,j} \sum_l \min(P_{il}, P_{jl})$, then $\tau(P(kr - r), kr)) \leq 1 - \min_{ij} \{\min(P_{ii'}, P_{ji'})\} \leq 1 - (T_{kr-1}/(\xi\bar{\omega}))^r$, $k \geq k_0$, thus $\alpha(P) \geq (T_{kr-1}/(\xi\bar{\omega}))^r$. According to theorem V.3.2 from [ISA76], a nonstationary Markov chain is weakly ergodic if $\sum_{k=k_0}^{\infty} \alpha(P(kr - r, r))) = \infty$. To prove this fact, it is sufficient to show, that $\sum_{k=k_0}^{\infty} (T_{kr-1}/\bar{\omega})^r = \infty$. Let $T_m = m^{-1/r}$, then $\sum_{k=k_0}^{\infty} (T_{kr-1}/\bar{\omega})^r = \sum_{k=k_0}^{\infty} (kr - 1)^{-1/r}/\bar{\omega})^r = (1/\bar{\omega})^r \sum_{k=k_0}^{\infty} 1/(kr - 1) \geq (1/\bar{\omega})^r \sum_{k=k_0}^{\infty} 1/(kr) = \infty$. Therefore, the LD automaton is weakly ergodic.

CONJECTURE A nonstationary Markov chain induced by the LD automaton is strongly ergodic.

EXAMPLE Suppose that the solution space consists of four points, 1, 2, 3, and 4 with the corresponding values of the objective function F_1, F_2, F_3, and F_4, and $F_1 = 0$, $F_2 = 1$, $F_3 = 2$, $F_4 = 2$. Let metric be defined in such way, that only two points belong to the neighborhood of each point, that is $N(1) = \{3, 4\}$, $N(2) = \{3, 4\}$, $N(3) = \{1, 2\}$, and $N(4) = \{1, 2\}$. Then the transition probability matrix $P(T)$ has the following form:

$$P(T) = \begin{bmatrix} 1 - T/2 & 0 & T/4 & T/4 \\ 0 & 1 - T & T/2 & T/2 \\ 1/2 & 1/2 & 0 & 0 \\ 1/2 & 1/2 & 0 & 0 \end{bmatrix}$$

This solution space has two minima, one of which is the global minimum and another is the local minimum. By solving $\pi \mathbf{P} = \pi$, and $\pi_1 + \pi_2 + \pi_3 + \pi_4 = 1$, we get the probability vector $\pi(T) = (2/(3+2T), 1/(3+2T), T/(3+2), T/(3+2T)) \rightarrow (2/3, 1/3, 0, 0)$, as $T \rightarrow 0$. From this example it follows that the LD automaton converged to the set of local minima, and probabilities are proportional to the depth of the cups.

8. TEST BED AND PARAMETERS OF PROCEDURES

The choice of a test bed for this work was greatly influenced by the recent paper by D. Johnson et al., Optimization by simulated annealing: An experimental evaluation (Part 1) [JOH87]. In this study the problem of partitioning the vertices of a graph into two equal-size sets in order to minimize the number of edges with endpoints in both sets was considered as the "arena" for testing the relative performance of SA. It was mentioned there that the SA demonstrated the best performance for this class of problems among the other combinatorial problems tried. The SA algorithm was tested on two different types of graphs: random graphs and geometric graphs. Geometric graphs are characterized by clustering of connected vertices and, therefore, represent an example of a problem with structure. We used the same two types of graphs and tried to retain the same values of parameters used for their generation.

The problem of graph partitioning formulated in [JOH87] was as follows: Given a graph $G = (V, E)$, where V is a set of vertices, and E is a set of edges, there are two sets V_1 and V_2 such that $V = V_1 \cup V_2$ and $V_1 \cap V_2 = \varnothing$. Two partitions are said to be neighbors if one can be obtained from the other by moving one vertex from one set to the other. The objective function is characterized as follows: $C(V_1, V_2) = |\{(m, v) \in E\colon m \in V_1, v \in V_2\}| + \beta(|V_1| - |V_2|)^2$, where m and v are two distinct vertices of the graph; β is the imbalance factor to penalize the unequal-sized partitions. In our case, β is 0.05. This approach makes it possible to use the imbalanced partitions as temporary solutions. Finally, if the solution is imbalanced, an heuristic is applied to make it balanced with a minimal increase in the cost (decrease is also possible).

Random graphs were generated by using two parameters, n_v and p, where $p, 0 < p < 1$, specifies the probability that any given pair of vertices constitute an edge, and n_v is the number of vertices. A formula $d = p(n_v - 1)$ defines the expected average degree of vertex. In our experiments the expected average degree of random graphs was 10.

The second type of graph has some inherent structure. These geometric graphs are characterized by two parameters n_v and d_v; d_v is the Euclidean distance between any given pair of vertices. To generate these graphs, $2n_v$ independent random numbers were selected from the uniform interval (0, 1) and were interpreted as coordinates of n_v points within the unit square. These points represent the vertices of the graph. An edge is added between any pair of vertices if and only if the Euclidean distance is less than or equal to d_v, and the expected average degree of vertex is defined as $d = n_v \pi d_v^2$. In our experiment, the expected average degree of vertex for the geometric graphs was also 10.

8.1 Parameters of the SA Algorithm

All parameters of SA were selected as in [JOH87]. Four graph sizes were selected 124, 250, 500, and 1000 vertices. Five different graphs of each dimension were randomly generated, that is, 20 different random graphs and 20 different geometric graphs with the average vertex degree equal to 10. Fifty runs with different random starts were performed for each graph of dimensions 124, 250, and 500. Twenty runs with different random starts were performed for graphs with 1000 vertices. Thus, 1700 runs for each algorithm in consideration were executed on random and geometric graphs. This constitutes 3400 runs for the two algorithms discussed in this paper. The initial probability of accepting a "bad" move is 0.4 and the initial parameter T_0 ("temperature") was determined from this condition. The cooling factor α was 0.95. The number of iterations with the same parameter T, called an "epoch length," was $16n_v$, where n_v is the size of the neighborhood. The process is terminated when the objective function did not decrease during the last five temperatures and the percentage of accepted moves did not exceed 2%. At this point the process is considered "frozen," and the best value of the objective function achieved during the run and time spent are registered.

8.2 Parameters of the LD Algorithm

There are two parameters in the LD algorithm: the decrease factor α and the initial probability of accepting a "bad" move. All parameters were chosen the same as for the SA algorithm; α was 0.95 and the acceptance probability was 0.4. The "epoch" length was $16n_v$, where n_v is the size of the neighborhood. Run time was the same as for SA.

8.3 Environment

All algorithms were coded in C and executed on the Appolo Domain/IC System SR9.5 network. The node type was DN4000 with the 68881 floating point accelerator and 8.0 Mb of main memory.

Table 1 Experimental Data*

	Random graphs		Geometric Graphs	
	Simulated annealing	Linear difference	Simulated annealing	Linear difference
124 vertices				
Average runtime	13	13	10	10
Average initial cut-size	310	310	266	266
Average cut-size achieved	188	177	25	19
Average no. of probes	43223	53940	32323	41220
Average no. of probes for reaching min. cut-size	32349	46044	21383	24712
Average time for reaching min. cut-size	8	10	5	5
250 vertices				
Average runtime	39	39	29	29
Average initial cut-size	634	634	554	554
Average cut-size achieved	364	344	45	39
Average no. of probes	132930	155388	95087	111708
Average no. of probes for reaching min. cut-size	111048	140035	72987	90005
Average time for reaching min. cut-size	31	34	21	22
500 vertices				
Average runtime	116	116	78	78
Average initial cut-size	1281	1281	1180	1180
Average cut-size achieved	714	689	75	76
Average no. of probes	355512	397704	245532	369324
Average no. of probes for reaching min. cut-size	311448	370434	201715	251765
Average time for reaching min. cut-size	100	107	62	69
1000 vertices				
Average runtime	263	263	207	207
Average initial cut-size	2530	2530	2380	2380
Average cut-size achieved	1410	1368	176	137
Average no. of probes	770336	834330	585993	689700
Average no. of probes for reaching min. cut-size	682264	798662	498016	640400
Average time for reaching min. cut-size	229	251	172	191

*Time units are seconds.

9. EXPERIMENTAL RESULTS

Experimental results for two algorithms described in the previous text are presented in Table 1. Table 1 contains results from the application of the two algorithms to random graphs and geometric graphs of increasing dimensions. This provides information on the average cut-size achieved for each of the graphs sizes by each algorithm, the average number of probes made by an algorithm during the run, the average time for reaching the minimal cost, etc.

9.1 Discussion

Our results on cut sizes obtained by the SA algorithm are reasonably close to those reported by Johnson et al. For SA on one graph of each dimension 124, 250, 500, and 1000 the best cuts found [JOH87] were respectively 178, 357, 628, and 1367. In our experiments the best cuts found by SA for the same dimension and expected average vertex degree were respectively 170, 338, 661, and 1306. One can see that these differences are $\sim 5\%$ of minimal value. This difference is due to the randomized procedure of graph generation.

On random graphs the SA algorithm lost to the LD algorithm for all dimensions.

On geometric graphs the LD was better than SA for three dimensions out of four (124, 250, 1000).

10. CONCLUSIONS

1. The SA algorithm belongs to the class of S-type GH-stochastic asymptotically optimal Moore automata with variable structure.
2. Other automata from the same class can be used for combinatorial optimization.
3. Performance characteristics of other automata are not inferior to those of SA.

REFERENCES

[BUS58] Bush, R., and F. Mosteller, *Stochastic Models for Learning* (Wiley, New York, 1958).

[CAM88] Cameron, P. J., M. Deza, and D. Frankl, Intersection theorems in permutation groups, *Combinatorica* 8 (3) (1988) 249–260.

[COH80] Cohen, G., and M. Deza, Some metrical problems on S_n, *Ann. Discrete Math.*, Vol. 8, 1980, 211–219.

[ISA76] Isaacson, D. L., and R. W. Madsen, *Markov Chains Theory and Applications* (John Wiley & Sons, New York, 1976).

[JOH87] Johnson, D., C. Aragon, L. McGeoch, and C. Schevon, Optimization by simulated annealing: Experimental evaluation (Part 1), Preprint.

[KIR83] Kirkpatrick, S., C. D. Gelatt, Jr. and M. P. Vecchi, Optimization by simulated annealing, *Science*, 220 (1983) 671–680.

[LAK73] Lakshmivarahan, S., and M. Thathachar, Absolutely expedient learning algorithms for stochastic automata, *IEEE Trans. on Systems, Man and Cybernetics*, SMC-3 (1973) 281–286.

[McM66] McMurthy, G., and K. Fu, A Variable Structure Automation Used as a Multinodal Search Technique, *IEEE Trans. Automat. Contr. AC-11* (1966) 379–387.

[MIT86] Mitra, D., F. Romeo, and A. Sangiovanni-Vincentelli, Convergence and finite time behavior of simulated annealing, *J. Adv. Appl. Probability* 18 (1986) 741–771.

[NAH85] Nahar, S., S. Sahni, and E. Shragowitz, Experiments with simulated annealing, *Proc. 22nd Des. Automation Conf.*, Las Vegas, June 1985, pp. 748–752.

[NAR74] Narendra, K., and M. Thathachar, Learning automata—a survey, *IEEE Trans. on Systems, Man and Cybernetics*, SMC-4 (1974) 323–334.

[REI65] Reiter, S., and G. Sherman, Discrete optimizing, *J. Soc. Ind. Appl. Math.* Vol. 13, 3 (1965) 864–889.

[REI66] Reiter, S., and D. Rice, Discrete optimizing solution procedures for linear and nonlinear integer programming problems, *Manage. Sci.* 11 (1966) 829–850.

[ROM84] Romeo, F., and A. Sangiovanni-Vincentelli, Probabilistic Hill Climbing Algorithms: Properties and Applications, Report No. UCB/ERL M84/34, Electronics Research Laboratory, University of California (1984).

[SER81] Sergienko, I., and M. Kaspschitzkaya, *Models and Methods for Solving Combinatorial Problem on Computers* (Naukova Dumka, Kiev, 1981).

[SHA69] Shapiro, I., and K. Narendra, Use of stochastic automata for parameter self-optimization with multimodal performance criteria, *IEEE Trans. System Science and Cybernetics*, SSC-5, 4(1969) 352–360.

[TSE61] Tsetlin, M., On the behavior of finite automata in random media, *Automatika i Telemechanika*, vol. 22, 10 (1961) 1345–1354.

[TSE69] Tsetlin, M., *Automation Theory and Modeling of Biological Systems* (Nauka, Moscow, 1969).

[TZ77] Tzypkin, Y., and A. Poznyak, Learning automata, *J. of Cybernetics and Information Science*, ASC, 1, #(2–4) (1977) 128–160.

[VAR63] Varshavskii, V., and I. Vorontsova, On the behavior of stochastic automata with a variable structure, *Automatika i Telemechanika*, vol. 24, 3 (1963) 353–360.

[VAR73] Varshavskii, V., *Collective Behavior of Automata* (Nauka, Moscow, 1973)

[VOR65] Vorontsova, I., Algorithms for changing stochastic automata transition probabilities, *Problemi Peredachi Informatzii* vol. 1, 3(1965) 122–126.

30

Solution of Large GSPN Models

GIANFRANCO CIARDO Software Productivity Consortium, Herndon, Virginia

KISHOR S. TRIVEDI Department of Computer Science, Duke University, Durham, North Carolina

ABSTRACT

Generalized stochastic Petri nets (GSPNs) can be effectively used to represent many systems in a compact way. Efficient algorithms for the translation of a GSPN into its underlying continuous-time Markov chain (CTMC) are known and have been implemented in a number of software packages. While GSPNs relieve the modeler from the cumbersome and error-prone task of building and inputting the CTMC by hand, their analytical tractability is still limited by the combinatorial growth of the underlying CTMC.

In order to avoid construction and solution of a large CTMC, we propose decomposing the GSPN into a set of subnets and separately solving individual subnets. Dependence among the subnets requires that, after solving each

---

This work was supported in part by the U.S. Office of Naval Research under contract N3014-88-K-0623

subnet, certain quantities be *exported* to other subnets. A fixed-point iteration is then used over the exported quantities.

We discuss ways of decomposing a net into subnets, the type of quantities that need to be exchanged between subnets, and the convergence of the fixed-point iterative schemes.

1. INTRODUCTION

Owing to the increasing complexity of computer and communication systems, there is an acute need for cost-effective techniques and tools for their performance and dependability evaluation. Markov and Markov reward models are often used for such evaluations. Unfortunately, the state spaces of Markov models are frequently unmanageable in practice.

Generalized stochastic Petri nets (GSPNs) [2] can be used to alleviate the problem of generation and specification of a large Markov model. GSPNs are powerful and compact in model specification, and efficient algorithms for automatic generation of the underlying Markov model are known and have been implemented [3,7,9]. Further extension to GSPN-reward models has also been proposed so that a powerful method of specification and generation of Markov reward models exists [9]. Nevertheless, the combinatorial growth of their state space (reachability graph) constitutes a major limitation to applicability of GSPN (-reward) models in real-life problems.

If we can identify states with low probability, we can avoid the generation of such states. Such *state truncation* methods are an integral part of many reliability [6] and availability analysis tools [11]. Enabling functions can be used to specify state truncation in GSPNs [9]. The error resulting from state truncation has also been analyzed [13,16].

An alternative way of avoiding the generation and solution of a large one-level model is to decompose the GSPN into nearly independent subnets. The time for the generation, the storage space, and the solution time are reduced drastically in this way. However, since the subnets are dependent, their solutions need to iteratively communicate. Such fixed-point iterative schemes have been used to solve non-product-form queuing networks [10,14]. The use of fixed-point iteration to solve large GSPNs is believed to be new, however.

Three important issues arise in this connection: how to decompose a GSPN into subnets, kinds of quantities that need to be communicated from one subnet solution to another subnet, and the convergence behavior of the iterative schemes. We do not propose an automatic method of decomposing GSPNs. Instead, we present four common types of GSPN structures that can be decomposed. We show that the probability that the subnet is in a desired state, the average rate at which a given condition goes off, and the expected time until the subnet satisfies a given condition are three quantities that suffice for exporting in the iterative schemes for the four GSPN structure types that we define. Questions of convergence of the iterative schemes are also explored but more work is required on this topic.

In Section 2, we recall the GSPN model and its solution, and in Section 3, we introduce the quantities needed in our decomposition approach. Finally, in Section 4, we show some nearly independent GSPN structures and we describe how to decompose such GSPNs.

2. THE GSPN MODEL

A Petri net (PN) [19,20] is a directed bipartite graph with two sets of nodes, *places* and *transitions*. *Input arcs* connect a place to a transition, *output arcs* connect a transition to a place. A multiplicity (positive integer) may label an arc. The *input* (*output*) *bag* for a transition is the bag [19] constituted by the input (output) arcs, considered with their multiplicity. Each place may contain any number of *tokens*. A *marking* is a bag representing the configuration of tokens in the places of the PN—it is the "state" of the PN.

A transition is *enabled* if its input bag is a subbag of the (current) marking. When a transition is enabled, it can *fire*, leading the PN into a different marking, obtained by subtracting its input bag from and adding its output bag to the current marking. A marking is *reachable* if it is obtained by a sequence of firings starting in the *initial marking*. The *reachability set* (*graph*) is the set (graph) of all the reachable markings (connected by arcs representing the transition firings). A set S of transitions is a *conflicting transition set* in a marking m if the contemporary firing of all the transitions of S is impossible in m, or, in other words, if the sum of the input bag of the transitions in S is not a subbag of m.

A GSPN [2] is a PN where each transition has an associated *firing time*, which can be zero (*immediate transition*) or exponentially distributed with a parameter dependent on the marking (*timed transition*). If several conflicting immediate transitions are enabled in a marking, a *firing probability* is specified for each of them. A firing probability for timed transitions does not need to be specified, since we assume a *race model*: the first firing time to elapse determines which transition fires first; contemporary elapsing of firing times has probability zero.

If at least one immediate transition is enabled, the marking is said to be *vanishing*, otherwise the marking is said to be *tangible*. Since the firing time of an immediate transition is zero, a GSPN does not remain in a particular vanishing marking for any length of time. In other words, the probability of finding the GSPN in a vanishing marking is zero.

A GSPN describes an underlying stochastic process, captured by the "extended reachability graph" (ERG), a reachability graph with additional stochastic information on the arcs. The ERG (hence the GSPN) is reducible to a CTMC by *eliminating the vanishing markings* [3,7,8]. The size of the ERG is usually very large; it represents the main obstacle to the analysis of GSPNs. The elimination of the vanishing markings reduces the number of states, but the number of nonzero entries in the matrix representation may increase [8].

Examples of output measures obtained from the analysis of a GSPN are the expected number (or probability distribution) of tokens in a given place or the throughput of a given transition, in steady state. They correspond to reward functions on the markings or on the transitions among markings, and they are easily computed after having solved for the steady-state probability of each marking.

3. NOTATION

In the following, $\#_{\mathcal{A}}(p,m)$ and $\mu_{\mathcal{A}}(t,m)$ indicate the number of tokens in place p and the rate of transition t when GSPN $\mathcal{A}$ is in marking m, respectively (m is omitted when no ambiguity can arise).

Note that GSPNs restrict timed transitions to have exponentially distributed firing times. Nevertheless, we can synthesize a timed transition with any positive coefficient of variation using a subnet [4,9,15]. For this reason, we will often say that the firing time of transition t has mean T_t and coefficient of variation CV_t.

Observing the marking of a GSPN $\mathcal{A}$, we can say that *condition c is ON* or *OFF*. For example, c could be "$\#_{\mathcal{A}}(p_7) = \#_{\mathcal{A}}(p_9)$." The following quantities can be defined:

$$\lambda_{\mathcal{A}}(c) = \text{steady-state average rate at which the condition } c \text{ goes } OFF \text{ given that it is } ON$$

$$PROB_{\mathcal{A}}(c) = \Pr\{\text{condition is } ON \text{ in steady state}\}$$

$$WAIT_{\mathcal{A}}(c) = E[\text{time until condition } c \text{ is } ON \text{ starting from steady state}]$$

Partition the set of states $\mathcal{S}$ of the CTMC underlying $\mathcal{A}$ (they correspond to the set of tangible markings) into $\mathcal{S}_c$ and $\mathcal{S}_{\bar{c}}$, where condition c is respectively *ON* or *OFF*. If $\underline{\pi}$ is the steady-state probability vector of the CTMC, and $Q = [q_{i,j}]$ is the infinitesimal generator matrix of the CTMC:

$$\lambda_{\mathcal{A}}(c) = \frac{\sum_{i \in \mathcal{S}_c, j \in \mathcal{S}_{\bar{c}}} \pi_i q_{i,j}}{\sum_{k \in \mathcal{S}_c} \pi_k}$$

and

$$PROB_{\mathcal{A}}(c) = \sum_{k \in \mathcal{S}_c} \pi_k = \frac{\lambda_{\mathcal{A}}(\bar{c})}{\lambda_{\mathcal{A}}(c) + \lambda_{\mathcal{A}}(\bar{c})}$$

where the last equality is proved using the relation

$$\sum_{i \in \mathcal{S}_c, j \in \mathcal{S}_{\bar{c}}} \pi_i q_{i,j} = \sum_{i \in \mathcal{S}_{\bar{c}}, j \in \mathcal{S}_c} \pi_i q_{i,j}$$

$WAIT_{\mathcal{A}}(c)$ is the mean time to absorption (MTA) when the markings in $\mathcal{S}_c$ are considered absorbing:

$$WAIT_{\mathcal{A}}(c) = \sum_{k \in \mathcal{S}_{\bar{c}}} x_k^{\bar{c}} \qquad \text{where} \quad \underline{x}^{\bar{c}} \text{ satisfies } \underline{x}^{\bar{c}} Q^{\bar{c}} = -\underline{\pi}^{\bar{c}}$$

($\underline{x}^{\bar{c}}$ represents the expected time spent in each nonabsorbing tangible marking before absorption, $Q^{\bar{c}}$ is the matrix obtained from Q by eliminating the rows and the columns corresponding to $\mathcal{S}_c$, and $\underline{\pi}^{\bar{c}}$ is the restriction of $\underline{\pi}$ to $\mathcal{S}_{\bar{c}}$).

These quantities are illustrated by means of the CTMC of Figure 1; the sojourn time in state i is $\sim EXPO(1/i)$, and its expected value is i. The steady-state probabilities are

$$\pi_1 = 0.1 \qquad \pi_2 = 0.2 \qquad \pi_3 = 0.3 \qquad \pi_4 = 0.4$$

If $\mathcal{S}_c = \{1, 2\}$ and $\mathcal{S}_{\bar{c}} = \{3, 4\}$, then

$$Q^{\bar{c}} = \left[\begin{array}{c|c} -1/3 & 1/3 \\ \hline 0 & -1/4 \end{array}\right] \qquad \underline{\pi}^{\bar{c}} = [0.3 \mid 0.4] \qquad \underline{x}^{\bar{c}} = [0.9 \mid 2.8]$$

$$Q^{c} = \left[\begin{array}{c|c} -1/1 & 1/1 \\ \hline 0 & -1/2 \end{array}\right] \qquad \underline{\pi}^{c} = [0.1 \mid 0.2] \qquad \underline{x}^{c} = [0.1 \mid 0.6]$$

and

$$\lambda(c) = \frac{0.2 \cdot 1/2}{0.1 + 0.2} = \frac{1}{3}$$

$$\lambda(\bar{c}) = \frac{0.4 \cdot 1/4}{0.3 + 0.4} = \frac{1}{7}$$

$$PROB(c) = 0.1 + 0.2 = 0.3$$

$$WAIT(c) = 0.9 + 2.8 = 3.7$$

$$WAIT(\bar{c}) = 0.1 + 0.6 = 0.7$$

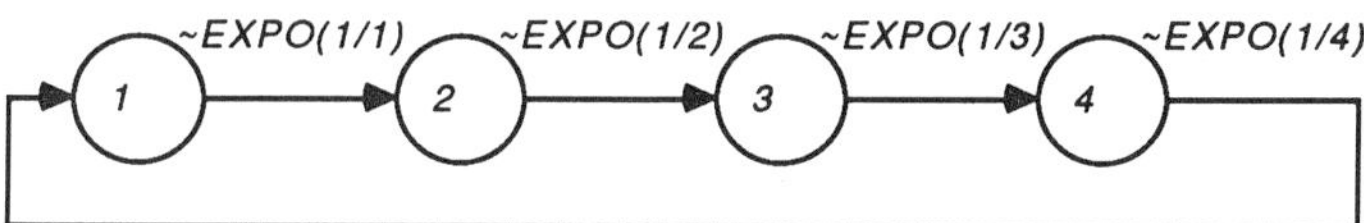

Figure 1 The computation of λ, *PROB*, and *WAIT*.

4. DECOMPOSITION BASED ON NEAR-INDEPENDENCE

Most systems can be logically decomposed according to criteria such as timing, functionality, or physical connectivity. The corresponding subsystems may not be studied in complete isolation; nevertheless, they are often largely autonomous and their interactions are limited in number and type. The corresponding GSPN models reveal this *near-independence* by exhibiting certain structures, which we use as a guide in our decomposition approach. The following steps are needed:

1. *Decomposition.* Given the initial GSPN $\mathcal{A}$, derive M subnets $\mathcal{A}_* = \{\mathcal{A}_1, \ldots, \mathcal{A}_M\}$.
2. *Imports.* Each subnet $\mathcal{A}_i$ may *import* certain quantities from other subnets and, in turn, may *export* quantities to other subnets. If $\mathcal{A}_i$ imports a quantity from $\mathcal{A}_j$, we write $\mathcal{A}_j \succ \mathcal{A}_i$. We denote the transitive closure of this "import relation" with the symbol $\succ\!\!\succ$. Often the import relation (or graph) is cyclic: both $\mathcal{A}_i \succ\!\!\succ \mathcal{A}_j$ and $\mathcal{A}_j \succ\!\!\succ \mathcal{A}_i$ hold. Even the case $\mathcal{A}_i \succ \mathcal{A}_i$ (cycle of length one) may arise.
3. *Partial order.* If the import graph is a DAG (directed acyclic graph), it defines a partial order on the subnets: if $\mathcal{A}_i \succ\!\!\succ \mathcal{A}_j$, $\mathcal{A}_i$ must be solved before $\mathcal{A}_j$. If neither $\mathcal{A}_i \succ\!\!\succ \mathcal{A}_j$ nor $\mathcal{A}_j \succ\!\!\succ \mathcal{A}_i$, $\mathcal{A}_i$ and $\mathcal{A}_j$ can be solved in any order, or even concurrently.
4. *Initial guesses.* If the import graph contains strongly connected components, they must be "opened" by specifying the order for the solution of the subnets in them. For some subnet, though, the imports are not available, so initial guesses must be provided.
5. *Fixed-point iteration.* After each subnet has been solved once, new values for the initial guesses are available; the iteration continues until the imports remain stable between successive iterations.
6. *Composition.* The results from solving the subnets are combined to compute the desired results for the original GSPN.

In the following, we present four GSPN structures that exhibit near-independence, and we show how to decompose them. Under certain conditions, exact results are obtained. Intuitively, we can extrapolate and argue that, when the conditions are "almost met," the results are "almost exact." This is analogous to applying Norton's theorem to queuing networks even when the product-form requirements are not satisfied.

4.1 Processor Sharing

Consider a GSPN $\mathcal{A}$ consisting of M disconnected subnets $\mathcal{A}_1, \ldots, \mathcal{A}_M$. The only interaction among these subnets occurs when they issue requests to a pool of K indistinguishable resources, which are assigned in a "processor-sharing" fashion. If $N \leq K$ requests for them exist, each request is assigned one resource; if $N > K$, the K resources are simultaneously shared among the N requests, so each request is assigned a fraction N/K of a resource. The resources do not need to be represented explicitly in the GSPN; only the definition of the rate of the transitions using them is affected by K. Let L_i denote the number of transitions using the resources in subnet $\mathcal{A}_i$.

Assume that, if place $p_{i,j}$ $(1 \leq i \leq M, 1 \leq j \leq L_i)$ in $\mathcal{A}_i$ contains r tokens, transition $t_{i,j}$, also in $\mathcal{A}_i$, requests r resources, then its rate is

$$\mu_{\mathcal{A}}(t_{i,j}) = \mu_{i,j} \cdot \#_{\mathcal{A}}(p_{i,j}) \cdot \frac{\min\left\{K, \left(\sum_{1 \leq l \leq M, 1 \leq k \leq L_l} \#_{\mathcal{A}}(p_{l,k})\right)\right\}}{\sum_{1 \leq l \leq M, 1 \leq k \leq L_l} \#_{\mathcal{A}}(p_{l,k})} \tag{1}$$

for some constant $\mu_{i,j}$.

It is natural to separate $\mathcal{A}$ into its M components. When solving $\mathcal{A}_i$, though, we need to know $ER[j]$, the expected number of requests that $\mathcal{A}_j$ $(j \neq i)$ makes on the resources, so the import graph is a complete graph. The following algorithm can then be used:

Algorithm Processor-Sharing

for $i = 1$ to M do
    $ER[i] = 0$
repeat
    $error = 0$
    for $i = 1$ to M do
        "solve $\mathcal{A}_i$ using $ER[l]$ $(\forall l, l \neq i)$"
        $x = \sum_{j=1}^{L_i} E[\#_{\mathcal{A}_i}(p_{i,j})]$
        $error = \max\left\{error, \frac{|ER[i] - x|}{|x|}\right\}$
        $ER[i] = x$
until $error < \epsilon$

The statement "solve $\mathcal{A}_i$ using $ER[l]$ $(\forall l, l \neq i)$" means that, when solving $\mathcal{A}_i$, the rate for transition $t_{i,j}$ is set to

$$\mu_{\mathcal{A}_i}(t_{i,j}) = \mu_{i,j} \cdot \#_{\mathcal{A}_i}(p_{i,j}) \cdot \frac{\min\left\{K, \left(\sum_{k=1}^{L_i} \#_{\mathcal{A}_i}(p_{i,k}) + \sum_{l=1, l \neq i}^{M} ER[l]\right)\right\}}{\sum_{k=1}^{L_i} \#_{\mathcal{A}_i}(p_{i,k}) + \sum_{l=1, l \neq i}^{M} ER[l]} \tag{2}$$

The denominator in the fraction is the total number of requests to the K resources "to the best of $\mathcal{A}_i$'s knowledge." Instead of using the averages $ER[i]$, we could use the distributions

$$PR[i,n_i] = PROB\left(\sum_{j=1}^{L_i} \#(p_{i,j}) = n_i\right) \qquad (0 \le n_i \le N_i)$$

where N_i is the maximum number of requests that $\mathcal{A}_i$ may have active at the same time, a finite value since the reachability set is finite. We can then define

$$\mu_{\mathcal{A}_i}(t_{i,j}) = \mu_{i,j} \cdot \#_{\mathcal{A}_i}(p_{i,j}) \cdot \sum_{n_1=0}^{N_1} \cdots \sum_{n_{i-1}=0}^{N_{i-1}} \sum_{n_{i+1}=0}^{N_{i+1}} \cdots$$

$$\sum_{n_M=0}^{N_M} \left(\prod_{l=1,l\neq i}^{M} PR[l,n_l] \right) \frac{\min\left\{K, \left(\sum_{k=1}^{L_i} \#_{\mathcal{A}_i}(p_{i,k}) + \sum_{l=1,l\neq i}^{M} n_l\right)\right\}}{\sum_{k=1}^{L_i} \#_{\mathcal{A}_i}(p_{i,k}) + \sum_{l=1,l\neq i}^{M} n_l} \tag{3}$$

which is closer to equation (1) than equation (2) is, because it correctly computes the expectation of the fraction in equation (1). Equation (2) computes the expectations of its components and then composes their values, introducing an additional error, since $E[1/X]$ is different from $1/E[X]$.

We suggest using equation (3) instead of equation (2) whenever possible, that is, whenever $\prod_{i=1}^{M} N_i$ is not too large. Note that $ER[i]$ is a *PROB*-type quantity.

Convergence of the Iteration

Algorithm processor-sharing performs an iteration of the form $\underline{x}^{(k+1)} = \underline{f}(\underline{x}^{(k)})$, with $\underline{f}$: $\mathbb{R}^M \to \mathbb{R}^M$, and initial guess $\underline{0}$. The convergence of the iteration is related to the existence and uniqueness of a *fixed point*, a vector $\underline{x}^*$ such that $\underline{x}^* = \underline{f}(\underline{x}^*)$.

Conditions for the existence and uniqueness of the fixed point are given by the Contraction Mapping Theorem [18]:

> Let $\underline{a} \in \mathbb{R}^M$, $\underline{b} \in \mathbb{R}^M$, and define $D = \{\underline{x} : \underline{a} \ge \underline{x} \ge \underline{b}\} \subseteq \mathbb{R}^M$. If $\underline{f}$ is continuous with continuous first partial derivatives and if $\forall \underline{x} \in D, \underline{f}(\underline{x}) \in D$, then there exists a fixed point $\underline{x}^* \in D$ satisfying $\underline{x}^* = \underline{f}(\underline{x}^*)$.
>
> If, in addition, $\exists c < 1$ such that $\forall \underline{x} \in D$, $\forall \underline{y} \in D$, $\|\underline{f}(\underline{x}) - \underline{f}(\underline{y})\| \le c \cdot \|\underline{x} - \underline{y}\|$, then the fixed point is unique.

We need to determine when a given GSPN corresponds to a function $\underline{f}$ having these properties. The conditions for existence are always met in our decomposition schema if the underlying reachability graphs are finite:

The imports are bounded: the probability of a certain event is always between zero and one, the expected number of tokens in a given place is always between zero and a maximum, the expected throughput on a certain path is always between zero and the expected firing rate of the slowest timed transition on the path (assume that the firing rate of at least one timed transition on the path is not affected by the imports).

The exports are continuous functions of the imports, since the rates of the underlying CTMCs are continuous functions of the imports and the exports are continuous functions of those rates; the same applies for the partial derivatives.

The condition for uniqueness is instead harder to prove (it is sufficient, not necessary). An even stricter condition is

$$\exists c < 1, \forall i, \forall j, \left| \frac{\partial f_i(\underline{x})}{\partial x_j} \right| \leq \frac{c}{M}$$

which is easier to test, but even more difficult to satisfy in practice.

In our "processor-sharing" algorithm, for example, we should prove that

$$\forall i, \forall l \qquad \left| \frac{\partial \sum_{j=1}^{L_i} E[\#_{\mathcal{A}_i}(p_{i,j})]}{\partial ER[l]} \right| \leq \frac{c}{M} \tag{4}$$

for some $c < 1$, but this is not true in general. Given the meaning of $ER[l]$, we can easily argue that, normally, increasing the number of available resources to subnet $\mathcal{A}_i$ by a factor ν (decreasing each $ER[l]$) can only decrease the number of steady-state requests to the resources by a factor ν', $1 \leq \nu' < \nu$. $\nu' = 1$ can only be achieved when the number of request is independent of the availability: trivially, when all the transitions in $\mathcal{A}_i$ need the resources and $\mathcal{A}_i$ is conservative. $\nu' = \nu$ would imply that decreasing the service time does not increase the throughput; in a closed system under a work-conserving policy, this can only happen, in the limit, when some other transition not requesting the resources is the bottleneck. Therefore, a "well-behaved" $\mathcal{A}_i$ satisfies

$$\forall i, \forall l \qquad \left| \frac{\partial \sum_{j=1}^{L_i} E[\#_{\mathcal{A}_i}(p_{i,j})]}{\partial ER[l]} \right| \leq c$$

for some $c < 1$, but not necessarily equation (4).

For $M = 2$, we can prove convergence when $\mathcal{A}_1$ and $\mathcal{A}_2$ are well behaved:

$$\left| \frac{df_1(x_2)}{dx_2} \right| = \left| \frac{\partial \sum_{j=1}^{L_1} E[\#_{\mathcal{A}_1}(p_{1,j})]}{\partial ER[2]} \right| \leq c$$

and

$$\left| \frac{df_2(x_1)}{dx_1} \right| = \left| \frac{\partial \sum_{j=1}^{L_2} E[\#_{\mathcal{A}_2}(p_{2,j})]}{\partial ER[1]} \right| \leq c$$

for some constant c (recall that $\forall i, 1 \le i \le M, \partial f_i/\partial x_i = 0$). The system of equations can be reduced to a single fixed-point equation in one variable:

$$x_1 = f_1(x_2) = f_1(f_2(x_1)) = g_1(x_1)$$

Then

$$\frac{dg_1(x_1)}{dx_1} = \frac{df_1(x_1)}{dx_2}\frac{df_2(x_1)}{dx_1} \le c^2 < c$$

This condition guarantees convergence of the iteration $x_1^{(k+1)} = g_1(x_1^{(k)})$.

THEOREM 1 Algorithm processor-sharing gives exact results if

$$\forall i, 1 \le i \le M \qquad \text{Var}\left(\sum_{j=1}^{L_i} \#_{\mathcal{A}_i}(p_{ij})\right) = 0$$

The proof is trivial: the partial derivatives in equation (4) are zero when $\sum_{j=1}^{L_i} \#_{\mathcal{A}_i}(p_{ij})$ is constant so convergence to a unique solution is guaranteed. Then the rates of the transitions using the resource in the exact and approximate nets are

$$\mu_{\mathcal{A}}(t_{ij}) = \mu_{\mathcal{A}_i}(t_{ij}) = \mu_{ij} \cdot \#_{\mathcal{A}_i}(p_{ij}) \cdot \frac{\min\{K, R\}}{R}$$

where $R = \sum_{i=1}^{M}\sum_{j=1}^{L_i} \#_{\mathcal{A}_i}(p_{ij})$ is a constant.

If the requests are really constant, algorithm processor-sharing requires two iterations. We could obtain the same results by computing R first and then solving each $\mathcal{A}_i$ once in isolation. Theorem 1, though, suggests that a high variance in the number of requests affects negatively both the quality of the approximation and the rate of convergence.

Empirical Results

In general, several resource types can be present [5,12]; the GSPN we consider for our numerical experiments (Figure 2) has two resources, one of type a and one of type b. Transitions t_3, t_4, u_2, and u_3 need resource a; transitions t_1 and u_4 need resource b. The remaining transitions, t_2 and u_1, do not share resources. The initial number of tokens N_p (in p_1) and N_q (in q_1) is set to 1, 2, and 5 independently (9 cases). The imports to $\mathcal{A}_1$ from $\mathcal{A}_2$ are

$$ER_a[2] = E[\#(q_2) + \#(q_3)] \qquad \text{and} \qquad ER_b[2] = E[\min\{\#(q_4), \#(q_5)\}]$$

while the imports to $\mathcal{A}_2$ from $\mathcal{A}_1$ are

$$ER_a[1] = E[\#(p_2) + \#(p_3)] \qquad \text{and} \qquad ER_b[1] = E[\#(p_1)]$$

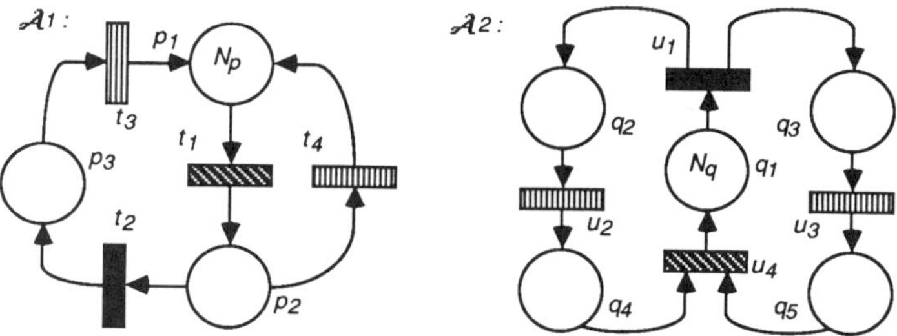

Figure 2 Two types of resources.

It is assumed that the number of "requests" waiting at u_4 is counted as the minimum of the number of tokens in q_4 and q_5; any other function of the marking of $\mathcal{A}_2$ could be used.

We choose to start the iteration with $\mathcal{A}_1$, so initial guesses are needed for $ER_a[2]$ and $ER_b[2]$. Using values between 0 and 10, well beyond the range of "reasonable" guesses, the same results are obtained. In all the cases, convergence ($\epsilon < 10^{-4}$) is achieved within four to 10 iterations, depending on the initial marking. The following rates are used in $\mathcal{A}$, $\mathcal{A}_1$, and $\mathcal{A}_2$:

$$\mu_{\mathcal{A}}(t_1) = \frac{1.0 \cdot \#(p_1)}{\#(p_1) + \min\{\#(q_4), \#(q_5)\}}$$

$$\mu_{\mathcal{A}_1}(t_1) = \frac{1.0 \cdot \#(p_1)}{\#(p_1) + ER_b[2]}$$

$$\mu_{\mathcal{A}}(t_2) = 2.0$$

$$\mu_{\mathcal{A}_1}(t_2) = 2.0$$

$$\mu_{\mathcal{A}}(t_3) = \frac{3.0 \cdot \#(p_3)}{\#(p_2) + \#(p_3) + \#(q_2) + \#(q_3)}$$

$$\mu_{\mathcal{A}_1}(t_3) = \frac{3.0 \cdot \#(p_3)}{\#(p_2) + \#(p_3) + ER_a[2]}$$

$$\mu_{\mathcal{A}}(t_4) = \frac{4.0 \cdot \#(p_2)}{\#(p_2) + \#(p_3) + \#(q_2) + \#(q_3)}$$

$$\mu_{\mathcal{A}_1}(t_4) = \frac{4.0 \cdot \#(p_2)}{\#(p_2) + \#(p_3) + ER_a[2]}$$

$$\mu_{\mathcal{A}}(u_1) = 1.0$$

$$\mu_{\mathcal{A}_2}(u_1) = 1.0$$

$$\mu_{\mathcal{A}}(u_2) = \frac{2.0 \cdot \#(q_2)}{\#(p_2) + \#(p_3) + \#(q_2) + \#(q_3)}$$

$$\mu_{A_2}(u_2) = \frac{2.0 \cdot \#(q_2)}{\#(q_2) + \#(q_3) + ER_a[1]}$$

$$\mu_{\mathcal{A}}(u_3) = \frac{3.0 \cdot \#(q_3)}{\#(p_2) + \#(p_3) + \#(q_2) + \#(q_3)}$$

$$\mu_{A_2}(u_3) = \frac{3.0 \cdot \#(q_3)}{\#(q_2) + \#(q_3) + ER_a[1]}$$

$$\mu_{\mathcal{A}}(u_4) = \frac{4.0 \cdot \min\{\#(q_4), \#(q_5)\}}{\#(p_1) + \min\{\#(q_4), \#(q_5)\}}$$

$$\mu_{A_2}(u_4) = \frac{4.0 \cdot \min\{\#(q_4), \#(q_5)\}}{\min\{\#(q_4), \#(q_5)\} + ER_b[1]}$$

Table 1 shows the relative error on the estimate of $ER_a[1], ER_b[1], ER_a[2]$, and $ER_b[2]$ as a function of the initial marking.

The quantities $\mu_{i,j}$ are related to the amount of work each token enabling transition $t_{i,j}$ requests from the resources. It is possible to consider a GSPN where only "the first token in line at $t_{i,j}$" can obtain the resource, by substituting $\#(p_{i,j})$ with the indicator $\mathbf{1}\{\#(p_{i,j}) > 0\}$. We can express the same concept by saying that the weight of transition $t_{i,j}$ is $\#(p_{i,j})$ or $\mathbf{1}\{\#(p_{i,j}) > 0\}$, respectively. In general, we can give an arbitrary positive marking-dependent weight to each transition, resulting in a different allocation of the resource(s). A zero weight would disable the transition whenever another transition with a positive weight is requesting the same resource. This corresponds to a priority policy, not a processor-sharing policy.

Table 1 Percent Errors for Shared Resources

		% Error on			
N_p	N_q	$ER_a[1]$	$ER_b[1]$	$ER_a[2]$	$ER_b[2]$
1	1	+1.51	−0.57	+0.51	−2.50
1	2	+3.08	−1.53	+0.87	−2.83
1	5	+8.20	−6.88	+2.15	−3.60
2	1	+0.06	−0.01	+0.89	−2.66
2	2	+0.99	−0.36	+1.33	−3.22
2	5	+6.82	−4.73	+3.41	−4.97
5	1	−0.96	+0.11	+0.89	−1.43
5	2	−1.90	+0.30	+1.41	−2.05
5	5	+0.51	−0.18	+3.99	−3.94

4.2 One-Way Condition Testing

Consider the GSPN $\mathcal{A}$ in Figure 3(a). Transition t_1 (timed or immediate) is enabled only when $\#(p_1) > 0$ and $\#(p_2) > 0$, but its firing does not remove the token in p_1. This GSPN is more general than it may seem at first, since the tokens in p_1 and p_2 may represent arbitrarily complex conditions in $\mathcal{X}_1$ and $\mathcal{X}_2$, respectively; we only assume that no other marking dependency exists between any transition in $\mathcal{X}_2$ and the marking of $\mathcal{X}_1$, or any transition in $\mathcal{X}_1$ and the marking of $\mathcal{X}_2$.

We can define $\mathcal{A}_1$ and $\mathcal{A}_2$ such that $\mathcal{A}_1 \succ \mathcal{A}_2$ and $\mathcal{A}_2 \not\succ \mathcal{A}_1$ [Figure 3(b)]. When subnet $\mathcal{A}_1$ is solved in isolation, the results exactly match the ones for $\mathcal{X}_1$ and p_1 that would be obtained from $\mathcal{A}$. This is a general property of our approach: any node without entering arcs in the import graph corresponds to a subnet whose results are "exact" (if the decomposition is done correctly).

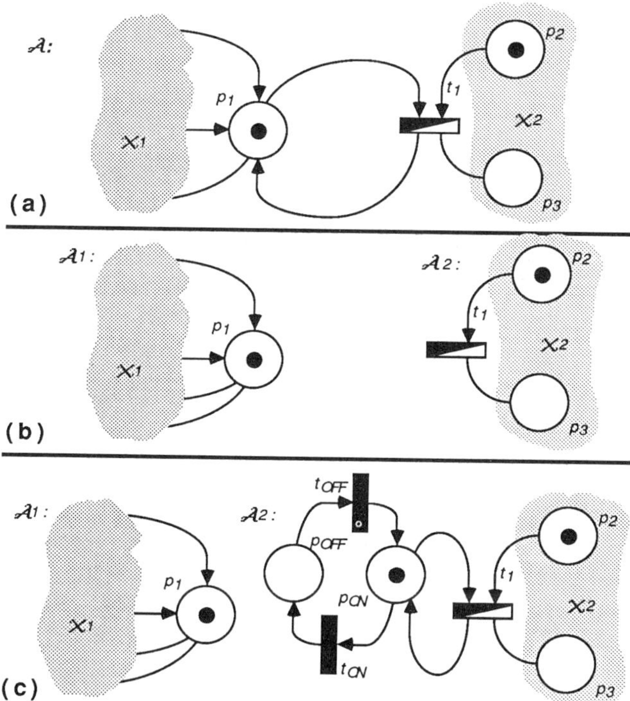

Figure 3 One-way condition testing.

Rarefaction and Delay

The definition of $\mathcal{X}_2$ in $\mathcal{A}_2$ is the same as in $\mathcal{A}$, but the timing of t_1 is different. If t_1 is a timed transition, it should be slowed down, since t_1 is enabled in $\mathcal{A}_2$ when $\#(p_1) = 0$ and $\#(p_2) > 0$, but it would not be enabled in $\mathcal{A}$. We can set

$$\mu_{\mathcal{A}_2}(t_1, m^{(2)}) = \mu_{\mathcal{A}}(t_1, m) \cdot PROB_{\mathcal{A}_1}(\#(p1) > 0)$$

where $m^{(2)}$ is the projection of m to $P^{\mathcal{A}_2}$ and $PROB_{\mathcal{A}_1}(\#(p_1) > 0)$ is computed from $\mathcal{A}_1$ in isolation. The definition is consistent because we required the rate of t_1 to depend only on the marking of $\mathcal{X}_2$. This method constitutes a *rarefaction* of the rate of t_1; the firing time for t_1 is interrupted every time the token is taken away from p_1. This method is analogous to the "shadow" CPU method proposed by Sevcik in another context [21].

THEOREM 2 Consider GSPN $\mathcal{A}$ in Figure 3(a) and its decomposition using rarefaction [Figure 3(b)]. Assume that the transitions in $\mathcal{X}_1$ can be speeded up by a factor σ and that $\mathcal{A}_1$ has an underlying ergodic CTMC of finite size k described by a transition rate matrix $Q^{(1)}(\sigma) = \sigma Q^{(1)}$. Then

$$\lim_{\sigma \to \infty} PROB_{\mathcal{A}}(c) = PROB_{\mathcal{A}_2}(c)$$

for any condition c on the marking of $\mathcal{X}_2$.

This theorem is proved in [8]; by continuity, it guarantees that, if the activities in $\mathcal{A}_1$ are "fast," the approximation obtained for $\mathcal{A}_2$ is good. If the transitions in $\mathcal{A}_1$ are fast, $\lambda_{\mathcal{A}_1}(\#(p_1) > 0)$ and $\lambda_{\mathcal{A}_1}(\#(p_1) = 0)$ are both fast. We then say that c is a "fast changing condition." If the condition changes slowly, the approximation is poor; furthermore, rarefaction cannot be applied if t_1 is an immediate transition. The two problems have the same origin; if t_1 is fast compared to the condition, the major effect of p_1 on t_1 does not lie in the repeated preemption of t_1, but in the time t_1 has to "wait" for a token to arrive in p_1 if none is there; once a token arrives in p_1, t_1 most likely fires without being preempted. The estimation of this *delay* τ is difficult; even if $\mathcal{A}_2$ does not affect $\mathcal{A}_1$, the condition is not tested (a token does not arrive in p_2) at "random" times for $\mathcal{A}_1$. There can be some synchronization between $\mathcal{A}_1$ and $\mathcal{A}_2$, even if it is only $\mathcal{A}_2$ to be affected by it. Unfortunately, if we do not have better information, we can only assume random incidence in $\mathcal{A}_1$ and set the expected delay τ to $WAIT_{\mathcal{A}_1}(\#(p_1) > 0)$. The rate for t_1 in $\mathcal{A}_2$ is then

$$\mu_{\mathcal{A}_2}(t_1, m^{(2)}) = \left(WAIT_{\mathcal{A}_1}(\#(p_1) > 0) + \frac{1}{\mu_{\mathcal{A}}(t_1, m)} \right)^{-1}$$

We can compose rarefaction and delay in a single heuristic, (*rar-del*),

$$\mu_{\mathcal{A}_2}(t_1,m^{(2)}) = \left(WAIT_{\mathcal{A}_1}(\#(p_1) > 0) + \frac{1}{\mu_{\mathcal{A}}(t_1,m)PROB_{\mathcal{A}_1}(\#(p_1) > 0)} \right)^{-1}$$

which appears to give the best results, since the delay $WAIT_{\mathcal{A}_1}(\#(p_1) > 0)$ becomes negligible and the rarefaction $PROB_{\mathcal{A}_1}(\#(p_1) > 0)$ determines the rate when the condition is fast changing, while the opposite is true if the condition is slow.

Compression

Rar-del leads to an approximation because $\mathcal{A}_1$ "may not be in steady state" when t_1 tests p_1. We can take a different approach and *compress* the behavior of $\mathcal{A}_1$ into a small subnet of $\mathcal{A}_2$ [Figure 3(c)] without changing the rate of t_1. The rates for t_{ON} and t_{OFF} are

$$\mu_{ON} = \lambda_{\mathcal{A}_1}(\#(p_1) > 0) \qquad \text{and} \qquad \mu_{OFF} = \lambda_{\mathcal{A}_1}(\#(p_1) = 0)$$

respectively.

THEOREM 3 Compression provides exact results if (i) condition "$\#(p_1) > 0$" is fast changing (in the limit), or (ii) the uninterrupted durations of time X_{ON} and X_{OFF} where the condition remains true or false in $\mathcal{A}_1$ are exponentially distributed.

While this theorem is easily proved [8] (given Theorem 2), its implications are of practical relevance: it suggests that the approximation of compression is acceptable not only if the condition is fast changing, but also if the coefficient of variation of X_{ON} and X_{OFF} is close to one.

The cost of compression can be high, though, since the number of markings for $\mathcal{A}_2$ is twice as large: for every marking of $\mathcal{X}_2$, a token is either in p_{ON} or in p_{OFF}. If n conditions had to be tested, 2^n as many markings would be generated, so compression continues to be plagued by some of the exponential growth present in the solution of the original GSPN $\mathcal{A}$. On the other hand, $\mathcal{A}_1$ could be studied in even greater detail and the firing time distributions for t_{ON} and t_{OFF} could match the first few moments, rather than the first moment only. In the GSPNs, this requires a more complex subnet to represent X_{ON} and X_{OFF}, but it provides a closer approximation.

It is interesting to compute the expected time $E[X]$ before the token in p_2 reaches p_3 in Figure 3(c), assuming steady state in the *ON*/*OFF* subnet. Define

$$\beta = \frac{\mu_1}{\mu_{ON} + \mu_1} \qquad \text{and} \qquad 1 - \beta = \frac{\mu_{ON}}{\mu_{ON} + \mu_1}$$

Then

$$E[X] = \underbrace{(1-\alpha)\frac{1}{\mu_{OFF}}}_{\text{time before } \#(p_{ON})=1}$$

$$+ \underbrace{\frac{1}{\mu_1 + \mu_{ON}} + \sum_{i=0}^{\infty} i \left(\frac{1}{\mu_{OFF}} + \frac{1}{\mu_1 + \mu_{ON}} \right) \beta (1-\beta)^i}_{\text{time before firing accounting for preemption}}$$

$$= \frac{1-\alpha}{\mu_{OFF}} + \frac{1}{\mu_1 + \mu_{ON}} + \left(\frac{1}{\mu_1 + \mu_{ON}} + \frac{1}{\mu_{OFF}} \right) \frac{1-\beta}{\beta}$$

$$= \frac{1-\alpha}{\mu_{OFF}} + \frac{1}{\mu_1 \alpha}$$

$(1-\alpha)/\mu_{OFF}$ corresponds to τ in the rar-del method, but their values coincide only if the *ON/OFF* periods are exponentially distributed in the original GSPN.

Rar-del and compression result in the same approximation if we can assume random incidence and if the results are dependent on $E[X]$ but not on the distribution of X. The second requirement is often met; for example, it is met if $\mathcal{X}_2$ blocks while waiting for the firing of t_1.

The first requirement, though, can be met only if $\mathcal{A}_2$ corresponds to an open system, with an infinite state space: assume that the arrival times of tokens in p_2, $\{A_1, A_2, \ldots\}$, are random incidence times for $\mathcal{A}_1$,

$$\pi_i^{(1)}(A_n) = \Pr\{\mathcal{A}_1 \text{ is in marking } i \text{ at time } A_n\} = \Pr\{\mathcal{A}_1 \text{ is in state } i\} = \pi_i^{(1)}$$

The departure times of tokens from p_2, $\{D_1, D_2, \ldots\}$, are dependent on the marking of $\mathcal{A}_1$, since

$$\Pr\{D_n \in [x, x+\delta)\} = 0 \Longleftrightarrow p_1 \text{ is empty during } [x, x+\delta)$$

Hence, the departure of tokens from p_2, or the number of tokens in p_2, must not affect the arrival of further tokens in p_2: p_2 must be unbounded. The result is not completely negative, since it suggests that the presence of a large number of tokens in $\mathcal{A}_2$ (approximating an open system) may nearly achieve random incidence.

Empirical Results

We compute the throughput ϕ_1 of transition t_1 in the GSPN of Figure 3. We vary the *OFF* and *ON* periods of condition "$\#(p_1) = 1$" (T_a, CV_a and T_b, CV_b), the time for a token to go from p_3 to p_2 (T_c, CV_c), and the time for t_1 (T_1). Since the results correspond to a seven-dimensional parameter space, we show only some of the percent errors in Table 2, where

$$\text{“percent error”} = \frac{\text{“approximate value”} - \text{“exact value”}}{\text{“exact value”}} \times 100$$

The following trends can be observed:

Rarefaction is often better than delay (delay may be completely off), but both are almost always worse than rar-del.

Increasing T_c or T_1 improves the approximation.

Increasing T_a improves the approximation for compression, but worsens it for rar-del.

Increasing T_b worsens the approximation, especially for rar-del.

Increasing CV_a and CV_b worsens the approximation for rar-del (with a few exceptions).

Small or large CV_a and CV_b have a similar effect in absolute value on the approximation for compression, but with opposite signs; this is intuitive,

Table 2 Percent Error for One-Way Condition Testing

							% Error on ϕ_1			
T_a	CV_a	T_b	CV_b	T_c	CV_c	T_1	Rarefaction	Delay	Rar-del	Compression
1	0.1	1	0.1	1	1.0	1	+4.60	+37.93	−4.18	−5.86
1	1.0	1	1.0	1	1.0	1	+11.11	+33.33	−4.76	0.00
1	10.0	1	10.0	1	1.0	1	+17.86	−49.49	−55.80	+6.07
1	0.1	1	0.1	20	1.0	20	+0.09	+49.12	−0.36	−0.71
1	1.0	1	1.0	20	1.0	20	+0.81	+49.35	−0.02	0.00
1	10.0	1	10.0	20	1.0	20	+7.55	+43.40	−0.72	+6.68
1	0.1	20	0.1	1	1.0	1	+34.51	+137.20	−8.88	−6.81
1	1.0	20	1.0	1	1.0	1	+44.35	+50.88	−22.64	0.00
1	10.0	20	10.0	1	1.0	1	+53.43	−82.46	−84.11	+6.29
20	0.1	1	0.1	1	1.0	1	+0.89	+2.07	−0.38	−0.30
20	1.0	1	1.0	1	1.0	1	+1.19	+1.31	−1.11	0.00
20	10.0	1	10.0	1	1.0	1	+1.57	−15.92	−17.58	+0.37
20	0.1	20	0.1	1	1.0	1	+30.08	−47.97	−54.09	−0.17
20	1.0	20	1.0	1	1.0	1	+30.30	−67.42	−69.93	0.00
20	10.0	20	10.0	1	1.0	1	+30.49	−96.16	−96.20	+0.15
1	0.1	20	0.1	1	0.1	20	+2.27	+1267.88	−0.21	−0.57
1	10.0	20	10.0	20	0.1	1	+115.14	−58.29	−61.89	+46.90

since the exponential assumption in compression tends to overestimate the waiting time if the variance is small and underestimate it if the variance is large; of course, the error for compression is zero when $CV_a = CV_b = 1$.

The maximum error for compression is 47%; the maximum error for rar-del is 96%, which should be considered unacceptable.

Compression often provides a better approximation than rar-del.

In conclusion, we can say that the approximation obtained from rar-del is acceptable when the condition changing is not much slower than the frequency at which the condition itself is tested, while it can be extremely good when the condition is fast changing. The values for the T's and the CV's were chosen to stress our methods. We also ran the same GSPN with $T_a, T_b, T_c, T_1 \in \{1,4\}$ and $CV_a, CV_b, CV_c \in \{0.5, 1, 2\}$, obtaining better approximations (the worst error was 47% for rar-del and 13% for compression).

4.3 Rendezvous

The processor-sharing policy considered in Section 4.1 is an abstraction useful in simplifying the modeling process. In reality, a resource can usually service only one request at a time. If the length of each service portion is large—possibly covering the whole service requirement—the processor-sharing approximation may be inappropriate.

Consider Figure 4, where subnet $\mathcal{X}_i$ $(1 \le i \le M)$ occasionally requests the token in p_0. If the token is available, it is immediately taken by the firing of u_i; alternatively, the token may be in use by some other $\mathcal{X}_j$ $(j \neq i)$ or it may be somewhere else in $\mathcal{X}_0$ (for bookkeeping reasons). y_i represents the usage of the token by $\mathcal{X}_i$; z_i represents the delay between the completion of

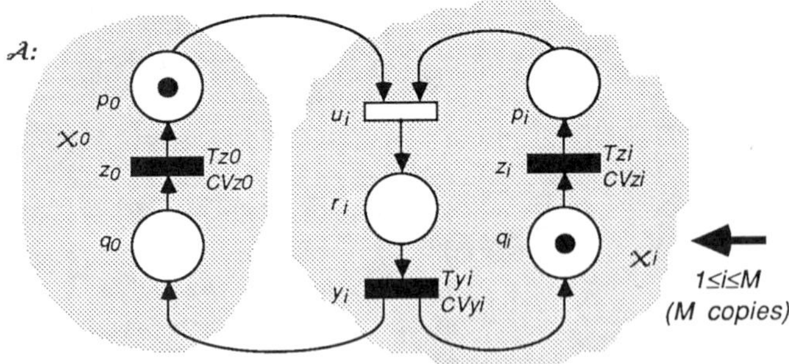

Figure 4 Rendezvous.

a request and the submission of the next one; z_0 represents the bookkeeping time between requests.

If K resources exist, $\#(p_0) = K$ initially. The subnets corresponding to z_0, y_i, and z_i may be arbitrary; we only require that the number of tokens representing requests and resources is preserved. For example, other activities could be progressing (tokens could be moving) in $\mathcal{X}_i$ even if a token is waiting in p_i.

The GSPN in Figure 4 corresponds also to the Ada [1] *rendezvous*. Assume that M producer tasks, $\mathcal{X}_1, \ldots, \mathcal{X}_M$, and one consumer task $\mathcal{X}_0$ exist. When producer $\mathcal{X}_i$ is ready to send out its product, it makes an entry call to $\mathcal{X}_0$. Only one token is circulating in each $\mathcal{X}_i$, and the portion of code represented by y_i corresponds to the operations performed during the rendezvous. We use this second interpretation in our discussion.

The Decomposition of the Rendezvous

The state space of Figure 4 is large because the $M + 1$ tasks are almost independent. If the subnets corresponding to z_i and y_i originate $N_i(\geq 2)$ and $L_i(\geq 1)$ markings, respectively, the total number of markings is

$$\underbrace{\prod_{j=0}^{M} N_j}_{\text{no rendezvous}} + \underbrace{\sum_{i=1}^{M} \left(L_i \prod_{j=1, j \neq i}^{M} N_j \right)}_{\mathcal{X}_i \text{ in rendezvous}} \tag{5}$$

We can solve each $\mathcal{X}_i$ $(1 \leq i \leq M)$ individually by estimating the amount of time each of them waits for a rendezvous (delay at u_i), but the solution of $\mathcal{X}_0$ is more complex.

Our decomposition makes use of the subnets in Figure 5, where only the value of T_{x_i} and T_{w_i} must be determined. When producer $\mathcal{X}_i$ makes an entry call to $\mathcal{X}_0$, it deposits a token in p_i. $\mathcal{A}_i$ is used to estimate the expected amount of time before $\mathcal{X}_0$ accepts the call. $\mathcal{A}_i$ represents the possible states of $\mathcal{X}_0$ when the token arrives in p_i. $\mathcal{X}_0$ may be executing locally, in q_0; it may be in rendezvous with $\mathcal{X}_j$ $(j \neq i)$, in r_j; or it may be waiting for a call, in p_0. The value of T_{x_j} represents the time elapsing from the arrival of a token in p_0 to the arrival of a token in p_j, when $\mathcal{X}_i$ is executing locally: it is estimated from $\mathcal{B}_j$ as

$$WAIT_{\mathcal{B}_j}(\#(p_j) = 1)$$

From $\mathcal{A}_i$, we can compute

$$WAIT_{\mathcal{A}_i}(\#(p_0) = 1)$$

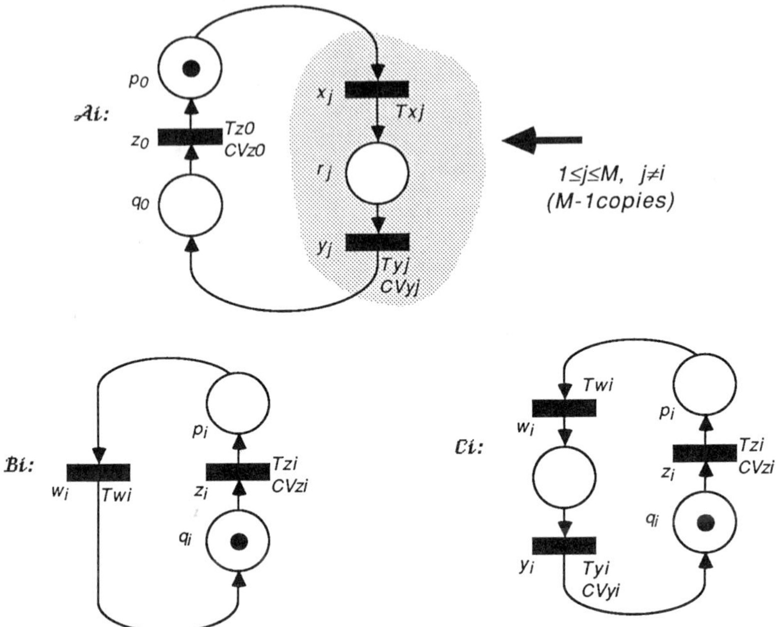

Figure 5 The decomposition of the rendezvous.

representing only the time until a token arrives in p_0 when $\mathcal{X}_i$ is waiting in p_i; once the token arrives in p_0, other producers may be waiting for it in addition to $\mathcal{X}_i$, so the waiting of $\mathcal{X}_i$ may not be over; $\mathcal{X}_0$ may decide to rendezvous with other producers before satisfying $\mathcal{X}_i$. If M is sufficiently large, we can assume that N, the number of times $\mathcal{X}_i$ is not chosen, has a modified geometric distribution with parameter

$$\alpha = \frac{1}{1 + \sum_{j=1,j\neq i}^{M} PROB_{\mathcal{C}_j}(\#(p_j) = 1)}$$

since $\mathcal{X}_i$ is waiting with probability one while producer $\mathcal{X}_j$ $(j \neq i)$ is waiting with probability $PROB_{\mathcal{C}_j}(\#(p_j) = 1)$, if we assume it is in steady state. $\mathcal{B}_i$ and $\mathcal{C}_i$ need to know the expected total time before $\mathcal{X}_i$ can rendezvous, T_{w_i}, computed as

$$\underbrace{WAIT_{\mathcal{A}_i}(\#(p_0) = 1)}_{E[\text{time before } \#\mathcal{A}_i(p_0)=1]} + \underbrace{\frac{1-\alpha}{\alpha}}_{E[N]} \cdot \underbrace{\frac{\sum_{j=1,j\neq i}^{M} PROB_{\mathcal{C}_j}(\#(p_j) = 1)(T_{y_j} + T_{z_0})}{\sum_{j=1,j\neq i}^{M} PROB_{\mathcal{C}_j}(\#(p_j) = 1)}}_{E[\text{length of a rendezvous with } \mathcal{A}_j]}$$

$$= WAIT_{\mathcal{A}_i}(\#(p_0) = 1) + \sum_{j=1, j \neq i}^{M} PROB_{\mathcal{C}_j}(\#(p_j) = 1)(T_{y_j} + T_{z_0})$$

We use $PROB_{\mathcal{C}_j}(\#(p_j) = 1)$, the probability of $\mathcal{X}_j$ being waiting in $\mathcal{C}_j$, not in $\mathcal{B}_j$. The difference between the two is the presence of y_j, corresponding to the rendezvous of $\mathcal{X}_j$ with $\mathcal{X}_0$. We use $\mathcal{B}_j$ to obtain $WAIT_{\mathcal{B}_j}(\#(p_j) = 1)$, needed by $\mathcal{A}_i$, because $\mathcal{X}_j$ cannot be in rendezvous with $\mathcal{X}_0$ when a token is in p_0; now the rendezvous is possible, so we use $\mathcal{C}_j$.

The export of $\mathcal{A}_i$ is $WAIT_{\mathcal{A}_i}(\#(p_0) = 1)$; it is needed by $\mathcal{B}_i$ and $\mathcal{C}_i$. The export of $\mathcal{B}_i$ is $WAIT_{\mathcal{B}_i}(\#(p_i) = 1)$; it is needed by $\mathcal{A}_j$ $(j \neq i)$. The export of $\mathcal{C}_i$ is $PROB_{\mathcal{C}_i}(\#(p_i) = 1)$; it is needed by $\mathcal{B}_j$ and $\mathcal{C}_j$ $(j \neq i)$. If results for $\mathcal{X}_0$ are sought, such as its throughput or the probability that it is waiting, we can define an additional GSPN $\mathcal{D}$, similar to $\mathcal{A}_i$, but without the condition $j \neq i$ (M copies instead of $M - 1$). $\mathcal{D}$ can be used to compute these results after the convergence has been achieved.

The number of markings in the subnets of Figure 5 using the quantities N_i and L_i with the same meaning as in equation (5) is

$$\mathcal{A}_i: \quad N_0 + \sum_{j=1, j \neq i}^{M} L_j \text{ markings}$$

$$\mathcal{B}_i: \quad N_i \text{ markings}$$

$$\mathcal{C}_i: \quad N_i + L_i \text{ markings}$$

The import graph has cycles, so iteration is needed; the total number of markings considered at each iteration is

$$\sum_{i=1}^{M} \left(\left(N_0 + \sum_{j=1, j \neq i}^{M} L_j \right) + (N_i) + (N_i + L_i) \right)$$

$$= 2 \sum_{i=1}^{M} N_i + M \left(N_0 + \sum_{i=1}^{M} L_i \right)$$

We conclude this section by observing that our decomposition of the rendezvous is related to the technique used for the solution of the stochastic rendezvous networks (SRNs) [23,24]. SRNs assume certain task structures, so it is possible to derive automatically a set of nonlinear equations from the stochastic description of the task system. The nonlinear system is then solved iteratively. In our approach, no structure is assumed, so it is possible to model more closely general control flow patterns and activities with nonexponentially distributed durations. For the same reason, though, the explicit derivation of the nonlinear equations becomes infeasible in practice (it would require the symbolic solution of the MC underlying a subnet). The evaluation of a nonlinear function in the solution of the SRNs corresponds, in our approach, to the solution of a linear system (the steady-state solution of $\mathcal{A}_i$, for example).

Empirical Results

We studied the GSPN of Figure 4 using the following parameters:

$$\begin{array}{ll} M = 5 & \\ T_{z_0} \in \{0.1, 1, 10\} & CV_{z_0} = 1 \\ T_{z_1} \in \{1, 10, 100\} & CV_{z_1} \in \{0.5, 1, 2\} \\ T_{y_1} \in \{0.1, 1\} & CV_{y_1} = 1 \\ T_{z_2} = T_{z_3} = T_{z_4} = T_{z_5} \in \{1, 10, 100\} & CV_{z_2} = CV_{z_3} = CV_{z_4} = CV_{z_5} = 1 \\ T_{y_2} = T_{y_3} = T_{y_4} = T_{y_5} = 1 & CV_{y_2} = CV_{y_3} = CV_{y_4} = CV_{y_5} = 1 \end{array}$$

for a total of $3^4 \cdot 2^1 = 162$ experiments, both exactly and with our decomposition scheme. Table 3 contains the percent error on the throughputs ϕ_0, ϕ_1, and ϕ_2 of tasks $\mathcal{X}_0$, $\mathcal{X}_1$, and $\mathcal{X}_2$, for a selected set of input parameter choices (the throughput of tasks $\mathcal{X}_3$, $\mathcal{X}_4$, and $\mathcal{X}_5$ is the same as that of $\mathcal{X}_2$).

As a general trend, the error for ϕ_i is reduced if the task spends more time locally (if T_{z_i} is increased). The dependency on the coefficient of variation is

Table 3 Percent Errors for Rendezvous

z_0	z_1		y_1	z_2	% Error on		
T	T	CV	T	T	ϕ_0	ϕ_1	ϕ_2
0.1	1	0.5	0.1	1	+2.5949	−3.9217	−7.8481
0.1	1	1.0	0.1	1	−3.2639	−5.3752	−8.1759
0.1	1	2.0	0.1	1	−12.1725	−6.8787	−8.7160
10	1	0.5	0.1	1	+0.4062	+1.0295	+0.6990
10	1	1.0	0.1	1	−0.0170	+1.0202	+0.6920
10	1	2.0	0.1	1	−0.7617	+1.0077	+0.6784
0.1	1	1.0	0.1	10	+2.7687	−9.8101	+0.1363
0.1	1	1.0	0.1	100	+0.0004	+0.0004	−0.0521
0.1	10	1.0	0.1	1	−10.2173	−0.5076	−9.4112
0.1	100	1.0	0.1	1	−5.6992	−0.0229	−9.3700

more complex. It appears that the error for ϕ_1 increases as CV_{z_1} increases, but only if $T_{z_0} < T_{z_1}$; the opposite happens if $T_{z_0} > T_{z_1}$ (although less noticeably).

Ordinarily, T_{z_0} is smaller than T_{z_i} $(1 \leq i \leq M)$, since $\mathcal{X}_0$ needs to "keep up" with requests from the M producers. T_{y_i} is small for the same reason, even in comparison to T_{z_0}. Under these conditions, the maximum error for the throughput of any producer is always less than 10%, often much less than 1%. Unfortunately, the error for the throughput of $\mathcal{X}_0$ is larger, up to 35%, although in most of the cases it is well below 10%.

4.4 Producers and Consumers

Consider Figure 6. If $N_1 = N_2 = 1$, we could consider this GSPN as the representation of the following code:

```
repeat
        execute X0
        parbegin
                execute X1; execute X2
        parend
forever
```

Noticeably, the same structure could also represent a rendezvous among two tasks, with $\mathcal{X}_0$ corresponding to the portion of code in the rendezvous. If N_i $(i = 1, 2)$ tokens are allowed to flow in $\mathcal{X}_i$, the GSPN could describe N_1 producers tasks and N_2 consumer tasks. Producers ready to pass their product to a consumer wait in q_1; one of them synchronizes with the first consumer ready to receive a product (first token to arrive in q_2). Analogously, when

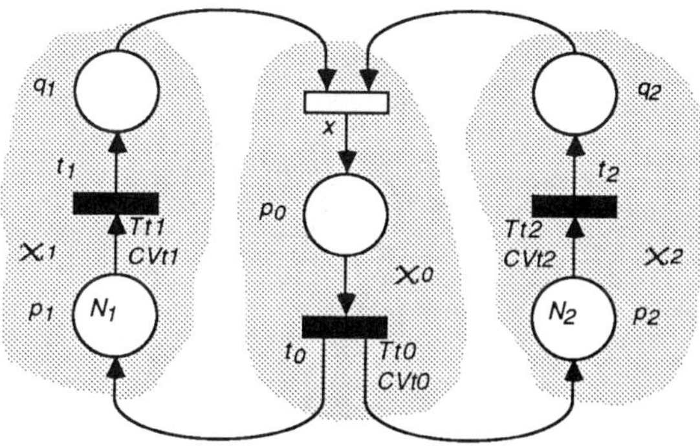

Figure 6 Multiple producers and consumers.

several consumers are waiting in q_2, one of them synchronizes with the first producer to arrive in q_1.

This GSPN structure could also represent a system where N_1 "parts of type 1" and N_2 "parts of type 2" circulate. Two parts, one of each type, can be assembled together, to form a part of type 0. When a part of type 0 leaves the system, two parts, one of type 1 and one of type 2, are added to the system.

Some preliminary observations are useful, before discussing the decomposition of this GSPN. First, $\#(p_0) \leq \min\{N_1, N_2\}$ at any time; the decomposition must guarantee this constraint. Then, the type of marking dependence for t_0, t_1, and t_2 must be specified. Typically, their speed depends at most on the respective input place. If the GSPN represents software tasks running on different processors, an infinite server behavior may be appropriate. If the GSPN represents a flexible manufacturing system where the machines (transitions) can only operate on one part at a time, a single server behavior is appropriate. Finally, the throughput ϕ of t_0, t_1, and t_2 is the same; this should hold true in the corresponding results from decomposition as well.

The decomposition of $\mathcal{A}$ requires two subnets, corresponding to the left and central portion, and to the right and central portion respectively (Fig. 7). The inhibitor arcs from p_0 to x_1 (in $\mathcal{A}_1$) and from p_0 to x_2 (in $\mathcal{A}_2$) ensure that $\#(p_0) \leq \min\{N_1, N_2\}$. Only the values T_{x_1} and T_{x_2} must be determined; the corresponding transitions x_1 and x_2 are of the single server type since, in $\mathcal{A}$, at least one among q_1 and q_2 has only one token when x is enabled. T_{x_1} represents the amount of time a token in q_1 must wait before a token arrives in q_2, hence we can define

$$T_{x_1} = WAIT_{\mathcal{A}_2}(\#(q_2) > 0)$$

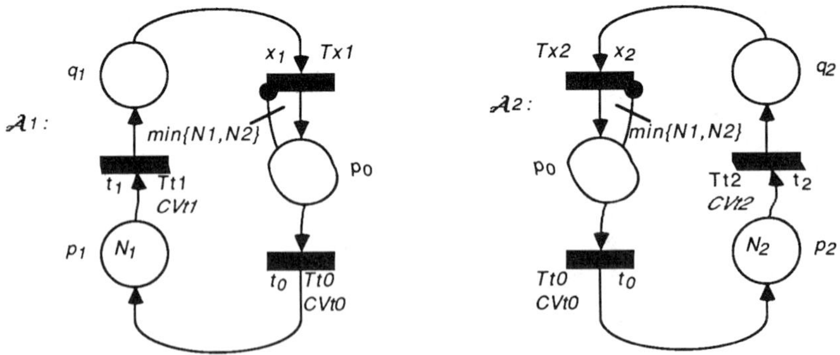

Figure 7 Decomposition of producers and consumers.

If we consider $\mathcal{A}_1$, though, the knowledge of how many tokens are in p_0 affects the estimation of the time a token has to wait in p_1. Assuming that $N_1 > N_2$,

$$\#(p_0) = N_2 \Longrightarrow \#(p_2) + \#(q_2) = 0$$

and

$$\#(p_0) = 0 \Longrightarrow \#(p_2) + \#(q_2) = N_2$$

In the former case, the probability that a token is in q_2 is zero, and in the latter it is the highest; even when q_2 is empty, the expected time before a token arrives in q_2 is smaller in the latter case. We must then compute

$$W_1[i] = WAIT_{\mathcal{A}_1}(\#(q_1) > 0 \mid \#(p_0) = i) \qquad 0 \leq i < \min\{N_1, N_2\}$$
$$W_2[i] = WAIT_{\mathcal{A}_2}(\#(q_2) > 0 \mid \#(p_0) = i) \qquad 0 \leq i < \min\{N_1, N_2\}$$

The quantity $WAIT(c \mid b)$ is analogous to $WAIT(c)$; the only difference is in the initial probability vector for the computation of the MTTA. Instead of using the steady-state probability vector $\underline{\pi}$, the "conditioned (on b)" steady-state probability vector $\underline{\omega}$ is used:

$$\omega_i = \begin{cases} \dfrac{\pi_i}{\sum_{j \in \mathcal{S}_b} \pi_j} & \text{if } i \in \mathcal{S}_b \\ 0 & \text{otherwise} \end{cases}$$

where $\mathcal{S}_b$ is the set of CTMC states in which condition b holds. An analogous definition can be given for $\lambda(c \mid b)$ and $PROB(c \mid b)$.

The rate of x_1 and x_2 is then dependent on the marking of p_0:

$$\mu_{\mathcal{A}_1}(x_1, m^{(1)}) = \frac{1}{W_2[\#_{\mathcal{A}_1}(p_0, m^{(1)})]}$$

and

$$\mu_{\mathcal{A}_2}(x_2, m^{(2)}) = \frac{1}{W_1[\#_{\mathcal{A}_2}(p_0, m^{(2)})]}$$

We are using the events "$\#_{\mathcal{A}_1}(p_0, m^{(1)}) = i$" and "$\#_{\mathcal{A}_2}(p_0, m^{(2)}) = i$" as if they were identical; in reality, they do indeed correspond to the event "$\#_{\mathcal{A}}(p_0, m) = i$" in the original system, but their probability, computed in $\mathcal{A}_1$ and $\mathcal{A}_2$ respectively, is not exactly the same.

Empirical Results

We studied $\mathcal{A}$ both exactly and approximately, with the following parameters:

$$N_1, N_2 \in \{1, 2, 3, 5, 10, 20\}$$
$$T_{t_0} = 2 \qquad T_{t_1} = 4 \qquad T_{t_2} = 6$$
$$CV_{t_0} = CV_{t_1} = CV_{t_2} = 1$$

assuming either a single server or an infinite server behavior for t_0, t_1, and t_2.

The results from our experiments were generally good; exact results were obtained for $N = \min\{N_1, N_2\} \in \{10, 20\}$. At the same time, though, a problem became apparent: the throughputs ϕ_1 and ϕ_2 estimated from $\mathcal{A}_1$ and $\mathcal{A}_2$ were somewhat different for small values of N. To make things worse, increasing only N_2 without increasing N_1 worsened the approximation for ϕ_1 with single server behavior, for ϕ_2 with infinite server behavior (Table 4).

A closer examination of the behavior of $\mathcal{A}$ provides the following explanation. The throughput ϕ is mostly determined by the "bottleneck" subnet, the slowest subnet in producing tokens for the synchronization in x. With a single server behavior, the bottleneck is $\mathcal{X}_2$ independent of N_1 and N_2 because $T_{t_1} < T_{t_2}$; ϕ_2 should be given more credibility. With an infinite server behavior, the bottleneck is determined by the relative value of N_1, N_2, T_{t_1}, and T_{t_2}, so it may be either $\mathcal{X}_1$ or $\mathcal{X}_2$. In general, we would like to characterize the "level of credibility" of ϕ_1 and ϕ_2. If $\mathcal{X}_1$ is the bottleneck, q_1 is almost always empty, because q_2 almost always contains tokens waiting to synchronize. Hence, both $\alpha_2 = PROB_{\mathcal{A}_2}(\#(q_2) > 0)$ and $(1 - \alpha_1) = PROB_{\mathcal{A}_1}(\#(q_1) = 0)$ are measures of the credibility of ϕ_1, while $\alpha_1 = PROB_{\mathcal{A}_1}(\#(q_1) > 0)$ and $(1 - \alpha_2) = PROB_{\mathcal{A}_2}(\#(q_2) = 0)$ are measures of the credibility of ϕ_2.

We are then faced with the determination of a function of ϕ_1, ϕ_2, α_1, and α_2 that defines our "best estimate" for the value of ϕ in the original GSPN $\mathcal{A}$. Reasonable heuristics are:

$$\phi = \begin{cases} \phi_1 & \text{if } \alpha_1 < \alpha_2 \\ \phi_2 & \text{otherwise} \end{cases}$$

Table 4 Percent Error for ϕ_1 and ϕ_2

		% Error on			
		(Single server)		(Infinite server)	
N_1	N_2	ϕ_1	ϕ_2	ϕ_1	ϕ_2
2	1	−0.0459	−2.1959	−2.7305	−0.2681
2	2	−5.1576	+3.7415	−2.7286	+1.0120
2	3	−9.6607	+5.6083	−2.7434	+4.3160
2	5	−13.6066	+4.0861	−0.4488	+8.0875
2	10	−16.6002	+0.9223	−0.0003	+10.5243
2	20	−17.3241	+0.0468	0.0000	+10.5264

Table 5 Percent Error for Producers and Consumers

		% Error on ϕ				% Error on ϕ	
N_1	N_2	Single server	Infinite server	N_1	N_2	Single server	Infinite server
1	1	−4.68151910	−4.68151910	5	1	−0.10961701	(a)
1	2	−5.43706799	−3.29045047	5	2	+0.84707453	−0.01356777
1	3	−7.69399494	−0.51873842	5	3	+1.23270087	+0.05960822
1	5	−10.38955720	−0.00420982	5	5	+0.77087544	+1.54073459
1	10	−12.75141050	(a)	5	10	−0.02343681	+0.51067213
1	20	−13.50198370	(b)	5	20	−0.15906097	+0.00000431
2	1	−1.99955510	−0.32558514	10	1	−0.00079990	(a)
2	2	+1.84500968	+0.06310237	10	2	+0.05458842	+0.00000002
2	3	+1.53514924	−0.02764103	10	3	+0.12527471	+0.00001590
2	5	−1.05397081	−0.12551209	10	5	+0.10901359	−0.00010499
2	10	−4.23451512	−0.00023199	10	10	+0.01646560	+1.15327783
2	20	−5.06603999	0.00000000	10	20	−0.00094632	+0.14086292
3	1	−0.77704006	−0.01574284	20	1	(a)	(b)
3	2	+2.02646846	+0.07385477	20	2	(a)	(b)
3	3	+2.41780221	+1.13233620	20	3	+0.00123075	(a)
3	5	+0.85192458	+1.17413966	20	5	+0.00180473	(a)
3	10	−1.09012956	−0.00662221	20	10	+0.00059958	+0.00002191
3	20	−1.45303755	(a)	20	20	−0.00000043	+0.46598564

$$\phi = \begin{cases} (1-\alpha_1)\phi_1 + \alpha_1\phi_2 & \text{if } \alpha_1 < \alpha_2 \\ \alpha_2\phi_1 + (1-\alpha_2)\phi_2 & \text{otherwise} \end{cases}$$

$$\phi = \frac{(1-\alpha_1+\alpha_2)\phi_1 + (1-\alpha_2+\alpha_1)\phi_2}{2}$$

$$\phi = \frac{(1-\alpha_1)\alpha_2\phi_1 + (1-\alpha_2)\alpha_1\phi_2}{(1-\alpha_1)\alpha_2 + (1-\alpha_2)\alpha_1}$$

$$\phi = \frac{(1-\alpha_1)\phi_1 + (1-\alpha_2)\phi_2}{(1-\alpha_1) + (1-\alpha_2)}$$

$$\phi = \frac{\alpha_2\phi_1 + \alpha_1\phi_2}{\alpha_1 + \alpha_2}$$

All are satisfactory, but no one method proves to be the best in all cases. The quantities ϕ_1 and ϕ_2 are almost always bounds for the exact value of ϕ, so it seems advantageous to use both, hoping that errors of opposite sign cancel each other (the first heuristic is often the worst indeed). The percent errors using the last heuristic are presented in Table 5. An entry "(a)" signifies that the error is zero given the precision of the machine (double-precision floating-point numbers are used). An entry "(b)" signifies that the solution method breaks down because the computation of α_1 or α_2 generates an underflow; while this is unpleasant as far as the tool used (SPNP) is concerned, the corresponding error would be zero.

5. GENERAL REMARKS ON THE DECOMPOSITION APPROACH

Although we presented only a handful of GSPN structures ([8] contains a few more), the class of GSPNs covered by them is large. In addition, different structures can be combined in the same GSPN (e.g., the branches of a parbegin–parend may require common resources), or the same structure may be applied recursively (e.g., "nested rendezvous": the code within a rendezvous between two tasks may contain an entry call to a third task).

On the other hand, we cannot claim to have addressed all the "interesting" GSPNs. Sometimes a GSPN has a completely new structure, but it is possible to derive an approach similar to the one we proposed, extending the class of GSPNs addressed by our decomposition framework. Other times, the GSPN is simply too intricate to lend itself to any decomposition. We believe that such GSPNs are rare in practice, because even the most complex system is usually built in a structured way and allows only limited interactions among subsystems.

We stress that our decomposition approach iteratively modifies the CTMC itself. Only two primitives are needed to define the imports: $\lambda_{\mathcal{A}}$ and $WAIT$ ($PROB$ is derived from $\lambda_{\mathcal{A}}$).

The reduction of the state space when applying our procedure is large, but a fair comparison of the "speedup" must take into account the number of iterations needed at the decomposition level and the behavior of the numerical methods, also expressed in number of iterations. If K iterations are needed at the decomposition level, if M subnets need to be analyzed at each iteration, and if, at the kth iteration ($1 \leq k \leq K$), subnet $\mathcal{A}_m$ ($1 \leq m \leq M$) generates a CTMC $Q_m^{(k)}$ with $\eta_m^{(k)}$ nonzero entries requiring $n_m^{(k)}$ iterations of the numerical method for its solution, the total cost is $\sum_{m=1}^{M} \sum_{k=1}^{K} \eta_m^{(k)} \cdot n_m^{(k)}$.

Two observations can be made at this point. The subnets $\mathcal{A}_m$ are often simple. If they correspond to a product-form queuing network or to a combinatorial model, for example, there is no need to solve them with standard GSPN solution methods. Sometimes the solution of a subnet can be written down in closed form.

Often, the measure(s) to be exported require a Markov analysis even if the model is simple (e.g., MTTA for some given initial probability vector), but the efficiency of the analysis might still be improved. The reachability graph for $\mathcal{A}_m$ must be generated only once, since it does not change from iteration to iteration. Even the differences between $Q_m^{(k)}$ and $Q_m^{(k+1)}$ are usually minor, restricted to a small change in some of the rates; in particular, the pattern of zero and nonzero entries does not change, so $\eta_m^{(k)} = \eta_m^{(k+1)}$. It is possible to exploit this similarity in two ways. First, $Q_m^{(\cdot)}$ should be generated once, leaving "symbolic" entries in correspondence to the imports for $\mathcal{A}_m$. At step k, $Q_m^{(k)}$ could then be generated from $Q_m^{(\cdot)}$ by binding the symbolic entries using the current values of the imports.

Furthermore, it is arguable that $\underline{\pi}_m^{(k)}$, the numerical steady-state solution of $\underline{\pi}_m^{(k)} Q_m^{(k)} = 0$, is only slightly different from $\underline{\pi}_m^{(k+1)}$, the solution of $\underline{\pi}_m^{(k+1)} Q_m^{(k+1)} = 0$, since $Q_m^{(k)}$ is close to $Q_m^{(k+1)}$. If we use $\underline{\pi}_m^{(k)}$ as the initial guess when solving $\underline{\pi}_m^{(k+1)} Q_m^{(k+1)} = 0$, few iterations are needed, especially when convergence at the decomposition level is nearly achieved. While the first improvement requires the ability to store the CTMC in a restricted symbolic format (as is done in SHARPE), the second improvement only requires storing a set of M probability vectors, from the kth to the $(k + 1)$th decomposition-level iteration.

6. FUTURE WORK

Two main efforts are needed to further our study. On a practical level, a GSPN tool that effectively supports decomposition is needed. At a minimum, this implies the ability of dealing with several GSPNs simultaneously exporting/importing data among them. With this tool, we hope to gain more experience on the effectiveness and quality of the approximation and on the typical structures arising in large, real-life problems.

On the theoretical side, two important issues remain open: the computation of bounds on the approximation, and a general theory on the convergence behavior when iterations are required.

REFERENCES

[1] *Reference Manual for the Ada Programming Language*. Technical Report ANSI/MIL-STD-1815A-1983, U.S. Department of Defense, Feb. 1983.

[2] M. Ajmone Marsan, G. Balbo, and G. Conte. A class of generalized stochastic Petri nets for the performance evaluation of multiprocessor systems. *ACM Trans. Comput. Systems*, 2(2):93–122, May 1984.

[3] M. Ajmone Marsan, G. Balbo, G. Ciardo, and G. Conte. A software tool for the automatic analysis of generalized stochastic Petri net models. In D. Potier, editor, *Modelling Techniques and Tools for Performance Analysis*, pages 155–170, INRIA, Elsevier Science Publishers B.V. (North Holland), 1985.

[4] M. Ajmone Marsan, G. Balbo, A. Bobbio, G. Chiola, G. Conte, and A. Cumani. On Petri nets with stochastic timing. In *Proceedings of the International Workshop on Timed Petri Nets*, Torino, Italy, July 1985.

[5] A. Blakemore and G. Schebella. Tools for analyzing dynamic properties of system and software designs. Aug. 1989. *Proceedings of the IFIP XI Congress*, San Francisco, Calif.

[6] M. Boyd, M. Veeraraghavan, J. Dugan, and K. Trivedi. An approach to solving large reliability models. In *Proceedings of IEEE/AIAA DASC Symposium*, San Diego, 1988.

[7] G. Chiola. *GreatSPN Users' Manual.* Technical Report, Dipartimento di Informatica, Università degli Studi di Torino, Torino, Italy, 1987.

[8] G. Ciardo. Analysis of large stochastic Petri net models. Ph.D. thesis, Duke University, Durham, NC, USA, 1989.

[9] G. Ciardo, J. Muppala, and K. S. Trivedi. SPNP: stochastic Petri net package. In *Proceedings of the Third International Workshop on Petri Nets and Performance Models* (PNPM89), Kyoto, Japan, Dec. 1989.

[10] E. de Souza e Silva and S. Lavenberg. A perspective on iterative methods for the approximate analysis of closed queueing networks. In *Mathematical Computer Performance and Reliability*, G. Iazeolla, P. J. Courtois, and A. Hordijk (eds.), Elsevier Science Publishers B.V. (North-Holland), 1984.

[11] A. Goyal and S. Lavenberg. Modeling and analysis of computer system availability. *IBM J. Res. Develop.*, 1988.

[12] M. Holliday and M. Vernon. A generalized timed Petri net model for performance analysis. In *Proceedings of the International Workshop on Timed Petri Nets*, Torino, Italy, July 1985.

[13] V. O. Li and J. A. Silvester. Performance analysis of networks with unreliable components. *IEEE Trans. Commun.*, COM-32(10):1105–1110, October 1984.

[14] I. Mitrani. Fixed-Point approximations for distributed systems. In *Mathematical Computer Performance and Reliability*, G. Iazeolla, P. J.

Courtois, and A. Hordijk (eds.), Elsevier Science Publishers B.V. (North-Holland), 1984.

[15] M. K. Molloy. On the integration of delay and throughput measures in distributed processing models. Ph.D. thesis, University of California, Los Angeles, 1981.

[16] R. R. Muntz, E. de Souza e Silva, and A. Goyal. Bounding availability of repairable computer systems. In *Proceedings of 1989 ACM SIGMETRICS and PERFORMANCE '89 International Conference on Measurement and Modeling of Computer Systems*, Berkeley, Calif., May 1989.

[17] S. Natkin. Reseaux de Petri stochastiques. These de Docteur Ingeneur, CNAM-Paris, Paris, June 1980.

[18] J. M. Ortega. *Numerical Analysis—A Second Course*. Academic Press, New York, 1972.

[19] J. L. Peterson. *Petri Net Theory and the Modeling of Systems*. Prentice-Hall, Englewood Cliffs, N.J., 1981.

[20] C. Petri. Kommunikation mit Automaten. Ph.D. thesis, University of Bonn, Bonn, West Germany, 1962.

[21] K. C. Sevcik. Priority scheduling disciplines in queueing network models of computer systems. In *1977 IFIP Congress Proceedings*, 1977.

[22] R. S. Varga. *Matrix Iterative Analysis*. Prentice-Hall, Englewood Cliffs, N.J., 1962.

[23] C. M. Woodside. Throughput calculation for basic stochastic rendezvous networks. *Perform. Eval.*, (9):143–160, 1988/89.

[24] C. M. Woodside et al. An active-server model for the performance of parallel programs written using rendezvous. *J. Systems Software*, 6(1 & 2):125–131, May 1986.

31

On Bounds for Token Probabilities in a Class of Generalized Stochastic Petri Nets

S. M. R. ISLAM* **and H. H. AMMAR†** Department of Electrical and Computer Engineering, Clarkson University, Potsdam, New York

ABSTRACT

In this paper we present methods to compute tight bounds for steady-state token probabilities of a class of generalized stochastic Petri net (GSPN) models. Such bounds also give a better estimate of the error produced when decompositions and aggregations are used to compute the various performance measures.

First, we describe a method to compute the best lower and upper bounds for conditional token probabilities of a class of GSPN subnets when only the subnet is considered. We show that such bounds can be improved if additional information about other subnets are available. We then extend the technique and outline an algorithm to compute the bounds for error due to aggregation and decomposition at the GSPN level. An example is also presented to illustrate the technique and algorithm.

*Current affiliation: IBM Corporation, Boca Raton, Florida.

†Current Affiliation: West Virginia University, Morgantown, West Virginia

1. INTRODUCTION

The generalized stochastic Petri net (GSPN) [7] models are powerful stochastic tools that have been used extensively for modeling a wide range of applications from flexible manufacturing systems to performability evaluation of parallel and distributed computers. They are particularly suitable for representing activities such as synchronization, concurrency, and communication in a system and therefore are applicable to many areas in the field of computer and information processing. However, the analysis becomes formidable when enough details are introduced into such models. Additionally, the presence of orders of magnitude difference in the duration of activities represented by timed transitions in a model leads to numerical instability.

The technique of time scale decomposition (TSD) of GSPNs developed in [1, 16] and [17] offers a step in the right direction for a solution of the above fundamental problems. It is applicable to systems containing activities whose durations differ by several orders of magnitude. The given model can be decomposed into a hierarchical sequence of aggregated subnets each of which is valid at a certain time scale. Such a decomposition is equivalent to the hierarchical aggregation of the underlying Markov process. The error of aggregation is in the order of the maximum degree of coupling between aggregates, which can be evaluated directly at the GSPN level.

The order of error due to aggregation sometimes does not give meaningful information; rather, a more informative description of the error is needed. A bound for the performance matrices will certainly be more informative than the order of error present in the matrices. Moreover, in certain situations, the designer is interested in a performance measure that depends mainly on the activities and resources modeled in a subnet and is weakly affected by the behavior of the other subnets in the GSPN model of the system. In such instances, it is of extreme practical importance to restrict the analysis to a subnet and to obtain sufficiently tight bounds for the token probabilities in that subnet.

In this paper, we are presenting techniques to compute these bounds directly at the GSPN level. After presenting necessary propositions, we extend our technique to compute bounds for error due to aggregation and decomposition at the GSPN level. We also present a simple example to illustrate the technique.

2. DESCRIPTION OF PREVIOUS WORK

Previous works in this area were mainly focused on computing bounds for delays and throughputs of GSPN models. Since the state space of GSPNs tends to explode with the introduction of more tokens in a net, bounds for throughputs that do not require enumeration of the state-space significantly increase the usability of GSPN models. The analysis of delays in a net is based either on the computation of visit counts before a trapping state is reached or using Little's result and flow balance.

Molloy [2] noted that if the average token flow in a network at steady state is conserved, a series of flow balance equations can be developed. The boundary of feasible solutions for these balance equations defines the bounds of token flow rate through the net. On the other hand, Bruell and Ghanta [3] developed algorithms for computing upper and lower bounds for a restricted class of GSPNs. The upper and lower bounds of delays are computed hierarchically, estimating maximum and minimum time of the path followed by the process in the network when there are several parallel paths available.

The work that we are presenting here differs from the previous work in two distinct ways. First, we are trying to find conditional token probability distribution in a subnet without solving the whole net. Since exact token probabilities cannot be determined without the knowledge of the whole net, we are interested in their bounds. Second, we know that aggregation and decomposition at the GSPN level introduce errors in the token probabilities; therefore, we are interested in computing the bounds for this error. The studies that were done in the spirit of bounds for probabilities were restricted at the Markov chain level. The most noticeable contributions were by Courtois and Semal [4,5]. They developed expressions for bounds on conditional steady-state probabilities associated with the submatrix of a large Markov chain. They also presented techniques that can obtain tighter bounds with the expense of more computations. Stewart [6] also presented techniques for computing *a priori* error bounds due to aggregation and decomposition. These bounds are in terms of the norms of the probability vectors. Our work extends these techniques to the GSPN level. Besides other obvious benefits [1,17], it is much easier to identify the subnet of interest at the GSPN level. For the sake of completeness, we will now briefly describe GSPNs.

2.1 Description and Analysis of GSPNs

We start our discussion with the concept of standard Petri nets (PNs) [8–10]. It is then extended from standard Petri nets to stochastic Petri nets [11] and later to generalized stochastic Petri nets [7]. A Petri net (PN) [8–10] comprises a set of places P, a set of transitions T, and a set of directed arcs A. Arcs connect transitions to places and places to transitions. The direction of an arc is represented by an arrow at the head of an arc. Places may contain tokens, which are drawn as black dots. The marking, or the state of a PN, is defined by the number of tokens contained in each place and is denoted by a vector M, whose ith component represents the number of tokens in place p_i, $[M]_{p_i} = m_i$. The construction of a PN model requires the specification of the initial marking M_1.

A place p_i is an input to a transition t_j if there exists at least one arc from the place to the transition. It is denoted by $I(p_i,t_j) > 0$, and the arc is called the input arc of transition t_j. A place p_k is an output from a transition t_j if there exists at least one arc from the transition to the place. It is denoted by $O(t_j,p_k) > 0$, and the arc is called the output arc of transition t_j. A transition is enabled when each of its input places contains at least as many tokens in it as the multiplicity of arcs from the place to the transition. An enabled transition can fire. The firing of a transition removes all enabling tokens from its input places and deposits tokens to its output places. The numbers of tokens deposited are equal to the multiplicity of output arcs. Clearly, the firing of a transition modifies the distribution of tokens in places and thus produces a new marking of the PN.

For a given initial marking M_1, the reachability set S is defined as the set of all markings that can be reached from M_1 by a sequence of transition firings. If the PN is such that, for any possible marking, the number of tokens in any place is less than or equal to k, for some integer k, then the PN is said to be *k-bounded*. Also, if for any marking $M \in S$, and for any transition $t_i \in T$, there exists a transition firing sequence starting from M and ending on a marking M' at which t_i is enabled, the PN is said to be *live*.

In a stochastic Petri net (SPN)[11], the transition firing time is an exponentially distributed random variable. A formal definition of SPN = (P,T,I,O,R) with an initial marking M_1 is thus the following:

$$P = \{p_1,p_2,\ldots,p_n\} \text{ is a set of places}$$
$$T = \{t_1,t_2,\ldots,t_m\} \text{ is a set of transitions}$$
$$A \subset \{P \times T\} \cup \{T \times P\}$$
$$R = \{r_1,r_2,\ldots,r_m\} \text{ is a set of firing rates}$$

such that r_j is the rate of exponential distribution associated with the firing time of the transition t_j; r_j is said to be marking dependent when it is a function of current marking M_i.

In a GSPN, the set of transitions T comprises two different classes of transitions. They are immediate (drawn as thin bars) and timed transitions (drawn as boxes). Immediate transitions fire in zero time once they are enabled. Timed transitions fire after a random, exponentially distributed enabling time. As mentioned earlier, GSPNs are logical extensions of SPNs, where immediate transitions represent transitions with infinite firing rate. Exactly as with SPNs, several transitions may be simultaneously enabled by a marking in a GSPN. The enabled transition fires with a probability

$$\frac{r_i}{\sum_{k \in H} r_k}$$

where the set of enabled transitions comprises H, and r_i is the firing rate of the transition that fires. Clearly, if H comprises one immediate transition and several timed transitions, then the immediate transition will fire with probability 1. However, for any marking at which several immediate transitions are enabled, they fire with equal probability unless a probability distribution over the enabled transitions selection is specified. This probability distribution is called the random switching distribution, and the set of enabled immediate transitions is said to form a random switch.

The analysis of GSPNs is based on characterizing the underlying process as a stochastically discontinuous Markov process (SDMP) [15,17]. The analysis of this process is based on the concept of state aggregation and can be divided into two steps. First, a discrete parameter Markov chain (MC) is obtained by considering only immediate transitions in the reachability graph of the GSPN. This MC determines a partition of the state space in terms of ergodic classes and transient states. Each ergodic class consists of one marking or several instantaneous markings. From the analysis of this MC we can determine the trapping probabilities from the transient state to the ergodic classes as well as the steady-state probabilities of the states within each ergodic class. From these ergodic classes a continuous parameter MC is obtained, the state space of which is defined on the set of ergodic classes. The steady-state probabilities of the ergodic classes can then be evaluated by analyzing the aggregated MC.

In the next section we developed some essential definitions and present two important propositions.

3. BOUNDS FOR CONDITIONAL TOKEN PROBABILITY DISTRIBUTIONS IN A SUBNET

In this section we will present bounds for conditional token probabilities in a subnet. We will derive these bounds from probabilistic arguments. Let us consider a subnet S_i. Further, assume that a process enters into this subnet through some initial marking. Since we are considering live GSPNs, after spending some time in the subnet, the process will exit from the subnet through some marking. If we can estimate the relative visit ratios of this process in various markings of the subnet, we can estimate the steady-state probabilities of these markings. However, the relative visit ratios depend on the initial marking of the subnet. Clearly, if this initial marking is known, the conditional steady-state probabilities of all markings in the subnet can be determined exactly. However, in the absence of such information, considering each marking of the subnet as an initial marking, we can estimate the conditional steady-state probabilities of markings. We can then take the maximum and minimum from these estimates to get a bound for the conditional steady-state probabilities of markings in the subnet.

We will now formally present a proposition based on the above argument. Before we can start, we must first present some definitions.

DEFINITION 3.1 *A subnet.* Let us consider the GSPN = (P,T,A,R) as defined in [1]. We further assume that the set of transitions T is composed of a set of slow transitions T_s and a set of fast transitions T_f. Then removing all slow transition $t_{is} \in T_s$ together with their input and output arcs will split the given GSPN into a set of subnets,

$$\{P_1, T_{1f}, A_1, R_1\}, \{P_2, T_{2f}, A_2, R_2\}, \ldots, \{P_i, T_{if}, A_i, R_i\}, \ldots, \{P_n, T_{nf}, A_n, R_n\}$$

where $\cup_{i=1}^{n} P_i = P$ and all $p_i \in P_i$ are connected.

The existence of more than one subnet is symptomatic of singularly perturbed SPNs as given in proposition 3 of [17]. If there is only one subnet, the SPN becomes regularly perturbed and solving the net at fast time scale will give the desired solution.

DEFINITION 3.2 *The input and output places and transitions of a subnet.* A place $p_i \in P_i$ in subnet i is said to be a member of the set of input places (IP_i) of subnet i (i.e., $p_i \in IP_i$), if there exists a slow transition $t_{is} \in T_s$ with p_i as an *output place* such that $O(t_{is}, p_i) \neq 0$. Similarly, a place $p_j \in P_i$ in subnet i is said to be a member of the set of output places (OP_i) of subnet i, (i.e., $p_i \in OP_i$) if there exists a slow transition $t_{ks} \in T_s$ with p_j as an *input place* such that $I(p_j, t_{ks}) \neq 0$. Moreover, the set of input (output) transitions of subnet i is defined as the set of all slow transitions with an output (input) place in IP_i (OP_i).

DEFINITION 3.3 *Stationary and non-stationary resource tokens.* A token is said to be a stationary resource token (SRT) if the token represents a resource and is confined within a subnet. That is, it does not go out of a subnet or comes into a subnet from the outside. On the other hand, a token is said to be a non-stationary resource token (NSRT) if the token representing a resource moves between subnets.

An example of an SRT is a token representing the availability of a resource local to a particular subnet. We will assume in the sequel that the initial marking of a subnet [1] is determined mainly by the distribution of NSRT's in the subnet.

DEFINITION 3.4 *Local markings of a subnet.* The local markings of a subnet i are defined as the markings defined only in terms of places $p_i \in P_i$ of the subnet i. Additionally, the markings are reached from the initial marking of the subnet by a transition firing sequence of transitions in the subnet only. A local marking j in a subnet i will be denoted by M_i^j.

DEFINITION 3.5 *The locality property of a subnet.* A subnet i is said to satisfy the locality property, if the firing probability or the firing rate of any of its transitions depends only on the local markings of the subnet. All subnets in this paper satisfy the locality property.

DEFINITION 3.6 *Recurrent subnets.* A subnet i is said to be recurrent if for some initial marking of the subnet, a local marking can be reached from any other local marking by a finite sequence of transition firings in the subnet only.

We now present the following proposition for the conditional bounds for steady-state token probabilities of a recurrent subnet.

PROPOSITION 3.1 Let us assume that S_i is a recurrent subnet with input place set IP_i and output place set OP_i. Additionally, the rates of slow transitions connected to OP_i are known. With no additional information, the tightest bounds for the conditional steady-state probabilities of local markings are given by the maximum and minimum of columns of matrix $\mathbf{T}$, where $\mathbf{T}$ is given by

$$\mathbf{T} = \Lambda^{-1}\overline{(\mathbf{I} - \mathbf{P}_i)^{-1}}$$

$\mathbf{P}_i$ is a substochastic matrix defined by the local markings of subnet i and $\overline{(\mathbf{I} - \mathbf{P}_i)^{-1}}$ is obtained by deleting the rows of $(\mathbf{I} - \mathbf{P}_i)^{-1}$ which are not the initial markings of subnet i. The elements of the diagonal matrix Λ are the row sums of $\overline{(\mathbf{I} - \mathbf{P}_i)^{-1}}$.

Proof. We perform the following modification to the GSPN.

Replace all subnets except S_i with a single place p_{ab}.
Remove all slow transitions that are connected to the input place set.

Since, we are assuming recurrent subnets there will be at least one marking in the subnet where a slow transition connected to the output place set will be enabled. Let us further assume that $M_1(i)$ is the initial marking of subnet i defined by the maximum number of NSRT and stationary resource tokens in the subnet i. Let $\mathbf{Z}$ be the transition probability matrix of the modified subnet including place p_{ab}. Therefore, the Markov chain defined by markings of this modified subnet will be transient with a set of absorbing states. These absorbing states are entered through the firing of slow transitions.

Let V_{ij} be the average number of visits to a marking j, starting at a marking i, before exiting to some absorbing marking $M_{ab}(i)$. Therefore V_{ij} [14] will be

$$V_{ij} = \delta_{ij} + \sum_{k=1} z_{ik} V_{ik}$$

where

$$\delta_{ij} = \begin{cases} 1 & \text{if } i = j \\ 0 & \text{otherwise} \end{cases}$$

The steady-state probability of marking j, given that starting at marking i, will be

$$v_{ij} = \frac{V_{ij}}{\sum_{l \in RG_{S_i}} V_{il}}$$

Writing all equations expressing V_{ij}s, we have the following matrix relation:

$$\mathbf{V} = \mathbf{C} + \mathbf{VP}_i$$

where $\mathbf{P}_i$ is the substochastic matrix (fundamental matrix) defined on all but absorbing markings of subnet i and $\mathbf{C}$ is a diagonal matrix of δ_{ij}. Therefore, $\mathbf{V}$ is given as

$$\mathbf{V} = \mathbf{C}(\mathbf{I} - \mathbf{P}_i)^{-1}$$

If we normalize $\mathbf{V}$ such that the row sums of $\mathbf{V}$ are equal to 1, the columns of $\mathbf{V}$ gives the maximum and minimum conditional probability of markings in subnet i. Clearly, $\mathbf{T}$ follows from $\mathbf{V}$ by removing the rows which do not correspond to initial marking and performing a row normalization. From [5], we can see that the row vectors of $\overline{\mathbf{\Lambda}^{-1}(\mathbf{I} - \mathbf{P}_i)^{-1}}$ form vertices of a polyhedron. Therefore, without additional information, the maximum and minimum of columns of $\mathbf{T}$ will give the bounds for the probability vector.

From the above proposition, it is clear that a tighter bound can be formulated if we know how we are entering into the subnet i—that is, the set of the initial marking(s) of the subnet. In this proposition, we are considering all initial markings are equally likely. If we know exactly how we are entering into the subnet, we can find the conditional steady-state marking probabilities exactly. Now, if we have further information from the rest of the network such that we can determine more preciously how we are entering into the subnet, we can formulate a better bound for the conditional steady-state marking probabilities. We present our second proposition that will enable us to formulate a tighter bound with a small extra cost for the information and computation.

PROPOSITION 3.2 Without loss of generality, let us assume that we have subnets S_{ji}, j: 1, ... which are connected by some transitions to subnet S_i

where S_i is defined in Proposition 3.1. Let us further assume that IP_{ji}, j: 1, ... are the set of input places of these subnets. With no additional information, the bounds for the conditional steady-state probability, $\hat{\pi}_i$, of local markings of subnet i are given by the maximum and minimum in each of the columns of the matrix $\mathbf{T}_{ji}$,

$$\min_j(\mathbf{T}_{ji}) \leq \hat{\pi}_i \leq \max_j(\mathbf{T}_{ji})$$

where

$$\mathbf{T}_{ji} = \Lambda_j^{-1}\overline{(\mathbf{I}-\mathbf{P}_{ji})^{-1}}\Delta_{ji}\overline{(\mathbf{I}-\mathbf{P}_i)^{-1}}$$

Proof. Let us consider the subnet S_{ji}. The initial markings of this subnet are the markings by which a process enters into the subnet. Clearly, the set of initial markings are defined by NSRTs that flow into the subnet. The process stays in the subnet S_{ji} for some time before making the final jump to an initial marking of the subnet S_i. We are interested in finding the trapping probabilities from the initial markings of subnet S_{ji} to the initial markings of subnet S_i. Let $\mathbf{P}_{ji}$ be a substochastic matrix defining the transition probabilities among all but the absorbing markings (absorbing marking of S_{ji} is defined in the same way we have defined absorbing markings of S_i; see Proposition 3.1). Clearly, the k-step transition probabilities are given by $\mathbf{P}_{ji}^k$. Therefore, the probability that a process enters into subnet S_{ji} through some initial marking of S_{ji} and after making k jumps within S_{ji} exits to an initial marking of S_i is given by $\mathbf{P}_{ji}^k\Delta_{ji}$, where Δ_{ji} is a rectangular nonnegative matrix that defines the transition probabilities between markings of subnet S_{ji} and initial markings of subnet S_i. These transition probabilities can be easily determined by examining the subnet S_{ji} and its interface with subnet S_i. Now, the overall trapping probabilities from the markings of S_{ji} to the initial markings of S_i are given by

$$\sum_k \mathbf{P}_{ji}^k\Delta_{ji} = (\mathbf{I}-\mathbf{P}_{ji})^{-1}\Delta_{ji}$$

Since, we are only interested in finding the trapping probabilities from the initial marking of S_{ji} to the initial markings of S_i, $\overline{(\mathbf{I}-\mathbf{P}_{ji})^{-1}}$ is formed by deleting the rows in $(\mathbf{I}-\mathbf{P}_{ji})^{-1}$ that are not initial markings of S_{ji}.

Clearly, $\mathbf{T}_{ji}$, the conditional steady-state probabilities of markings in S_i when the system is entered through some initial marking of S_{ji}, is given by

$$\mathbf{T}_{ji} = \Lambda_j^{-1}\overline{(\mathbf{I}-\mathbf{P}_{ji})^{-1}}\Delta_{ji}\overline{(\mathbf{I}-\mathbf{P}_i)^{-1}}$$

where Λ_j is a diagonal matrix such that the row sum of $\mathbf{T}_{ji}$ is one. Therefore, the columns of the matrix $\mathbf{T}_{ji}$ will give the maximum and minimum of the conditional steady-state marking probabilities. We must try this on all possible subnet S_{ji}, j: 1, ... to get a bound on the conditional probability vector.

Therefore, we can express the kth element of conditional probability vector $[\hat{\pi}_i]_k$ as follows:

$$\min_{j,l}[\mathbf{T}_{ji}]_{lk} \leq [\hat{\pi}_i]_k \leq \max_{j,l}[\mathbf{T}_{ji}]_{lk}$$

where $[\mathbf{T}_{ji}]_{lk}$ is the lth element in the kth column of $\mathbf{T}_{ji}$.

Now, we have to show that $\mathbf{T}_{ji}$ gives better bounds. Since $(\mathbf{I} - \mathbf{P}_{ji})^{-1}$ and $(\mathbf{I} - \mathbf{P}_i)^{-1}$ are nonnegative matrices, the polyhedron determined by the normalized rows of $(\mathbf{I}-\mathbf{P}_i)^{-1}$ contains the polyhedron determined by the normalized rows of their product (see [4], Lemma 1.) Clearly, Proposition 3.2 provides a tighter bound in exchange for more computation.

Although Proposition 3.2 provides better bounds than Proposition 3.1, we can see that Proposition 3.1 is a special case of Proposition 3.2 where the number of additional subnets S_{ji} is zero. In Proposition 3.2 we are assuming that the initial markings of subnets S_{ji} are equally likely. However, if we can determine the trapping probabilities into these initial markings from some additional subnets, we can estimate an even tighter bound. We can repeat this process to get increasingly tighter bounds with an increase in the computation. If we are able to know all subnets, we can uniquely determine the initial marking and we will have the exact solution.

3.1 Initial Markings of the Subnet

From the propositions we clearly see that one of the important steps in computing bounds for conditional token probability distribution is the determination of the set of local markings of the subnet and associate initial markings. Since a subnet can only be entered through depositing tokens (NSRTs) in the input places, the NSRTs in the input places clearly define the initial markings of the subnet. We assume that the number and type of resources represented by the NSRTs that are available to the subnet are known. This enables us to determine the set of initial markings of the subnet without knowledge of the rest of the subnet. We will illustrate this with a simple example. Figure 2 show a GSPN representation of a buffer multiprocessor system and its three subnets will be shown in Figure 3. Let us assume that we are interested in resource utilizations in the subnet S_1. From Table 1, we know that places p_1 and p_2 represent a number of available buffers and processors respectively in the subnet. Therefore, if the maximum number of available buffers and processors to the subnet are m and k respectively, then the set of initial markings of subnet S_i will be given by tokens in places p_3, p_4 and in input places p_1 and p_2 where tokens in places p_1 and p_2 are given by

$$p_1 = i, p_2 = j: i = 1,\ldots,m; j = 1,\ldots,k$$

Table 1 Description of the GSPN Modeling Buffer–Multiprocessor Subsystem

p_1	available buffers
p_2	available processors
p_3	task in a buffer waiting for a free processor
p_4	task executing in a processor
p_5	failed buffers
p_6	failed processors
t_1	task arrival with rate λ
t_2	completion of task in processor with rate $m_2\mu$
t_3	available buffer failure with rate $m_1 f_b$
t_4	buffer repair with rate r_b
t_5	available processor failure with rate $m_2 f_p$
t_6	processor repair with rate r_p
i_1	if a processor is available and a task is waiting in the buffer then start processing of that task and free the buffer

In our example, the set of markings generated by each of these initial markings are disjoint. In the case of disjoint sets the fundamental matrix of subnet S_1, $\mathbf{P}_i$, is given by

$$\mathbf{P}_i = \text{diag}\,\{\mathbf{P}_{i1}, \mathbf{P}_{i2}, \ldots\}$$

where fundamental matrices $\mathbf{P}_{i1}, \mathbf{P}_{i2}, \ldots$ are formed by the disjoint markings. Hence, $(\mathbf{I} - \mathbf{P}_i)^{-1}$ can be computed readily as follows:

$$(\mathbf{I} - \mathbf{P}_i)^{-1} = \text{diag}\,\{(\mathbf{I} - \mathbf{P}_{i1})^{-1}, (\mathbf{I} - \mathbf{P}_{i2})^{-1}, \ldots\}$$

Note that in this case the search for maximum and minimum of columns of the matrix $\mathbf{T}$ must be restricted within the blocks formed by the disjoint markings.

4. BOUNDED AGGREGATION

It is known that aggregation in a near-completely decomposable system introduces error. The criteria of the order of this error has also been established in [12]. As mentioned earlier, it is more desirable to get bounds in the range of magnitude of this error rather than the order. In the next section, we will show an example of such an aggregation at the GSPN level and will obtain bounds for the steady-state probability distribution. The bounded aggregation is done as follows:

First determine the reachability graph of the aggregated GSPN [1].
Find the bounds for the conditional token probabilities for each of the subnets for each aggregated global marking.
Since the transition firing rate in the aggregated GSPN depends on the token probability distribution within the subnet, we can compute the bounds for the rates from the results of the previous step.
Find the generator matrix that gives the maximum and minimum of the steady-state probabilities of the aggregated GSPN.
Solve the generator matrices to compute bounds for the probabilities of aggregated markings.
From the bounds for probability of aggregated markings and the bounds for conditional token probability in the subnet, the bounds for token probabilities can be determined.

We present an example that illustrates the technique of computing both bounds for the conditional steady-state probability of a subnet and the bounds for steady-state probabilities in a GSPN when solved using aggregation and decomposition.

5. AN EXAMPLE OF A BUFFER–MULTIPROCESSOR SUBSYSTEM

Let us consider a buffer–multiprocessor subsystem consisting of n processors and a pool of buffers as shown in Figure 1 [1, 13]. Tasks are arriving as a Poisson process with the rate λ. If the buffer pool is empty, they are rejected; otherwise they are placed in buffers to be served by the first available free processor. As soon as a task residing in a buffer gets a free processor, that buffer is released and can be used to hold new tasks. We also assume that the processing times are independent and exponentially distributed with each processor having a processing rate μ.

Faults occur in both buffers and processors as a Poisson process with rate f_b and f_p respectively. For the simplicity and to show the general technique we assume that faults in both buffers and processors occur only when they are waiting for a task. Repairs are performed on both buffers (Poisson process with rate r_b) and processors (Poisson process with rate r_p) one at a time (single server).

Figure 2 shows $S = (P,T,A,R)$, the GSPN representation of the system in Figure 1, where

$$P = \{p_1,p_2,p_3,p_4,p_5,p_6\}$$
$$T = \{t_1,t_2,t_3,t_4,t_5,t_6,i_1\}$$
$$A \subset \{P \times T\} \cup \{T \times P\}$$

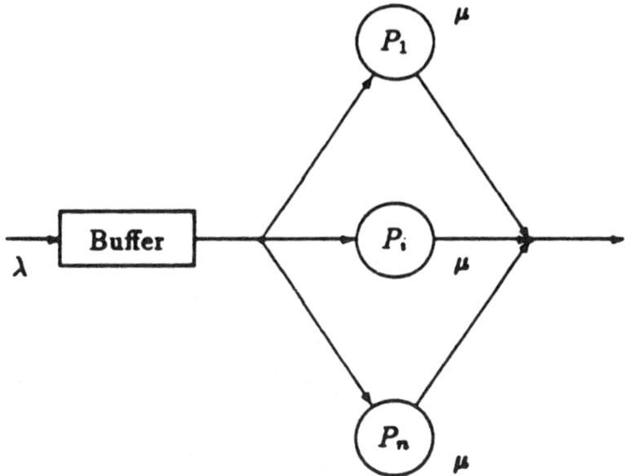

Figure 1 Buffer–multiprocessor subsystem.

$$R = \{r_1, r_2(m), r_3(m), r_4, r_5(m), r_6\}$$
$$M_i = (m_1, m_2, m_3, m_4, m_5, m_6)$$
$$\begin{aligned} r_2(m) &= m_4 r_2 & r_2 &= \mu \\ r_3(m) &= m_1 r_3 & r_3 &= f_b \\ r_5(m) &= m_2 r_5 & r_4 &= r_b \\ r_1 &= \rho\mu & r_5 &= f_p \\ & & r_6 &= r_p \end{aligned}$$

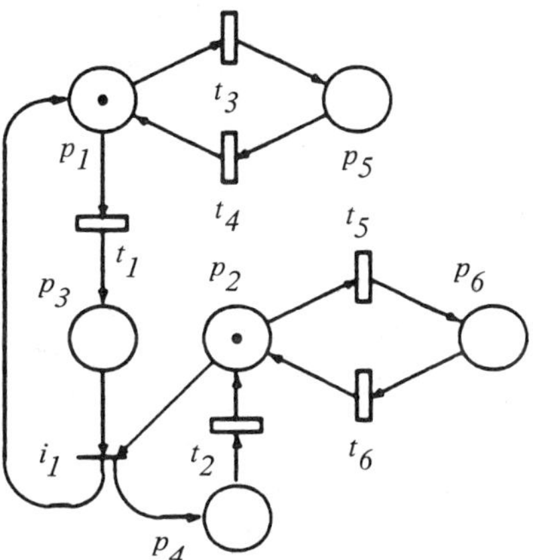

Figure 2 GSPN representation of a buffer–multiprocessor subsystem.

A description of the elements of the sets P and T is given in Table 1. M_i is some marking in the reachability set of S. The initial marking is $M_1 = (1,1,0,0,0,0)$ (i.e., one fault free buffer and one fault free processor to begin with). We will now solve the given GSPN model with the following parameter values [1]: $\mu = 1$, $\rho = 0.8, f_p = 10^{-4}, f_b = 10^{-4}, r_p = 10f_p$, and $r_b = 10f_b$, where $\rho = \lambda/\mu$. We will briefly show how the aggregated GSPN is obtained. A detailed description of this procedure is given in [1].

When modeling a system using GSPNs, transitions correspond to activities taking place in the system. Therefore, activities such as job arrival, job completion, failure and repair of processor and buffer are represented by timed transitions. Since the failure and repair rates are normally orders of magnitude smaller than the rate of the activities such as job arrival and completion, the sets of transitions T_s and T_f can be easily defined. From the given numerical values, clearly $T_s = \{t_3, t_4, t_5, t_6\}$ and $T_f = \{t_1, t_2, i_1\}$. At the fast time scale the failure and repair activities are so far apart in time that the system can be assumed fault free. Removing transitions in T_s decomposes the network (as shown in Figure 3) into the three subnets $S_1 = \{P_1, T_1, A_1, R_1\}, S_2 = \{P_2, T_2, A_2, R_2\}$ and $S_3 = \{P_3, T_3, A_3, R_3\}$, where

$$P_1 = \{p_1, p_2, p_3, p_4\}$$
$$T_1 = \{t_1, t_2, i_1\}$$
$$A_1 \subset \{P_1 \times T_1\} \cup \{T_1 \times P_1\}$$
$$R_1 = \{r_1, r_2(m)\}$$
$$M_1^1 = (1,1)$$
$$r_2(m) = m_2 r_2$$

$P_2 = \{p_5\}$	$P_3 = \{p_6\}$
$T_2 = \varnothing$	$T_3 = \varnothing$
$A_2 = \varnothing$	$A_3 = \varnothing$
$R_2 = \varnothing$	$R_3 = \varnothing$
$M_2^1 = (0)$	$M_3^1 = (0)$

To obtain the aggregated GSPN (AGSPN) each fast subnet i is aggregated into a single place p_{ai}, $i = 1, 2, 3$, as shown in Figure 4 where M_i^1 represents initial marking of these subnets. From the set of slow transitions T_s we define the set of transitions T_a of the aggregated net such that for each transition in T_s with an input place in subnet i and an output place in subnet j, we define a transition in T_a with with an input place p_{ai} and an output place p_{aj}. This completes the structure of the AGSPN as shown in Figure 4.

To complete the definition of the aggregated net we must define the initial marking and the rates of timed transitions. The initial marking of the AGSPN

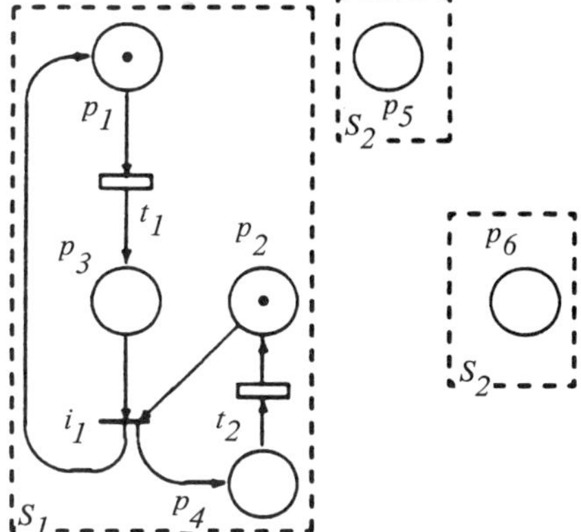

Figure 3 Subnets in the GSPN model of a buffer–multiprocessor subsystem.

is determined by the NSRTs in input places of each subnet. Clearly, $M_1^1 = (1,1)$, is the initial marking of subnet 1. Similarly $M_2^1 = (0)$ and $M_3^1 = (0)$. Hence the initial marking of the given AGSPN is $M^1 = ((1,1),0,0)$.

The rates of transitions in T_a are marking dependent. They are determined from the rates of transitions in T_s and the steady-state probability distribution of local markings in each fast subnet. These probability distributions are obtained by analyzing each subnet in isolation with a local initial marking determined from the current marking of the aggregated net. To analyze this aggregated net, we must create its reachability graph RG.

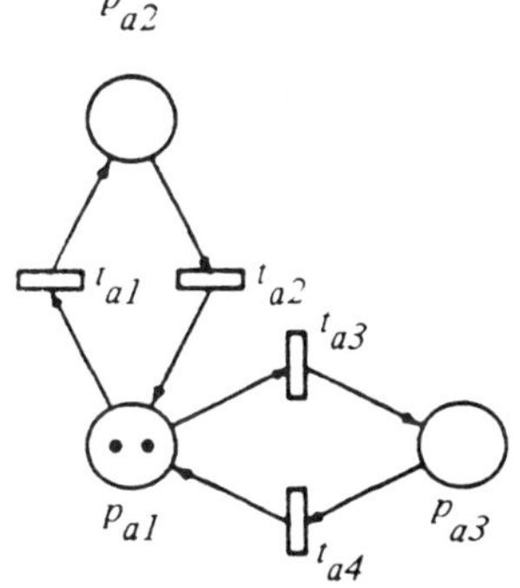

Figure 4 Aggregated subnets of the GSPN model of a buffer–multiprocessor subsystem.

The current marking is M^1. Therefore RG $= \{M^1\}$. Let the probability of finding j tokens in place p_i of subnet S_1 be denoted by $\Pr[p_i(j)]$ and the marking-dependent firing rates are denoted by $r'_{a_1}(M^1), \ldots, r'_{a_4}(M^1)$. Therefore,

$$r'_{a_1}(M^1) = r_3 \Pr[p_1(1)]$$
$$r'_{a_2}(M^1) = 0$$
$$r'_{a_3}(M^1) = r_5 \Pr[p_2(2)]$$
$$r'_{a_4}(M^1) = 0$$

In marking M^1 both transitions t_{1s} and t_{3s} can fire. Firing of t_{1s} leads to a new marking $M^2 = ((0,1),1,0)$. At M^2 we have

$$r'_{a_1}(M^2) = 0$$
$$r'_{a_2}(M^2) = r_4$$
$$r'_{a_3}(M^2) = r_5$$
$$r'_{a_4}(M^2) = 0$$

and only t_{2s} and t_{3s} can fire. Firing of t_{2s} brings the net back to the old marking M^1. Firing of t_{3s} leads the net to a new marking $M^3 = ((0,0),1,1)$. At M^3,

$$r'_{a_1}(M^3) = 0$$
$$r'_{a_2}(M^3) = r_4$$
$$r'_{a_3}(M^3) = 0$$
$$r'_{a_4}(M^3) = r_6$$

and only t_{2s} and t_{4s} can fire. Firing of t_{4s} will bring the net to a old marking M^2, and firing of t_{2s} will take the net to a new marking $M^4 = ((1,0),0,1)$. Note that firing of t_{3s} in M^1 will also lead to this marking. Therefore,

$$r'_{a_1}(M^4) = 0$$
$$r'_{a_2}(M^4) = 0$$
$$r'_{a_3}(M^4) = 0$$
$$r'_{a_4}(M^4) = r_6$$

Clearly, only t_{4s} can fire which will bring the net to a old marking M^1. Furthermore, there is no transition that can lead the net to a new marking. Therefore, the reachability set $RG = \{M^1, M^2, M^3, M^4\}$ is complete. The complete reachability graph is shown in Figure 5.

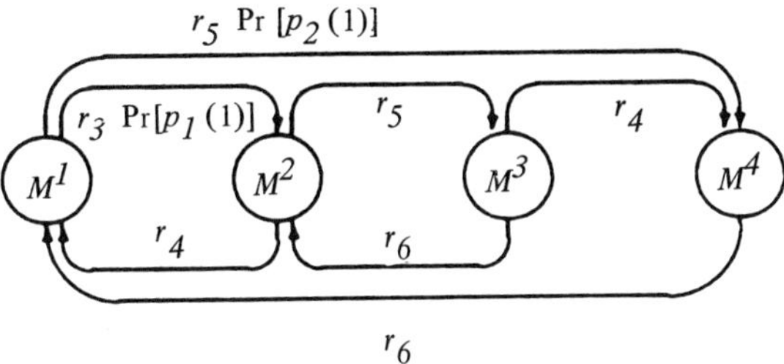

Figure 5 Reachability graph of the AGSPN.

We are interested in finding the bounds for resource utilization in the buffer–multiprocessor system as shown in Figure 1. Clearly, this depends on the bounds for Pr$[p_1(1)]$ and Pr$[p_2(1)]$ when the initial marking of S_1 is (1, 1). These give the bounds for the transition rates $r'_{a_1}(M^1)$ and $r'_{a_3}(M^1)$ which, in turn, define the maximum generator matrix $\mathbf{Q}_{\max}$ and minimum generator matrix $\mathbf{Q}_{\min}$.

We will now determine the bounds for Pr$[p_1(1)]$ and Pr$[p_2(1)]$. The local markings M_1^1, M_1^2, and M_1^3 of subnet S_1 for the given initial marking are as follows:

	p_1	p_2	p_3	p_4
M_1^1	1	1	0	0
M_1^2	1	0	0	1
M_1^3	0	0	1	1

In marking M_1^1 slow transitions t_3 and t_5 are enabled. Similarly, in marking M_1^2 only one slow transition, t_3, is enabled. However, in marking M_1^3 no such slow transition is enabled. Now, we can easily determine the fundamental matrix $\mathbf{P}_1$ as follows:

$$\mathbf{P}_1 = \begin{bmatrix} 0.555469 & 0.444420 & 0.0 \\ 0.555525 & 0.0 & 0.444420 \\ 0.0 & 0.555556 & 0.444444 \end{bmatrix}$$

Clearly,

$$(\mathbf{I} - \mathbf{P}_1)^{-1} = \begin{bmatrix} 6451.794 & 5160.924 & 4128.509 \\ 6451.155 & 5162.213 & 4129.540 \\ 6451.155 & 5162.213 & 4131.340 \end{bmatrix}$$

Normalizing the rows we get

$$\mathbf{T} = \begin{bmatrix} 0.409866 & 0.327860 & 0.266227 \\ 0.409782 & 0.327907 & 0.262311 \\ 0.409735 & 0.327870 & 0.262395 \end{bmatrix}$$

A close examination of the subnet S_1 will show that each local marking can be the initial marking through which the subnet can be entered. Therefore, the bounds for the $\Pr[p_1(1)]$ and $\Pr[p_2(1)]$ are given as follows:

$$\min\left\{\Pr[M_1^1] + \Pr[M_1^2]\right\} \leq \Pr[p_1(1)] \leq \max\left\{\Pr[M_1^1] + \Pr[M_1^2]\right\}$$

$$\min\left\{\Pr[M_1^1]\right\} \leq \Pr[p_2(1)] \leq \max\left\{\Pr[M_1^1]\right\}$$

Therefore,

$0.737595 \leq \Pr[p_1(1)] \leq 0.737773$

$0.409735 \leq \Pr[p_2(1)] \leq 0.409866$

We are interested in the bounds for the aggregated marking M^1. Therefore, the maximum generator matrix $\mathbf{Q}_{\max}$ and the minimum generator matrix $\mathbf{Q}_{\min}$ of the AGSPN of Figure 4 are given by

$$\mathbf{Q}_{\max} = \begin{bmatrix} -1.147609 \times 10^{-4} & 0.737773 \times 10^{-4} & 0.0 & 0.409866 \times 10^{-4} \\ 10^{-3} & -1.1 \times 10^{-3} & 10^{-4} & 0.0 \\ 0.0 & 10^{-3} & -2.0 \times 10^{-3} & 10^{-3} \\ 10^{-3} & 0.0 & 0.0 & -10^{-3} \end{bmatrix}$$

and

$$\mathbf{Q}_{\min} = \begin{bmatrix} -1.147330 \times 10^{-4} & 0.737595 \times 10^{-4} & 0.0 & 0.409735 \times 10^{-4} \\ 10^{-3} & -1.1 \times 10^{-3} & 10^{-4} & 0.0 \\ 0.0 & 10^{-3} & -2.0 \times 10^{-3} & 10^{-3} \\ 10^{-3} & 0.0 & 0.0 & -10^{-3} \end{bmatrix}$$

Solving these two matrices we get the bounds for the steady-state probability vector $\pi_a = [M^1, M^2, M^3, M^4]^T$ as follows:

$$\begin{bmatrix} 0.894235 \\ 0.062808 \\ 0.003140 \\ 0.039790 \end{bmatrix} \leq \pi_a \leq \begin{bmatrix} 0.894258 \\ 0.062832 \\ 0.003141 \\ 0.039794 \end{bmatrix}$$

Therefore, the bounds for the utilization of the processor and buffer, U_p and U_b, are given by

$$\min\left\{\Pr[M^1]\right\} \min\left\{\Pr[p_4(1)]\right\} \leq U_p \leq \max\left\{\Pr[M^1]\right\} \max\left\{\Pr[p_4(1)]\right\}$$

$$\min\{\Pr[M^1]\}\min\{(1-\Pr[p_1(1)])\} \leq U_b \leq \max\{\Pr[M^1]\}\max\{(1-\Pr[p_1(1)])\}$$

Solving, we get

$$0.527718 \leq U_p(0.527784) \leq 0.527883$$
$$0.340259 \leq U_b(0.340270) \leq 0.340417$$

The numbers within braces with U_p and U_b are from the exact solution. Clearly, we have a fairly tight bound for the resource utilization. This is because there is an order of magnitude difference in the rates of slow and fast transitions and the error due to aggregation is clearly seen to be the equal to the order of maximum degree of coupling between the subnets.

6. CONCLUSION

In this paper, we have presented techniques of estimating the bounds for the error due to aggregation where the analysis is done at the GSPN level as opposed to the Markov chain level. We have also presented techniques to compute, with various degrees of accuracy and complexity, the bounds for the conditional token probabilities of a class of GSPN models when only part of the net, called the subnet, is available. This is extremely important when the whole net is prohibitively large and the measure of performance that we are interested in depends mainly on the activities and resources modeled in the subnet and weakly affected by the activities in the rest of the net. To illustrate the overall technique, the bounds for various resource utilizations are determined using a simplified example of a buffer–multiprocessor system.

REFERENCES

[1] H. H. Ammar and S. M. R. Islam, Time scale decomposition of a class of generalized stochastic Petri net models, *IEEE Trans. Software Eng.*, vol. SE-15, no. 6, pp. 809–820, June 1989.

[2] M. K. Molloy, Fast bound for stochastic Petri nets, in *Proc. Int. Workshop on Timed Petri Nets*, Torino, Italy, July 1985.

[3] S. C. Bruell and S. Ghanta, Throughput bounds for generalized stochastic Petri nets, in *Proc. Int. Workshop on Timed Petri Nets*, Torino, Italy, July 1985.

[4] P. J. Courtois and P. Semal, Computable bounds for conditional steady-state probabilities in large Markov chains and queuing models, *IEEE J. Selected Areas Commun.*, vol. SAC-4, no. 6. Sept. 1986.

[5] P. J. Courtois and P. Semal, Bounds for the positive eigenvectors of nonnegative matrices and for their approximation by decomposition, *J. ACM*, vol. 31, no. 4, Oct. 1984.

[6] G. W. Stewart,Computable error bounds for aggregated Markov chain, *J. ACM*, vol. 30, no. 2, April 1983.

[7] M. A. Marsan, G. Balbo, and G. Conte, A class of generalized stochastic Petri nets for the performance analysis of multiprocessor systems, *ACM Trans. Comput. Systems*, vol. 2, no. 2, May 1984.

[8] C. Petri, Communication with Automata, Technical Report, RADC-TR-65-377, Jan. 1966, vol. 1, RADC, Griffiths Air Force Base, N.Y.

[9] J. L. Peterson, *Petri Net Theory and the Modeling of Systems*, Englewood Cliffs, N.J.: Prentice-Hall, 1981.

[10] J. L. Peterson, Petri nets, *ACM Comput. Surveys*, vol. 9, no. 3, Sept. 1977.

[11] M. K. Molloy, Performance analysis using stochastic Petri nets, *IEEE Trans. Comput.* vol. C-31, no. 9, Sept. 1982.

[12] P. J. Courtois, *Decomposability: Queuing and Computer Systems Applications*, New York: Academic Press, 1977.

[13] J. F. Meyer, A. Movaghar, and W. H. Sanders, Stochastic activity networks, structures, behavior, and application, in *Proc. Int. Workshop on Timed Petri Nets*, Torino, Italy, July 1985.

[14] K. S. Trivedi, *Probability and Statistics with Reliability, Queuing, and Computer Science Applications*, Englewood Cliffs, N.J.: Prentice-Hall, 1982.

[15] M. Goderch, A. S. Willsky, S. S. Sastry, and D. A Cantanon, Hierarchical aggregation of singularly perturbed finite state Markov process, *Stochastic*, vol. 8, 1983.

[16] H. H. Ammar and R. W. Liu, Analysis of the generalized stochastic Petri nets by state aggregation, in *Proc. Int. Workshop on Timed Petri Nets*, Torino, Italy, July 1985.

[17] H. H. Ammar, Y. F. Huang and R. W. Liu, Hierarchical models for systems reliability maintainability, and availability, *IEEE Trans. Circuits Systems*, vol. CAS-34, no. 6, June 1987.

32

The Importance of Bias Terms for Error Bounds and Comparison Results

NICO M. VAN DIJK Department of Economical Sciences and Econometrics, Free University, Amsterdam, The Netherlands

ABSTRACT

Some recent results are surveyed and unified showing that bias terms of Markov reward structures are essential to conclude error bound or monotonicity results for steady-state measures when dealing with

perturbations
finite truncations
system modifications, or
system comparisons (bounds)

The results are illustrated by a non-product-form queuing network example with practical phenomena such as blocking, overflowing, and breakdowns. Monotonicity results and explicit error bounds are hereby obtained.

1. INTRODUCTION

Markov chains are known to be a powerful modeling tool for a variety of practical situations. Applications are found in telecommunication (queuing networks, broadcasting, satellite communication), computer performance evaluation (computer networks, parallel programming, store and forward buffering), manufacturing (assembly lines, material handling systems), reliability (maintenance, breakdown analysis), and inventory theory, as well as combinations of these as performability analysis of computer systems with breakdowns, error detections or fault tolerance. Modeling of realistic situations, however, may itself already introduce inaccuracies by making simplifying assumptions and by using estimated parameter values.

Traditionally, steady-state behavior is the prime interest, such as to evaluate throughput, system efficiency, mean workload, response time, or system availability. But transient analysis also becomes more and more important. As closed-form expressions are available only in a limited number of situations (such as a Jackson network without blocking, a finite reversible material handling system, or a simple parallel computer system with Poissonian input), numerical or approximate computations are to be performed. These, however, easily become astronomic and usually encounter unspecified imprecisions. *A priori* error bounds for the accuracy or order of accuracy are thus of interest.

Generally, different types of "approximate" results can hereby be involved. Let us discuss the basic ones in more detail.

1.1 Perturbation

Markov (reward) chains are usually studied under the assumption that the one-step transition probabilities (and rewards) have 100% known values. In practice, in contrast, these are usually to be obtained from statistical estimation, such as by interval estimates for the arrival and service rates in a queuing system. Inaccuracies in parameter values or the input data of Markov chains are hereby naturally encountered. Alternatively, by studying the effect of perturbations in the input data (parameters) of a Markov (reward) process, qualitative or sensitivity results with respect to essential system parameters can be obtained.

In [9] and [14] perturbation results are given for stationary probabilities of finite Markov chains. The accuracy of these results was characterized by the so-called fundamental matrix. In [5] and [43] error bounds for approximate undiscounted finite-horizon and discounted infinite-horizon Markov reward structures are derived. The order of these bounds, however, did not allow a limiting result for the average reward case. An extension to the average reward case was therefore developed in [38]. Herein, the role of bias-terms, which are related to fundamental matrices, was made explicit. One-dimensional or random walk type queuing applications could be dealt with by bounding these bias terms based on mean first-passage time expressions for random walks. In [30] these results were generalized to unbounded reward structures and a simple two-dimensional queuing application.

1.2 State-Space Truncation

Practical situations often encounter large or infinite state spaces (e.g., an infinite server queue, an open queuing network, or a maintenance system with an *a priori* unbounded lifetime). For computational purposes truncation of the state space is then required.

Though the technique of state-space truncation is a common feature in practice, theoretical support in terms of orders of accuracy or rates of convergence hardly seems available. Convergence proofs as the truncation size tends to infinity have already been investigated in the early 1950s by Savymsakov and were crystallized most notably by Seneta (cf. [15], [16]) with reference to private communications with Kendall (cf. [15]). A detailed study of these convergence results as well as an extensive list of related literature can be found in [17]. In this latter reference, also simple error bounds are provided (cf. theorem 6.4 and its corollary, p. 215), but these are just robust bounds, which do not secure an order of accuracy.

A priori truncation error bounds not only provide a measure of inaccuracy for fixed truncations but also lead to a stop criterion or a convergence rate when approximating an infinite system by a sequence of finite systems, such as for optimization purposes.

Conversely, one might wish to know the inaccuracy of modeling finite systems by infinite analogues in order to obtain simple performance bounds (cf. [32], [33], [36]). Finally, as a special application, methods based on bounded transition rate functions, such as the uniformization method, are not generally applicable to infinite systems (e.g., an infinite server queue). By providing state-space truncation error bounds such methods can be made applicable in an approximate manner (cf. [35]).

1.3 Modifications

Explicit steady-state expressions are usually obtainable only under strong assumptions that are often unrealistic. In service systems, for example, inputs are frequently assumed to be Poissonian merely for convenient modeling, while in practice inputs are more likely to be of a large finite source and state-dependent nature.

Product-form expressions for queuing networks are typically violated by practical phenomena such as blocking, dynamic routing, breakdowns, and priorities. By slightly modifying the system protocols or rather the underlying transition structure, conditions can be met from which product-form "approximations" can be concluded. Simple performance estimates based on product form modifications have so been established for various non-product-form systems (cf. [28], [31], [37], [39], [40]). *A priori* analytic error bounds for these estimates, though, have as yet not been provided.

1.4 Comparison Results

A system can be compared under different situations, for example, (1) to investigate the effect of changing certain system parameters such as a storage or service capacity (cf. [1], [2], [17], [18], [19], [20], [21], [22], [23], [24], [25]), (2) to determine a better protocol such as for dynamic job or server allocation (cf. [3], [17], [39], [42]), or (3) to conclude that a specific transition modification leads to a performance bound (cf. [21], [28], [31], [37], [39], [40], [46]).

Established comparison, or relatedly monotonicity, proof techniques as the one-step comparison technique employed in [6], [23], [44], and [45], and the related sample path technique in [18], [19], [20], [22], [25], [39], however, do not generally apply (cf. [40]). The Markov reward proof technique that will be illustrated in this paper has already proven to be a fruitful addition to this end (cf. [28], [29], [32], [33], [36], [37], [40]).

All of the above different types of "approximations" thus come down to some kind of modification or perturbation of the transition structure and/or a truncation or extension of the state space. This paper, therefore, will provide a tool to conclude error bounds for comparing an original and modified system. Particularly, error bound or comparison results can hereby be concluded when dealing with perturbations; finite truncations; system modifications, or system comparisons (bounds).

The key-step to these results turns out to be one and the same: an estimation of the so-called bias terms for the specific required Markov reward structure in order. To this end, recent results from [30], [34], [38], and [40] are surveyed in a unifying manner merely to highlight the importance of these bias terms and the various ways that they can be exploited for different purposes.

Bias terms (or fundamental matrices) are common knowledge in Markov decision theory as the key factor to determine average optimal policies (cf. [11], [26]). They are also known to be directly related to mean first passage times, which play a key role in the conditioning and convergence of numerical procedures to solve steady-state equations (cf. [7], [9], [14]). Unfortunately, explicit expressions of passage times can be obtained in only a limited number of simple situations such as random walks (cf. [7]).

Most essential, in contrast, is that in concrete situations one can quite frequently derive explicit analytic bounds for bias terms by employing an inductive Markov reward proof technique. The steps can become rather technical and complicated when a large number of different types of transitions are possible. For many natural and simple multidimensional transition structures though, such as most typically queuing networks, it has already proven to be most fruitful (cf. [32], [36]). This paper also aims to illustrate this important aspect of bias-terms.

A pilot example will therefore be analyzed in detail. This example concerns a simple but multidimensional queuing network, which includes typical practical features mentioned earlier: finite capacities, overflow, and breakdowns. Both comparison results and error bounds will be established for a perturbation, a state-space truncation, and a system modification. The verification of the necessary conditions and the importance of bias-terms will hereby be illustrated.

2. GENERAL RESULTS

This section will provide the key theorem from which error bound and monotonicity results of steady-state measures of Markov chains can be concluded. The theorem itself is presented in Section 2.1. Its conditions will be discussed in Section 2.2. A transformation to continuous-time Markov chains is given in Section 2.3, while in Section 2.4 the results are particularized to queuing networks.

2.1 Key Theorem

Consider a Markov chain $\{X_t, t = 0, 1, 2, \ldots\}$ with states ordered at $N = \{1, 2, \ldots\}$ and one-step transition probability matrix $P = (p(i,j))$. Without loss of generality assume that this Markov chain is irreducible as some set S.

Let $\{\bar{X}_t, t = 0, 1, 2, \ldots\}$ be another Markov chain with states also ordered at $N = \{1, 2, \ldots\}$ and one-step transition probability matrix $\bar{P} = (\bar{p}(i,j))$. Again, without loss of generality assume that this Markov chain is irreducible at some set $\bar{S}$. We impose the condition:

$$\bar{S} \subset S \tag{1}$$

From now on, we use the overbar symbol for an expression concerning the second chain and the overbar in parentheses to indicate that an expression is to be read for both chains. Let $\overset{(-)}{T}$ and $\{\overset{(-)}{T}_t \mid t = 0, 1, 2, \ldots\}$ be operators on arbitrary functions f as defined by

$$\begin{aligned}
{}^{(\bar{T})}f(i) &= \sum_j {}^{(\bar{p})}_{ij} f(j) \\
{}^{(\bar{T})}_{t+1} &= {}^{(\bar{T})(\bar{T})}_{t} \\
{}^{(\bar{T})}_{0} &= I
\end{aligned} \tag{2}$$

and for a given function r define functions ${}^{(\bar{V})}_N$, $N = 0, 1, 2, \ldots$, by

$$ {}^{(\bar{V})}_N = \sum_{t=0}^{N-1} {}^{(\bar{T})}_t {}^{(\bar{r})} \tag{3}$$

In words, ${}^{(\bar{V})}_N(i)$ represents the total expected reward over N periods when starting in state i at time $t = 0$ and receiving a one-step reward $r(X_t)$ at time $t = 0, 1, \ldots, N-1$. Then for any given state $l \in \bar{S}$,

$$ {}^{(\bar{g})} = \lim_{n\to\infty} \frac{1}{N} {}^{(\bar{V})}_N(l) \tag{4}$$

is the expected average reward provided this limit exists. The following key theorem can now be proven. It provides a pair of conditions from which an error bound or a comparison of the performance of both chains can be concluded.

THEOREM 2.1 (ERROR BOUND) Suppose that for some nonnegative function Φ, some initial state $l \in \bar{S}$, some constants $\epsilon, \delta, \beta > 0$, all $i \in \bar{S}$ and $t \geq 0$:

$$\left| \sum_j [\bar{p}(i,j) - p(i,j)][V_t(j) - V_t(i)] \right| \leq \epsilon \Phi(i) \tag{5}$$

$$|(\bar{r} - r)(i)| \leq \delta \Phi(i) \tag{6}$$

$$\bar{T}_t \Phi(l) \leq \beta \tag{7}$$

Then

$$|\bar{g} - g| \leq [\epsilon + \delta]\beta \tag{8}$$

Proof. As for all t:

$$ {}^{(\bar{V})}_{t+1} = {}^{(\bar{r})} + {}^{(\bar{T})(\bar{V})}_{t} \tag{9}$$

by virtue of (3), while the transition probabilities $\bar{p}(\cdot,\cdot)$ remain restricted to $\bar{S} \subset S$, for arbitrary $l \in \bar{S}$ we can write:

$$(\bar{V}_N - V_N)(l) = (\bar{r} - r)(l) + (\bar{T}\bar{V}_{N-1} - TV_{N-1})(l)$$

$$= (\bar{r} - r)(l) + (\bar{T} - T)V_{N-1}(l) + \bar{T}(\bar{V}_{N-1} - V_{N-1})(l)$$

$$= \sum_{t=0}^{N-1} \bar{T}_t([\bar{r} - r] + [(\bar{T} - T)V_{N-t-1}])(l) + \bar{T}_N(\bar{V}_0 - V_0)(l) \tag{10}$$

where the latter equality follows by iteration. Now note that the last term in the last right-hand side is equal to 0 as $\bar{V}_0(\cdot) = V_0(\cdot) = 0$ by definition. Further, for any s and i:

$$(\bar{T} - T)V_s(i) = \sum_j [\bar{p}(i,j) - p(i,j)]V_s(j)$$

$$= \sum_j [\bar{p}(i,j) - p(i,j)][V_s(j) - V_s(i)] \tag{11}$$

By substituting equation (11) in equation (10), taking absolute values, and noting that $\bar{T}_t$ is a monotone operator for all $t \geq 0$, we obtain from equations (5), (6), (7), and (10):

$$|(\bar{V}_N - V_N)(l)| \leq [\delta + \epsilon| \sum_{t=0}^{N-1} \bar{T}_t \Phi(l) \leq [\delta + \epsilon]\beta N \tag{12}$$

Applying equation (4) completes the proof.

REMARK 2.1 Clearly, the conditions of equations (5), (6), and (7) could have been combined in one bounding condition that can be applied directly to equation (10). The present slightly more restrictive conditions are preferred as they appear more practical for verification. In the next theorem, however, which concentrates only on monotonicity, the combination of equations (5) and (6) has proven to be most essential in applications (cf. [28], [37], [40]).

THEOREM 2.2 (COMPARISON RESULT) Suppose that for all $i \in \bar{S}$ and $t \geq 0$:

$$[\bar{r} - r](i) + \sum_j [\bar{p}(i,j) - p(i,j)][V_t(j) - V_t(i)] \geq 0 \ (\leq 0) \tag{13}$$

Then

$$\bar{g} \geq g \ (\bar{g} \leq g) \tag{14}$$

Proof. This follows directly from substituting equation (11) in equation (10) and noting that the operators $\bar{T}_t$ are monotone (i.e., $T_t f \geq \bar{T}_t g$ whenever $f \geq g$ componentwise).

REMARK 2.2 (IMPORTANCE OF BIAS-TERMS) The crucial step for the above theorems is the simple relation (11). This step enables one to transform conditions upon $V_t(\cdot)$ in so-called bias-terms: $V_t(j) - V_t(i)$. While $V_t(\cdot)$ generally grows linearly in t, bias terms for given i and j are generally bounded uniformly in t. More precisely, when $r(\cdot)$ is bounded, say $|r(i)| \leq B$ for all i, by simple Markov reward arguments (cf. [38]) one proves

$$|V_t(j) - V_t(i)| \leq 2B \min[R_{ij}, R_{ji}] \tag{15}$$

where R_{ij} is the expected number of steps (mean first passage time; see [7]) to reach state j out of state i. A similar though more technical result in terms of these times can be given also for unbounded rewards (cf. [30]). Most essentially, however, closed-form expressions or simple bounds for mean first passage times seem to be limited to simple one-dimensional random walks (cf. [7]). In the next section, therefore, we will illustrate how estimates for bias terms can be derived by inductive Markov reward arguments or more precisely by employing equation (9). Most notably, multidimensional applications such as queuing networks (also see step 1 in Section 2.2) can hereby be dealt with.

REMARK 2.3 (UNBOUNDED REWARDS) Note that no conditions are imposed upon the one-step reward function $r(\cdot)$ other than that we implicitly assume the average rewards g and $\bar{g}$ to be well defined. Particularly, unbounded rewards are allowed. For instance, by using a linear one-step reward function $r(i) = i$ we can compute a mean queue length of an infinite system. As a special example of the bounded case, g represents the steady state probability of a set G if we choose

$$r(i) = \begin{cases} 1 & \text{for } i \in G \\ 0 & \text{otherwise} \end{cases}$$

REMARK 2.4 (STATE LABELING) For expository convenience the states were labeled in a countable manner. Clearly, for multidimensional applications such as queuing networks we can always label the states in an appropriate one-dimensional manner. But more conveniently, the results of this section can directly be reread with multidimensional states by simply identifying a state with symbols i and j. In Section 2.5 this will be applied for queuing networks.

REMARK 2.5 (TRANSIENT RESULTS) In analogy with [28], both Theorem 2.1 and Theorem 2.2 can be extended to transient analysis of reward structures up to an exit time.

2.2 Discussion of Conditions

In order to give more insight in the conditions and their possible verification, this section briefly discusses the steps involved for applying Theorem 2.1.

First of all, one must typically think of either β or $[\delta + \epsilon]$ to be small. For convenience, let $\Delta(i,j) = \bar{p}(i,j) - p(i,j)$.

Step 1 (Bounded bias terms)

As the first and most essential step, one has to find estimates (bounds) B_{ij} for specific i and j such that for all s:

$$|V_s(j) - V_s(i)| \leq B_{ij} \tag{16}$$

Conveniently, in condition (5) this is required only for all i,j with

$$|\Delta(i,j)| > 0 \tag{17}$$

For example, with i the number of jobs in a discrete-time birth-death queuing system we only need to consider $j = i - 1$ and $j = i + 1$.

To verify equation (16), an inductive Markov reward technique based on equation (9) can be applied where T is to be written out in the transition probabilities as per equation (2). The structure of these probabilities and an induction hypothesis upon V_t and the reward $r(\cdot)$ together must prove equation (16) also for $s = t + 1$. This key step is conceptually straightforward but can be rather complex in concrete multidimensional applications.

Step 2 (Transition Differences)

Secondly, one has to find out whether the differences in the transition structure $\Delta(\cdot,\cdot)$ are small or just bounded up to a state-dependent scaling function $\Phi(\cdot)$. For illustration, we think of $\Phi(\cdot) = 1$ and present the following examples.

EXAMPLE 1 Consider a discrete-time birth-death model representing a discrete-time single-server queue with arrival probability α and service probability γ per step [i.e., with probability $\alpha(1-\gamma)$ a job arrives but no job leaves, while with probability $\gamma(1 - \alpha)$ a job leaves but no job arrives] and consider the same model with arrival probability $\alpha \pm \tau$ (perturbation) as resulting from a statistical confidence interval for estimating α, where τ is to be thought of as being small. Then with i the number of jobs present,

$$|\Delta(i,i + 1)| \leq \tau$$

EXAMPLE 2 Consider the model as in the example above but now with rejection of arrivals (state-space truncation) if upon arrival the number i of jobs present is equal to some limit L. Then with $1(A)$ an indicator function of event A,

$$|\Delta(i, i+1)| \leq \alpha 1(i = L)$$

Step 3 (Bounding Function Φ)

By comparing the transition structures, candidates for an appropriate bounding function $\Phi(\cdot)$ may come up naturally. Here one may typically think of polynomial type functions. In Example 2 above, for instance, condition (5) will be satisfied with some constant B resulting from equation (16) and

$$\epsilon = \begin{cases} \alpha B/L & \text{if } \Phi(i) = i \\ \alpha B & \text{if } \Phi(i) = 1(i = L) \end{cases}$$

Step 4 (Stability)

Which option of $\Phi(\cdot)$ is appropriate will eventually depend on whether we can easily verify equation (7), requiring that its expected value over time remains bounded (stability) by either a small or just a finite number. As illustration, again for Example 2 from above, analogously to the standard continuous-time single-server queue one easily shows with $\lambda = \alpha(1-\gamma)$ and $\mu = \gamma(1-\alpha)$:

$$\beta = \begin{cases} \beta_1 = 1 & \text{if } \Phi(i) = 1 \\ \beta_2 = (\lambda/\mu)^L[\sum_{k=0}^{L}(\lambda/\mu)^k]^{-1} & \text{if } \Phi(i) = 1(i = L) \\ \beta_3 = \sum_{k=0}^{L} k(\lambda/\mu)^k[\sum_{k=0}^{L}(\lambda/\mu)^k]^{-1} & \text{if } \Phi(i) = i \end{cases}$$

Summarizing, Theorem 2.1 can thus be applicable in the following twofold manners, provided the bias terms can be estimated as per equation (16):

1. By showing that the impact of the difference Δ in the transition structures on the bias terms is sufficiently small, such as for Example 1 with $\epsilon = \tau B$ and $\beta = \beta_1 = 1$ by using $\Phi(i) = 1$, or Example 2 with $\epsilon = \alpha B/L$ and $\beta = \beta_3$ by using $\Phi(i) = i$.
2. By showing that the expected value of the scaling function or the probability of being in states where this difference is significant, is sufficiently small, such as for Example 2 with $\epsilon = \alpha B$ and $\beta = \beta_2$ by using $\Phi(i) = 1(i = L)$.

2.3 A Special Case: Truncation

To illustrate how truncations are covered by the results of Section 2.1, consider the special case that for some $L < \infty$:

$$\begin{aligned}
&\bar{p}(i,j) = 0 && j > L,\ i \le L \\
&\bar{p}(i,j) = p(i,j), && j \ne t[i],\ j \le L,\ i \le L \\
&\bar{p}(i,t[i]) = p(i,t[i]) + \sum_{j>L} p(i,j) && i \le L
\end{aligned} \tag{18}$$

where $t[i] \le L$ is some given "state of truncation" for any $i \le L$. In words that is, all transitions of the original matrix $p(i,j)$ out of state i beyond a certain threshold L are reflected to one and the same state $t[i]$. Condition (5) now simply reduces to:

$$\left| \sum_{j>L} p(i,j)[V_t(j) - V_t(t[i])] \right| \le \epsilon\Phi(i) \tag{19}$$

The flexibility in choosing different absorption states $t[i]$ for different states i will be convenient for multidimensional applications.

REMARK 2.6 (OTHER TRUNCATIONS) The truncation (18) is a natural one as it corresponds to the original model as long as the truncation limit L is not exceeded. Clearly, similar conditions can be provided for other types of truncations. For example, rather than letting a transition $i \to j$ for all $j > L$ transform into one and the same state $t[i]$, we can also let it transform into different states in a randomized manner.

2.4 Continuous-Time Markov Chains

Various Markov chain applications, most notably in queuing, are of a continuous- rather than discrete-time nature. In order to apply the above results the standard uniformization technique (cf. [12], p. 261, [26], p. 110) can then be applied provided the transition rates are uniformly bounded. More precisely, let $\{\overset{(-)}{X}_t \mid t = 0, 1, 2, \ldots\}$ be a continuous-time irreducible Markov chain at some set $\overset{(-)}{S}$ with transition rates $\overset{(-)}{q}(i,j)$ such that for some $Q < \infty$ and all $i \in \overset{(-)}{S}$:

$$\sum_{j \ne i} \overset{(-)}{q}(i,j) \le Q \tag{20}$$

Let $\overset{(-)}{R}(i)$ be a reward rate, that is, the reward per unit of time when the chain is in state i and $\overset{(-)}{G}$ the corresponding average expected reward (per unit of time). The results of Section 2.1 then apply by substituting

$$\left.\begin{aligned}
&g = \overset{(-)}{G}/Q \\
&\overset{(-)}{r} = \overset{(-)}{R}/Q \\
&\overset{(-)}{p}(i,j) = \overset{(-)}{q}(i,j)/Q \qquad (j \neq i) \\
&\overset{(-)}{p}(i,j) = \overset{(-)}{q}(i,j)/Q + [1 - \sum_{j \neq i} \overset{(-)}{q}(i,j)/Q]
\end{aligned}\right\} \tag{21}$$

The conditions (5) and (6) are then natural to be satisfied with $\delta = \alpha/Q$ and $\epsilon = \gamma/Q$, where α and γ are to be thought of as small, so that from equation (8):

$$|\bar{G} - G| \leq [\alpha + \gamma]\beta \tag{22}$$

REMARK 2.7 For the unbounded case, that is, without equation (20), an approximate uniformization can be applied as developed in [35]. The details are rather technical.

2.5 Special Application: Queuing Networks

As queuing networks are an application area of significant interest while both a multidimensional (Remarks 2.2 and 2.4) and continuous-time (Section 2.4) structure are to be taken into account, this subsection will particularize the results of Section 2.1 to queuing networks.

Model

Consider an arbitrary open or closed single-class exponential queuing network with N service stations (hereafter called the original model), such as illustrated below.

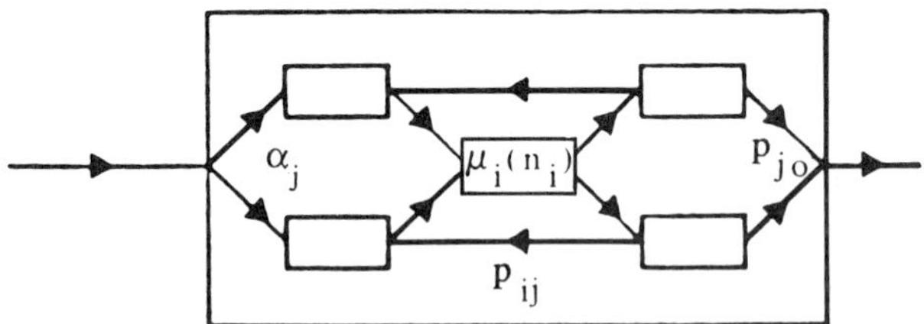

The state of the network is described by $\bar{n} = (n_1, \ldots, n_N)$ where n_i is the number of jobs at station i, $i = 1, \ldots, N$. By $n + e_i$ or $n - e_i$ we denote the state equal to $\bar{n}$ except for one job more respectively less at station i, where $\bar{n} - e_i = \bar{n}$ for $n_i = 0$, $i = 1, \ldots, N$ and where we also allow $i = 0$ with the convention that $\bar{n} + e_0 = \bar{n}$. Consequently, by $\bar{n} - e_i + e_j$ we denote the state equal to $\bar{n}$ with one job moved from station i into station j, where $i = 0$ corresponds to an external arrival at station j and $j = 0$ to a departure from the system at station i.

Let $q(\bar{n}, \bar{n} - e_i + e_j)$ be the transition rate for a change from state $\bar{n}$ in state $\bar{n} - e_i + e_j$, while transition rates for changes not of this form are assumed to be 0. For example, for a standard Jackson network we have

$$q(\bar{n}, \bar{n} - e_i + e_j) = \mu_i p_{ij}$$

with μ_i the service rate at station i and p_{ij} the routing probability from station i to j, while an additional capacity constraint N_j at station j yielding a reflective blocking (communication protocol) is parametrized by

$$q(\bar{n}, \bar{n} - e_i + e_j) = \mu_i p_{ij} 1\,(n_j < N_j)$$

Assumptions

1. The underlying Markov jump process is irreducible at some set S of admissible states $\bar{n}$ with a unique stationary distribution $\pi(\cdot)$.
2. We can choose a finite Q such that

$$Q \geq \sup_{\bar{n} \in S} \sum_{i,j} q(\bar{n}, \bar{n} - e_i + e_j)$$

3. For given reward rate $R(\bar{n})$ the measure G is well defined by

$$G = \sum_{\bar{n}} \pi(\bar{n}) R(\bar{n})$$

Approximate Model

Now consider a modified version of the single class exponential queuing network (hereafter called the modified model) with a description as above, but with $q(\bar{n}, \bar{n} - e_i + e_j)$ replaced by $\bar{q}(\bar{n}, \bar{n} - e_i + e_j)$, the assumptions 1,2, and 3 adopted with S, π, r, and g replaced by $\bar{S}$, $\bar{\pi}$, $\bar{r}$, and $\bar{g}$, and $\bar{S} \subset S$.

Comparison Result

In order to compare the original and modified network, now define the functions $\overset{(-)}{V}_t$ and operators $\overset{(-)}{T}_t$ for functions $f\colon \overset{(-)}{S} \to R_1$ by

$$\overset{(-)}{V}_{t+1}(\bar{n}) = \overset{(-)}{R}(\bar{n}) Q^{-1} + \overset{(-)}{T}\overset{(-)}{V}_t(\bar{n})$$

and

$$\overset{(-)}{T}_0 f(\bar{n}) = f(\bar{n}) \qquad \overset{(-)}{T}_{t+1} f(\bar{n}) = \overset{(-)}{T}\overset{(-)}{T}_t f(\bar{n})$$

where

$$ {}^{(\bar{T})}f(\bar{n}) = \sum_{i \neq j=0}^{N} [{}^{(\bar{q})}(\bar{n}, \bar{n} - e_i + e_j)Q^{-1}]f(\bar{n} - e_i + e_j) \tag{23} $$

$$ + [{}^{(\bar{q})}(\bar{n}, \bar{n})Q^{-1} + 1 - \sum_{i \neq j=0}^{N} {}^{(\bar{q})}(\bar{n}, \bar{n} - e_i + e_j)Q^{-1}]f(\bar{n}) $$

The following application of Theorem 2.1 then provides an error bound for the difference $|\bar{G} - G|$ without having to compute the stationary distributions ${}^{(\bar{\pi})}$. Herein, let

$$ \Delta(\bar{n}, \bar{n} - e_i + e_j) = \bar{q}(\bar{n}, \bar{n} - e_i + e_j) - q((\bar{n}, \bar{n} - e_i + e_j) \qquad (\bar{n} \in \bar{S}) \tag{24} $$

THEOREM 2.3 (ERROR BOUND) Suppose that for some nonnegative function Φ, some initial state $\bar{l} \in \bar{S}$, constants α, γ, $\beta > 0$, all $\bar{n} \in \bar{S}$ and $t \geq 0$:

$$ \left| \sum_{i,j=0}^{N} \Delta(\bar{n}, \bar{n} - e_i + e_j)[V_t(\bar{n} - e_i + e_j) - V_t(\bar{n}) \right| \leq \alpha\Phi(\bar{n}) \tag{25} $$

$$ |\bar{R}(\bar{n}) - R(\bar{n})| \leq \gamma\Phi(\bar{n}) \tag{26} $$

$$ \bar{T}_t\Phi(\bar{l}) \leq \beta \tag{27} $$

Then

$$ |\bar{G} - G| \leq \beta[\alpha + \gamma] \tag{28} $$

Proof. Directly by applying Theorem 2.1 with the substitutions (21), $\delta = \alpha/Q$, $\epsilon = \gamma/Q$ and noting that the difference $\Delta(\bar{n}, \bar{n})$ in the transition rates for a change form a state $\bar{n}$ in itself vanishes in equation (5).

In a similar manner, we conclude from Theorem 2.2:

THEOREM 2.4 (MONOTONICITY) Suppose that for all $\bar{n} \in \bar{S}$ and $t \geq 0$:

$$ [\bar{R} - R](\bar{n}) + \sum_{i\,j=0}^{N} \Delta(\bar{n}, \bar{n} - e_i + e_j)[V_t(\bar{n} - e_i + e_j) - V_t(\bar{n})] \geq 0 \; (\leq 0) \tag{29} $$

Then

$$ \bar{G} \geq G \qquad (\bar{G} \leq G) \tag{30} $$

3. A PILOT EXAMPLE: A BREAKDOWN OVERFLOW MODEL

This section will contain some concrete applications of the results. In particular, the estimation and importance of bias terms will hereby be illustrated.

To this end, a special queuing network example will be analyzed. Though relatively simple, this example has no closed product-form expression and can be seen as a pilot example of practical aspects such as a multidimensional transition structure, finite constraints and blocking, dynamic (overflow) routing, and breakdowns.

3.1 Model

Description

Consider a queuing network of two parallel service stations with storage capacities for at most N_1 and N_2 jobs respectively, where N_1 or N_2 can be infinite. Jobs arrive at the system according to a Poisson process at rate λ. An arriving job first attempts to enter station 1. If station 1 is saturated ($n_1 = N_1$) it routes to station 2. If also this station is saturated ($n_1 = N_1$ and $n_2 = N_2$) it is lost. Further, jobs at station 2 can never switch (back) to station 1. The service rate at station i is $\mu_i(n_i)$ when n_i jobs are present, where we assume $\mu_i(\cdot)$ to be nondecreasing and bounded, and where $\mu_i(0) = 0, i = 1, 2$.

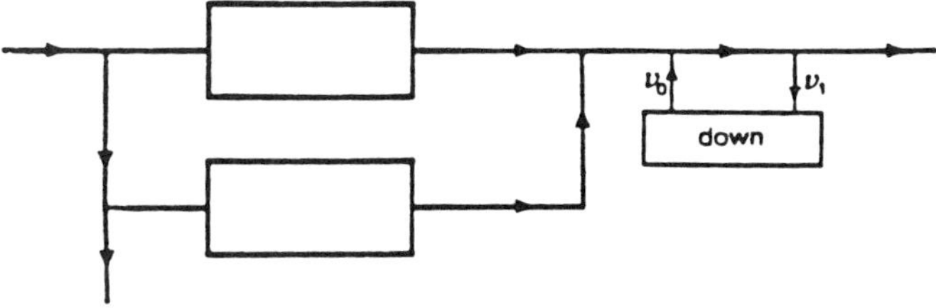

As a special complication, furthermore, the output channel is subject to breakdowns, which render it inoperative for a period of time regardless of whether jobs are present or not. More precisely, this channel is alternatively "up" and "down" for exponential periods with parameters ν_1 and ν_0 respectively. When this channel is down, jobs completing a service are prohibited from leaving the system and have to remain at their station. [This latter assumption corresponds to the so-called communication protocol and can be interpreted as if blocked jobs (messages) are to be reserved (retransmitted), or, as if the servicing at the stations is stopped as long as the channel is down.]

Background

The system under consideration has no product-form solution for the steady-state distribution (necessary partial balance conditions to this end are easily shown to fail). Only some explicit marginal results are available.

1. For the pure overflow case, that is, with $\nu_1 = 0$ so that breakdowns never occur, the overflow stream is known to be hyperexponential (cf. [41]). The overflow station 2 separately can then be analyzed as a $G/M/N_2$ system. However, our performance of interest may also depend on station 1.
2. For the simple single-station case, that is, assuming either $N_1 = \infty$ or $N_2 = 0$, this example is known as an "independent breakdown" model for which, despite its simplicity and independence, a closed-form expression seems to be available only for the generating functions (cf. [8], p. 103).

Parametrization

Let the state $[\bar{n}, \theta]$ with $\bar{n} = (n_1, n_2)$ denote the number of jobs n_1 and n_2 at stations 1 and 2 while $\theta = 1$ or $\theta = 0$ depending upon whether the system (output channel) is "up" or "down," respectively. Set $n = n_1 + n_2$. Further, throughout let $1_{\{A\}}$ denote an indicator of an event A, that is, $1_{\{A\}} = 1$ if A is satisfied and $1_{\{A\}} = 0$ otherwise.

Note that the results of Section 2.4 can be adopted directly with $\bar{n}$ replaced by $[\bar{n}, \theta]$. From here onward this will be applied throughout without further mentioning.

The transition rates of the above model are now given by

$$q([\bar{n}, \theta], [\bar{n}', t\theta']) = \begin{cases} \lambda 1_{\{n_1 < N_1\}} & n_1' = n_1 + 1,\ n_2' = n_2,\ \theta' = \theta = 0, 1 \\ \lambda 1_{\{n_1 = N_1, n_2 < N_2\}} & n_2' = n_2 + 1,\ n_1' = n_1,\ \theta' = \theta = 0, 1 \\ \mu_1(n_1) 1_{\{\theta = 1\}} & n_1' = n_1 - 1,\ n_2' = n_2,\ \theta' = 1 \\ \mu_2(n_2) 1_{\{\theta = 1\}} & n_2' = n_2 - 1,\ n_1 = n_1,\ \theta' = 1 \\ \nu_0 1_{\{\theta = 0\}} & n_1' = n_1,\ n_2' = n_2,\ \theta' = 1 \\ \nu_1 1_{\{\theta = 1\}} & n_1' = n_1,\ n_2' = n_2,\ \theta' = 0 \end{cases} \tag{31}$$

Clearly, the corresponding Markov chain is irreducible at

$$S = \{[\bar{n}, \theta] \mid 0 \le n_1 \le N_1 \qquad 0 \le n_2 \le N_2, \theta = 0, 1\} \tag{32}$$

and by assumption we can choose a finite Q such that

$$Q \ge \lambda + \nu_0 + \nu_1 + \mu_1(n_1) + \nu_2(n_2) \qquad ([\bar{n}, \theta] \in S) \tag{33}$$

Consider the possible reward rates R:

$$R = R_s([\bar{n}, \theta]) = 1_{\{\theta=1\}}[\mu_1(n_1) + \mu_2(n_2)] \tag{34}$$
$$R = R_j([\bar{n}, \theta]) = 1_{\{\theta=1\}}\mu_j(n_j) \qquad (j = 1, 2)$$

Then G, the average reward as according to Section 2.4, is the actual "throughput" of the system (R_s) or, of station $j(R_j)$. We wish to compare this throughput G with values $\bar{G}$ and estimate

$$|\bar{G} - G|$$

for values $\bar{G}$ in various "approximate" situations. The key step to this end is the estimation of the bias terms as established in the next section.

3.2 Estimation of Bias Terms

Let $V_t(\cdot)$ be defined as according to equation (23) with R of any form (34) and define

$$\Delta_i V_t([\bar{n}, \theta]) = V_t([\bar{n} + e_i, \theta]) - V_t([\bar{n}, \theta]) \tag{35}$$

The following lemma then provides estimates for the bias terms that will appear to be essential for the application of both Theorem 2.1 and Theorem 2.2 in various situations.

LEMMA 3.1 For both $i = 1, 2$, all $[\bar{n} + e_i, \theta] \in S$ and all $t \geq 0$, we have

$$0 \leq \Delta_i V_t([\bar{n}, \theta]) \leq 1 \tag{36}$$

Proof. We will give the proof for the case $R = R_s$. The proof for the case $R = R_j$ is almost identical and left as a remark (see Remark 3.1 below).

The proof will be given by induction to t. Clearly, equation (36) holds for $t = 0$ as $V_0(\cdot) = 0$ by definition. Suppose that equation (6) holds for $t \leq m$. Then by virtue of equations (23) and (31) and for convenience writing $h = Q^{-1}$, we obtain for any $[\bar{n} + e_i, \theta] \in S$:

$$\begin{aligned}
\Delta_i V_{m+1}([\bar{n}, \theta]) = {} & \{h\,[\mu_1(n_1) + \mu_2(n_2) + [\mu_i(n_i + 1) - \mu_i(n_i)]]\,1_{\{\theta=1\}} \\
& + h\lambda 1_{\{i=1\}} 1_{\{n_1+1<N_1\}} V_m([\bar{n} + e_1 + e_1, \theta]) \\
& + h\lambda 1_{\{i=1\}} 1_{\{n_1+1=N_1, n_2<N_2\}} V_m([\bar{n} + e_1 + e_2, \theta]) \\
& + h\lambda 1_{\{i=1\}} 1_{\{n_1+1=N_1, n_2=N_2\}} V_m([\bar{n} + e_1, \theta]) \\
& + h\lambda_{\{i=2\}} 1_{\{n_1<N_1\}} V_m([\bar{n} + e_2 + e_1, \theta]) \\
& + h\lambda 1_{\{i=2\}} 1_{\{n_1=N_1, n_2+1<N_2\}} V_m([\bar{n} + e_2 + e_2, \theta]) \\
& + h\lambda 1_{\{i=2\}} 1_{\{n_1=N_1, n_2+1=N_2\}} V_m([\bar{n} + e_2, \theta]) \\
& + h\nu_0 1_{\{\theta=0\}} V_m([\bar{n} + e_i, 1]) + h\nu_1 1_{\{\theta=1\}} V_m([\bar{n} + e_i, 0])
\end{aligned}$$

$$
\begin{aligned}
&+ h\mu_1(n_1)1_{\{\theta=1\}}V_m([\bar{n}+e_i-e_1,1]) \\
&+ h\mu_2(n_2)1_{\{\theta=1\}}V_m([\bar{n}+e_i-e_2,1]) \\
&+ h[\mu_i(n_i+1)-\mu_i(n_i)]1_{\{\theta=1\}}V_m([\bar{n}+e_i-e_i,1]) \\
&+ [1-h\lambda-h\nu_\theta-h1_{\{\theta=1\}}[\mu_1(n_1)+\mu_2(n_2)+\mu_1(n_i+1) \\
&- \mu_i(n_i)]]V_m[\bar{n}+e_i,\theta])\} \\
&- \{h[\mu_1(n_1)+\mu_2(n_2)]1_{\{\theta=1\}} \\
&+ h\lambda 1_{\{i=1\}}1_{\{n_1+1<N_1\}}V_m([\bar{n}+e_1,\theta]) \\
&+ h\lambda 1_{\{i=1\}}1_{\{n_1+1=N_1,n_2<N_2\}}V_m([\bar{n}+e_1,\theta) \\
&+ h\lambda 1_{\{i=1\}}1_{\{n_1+1=N_1,n_2=N_2\}}V_m([\bar{n}+e_1,\theta]) \\
&+ h\lambda 1_{\{i=2\}}1_{\{n_1<N_1\}}V_m([\bar{n}+e_1,\theta]) \\
&+ h\lambda 1_{\{i=2\}}1_{\{n_1=N_1,n_2+1<N_2\}}V_m([\bar{n}+e_2,\theta]) \\
&+ h\lambda 1_{\{i=2\}}1_{\{n_1=N_1,n_2+1=N_2\}}V_m([\bar{n}+e_2,\theta]) \\
&+ h\nu_0 1_{\{\theta=0\}}V_m([\bar{n},1]) + h\nu_1 1_{\{\theta=1\}}V_m([\bar{n},0]) \\
&+ h\mu_1(n_1)1_{\{\theta=1\}}V_m([\bar{n}-e_1,1]) + h\mu_2(n_2)1_{\{\theta=1\}}V_m([\bar{n}-e_2,1]) \\
&+ h[\mu_i(n_i+1)-\mu_i(n_i)]1_{\{\theta=1\}}V_m(\bar{n},1) \\
&+ [1-h\lambda-h\nu_\theta-h1_{\{\theta=1\}}[\mu_1(n_1)+\mu_2(n_2)+[\mu_i(n_i+1) \\
&- \mu_i(n_i)]]V_m([\bar{n},\theta]\} \\
= {}& h[\mu_i(n_i+1)-\mu_i(n_i)]1_{\{\theta=1\}} \\
&+ h\lambda 1_{\{i=1\}}1_{\{n_1+1<N_1\}}\Delta_1 V_m([\bar{n}+e_1,\theta]) \\
&+ h\lambda 1_{\{i=1\}}1_{\{n_1+1=N_1,n_2<N_2\}}\Delta_2 V_m([\bar{n}+e_1,\theta]) \\
&+ h\lambda 1_{\{i=1\}}1_{\{n_1+1=N_1,n_2=N_2\}}\,[V_m([\bar{n}+e_1,\theta])-V_m([\bar{n}+e_1,\theta])] \\
&+ h\lambda 1_{\{i=2\}}1_{\{n_1<N_1\}}\Delta_2 V_m([\bar{n}+e_1,\theta]) \\
&+ h\lambda 1_{\{i=2\}}1_{\{n_1=N_1,n_2+1<N_2\}}\Delta_2 V_m([\bar{n}+e_2,\theta]) \\
&+ h\lambda 1_{\{i=2\}}1_{\{n_1=N_1,n_2+1=N_2\}}\,[V_m([\bar{n}+e_2,\theta])-V_m([\bar{n}+e_2,\theta])] \\
&+ h\nu_0 1_{\{\theta=0\}}\Delta_i V_m([\bar{n},1]) + h\nu_1 1_{\{\theta=1\}}\Delta_i V_m([\bar{n},0]) \\
&+ h\mu_1(n_1)1_{\{\theta=1\}}\Delta_i V_m([\bar{n}-e_i,1]) \\
&+ h\mu_2(n_2)1_{\{\theta=1\}}\Delta_i V_m([\bar{n}-e_2,\theta]) \\
&+ h1_{\{\theta=1\}}[\mu_i(n_i+1)-\mu_i(n_i)]\,[V_m([\bar{n},1])-V_m([\bar{n},1])] \\
&+ [1-h\lambda-h\nu_\theta-h1_{\{\theta=1\}}[\mu_1(n_1)+\mu_2(n_2)+\mu_i(n_i+1) \\
&- \mu_i(n_i)]]\Delta_i V_m([\bar{n},\theta])
\end{aligned} \tag{37}
$$

First of all, note that the fourth, seventh, and next to last term in the last expression are equal to 0, while the coefficient of the last term is nonnegative by virtue of equation (33) and $h = Q^{-1}$. By substituting the induction hypothesis $\Delta_i V_m(\cdot) \geq 0$ and recalling that $\mu_i(\cdot)$ is nondecreasing we hereby conclude that $\Delta_i V_{m+1}(\cdot) \geq 0$. To conclude the upper estimate $\Delta_i V_{m+1}(\cdot) \leq 1$, recall that the next to last term is equal to 0 while its coefficient is exactly equal to the additional nonnegative first term $h 1_{\{\theta=1\}}[\mu_i(n_i + 1) - \mu_i(n_i)]$. By substituting the hypothesis $\Delta_i V_m(\cdot) \leq 1$ and using equation (33) again where $h = Q^{-1}$, we hereby verify also

$$\Delta_i V_{m+1}(\cdot) \leq 1$$

REMARK 3.1 For the case $R = R_j$ the only difference is the first term in the right hand side of expression (37). Here it would become $h 1_{\{j=i\}}[\mu_i(n_i + 1) - \mu_i(n_i)]1_{\{\theta=1\}}$. For $j = i$ the arguments thus remain. For $j \neq i$, substitution of the induction hypothesis (36) for $t = m$ and relation (33) directly leads to equation (36) for $t = m + 1$.

3.3 Application 1: Perturbation

Reconsider the model of Section 3.1 with arrival rate $\bar{\lambda}$ instead of λ. Let all quantities of Section 3.1 be defined accordingly with an overbar symbol where we assume that $\bar{Q} = Q$ also satisfies relation (33) with λ replaced by $\bar{\lambda}$ (note that this can always be established) and where we use the same reward $\bar{R} = R$.

In order to determine the effect of the perturbation $\lambda \rightarrow \bar{\lambda}$ we will verify the conditions of Theorems 2.3 and 2.4. To this end, we conclude from equations (24) and (31) and with $\Delta([\bar{n}, \theta], [\bar{n} - e_i + e_j, \theta']) = 0$ if $\theta' \neq \theta$ and $i \neq j$:

$$\sum_{\substack{\theta'=0,1 \\ i,j=0,1,2}} \Delta([\bar{n}, \theta], [\bar{n} - e_i + e_j, \theta'])\,[V_t([\bar{n} - e_i + e_j, \theta']) - V_t([\bar{n}, \theta])]$$

$$= [\bar{\lambda} - \lambda]\{1_{\{n_1 < N_1\}}\,[V_t([\bar{n} + e_1, \theta]) - V_t([\bar{n}, \theta])] + 1_{\{n_1 = N_1, n_2 < N_2\}}\,[V_t([\bar{n} + e_2, \theta]) - V_t([\bar{n}, \theta])]\} \tag{38}$$

Choosing

$$\Phi([\bar{n}, \theta]) = 1 \qquad ([\bar{n}, \theta] \in S) \tag{39}$$

so that equation (27) is satisfied with $\beta = 1$ for any initial state $\bar{l} = [\bar{n}, \theta] \in S$, and recalling that $\bar{R} = R$, by substituting equation (8) in equation (25) and (29), using Lemma 3.1 and applying Theorems 2.3 and 2.4, we conclude:

RESULT 3.1 With $\overset{(-)}{G}$ the total throughput of the system:

$$\bar{G} \geq G \quad (\bar{G} \leq G) \qquad \text{if} \quad \bar{\lambda} \geq \lambda \qquad (\bar{\lambda} \leq \lambda) \tag{40}$$

and

$$|\bar{G} - G| \leq |\bar{\lambda} - \lambda| \tag{41}$$

REMARK 3.2 Equations (40) and also (41) may seem trivial. However, as per counterintuitive examples such as in [28] and [40], the actual throughput can sometimes be increased (decreased) by decreasing (increasing) arrival intensities at particular periods.

3.4 Application 2: Finite Truncation

Assume that the second (or overflow) station is an infinite-server queue, that is, $N_2 = \infty$ and $\mu_2(n_2) = n_2$. We are then dealing with an infinite system. In order to apply methods (such as numerical) for finite systems, consider the approximate truncated model with $N_2 = L$ [and $\mu_2(n_2) = n_2$], where L is some fixed finite number. Further, in this case consider the reward rates $R = R_j, j = 1, 2$ to evaluate the individual station throughputs.

Thus, let all expressions of Section 3.1 with an overbar correspond to the truncated model with $N_2 = L$ while without overbar to $N_2 = \infty$, and use $R = \bar{R} = R_j$ for $j = 1$ or $j = 2$. First of all, note that

$$\bar{S} = \{[\bar{n}, \theta] \mid 0 \leq n_1 \leq N_1, 0 \leq n_2 \leq L, \theta = 0, 1\} \subset S$$

By comparing the transition rates of equation (31) for $N_2 = \infty$ (q-model) and $N_2 = L$ ($\bar{q}$-model), we conclude from equation (24) and (31) that for $[\bar{n}, \theta] \in S$:

$$\sum_{\substack{\theta'=0,1 \\ i,j=0,1,2}} \Delta([\bar{n}, \theta], [\bar{n} - e_i + e_j, \theta'])[V_t([\bar{n} - e_i + e_j, \theta']) - V_t([\bar{n}, \theta])]$$
$$= \lambda 1_{\{n_1 = N_1, n_2 = L\}} [V_t([\bar{n} + e_2, \theta]) - V_t([\bar{n}, \theta])] \tag{42}$$

With

$$\Phi([\bar{n}, \theta]) = 1_{\{n_1 = N_1, n_2 = L\}} \qquad ([\bar{n}, \theta] \in S) \tag{43}$$

and Lemma 3.1 we thus conclude that equation (25) is satisfied with $\alpha = \lambda$. As $\bar{R} = R$ at $\bar{S}$, Theorem 2.4 now directly formalizes the intuitively obvious result

$$\bar{G} \leq G \qquad (R = \bar{R} = R_j, j = 1, 2) \tag{44}$$

LEMMA 3.2 With $\bar{0} = (0, 0)$ and $\tau = \mu_2(1 + \nu_1/\nu_0)$, we have for all t

$$\bar{T}_t \Phi([\bar{0}, 1]) \leq (\lambda\tau)^L / L! \tag{45}$$

Proof. This will only be given in heuristic steps. The details can be formalized but are rather technical and require application of Theorem 2.2 itself and an inductive proof technique as in the proof of Lemma 3.1.

Step 1: Clearly, by directly routing all jobs to the second station, the steady-state probability of L jobs at the second station is estimated from above by that for at least L jobs in an $M/M/\infty/\infty$ system with appropriate service rate τ so as to take into account the breakdowns. This tail probability in turn is estimated from above by the right-hand side of equation (45).

Step 2: Similarly to Lemma 3.1.2 in [30] and based upon the special initial state $[\bar{O}, 1]$, one can show that

$$\bar{T}_t\Phi([\bar{O}, 1])$$

is nondecreasing in t. Combination of steps 1 and 2 completes the proof

Recalling that $R = \bar{R} = R_j$ for $j = 1$, or $j = 2$, we now obtain from Theorem 2.3, Lemma 3.1, and equation (42):

RESULT 3.2 With $\tau = \mu_2(1 + \nu_1/\nu_0)$, we have

$$|\bar{G} = G| \leq \lambda(\lambda\tau)^L/L! \tag{46}$$

REMARK 3.3 Note here the special role of the initial state $\bar{I} = [\bar{O}, 1]$ and scaling function Φ.

3.5 Application 3: Modification (Simple Throughput Bound)

Reconsider the model of Section 3.1 in which the Poisson arrivals are replaced by a finite source input of M sources with exponential parameter γ per source. More precisely, when n jobs are present in the system, a job will arrive with arrival intensity $(M - n)\gamma$. Here it is quite natural to assume that M is large while γ is small. Further, as in Section 3.4, let the second (overflow) station be an infinite-server station, that is, $N_2 = \infty$ and $\mu_2(n_2) = n_2$ while also $\mu_1(n_1) = n_1$ is assumed. The total number of jobs in the system, however, can never exceed M.

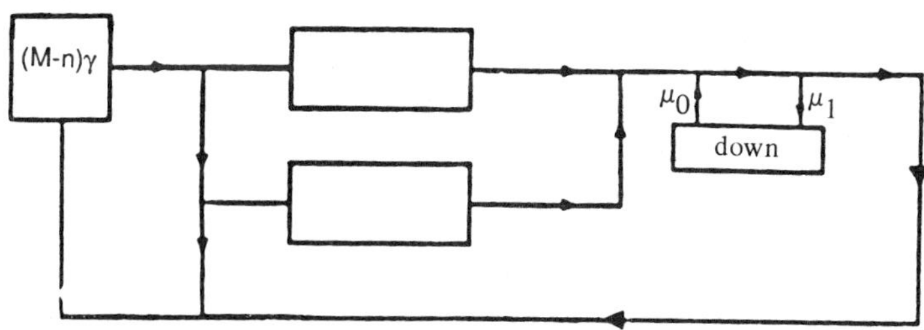

The throughput $\bar{\lambda}$ of this system cannot be calculated easily. To this end, we will approximate this finite system by the infinite Poisson arrival system of Section 3.1 with $N_2 = \infty$, $\mu_2(n_2) = n_2$, and

$$\lambda = \gamma M \tag{47}$$

Note that this "approximation" involves in fact both a truncation (or equivalently an infinite extension) and a state-dependent perturbation [arrival rates $(M - n)\gamma$ as opposed to λ].

We aim to show that λ is a simple upper bound and calculate an error bound on its accuracy. To this end, denote the expressions of Section 3.1 for the finite source system with an over bar and for the infinite system (as in Section 3.1 itself) without. Hence,

$$\bar{S} = \{[\bar{n}, \theta] \mid 0 \leq n_1 \leq N_1, 0 \leq n_2 \leq N_2; n_1 + n_2 \leq M, \theta = 0, 1\} \subset S$$

Further, use $R = \bar{R} = R_s$ given by equation (34). From the difference in the arrival intensities and with $\gamma = \lambda/M$ as by equation (47) and $n = n_1 + n_2$, we obtain for $[\bar{n}, \theta] \in \bar{S}$:

$$\begin{aligned} &\sum_{\substack{\theta'=0,1 \\ i,j=0,1,2}} \Delta([\bar{n}, \theta], [\bar{n} - e_i + e_j, \theta'])[V_t([\bar{n} - e_i + e_j, \theta']) - V_t([\bar{n}, \theta])] \\ &= -n[\lambda/M]\{1_{\{n_1 < N_1\}}[V_t[\bar{n} + e_1, \theta]) - V_t([\bar{n}, \theta])] \\ &\quad + 1_{\{n_1 = N_1\}}[V_t([\bar{n} + e_2, \theta]) - V_t([\bar{n}, \theta])]\} \end{aligned} \tag{48}$$

choosing

$$\Phi([\bar{n}, \theta]) = n \qquad ([\bar{n}, \theta] \in S) \tag{49}$$

and applying Lemma 3.1 thus implies that equation (25) is satisfied with $\alpha = [\lambda/M]$. As $\bar{R} = R = R_s$, Theorem 2.3 can now be applied if we determine β.

LEMMA 3.3 With $\bar{0} = (\bar{0}, 0)$ and W the expected sojourn time of a job in the infinite model, we have

$$\bar{T}_t \Phi([\bar{0}, 1]) \leq \lambda W \tag{50}$$

Proof. This can be given similar to that of Lemma 3.2 in [32], based on showing that the left-hand side equation of (50) is nondecreasing in t (as Step 2 in Lemma 3.2), a bounding argument (as Step 1 in Lemma 3.1), but now for mean queue lengths and Little's result.

Recalling that $R = \bar{R} = R_s$ and noting that the standard sojourn time is estimated from above, we thus obtain from Theorems 2.3 and 2.4, relation (48), Lemma 3.1, and Lemma 3.2

RESULT 3.3

$$0 \le \lambda - \bar{\lambda} \le [\lambda^2/M] \max[\mu_1^{-1}, \mu_2^{-1}[1 + \nu_1/\nu_0]$$

REMARK 3.4 As in Section 3.4, again note the special role of the initial state $\bar{l} = [\bar{0}, 1]$ and scaling function Φ. Particularly, observe that the scaling function is unbounded.

4. EVALUATION

Various types of "approximate" modeling may arise when analyzing a Markov chain model such as for a queuing network. Most notably, inaccuracies in system input data (perturbations), finite approximations of infinite systems (truncations), simplifying transition assumptions to obtain simple bounds (modification), or comparison of system analogs under different parameters or protocols (comparisons) can be involved. A tool has been provided by which error bounds or comparison results for "approximate" modeling (numerical, bounding, approximating) can be concluded. The key step is the estimation (bounding) of so-called bias terms for Markov reward structures. To this end, an inductive Markov reward technique can be employed. This technique applies to multidimensional structures such as most notably queuing networks with practical phenomena such as blocking, dynamic routing, machine failures, and job priorities. Further application of this "bias-term" tool seems promising.

REFERENCES

[1] Adan, I. J. B. F., and Wal, J. van der (1989), Monotonicity of the throughput of a closed queueing network in the number of jobs, Operation Res. 37, 953-957.

[2] Adan, I. J. B. F., and Wal, J. van der (1987), Monotonicity of the throughput of an open queueing network in the interarrival and service times, Memorandum COSOR 87-05, Department of Mathematics and Computer Science, Eindhoven, University of Technology.

[3] Adan, I. J. B. F., and Wal, J. van der (1989), Monotonicity of the throughput in single server production and assembly networks with respect to the buffer sizes, in "Queueing Networks with Blocking" (eds. H. G. Perros and T. Altiok) North-Holland, 345–356.

[4] Barboúr, A. D. (1982), Generalized semi-Markov schemes and open queueing networks, *J. Appl. Prob. 19*, 469–474.

[5] Hinderer, K. (1978), On approximate solutions of finite-stage dynamic programs. *In Dynamic Programming and Its Applications*, ed. M. L. Puterman, Academic Press, New York, 289–318.

[6] Hordijk, A., and Ridder, A. (1987), Stochastic inequalities for an overflow model, J. Appl. Prob. 24, 696–708.

[7] Kemeny, J. G., Snell, J. L., and Knapp, A. W. (1966), *Denumerable Markov Chains*, Van Nostrand, Princeton, N.J.

[8] Jaiswal, N. K. (1968), *Priority Queues*, Academic Press, New York.

[9] Meyer, C. D., Jr. (1980), The condition of a finite Markov chain and perturbation bounds for the limiting probabilities, *SIAM J. Alg. Disc. Math.* 1, 273–283.

[10] Rohlicek, J. R., and Willsky, A. S. (1988), The reduction of perturbed Markov generators: an algorithm exposing the role of transient states, *J. ACM* 35, 675–696.

[11] Ross, S. M. (1970), *Applied Probability Models with Optimization Application*, Holden–Day, San Francisco.

[12] Ross, S. M. (1984), *Introduction to Probability Models*, Academic Press, New York.

[13] Ross, S. M. (1987), Approximating transition probabilities and mean occupation times in continuous-time Markov chains, *P.E.I.S.* 1, 251–264.

[14] Schweitzer, P. J. (1968), Perturbation theory and finite Markov chains, *J. Appl. Prob.* 5, 401–413.

[15] Seneta, E. (1967), Finite approximations to infinite non-negative matrices, *Proc. Cambridge, Phil. Soc.* 63, 983–992.

[16] Seneta, E. (1968), The principles of truncations in applied probability, *Comm. Math. Univ. Carolina* 9, 533–539.

[17] Seneta, E. (1980), *Non-negative Matrices and Markov Chains*, Springer Verlag, New York.

[18] Shanthikumar, J. G., and Yao, D. D. (1986), The effect of increasing service rates in closed queueing network, *J. Appl. Prob.* 23, 474–483.

[19] Shanthikumar, J. G., and Yao, D. D. (1987), Stochastic monotonicity of the queue lengths in closed queueing networks, Research Report, University of California, Berkeley, *Operations Res.* 35, 583–588.

[20] Shanthikumar, J. G., and Yao, D. D. (1987), General queueing networks: Representation and stochastic monotonicity, *Proc. 26th IEEE Conference on Decision and Control*, 1084–1087.

[21] Shanthikumar, J. G., and Yao, D. D. (1988), Throughput bounds for closed queueing networks with queue-dependent service rates, *Performance Eval.* 9, 69–78.

[22] Shanthikumar, J. G., and Yao, D. D. (1988), Monotonicity properties in cyclic queueing networks with finite buffers, *Proc. First Int. Workshop on Queueing Networks with Blocking*, North Carolina, May 1988.

[23] Stoyan, D. (1983), *Comparison Method for Queues and Other Stochastic Models*, Wiley, New York.

[24] Suri, R. (1985), A concept of monotonicity and its characterization for closed queueing networks, *Operations Res.* 33, 606–624.

[25] Tsoucas, P., and Walrand, J. (1989), Monotonicity of throughput in non-Markovian networks, *J. Appl. Prob.* 26, 134–141.

[26] Tijms, H. C. (1986), *Stochastic Modelling and Analysis; A Computational Approach*, Wiley, New York.

[27] Tijms, H. C., and Eikeboom, A. M. (1986), A simple technique in Markovian control with applications to resource allocation, *Operations Res. Lett.* 1, 25–32.

[28] Van Dijk, N. M. (1988), Simple bounds for queueing systems with breakdowns, *Performance Eval.* 8, 117–128.

[29] Van Dijk, N. M. (1988), A formal proof for the insensitivity of simple bounds for finite multi-server non-exponential tandem queues based on monotonicity results, *Stochastic Proc. Appl.* 27, 261–277.

[30] Van Dijk, N. M. (1988), Perturbation theory for unbounded Markov reward processes with applications to queueing, *Adv. Appl. Prob.* 20, 99–111.

[31] Van Dijk, N. M. (1989), A simple bounding methodology for non-product-form queueing networks with blocking, in "Queueing Networks with Blocking" (eds. H. G. Perros and T. Altiok) North-Holland, 3–17.

[32] Van Dijk, N. M. (1989), A simple throughput bound for large closed queueing networks with finite capacities, *Perform. Eval.* 10, 153–167.

[33] Van Dijk, N. M. (1990), Analytic error bounds for approximations of queueing networks with an application to alternate routing, *J. Austr. Math. Soc.* Ser. B 31, 241–258.

[34] Van Dijk, N. M. (1989), Truncation of Markov chains with applications to queueing, Operations Res. (To appear).

[35] Van Dijk, N. M. (1988), Approximate uniformization for continuous-time Markov chains with an application to performability analysis, Research report 54, Free University, Amsterdam, Submitted.

[36] Van Dijk, N. M. (1988), Error bounds for comparing open and closed queueing networks with an application to performability analysis, *J. ACM.* (To appear)

[37] Van Dijk, N. M., and Lamond, B. F. (1988), Bounds for the call congestion of finite single-server exponential tandem queues, *Operations Res.* 36, 470–477.

[38] Van Dijk, N. M., and Puterman, N. L. (1988), Perturbation theory for Markov reward processes with applications to queueing systems, *Adv. Appl. Prob.* 20, 79–89.

[39] Van Dijk, N. M., Tsoucas, P., and Walrand, J. (1988), Simple bounds and monotonicity of the call congestion of finite multiserver delay systems, *Probability in the Eng. Info. Sci.* 2, 129–138.

[40] Van Dijk, N. M., and Van der Wal, J. (1989), Simple bounds and monotonicity results for multi-server exponential tandem queues, *Queueing Systems* 4, 1–16.

[41] Van Doorn, E. A. (1984), On the overflow process from a finite Markovian queue, *Perfor. Eval.* 4, 233–240.

[42] De Waal, P. R., and Van Dijk, N. M. (1988), Monotonicity of performance measures in a processor sharing queue, Perform. Eval. (To appear.)

[43] Whitt, W. (1978), Approximations of dynamic programs I, *Math. Operations Res.* 3, 231–243.

[44] Whitt, W. (1981), Comparing counting processes and queues, *Adv. Appl. Prob.* 13, 207–220.

[45] Whitt, W. (1986), Stochastic comparisons for non-Markov processes, *Math. Operations Res.* 11 (4), 608–618.

[46] Yoon, B. S., and Shanthikumar, B. S. (1989), Bounds and approximations for the transient behavior of continuous time Markov chains, *P.E.I.S.* 3, 175–198.

Poster Papers

33

Numerical Solution of Markov Reward Models Using Laguerre Functions

S. M. R. ISLAM* **and H. H. AMMAR**† Department of Electrical and Computer Engineering, Clarkson University, Potsdam, New York

Performability analysis is an important and growing research area. One of the bases of performability analysis is the Markov reward model. However, it is extremely difficult to solve these models, and no closed-form solution exists for large repairable systems. As a result, numerical solution techniques are adapted to solve such systems.‡ The bulk of the work needed when solving Markov reward models is in the inverse of the transform equation. We use the Laguerre functions to invert this transform equation. However, as with any numerical technique, it requires careful manipulations for a meaningful result. We present a new algorithm to compute the moments and distribution of performability of repairable systems that can be described

*Current affiliation: IBM Corporation, Boca Raton, Florida

†Current affiliation: West Virginia University, Morgantown, West Virginia

‡S. M. R. Islam and H. H. Ammar, Performability of the hypercube, *IEEE Trans. Reliability*, Dec. 1989.

by cyclic Markov reward models. We also discuss aspects of the numerical computations involved such as error estimation and series convergence.

The algorithm presented computes both performability distribution and its higher moments in $O(n^3)$. As mentioned earlier, we use Laguerre functions to numerically invert the transform equation. A brief description of the algorithm is as follows:

Let X_0 be the initial state and $Y(t)$ be the accumulated reward until time t. That is,

$$S_i(x,t) = \text{Prob}[Y(t) \leq x \mid X_0 = i] \qquad i = 1,\dots,n$$

We then have

$$\mathbf{S}^{+*}(\phi,\delta) = (\delta\mathbf{I} + \phi\mathbf{F} - \mathbf{Q})^{-1}e$$

where $S_i^{+*}(\phi,\delta)$ is the Laplace–Stieltjes transform (shown by $+$) of $S_i(x,t)$ in x variable and the Laplace transform (shown by $*$) in t variable and defines

$$\mathbf{S}^{+*}(\phi,\delta) = [S_1^{+*}(\phi,\delta),\dots,S_n^{+*}(\phi,\delta)]^T$$

where $\mathbf{e}$ is a column vector of 1. In addition, $\mathbf{Q}$ is the Markov generator matrix of the reward model and $\mathbf{F}$ is a diagonal matrix of nonnegative reward rates obtained from the reward structure. The performability density $s(x,t) = (d/dx)\mathrm{S}(x,t)$, distribution $\mathrm{S}(x,t)$, and ith moment $\mathbf{M}^i(t)$ are given by

$$\mathrm{s}(x,t) = e^{ct}\sum_m^{\infty}\sum_k^{\infty} \mathrm{s}_{mk}^{++} l_{mk}(x,t)$$

$$\mathrm{S}(x,t) = \int_0^x \mathrm{s}(x',t)\,dx' = 1 - \left(e^{ct}\sum_m^{\infty}\sum_k^{\infty} \mathrm{S}_{mk}^{++} l_{mk}(x,t)\right)$$

$$\mathbf{M}^i(t) = \int_0^{\infty} x^i \mathrm{s}(x,t)\,dx = e^{ct}\sum_{m=0}^{\infty}\sum_{k=0}^{\infty}\sum_{j=0}^{k}(-1)^m \mathrm{S}_{mj}^i l_k(t)$$

where the Laguerre coefficient $\mathrm{s}_{mk}^{++} = \mathbf{E}_{mk}\mathbf{D}$ and $n \times n$ matrix (for $m, k > 0$)

$$\mathbf{E}_{mk} = \mathbf{E}_{m-1,k-1} + (\mathbf{E}_{m-1,k} - \mathbf{E}_{m-1,k-1})\mathbf{A} + (\mathbf{E}_{m,k-1} - \mathbf{E}_{m-1,k-1})\mathbf{B}$$

Additionally, $\mathbf{E}_{00} = \mathbf{I}$, $\mathbf{E}_{10} = \mathbf{A}$, and $\mathbf{E}_{01} = \mathbf{B}$.

The $n \times n$ matrices $\mathbf{A}$, and $\mathbf{B}$ and $n \times 1$ matrix $\mathbf{D}$ are given as $(\mathbf{I}+\mathbf{F}-2\hat{\mathbf{Q}})^{-1} = \mathbf{Z}$, $2\mathbf{Ze} = \mathbf{D}$, $\mathbf{Z}(\mathbf{I}-\mathbf{F}-2\hat{Q}) = \mathbf{A}$, $\mathbf{Z}(-\mathbf{I}+\mathbf{F}-2\hat{\mathbf{Q}}) = \mathbf{B}$, and $\mathbf{Z}(\mathbf{I}+\mathbf{F}+2\hat{\mathbf{Q}}) = \mathbf{C}$, where $\hat{\mathbf{Q}} = \mathbf{Q} - c\mathbf{I}$ and $\mathbf{A} + \mathbf{B} + \mathbf{C} = \mathbf{I}$.

Additionally, $\mathrm{S}_{mk}^{++} = 2\sum_{i=m+1}^{\infty}\sum_{j=0}^{k}(-1)^{m+i}s_{ij}^{\#}$ and $s_{mk}^{\#} = (s_{mk}^{++} - s_{m,k-1}^{++} - s_{m-1,k}^{++} + s_{m-1,k-1}^{++})$. Furthermore, $l_{mk}(x,t) = l_m(x)l_k(t)$, $l_{k+1}(x) = (1/(k+1))[(2k+1-x)l_k(x) - kl_{k-1}(x)]$ and $l_0(x) = \exp(-x/2)$. the recursive function

$\mathbf{S}^i_{mk}$, $i > 0$ is given by $\mathbf{S}^i_{mk} = -[(m+1)\mathbf{S}^{i-1}_{m+1,k} - 2m\mathbf{S}^{i-1}_{m,k} + (m-1)\mathbf{S}^{i-1}_{m-1,k}]$, and the initial condition is $\mathbf{S}^0_{mn} = \mathbf{s}^{\#}_{mn}$.

1. THE ALGORITHM

Once the matrices $\mathbf{E}_{mk}$ are known, the algorithm to compute various performability measures is straightforward. The structure of the overall algorithm is as follows:

Phase A Compute and store column vectors $\mathbf{s}^{++}_{mn}$. This is the essential phase required during computation of all performability measures. This is also the most expensive to compute ($O(n^3)$)). However, it is only computed once for a given model.

Step 1	Create matrices **A**, **B**, and **D**.	$O(n^3)$
Step 2	Create matrices $\mathbf{E}_{mk}, m = 0, \ldots, m_{\max}; k = 0, \ldots, k_{\max}$.	$O(n^3)$
Step 3	Create column vectors $s^{++}_{mk} == \mathbf{E}_{mk}\mathbf{D}$ and save.	$O(n^2)$

Phase B In this phase, computations are performed to compute different performability measures in no greater than $O(n)$. Here we will show steps for computing $M^i(t)$. Steps for other performability measures are straightforward.

```
for1 (k up to k_max) do
    M^i(t) = M^i(t) + l_k(t)S^t                O(n)
    for2 (j up to k and m up to m_max) do
        S^t = S^t + (-1)^m S^i_mj              O(n)
    end for2
end for1
```

2. NUMERICAL COMPUTATIONS

The most important and sensitive part of the computation is the evaluation of the Laguerre coefficients $\mathbf{s}^{\#}_{mk}$, which forms a rapidly decreasing sequence, and experience shows that the accuracy of the sum of an infinite series when approximated by a finite one directly depends on these coefficients. Since $\mathbf{M}^0(t) = \mathbf{e}$, it can be used as an important indicator function to monitor the integrity of the model and solution process. Convergence of $\hat{\mathbf{M}}^0(t)$ to something other than **e** indicates some problems either in the model description (such as the row sums of the matrix **Q** not zero) or in the solution process. We note that multiplying the s(x,t) by $e^{-ct}(c \geq 0)$ is equivalent to subtracting

c from diagonal elements of the matrix $\mathbf{Q}$. A proper choice of c is very important for the convergence of the infinite series, particularly for the higher moments; c must be chosen such that $\int_0^\infty e^{-ct} \int_0^\infty s_i(x,t)\,dx\,dt < \infty$, and $\int_0^\infty e^{-2ct} \int_0^\infty \{s_i(x,t)\}^2\,dx\,dt < \infty$.

3. ADVANTAGES AND LIMITATIONS

The algorithms computes both the moments and distribution in $O(n^3)$ The other $O(n^3)$ algorithm* that computes the performability distribution of a repairable system does not compute the higher moments. Briefly, this algorithm* also requires determination of eigenvalues, transformations of matrices, and solution of linear systems, and must be done in terms of complex numbers. The algorithm presented here requires only one matrix (which is always nonsingular) inversion and lots of matrix multiplication and additions (all real), and is therefore conceptually simpler and easy to automate. However, the algorithm presented, in general, requires more storage than the other algorithm.

*R. Smith, K. S. Trivedi, and A. V. Ramesh, Performability analysis: measures, an algorithm, and a case study, *IEEE Trans. Comput.* vol. C-37, no. 4, April 1988.

34

The Analysis of a Manufacturing Unit by Sparse Matrix Techniques*

VINCENT A. BARKER Institute for Numerical Analysis, Technical University of Denmark, Lyngby, Denmark

BO FRIIS NIELSEN Institute of Mathematical Statistics and Operations Research, Technical University of Denmark, Lyngby, Denmark

1. INTRODUCTION

In the modeling and analysis of flexible manufacturing systems (FMS) [3,4], queuing network theory has recently received much attention. For example, models based on BCMP-type networks have been considered [2,7]. Since many systems, however, do not have product-form solutions [8], work has been done to reformulate the problem in terms of product-form models [9]. Further, a number of papers have dealt with the approximate solution of continuous-time Markov chain models of queuing networks—for example, using techniques based on aggregation [6].

In this study we analyze a specific production unit model whose Markov chain is not known to have a closed-form solution. Since the infinitesimal

*This work has been supported in part by the Danish Natural Science Research Council and UNI-C, the Danish Computing Center for Research and Education at the Technical University of Denmark.

generator matrix turns out to be very large and sparse, we solve the steady-state equations iteratively using NSPCG, a program package from the Center for Numerical Analysis at the University of Texas at Austin.

2. MODEL DESCRIPTION

We consider a small production unit, with a single AGV (automated guided vehicle) [3] as transportation equipment, and a small number of machines, say one or two, with limited buffer capacity. We allow two different types of parts to enter the system. It is assumed that the retrial strategy makes the overall arrival process Poissonian: that is, parts that cannot enter the system due to congestion wait a random time before trying again. Each machine has between zero and three input buffer positions and a single output buffer position. This model makes it possible to investigate a number of production unit properties. Of primary interest is the effect of the limited buffer capacity, which can lead to the blocking of machine units and thus reduced efficiency.

3. THE GENERATOR MATRIX

A problem in the analysis of non-product-form networks is the construction of the generator matrix. Since the Markov chain does not have a simple form, we need an algorithm to select feasible states and define transition intensities. It is appropriate to define the state space multidimensionally. Thus for each machine position there are indicators of occupancy. For example, the output buffer position can be vacant or occupied by a customer of a specific class. An additional state describes the status of the AGV. A number of rules relate the admissible values of the individual elements of the state vector. However, some element combinations that satisfy these rules cannot be reached from any other admissible state, implying that the generator matrix is reducible. Our approach has been first to construct the reducible matrix and then to find the irreducible submatrix, using for this purpose the Tarjan algorithm as implemented in the Harwell Subroutine Library.

4. NUMERICAL ASPECTS

After some simple transformations we obtain a system $Ax = 0$, where A is a sparse, singular, irreducible M-matrix with zero column sums, and x is a transformation of the steady-state probability vector. Many iterative methods are applicable to this problem [1]. We have used NSPCG, a package from the University of Texas at Austin that implements a number of state-of-the-art iterative methods for sparse linear systems [5]. Our calculations have been performed on the AMDAHL VP1100 at the UNI-C Computing Center in Denmark. Sample result: Using the ORTHOMIN method with preconditioning we were able to solve a system of order 69805 in 34 seconds with a relative residual error reduction of 0.41×10^{-9}.

REFERENCES

[1] Barker, V. A.: Numerical solution of sparse singular systems of equations arising from ergodic Markov chains. *Stochastic Mod.*, Vol. 5 (1989), pp. 335–381.

[2] Baskett, F., Chandy, K. M., Muntz, R. R., and Palacios, F. G.: Open, closed and mixed networks of queues with different classes of customers. *J. ACM*, Vol. 22 (1975), N. 2, pp. 248–260.

[3] Buzacott, J. A., and Yao, D. D.: On queueing network models of flexible manufacturing systems. *Queueing Systems*, Vol. 1 (1986), N. 7, pp. 5–27.

[4] Dallery, Y.: On modelling flexible manufacturing systems using closed queueing networks. *Large Scale Systems*, Vol. 11 (1986), pp. 109–119.

[5] Oppe, T. C., Joubert, W. D., and Kincaid, D. R.: *NSPCG User's guide, Version 1.0*, Report CNA-216, Center for Numerical Analysis, University of Texas at Austin, Austin, Tex., 1988.

[6] Schweitzer, P. J.: Aggregate modelling of queueing networks. *Proc. ITC 12*, Torino, Italy, June 1988, pp. 4.1B.1–4.1B.7.

[7] Vliet, M., and Wassenhove, L. N.: *Operational Research Techniques for Flexible Manufacturing Systems*. Tutorial papers in Operational Research, Shahani & Stainton, eds., OR31, Southampton, 1989 pp. 91–119.

[8] Yao, D. D.: Queueing Models of Flexible Manufacturing Systems. Ph.D. dissertation, University of Toronto, 1983.

[9] Yao, D. D., and Buzacott, J. A.: Modelling a class of flexible manufacturing systems with reversible routing. *Operations Res.*, Vol. 35 (1986), pp. 87–93.

35

Approximating the Stationary Distribution of an Infinite Stochastic Matrix

DANIEL P. HEYMAN Bellcore, Red Bank, New Jersey

EXTENDED ABSTRACT

We are given a Markov chain with states 0, 1, 2, ... We want to get a numerical approximation of the steady-state balance equations. To do this, we truncate the chain, keeping the first $n + 1$ states, make the resulting matrix stochastic in some convenient way, and solve the finite system. The purpose of this paper is to provide some sufficient conditions that imply that as $n + 1$ tends to infinity, the stationary distributions of the truncated chains converge to the stationary distribution of the given chain. Our approach is completely probabilistic, and our conditions are given in probabilistic terms. We illustrate how to verify these conditions with several examples.

We number the states 0, 1, 2, ... Let $\{X_\infty(m), m = 0, 1, \ldots\}$ be the original chain, and $\{X_n(m), m = 0, 1, \ldots\}$ be the truncated chain, and let P_∞ and P_n be the transition matrices. We assume that P_n is formed from the $n \times n$ northwest corner of P_∞ and an augmentation rule that makes P_n stochastic. The key idea is to define X_∞ and X_n on the same probability space by defining them in terms of a sequence of i.i.d. uniform random variables, just as in a Monte

Carlo simulation. When this is done, we show that a theorem in Heyman and Whitt (1989) can be applied. Let C_n be the first-passage time from state zero to state zero in process X_n, including $n = \infty$. The theorem is the following.

THEOREM If there is a positive random variable Z with finite expected value such that $C_n \leq C_\infty + Z$, then the stationary distribution of X_n converges to the stationary distribution of X_∞ as $n \to \infty$.

We apply this condition to rederive five results from the literature. (1) Augment the first column kept for any transition matrix; any augmentation for (2) Markov matrices (one row bounded away from zero) and (3) stochastically monotone matrices; (4) augment the last column kept for upper-Hessenberg matrices; and (5) spread the truncated probability arbitrarily over an (essentially) finite set of states for lower-Hessenberg matrices.

REFERENCES

[1] Heyman, D. P., and W. Whitt (1989), Limits of queues as the waiting room grows, *Queueing Theory Appl.*, vol. 5, pp. 381–392.

[2] Heyman, D. P. (1991), Approximating the stationary distribution of an infinite stochastic matrix, *J. Appl. Prob.*, in press.

36

Transient Analysis of the M(t)/M(t)/1 Queue*

JI ZHANG Department of Electrical Engineering, University of Hawaii at Manoa, Honolulu, Hawaii

EDWARD J. COYLE School of Electrical Engineering, Purdue University, West Lafayette, Indiana

Extensive work has been done to study general time-dependent queuing systems [1–3]; see [4] for a survey. The subject of study of this paper, and of many other papers, is the time-dependent $M/M/1$ queue, which is also called the $M(t)/M(t)/1$ queue. The goal is to find the solution to the following set of difference-differential equations:

$$\dot{\pi}_0(t) = -\lambda(t)\pi_0(t) + \mu(t)\pi_1(t) \tag{1}$$

$$\dot{\pi}_n(t) = -(\lambda(t) + \mu(t))\pi_n(t) + \mu(t)\pi_{n+1}(t) + \lambda(t)\pi_{n-1}(t) \qquad n \geq 1 \tag{2}$$

where $\pi_n(t)$ is the probability that n customers are in the queue at time t.

Usually the key to the solution is how the boundary values $\pi_0(t)$ and $\pi_1(t)$ are obtained. Most other approaches try to approximate the boundary probability $\pi_0(t)$ [5–7]. In this paper we describe a method to determine $\pi_0(t)$

*This research was supported by NSF under grant CDS-8803017 to the ERC in Intelligent Manufacturing Systems at Purdue University, and by a grant from the AT&T Foundation.

exactly through the use of integral equations. To this end, another approach [8] exists to solve equations (1) and (2) that also yields integral equations. The difficulty in this approach, however, is that the boundary probability function $\pi_0(t)$ can not be found explicitly from the solution of the integral equation. In our method, the unknown function in the integral equation is $\pi_0(t)$, not the entire distribution. This is an important advantage because the expected number of busy customers $N(t)$ can be found directly from $\pi_0(t)$. In addition, the method used in this new technique is more general and can be extended to the analysis of more general time-dependent queues.

Assume, for simplicity, that at $t = 0$ the queue is empty: $\pi_0(0) = 1$. Define $\pi(Z,t) \triangleq \sum_{n=1}^{+\infty} \pi_n(t)Z^n$; then $\pi(Z,0) = 0$. Equation (2) can then be transformed into the following two differential equations:

$$\dot{\pi}_0(t) = -\lambda(t)\pi_0(t) + \mu(t)\pi_1(t) \tag{3}$$

$$\dot{\pi}(Z,t) = \pi(Z,t)[\lambda(t)Z + \mu(t)Z^{-1} - (\lambda(t) + \mu(t))] + Z\lambda(t)\pi_0(t) - \mu(t)\pi_1(t) \tag{4}$$

The solution of equation (4) can be written in terms of $\pi_0(t)$, $\pi_1(t)$ via the following form of the convolution integral:

$$\pi(Z,t) = \int_0^t [\lambda(\eta)\pi_0(\eta) - \mu(\eta)\pi_1(\eta)]\Phi_Z(t,\eta)\,d\eta$$

$$\Phi_Z(t,\eta) = e^{\int_\eta^t [\lambda(\tau)Z + \mu(\tau)Z^{-1} - (\lambda(\tau)+\mu(\tau))]\,d\tau} \tag{5}$$

Note that $\pi(Z,t)$ satisfies $\frac{1}{2\pi j}\oint_{|Z|=1} \pi(Z,t)\,dZ = 0$. Apply this result to equation (5) and switch the order of the integrals:

$$\int_0^t [\lambda(\eta)\pi_0(\eta)R_1(t,\eta) - \mu(\eta)\pi_1(\eta)R_2(t,\eta)]\,d\eta = 0 \tag{6}$$

where

$$R_1(t,\eta) = \frac{e^{-a(t,\eta)-b(t,\eta)}}{a(t,\eta)^2} \sum_{n=2}^{+\infty} \frac{(a(t,\eta)b(t,\eta))^n}{n!(n-2)!}$$

$$R_2(t,\eta) = \frac{e^{-a(t,\eta)-b(t,\eta)}}{a(t,\eta)} \sum_{n=1}^{+\infty} \frac{(a(t,\eta)b(t,\eta))^n}{n!(n-1)!}$$

with $a(t,\eta) = \int_\eta^t \lambda(\tau)\,d\tau$ and $b(t,\eta) = \int_\eta^t \mu(\tau)\,d\tau$. By bringing equation (1) into equation (6), and by doing some manipulations, we can obtain the following theorem:

THEOREM $\pi_0(t)$ satisfies the following Volterra integral equation of the first kind:

$$\int_0^t \pi_0(\eta)\tilde{K}(t,\eta), d\eta = \tilde{g}(t) \tag{7}$$

where $\tilde{g}(t) = R_2(t,0)$ and $\tilde{K}(t,\eta) = \lambda(\eta)(R_1(t,\eta) - R_2(t,\eta)) + (\partial/\partial\eta)R_2(t,\eta)$.

However, the above equation is numerically sensitive. In other words, small error in $\tilde{g}(t)$ can result in large deviation in $\pi_0(t)$ from the true value. To avoid this undesired property, we change equation (7) into another numerically stable form by differentiating both sides with respect to t. The final form is a Volterra integral equation of the second kind.

$$\pi_0(t) = g(t) + \int_0^t \pi_0(\eta)K(t,\eta)\,d\eta \tag{8}$$

where $g(t)$ and $K(t,\eta)$ can be determined from $\lambda(t)$ and $\mu(t)$. Equation (8) has been shown to be very tractable numerically. The expected queue size, $N(t)$, is given by $N(t) = \int_0^t [\lambda(\tau) - \mu(\tau)(1 - \pi_0(\tau))]\,d\tau$. $\pi(Z,t)$ can also be expressed in terms of $\pi_0(t)$:

$$\pi(Z,t) = \Phi_Z(t,0) - \pi_0(t) + \int_0^t \pi_0(\eta)[\lambda(\eta)(1-Z) + \mu(\eta)(1-Z^{-1})]\Phi_Z(t,\eta)\,d\eta$$

The basic idea in this approach can be easily modified and extended to the transient, or time-dependent, analysis of other more sophisticated queuing systems for modeling multiaccess computer networks and voice/data integrated networks.

REFERENCES

[1] W. Grassmann, Transient solutions in Markovian queues, *Euro. J. Operations Res.*, 1, 1977, pp. 396–402.

[2] G. F. Newell, Queues with time-dependent arrival rates I, II, III, *J. Appl. Probability*, May 1968, pp. 436–451, 579–590, and 591–606.

[3] J. B. Keller, Time-dependent queues, *SIAM Rev.*, vol. 24, No. 4, October 1982, pp. 401–412.

[4] S. K. Tripathi and A. Duda, Time-dependent analysis of queueing systems, *INFOR*, vol. 24, no. 3, 1986.

[5] K. L. Rider, A simple approximation to the average queue size in the time-dependent $M/M/1$ queue, *J. ACM*, vol. 23, no. 2, April 1976, pp. 361–367.

[6] M. H. Rothkopf and S. S. Oren, A closure approximation for the Nonstationary $M/M/s$ queue, *Manage. Sci.*, vol. 25, no. 6, June 1979, pp. 522–534.

[7] W. A. Massey, Asymptotic analysis of the time dependent $M/M/1$ queue, *Math. Operations Res.*, vol. 10, no. 2, May 1985, pp. 305–327.

[8] A. B. Clarke, A waiting line process of Markov type, *Ann. Math. Statist.*, 27, 1956, pp. 452–459.

37

Experimental Results on Matrix-Analytical Solution Techniques: Extensions and Comparisons*

LEVENT GÜN† Department of Electrical Engineering, University of Maryland, College Park, Maryland

EXTENDED ABSTRACT

The matrix-analytic analysis of many stochastic models requires the solution of nonlinear matrix equations. In this paper, several iterative algorithms for solving these equations are reviewed and insights gained on their convergence behavior are reported. An extension to these algorithms based on *linear extrapolation* is proposed with a view toward extending the computational feasibility of the matrix-analytic techniques to even larger classes of problems.

*Full paper appeared in *Stochastic Models*, Special Issue on Computer Experimental Methods in Probability, Volume 5, pp. 662–682, 1989.
This work was supported partially through National Science Foundation grant ECS-83-51836 and partially through a grant from AT&T Bell Laboratories.
†Current affiliation: IBM Communications Systems, Research Triangle Park, North Carolina

This proposed extension has been observed to considerably improve the efficiency of the existing algorithms, especially when the convergence rates are slow.

Matrix-analytic techniques allow for the analysis of broad classes of applied probability problems. The major limitation in the applicability of the matrix-analytical techniques is the "curse of dimensionality," which may incur intolerably long CPU times in various iterative matrix computations. It is therefore crucial that efforts be made to construct efficient algorithms with improved rates of convergence properties, so that the solution of even larger classes of problems can be made feasible through the matrix-analytic techniques.

Central to the matrix-analytic approach is the entry-wise minimal nonnegative solution of nonlinear matrix equations of the form

$$G = \sum_{n=0}^{\infty} A_n G^n \qquad \text{and} \qquad R = \sum_{n=0}^{\infty} R^n A_n \tag{1}$$

Several iterative algorithms available in the literature for solving these matrix equations are summarized as noted below.* These algorithms are stopped when all the entries in the successive iterates of the G (or R) matrix are within a small number, say ϵ, of each other, that is,

$$\max_{1 \leq i,j \leq m} |(G_{k+1})_{ij} - (G_k)_{ij}| < \epsilon \tag{2}$$

Computational experience has shown that when the convergence rate is slow, even when the successive iterates satisfy equation (2), say at iteration K, the matrix G_K can be far away from the limiting G matrix, that is, $G_{ij} - (G_K)_{ij}$ can be as much as 10^2 to 10^4 times ϵ for some i and j depending on the convergence rate. However, the rows of G_K, when considered as vectors, were observed to be almost collinear with the corresponding rows of the G matrix. Therefore, the most recent iterates, G_K and G_{K-1}, can be used to obtain an approximation G_K^* to the *stochastic* matrix G by linear extrapolation, that is,

$$G_K^* = G_K + L[G_K - G_{K-1}] \tag{3}$$

where L is a *diagonal* matrix so that the matrix G_K^* is stochastic. Computational experience has shown that even when G_K is far away (relative to ϵ) from being a stochastic matrix, $G^* - k$ is typically much closer to G and use of G_K^* leads to much more accurate results in the calculation of the performance measures of interest.

These observations suggest that if G_K^* is the stochastic matrix obtained from the linear extrapolation of G_K and G_{k-1}, then for the same ϵ, the stopping criterion

$$\max_{1 \leq i,j \leq m} |(G^*_{k+1})_{ij} - (G^*_k)_{ij}| < \epsilon \tag{4}$$

should be satisfied at an iteration K^* that is much smaller than K, and $G^*_{K^*}$ should be closer to G then G_K. Indeed, the stochastic matrix $G^*_{K^*}$ was observed to be much closer to G than G_K in all the numerical examples performed, and the number of iterations K^* was much smaller than K especially when the convergence rate is slow.

Note that this proposed simple extension to the existing iterative algorithms is based on the fact that the matrix G is *stochastic* and does not apply to the computation of the R matrix except for the quadratic case, where

$$R = A_0(I - A_1 - A_0G)^{-1} \tag{5}$$

The results of the extensive numerical study presented in the reference cited earlier* lead to several recommendations, some of which are summarized here.

In solving for the G matrix in the $M/G/1$ paradigm of Neuts, extrapolating the iterates to a stochastic matrix at each step results in significant savings (up to an order of magnitude) in the number of iterations and CPU time. More importantly, such a modification in the existing algorithms avoids inaccuracies that result from large ϵ values (10^{-5} and larger) by providing direct information on the accuracy of the computed performance measures.

Therefore in the matrix-geometric case it is also recommended that the G matrix be iteratively computed using the extrapolation technique proposed here and that R be obtained in terms of G via equation (5).

When both the $M/G/1$ and the matrix-geometric approach can be used on a problem, the former is desirable from the point of view of accuracy and computational complexity except in that the latter approach may simplify the computation of the performance measures such as the queue length distributions.

38

Lumping in Markov Reward Processes

VICTOR F. NICOLA IBM Thomas J. Watson Research Center, Yorktown Heights, New York

Explicit conditions on the transition probabilities for lumping in a discrete-time Markov chain (DTMC) are well known and were given by Kemeny and Snell in 1960 [1]. They distinguish between "strong" lumpability for which the process is lumpable for any initial probability distribution on the states and "weak" lumpability for which the process is lumpable only for some initial probability distributions. In this paper, we obtain conditions for lumping in a continuous-time Markov reward process. We introduce the notion of "proportional dynamics" and give necessary and sufficient conditions for it to hold. We also show that proportional dynamics for a given measure is sufficient for weak lumpability for the same measure, it also implies unlumpability (i.e., unlumping after lumping.) We consider measures, such as the transient probabilities, the distribution of accumulated reward, the expected accumulated reward, and the instantaneous reward. The latter can be specialized to availability and reliability measures.

A Markov reward process is described by an underlying time-homogeneous continuous-time Markov chain (CTMC) $\{Z(t), t \geq 0\}$, where $Z(t)$ is the state of the process at time t. The state space S of the underlying CTMC is assumed finite, that is, $Z(t) \in S = \{1,2,\ldots,n\}$. A constant reward rate $R_i \geq 0$ is associated with state i. $\mathbf{R} = \{R_i, 1 \leq i \leq n\}$ is the diagonal reward matrix. Let Q_{ij}, $i \neq j$, $1 \leq ij \leq n$, be the infinitesimal transition rate from state i to state j, and $Q_{ii} = -\sum_{j=1, j\neq i}^{n} Q_{ij}$. Then $\mathbf{Q} = \{Q_{ij}, 1 \leq i,j \leq n\}$ is the generator matrix for the underlying CTMC, with row sums equal to zero. Consider the partition of the state space $S = \{S_1, S_2, \ldots, S_m\}$. We define the underlying lumped process $Z'(t)$ with a finite state space $S' = \{1,2,\ldots,m\}$, such that for any state i, $1 \leq i \leq m$, in the lumped process $Z'(t) = i$ if and only if $Z(t) \in S_i$, $\forall t \geq 0$. Conditions for lumping for a given partition of the underlying process $Z(t)$ are those under which the lumped process $Z'(t)$ is also a CTMC. The lumped Markov reward process can be similarly described, except that the state space of the underlying CTMC is now S', with a generator matrix $\mathbf{Q}' = \{Q'_{ij}, 1 \leq ij \leq m\}$ and a constant reward rate $R'_i \geq 0$ associated with state i. $\mathbf{R}' = \{R'_i, 1 \leq i \leq m\}$ is the diagonal reward matrix. $\mathbf{Q}'$ and $\mathbf{R}'$ are determined from the lumping conditions. We define the matrices $\mathbf{U}$ and $\mathbf{V}$ that are constructed from the given partition. $\mathbf{U} = \{U_{ij}, 1 \leq i \leq m, 1 \leq j \leq n\}$ is the "distributor" $m \times n$ matrix, such that the ith row is a probability vector restricted to states in S_i, that is, its components for states in S_i sum to one, and its components for all remaining states are 0. $\mathbf{V} = \{V_{ij}, 1 \leq i \leq n, 1 \leq j \leq m\}$ is the "collector" $n \times m$ matrix, such that the jth column is a vector with its components equal to one for states in S_j and 0 otherwise. It follows that $\mathbf{UV} = \mathbf{I}'$ (the identity matrix).

The term $\pi(t) = \{\Pr[Z(t) = i], 1 \leq i \leq n\}, t \geq 0$, is the transient probabilities vector of the underlying process $Z(t)$, $t \geq 0$. The accumulated reward up to time t is a random variable denoted by $Y(t)$. We define the matrix $\mathbf{Y}(x,t) = \{Y_{ij}(x,t)\}, 1 \leq i,j \leq n\}$, with the functions

$$Y_{ij}(x,t) = \frac{d}{dx} \Pr[Y(t) \leq x, Z(t) = j \mid Z(0) = i], \qquad 1 \leq ij \leq n$$

The term $\mathbf{y}(x,t) = \pi^T(0)\mathbf{Y}(x,t)$ is obtained by unconditioning on the initial probability distribution; $y'(x,t)$ is the corresponding vector in the lumped process. For lumping we must have $\mathbf{y}^T(x,t)\mathbf{V} = \mathbf{y}'^T(x,t)$, $\forall x,t \geq 0$.

1. STRONG LUMPABILITY

A Markov reward process is strongly lumpable for a given measure if for every initial probability vector $\pi(0)$, the lumped process is a Markov reward process giving the same measure.

THEOREM 1 A necessary and sufficient condition for strong lumpability for the distribution of accumulated reward in a Markov reward process is

$$\mathbf{VUQV} = \mathbf{QV} \qquad \text{and} \qquad \mathbf{VURV} = \mathbf{RV}$$

2. PROPORTIONAL DYNAMICS AND WEAK LUMPABILITY

If a Markov reward process is not lumpable for every initial probability vector, then it may be lumpable for some initial probability vector $\pi(0)$. In this case we say that the Markov reward process is weakly lumpable with respect to $\pi(0)$. We introduce the notion of proportional dynamics, which is sufficient for weak lumpability and unlumpability. For proportional dynamics we must have $\mathbf{y}^T(x,t)\mathbf{VU} = \mathbf{y}^T(x,t)$, $\forall x, t \geq 0$. Note that while lumping is defined for vector as well as scalar measures, proportional dynamics is defined only for vector measures, such as $\pi(t)$ and $\mathbf{y}(x,t)$.

THEOREM 2 A necessary and sufficient condition for proportional dynamics for the distribution of accumulated reward in a Markov reward process is

$$\mathbf{UQVU} = \mathbf{UQ} \qquad \text{and} \qquad \mathbf{URVU} = \mathbf{UR}$$

with $\mathbf{U}$ uniquely determined from $\pi^T(0)\mathbf{VU} = \pi^T(0)$. This condition is sufficient for weak lumpability.

The above theorems can be used to deduce conditions for strong and weak lumpability for other measures of interest [2]. In general, we have shown that lumping (respectively, proportional dynamics) a Markov reward process for a given measure requires at least lumping (respectively, proportional dynamics) its underlying CTMC, possibly with an additional condition on the reward rates associated with each state. For the distribution of accumulated reward, lumping (respectively, proportional dynamics) requires equal reward rates associated with states in the lumped (respectively, dynamically proportional) subset. For the instantaneous reward and the expected accumulated reward, the condition of equal reward rates is required for strong lumpability, however, there is no condition on the reward rates for proportional dynamics and weak lumpability.

REFERENCES

[1] Kemeny, J. G., and Snell, J. L. *Finite Markov Chains*, 2nd Edition, Springer-Verlag, New York, 1978.

[2] Nicola, V. F., Lumping in Markov Reward Processes, IBM Research Report RC 14719, Yorktown Heights, N.Y., 1989.

39

Accelerated Convergence through Extrapolation for the Discounted Return in Markov Chains

JEFFREY L. POPYACK Department of Mathematics and Computer Science, Drexel University, Philadelphia, Pennsylvania

EXTENDED ABSTRACT

We are concerned with methods for accelerating the convergence of iterative methods for computing the discounted return for infinite-horizon, finite-state Markov chains. We define the set of states as $\mathbf{S} = \{1, 2, \ldots, n\}$, transition matrix $P = \{p_{ij}\}$, cost vector $c = \{c_i\}$, and discount factor $\alpha \in (0, 1)$. The objective is to determine, for each state i, the long-range expected discounted cost. This cost, f_i, is determined by $f_i = \alpha \sum_{j \in S} p_{ij} f_j + c_i, i \in \mathbf{S}$. The existence of a unique f is guaranteed.

Let $v : \mathbf{S} \to \mathcal{R}^n$. Define $H(v) = \alpha P v + c$. We note that $H(f) = f$. Extrapolation methods are described by $v^{k+1} = H(v^k) + \Delta^k$, for some vector Δ^k. For instance, the L_∞ *norm reducing extrapolation*, or *Rowsum* extrapolation, chooses $\Delta^k = \underline{1}(\alpha/(1-\alpha))(L'^k + L''^k)/2$, where $L'^k = \min_{i \in S} \{H_i(v^k) - v_i^k\}$, $L''^k = \max_{i \in S} \{H_i(v^k) - v_i^k\}$, and $\underline{1} = (1, \ldots, 1)^\tau \in \mathcal{R}^n$. Below, we describe new extrapolations.

Error Sum Extrapolation (ESE)

We suppose $v^k = f + \Delta^k \underline{1} + \epsilon^k$, where Δ^k is a scalar which represents the major portion of the error between f and v^k, and ϵ^k is a vector whose components sum to 0, that is, $v_i^k = f_i + \Delta^k + \epsilon_i^k$, $\sum_{i \in S} \epsilon_i^k = 0$. A reasonable approximation for Δ^k in this circumstance is given by $\Delta'^k = -\sum_{i \in S}(H_i(v^k) - v_i^k)/n(1-\alpha)$, and we compute v_i^{k+1} as $v_i^{k+1} = H_i(v^k) - \alpha\Delta'^k$, $i \in \mathbf{S}$.

Absolute Error Extrapolation (AEE)

We suppose that, for each $i \in \mathbf{S}$, $|v_i^k - f_i| = \Delta^k + \epsilon_i^k$, where again Δ^k is a scalar and $\sum_{i \in S} \epsilon_i^k = 0$. Denote $A^k = \{i \in \mathbf{S}, | \, v_i^k \leq f_i\}$, $A'^k = \mathbf{S} - A^k$. We see that, for $i \in A^k$, $v_i^k = f_i - \Delta^k - \epsilon_i^k$, and for $i \notin A^k$, $v_i^k = f_i + \Delta^k + \epsilon_i^k$. Since it is not easily discernible whether $i \in A$ for given i, we approximate A^k with a computable set B^k: Denote $B^k = \{i \in \mathbf{S} \mid v_i^k \leq H_i(v^k)\}$, $B'^k = \mathbf{S} - B^k$, $d_i = \sum_{j \in A'} p_{ij} - \sum_{j \in A} p_{ij}$, $D = \sum_{j \in B} d_i - \sum_{j \in B'} d_i$. (We drop the k superscript from A, A', B, and B' without confusion). A good approximation for Δ^k is given by $\Delta'^k = \sum_{i \in S} |H_i(v^k) - v_i^k|/(n + \alpha D)$, and $v_i^{k+1} = H_i(v^k) - \alpha d_i \Delta'^k$, with the set B substituted for A. Computational experience shows that in practice, this extrapolation is effective. It is easily shown that when $A = S$ or $A' = S$, the result of AEE is identical to that of ESE.

It can be proved that the extrapolation ESE yields a sequence that converges to f. In the proof, we express ESE in the form of a stationary iteration $v^{k+1} = Rv^k + \gamma$. We show that f is a fixed point of the iteration, and $\rho(R) = \alpha|\lambda_2|$, where $\rho(R)$ is the special radius of R, and λ_2 is the subdominant eigenvalue of P.

A similar proof for the convergence of the extrapolation AEE has been elusive, since AEE represents a nonstationary iteration. In this case, we may express AEE in the form $v^{k+1} = R^k v^k + \gamma^k$. We are able to show that f is a fixed point. Further, we can show that $R^k = \alpha P Q^k$, where $\rho(Q^k) = 1$. These results suggest convergence, since $R^k = \alpha P Q^k$, $\alpha < 1$, $\rho(P) = 1$, $\rho(Q^k) = 1$, and $v^{k+1} - f = R^k v^k + \gamma^k - (R^k f + \gamma^k) = R^k(v^k - f)$. However, we do not know whether $\rho(R^k) < 1$. We further note that even the conditions $\{\rho(R^k) < 1\}$ are not sufficient to guarantee convergence to the fixed point of a nonstationary iteration.

The new extrapolations have been tested numerically and compared with eight other well known acceleration techniques: Jacobi, Gauss–Seidel, Gauss–Jacobi, minimum Gauss–Seidel row sum reordering, successive overrelaxation, variable successive overrelaxation, lower bound extrapolation, and L_∞ norm reducing extrapolation (Rowsum). The results suggest that AEE outperforms the other methods for problems with nonsparse transition matrices. In all test cases, AEE required less total computational effort than Rowsum, often considered the most robust extrapolation technique. All extrapolation methods, including Rowsum, were less predictable when the iteration matrix was sparse. For large test cases, AEE required less total computational effort than policy iteration. In fact, AEE seems to grow more powerful as n grows large. We further note that AEE does not seem to be adversely affected by values of α close to 1. In conclusion, the extrapolation AEE appears to be the most powerful method for accelerating convergence for nonsparse MDP's of large scale. Especially encouraging is the fact that this extrapolation has been developed for use in conjunction with a notably slow iterative method, namely pre-Jacobi.

40

Computing the Fundamental Matrix for a Reducible Markov Chain

THEODORE J. SHESKIN Industrial Engineering Department, Cleveland State University, Cleveland, Ohio

ABSTRACT

Many important quantities for a reducible Markov chain can be expressed in terms of the fundamental matrix. These quantities include (1) the expected time that the process is in each transient state for each possible transient starting state, (2) the mean time to absorption for each possible transient starting state, and (3) the probabilities of absorption for each possible transient starting state. In this paper a new matrix construction algorithm is presented for computing the fundamental matrix for any finite, reducible Markov chain. The algorithm contains three steps: construction, reduction, and multiplication. The algorithm requires about the same amount of storage and the same number of arithmetic operations as matrix inversion based on the LU decomposition.

To begin the construction step, arrange the transition probability matrix for a finite, reducible Markov chain in canonical form. The states are partitioned into M mutually exclusive closed communicating classes and a set of transient states. The submatrix S governs transitions among the transient states. Assume that S contains r states. Create three square, diagonal matrices, R, U, and T, of order r. Construct a composite matrix B of order $2r$ by adjoining matrices R, U, and T to S. Arrange the composite matrix in the form

$$B = \begin{bmatrix} T & U \\ R & S \end{bmatrix} \tag{1}$$

The three diagonal matrices are constructed in the following manner. Set an element on the main diagonal of T equal to the sum of the elements in the corresponding row of S, provided that the corresponding row sum of S is less than one. Otherwise, set the element on the diagonal of T equal to zero. Set every element on the main diagonal of U equal to one minus the corresponding element on the main diagonal of T. Set $R = U$.

The reduction step is almost identical to the one used in the author's Markov chain partitioning algorithm. The major difference is that the step begins when $n = 2r$. Reduction ends when $n = r$, indicating that the final reduced matrix, B_r, is of order r. The multiplication step is based on the result that when a matrix B is partitioned as in equation (1), then

$$B_r = T + U[I - S]^{-1}R \tag{2}$$

All of the diagonal elements of U and all of the diagonal elements of R are greater than zero. Therefore, each of these equal diagonal matrices is nonsingular and has an inverse. Solving equation (2) for the fundamental matrix,

$$N = [I - S]^{-1} = U^{-1}[B_r - T]R^{-1} \tag{3}$$

Software Demonstrations

41

MARKOV1: Software to Formulate and Solve Markov Chains

THOMAS R. BOWE and WILLIAM DAPKUS Decision System Associates, Woodside, California

1. SOFTWARE DESCRIPTION

MARKOV1 is an interactive software tool used to formulate and solve discrete space, continuous time Markov and semi-Markov models. Applications include:

Protection requirements for independent power producers.
Reliability worth (e.g., distribution system design–throwover schemes).

The software is menu-driven and allows easy model formulation in a spreadsheet-like environment. You can experiment with alternative formulations and see the results immediately. You may write formulas to estimate model inputs and build customized input screens.

MARKOV1 comes with example applications for reliability, availability, maintenance (Fig. 1), project planning and inventory control. You may build custom applications for use by others who may not have specific knowledge of Markov modeling methods.

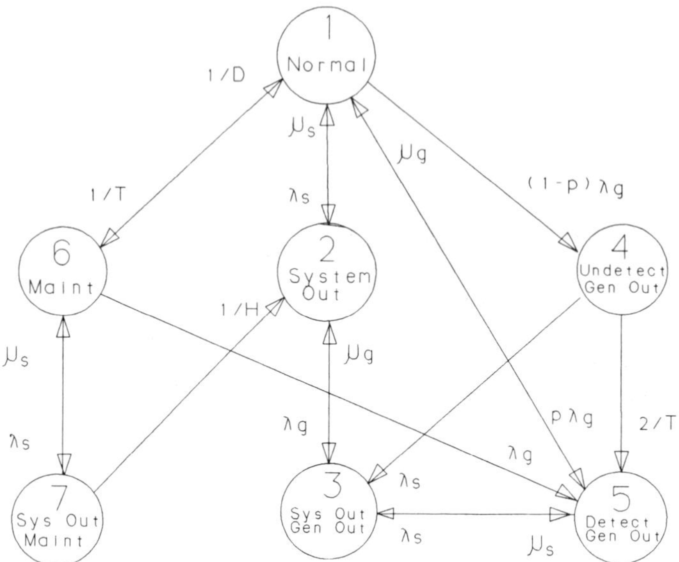

Figure 1 State representation for maintenance model.

MARKOV1 displays many statistics associated with Markov models, including steady-state probabilities, entrance rates, absorption and first passage probabilities and times, annual discounted costs, cost of first transitions, and transient behavior (Fig. 2). Automatic sensitivity analysis is available.

MARKOV1 is for engineers, analysts, researchers and students. MARKOV1 runs on personal computers that use the MS-DOS operating system. A math coprocessing chip (80x87) is recommended.

2. MARKOV1 SPECIFICATIONS: FEATURES AND PERFORMANCE

Features

Markov and semi-Markov models. Holding times may be constant, exponential or gamma distributed.

Trapping states.

Transient analysis.

User defined formulas with variables and math operators.

Markov reward process with linear and exponential cost curves, continuous time discounting, multiple alternatives, costs grouped into categories.

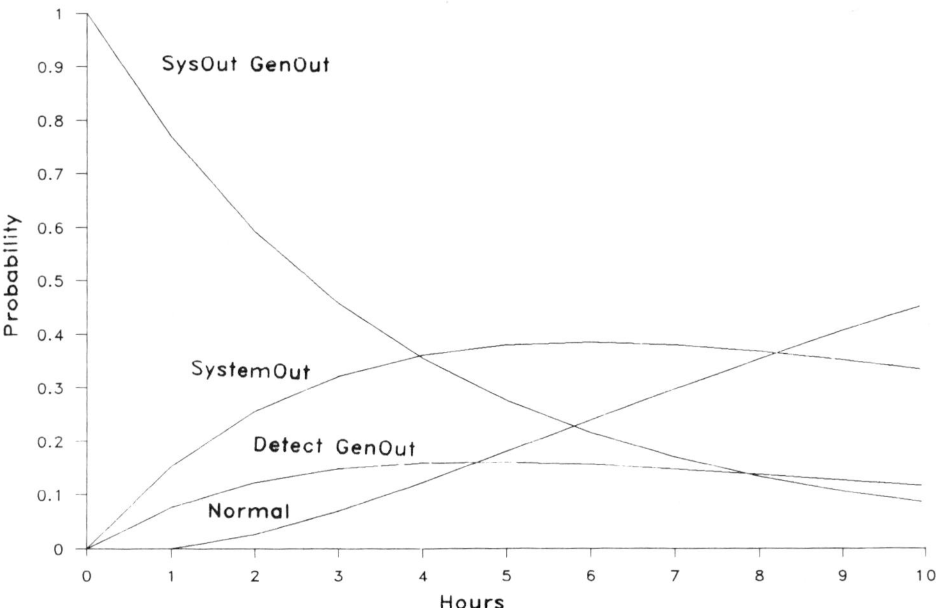

Figure 2 Transient behavior: starting in State 3, probability of being in each state.

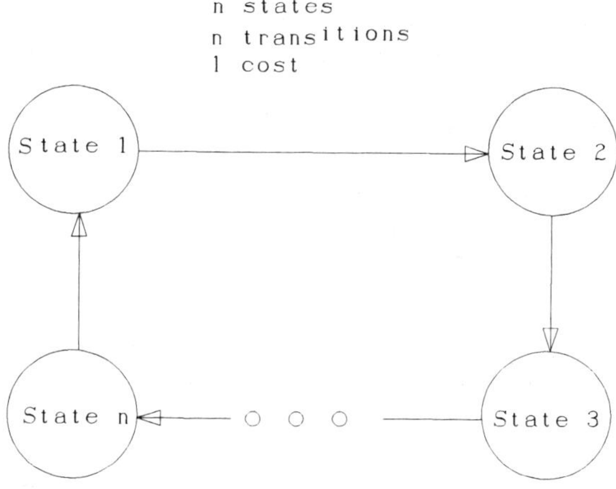

Figure 3 Benchmark model.

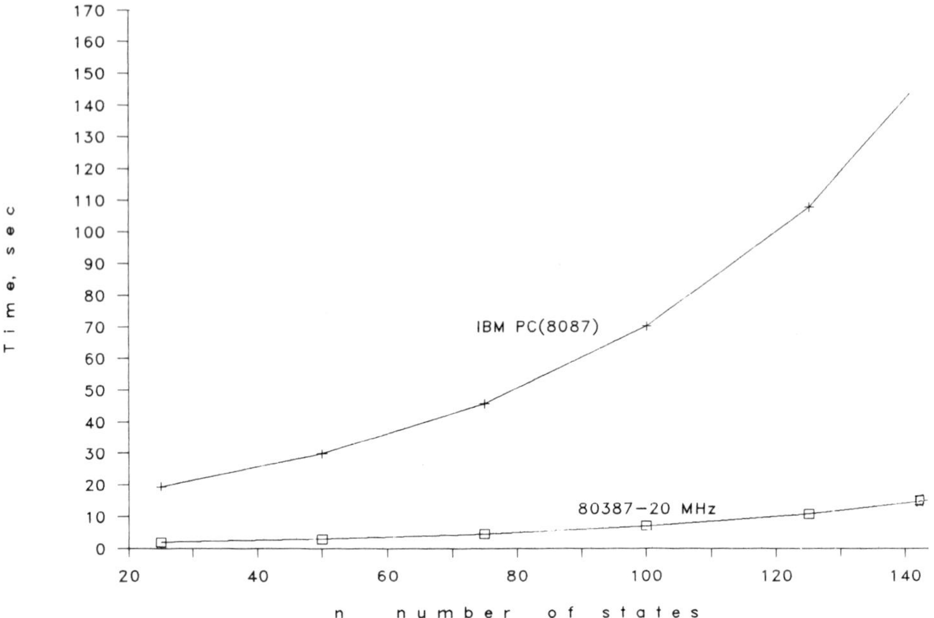

Figure 4 MARKOV1 computation time.

Statistics calculated: steady-state probabilities, entrance rates, absorption and first passage probabilities, expected annual costs, cost of first transition.

User-friendly interface, menu-driven, spreadsheet-like, interactive environment.

Tabular and graphical outputs to screen, printer or file.

Automatic sensitivity analysis.

Performance

Number of states, transitions, costs, and alternatives are only limited by computer memory. Personal computers with 640k allow models with up to 200 states.

Computation times depend on the number of states, transitions, and costs in your model. Figure 4 shows how computation time depends on the number of states in the simple benchmark model (Fig. 3).

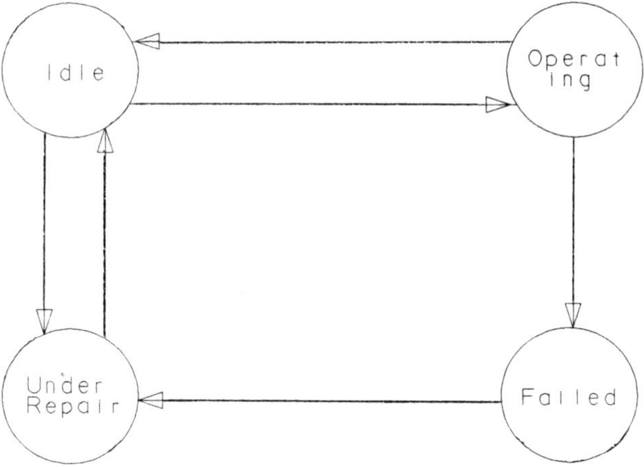

Figure 5 Generator state diagram.

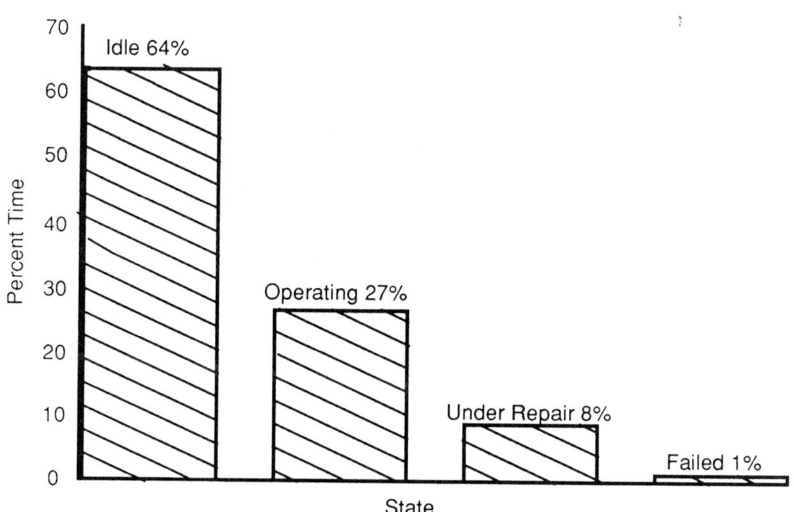

Figure 6 Generator availability.

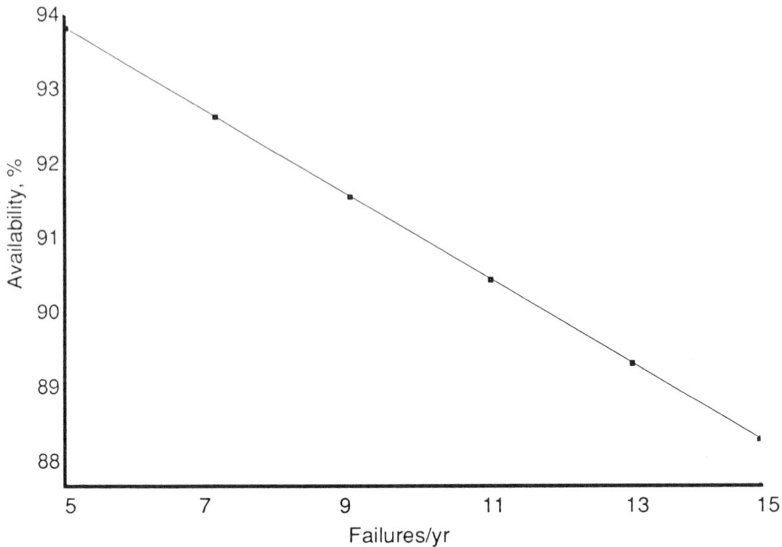

Figure 7 Sensitivity of generator availability to number of failures per year.

3. EXAMPLE APPLICATION: PREDICTING GENERATOR AVAILABILITY

1. Select a model. The first step is to select a model to fit the application. In this application we are interested in predicting generator availability (Fig. 5). We load the file GENAVL1 from disk.
2. Estimate parameters. The second step is to estimate parameters. This may require review of historical data and subjective judgment. These parameters are used in formulas for the transition probabilities and holding times.

Parameters

Events per year:	
Startups/yr	195
Normal shutdowns/yr	185
Failures/yr	10
Preventative maintenance/yr	5
Mean time:	
Idle	28
Operating	12

Failed	6
Repair	48

3. Calculate statistics. The third step uses MARKOV1 to calculate statistics for the application. We are interested in the percent of time the generator is available (e.g., idle or operating Fig. 6).
4. Determine sensitivity. You can use the automatic sensitivity capability of MARKOV1 to see how the results vary as a function of the parameters. Here, we see how availability depends on the annual number of failures of the generator (Fig. 7).

For more information contact Decision Systems Assoc. (DSA), 4244 Jeffferson Ave., Woodside, CA 94062, 415-851-7591 or 415-369-0501.

42

Queuing Network Analysis Package-2 (QNAP2)

KSHEMENDRA PAUL Techno-Sciences, Inc., Greenbelt, Maryland

QNAP2 is a portable modeling environment providing specific facilities for building, handling, and solving queuing network models. It is comprised of a collection of resolution algorithms, and of a common user interface for model description, analysis control, and result presentation. QNAP2 has both exact and approximate analytic solvers, Markovian solvers, and discrete event simulation. It is an extremely powerful package, which has been successfully used for problems of the type discussed here.

QNAP2 was developed at INRIA, a national research laboratory of France. Over 140 sites in Europe, Japan, and North America currently use QNAP2, and it is a fully supported commercial product. Techno-Sciences is the North American distributor.

Queuing network modeling is widely recognized as a powerful tool for design and analysis of discrete flow systems. QNAP2 is an interactive software package designed to facilitate the description and analysis of queuing network models for large and complex systems. QNAP2 is comprised of:

A collection of resolution algorithms based on discrete event simulation, exact and approximate analytical methods, and Markovian methods to solve for the steady-state solution of the network.
A high-level, easy-to-learn interactive language for model building, analysis control, and results editing.

1. QNAP LANGUAGE

The QNAP2 language is a high-level, object-oriented, interactive language. It includes two levels of specification:

A command language consisting of a set of parametrized commands. These commands correspond to the main functionalities of QNAP2: declarations of the objects and variables manipulated in a model; basic structure of the service stations; model analysis control; and interactive dialogue specification.
An algorithmic language, derived from SIMULA and PASCAL. The algorithmic language allows: service specifications involving complex mechanisms and the definition of specific interfaces tailored to the user's need.

The QNAP2 language is easy to learn and use. The elements of queuing network modeling are built in within the two levels of the language: the basic structure of a queuing network model can be simply defined at the level of the command language and through the specific parameters of the commands; the algorithmic language provides a large set of language features and specific procedures for the progressive enhancement of models.

The QNAP2 language allows the construction of multilevel models and the flexible combination of the QNAP2 solvers on the different modeling levels.

2. QNAP2 SOLVERS

QNAP2 is unique in that it offers access to three types of solvers through the same applicative language:

1. Analytical solvers
   a. Exact solvers based either on the well-known convolution algorithms (the QNAP2 implementation is more powerful than in any other known package); or on mean value analysis;
   b. Approximate solvers based either on the diffusion approximation or on iterative algorithms; or on various heuristics.
   Dispatching between the Analytical Solvers (exact and approximate ones) is automated on the basis of input data. It may be forced if needed.

2. A Markovian solver: This solver computes eigenvalues of the state transition matrix in order to obtain steady-state information. Numerical algorithms have been improved to decrease computation time. This solver provides exact solutions for models involving complex synchronizations.
3. An event-driven simulator: The simulator supports the detailed descriptions of complex structures, algorithms, and synchronization mechanisms by mean of the extended features of the QNAP2 language (semaphores, resources, flags to synchronize processes, object attributes to describe properties of these objects). It provides the user with the standard results and with confidence interval analysis.

3. MODELING FRAMEWORK

The modeling framework supported by QNAP2 is the representation of a system as a network of stations (nodes) through which customers (packets, messages, calls) circulate. A station can consist of one or several servers serving a single input queue. The station represents the physical or logical processing facilities of the system, and intuitively one can see that they might map onto the functional modules of the communications node. Each station has a service function, which can be either a probabilistic model of the processing delay, or an algorithmic calculation. An example of the latter might be a shortest path calculation for dynamic routing.

QNAP2 does not provide a large set of predefined modeling mechanisms (i.e., memory allocation, polling, dynamic queue selection), which, experience shows, is seldom exactly what is needed to handle whatever specific problem is under investigation. Rather, QNAP2 offers a limited set of standard primitives and mechanisms (i.e., time delays, synchronization operations with other customers, customer generation) and a powerful language, which allow the user to build almost any complex mechanism.

4. QNAP2 USAGE

Three possible uses are relevant:

Modeling workbench: for efficient and progressive problem analysis relying on the conciseness of the language and the spectrum of methods which are directly accessible.

Modeling environment: for the development of large and complex modeling studies, specific features enable the user to build large models involving repetitive or embedded structures. Also, dedicated hybrid solution techniques can

be defined for improved efficiency of the analysis and, if the model is used intensively, a tailored user interface can be added to the model for parameter input and result output.
Modeling testbench: for research in queuing network modeling; as indicated above, new solvers can be implemented and tested against existing solvers.

5. PORTABILITY

QNAP2 is coded in a subset of FORTRAN 77 and is thus fully portable, and is supported on virtually every major computer system.

43

MARCA: Markov Chain Analyzer

WILLIAM J. STEWART Department of Computer Science, North Carolina State University, Raleigh, North Carolina

MARCA (MARkov Chain Analyzer) is a software package designed to facilitate the generation of large Markov chains (the state space and the transition matrix); to analyze the structure of the chain, in particular to determine the extent to which the chain may be decomposed; and to compute both the transient distribution at any time t, and the stationary probability distribution.

A state of any Markov chain may be represented as a vector of length l. In MARCA, each component of this state descriptor vector is referred to as a bucket. Each bucket contains a number of balls that is equal to the numeric value of the corresponding vector component. The evolution of the Markov chain is represented by the movement of balls among the buckets. The movement of a ball from one bucket to another is called a *transition* from a *source bucket* to a *destination bucket*, and the *rate of transition* is defined as the rate at which the source bucket loses balls to the destination bucket. The rates will obviously vary from system to system and from bucket to bucket. A transition of particular interest is an *instantaneous transition*, which means, as its name implies, that the actual movement takes no time at all. Such transitions often serve to implement scheduling policies.

1. GENERATION OF SYSTEM STATES AND TRANSITION MATRIX

As the states of the Markov chain are generated, they are stored in a one-dimensional array called *LIST*. Simultaneously, the matrix of transition rates is also formed and stored in compact form. The list of states is built as follows. An initial feasible state suppled by the user is examined to determine which states it can reach in a single step transition. For each destination state, *newstate*, a subroutine written by the user and called *RATE* is used to determine the rate at which this transition occurs. If the transition is zero or negative, the destination state is ignored. On the other hand, if the transition rate is strictly positive, a second user written subroutine called *INSTANT* is called to see if an instantaneous transition(s) may be made from *newstate*. If so, the result of the instantaneous transition becomes *newstate*, so that it is now a single step transition from *currentstate*. The list of states is now searched to determine if it already contains *newstate*. If *newstate* is not in the list of states, then it is added to the end of *LIST* and a pointer, *LIMIT*, which indicates the number of states in the list, is incremented by one. The rate of transition obtained by *RATE* and its position in the transition matrix are now stored in compact form.

When all destination states that emanate from *currentstate* have been so treated, the diagonal element of the matrix is computed as the negated sum of the rates and is also stored. The next state in *LIST*, which is marked by a variable *MARKER*, becomes *currentstate* and the process repeated. When the value of *MARKER* exceeds that of *LIMIT* all the states will have been generated.

2. NEAR DECOMPOSABILITY

After the states of the system and the matrix of transition rates have been generated, MARCA tests the system for near-decomposability. This is accomplished by temporarily setting to zero those nonzero elements of the stochastic transition probability matrix that are less than some threshold, *gamma*, and then applying an algorithm to determine the strongly connected components of the resulting matrix. MARCA performs this for a range of values of *gamma* and produces a table that displays the number of disjoint groups, their average size, and standard deviation. If the number of groups is not large (less than 50), it will also list the size of each group. This information may be useful in choosing a particular solution method.

3. COMPUTATION OF STATIONARY DISTRIBUTIONS

Currently in MARCA there is a variety of different types of algorithms for determining the stationary probability vector: direct methods, fixed-point iteration methods, subspace iteration methods, and combination of these. A particular method must be chosen by the user. In future implementations, it is envisaged that this choice could be made automatically. The current set of algorithms available is:

SOR	Successive overrelaxation
SSOR	Symmetric SOR
POWER	Power iteration
ARNOLDI	The method of Arnoldi
GE	Gaussian elimination (compacted form)
PREC. GMRES	Preconditioned generalized minimum residual
PREC. ARNOLDI	Preconditioned Arnoldi
SOR-GMRES	GMRES preconditioned with SOR
SSOR-GMRES	GMRES preconditioned with SSOR
PREC. POWER	Preconditioned fixed-point iterations
CGS	Conjugate gradients squared
NCD-SOLVER	Iterative aggregation/disaggregation

Three different incomplete factorizations are available and may be used as preconditioners. The first is called ILU0 and is an incomplete factorization in which the structure of the LU decomposition is forced to be identical to the structure of the original infinitesimal generator matrix. The second is called ILUTH and in this decomposition only elements that are larger than a specified threshold are kept. The third is ILUK and here only a fixed number (K) of elements are kept per row.

Some of these methods are included because this is an ongoing research project. For example, in many cases we have found that the method of conjugate gradient squared failed to perform satisfactorily. GMRES preconditioned with SOR or SSOR generally performs less well than when it is preconditioned with one of the preconditioners ILU0, ILUTH, or ILUK. The Power method is for comparison (and historical) purposes only.

When the stationary probability vector has been generated, MARCA determines the marginal distributions for each of the buckets of the model as well as the mean number of balls in each bucket and the standard deviation. A facility is also included to provide the user with access to the stationary distributions to compute model characteristics other than the marginal probabilities just described.

4. TRANSIENT DISTRIBUTIONS

The numerical method used to compute the transient distributions is the randomization method. The user is requested to input a time at which the probability distribution is required, a beginning state and an accuracy parameter. The transient distribution is computed at that time and from it, the marginal probabilities. The user may request the probabilities of different states, or the transient solution at a different point of time.

44

Q-Lib: A Software Package for Queuing Models

H. C. TIJMS Department of Econometrics, Free University, Amsterdam, The Netherlands

We demonstrate an IBM-PC version of a software package developed at the Vrije University, Amsterdam, in cooperation with M. H. van Hoorn and L. P. Seelen. The package is organized as an iterative program consisting of various queuing models:

The infinite-capacity $GI/G/c$ queue.
The finite-capacity $GI/G/c$ queue.
The machine repair model.

The specific interarrival and service time distributions covered by the package include deterministic, generalized Erlangian, and hyperexponential distributions. The package is fast and uses special-purpose iterative algorithms (GAUSS–Seidel, aggregation/disaggregation) for the numerical solution of large-scale Markov chain equations. The computing times are relatively short even for a large number of servers.

45

SPNP: The Stochastic Petri Net Package

GIANFRANCO CIARDO Software Productivity Consortium, Herndon, Virginia

JOGESH MUPPALA and KISHOR S. TRIVEDI Department of Computer Science, Duke University, Durham, North Carolina

The stochastic Petri net package (SPNP) is a versatile modeling tool for the solution of generalized stochastic Petri nets (GSPN). The GSPN models are described in the input language for SPNP called CSPL (C-based SPN language). The CSPL is an extension of the C programming language with additional constructs that facilitate easy description of GSPN models. The full power and generality of C is available, but a working knowledge of C is sufficient to use SPNP effectively.

The GSPN models specified to SPNP are actually "GSPN reward models" that are based on the "Markov reward models" paradigm. This provides a powerful modeling environment for:

Dependability (reliability, availability, safety) analysis.
Performance analysis.
Performability modeling.

A number of important Petri net constructs like marking dependency, variable cardinality arc, and enabling functions facilitate the construction of models for complex systems. The package also allows logical analysis on the

Petri net whereby any general assertions defined on the Petri net are checked for each marking of the net. The GSPN may be solved to obtain either steady-state measures or transient measures. The package allows the specification of custom measures, although a standard set of measures is available. The measures are defined in terms of reward rates associated with the markings and/or rewards associated with the transitions of the GSPN. Sensitivity analysis allows the user to evaluate the effect of changes in an input parameter on the output measures. This is useful in system optimization and bottleneck analysis.

46

SHARPE: Symbolic Hierarchical Automated Reliability and Performance Evaluator

ROBIN A. SAHNER Urbana, Illinois

KISHOR S. TRIVEDI Department of Computer Science, Duke University, Durham, North Carolina

SHARPE (symbolic hierarchical automated reliability and performance evaluator) is a hierarchical modeling tool for analyzing a class of performance, reliability/availability (dependability), and combined performance/reliability/availability (performability) models. It provides a specification language and solution methods for the following model types (the words in the parentheses represent the class of models for which the model type is applicable):

1. Series-parallel reliability block diagrams (dependability).
2. Fault trees (reliability).
3. Reliability graphs (reliability).
4. Acyclic Markov chains (dependability, performance, performability).
5. Irreducible cyclic Markov chains (dependability, performance, performability).
6. Cyclic Markov chains with absorbing states (dependability, performance, performability).
7. Acyclic semi-Markov chains (dependability, performance, performability).

8. Irreducible cyclic semi-Markov chains (dependability, performance, performability).
9. Series-parallel directed (acyclic) graphs (dependability, performance).
10. Product-form single chain (closed) queuing networks (dependability, performance).
11. Generalized stochastic Petri nets (dependability, performance, performability).
12. PMS (processor–memory–switch) models (dependability).

The Markov models can have reward rates associated with the states of the Markov chains allowing the specification and solution of "Markov reward models." The output measures obtained from the models are either steady-state or transient depending on the model type. Typical output measures include the cumulative distribution functions including mean and variance for most of the model types, state probabilities (Markov models), and specialized measures like the expected throughput, utilization, queue length (PFQN) and expected number of tokens in a place, and the expected throughput of a transition (GSPN).

The output distribution functions are symbolic in time t and they can thus be used as parameters to other submodels. These are represented in terms of exponential polynomials. Scalar values such as probabilities or the mean and variance of a distribution can also be used as parameters to other submodels. This facilitates the hierarchical combination of different model types to derive an overall model for a system.

SHARPE is written in C, and versions are available for VMS (Vax) and UNIX (Vax, Sun). The object code is about 600 kbytes on SUN/UNIX.

Index